2011 18th European Microelectronics & Packaging Conference

(EMPC 2011)

Brighton, United Kingdom
12-15 September 2011

IEEE Catalog Number: CFP1154H-PRT
ISBN: 978-1-4673-0694-2

Copyright © 2011, International Microelectronics and Packaging Society
All Rights Reserved

***This publication is a representation of what appears in the IEEE
Digital Libraries. Some format issues inherent in the e-media version may
also appear in this print version.*

IEEE Catalog Number: CFP1154H-PRT
ISBN 13:   978-1-4673-0694-2

Additional Copies of This Publication Are Available From:

Curran Associates, Inc
57 Morehouse Lane
Red Hook, NY 12571 USA
Phone: (845) 758-0400
Fax: (845) 758-2633
E-mail: curran@proceedings.com
Web: www.proceedings.com

2011 18th European Microelectronics & Packaging Conference

(EMPC 2011)

Brighton, United Kingdom
12-15 September 2011

IEEE Catalog Number: CFP1154H-PRT
ISBN: 978-1-4673-0694-2

Keynotes

'A Centre of Excellence to improve the Quality of Life and Human Health', 1
Prof. Radimir Vrba, CEITEC, Central European Institute of Technology, Brno, Czech Republic

'Future challenges and trends - Highlights of iNEMI 2011 Roadmap' 9
Bill Bader, CEO of iNEMI, USA

TuA1: Packaging for Medical Applications

'Fabrication of a Novel Biochip Multi-Electrode Array Using Polyester Insulated Electrodes with 18
Micro-Well Features for Cardiomyocyte Analysis'
Olivia Flaherty, Loughborough Univ., UK

'Innovative Manufacturing and 3-Dimensional Packaging Methods of Ultrasonic Transducers for 24
Medical Applications'
Jack Hoyd-Gigg Ng, Heriot Watt Univ., UK

'Evaluation of Co-fired Platinum /Alumina High Density Feed-through for Implantable 29
Neurostimulator Applications'
William Kinzy Jones, Florida International Univ., USA

TuA2: Substrate Technologies

'Co-Firing of LTCC Modules with Ag-electrodes and embedded Ferrite layers' 34
Hamid Naghibzadeh, Federal Materials Research Inst., Germany

'Embedded RFID TAG inside PCB board to improve supply chain management' 40
Julien Viret, Loughborough Univ., UK

'Manufacture of a 3-D Package Using Low Temperature Co-fired Ceramic Technology' 45
Yves Lacrotte, Heriot Watt Univ., UK

TuA – Display Presentations

'Non-Destructive X-Ray Mapping of Strain & Warpage of Die in Packaged Chips' 53
Patrick McNally, Dublin City Univ, Ireland

'Design, modelling and fabrication of novel MEMS structure utilizing carbon nanotubes' 61
Jiri Haze, Brno Univ of Tech, Czech Republic

'Inverse Modelling of the Button Shear Test' 66
Ilko Schmadlak, Freescale Semiconductor, USA

'Changes in Water Absorption and Modulus of Elasticity of Flexible PCB Materials in High RH Testing' 70
Sanna Maria Lahokallio, Tampere Univ, Finland

'LTCC-based capacitive pressure sensor in harsh environment'   76
Darko Belavic, HIPOT, Slovenia

TP1: Advanced Packaging Applications

'A Comprehensive Packaging Solution For Next Generation IC Substrates'   80
Stephen Kenny, Atotech, Germany

'Modular Microelectronics by System-in-Packages with Embedded Components'   84
Andreas Ostmann, Fraunhofer IZM, Germany

'Packaging Challenges of Smaller Power MOSFETs & Schottky Diodes for the Automotive Sector'   89
Kevin J Keller, OnSemi, USA

'Application of an Angular Exposure System to Fabricate True-Chip-Size Packages for SAW Devices'   94
Barbara Monika L'huillier, SUSS Micro Tec, Germany

TP2: Manufacturing Technologies

'Silver-Palladium Pastes for Aluminium Nitride Applications'   100
Kathrin Reinhardt, Fraunhofer IKTS, Germany

'Miniaturization: Solder Paste Attributes for Maximizing the Print & Reflow Manufacturing Process   106
Window'
Karthik Vijay, Indium Corp., UK

'Inkjet Printing of Electrical Vias'   116
Werner Zapka, Xaar, Sweden

'Copper Wire Bonding Experiences from a Manufacturing Perspective'   120
Bernd Appelt, ASE Group, USA

TuP – Display Presentations

'High speed bending of 2nd level interconnects on printed circuit boards for automotive electronics'   124
Marcel Kouters, TNO, The Netherlands

'Reliable hermetic MEMS chip-scale packaging'   131
Guido Spinola Durante, CSEM, Switzerland

'High resolution microstructural investigation of lead-free Al-Ge & Al-Ge-Cu alloys for high temp   137
silicon die attach'
Sandy Klengel, Fraunhofer IWMH, Germany

'Reliability of ACA joined thinned chips on rigid substrates under humid conditions'   142
Laura Katriina Frisk, Tampere Univ, Finland

'BGA Lifetime Prediction in JEDEC Drop Tests Accounting for Copper Trace Routing Effects'   149
Frank Kraemer, Saarland Univ, Germany

'Electronic Packaging for Bio-Diagnostic Microfluidics Application'   156
Mark Shaw, ST Microelectronics, Italy

IMAPS-Europe Electronics Business Council Session

'Manufacturing business decisions: design and manufacture of new generation circuits'   164
Michael A Lane, President, Struthers Dunn, South Carolina, USA

'A Smarter Supply Chain – End to End Quality Management'   170
Dr Kitty Pearsall, DE, ISC Procurement Engineering Organisation, IBM, USA

'A Cross Supply Chain Collaborative Approach to Addressing the   175
Technical Challenges of Today's Electronics Manufacturing Industry'
Brian Smith & Marshall Andrews, High Density Packaging User Group

WA1: Joint Behaviour
'Computational Parametric Study on the Strain Hardening Effect of Lead-free Solder Joints in Board   181
Level Mechanical Drop Tests'
Jiang Tong, Univ. of Science & Tech, Hong Kong

'Reliability Model for Assessment of Lifetime of Lead-Free Solder Joints'   187
Ivan Szendiuch, Brno Univ. of Technology, Czech Republic

'Modelling and Experimental Measurement of Multiple Joint Lead Free Solder Interconnects   192
Subjected to Low Cycle Mechanical Fatigue',
Elisha Kamara, Univ. Greenwich, London, UK

'Effect of heating method on microstructure of Sn-3.0Ag-0.5Cu solder on Cu substrate'   198
Hiroshi Nishikawa, Osaka Univ., Japan

WA2: Physico-Chemical Investigations
'Electromigration Study of Nano Al Doped Lead- Free Sn-58Bi on Cu and Au/Ni/Cu Ball Grid Array   202
Packages'
Shafiq Ismathullakhan, City Univ., Hong Kong

'Characterisation of ion transport during electroplating of high aspect ratio microvia by megasonic   209
agitation'
Suzanne Costello, Heriot Watt Univ., UK

'Electrical Properties of an Isotropic Conductive Adhesive Filled with Silver Coated Polymer Spheres'   216
Shiwani Jain, Loughborough Univ., UK

'An Investigation of Lead-free Thick-film Resistors on LTCC Substrates – Preliminary Results'   223
Marko Hrovat, Josef Stefan Inst., Slovenia

WA3: Reliability of Microstructures
'Characterization of intermetallic compounds in Cu-Al ball bonds'   229
Marcel Kouters, TNO, Netherlands

'Effects of additional Co atoms on microstructural evolution in Sn-Ag-Bi-In solder under current   236
stressing'
Katsuaki Suganuma, Osaka Univ., Japan

'Influence of Ti- surface contamination on reliability of Al wire bonds'   240
Golta Khatibi, University of Vienna, Austria

'Polymer cored BGA ball reliability optimisation'   247
David C Whalley, Loughborough, Univ., UK

WA4: Electrical & Structural Modelling
'Power Loss due to Periodic Structures in High-Speed Packages and PCBs'   252
Paul Huray, Intel Corporation, USA

'DoE Simulations and Measurements with the microDAC Stress Chip for Material and Package   260
Investigations'
Florian Schindler-Saefkow, Fraunhofer ENAS, Germany

'Advancements in Fracture and Failure Simulation for Electronic Package Applications'   265
David Reid, SIMULIA, Providence, USA

WA5: Substrate Interconnections
'New Improvements in Thermal Management: Thick Print Copper Thick Film Replacement for DBC'   270
Tracey Smolinsky, Heraeus Materials Technology, USA

'Fine-line on LTCC-Substrates for 60 GHz Line Coupled Filters'   278
Jens Müller, TU Ilmenau, Germany

'Die-bonding by using thin Ag flakes'   283
Soichi Sakamoto, Osaka Univ, Japan

'Novel Approaches to create low cost Through Silicon Vias'   288
Jan Eite Bullema, TNO, Netherlands

WA6: Reliability of Interconnects
'Acceleration factors of combined reliability tests of lead-free interconnections'   293
Przemyslaw Matkowski, Wroclaw Univ of Technology, Poland

'Thermo-Mechanical Fatigue Life Evaluation for SnPb and SnAg Solders'   301
Noritake Hiyoshi, Ishikawa National College of Tech, Japan

'Effects of microstructure on creep deformation of Sn-3.5Ag alloys'   308
Jin Yu, Kaist, South Korea

'Characterization of 300mm Large Scale embedded Wafer Level BGA'   313
Seung Wook Yoon, STATS ChipPAC, Singapore

WA – Display Presentations
'Advanced Thermal Materials for Heat Sinks in Microelectronics '   318
Mathias Ekpu, Univ Greenwich, UK

'Emerging Nanotech-based Thermal Interface Materials for Automotive Electronic Control Units'.   326
Kenny Otiaba, Univ Greenwich, UK

'3D LTCC & Flexible Interconnects Using Galvanized Layers'   334
Michal Nicák, Brno Univ, Czech Rep.

'Solder TIMs for Improved Thermal Management'   338
Karthik Vijay, Indium Corp. UK

WP1: Nano Technology

'Low-Pressure Sintering of Ag Micro and Nano particles for a High Temp Stable Pick & Place Die   346
Attach'
Julian Kähler, TU Braunschweig, Germany

'Low-temp, photonic approach to sintering ink-jet printed conductive microstructures containing   353
nano silver particles'
Tomasz Falat, Wroclaw Univ of Technology, Pl

'Influence of different type protective layer on silver metallic nano-particles for Ink-Jet printing   357
technique'
Andrzej Moscicki, Amepox Microelectronics, Poland

'Additive Photolithography Based Process for Metal Patterning Using Chemical Reduction on Surface   363
Modified Polyimide'
David Ewan Watson, Heriot Watt Univ., UK

WP2: Bonding Technology

'Copper Makes A Lot Of Cents'   370
Chris Flowers, CSR, Cambridge, UK

'Current Industry Adoption of Fine-Pitch Cu Wire Bonding'   376
Grace O'Malley, iNEMI, Ireland

'Use of Harsh Wire Bonding to Evaluate Various Bond Pad Structures'   380
Stevan G Hunter, OnSemi, USA

'Flip-chip bonding of thin Si dies onto PET foils: possibilities and applications'   388
Jeroen van den Brand, TNO, Netherlands

WP3: Reliability Testing & Prognostics

'Effects of different combinations of environmental test on the reliability of UHF RFID tags'   394
Kirsi Saarinen, Tampere Univ, Finland

'A Dynamic Bending Method in CSP Package Validation for Portable Electronics'   399
Jeffrey ChangBing Lee, Integrated Service Technology Inc., Taiwan

'Data analysis techniques for real-time prognostic and health management of semiconductor   405
devices'
Thamotharampillai Sutharssan, Univ., Greenwich, London, UK

WP – Display Presentations

'Planar Thick Film Inductor Characterization'   412
Jiri Pulec , Brno Univ, Czech Rep.

'Laser Patterning of Thin Films for Luminescence Applications'   416
Thomas Höche, Fraunhofer IWMH

'Effect of Substrates Surface Condition on the Morphology Silver Patterns Formed by Inkjet Printing'   421
Zhaoting Xiong, Loughborough Univ, UK

Embossed ceramic reflectors with nano-dispersive coatings for compact optoelectronic systems   425
Wolfgang Buss, Fraunhofer IOF

Dispensing process with additional ultrasonic energy   430
Martin Bursik, Brno Univ, Czech Rep.

3D TSV Middle-end Processes and Assembly/Packaging Technology   434
Seung Wook Yoon, STATS ChipPAC, Singapore

WP4: Photo-voltaics & LED Packaging
'Ultra Fine Line Print Process Development for Silicon Solar Cells'   440
Tom Falcon, DEK, UK

Mechanical Problems of Manufacturing Processes for Photovoltaic Modules'   445
Steffen Wiese, Saarland Univ., Germany

'Development of an intelligent integrated LED system-in-package'   451
Alexander Gielen, TNO, Netherlands

WP5: MEMS Packaging
'Multichip MEMS Sensor Packaging'   458
Mark Shaw, ST Microelectronics, Italy

'Miniaturised Integrated Hybrid Microsystem Assemblies for Harsh Environment Applications'   466
Stephen Riches, GE Aviation Systems, UK

'Requirements for microfluidic sensors based on ceramic technologies'   470
Samuel Hildebrandt, TU Dresden, Germany

ThA1: Eco Energy Technology & Reliability
'Multilayer technology as integration system for ceramic micro fuel cells'   474
Adrian Goldberg, Fraunhofer IKTS, Germany

'The i-Module Approach: Towards Improved Performance and Reliability of Photovoltaic Modules'   480
Jonathan Govaerts, IMEC, Belgium

'Improved Testing of Soldered Busbar Interconnects on Silicon Solar Cells'   485
Robert Klengel, Fraunhofer IWMH, Germany

'Solidification Processes of SnCu and SnAgCu solder alloys and interface reactions to charactise solar   490
cell interconnections processes'
Sebastian Schindler, Fraunhofer CSP, Germany

ThA2: Embedded Packaging Technology
'3D Stacking Approaches for Mould Embedded Packages'   494
Tanja Braun, Fraunhofer IZM, Germany

'Chip embedding technology developments leading to the emergence of miniaturized system-in-   502
packages'
Dionysios Manessis, Fraunhofer IZM, Germany

'Technology development for a low-cost, roll-to-roll chip embedding on PET foils'   510
Maarten Cauwe, IMEC, Belgium

ThA3: Special Applications Technologies
'Novel Approach for Integrating Electronics into Textiles at Room Temperature using a Force-Fit   516
Interconnection'
Erik Simon, Fraunhofer IZM, Germany

'Microsoldering with short-pulsed IR Laser for Textile Fabrics'   523
Hartmann Hieber, ICR, Germany

'High Temperature Electronics for Harsh Environments'   528
Bob Hunt, Strategic Technologist, CMAC MicroTechnology, Great Yarmouth, UK

ThA4: Advanced Packaging
'Fine Pitch Flip-Chip Chip-Scale Packaging'   533
Bernd Appelt, ASE Group, USA

'Copper Wire Bond investigation on Multiple Surface Finishes - Enabling Wire Bond Packages without   537
Gold'
Nigel White, Atotech, Germany

'Low Temperature Wafer Bonding Technologies'   543
Marco Haubold, Fraunhofer ENAS, Germany

Additional Papers

'Effect of Thermal Shock Conditions on Reliability of Chip Ceramic Capacitors   551
Alexander Teverovsky, Dell Services Fed Gov

CEITEC - Centre of Excellence in Life Sciences, Advanced Materials and Technologies to Improve the Quality of Life and Human Health

Radimir Vrba

Brno University of Technology, CEITEC - Central European Institute of Technology,
Technicka 3058/10, CZ-61600 Brno, Czech Republic

Phone +420 541146161, Fax +420541146298 and E-mail vrba@ceitec.vutbr.cz

Abstract

CEITEC - Central European Institute of Technology is a project to develop a European centre of scientific excellence in the fields of life sciences and advanced materials and technologies, which is aiming to set up a centre with state-of-the-art infrastructure and conditions for best scientific workers. It was prepared by four universities and two research institutes in Brno, Czech Republic. CEITEC creates new working positions, ensures conditions for research excellence and enables effective cooperation with innovative companies and institutes.

Key words: advanced materials, technologies, life sciences, scientific excellence, infrastructure

Introduction

Central European Institute of Technology was prepared as a project of a European centre of scientific excellence since 2007 year to compete in 2010 for funding provided by Operational Programmes of the European Commission for the period of 2011 up to 2015 year. The project was prepared by six academic and research partners interdisciplinary with full synergy in the fields of life sciences and advanced materials and technologies, aiming to set up a centre with state-of-the-art infrastructure and promising conditions for best scientific workers. The project was evaluated as the best one by the Ministry of Education, Youth and Sports of the Czech Republic and particularly by the European Commission International Approval Board.

CEITEC will be located in of Brno in the Czech Republic, both buildings will be constructed before the end of 2013 and fully equipped by scientific instruments a technologies before the end of 2014.

The CEITEC projects will end in 2015 and after this, in 2016 to 2021, in the period of sustainability will run according to European Commission rules for operational programme funding.

The CEITEC centre creates gradually new working positions for up to 550 researchers recruited between 2011 and 2015 year. The centre ensures conditions for research excellence and enables effective cooperation with innovative companies and institutes in 7 research programmes shared by 54 research groups with 10 core facilities and 25,000 m^2 of new laboratories. The project with full budget of more than 200 million euro started in the beginning of 2011 year.

Common Research Objectives

The targeted cooperation within the research programmes is assured by common research objectives. They define the general scientific orientation of the centre and reflect synergic

Fig. 1: CEITEC Buildings – advanced materials and technologies *(links)*, **life sciences** *(right, oval)*

integration in the field of life sciences and advanced materials and technologies:

- To understand the mechanisms of the genesis and spreading of important diseases, methods of their prevention, early diagnostics and therapy.
- To utilise plant systems as renewable sources of materials and biologically active compounds.
- To develop advanced materials and functional nanostructures for medicine, energy and information and communication technologies.
- To utilise information and communication technologies for biomedicine.

Research Programmes

CEITEC is based on the synergy of seven research programmes headed by the research programme coordinator each. Two of seven research programmes belong to advanced materials and technologies and five programmes represent most part of life sciences branch.

RP1 ADVANCED NANOTECHNOLOGIES AND MICROTECHNOLOGIES. The research programme is aimed at nanotechnologies and nanoscience covering materials and functional structures suitable for nanoelectronics and nanophotonics, generally. It will address both the preparation and characterisation of nanostructures. Attention will be paid to the research of 2D – 0D nanostructures. The research will consider semiconductor nanostructures, metallic and magnetic nanostructures, nanotubes and nanowires, supramolecular structures and novel nanoelectronic circuits (More-than-Moore), etc. The interconnection of nanostructures with peripheries and special micro-circuits will make it possible to utilise the properties of nanostructures in various applications, for instance sensors and actuators.

RP2 ADVANCED MATERIALS. The research of advanced materials covers the synthesis of materials, and the analysis of the structures and properties of advanced materials. The aim of the research is to develop novel materials of complex properties and propose novel application areas for these materials. The research is focussed on advanced ceramic materials, advanced polymeric materials and composites, and advanced metallic materials.

RP3 STRUCTURAL BIOLOGY. The research programme will integrate the three dimensional structural information describing large macromolecular assemblies - proteins, nucleic acids and their complexes - into functional contexts in order to gain an understanding of the vital processes at the cellular level. It aims to achieve European competitiveness, to stimulate regional development, and to facilitate biomedical research and biotechnologies. At the application level, its results will facilitate developments of next-generation diagnostic and therapeutic strategies for the treatment of human diseases and solution for health.

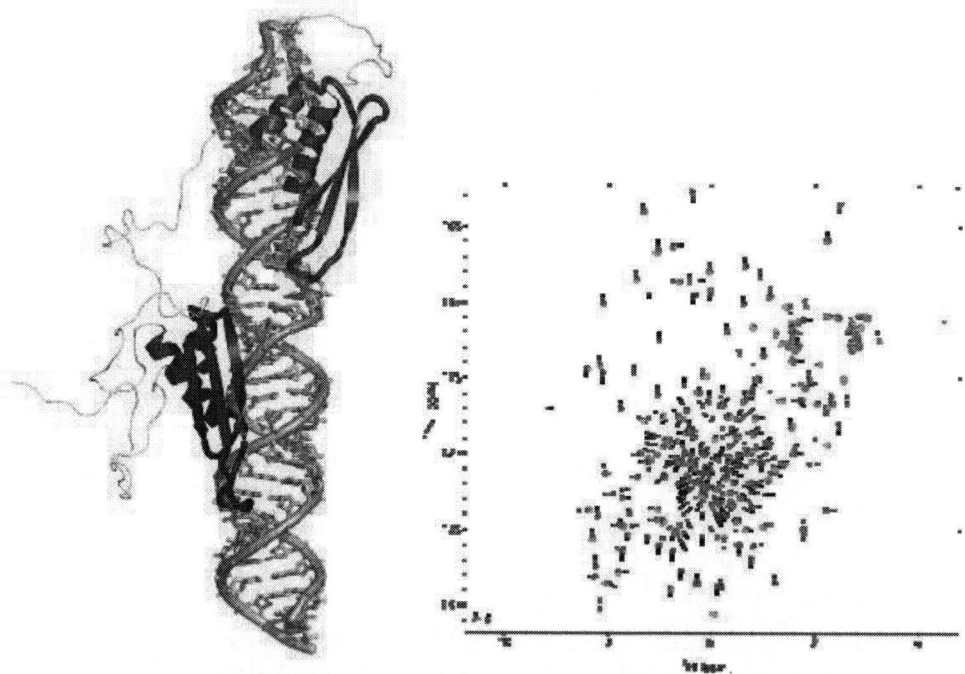

Fig. 2: Example of NMR spectroscopy investigation of RNA sequence- and structure-dependent recognition by proteins in Structural Biology research programme

RP4 GENOMICS AND PROTEOMICS OF PLANT SYSTEMS. The programme is focussed on the systematic analysis of genes and complete genomes, the subcellular distribution of gene products and a detailed description of the structure, function and evolution of chromosomes and their elements to promote an understanding of the key processes regulating metabolism, growth and differentiation. In addition, we aim to elucidate the molecular nature of the developmental decisions and response of organisms to the environment. To achieve this, detailed knowledge of cellular signalling pathways, protein composition, their subcellular trafficking and distribution is required in addition to genomic data.

RP5 MOLECULAR MEDICINE. The research programme aims to improve diagnostics, therapy, and prevent important human diseases. The main diseases studied will be cancer; neuromuscular, neurodegenerative and metabolic diseases; infectious diseases; and quantitative and qualitative immune defects. The project aims to bring novel findings in the pathogenesis of those diseases and produce experimental strategies for their treatment and prevention.

medicina"). Both research programmes (5 Molecular Medicine and 7 Molecular Veterinary Medicine) represent parallel lines approaching the same issues (pathogen genomics, molecular mechanisms of various types of diseases, host genomics) as well as diseases shared by animals and humans and/or transmitted between them.

RP1 Advanced Nano and Microtechnology Programme in More Details

In the field of Surface Science, Thin Film Physics, and Nanotechnology activities dealing with a large variety of problems namely the modification of surfaces and etching microstructures / nanostructures with ion beams, development and application of ion beam assisted deposition technique for the preparation of metal, oxide and nitride thin films and multilayers, development and application of ultravacuum equipment for ion assisted molecular beam deposition, such as Ga, GaN, development and application of in situ analytical and microscopic methods working under Ultra High Vacuum (UHV) conditions – Scanning Probe Microscopy (SPM), Secondary Ion Mass Spectrometry (SIMS), Time-of-Flight Low-Energy

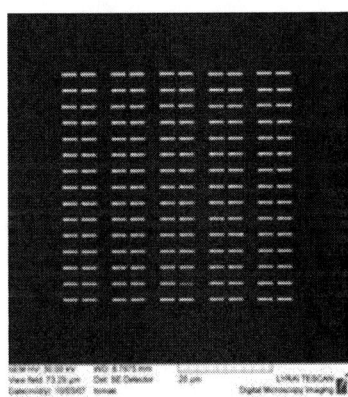 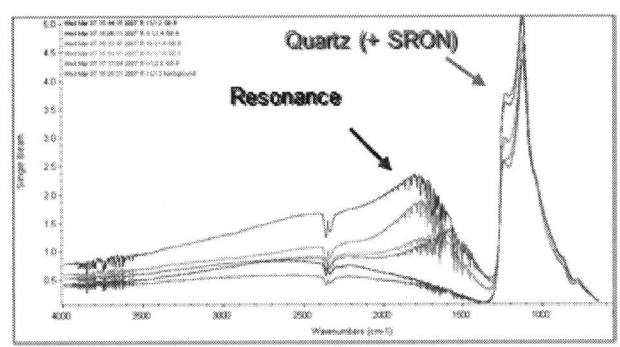

Fig. 3: An array of plasmonic antennas fabricated by EBL and corresponding FTIR response [1]

RP6 BRAIN AND MIND RESEARCH. The research programme will be established to promote a collaborative theoretical, experimental, and clinical study of the brain from the molecular to the behavioural and cognitive levels. Interdisciplinary research will be completed in the fields of neurobiology, neuropsychopharmacology, functional neuroanatomy, neurophysiology, neuroimaging, neuropsychology, neurology, psychiatry, and computational neuroscience.

RP7 MOLECULAR VETERINARY MEDICINE. Important biological processes and diseases will be studied in selected animal models. In terms of the philosophical approaches used, links to the research programme 5 Molecular Medicine are especially important ("una vita, una sanitas, una

Ion Scattering (TOF-LEIS), X-Ray Photoelectron Spectroscopy (XPS), Low Energy Electron Diffraction (LEED) and Thermal Desorption Spectroscopy (TDS), and the study of the interaction of SPM local probes with surfaces. The expertise in nanostructured materials comprises the fabrication of nanostructures by Focused Ion Beam (FIB) and other lithographic methods, the application of SPM for the fabrication of nanostructures, e.g. by Local Anodic Oxidation, guided growth of metallic nanostructures on the surfaces patterned by focused ion beam or Atomic Force Microscopy (AFM), and the study of the integral and local electric, magnetic and photonic properties of nanostructures for plasmonics [1] and spintronics [2] with potential applications in (bio)sensing.

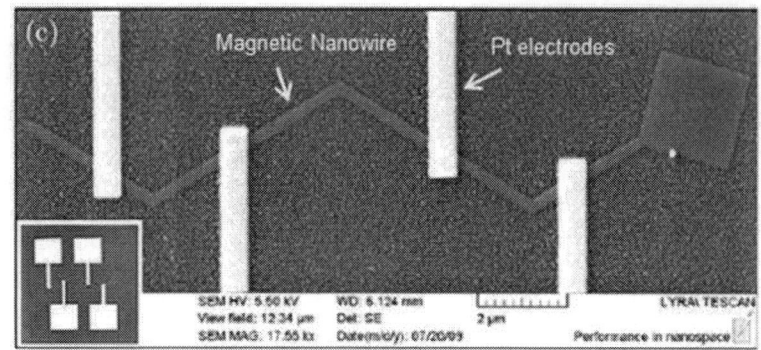

Fig. 4: A magnetic (FeNi/Cu/Co) nanowire and Pt electrodes fabricated by FIB for a study of the motion of magnetic domain walls [2]

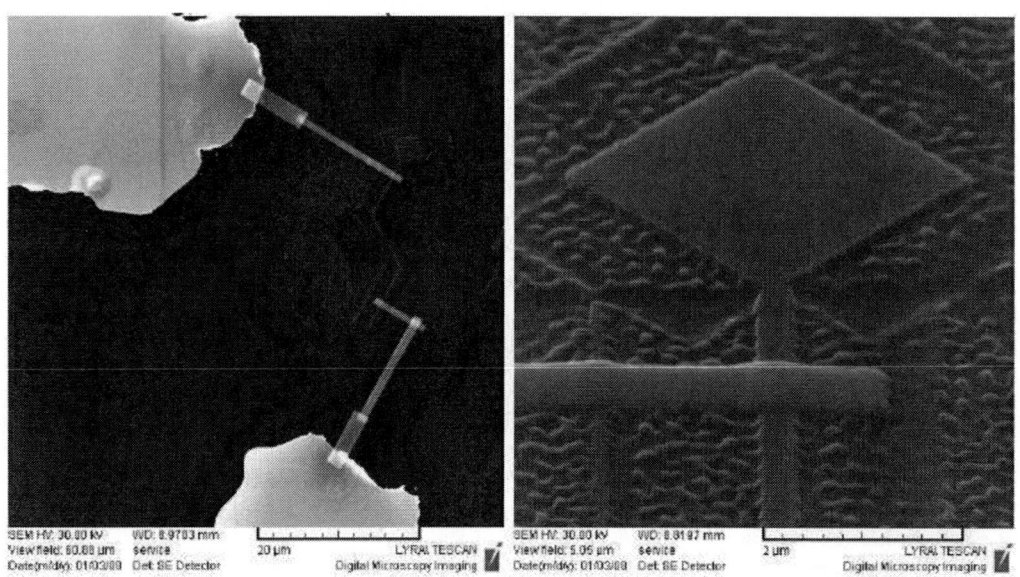

Fig. 5: Spin valve nanowires

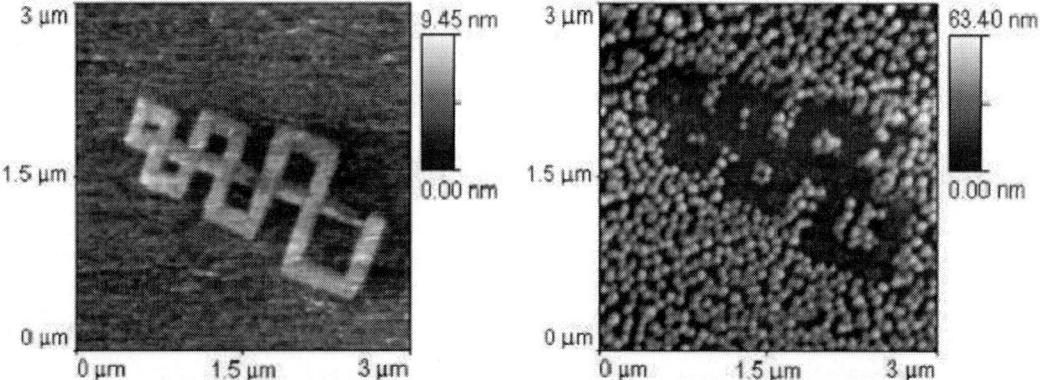

Fig. 6: Selective growth. AFM image (contact mode) of a zig-zag structure made by LAO *(left)*, the same after deposition of 8 ML of Ga at 350 °C *(right)*

Analytical methods are focused on studies of selected materials and layered structures, in particular their optical response and structural properties. This is aimed at metals, semiconductors and insulators that are of interest either independently, or as constituents of layered systems. Optical spectroscopy in a wide range (from far-infrared to ultraviolet) is used to study vibrational and electronic states and their interactions, e.g. the changes of the optical response with temperature. The expertise in nanostructured materials comprises the electronic structure of low-dimensional III-V

methods for selective growths, etc. Sonochemical methods, laser induced assembling, synthesis of inorganic-organic complex materials, coordination and modified polymers and supramolecular entities, chemical and electrochemical growth. All the developed methods will be utilized directly for fabrication of nanostructures, advanced planar materials and devices specified in Part 2.

2. Investigation of the functional properties of nanostructures: Specification and optimization of the functional properties of nanostructures for nanoelectronics, nanophotonics and (bio)sensing

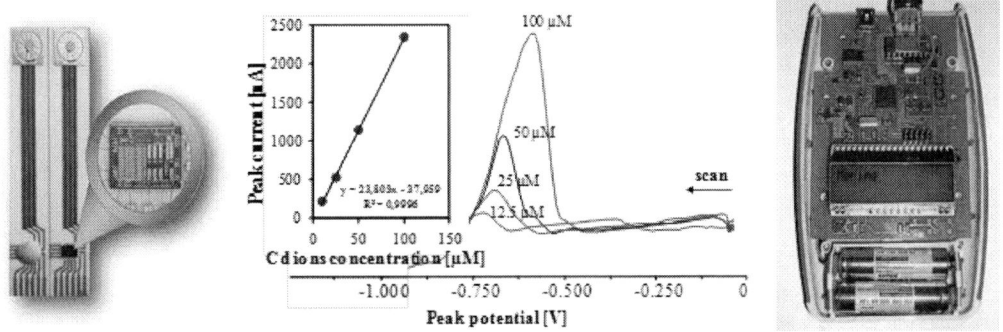

Fig. 7: Electronic chips for biosensing - CMOS technology and mixed mode electronics

structures, from quantum wells to self-organized strained quantum dots. In addition, nanostructured oxide materials (including superconducting and magnetic components) belong to a long-term activity [3].

Activities aimed at sensors and submicron electronic system design are dealing with the design of nanostructured electrodes for chemical, biochemical and medical sensing [4]. The expertise comprises chemical wet sensor- and gas sensor development and integration of sensors into microsystems. It includes research and development of microflow systems with sensitive transducers and lab-on-chips. The expertise also comprises intermetallic and heteromaterial system connections.

RP1 Advanced Nano and Microtechnology Programme Research Plans till 2015 year

The research activities may be characterized as follows:

1. Fabrication of nanostructures by "bottom-up" methods and by "top-down" methods (nanolithography): Development of methods on nanostructure fabrication: planar physical and plasmochemical methods using Electron-Beam Laser (EBL), UV lithography, Focused Ion Beam (FIB), Scanning Probe Microscopy (SPM) lithography, imprint technology. Molecular Beam Epitaxy (MBE), Chemical Vapor Deposition (CVD), Atomic Layer Deposition (ALD), Plasma Enhanced Chemical Vapor Deposition (PECVD), hybrid

prepared in Part 1, their correlation with geometrical/structural parameters of nanostructures and operational parameters. Novel and unique properties of nanostructures not observable at conventional materials and microstructures open the ways for qualitatively new applications.

3. The development of submicron devices and nanostructures: Development of the methods and techniques for implementation of outputs from Part 1 to 2 into higher functional integrated systems, special electronic circuits on a chip, nano- and micro-electromechanical systems (MEMS/NEMS) and sensors and advanced materials for them. Development of libraries for the design of integrated systems and their fabrication using CEITEC planar technologies, specifying new techniques and methods of packaging and testing of nanostructures and devices. Nanostructures implemented into these systems will enhance the performance and efficiency of nowadays devices and systems and widen the areas of their applications (e.g. biosensors, biochips, biorecognition tranducers).

4. The development of analytical and measurement methods: Development of the techniques and methodologies for microscopy, analysis and metrology of nanomaterials / nanostructures, and for diagnostics of their properties - new techniques of nanometrology by SPM, optical methods, combination of more techniques (SEM, AFM, etc.). This will be used as supportive activities for Parts 1 to 3 and

characterisation of nano- and micro-structures generally.

RP2 Advanced Materials Programme in More Details

The research of advanced materials covers the synthesis of materials, and the analysis of the structure and properties of advanced materials. The aim of the research is to develop novel materials of complex properties and propose novel application areas for these materials. The research is focused on advanced ceramic materials, advanced polymeric materials and composites, and advanced metallic materials.

Research in advanced ceramic materials will be aimed to the preparation of precursors of advanced ceramic materials and composites using modern advanced methods of inorganic ceramic powder synthesis and surface or bulk modifications of ceramic nanoparticles. By means of application of novel ceramic shaping and sintering methods and using advanced ceramic precursors will be developed new heterogeneous functionally graded and nanostructural ceramic materials. Characterisation of composition, structure and properties of advanced ceramic materials, modelling of the structure-property-function relationships and

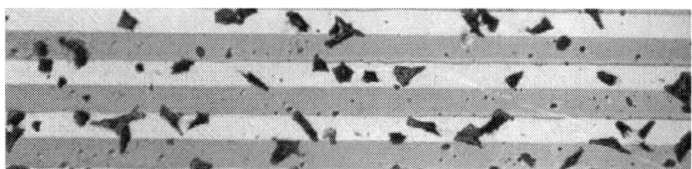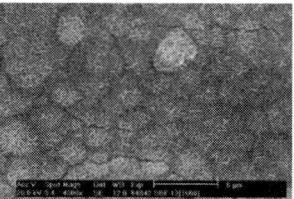

Fig. 8: Bioactivity of zirconia-alumina ceramics

Overall goal is to establish an equipment and personnel infrastructure further enhancing excellence in the research of advanced (polymeric, ceramic, metallic and composite) materials and their applications in various industrial segments, medicine and services. The main effort will be devoted to investigating advanced methods of preparing multifunctional homogeneous and heterogeneous advanced materials, characterizing their structure on various dimensional scales, quantifying structure-property-function relationships on the various structural levels and developing procedures for engineering properties of this class of materials in the process of their preparation.

testing of ceramic materials from the view of potential applications will be carried out. Research in this area of advanced ceramic materials will focuse to (a) synthesis of ceramic powdered materials, (b) consolidation, shaping and sintering of ceramic materials, (c) physico-chemical properties of ceramic materials, (d) electroanalytical diagnostic of materials (ceramics, polymers, and composites will be tested by electrical and electrochemical methods EIS, CV, LSV, DPV).

Research in materials for sensors and technological processes control systems is aimed to the advanced robotic, control and sensor systems using new materials. New robotic systems in the

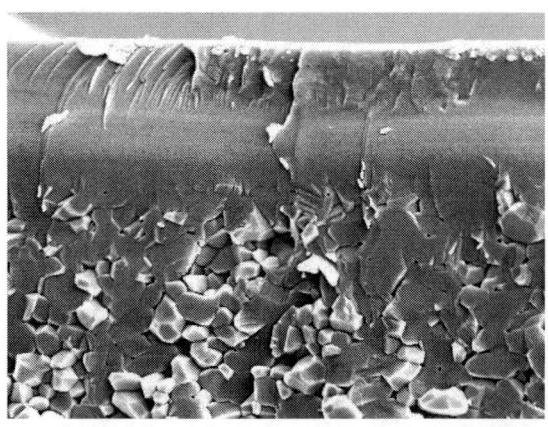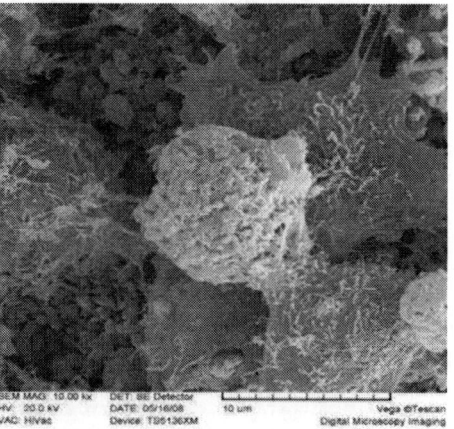

**Fig. 9: Bioinert ceramic implant with bioactive surface. Nanostructured coating on the surface of implant *(left)*, growth of bone tissue on the surface of implant activated by nanostructured surface *(right)*

area of rescue systems and life-science applications will be developed (e.g. intelligent service and prosthetics aids). New approaches inspired by the nature will be used in the development works. Control technologies for technological processes with critical reliability and safety will be studied. Particular results of the control technologies development will be also applicable beyond the scope of robotics systems, especially applications in control of traction drives for ecological transportation drives will be considered. The topic includes also smart sensors design research based on new materials. The smart sensors are expected to provide necessary process data gathered from technical and biological systems interacting with the controlled robotic systems. The particulate research activities include: (a) intelligent sensors using new materials, (b) robotic systems for special and hazardous environment, (c) advanced control systems, (d) instrumentation, communication and special electronics for technological processes.

Research in advanced polymers and composites will focus on advanced methods of preparing multifunctional homogeneous and heterogeneous polymeric materials (biomaterials, electro materials and structural materials) to characterize their structure on various dimensional scales, to quantify structure-property-function relationships on the various structural levels and to develop procedures for engineering the properties of this class of materials in the process of their preparation. The particulate research activities include: (a) preparation of advanced multifunctional polymeric and biopolymeric materials, (b) quantification of the structure-property-function relationships in bulk materials, (c) mechanisms of polymer degradation, (d) electrical properties and life-time prediction by means of accelerated ageing, (e) computer modelling and simulations.

Research in advanced metallic materials concentrates to an extensive investigation of the properties (predominantly mechanical ones, but in specific cases also magnetic, electrical and other ones) of the selected advanced materials in relation to their microstructure will be carried out. Another line of activity will be the study of the mechanisms of degradation processes in advanced materials under conditions simulating service conditions. The particulate research activities include: (a) mechanical properties of advanced metallic materials, (b) microstructure, diffusion and thermodynamics of advanced metallic materials, (c) multiaxial fatigue of advanced metallic materials.

Research in structure and phase analysis of advanced materials is represented by a core facility Structural Analysis Laboratory, which will be equipped with top-class instruments for transmission and scanning microscopy, microanalysis, and X-ray diffraction analysis, with the aim of providing for such work for which it would not be economic and meaningful to build specialized laboratories. Simultaneously, these specialized services will also be offered to interested parties outside CEITEC centres. The research will proceed in close relation to the activities of previous teams. Its priority will therefore be to focus on the study of the microstructure, submicrostructure and local chemical analysis of new advanced ceramic and polymer materials and composites based on these materials. Another research area will be composites based on metallic matrices (Ni-SiO2 and other types) and materials with ultra-fine grain obtained via SPD (Severe Plastic Deformation), in particular by the ECAP method.

RP2 Advanced Materials Programme Research Plans till 2015 year

The research activities may be characterized as follows:

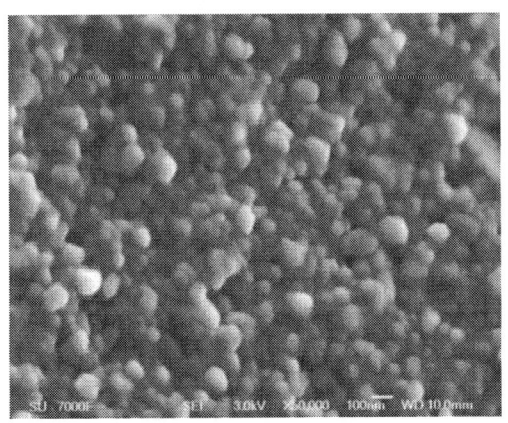 

Fig.10: Bulk BaTiO$_3$ and SrTiO$_3$ (relative density 98 %, grain size 80 nm) (nano)ceramics and composites *(left)*, synthesis and sintering of dielectric core SiO$_2$ and magnetic shell Ni ferrite ceramics *(right)*

1. Biomaterials: The development of novel composite biomaterials which can induce the growth of connective tissue on the surface of implants and therefore accelerate healing and improve the strength and biological stability of the implant-tissue connection (ceramic and metallic materials for replacement of soft and hard tissues, materials for orthopaedic devices).

2. Materials for energetics, communication and ecology: The development of novel composite materials with a functionally graded structure for the improvement of the efficiency and lifetime of components and devices for energetics, communication and control technologies (conductive ceramic and polymer materials for electrodes, novel actuators, sensor components, control and instrumentation systems for technological processes, catalyst for the decomposition of gaseous pollutants, biopolymers and precursors from plants and plant residues).

3. Structural materials: The development of novel polymeric, metallic and ceramic composites with excellent mechanical and thermal properties for structural applications (transparent ceramic materials; thermally and chemically resistant ceramic composite materials; impact-resistant ceramic composites; polymer multifunctional composites for high-tech engineering applications).

Conclusions

The main benefit of the CEITEC project is the unique combination of life and material sciences. The core of the project consists of seven research programmes, bringing together research groups working in specific and progressive scientific fields and disciplines. Vision of the centre is "The establishment of a Centre of Excellence conducting research to improve the Quality of Life and Human Health". We are creating a supra-regional centre of scientific excellence, whose results will try to be comparable to those of top centres and will thus contribute to strengthening the position of Brno as a recognised European scientific centre. We strive to be completely on a par with the world's leading institutions of this kind, taking advantage of the unique opportunities arising from synergies between life and traditional sciences and collecting well known and skilled scientists.

Acknowledgements

This work was supported by the project "CEITEC - Central European Institute of Technology" (CZ.1.05/1.1.00/02.0068) funded by European Regional Development Fund in the frame of Research and Development for Innovations Operational Programme in No. 1 Priority Axis.

Special thanks to my colleagues Tomáš Šikola, Jaroslav Cihlář and Jan Neuman willingly assisting with this paper manuscript preparation.

References

[1] T. Šikola, R. D. Kekatpure, E. S. Barnard, J. S. White, P. Van Dorpe , L. Břínek, O. Tomanec, J. Zlámal, D. Lei, Y. Sonnefraud, S. A. Maier, J. Humlíček, M. L. Brongersma, "Mid – IR plasmonic antennas on silicon-rich oxinitride absorbing substrates: nonlinear scaling of resonance wavelengths with antenna length", Appl. Phys. Lett. 95 (2009), 253109.

[2] M. Urbánek, V. Uhlíř, P. Bábor, E. Kolíbalová, T. Hrnčíř, J. Spousta, T. Šikola, "Focused ion beam fabrication of spintronic nanostructures: an optimization of the milling process", Nanotechnology 21 (14), (2010), 145304.

[3] P. Klenovský, V. Křápek, D. Munzar, J. Humlíček, "Electronic structure of InAs quantum dots with GaAsSb strain reducing layer: Localization of holes and its effect on the optical properties", Appl. Phys. Lett. 97 (2010), 203107.

[4] O. Zitka, H. Skutková, H. Kryštofová, J. Hubálek, et al., "Rapid and Ultrasensitive Method for Determination of Phytochelatin(2) using High Performance Liquid Chromatography with Electrochemical Detection", Intern. Journ. of Electrochem. Science 6 (2011), 1367.

[5] "CEITEC - Central European Institute of Technology", CZ.1.05/1.1.00/02.0068 project, Research and Development for Innovations Operational Programme Priority Axis No. 1, Application Form, 2010, Masaryk University, Brno University of Technology.

[6] K. Maca, V. Pouchly, Z. Shen: Two-step Sintering and Spark Plasma Sintering of Al_2O_3, ZrO_2 and $SrTiO_3$ Ceramics, Integrated Ferroelectrics 99(1), p. 114-124, 2008

[7] J. Cihlar, M. Trunec, V. Sida, "Ceramic Femoral Component of Knee Prothesis", Bioceramics 11 (1998), 69

[8] M. Trunec, K. Maca, Z. Shen, "Warm Pressing of Zirconia Nanoparticles by the Spark Plasma Sintering Technique", Scripta Materialia, 59 (2008), 23

[9] M. Trunec, K. Maca, "Compaction and Pressureless Sintering of Zirconia Nanoparticles", J. Am. Ceram. Soc., 90 (2007) 2735

[10] J. Cihlar and M. Trunec," Injection Moulded Hydroxyapatite Bioceramics", Biomaterials 17 (1996), 1905

[11] M. Trunec, J. Cihlar, S. Diethelm, J. Van Herle, "Tubular La0.7Ca0.3Fe0.85Co0.15O3-δ perovskite membranes, Part I: Preparation and Properties", J. Am. Ceram. Soc., 89 (2006), 949

HIGHLIGHTS OF 2011 iNEMI TECHNOLOGY ROADMAP

Bill Bader, Chuck Richardson, Robert C. Pfahl, Jr. and Grace O'Malley

International Electronics Manufacturing Initiative (iNEMI)

Herndon, Virginia, USA

Bill.bader@inemi.org; chuck.richardson@inemi.org, bob.pfahl@inemi.org; gomalley@inemi.org

ABSTRACT

iNEMI released its 2011 Technology Roadmap to industry on March 29, 2011. The Roadmap covers six product segments and 21 technology areas pertinent to electronics manufacturing. Every two years iNEMI maps future manufacturing technology needs of the global electronics industry for the next ten years. It discusses the major business and technology issues, paradigm shifts, emerging technologies and markets, technology gaps, and identified needs in each of six product segments: aerospace/defense, automotive, consumer/portable, medical, Netcom, and office/large business systems. This paper highlights two product sector chapters (Automotive, Medical) and three technology chapters (MEMS, Packaging, and Environmentally Conscious Electronics).

INTRODUCTION

The iNEMI roadmap charts future opportunities and challenges for the electronics manufacturing industry. Our widely utilized roadmaps influence Research & Development (R&D) investments and technology deployment around the world.

Updated every two years, the roadmap sets direction for technology development and deployment by predicting future packaging, component and infrastructure needs and describing critical technical and business elements required to support industry growth. This information provides the basis for collaborative projects undertaken by our members (via iNEMI's Technology Plan Document) as well as research priority focus for the global electronics industry (as identified in iNEMI's Research Priorities Document).

ROADMAP PROCESS:

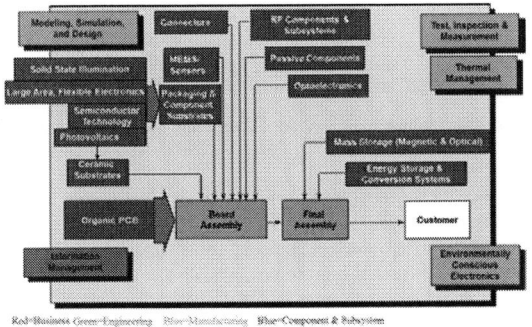

Red=Business Green=Engineering Blue=Manufacturing Blue=Component & Subsystem

Figure1 Twenty-one TWGs

The 2011 Roadmaps were developed by twenty-one Technology Working Groups, in response to inputs from representatives of OEMs in six Product Emulator Groups. These are; Automotive, Medical, Consumer / Portable, Office / Large Business Systems, Netcom - Network / Datacom / Telecom, Aerospace / Defense.

The leading electronic systems manufacturers are basing their strategic planning for new products on the assumption that the electronics infrastructure will develop and implement the technology to meet these key drivers.

SITUATION ANALYSIS

BUSINESS:

The electronics industry has done better than most in seeing solid growing demand in many of the market segments – in spite of the economic conditions – as consumers continue to demand creative new products that can improve productivity, facilitate ubiquitous communication and entertain.

Rapid consumer product lifecycles quickly drive premium pricing towards commodity levels with only the most creative products enjoying a relatively longer period of healthy margins at the OEM level.

Asia continues to enjoy a very high percentage of electronics manufacturing. Security concerns, rising transportation costs, and an increasing emphasis on sustainable business practices have recently had the effect of slowing down this movement in some segments.

REGULATORY:

Regulatory requirements continue to expand on a global basis both in terms of new directives as well as expansion of existing laws and guidelines. We are now seeing an explosion of new requirements from many regional, national, and local governments. Industry is struggling to keep up with the ever expanding portfolio of regulations.

• To meet regional legislative requirements, manufacturers must remove environmental "Materials of

Concern". The list of banned materials is expanding with no end in sight. Harmonization of global requirements is a major challenge.

• "High Reliability" product manufacturers are vulnerable as they are being pressured into using new materials for which reliability may not be well understood.

• Determination of carbon footprint is an expanding requirement on industry with significant challenges remaining to deploy consistent methodologies.

MARKET:

The boundaries among computers, communications and entertainment products have blurred. Flat panel displays are the norm for virtually all applications with touch screen becoming more prevalent in a number of product categories. Wireless products continue to proliferate, and this expansion is opening up new applications in a number of segments. Home and office functionality is being added to automotive products with growing concerns over driver distraction. The needs of the telecommunication and data communication infrastructures are converging. With the move to all digital communications and storage, we see the convergence of a number of markets:

• Medical-Consumer
• Automotive-Entertainment
• Communication-Entertainment
• Computing-Entertainment
• Computing-Security

TECHNOLOGY

Multi-core processors are now the norm for most computing applications. A consequence of the expected demise of the traditional scaling of semiconductors is the increased need for improved cooling and operating junction temperature reduction due to large leakage currents. The consumer's demand for thin multifunctional products has led to increased pressure on alternative high density packaging technologies. High-density three-dimensional (3D) packaging of complete functional blocks has become the major challenge in the industry.

• RF System-in-Package (SiP) applications have become the technology driver for small components, packaging, assembly processes, and high density substrates.

• The use of motion-gesture sensors in various consumer and portable devices will expand the MEMS gyroscope landscape (both 2D-axis and 3D-axis), and is expected to see an exponential growth.

• RFID finally will be replacing the barcodes and 2D identification and a number of OEMs will integrate to Information Technology (IT) systems for tracking Printed Circuit Boards (PCBs) and full system level products.

• Performance requirements such as increased bandwidth and lower power are driving the 3D ICs designed with through silicon vias.

Medical Product Sector

As the medical electronics market matures and develops, a wider group of companies has begun serving this growing market due to sustained growth, and higher margins than traditional electronics assembly markets. This growth has lead to the need for better understanding of the technical challenges associated with medical electronics manufacturing. This is reflected in the shift of focus within the electronics industry and within iNEMI to concentrate on the medical product sector. While this product sector encompasses traditional or widely known products such as; implantable medical devices, information technology used for patients' records, medical diagnostic tools, and monitoring devices, the medical electronics market is being currently fueled by an explosive growth in personal medical electronics. The shift towards home or patient centric health care has lead to a very rapid growth in personal healthcare monitoring, diagnostic and preventative medical electronics. Examples, such as glucose monitoring, pedometers, external portable defibrillators, thermometers, women's health products and others, are a few products readily available at a local pharmacy or department store. In addition, the drive towards patient records portability has spurred a growth in information technology in the medical sector as healthcare providers and insurance companies strive towards global accessibility of personal health records of patients.

Situation Analysis

Aging of the population is a world-wide phenomenon. The most dramatic growth is expected in many smaller and/or developing nations. For example, Latin America will exhibit a dramatic increase in the elderly, especially in countries such as Cuba, Puerto Rico, Chile, and Trinidad and Tobago. Over the next 3 decades, the aging index is expected to double or triple..

Market Size Prediction

In a similar fashion to the last roadmap cycle, iNEMI has asked Prismark to provide a range of direct support services related to the synthesis and production of its upcoming bi-annual roadmap.

• Prismark estimates that medical electronics equipment production totaled $76Bn in 2009, accounting for about 6% of the global electronics industry. This market is expected to continue to increase at an average rate of 5% per year to reach $103Bn in 2015. Other than military and aerospace electronics, medical electronics was the only sector not to experience a decline in 2009 compared to 2008, and actually grew by about 7% year-over-year.

Most medical electronics systems by value are produced in North America today. However, Asian countries, such as China and India, are the fastest growing markets for medical equipment, and leading medical electronics companies, such as Siemens, GE, and Philips, are increasing product design and assembly capabilities in these countries. One unique aspect of medical electronics produced in developing markets is that primarily, the products are being produced for local/regional use rather than for export. Some notable exceptions, however, are in Malaysia and Singapore, where contract assembly houses have sought and received certification to produce medical products. Most of those goods are external portable or sensory products (thermometers, blood pressure sensors and similar).

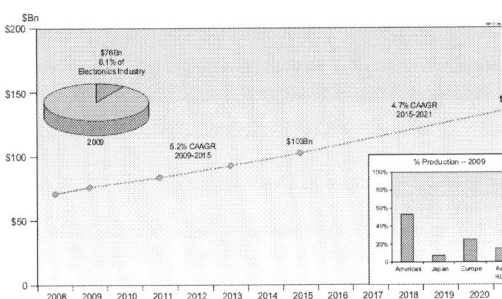

Figure 2: Prismark Partners estimate of medical market size.

Product Classification in Medical Sector

Equipment

- *Large Infrastructure Medical Equipment* includes products such as medical imaging systems (e.g., x-ray and MRI), IT equipment (e.g., picture archival communication systems PACs), and biochemical analysis equipment (e.g., lab instruments and DNA analyzers).

- *Small Stationary and Portable Medical Equipment* includes products such as patient monitoring systems that are used to measure and monitor patients' vital signs and other bodily functions. The segment also includes home diagnostics products such as blood pressure cuffs (including wireless), blood glucose meters, pulse oximeters, and biochemical analysis meters.).

- *Implantable Medical Equipment* includes major therapy devices such as pacemakers and implantable cardioverter defibrillators (ICDs), and the market is rapidly expanding beyond these systems. Devices such as neurostimulators (e.g., for Parkinson's disease) and drug pumps (e.g., for insulin release) are also being brought to market. Increased reliability, greater functionality, and miniaturization are the main technical drivers in this segment.

IT Sector

- Office Information Technology: The final category includes devices that store and manage patient data and move information for the medical community, as well as novel personal health devices that may be brought to the health care provider's office and health information is downloaded. These devices

may not have regulatory or form factor restrictions. The devices in this category include Picture Archival Communication Systems, (PACS), telemedicine devices, electronic health records carried by the patient, and computers in the medical offices, to name a few.

Critical Issues

There are three critical issues that impact the medical products group, that are unique to the sector, and are difficult to control, influence or affect. The first is the growing uncertainty of the Pb-free solder issues and long term availability of Sn-Pb components. The second and third are non technical issues, but are broadly impacting. They include the growing litigation in society that is expanding the overhead of regulatory organizations in the Americas, and the third is the high rate of cost escalation and hospital insolvency.

Automotive Product Sector

The main factor that distinguishes the Automotive Product Sector Emulator from the other iNEMI (International Electronics Manufacturing Initiative) Product Emulators is the environment in which the product must perform. The products must perform reliably in automobiles, light-duty, medium-duty, and heavy-duty trucks. Many of the attributes, such as cost, density, and components, overlap into the other emulators. Increasing density is important for these applications because of cost, size, and weight reductions. The assembly and manufacturing / test equipment requirements are also critical because of the reliability requirements.The challenge for the Automotive Product Sector Emulator is to adapt other emulators' technologies to meet the high temperature, environmental, and reliability requirements cost effectively.

Market Size Prediction

The automotive electronics industry is approximately 9% of global electronics production and was $114 billion in 2009. Growth is expected to increase about 8.9% per year through 2015. This growth rate is driven by a disastrous 2009 used as the base year, combined with an increase in average electronic module content per vehicle, and very strong unit car sales growth in Asia. In 2009, China surpassed the U.S. as the largest car market in the world.

Situation Analysis

The major trends driving the demand for increased electronics penetration in automobiles include:

- Stricter fuel economy and emissions mandates
- Legislated requirements for advanced safety systems, such as advanced airbags and on-board tire pressure monitoring
- Consumer demand for greater vehicle efficiencies driven by escalating global crude oil prices

- Consumer demand for greater safety, comfort, and convenience features
- Consumer demand for luxury features
- Growth of hybrid and electric vehicles

Given very rapid growth in local car production and the gradual shift to low cost production in Asia, Asia now accounts for the majority of production value of automotive electronics, with about 37% share in 2009. This region is followed by Europe with 31% production share, Japan with 17% share, and the Americas with 16% share.

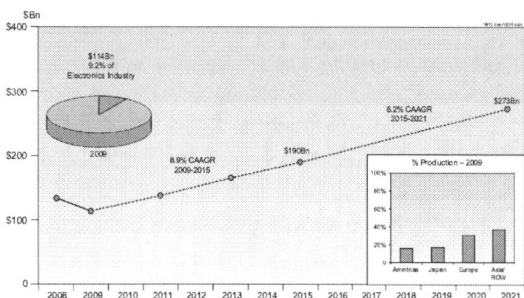

Figure 3: Prismark Partners' estimate of Automotive Electronics market size.

Embedded and non-embedded sensors are used in many automotive systems. Many of these sensors are MEMS devices. MEMS applications for the automotive sector have come a long way since the introduction of a micro-machined MAP sensor in 1979. An average car today consists of over 40 different sensors, ~30% of which are micro-machined. Application areas include engine management, passenger safety and comfort as well as environmental care. A new emerging application area is the office and entertainment sector, which will require devices such as integrated microphones and devices for optical signal handling. In addition, wireless systems such as direct tire pressure have been introduced in the last year.

MEM automotive application areas can be broadly classified as shown below,

- Integrated sensors for pressure (MAP, Fuel, Occupant Detection, Tire, Air Bags), acceleration, non contact temperature, airflow, fuel flow, angular rate (Electronic Stability Control, Roll Over) sensor.
- Gas/Chemical sensors for in-cabin air quality, monitoring exhaust gas composition and oil quality.
- Actuators/valves for fuel injection
- Optical/Infra-Red sensors for in-car LANs, HVAC control, Occupant Sensing, Night-Vision and in-vehicle displays.
- Polymer based sensors for humidity detection
- Radar based sensors for Back-up Aid, Blind Spot Detection, and Adaptive Cruise Control

Key Drivers: Cost, Size, Quality, Reliability

The key driver for the automotive sector is cost. All other drivers are assumed to be met and cost is often used to determine the company that wins..

Size is also a premium in the automotive environment. Space in the passenger compartment, engine compartment, on-engine, and on-transmission are limited. The supplier that offers a smaller box that is equal or better in cost has an advantage. Some higher end vehicles can have up to 100 electronic controllers in a vehicle.

Quality is a key metric that all automotive suppliers are measured on by using Assembly Plant Returns (APR). This metric is kept in parts per million (ppm). Assembly Plant Returns are returned to the supplier for root cause and corrective action.

Reliability is measured in warranty returns from the field. Warranty rates are calculated for 30, 60, 90, 180, and 360 days of exposure in the field. This metric is measured in Incidents per Thousand Vehicles (IPTV).

Critical (infrastructure) Issues

There are several paradigm shifts occurring now in the automotive sector. The first is the electrification of the vehicle.

The second paradigm shift is the connected vehicle. Internet access in the vehicle will grow to 62.3 million vehicles by 2016.

The third paradigm shift will be the advancement of safety systems. Systems like adaptive cruise control to maintain a safe distance. Systems that detect vehicles in blind spots, warn if a driver is falling asleep, help see at night and/or in foggy conditions. Systems that detect lane departures and systems like vehicle stability control that help the driver maintain his or her commanded path. Many of these systems already exist today. The key is to reduce their cost so that lower end vehicles can afford these systems.

MEMS

Similar to discrete sensor technologies, Micro-Electro-Mechanical Systems (MEMS) have an extremely diverse application set, ranging from physical to optical, chemical, and biological, as well as a diversity of materials and methods used to manufacture them. A first impression of MEMS would undoubtedly start with the theme of miniaturization for realizing ever-smaller sensors and actuators. However, where they truly stand apart is by the integration (also referred to as co-integration) of added functionalities: combining sensing and actuation operations with information processing, signal conditioning, built-in test, and communications. MEMS technology is a child or ancillary innovation of semiconductor electronics much as people have pointed toward the information technology being an ancillary innovation of semiconductor technology. However, MEMS diversity, including front end manufacturing techniques, back end manufacturing techniques, required testing procedures, diversified materials and the lack of a unit cell or standardized precursors such as the transistor and MOS or Bipolar technologies, make the MEMS manufacturing activity much more like a job shop than a High Volume Semiconductor facility.

MEMS are the second wave of micro manufacturing which emphasizes the mechanical nature of materials often used in semiconductor manufacturing. The three main manufacturing technologies for MEMS are: Bulk Micromachining, Sacrificial Surface Micromachining and HARM or High Aspect Ratio Micromachining. Both Bulk Micromachining and Sacrificial Surface Micromachining use materials similar in nature to semiconductor materials for the most part but HARM based products have exceptionally larger materials selection (including plastics).

Most of today's leading applications of MEMS are found in office automation (inkjet), firms that initiated their efforts in office automation and found other market avenues (TI's DMD) and the current application in the portable consumer electronics market, such as the mobile phone and electronic gaming systems, has resulted in new directions for growth in the industry. The recognition of the consumer's desire for MEMS manufacturers are responding to this demand by increasing functionality, adding more axes of inertial sensing, developing new devices such as micro speakers, and pico projectors, for example, and adding intelligence (information processing) and new interfaces (communications). This says nothing about the myriad of biomems devices both current and future that promise to reinvigorate the industry still again.

Market

The MEMS Market is diverse, with solutions for a wide variety of applications. The MEMS market is rebounding in 2010 after two years of recession in 2008 and 2009. The CAGR of the next 5 years for the MEMS industry is estimated to be close to 11% compared to 6% for the IC industry according to the IC insight MEMS 2010 report. Out of the 7B$ of 2010 MEMS market, more than half of it should come from consumer and automotive applications.

The Automotive industry sector is a very large market for MEMS and all cars on the market today are equipped with multiple sensors. Looking forward, more attention is being placed on new clusters of inertial sensors for electronic stability control (ESC) in an effort to avoid crashes. Automakers must include ESC as standard equipment on new U.S. cars by the 2012 model year.

Europe announced that it will also force ESC and Japan should shortly announce the same. ESC is making use of accelerometers and gyros.

Standardization

The global competition to manufacture high-volume innovative products with enhanced functions and performance enabled by MEMS/NEMS is intense because nations want to strengthen their economies and create new jobs for their citizens. International standards and their associated metrologies are significant enablers for success at all stages of MEMS/NEMS innovation - from research, development, initial deployment, high-volume commercialization, end of initial useful life, to recycling and disposal. Successful MEMS/NEMS innovation requires international standards based on the best of science and engineering.

The MEMS industry requires a high degree of customization in its manufacturing methods. Major product lines of the industry have traditionally been led by vertically integrated companies who operated in-house design, manufacturing, packaging, test, and distribution, and who could set their own internal standards. However, the recent growth in availability of MEMS foundries has occurred with a shift in the business model for new product development in order to lower costs. The "fabless" business model-approach includes developing relationships with companies who specialize in device and system design, foundries for manufacturing, packaging houses, testing, and distribution partners. This has renewed interest and discussion in the role for standards, especially the characterization and testing of materials and devices, to reach a common language for the buyer-seller interface in the marketplace.

The industry has adopted some existing standards from the semiconductor electronics industry. Beyond that, SEMI and ASTM have led the development of new standard protocols for MEMS device testing.

The MEMS Industry Group (MIG), NIST, SEMI, and IMEC, have partnered to organize a series of workshops to identify, prioritize, and roadmap standardization needs for the industry. This work is in progress, but there is consensus growing that material and device level test and calibration methods are an area of industrial need.

MEMS fabrication in general shares many commonalities with IC fabrication as both employ automated batch wafer processing in a semiconductor cleanroom environment. Accelerometer and gyroscope fabrication borrows many standard processing technologies and equipment for ICs, e.g., lithography, thin film deposition, and dry etching. However, it differs substantially from IC's when it comes to the formation and the subsequent release of free standing microstructures which are essential for inertial sensing. These microstructures are often quite large (100's of μm across to well over 1 mm) and relatively thin (from a few μm to 10's of μm). They are therefore very fragile and prone to mechanical damage prior to encapsulation. It often takes a tremendous time and effort to develop a viable MEMS fabrication platform.

Since most accelerometer and gyroscope die are encapsulated at the wafer level, they are not unlike regular IC die and for which standard assembly processes and plastic packages for ICs are usually applicable. What used to be "one product, one package" for MEMS is gradually standardizing into a few common QFN or LGA packages for accelerometers and gyroscopes.

A mechanical stimulus is typically required for accelerometer and gyroscope testing. This can get complicated and costly as the number of sensing axes and functionality increase. Many high end products, e.g., those for automotive applications, also require temperature testing. MEMS companies have initially come up with their own custom test solutions along with many innovative designs for the mechanical stimulus, mechanization, test fixture, and handler.

Unlike the IC model, there is currently no pure play MEMS foundry which offers a standardized, production proven, accelerometer or gyroscope process platform for open use. For the backend, there are well -established, semiconductor packaging houses with more than a decade of experience in the high volume assembly of MEMS devices. Package level testing, however, remains well guarded IP as it is very challenging to provide adequate test coverage at low cost for the increasingly sophisticated MEMS system-in-package products which are loaded with smart algorithms for the sensing of multiple physical parameters. As a result, Design for Testability has received utmost attention for these products. The cost of package level testing and the associated yield loss at this step can often be a significant cost differentiator among MEMS IDMs. The lack of test standardization or solution, however, remains a significant bottleneck for the MEMS industry.

Accelerometer and Gyroscope Future Needs

A future trend for consumer applications is toward sensor fusion, or the integration of multiple sensor components in the same package, e.g., an 11-DOF (Degree of Freedom) sensor system-in-package that combines a 3-axis accelerometer, 3-axis gyroscope, 3-axis magnetometer, altimeter, and temperature sensor, to perform sophisticated tasks of motion sensing and navigation (in conjunction with GPS).

Medical MEMS

MEMS being utilized in medical products are finding new applications everywhere in the health industry with rapid growth significantly past 2015.

The microsystem technologies market for healthcare applications will grow from $1.2 billion in 2009 to $4.5 billion in 2015, representing over 1 billion units per year in 2015. Meanwhile, wireless systems will exceed 50% market share..

MEMS enable dramatic new possibilities for detecting, analyzing, and manipulating biomaterials, from proteins to bacteria to blood.

MEMS combine silicon-based microelectronics with micromachining technology, making it possible to develop complete systems on a chip. Devices generally range from a millimeter down to 20 micrometers.

Lab on a chip and Related Microdevices:

- Biochemical assays
- Immunoassay: detect bacteria, viruses and cancers based on antigen-antibody reactions.
- Dielectrophoresis: detection of cancer cells and bacteria.

- Blood sample preparation: can crack cells to extract DNA.
- Cellular lab-on-a-chip for single-cell analysis.
- Ion channel screening (patch clamp)
- Testing the safety and efficacy of new drugs, as with lung on a chip

Some of the healthcare improvements promised to future generations are: new in-vitro diagnostic systems, new therapy strategies, genetic disease treatment, artificial pancreas, drug discovery processes, and others.

Regulation

The industry is in the early stages of discussions on the need to establish standard protocols and testing methodologies that can be used to help streamline the process for FDA approval of portable medical diagnosis and drug delivery systems. The types of testing required for these products must be finalized and agreed to by all participating companies. Class I, II, and III medical devices will need their own set of regulations and requirements.

PACKAGING

The pace of change in packaging technology today has accelerated to the highest rate in history. The penetration of electronics into virtually every segment of society is driving his demand. Communication, transportation, education, agriculture, entertainment, health care, environmental controls (heating and cooling), defense and research all rely heavily upon electronics today. This diversity of application and the never ending demand for both lower cost and higher performance cannot be achieved without major changes in architecture, materials and manufacturing processes. These new technologies include SiP, Wafer Level Packaging (WLP), wafer thinning and Through Silicon Vias (TSV) today. In the near future, we will see additional changes with the incorporation of nano-materials.

TECHNOLOGY CHALLENGES:

Reliability

Many factors determine the reliability of electronic components and systems. SiP products have higher thermal cycle count due to the use pattern of consumer electronics and greater mechanical stress due to vibrations and dropping for the same reason. The storage and use environments also have a wider range than many conventional electronic systems. Meeting the reliability requirements of future SiP components and systems will require tools and procedures that are not yet available.

Manufacturing Equipment and Processes

The shift in equipment for single chip packaging from dedicated hard tooling based equipment to more flexible manufacturing platforms is still in process. The drive for higher parallelism in package assembly, higher speed and other changes to reduce cost continues with increased importance in the current economic environment.

Packaging innovations such as WLP and SiP have specialized equipment requirements. New generations of equipment will be required for wafer level interconnect structures, specialized under bump metallurgy, TSV, and embedded wafer level structures.

Emerging SIP products require assembly equipment with greater versatility and precision. Assembly of SIP with a variety of IC types, optical devices, MEMS devices, and biochips on the same substrate will require substantial extension of current assembly equipment capability.

Materials

New requirements for improved reliability, performance, cost, and compatibility with new semiconductor materials are driving the development of a broad base of compliant materials for packaging applications.

The migration to "green" materials that are lead-free and halogen-free compatible are in full swing. Materials for the traditional wire bond and flip chip packages will have to be improved to meet lead-free, halogen-free, and low-κ /ultra low-κ requirement.

The development of new packaging technologies, such as SiP, WLP, embedded die and passives, and TSV, will call for innovations in design of materials and materials processing innovations beyond what is available today. WLP will require materials with improved or different properties as it evolves to meet new packaging applications. Different metallization systems for both redistribution traces and under bump metallization, as well as new dielectric polymers, are needed to meet the ever changing reliability requirements for portable electronic devices. The development of fanout WLP and embedded passives/actives will require new low-temperature embedding polymers and low-temperature cure redistribution layer polymers. TSVs will benefit from new dielectric insulators and conductive via filling media for improved low cost manufacturability. Integrated passive devices (IPDs) will also require better materials, with improved electrical properties, for both resistive and capacitive devices.

Warpage

The increasing package diameter, decreasing package thickness, smaller ball size required for the increasing pin count and higher reflow temperatures associated with lead free solder are resulting in warpage becoming the primary limiting factor for ball pitch and ball size in BGA packages.

Package Substrates

The package substrates are now the most expensive component in a typical package. Substrate cost will present an increasing challenge as the thermal density continues to increase and the pitch continues to decrease. The technology exists to meet the requirements but in many cases the cost of more advanced materials and finer pitch may be prohibitive.

Table 1: Package Substrate Parameters

Wafer Level Packaging

WLP is one of the most rapidly growing packaging technologies. Today WLP includes packages that are larger than the die with both fan-in and fan-out terminals. WLP enables decrease in size, increase in operating frequency and cost reduction compared to wire bonding or flip chip. WLPs with fan-in design today are typically for low I/O count and small die sizes. They are mainly being used in portable consumer markets where small size, thickness, and weight are an additional advantage to cost. WLP now incorporates many different structures to meet specific application targets.

Environmentally Conscious Electronics

Global environmental issues and the resulting stakeholder concerns are accelerating the demand for environmentally conscious electronics (ECE) products. Active public debate has been focusing on climate change, access to clean water, e-waste, hazardous materials, and conflict free materials sourcing. These issues are influencing policy makers, advocacy groups, and consumers to demand greater transparency from manufacturers on the environmental aspects of electronic products.

These concerns are emerging in new industry standards, legislation, and eco-labels that are expanding their scope to cover product environmental and sustainability performance throughout the entire product lifecycle. This includes extraction and manufacture of input raw materials, product manufacturing, transport and logistics, product use, and end of life management.

The Information and Communications Technology (ICT) industry has an unprecedented opportunity to enable new technologies to help solve these global environmental challenges. Part of the solution is to make electronics products more energy efficient and sustainable, but the key to the broader problem is to develop new technologies that can be deployed broadly across many sectors of the economy.

As in previous years, the 2011 ECE Roadmap is structured around five topic areas: Materials, Energy, Recycling, Eco-Design, & Sustainability

Key recommendations and issues raised in the five sections follow:

Materials:

The need for new and novel materials in emerging technologies runs counter to the growing global restrictions on materials use. An urgent need is a more consensus approach to technically sound assessment methodologies to quantify environmental impacts of materials and potential trade-offs of alternatives. A second urgent need is development of data structures and databases for the environmental aspects of materials and their alternatives.

Continued support is needed for solving remaining challenges for lead-free reliability and manufacturability issues, as well as the work on PVC alternatives and HFR – free leadership for printed circuit boards.

Industry must be more involved in policy making on material restrictions so that the regulatory community better understands the trade-offs inherent in material substitution.

Energy:

Harmonization of global energy efficiency standards and labeling remains a priority including closing gaps in metrics, measurement, and testing.

Support the iNEMI product carbon footprint position by promoting credible, workable methodologies and appropriate use of comparative data.

Support research and initiatives to increase the energy efficiency of electronic products, but more importantly, support development of new energy saving technologies that can improve efficiencies in many economic sectors.

Support energy efficiency and harmonized standards for data centers and cloud computing.

Recycling:

A patchwork approach to regulations, recycling methods, and product eco-design continues to hamper the development of a sustainable infrastructure with a viable recycled materials market for use in new products and other applications. Key recommendations include:

Establish new electronic applications for postconsumer blended plastics (e.g., housings for power supplies) and for recycled CRT leaded glass

Develop design guidelines for both product reuse and demanufacturing

Establish a forum to encourage communications with manufacturers, recyclers, NGOs, and policy makers to resolve barriers to recycling

Encourage efficiency, reduced costs and innovation in electronics recycling systems through competition and other market-based mechanisms

Eco-Design:

Provide more and earlier feedback to Eco-label organizations and regulators on new technologies, life-cycle thinking, scientific / test data and design cycles. A multi-stakeholder group that includes industry can

successfully develop accepted environmental standards based on harmonized principles

Continue supporting IEEE 1680 as a workable design standard, but support efforts to resolve the adversarial revision process.

Sustainability:

Electronics are a solution towards sustainability. While there is the responsibility to minimize the electronics industry's environmental impact from operations, products, and services, the compelling news is that the electronics industry has a particular opportunity to make society function more efficiently and easily and thereby mitigate society's impact on the environment. Solutions that enable other sectors to improve their environmental footprint make use of electronics and need to be implemented now to enable opportunities.

Brominated Flame Retardants (BFRs)

Taking their lead from both regulatory and NGO-driven market pressures, a number of consumer electronics OEMs have made public goals to limit the use of brominated flame retardants (BFRs) and polyvinyl chloride (PVC) from their products in 2009 and 2010. A few have gone even further, promising to remove all organic and inorganic halogens, a chemical group which includes fluorine, chlorine, bromine, iodine, and astatine. Despite the wide variety in goals, the term "halogen-free" is often used to refer to any type of halogen restrictions. However, there is much debate as to what level of reduction constitutes "halogen-free."

While there are over 75 BFRs, the ones most commonly used in electronics include decabromodiphenyl ether (deca-BDE) and tetrabromobisphenol A (TBBPA). Octa- and penta-BDEs are rarely (if ever) used in electronics. Deca-BDE is used in some television casings and TBBPA is widely used in circuit boards and other component casings. PVC is used in electronic cabling and plastic parts.

While many OEMs have already committed to removing BFRs and PVCs, discussion still surrounds exactly to what extent halogens need to be removed from electronics products. While the removal of BFRs and PVCs is the first of the environmental groups' goals, other flame retardant compounds have already been identified as future targets.

In the absence of other standards, many electronics manufacturers are applying the 900 ppm limits for chlorine and bromine in laminate materials to entire electronic products.

ROADMAP RESEARCH KEY TOPICS

DESIGN:

• Co-design of mechanical, thermal and electrical performance of the entire chip, package and associated heat removal structures.

• Simulation tools for nano devices and materials.

- Integrated design and simulation tools for RF modules and devices.

- Electronics-manufacturing simulation and modeling tools for the designer.

- Cost effective, improved thermal management.

- New capability to close the gap between chip and substrate interconnect density.

MANUFACTURING TECHNOLOGY:

- The development of new approaches to organic substrate fabrication that address needs for dramatic increases in density, reduced process variability, improved electrical performance and significant reductions in cost.

- Manufacturing processes for dealing with warpage and thin format products – Wafer, Package, PCB

- 3D Package Stacking Development - Assembly, Cooling, Reliability

- 3D TSV Commercialization by the development of the industry infrastructure as well as the supply chain.

- Low Temperature Assembly

MATERIALS DEVELOPMENT:

- A combination of materials and fabrication research to support the development of monolithically integrated optics and electronics.

- Low cost, higher thermal conductivity, packaging materials, such as adhesives, thermal pastes and thermal spreaders.

- Next generation of solder materials to replace the high cost/high temperature silver containing alloys.

- New interconnect technologies deploying nano-materials to support decreased pitch and increased interconnect frequencies.

- High-performance laminates that are competitively priced.

- • Reliability testing methodologies for new materials.

ENERGY AND THE ENVIRONMENT:

- Establish shared, peer reviewed, data bases for sustainability.

- Need to engage with policy makers and NGOs to establish life cycle basis for decisions.

- Need Materials Usage modeling to track material flows within and between nations.

- Support R&D to create a sustainable infrastructure and viable recycled materials market.

PARADIGM SHIFTS

Many of the Technology Working Groups identified paradigm shifts that are taking place now and potential paradigm shifts that might occur in the future. This information is critical for infrastructure providers to identify where non-linear changes may occur in the future. These changes provide both opportunities and risks for individual firms.

The need for continuous introduction of complex, multifunctional new products to address the converging markets (first identified in 2004) has continued to favor the development of functional, modular components or SiP (both 2-D and 3-D structures). This paradigm shift in the design approach increases the flexibility, shortens the product design cycle and places the test burden on the producers of the modules.

The standard platform movement that is developing in the telecom market is helping to address the disrupting and disaggregating of both the design chain and the supply chain. This movement could accelerate the introduction of new functions by making them easier to do. Telecom, computing, IT and military sectors are affected.

The board assembly roadmap is predicting another migration to lower temperature and lower cost lead-free solder materials in 2011-2017.

Other paradigm shifts include:

- Optical Interconnects at the backplane/board level will eventually be deployed – driven by increasing performance requirements and more cost effective solutions.

- Printed electronics moving into initial applications with many infrastructure challenges.

- Electronic component suppliers are looking to utilize embedded passive and active components, systems in package, systems on chip, or any other means to densely pack ICs with increased functionality.

- High density PCBs will use discrete devices only down to 0201 format.

- High Density Substrates will use embedded passives or 01005 discretes based on size constraints.

AKNOWLEGEMENT

The authors gratefully acknowledge the contributions of nearly 600 direct participants and hundreds more indirect participants in this cycle's roadmap. Many participants are members of the 13 collaborating organizations that contributed excerpts from their own roadmaps and helped to recruit participants from their membership.

Fabrication of a Novel Multi-Electrode Array (MEA) Biochip Using Polyester Insulated Electrodes with Microwell Features for Cardiomyocyte Analysis.

Olivia M Flaherty[1], Xiaoyun Cui[1], Divya Rajamohan[2], David Anderson[2], David Hutt[1], Chris Denning[2], Paul P Conway[1], Andrew A West[1].

[1]The Wolfson School of Mechanical and Manufacturing Engineering, Loughborough University, Loughborough, Leicestershire. LE11 3TU. UK.
[2]The Wolfson Centre for Stem Cells, Tissue Engineering & Modelling (STEM), Centre for Biomolecular Sciences, University of Nottingham, University Park, Nottinghamshire. NG7 2RD. UK.

O.M.Flaherty@lboro.ac.uk

Abstract

There is an increasing interest in Multi-Electrode Array (MEA) technology arising from a diversity of bio-scientific areas, which is especially strong in stem cell, drug discovery or safety pharmacology applications. There are limitations in the quantity of useful data captured during an MEA-based trial that can be attributed to reasons such as the time intensive nature of MEA application, the subsequent time-consuming and often complex data analysis, and the lack of standardisation between applications. MEAs that are more widely applicable to the diverse and demanding requirements of applications in research and industry are necessary to improve efficiencies and reduce costs. Existing electronics manufacturing approaches have been brought together here to realise a novel MEA biochip. Fabrication of this MEA biochip design that is suited to higher throughput screening of cardiomyocyte-based drug trials has been pursued. The MEA biochip has been successfully manufactured using approaches that are novel in this application domain. Results achieved have provided a device that has supported cell survival and attachment. Assessment of practical bioscientific protocols using this device compared to a standard model indicates that the design changes implemented in this setting could save users up to approximately £115 per experiment, in terms of labour cost savings alone.

Key words: *Cardiomyocyte, Melinex® Film, Excimer Laser, Micro-electrode array.*

Introduction

The development of extracellular electrophysiological recording devices, multi- (or micro-) electrode array (MEA) systems, is a rapidly evolving discipline across both research and industrial settings. The use of MEA technologies is not new; however in recent years as more interest has been focused towards the true scientific potential of these tools, new questions pertaining to how best to optimise the acquisition of the required data and how to best extract meaning from those data have been posed. This heightening level of interest comes from a diversity of bio-scientific areas but is especially strong in stem cell (SC) [1-3] and drug development and discovery [4-6] MEA applications. Recent evidence of this exploration and the subsequent development of new solutions can be reviewed in previous Proceedings of the MEA Meeting, held bi-annually in Reutlingen, Germany [7-9]. In essence, an MEA system is a tool for detecting and recording electrical signaling in living cells or tissue. The vast majority of electrogenic tissue type studies conducted using MEA technology utilise neural and cardiac tissues [6] but work pertaining to other electrogenic and non-electrogenic cell types has been attempted [10-13].

Limitations in the quantity of useful data that are captured during an MEA-based experiment can be attributed to a number of reasons: the time intensive nature of MEA system set-up and application [14], the consequent timely and complex data analysis [15] and, the lack of standardisation across disciplines [16] that has allowed for different approaches and protocols from one MEA user group to another. MEA systems are required that allow users to perform studies on larger scales [17], the result of which will be more data generation, that will thus require more advanced data processing, ideally within a shorter period of time.

As in-vitro recording using MEA systems continues to grow it has been widely recognised that the development of the MEA biochip has relied heavily upon the timely progression of suitable fabrication techniques [18]. Generally, MEAs are designed on a planar surface using conventional microelectronics or micro-electro-mechanical system (MEMS) technology; this involves thin film deposition and varying photolithography techniques [19].

Popular commercially available MEA biochips are typically manufactured using glass based substrates; these materials tend to be rigid and fragile. The microelectrodes embedded in the MEA base substrate are presently made from titanium nitride (TiN), platinum (Pt) or gold (Au), and are traditionally configured in an 8x8 grid array. One electrode is missing at each corner of the grid array meaning a total of 60 microelectrodes are present in a standard MEA biochip. The emergent standard geometry of the microelectrodes within the 8x8 arrangement is a diameter of 30µm spaced with an inter-electrode distance (pitch) of 200µm. Therefore the total area of the base substrate that is exploited as the MEA "workspace" is just 1.4mm^2. The workspace is insulated on standard MEA biochips by spin coating of a suitably biocompatible material. Commonly used insulation materials include silicon nitride and the epoxy-based photoresist SU-8, which are known to be chemically resistant and relatively impermeable to environmental moisture and ions [19, 20]. The microwell feature that is being incorporated into this biochip design does not require application of such an insulator as the electrodes, conductive channels and contact pads are manufactured on a different surface. This could potentially eliminate the current requirement to pre-treat the MEA workspace with a suitable attachment matrix prior to seeding the living cells.

The most commonly used polymers that could be applied to meet this design preference are polyester, polyimide, liquid crystal structures and polydimethylsiloxane. Polyester and polyimide in film form exhibit a combination of properties otherwise exhibited in a number of other polymer films either individually or in inferior combinations. Properties exhibited by these films include: low stress, low moisture uptake, high modulus and good ductility. These properties are ideally suited to biotechnology applications. A commercially available example of a polymer film suited specifically to cell growth is Melinex® by DuPont, which is a polyester film that has been treated specifically to enhance cell attachment [21]. The Excimer laser is a laser variant that can facilitate formation of micrometer scaled structures that is frequently used with these types of materials, where a high degree of design flexibility or complex patterning is required. The properties of the Excimer laser that make its use excellent in such circumstances are its high resolution, good precision and high coupling efficiency [22]. The authors of this paper identified the potential to use Melinex® film in unison with an Excimer laser to produce enhanced MEA biochip features.

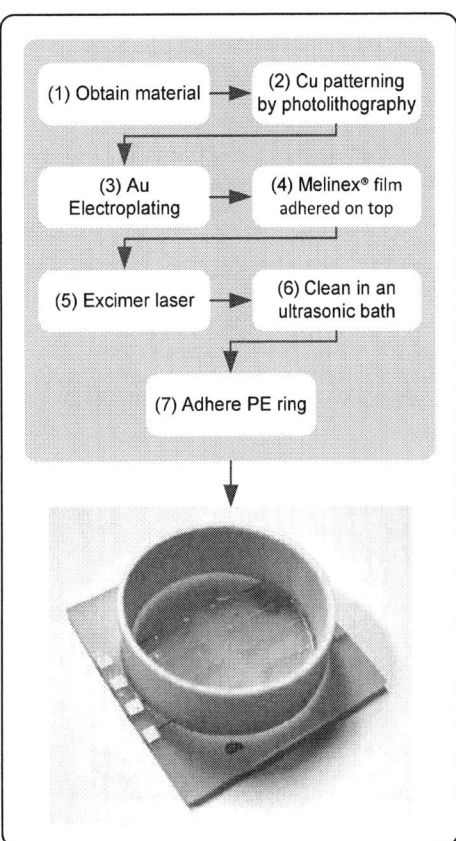

Figure 1: Manufacturing workflow schematic

The Manufacturing Workflow

Figure 1 demonstrates the novel MEA biochip manufacturing workflow. Standard FR4 substrate was used with a 35 micron copper foil coating (Mega Electronics Ltd, UK). A negative photoresist was applied over the surface using a laminator roller at 115°C. A pre-designed photomask defining electrode sites, conduction lines and contact pads was secured temporarily over the photoresist surface and the combined sheet exposed to UV light. The exposed areas of photoresist were then stripped away and the excess Cu etched away using a concentrated ferric chloride solution. Samples were then outsourced to an external company for Au coating (PMD Group, Coventry). The Au coating served to seal in the Cu, which is toxic to living cells, thus producing a cell-friendly contact surface on the 1mm diameter electrodes. The Au coating was approximately 1µm thick. The Au plated substrates were cleaned using an ultrasonic bath to remove any dust or debris that may have resulted from the manufacturing processes.

A cut-to-size disc of Melinex® film was adhered over the electrode sites in the same location as the future positioning of the media well. An Exitech Excimer laser was used to prepare

19

microwells directly over the electrodes with a diameter of 500μm. The Melinex® film was laser ablated until the underlying Au surface of the electrode was exposed; the depth of the microwells was therefore equal to the thickness of the Melinex® film, 125μm. Different size and shape microwells can be made by making various ablation masks, replacing the circular one used in this instance, and re-focusing the system. Finally, samples were cleaned again in an ultrasonic bath prior to the addition of a polyester (PE) ring, made from a cut-to-length standard industrially available PE tube. The PE ring was adhered directly onto the Melinex® surface using a non-toxic silicon-based adhesive, to serve as a media chamber around the cells when in situ.

Excimer Laser Drilling

The Excimer laser system utilised consisted of a short-pulse Lambda Physik Excimer laser source (KrF, 248nm) and an Aerotech positioning system. The laser application was tested using alternative substrates (i.e. glass and polyester coverslips) of equivalent thickness and the Melinex film demonstrated the best compatibility with the laser process.

Process Time

The process time required throughout the various manufacturing workflow stages was captured using internationally standardised CIMOSA (computer integrated manufacturing open system architecture) enterprise modelling techniques (ISO 19439:2006(E)). This modelling procedure was also engaged in to track consumables use and labour inputs throughout the product innovation process, allowing costs of manufacture to be monitored.

Figures 2 (A&B) present the critical stages of manufacture that differ from those used to produce commercially available MEA biochips; production of the MEA base substrate and production of the microwell features, where resource use and time in production were of specific interest to this work. This data was captured and iteratively reviewed as workflow amendments were made, to ensure every aspect of the manufacturing process was appropriately monitored and communicable.

Results

The following results were generated during the combined assessment of the novel manufacturing approach and the resulting microwell integrated MEA biochip. Assessment of material biocompatibility and the level of success in terms of the manufacturing combination with respect to in-use characteristics have been pursued.

The base substrate and microwell features were optically inspected for defects prior to use in live trials. A cross section micrograph was used to

A

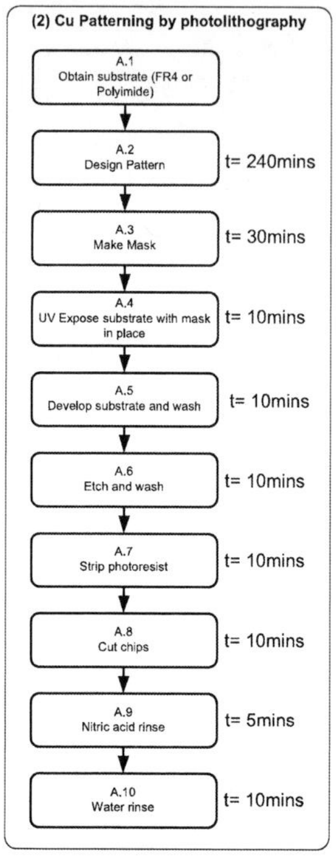

B

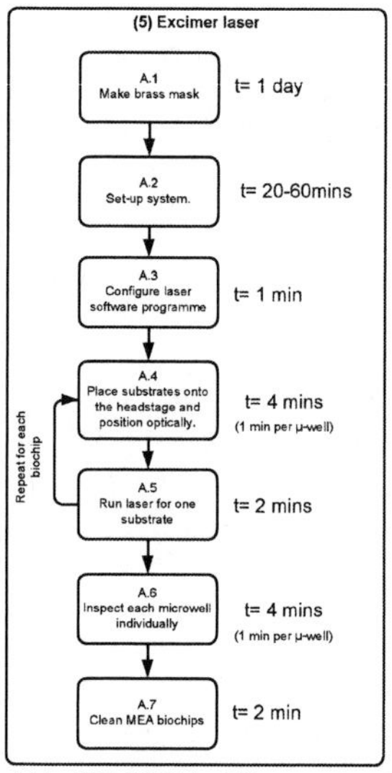

Figure 2: (A) CIMOSA (Computer Integrated Modeling Open System Architecture) activity model of workflow developed to produce base substrate and the associated labour time. (B) Activity model of the workflow and associated labour time for creation of microwell features at electrode sites.

view the microwell feature in closer detail. Figure 3 reveals that the Excimer laser accurately removes the Melinex® film to expose the underlying electrode surface. Conductivity measurements using a Hewlett Packard 4284A LCR multimeter also showed that the machined surface had good conductivity, implying that no insulator film remained at the bottom of the microwell.

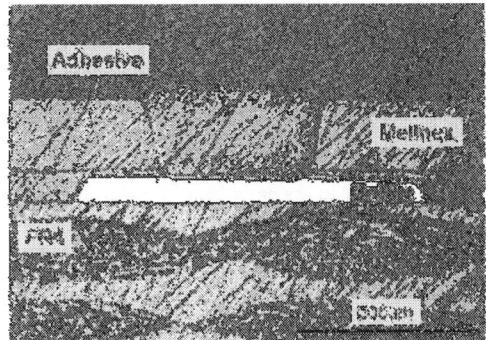

Figure 3: A cross-sectional micrograph image of the microwell feature.

The alignment of the microwell feature over the underlying electrode was variable across prototype samples manufactured. This was due to the relatively imprecise alignment of the samples on the alignment system's headstage; samples were positioned by hand in this instance. This is a manufacturing parameter that can be more reliably controlled in future prototyping to allow a greater degree of freedom in experimenting with microwell geometries, such as, for example, larger microwell diameters.

A

B

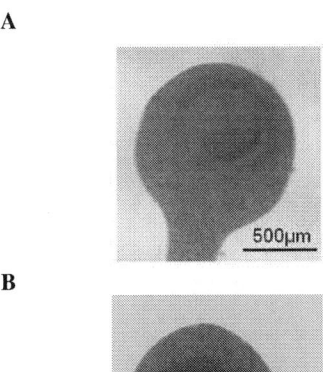

Figure 4: Optical micrographs of microwells under media conditions, demonstrating variation in position of microwells relative to the underlying electrode. A) Microwell containing a human Embryonic Stem Cell (hESC)-derived cardiomyocyte cluster with imprecise alignment. B) An empty microwell showing improved alignment.

Stem cell (SC) derived cardiomyocytes are still a relatively new cell source, and as such are under constant attention in terms of improving beating cluster reliability of production and control over inherent variability. Testing of this device was completed in batches due to constraints on the number of beating clusters that could be acquired for this work. Initial employment of this design was unsuccessful due to contamination. Manufacturing facilities used to produce the biochips were not in a clean room environment, so more careful attention was paid towards cleaning prior to hESC cardiomyocyte seeding; subsequent tests showed 100% cell survival.

Beating clusters seeded in MEA biochip prototypes showed varying successes in terms of cell attachment. This is thought to be due to both the movement of the MEA biochip from the culture hood to the incubator, the contraction movement of the cluster as it beats and the microwell geometry. A beater successfully attached inside a micro-well can be seen in Figure 4A. Other beating clusters that had been positioned directly over the microwells had moved and attached to positions in excess of 500µm from their initial position at the microwell centre. From these observations we can confirm the suitability of the materials selected in manufacturing and the approach used to realise the novel biochip design.

In addition to these observations it is noteworthy that responses from the end user's point of view were recorded, highlighting concerns pertaining specifically to the increased optical constraints of this design when compared to commercially available alternatives. No good signals from beating clusters have been recorded at the time of submission. A statistically significant number of tests have not been completed due to limitations in cell supply. The seeding of more beating clusters is required before results can be reported pertaining to the quality and reliability of the signal acquisition.

Quantitative information derived from deep level CIMOSA modeling of the bioscientific aspects (i.e. laboratory protocols) of MEA employment has also been completed. The level of detail in modelling across all project elements has made it possible to estimate where time savings are probable during biochip use in the target application domain. Data shown in Figure 5 indicates that the design changes implemented in this setting could save users approximately 8 hours of set-up and trial time. In this instance 8 channels (and therefore beating clusters) will be assessed in one test. If a research associate's hourly rate is broken down to approximately £14 per hour, an estimated saving of ~£115 can be made per experiment, when compared to if 8 commercially available standard biochips were employed sequentially, as is the currently the case in this application area.

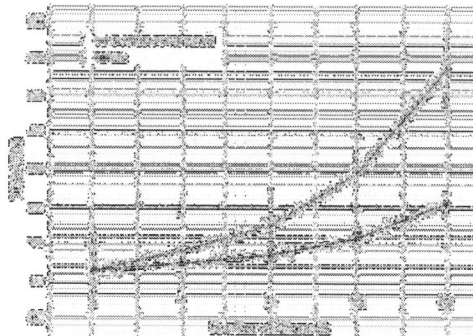

Figure 5: Projected labour saving for this design vs. commercially available standard, based on real-life timing data extracted and documented. LU = Loughborough University MEA biochip. Prototype biochips presented 8 electrode sites, providing an opportunity to capture the equivalent amount of data to 8 sequential trials performed using a standard commercially available biochip.

Conclusion

Biocompatibility of material choices and suitability of the manufacturing approach have been assured as a result of this preliminary work.

The manufacturing approach presented here has allowed a new MEA biochip design to be demonstrated that permits more recording sites per single media well. Consequently less time is spent on day-to-day culture maintenance, there is an increased ease of initial cell positioning due to the microwell features, and chemical treatments (such as drug candidates) can be administered to all of the cells in a trial at exactly the same time without the requirement of an expensive perfusion system. These advantages deliver increased convenience and consequential time and cost savings to user groups in this particular application.

Further attention is required with regard to the microwell dimensional optimization, optical inspection conveniences, and quality of the output signals.

Acknowledgements

This work was supported by the Engineering and Physical Sciences Research Council under grant awards EP/E002323/1 and EP/C534247/1. The authors would like to give special thanks to PMD Group, Coventry, UK for their expertise and efforts in the gold electroplating process.

References

[1] S. Cho and H. Thielecke, "Electrical characterisation of human mesenchymal stem cell growth on microelectrode", Microelectronic Engineering, Vol 85, pp1272-1274, 2008.

[2] M.M. Mahlstedt, D. Anderson, J.S. Sharp, R. McGilvray, M.D. Barbadillo Munoz, L.D. Buttery, M.R. Alexander, F.R.A.J. Rose and C. Denning, "Maintenance of Pluripotency in Human Embryonic Stem Cells Cultured on a Synthetic Substrate in Conditioned Medium", Biotechnology and Bioengineering, Vol 105, No.1, pp130-140, 2009.

[3] K. Banach, M.D. Halbach, P. Hu, J. Hescheler and U. Egert, "Development of electrical activity in cardiac myocyte aggregates derived from mouse embryonic stem cells." American J. of Physiology – Heart and Circulatory Physiology, Vol 284, H2114-H2123, 2003.

[4] C. Denning and D. Anderson, "Cardiomyocytes from human embryonic stem cells as predictors of cardiotoxicity", Drug Discovery Today, Vol 5, No. 4, pp223-232, 2008.

[5] S.E. Harding, N.N. Ali, M. Brito-Martins and J. Gorelik, "The human embryonic stem cell-derived cardiomyocyte as a pharmacological model", Pharmacology and Therapeutics, Vol 113, pp341-353, 2007.

[6] A. Stett, U. Egert, E. Guenther, F. Hofmann, T. Meyer, W. Nisch & H. Haemmerle, "Biological application of microelectrode arrays in drug discovery and basic research" Analytical & Bioanalytical Chemistry, Vol 377, pp.486-495, 2003.

[7] 5th Intl Meeting Substrate-Integrated Microelectrode Arrays, Germany, 2006. www.nmi.de/meameeting2008/MEA_2006.html

[8] 6th Intl Meeting on Substrate-Integrated Microelectrode Arrays, Germany, 2008 www.nmi.de/meameeting2008/index.php

[9] 7th Intl Meeting on Substrate-Integrated Microelectrode Arrays, Germany, 2010. www.nmi.de/meameeting2010/proceedings.php

[10] W.J. Lammers, A. al-Kais, S. Singh, K. Arafat, and T.Y. el-Sharkawy, "Multielectrode mapping of slow-wave activity in the isolated rabbit duodenum", Journal of Applied Physiology, Vol 74, No. 3, pp1454-1461, 1993.

[11] S. Nakayama, K. Shimono, H.-N. Liu, H. Jiko, N. Katayama, T. Tomita and K. Goto, "Pacemaker phase shift in the absence of neural activity in guinea-pig stomach: a microelectrode array study", J. of Physiology, Vol 576, No. 3, pp 727-738, 2006.

[12] C.-K. Yeung, J.K.-Y. Law, S.-W. Sam, S. Ingebrandt, H.-Y.A. Lau, J.A. Rudd and M. Chan, "The Use of Microelectrode Array (MEA) to Study Rat Peritoneal Mast Cell Activation", J. Pharmacological Science, Vol 107, pp201-212, 2008.

[13] Q. Liu, W. Ye, L. Xiao, L. Du, N. Hu and P. Wang, "Extracellular potentials recording intact olfactory epithelium by microelectrode array for a bioelectronic nose", Biosensors and Bioelectronics, Vol 25, pp2212-2217, 2010.

[14] G. Kang, J.-H. Lee, C.-S. Lee, and Y. Nam, "Agarose microwell based neuronal micro-circuit

arrays on microelectrode arrays for high throughput drug testing", Lab on a Chip, Vol 9, pp3236-3242, 2009.

[15] C. Bedard, H. Kroger, and A. Destexha, "Modeling Extracellular Field Potentials and the Frequency-Filtering Properties of Extracellular Space", Biophysical Journal, Vol 86, pp1829-1841, 2004.

[16] P. Watson, T. Jackson, G. Pitsilis, F. Gibson, J. Austin, M. Fletcher, B. Liang and P. Lord, "The CARMEN Neuroscience Server", Nottingham, UK, e-Science All Hands Meeting 2007, 1- 9- 2007.

[17] A.F.M. Johnstone, G.W. Gross, D.G. Weiss, O.H.-U. Schroeder, A. Gramowski, and T.J. Shafer, "Microelectrode arrays: A physiologically based neurotoxicity testing platform for the 21st Century", NeuroToxicity, Vol 31, pp331-350, 2010.

[18] M. Ferrari, ed., "BioMEMS and Biomedical Nanotechnology", Springer Science and Business Media LLC, New York, USA, 2006.

[19] S.A. Wilson, R.P.J. Jourdain, Q. Zhang, R.A. Dorey, C.R. Bowen, M. Willander, Q.U. Wahab, S.M. Al-hilli, O. Nur, E. Quandt, C. Johansson, E. Pagounis, M. Kohl, J. Matovic, B. Samel, W. van der Wijngaart, E.W.H. Jager, D. Carlsson, Z. Djinovic, M. Wegener, C. Moldovan, R. Iosub, E. Abad, M. Wendlandt, C. Rusu, and K. Persson, "New materials for micro-scale sensors and actuators: An engineering review", Materials Science and Engineering: R: Reports, Vol 56, No.1-6, pp1-129, 2007.

[20] L. Lorenzelli, B. Margesin, S. Martinoia, M.T. Tedesco, and M. Valle, "Bioelectrochemical signal monitoring of in-vitro cultured cells by means of an automated microsystem based on solid state sensor-array", Biosensors and Bioelectronics, Vol 18, No.5, pp621-626, 2003.

[21] T. Stieglitz, H. Beutel, R. Keller, M. Schuettler, and J.U. Meyer, "Flexible, polyimide-based neural interfaces", Proceedings of the Seventh International Conference on Microelectronics for Neural, Fuzzy and Bio-Inspired Systems, Micorneuro'99, pp112-119, 1999.

[22] H.G. Rubahn, ed., "Laser applications in surface science and technology", Wiley, New York, 1999

Innovative Manufacturing and 3-Dimensional Packaging Methods of Micro Ultrasonic Transducers for Medical Applications

Jack Hoyd-Gigg Ng[1], Robert T.Ssekitoleko[2], David Flynn[1], Robert W. Kay[1], Christine Démoré[3], Sandy Cochran[3] and Marc P.Y. Desmulliez[1]

[1]MIcroSystems Engineering Centre (MISEC), School of Engineering & Physical Sciences, Heriot-Watt University, Edinburgh EH14 4AS, Scotland, UK

[2]Department of Bioengineering, University of Strathclyde, Glasgow, G4 0NW, Scotland, UK

[3]Institute for Medical Science and Technology (IMSaT), University of Dundee, Dundee, DD2 1FD, Scotland, UK

+44 (0)131 451 3783, j.h.ng@hw.ac.uk

Abstract

This article presents the innovative manufacturing and 3-dimensional packaging of a miniaturized ultrasonic transducer for medical applications. A spirally rolled-up flexible circuit was employed for the interconnection of a linear transducer array to be used in a needle size footprint. Photolithography, precision dicing and low temperature conductive bonding are amongst the technologies under investigation for the realization of this high frequency ultrasound device.

Key words: Linear array transducer, high frequency ultrasound, 3D packaging.

Introduction

Operating frequency above 30MHz is required to achieve less than 100μm feature resolution for medical ultrasound high quality imaging. Such high frequency transducers have applications not only in diagnostic imaging but also in guiding surgical procedures. This article presents the development of a sensitive high frequency transducer with fine-scale element dimensions, housed in a package of an intervention tool such as a biopsy needle for inserting through the human skin to allow for easier *in vivo* pathology. Microfabrication of such devices are challenging due to the constraints implied by the construction of the ultrasound device. These constraints include the maximum processing temperature tolerated by the piezoelectric ceramic material, the acoustic impedance of the materials employed and the design of the package which might affect the ultrasonic beam forming and focusing.

For example, a linear transducer array operating at 30MHz, requires element pitch (electrode + gap) equal to the imaging wavelength, 50μm, thereby requiring definition of element electrodes with a width of 25μm or less. A high number of elements, for instance 128, are desired for image scans over a wide area. Fine pitch bonding and high density interconnection normally used in the microelectronics industry is required to manufacture the next generation of ultrasound devices.

Research in incorporating ultrasound transducers at the tip of surgery guidance needles [1] and catheters [2] has been carried out for some time. Single element transducers have a fixed focus and a limited image depth of field; beamforming for a movable focus and an enlarged depth of field can be achieved electronically by arrays of transducers [3]. Efforts in constructing arrays transducers at the tip of a catheter of 3 mm inner diameter have been carried out recently [4]. These types of arrays transducers require high density and small pitch interconnections. Furthermore, a new piezoelectric ceramic single crystal material, lead magnesium niobate – lead titanate (PMN-PT), is becoming more popular for higher sensitivity imaging owing to its properties of high piezoelectric coefficient and electromechanical coupling factor [5-6]. PMN-PT has a maximum processing temperature of 60°C without depoling which is important for its piezoelectric properties. This adds another obstacle to choosing the manufacturing and packaging technologies currently in use in industries including wire bonding and flip chip bonding which usually require temperatures of 120°C and above. Thus, the design of a manufacturing process incorporating fabrication technologies from different fields has been investigated that capitalizes on our previous work on SU-8/TiO₂ composite, flexible circuits, powder blasting, and precision dicing [7].

Reduction in the number of process steps and of the processing time has been taken into careful considerations in this new manufacturing and packaging design. The reduction of the footprint devoted to interconnects is enabled by a fan-out distribution layer crearted on a flexible susbstrate which is spirally rolled up within the needle. This technique replaces the conventional lateral stacking approach found in silicon IC chip packaging where Through Silicon Vias (TSVs) are created, filled and aligned for each layer in the stack [put a reference here about stacking of die]. The rolled-up flexible interconnection also relieves the ultra high aspect ratio of 67:1 that would be required if the connections were made in a planar manner along a 10 cm length needle. The device structure and the construction steps of the arrays transducer are first described in this article. The fabrication issues of this process are then discussed. A working prototype following the proposed construction process flow is presented in the final section.

Device Structure

The element electrodes are patterned on the surface of the active piezoelectric layer of the device, which comprises 55μm pillars of piezoelectric ceramic, lead zirconate titanate (PZT-5H), with a 90μm pitch and embedded in an epoxy matrix, forming a piezocomposite. PZT-5H was used instead of PMN-PT for the prototype due to cost considerations whilst new bonding methods for reducing the processing temperatures are still being investigated. The piezocomposite layer is sandwiched between an acoustic matching layer and an acoustic absorbing backing layer as shown in Figure 1. The matching layer allows the efficient acoustic energy transfer into the tissue. The backing layer is a dense, acoustically lossy material behind the piezoelectric layer that provides damping for the piezoelectric resonator and allows a short pulse to be transmitted forward from the array. It also prevents reflection from the bottom of the transducer as well as providing structural support. This is especially beneficial for providing resistance to mechanical and thermal shocks when high power ultrasound is used for therapeutic applications. The thickness of these acoustic layers and the element pitch size for 15, 25 and 50MHz operating frequencies considered for prototypes are listed in Table 1. These are based on PMN-PT active layer and a matching layer of 15% alumina filled epoxy by volume.

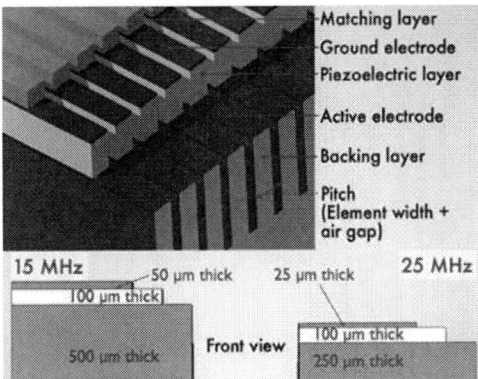

Figure 1: Structures of the 15MHz and 25MHz linear array transducers. The top figure shows a 3D view of the arrangement of the layers. The bottom figure shows a front view illustrating the thickness of the layers.

Table 1: Layer thickness and pitch size for the linear array transducer at 15, 25 and 50MHz frequencies.

Layer	15MHz	25MHz	50MHz
Piezocomposite (μm)	100	80	40
Matching (μm)	50	27	14
Backing (μm)	500	250	200
Pitch (element width + gap in μm)	100	60	30

In the current design, the matching layer, piezocomposite layer and backing layer adopt a staggered configuration as shown in Figure 1. The active electrodes are exposed on the top face of the backing layer and the ground electrodes are exposed on the top face of the piezocomposite layer. A fan-out flexible circuit is then attached to the active electrodes which, in turn, can be spirally rolled up along the long axis of the needle as shown in Figure 2. By using this configuration, only one or two strips of flexible circuit strips are needed for for connecting up to 128 elements in the linear array. If two strips are needed, they will be rolled up side by side forming an interlaced strand.

Since it is important to have a high level of control of beam quality and focus for the precise volume of tissue to be ablated or imaged, mechanical and electrical cross talks must be minimized. The air gaps in between the elements as shown in Figure 1 inhibit any undesired waves propagation in the piezocomposite layer generated from electric field spreading from individual element electrodes. Electrical cross talk can be further minimized by introducing an unconnected metal track in between each active connection in the high density fan-out flexible circuit strips.

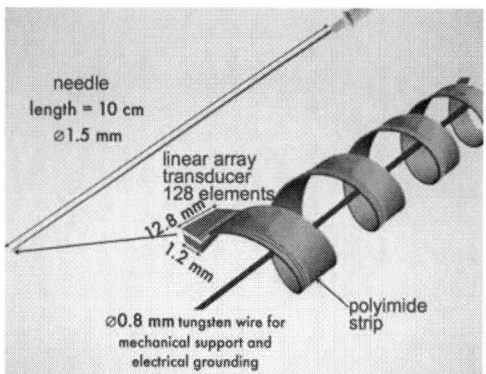

Figure 2: Schematic diagram illustrating the interconnection between the transducer at the tip to the other end of a needle by a spirally rolled-up polyimide based flexible circuit.

Fabrication Methods and Prototype

A. The layers in the transducer

A layer of 1-3 composite was manufactured by the common "dice and fill" technique whereby a Loadpoint Microace mechanical dicing saw is used to cut kerfs into a piece of bulk piezoceramic, the kerfs being subsequently backfilled with epoxy [8]. A slow spindle speed of 10,000 rpm and a low feed rate < 0.2 mm/s were used to prevent damage whilst dicing the fine pillars. Flatness in each of the matching layer, the piezoelectric layer and the backing layer is very important in preventing interference with wave propagation. A Logitech lapping and polishing machine was used to lap after each time epoxy was poured and cured. Figure 3 shows the piezocomposite layer after lapping.

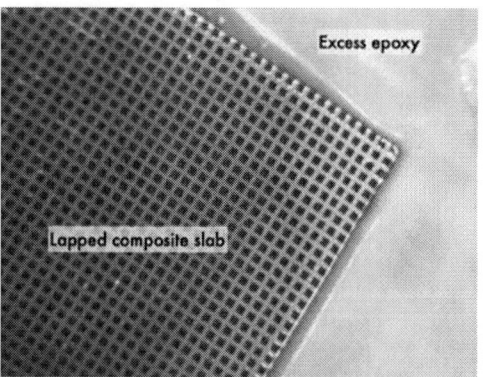

Figure 3: Image of a lapped composite slab consisting of diced ceramic pillars of 50 x 50μm and a kerf of 30μm which has been filled with epoxy.

For the active electrodes, a titanium layer was deposited onto the piezocomposite layer by physical vapour deposition as a first trial. A problem found was that the piezoelectric material became depoled by the temperature rise during the metal vapour evaporation. It was then decided that the electrodes would be formed on the backing layer instead as

shown in Figure 1. However this means that the backing layer would need to be bonded to the piezocomposite layer by another interlayer material.

Epoxy dispersed with alumina particles was casted on top of the piezocomposite layer to form the matching layer. Silver paint was used to form ground electrodes in between the matching layer and the piezocomposite layer. After lapping, the sample was released from the carrier and flipped over for casting the backing layer using epoxy dispersed with tungsten particles.

By making use of an interlayer bonding material, it was decided that the matching layer and the backing layer are to be cured and lapped on individual carriers separate from the piezocomposite layer. This allows higher curing temperatures that shorten curing time as well as saving time in flipping over the sample and preparing the layers in a serial manner. However, releasing the thin matching layer from a carrier requires delicate care.

The elements were defined by dicing after all the layers and the flexible circuits had been assembled.

B. Flexible circuits

Since the polyimide flexible circuit is directly bonded to the linear array elements, the line width and spacing of the copper tracks on the polyimide strip is a limiting factor for the pitch size possible on the transducer. Photolithography using the Ordyl dry film photoresist was applied to a copper clad polyimide laminate. The copper thickness was 18μm and the polyimide was 25μm. These thicknesses give fairly good integrity for rolling up the circuit. Cupric chloride was used as the etchant which provides a fast etch rate. Using this setup, moderate feature size as shown in Figure 4 can be produced reliably without defects. The produced line width and gap are 114μm and 157μm, respectively, from an original dimension of both 125μm at the design stage. These deviations can be attributed to the curing chemistry of the dry film photoresist which generally has a lower feature resolution than liquid photoresist, the development chemistry, the 40μm thick dry film, the thickness of the copper and the etch rate as well as additives and other conditions in the etching chemistry. Further experiments using the current setup indicate that a line width of 60 μm and a gap size of 65 μm are achievable.

Conventional wet etching of the copper reaches its limits when the line width and gap dimensions are reduced to 50μm in the design of a 15MHz transducer. , Figure 5 shows that some copper could not be etched in the narrow gaps due to insufficient transport of liquid etchant into the gaps. The copper edges of the traces also appear irregular. Other etching options for example oxygen plasma dry etching can be used to overcome the diffusion limit of wet etching.

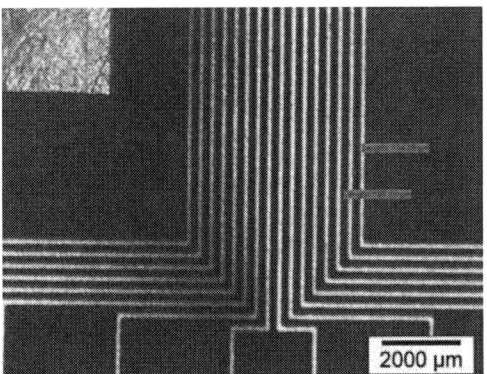

Figure 4: Microscope image of a flexible circuit fabricated by photolithography.

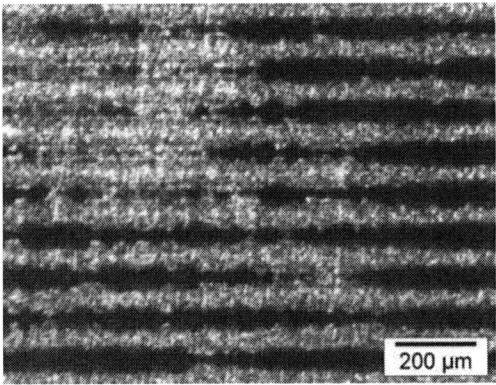

Figure 5: Microscope image of a flexible circuit with 50μm line width and gap afer copper etching.

C. Bonding

As already mentioned, an interlayer conductive bonding material is to be used to assemble the matching and the backing layers to the piezocomposite layer. Isotropic conductive adhesive (ICA) commonly used as a solder replacement can be blanket coated and partially cured on the matching and the backing layers individually first. Then the two layers can be bonded to sandwich the piezocomposite layer by a low temperature technique in order to avoid depoling of the piezoelectric material.

Anisotropic conductive film (ACF) has been considered for the bonding of the flexible circuit onto the backing layer with active electrodes exposed by the staircase arrangement as shown in Figure 1. ACFs achieve electrical conduction in the z-direction by deforming the metal particles dispersed in the film if the appropriate temperature and pressure are used. It is important that a minimum contact pitch area is provided to allow for sufficient amount of metal particles to be in contact within the contact area. Testing of a fine pitch type ACF for bonding is underway.

Alternatively, since dicing is carried out through the whole transducer assembly, which includes the flexible circuit, to define the elements by create air gaps, ICA could be used for the bonding.

D. First prototype

A prototype has been manufactured as presented in Figure 6. Excellent alignment was achieved between the copper tracks on the flexible circuit and the dicing blade for defining the elements as shown in Figure 7. Impedance analysis measurement reveal that the assembled transducer operates at 5.8 MHz as shown in Figure 8.

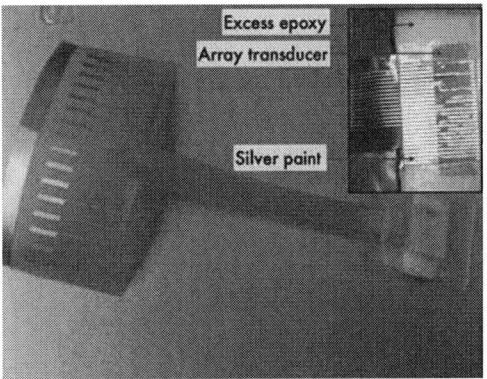

Figure 6: Photograph of the first prototype assembled. Inset shows the connection between the flexible circuit and the linear array transducer.

Further iterations of prototypes with higher element count, reduced pitch size (and therefore higher operating frequency) and extended fabrication details together with the results from the transducers in-water testing will be presented at the conference.

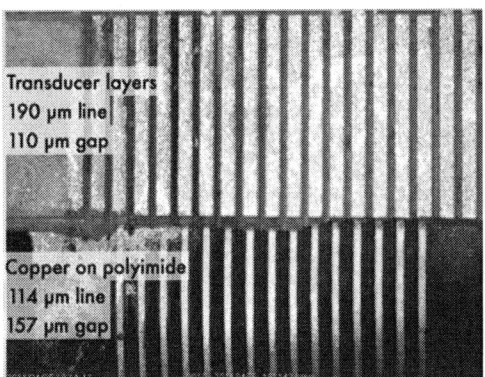

Figure 7: Image of the diced linear array aligned to the copper tracks on the flexible circuit.

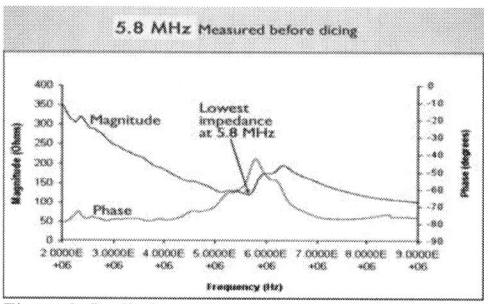

Figure 8: Preliminary impedance measurements.

Conclusion

A new approach in realizing a miniaturized linear array transducer has been presented which partially meets the challenges of interconnection in a very narrow and ultra high aspect ratio footprint with the potential of achieving high frequencies. The limiting factors for manufacturing small pitch arrays identified through the process flow are: (i) photolithography on flexible substrates, (ii) contact area allowed for bonding between the flexible circuits and the active electrodes, and (iii) rupture of the fine width element during dicing. In addition, the proposed processing steps are up-scalable and suitable for wafer-level manufacturing.

Acknowledgements

The authors of this article would like the financial support of the British Engineering and Physical Sciences Research Council (EPSRC) through the Innovative Research Manufacturing Research Centre (IMRC) programme. In particular the authors acknowledge the support of the ISPUD project from the James Watt Institute for High-Value Manufacturing, Edinburgh, UK.

References

[1] Q. Zhou *et al.*, "Design and fabrication of PZN-7%PT single crystal high frequency angled needle ultrasound transducers," IEEE Trans Ultrasonics, Ferroelectrics and Frequency Control, 55, pp. 1394-1399, 2008.

[2] M. Makuuchi, G. Torzill, and J. Machi, "History of intraoperative ultrasound," Ultrasound in Medicine and Biology, 24, pp. 1229-1242, 1998.

[3] J. M. Cannata, J. A. Williams, Q. Zhou, T. A. Ritter, and K. K. Shung, "Development of a 35-MHz piezo-composite ultrasound array for medical imaging," IEEE Trans Ultrasonics, Ferroelectrics and Frequency Control, 53, pp. 224-236, 2006.

[4] D. Stephens *et al.*, "Multifunctional catheters combining intracardiac ultrasound imaging and electrophysiology sensing," IEEE Trans Ultrasonics, Ferroelectrics and Frequency Control, 55, pp. 1570-1581, 2008.

[5] K. C. Cheng, H. L. W. Chan, C. L. Choy, Q. Yin, H. Luo, and Z. Yin, "Single crystal PMN-0.33 PT/epoxy 1-3 composites for ultrasonic transducer applications," IEEE Trans Ultrasonics, Ferroelectrics and Frequency Control, 50, pp. 1177-1183, 2003.

[6] X. Jiang, K. Snook, A. Cheng, W. S. Hackenberger, and X. Geng, "Micromachined PMN-PT single crystal composite transducers -- 15-75 MHz PC-MUT," IEEE International Ultrasonics Symposium, 2008.

[7] J.H.-G. Ng *et al.*, "Progress towards the development of novel fabrication and assembly methods for the next generation of ultrasonic transducers," Electronics System Integration Technologies Conference (ESTC), Berlin, September 13-16, pp. 179, 2010.

[8] K. A. K. H. P. Savakas, And R. E. Newnham, "PZT-epoxy piezoelectric transducers: A simplified fabrication procedure," Materials Research Bulletin, 16, pp. 677-680, 1981.

Evaluation of Cofired Platinum /Alumina High Density Feedthrough for Implantable Neurostimulator Applications

Ali Karbasi, W. Kinzy Jones

Department of Mechanical and Material Engineering, Florida International University, Miami, Fl, USA 33199

+1(305) 348-2345, jones@fiu.edu

Abstract

This work evaluated materials and process interactions in the development of a high density platinum / alumina feedthrough. Effect of Pt particle size, firing rates, firing temperature on the sintering, densification and secondary reactions are evaluated, as well as the effect of alumina additive to minimize the difference in expansion coefficient during cooling.

Key words: High density feedthrough, Biocompatible, Implantable neurostimulator, Hermetic device

Introduction

There is demand for increased input/output hermetic feedthroughs for application in implantable biomedical devices, especially for use in neurostimulating prosthetic devices such as, cochlear implants, muscular stimulators and retinal prosthesis. The hermetic feedthrough provides electrical signal transmission through the hermetic enclosure to appropriate electrode structures. Major classes of low I/O feedthroughs, such as cardiac pacemakers, where only 2-4 I/Os are required, are alumina ceramics brazed to a platinum wire pin. This structure has a long and high reliability history in implantable devices, due to biocompatibility of both platimun and alumina.

Strong bonding between metal and ceramic is the most important factor to develop a hermetic device. These bonds isolate electronic components and hermetically seal them when implanted in the body. The bonding mechanism in semiconductor type oxide (i.e. TiO_2, CeO_2, etc) and most of the metals is chemical in nature, especially group *VIII* elements which include the Platinum metal group [1]. However, in case of insulators oxide, like Al_2O_3 and SiO_2, both theoretical and experimental results demonstrate that the electronic interactions between metals and oxide surfaces in low temperature are dominated by either interfacial bonding or metal polarization effects [2]. Pt/Al_2O_3 bonds are particularly interesting because of the proven biocompatibility of the both materials [3] and their strong bonding [4]. The long-range interaction, such as space charge transfer, which was observed at Pt/TiO_2 interfaces, has been rarely observed, and strong bonding between platinum and alumina are believed to be physical in nature [4-6]. However, Several studies [1, 7-9] of platinum supported on alumina have reported changes in the catalytic performance in high temperature application, which is proposed to be caused by the interaction between a metal and an oxide; support oxide can alter the stability and morphology of the metal particles and promote compound formation in the Pt/Al_2O_3 system at high temperature, impacting the interaction range [5].

In addition to bonding, thermal expansion and shrinkage mismatch between materials must be tailored to insure the hermeticity of the device. Experimental results of the thermal expansion and shrinkage of the platinum [10] and alumina [11] indicate a large difference in their thermal expansion coefficient. Considering cofiring of the metal/ceramic multilayer structure, mismatched sintering kinetics between the metal and ceramics, in addition to the high value of thermal expansion difference on cooling, could generate undesirable defects including delamination, cracks, and camber in the final products [12, 13].

Physical properties such as melting point of nano particles are expected to be different than the bulk melting point because of the high ratio of surface area to volume for nano particles and the difference in the electronic structure of the nano particles [14, 15]. The general equation that determine the depression in melting point of nano sized particles are

$$\frac{T_m(r)}{T_m(\infty)} = 1 - \frac{4}{\rho_s L}\left\{\gamma_s - \gamma_l\left(\frac{\rho_s}{\rho_l}\right)^{2/3}\right\}\frac{1}{d}$$

where $T_m(r)$ and $T_m(\infty)$ are the melting points of particle of radius r and the bulk material respectively, ρ_s and ρ_l are the density of solid and liquid respectively, and γ_s and γ_l are the surface free energy of the solid and liquid. A 300 °C reduction in melting point for 5 nm gold particles has been observed.

Heating rate and firing atmosphere are very important to control the densification behavior and the interface properties of platinum and alumina. In the current paper, the effect of platinum particle size on the metallization behavior is studied as the first step. Then it focuses on the study of optimized firing condition (i.e. rate and atmosphere) to develop a hermetic biocompatible feedthrough. In addition to that, the effect of shrinkage controlling agent on densification of platinum is also studied to develop suitable conductive ink.

Experiments

Conventional high temperature cofired ceramics (HTCC) processing was used to develop the feedthrough [16]. 96% ceramic alumina tape (Maryland Ceramic & Steatite Co.) and three different platinum powder morphologies-nano powder Pt black, a spray dried and sintered Pt black platinum (Heraeus Inc. PM-110-10) and a spherical Pt (4 μm) from Alfa Asar were used as the primary materials. The inks were mixed on a Hoover Color Muller parallel plate mixer. The ceramic tape was punched with a Keko PAM mechanical punch machine and punched vias were filled with platinum ink by the means of a PTC 1000 vacuum assisted bladder filler. Filled tapes were stacked, isostatically laminated and fired up to the temperature range of 1350-1550 °C in different atmosphere and with different heating rate.

Shrinkage of platinum powder and ceramic alumina tape was measured by the Horizontal optical dilatometer Misura® ODLT. TA Q600 SDT machine was used to measure the colorimetric behavior of the platinum in high temperature. Phenom tabletop SEM was used to take a backscatter SEM image of the samples and JEOL JSM-6330F FE/SEM was used to capture the secondary electron image of the samples. The SEM images were analyzed by the use of ImageJ software to calculate the densification behavior of materials.

Results and Discussion

Initial Feedthrough Development

A cofired Pt feedthrough metallization was formulated for cofiring with a 96% alumina green tape to make the alumina substrate. Nanoscale Pt black was mixed with a micron size Pt (and alumina powder to control thermal expansion matching to the alumina). The nanoscale Pt melts below 1550°C, the firing point for the 96% alumina, greater than 200 °C below the melting point of Pt (1772 °C), providing liquid phase sintering of the micron size Pt. Figure 1 is a fracture microphotograph of a 250 μm Pt via structure showing excellent densification and adhesion in the cofired alumina structure. A 61 I/O 10 mil diameter Pt via feedthrough structure was fabricated and tested for hermeticity, exceeding 1×10^{-7} cc-He/sec.

Work then began to reduce the diameter of the feedthrough to 100 microns to increase the I/O count to greater than 200, at which point new phenomenon were observed due to the higher surface area to volume of the via. The platinum metallization exhibited complete melting in air firing, causing the added alumina for expansion control to float to the surface as a slag.

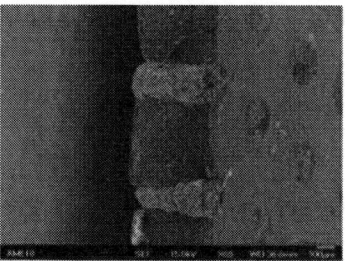

Figure 1- Hermetic cofired Pt Feedthrough

Effect of Platinum Particle size

Mixtures of the three types of platinum were evaluated to determine if the small diameter via would melt. In all cases, either combining the nano platinum with the spray dried pre-sintered nano Pt or the 4 μm spherical Pt particle or only the 4 μm spherical Pt particles, which should show no melting from particle size effects, melting occurred the results are showed in figure 2.

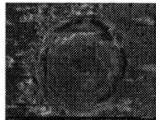  

Macro+ Nano Particle   Nano + Spherical Particle   Spherical Particle

Figure 2- Reaction kinetics of air firing different Platinum particle size powders

Effect of Heating Rate

Figure 3 showed the densification behavior of nano platinum and alumina tape with temperature with the heating rate of 5°C/min up to the temperature 1500 °C.

As it can be seen in about 1500 C the density of nano platinum increased about 35%, while alumina's was increased only about 17%. The high difference in densification and sintering kinetics of platinum and alumina could create defect like cracks and camber in the final multilayer assembly. However, densification behavior of the platinum and ceramic could be controlled by the heating rate. Basically higher densification can achieve by lowering the heating rate [17].

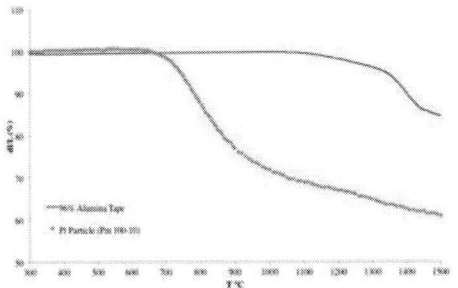

Figure 3. The nonisothermal densification behavior of platinum powder and alumina tape.

SEM images of feedthrough assembly fired in four different heating rates are showed in figure 4 and a summary of their densification results are also showed in table 1. Each measurement in the table 1 is the average of the 5 – 10 platinum via in one sample.

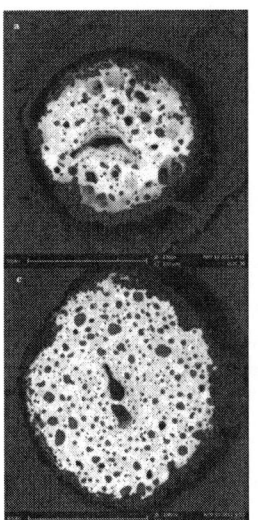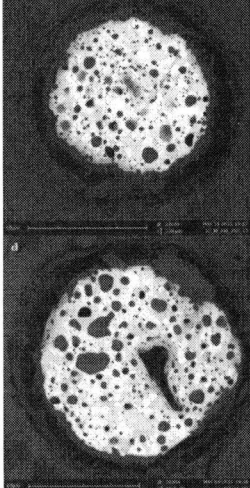

Figure 4. Backscatter SEM image of the platinum via fired in different heating rate. a) 0.2 °C/min, b) 1 °C/min, c) 5 °C/min and d) 20 °C/min.

As it can be seen in the pictures all of the four samples showed delamination of metal from ceramic and camber generation. Still, the results in table 1 showed that the minimum difference between platinum and alumina archived with the heating rate around 5 C/min.

Table 1. Densification behavior of platinum and alumina in different heating rate.

Heating rate (°C/min)	Platinum (µm)	Alumina (µm)	Difference (µm)
0.2	64.068	86.891	22.822
1	66.287	87.707	21.42
5	68.36	82.37	14.01
20	69.775	88.043	18.268

Diameter of platinum via metallization is decreased by decreasing the heating, while we cannot see such behavior for alumina. Decreasing the heating rate increases the shrinkage of the alumina. On the other hand the interaction of platinum and alumina can increase the adhesion force between them. The strong metal support interaction that was observed several times in the platinum supported catalyst, could describe this interaction. Hwang et. al. [18] and Luo et. al. [19] showed that platinum aluminate (i.e. $Pt_xAl_yO_z$) structure could be formed in the process of sintering Pt/Alumina mixture; in addition to that low heating rate can increase the diffusion of platinum to alumina and alumina to platinum and increase their physical adhesion. In result of such interaction, platinum tries to pull in alumina with itself. However the interaction force is not enough, which makes ceramic to tear apart in the interface area as it can be seen in the figure 4.

It should be noted, that the hole which can be seen in almost all of the samples could be attributed to the formation of volatile platinum oxide which is usually formed in temperatures above 650 C [20].

Effect of Atmosphere

Atmosphere condition is the most important factor to control the interface of platinum and alumina. In addition to that replacing air with neutral atmosphere, such as argon, could eliminate evaporation of platinum. SEM image of sample fired in the air and Ar atmosphere with heating rate of 5 C/min are showed in figure 5.

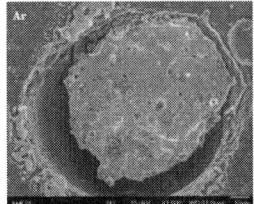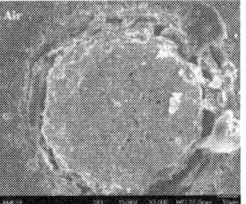

Figure 5. Secondary electron SEM image of feedthrough fired in air and argon atmosphere.

As it can be seen that platinum densified in Ar more than air. This behavior of platinum could be attributed to the decomposition of platinum dioxide and desorption of oxygen from the bulk of platinum. This oxygen layer is adsorbed on the layer of platinum below the surface and it is almost inert in temperature below 800 C [21]. The results of DSC experiments of sintering of platinum are shown in figure 6. The colorimetric experiment followed the same condition was used in firing of feedthrough. As it can be seen, the exothermic reaction of decomposition occurred around 1000 – 1200 C. Heat generation in neutral atmosphere is more than air, which could result in local super heating in sample and increase the shrinkage of the platinum.

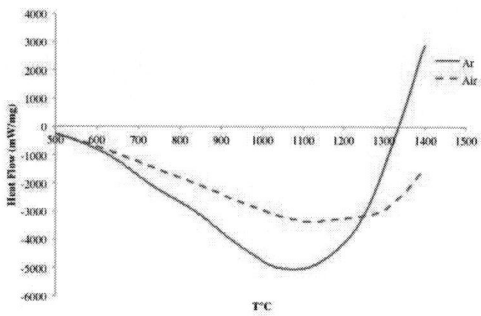

Figure 6. DSC analyses of the platinum powder in air and argon.

Reaction of Pt/Al$_2$O$_3$ could be enhanced in reduced atmosphere [6,22]. Hydrogen can reduce alumina and Pt$_3$Al is the major product of this reaction. However, pure reduced atmosphere can only worsen the shrinkage situation by increasing the reduction reaction of platinum and increase the shrinkage of platinum [23]

Shrinkage control agent

Desired shrinkage and thermal expansion of the conductive ink could also be achieved by optimizing the formulation of the ink for via filling. Low thermal expansion materials, such metal oxides (e.g. alumina) can be used to both control the sintering of platinum particles and adjust the overall expansion coefficient of the composite metallization to match that of the alumina. Result of application of 20% volume faction of Al$_2$O$_3$ in the platinum ink is shown in figure 7. All the samples are fired with the 5 C/min heating rate

Metal oxides are very stable compounds with low thermal expansion, which can reduce the amount of shrinkage of platinum. However as it can be seen in the figure 7, their low density causes separation from the body of platinum ink in high temperature and segregation to the surface of the assembly, especially with melting of the platinum metallization.

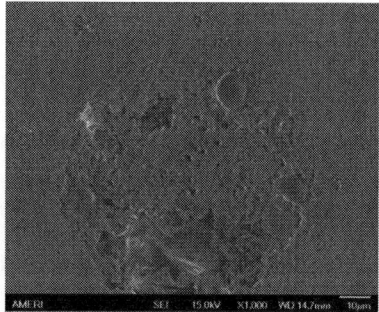

Figure 7. Platinum via with, 20% Al$_2$O$_3$

Conclusion

The results of firing of feedthrough assembly in different condition showed that air is not appropriate environment for firing platinum/ alumina combination. Mixture of hydrogen and argon mixture is the best possible environment for firing the feedthrough. In addition to that heating rate is also important in this process. Best results archived by heating rate about 5-10 C/min. Introducing tungsten to the formulation of the platinum ink also could improve the sintering kinetics of platinum and improve the properties of feedthrough assembly.

Acknowledgements

The authors acknowledge Expert System Solutions and Haiku Tech for measuring the densification behavior with optical dilatometry experiments and the support of the Advanced Materials Engineering Research Institute (AMERI) at FIU for SEM and DSC analysis. Funding is acknowledged from MIT Grant 800000195.

References

1. Tauster SJ, Fung SC, Baker RTK, Horsley JA. Strong interactions in supported-metal catalysts. Science. 1981;211(4487):1121.
2. Fu Q, Wagner T. Interaction of nanostructured metal overlayers with oxide surfaces. Surface Science Reports. 2007;62(11):431-98.
3. Merrill DR, Bikson M, Jefferys JGR. Electrical stimulation of excitable tissue: design of efficacious and safe protocols. Journal of Neuroscience Methods. 2005;141(2):171-98.
4. Allen RV, Borbidge WE. Solid state metal-ceramic bonding of platinum to alumina. Journal of Materials Science 1983;18:2835-43.
5. Panfilov P, Bochegov A, Yermakov A. The Transition Layer in Platinum-Alumina. Platinum Metals Review. 2004;48(2):47-55.
6. Suppel KP, Forrester JS, Suaning GJ, Kisi EH. A Study of the Platinum/Alumina Interface. Advances in Science and Technology. 2006;45:1417-22.
7. Ruckenstein E, Hu XD. Mechanism of redispersion of supported metal catalysts in oxidative atmospheres. Langmuir. 1985;1(6):756-60.
8. Andersson S, Brühwiler PA, Sandell A, Frank M, Libuda J, Giertz A, et al. Metal-oxide interaction for metal clusters on a metal-supported thin alumina film. Surface Science. 1999;442(1):L964-L70.
9. Linsmeier C, Taglauer E. Strong metal-support interactions on rhodium model catalysts. Applied Catalysis A: General. 2010;Article in Press.
10. Arblaster JW. Crystallographic properties of platinum. Platinum Metals Review. 1997;41:12-20.

11.	Huntz AM, MarÈchal L, Lesage B, Molins R. Thermal expansion coefficient of alumina films developed by oxidation of a FeCrAl alloy determined by a deflection technique. Applied Surface Science. 2006;252(22):7781-7.

12.	Jean J, Chang C. Camber development during cofiring Ag-based low-dielectric-constant ceramic package. Journal of Materials Research. 1997;12(10):2743-50.

13.	Chang C-R, Jean J-H. Effects of Silver-Paste Formulation on Camber Development during the Cofiring of a Silver-Based, Low-Temperature-Cofired Ceramic Package. Journal of the American Ceramic Society. 1998;81(11):2805-14.

14. Buffat, P. and Borel, J.P. ," Size effect on melting temperature of gold particles", Physical Review A, Vol. 13, No. 6, June 1976.

15. Cortie, M.B, " The Weird World of Nanoscale Gold", Gold Bulletin 2004 • 37/1–2.

16.	Gangqiang W, Folk EC, Barlow F, Elshabini A. Fabrication of microvias for multilayer LTCC substrates. Electronics Packaging Manufacturing, IEEE Transactions on. 2006;29(1):32-41.

17.	Bernard-Granger G, Guizard C, Addad A. Sintering of an ultra pure α-alumina powder: I. Densification, grain growth and sintering path. Journal of Materials Science. 2007;42(15):6316-24.

18.	Hwang CP, Yeh CT. Platinum-oxide species formed by oxidation of platinum crystallites supported on alumina. Journal of Molecular Catalysis A: Chemical. 1996;112(2):295-302.

19.	Luo M-F, Ten M-H, Wang C-C, Lin W-R, Ho C-Y, Chang B-W, et al. Temperature-Dependent Oxidation of Pt Nanoclusters on a Thin Film of Al2O3 on NiAl(100). The Journal of Physical Chemistry C. 2009;113(28):12419-26.

20.	Jehn H. High temperature behaviour of platinum group metals in oxidizing atmospheres. Journal of the Less Common Metals. 1984;100:321-39.

21.	Gland JL, Sexton BA, Fisher GB. Oxygen interactions with the Pt(111) surface. Surface Science. 1980;95(2-3):587-602.

22.	Klomp J. Ceramic-metal reactions and their effect on the interface microstructure. Ceramic microstructures '86: Role of interfaces; Proceedings of the International Materials Symposium; 28-31 July 1986; Berkeley, CA; UNITED STATES1987. p. 307-17.

23.	Hassan SA. Kinetics of sintering of unsupported platinum catalyst in nitrogen, oxygen and hydrogen atmospheres. Journal of Applied Chemistry and Biotechnology. 1974;24(9):497-503.

Co-Firing of LTCC Modules with Embedded Ferrite Layers

H. Naghib-zadeh [1], T. Rabe [1], J. Toepfer [2], R. Karmazin [3]

1. Federal Institute for Material Research and Testing, Berlin, Germany,

2. University of Applied Science Jena, Germany,

3. Siemens AG, Munich, Germany

+493081044442, Hamid.naghib-zadeh@bam.de

Abstract

Further miniaturization of electronic packaging calls for integration of magnetic functional components into LTCC modules. For integration of magnetic function into LTCC, low fired MnZn- and NiCuZn-ferrites which can be fully densified at the standard LTCC sintering temperature of 900 °C were developed. To co-fire these ferrite tapes with dielectric tapes the sintering shrinkage and the coefficient of thermal expansion of ferrite and dielectric tapes must be matched. For each ferrite material a new LTCC dielectric material was designed. The embedded ferrite tapes into new LTCC dielectric tapes can be sufficiently densified during co-firing at 900 °C without any cracking. Compared to separately sintered ferrites the permeability of embedded ferrite tapes is reduced. For embedded NiCuZn ferrites permeabilities between 230 and 570 (at 2 MHz) according to the thickness of the embedded ferrite layer were measured. For embedded MnZn ferrites a permeability of 300 was measured.

Key words: LTCC, co-firing, ferrite

Introduction

LTCC (Low temperature co-fired ceramics) are preferably used as 3D wiring circuit board in microelectronics where harsh environmental conditions and high reliability are demanded [1]. Realization of highly integrated and multi-functional LTCC modules calls for incorporating passive components such as resistors, capacitors, inductors, and other functional parts in LTCC multilayer.

Specially, integration of ferrite tapes into LTCC dielectric tapes enables many new applications due to effective integration of magnetic function into LTCC-modules, e.g. for transducers, antennas or circulators.

Standard sintering temperatures of LTCC are below 900 °C which allows the use of low resistivity and low cost silver conductors. However, ferrite materials require sintering temperatures higher than 900 °C. Hence, for co-firing of ferrite tapes into LTCC dielectric tapes, low fired ferrite materials are required. In this work, we used newly developed low fired MnZn- and NiCuZn-ferrites [2, 3, 4].

To reach a defect-free co-firing of sandwiched LTCC laminates consisting of ferrite tapes and standard LTCC tapes the shrinkage behaviour and the coefficient of thermal expansion of both components must be matched [5]. Furthermore the diffusion processes at the interfaces have to be controlled to have a good bonding without a large extent of infiltration [1].

In this work, we exhibit successful manufacturing of sandwiched LTCC laminates consisting of ferrite tapes and adapted LTCC dielectric tapes. In addition we discuss microstructure and magnetic properties of embedded ferrites.

Experimental

Two low fired NiCuZn and MnZn-ferrite tapes were used for co-firing with LTCC dielectric tapes. NiCuZn-ferrite contains 0.5 wt% Bi_2O_3 as sintering aid and was developed by University of Applied Science Jena [2, 3]. MnZn-ferrite contains a few volume % Bi_2O_3-B_2O_3-SiO_2-ZnO-glass as sintering aid and was developed by Siemens AG in Munich [4]. Both ferrite materials can be fully densified at 900 °C.

Two new tailored dielectric tapes BAM[474] and BAM[562] were used for co-firing with ferrite tapes. The dielectric tapes are typical glass ceramic composites consisting 70 vol. % glass filled with corundum and quartz.

Lamination was performed at 75°C and 25 MPa for 10 minutes using isostatic press (KECO Equipment Ltd., Slovenia). Debindering of laminates was carried out at 500 °C in air. Sintering of laminates containing NiCuZn-ferrites was carried out in air at

900 °C for 2 hours, while for sintering of laminates containing MnZn-ferrites nitrogen atmosphere must be used with dwell time of 8 hours at 900 °C.

The coefficient of thermal expansion (CTE) and the shrinkage behaviours of laminates were measured using a dilatometer DIL802 (Bähr GmbH, Hüllhorst, Germany). The heating rates for dilatometric measurements were 5 K/min. Bulk density and open porosity were determined by 'Archimedes' method. XRD measurements were processed with PW 1710 powder diffractometer (Philips, Eindhoven, Netherlands). The microstructure of the samples was characterized by scanning electron microscopy (SEM, ZEISS Gemini Supra 40, Oberkochen, Germany).

The complex initial permeability was determined in tape ring cores by impedance analysis (HP 4291A, Santa Clara, CA, USA) between 1 MHz and 1.8 GHz.

Results and Discussion

Main properties of sintered monolithic laminates of ferrite and dielectric materials are summarized in Table 1. All monolithic laminates were found to reach complete densification after sintering at 900 °C. Low fired ferrite tapes exhibit remarkably high permeabilities despite addition of sintering aids.

Table 1: Main properties of sintered materials.

Material	Sintering density [g/cm^3]	ε_r at 100 MHz	μ_i at 2 MHz
BAM[474]	2.7	5.5	
BAM[562]	2.8	5.9	
NiCuZn-ferrite	5.2		800
MnZn-ferrite	5.5		600

Co-firing of NiCuZn-ferrite in LTCC

Figure 1 shows the linear shrinkage of the NiCuZn-ferrite compared to BAM[474] dielectric tapes. It is obvious that there is a good adaptation of

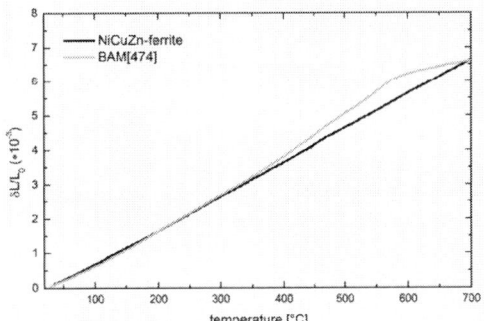

Figure 2: CTE curves of NiCuZn-ferrite and BAM[474] dielectric material

shrinkage curves of both materials. However, residual differences between the shrinkage curves can induce constrained sintering which cause shrinkage differences in lateral and vertical direction. As shown in figure 2, the CTE curves of ferrite and dielectric materials are very similar and this is an important requirement for a successful co-firing, i.e. the cooling process without cracks.

Co-firing of NiCuZn-ferrite tapes with BAM[474] dielectric tapes was reported in previous paper [1]. Defect-free sandwiched laminates could be manufactured. No open porosity could be measured after co-firing at 900 °C for 2 h.

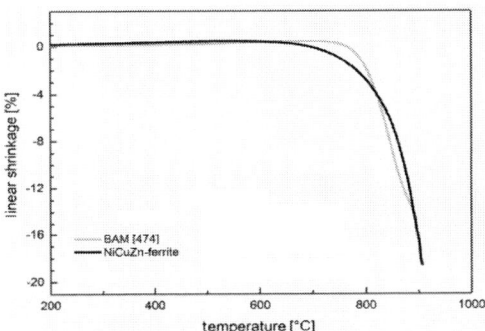

Figure 1: Shrinkage curves of NiCuZn-ferrite and BAM[474] dielectric material

Figure 3: SEM micrograph of fracture surface of monolithic NiCuZn-ferrite (top) and NiCuZn-ferrite co-sintered between BAM[474] dielectric tapes (bottom)

It was shown that growth of ferrite grains is hindered near the interface with dielectric tapes while monolithic ferrite exhibited a coarse-grained microstructure with grain size up to 60 μm in the whole volume of the sample (see figure 3). These changes in microstructure probably caused the reduction of the permeability for embedded ferrite.

In this study we changed the thickness of embedded ferrite layer and investigated the impact of these changes on the ferrit microstructure and permeability. Sandwiched laminates with ferrite thicknesses of 370, 550, 940 and 1540 μm and BAM[474] dielectric layer on top and bottom of laminates were manufactured.

No cracks or delamination were found in the cross section of the sandwiched laminates as shown in figure 4. Figure 5 demonstrates SEM micrographs of the fracture surface of sandwiched laminates. The sample with 1540 μm thick ferrite exhibits close-packed large ferrite grains in the middle of the ferrite layer and a fine-grained microstructure with average grain size of 1 μm near the interface with BAM[474] (see figure 6).

The large grains in the middle of ferrite layer are no more close-packed for 940, 550 and 370 μm thick embedded ferrites. Furthermore, with decreasing thickness of the embedded ferrite, the volume fraction of large ferrite grains decreases.

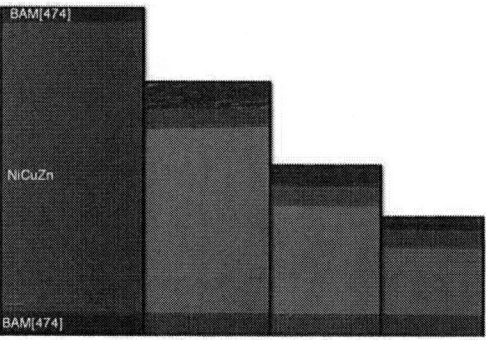

Figure 4: SEM micrograph of the sandwiched laminate NiCuZn/BAM[474] with different thicknesses of ferrite layer.

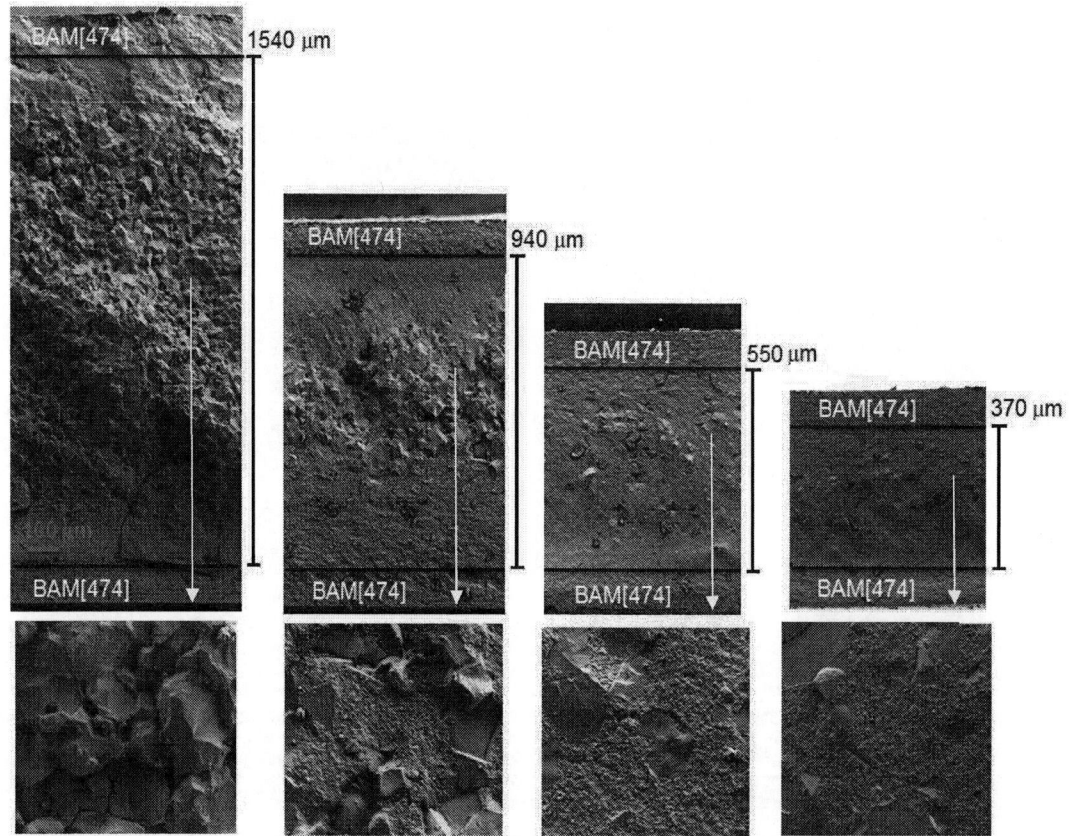

Figure 5: SEM micrograph of fracture surface of the sandwiched laminate NiCuZn/BAM[474] with different thicknesses of ferrite layer (top) and the microstructure in the middle of the ferrite layer (bottom).

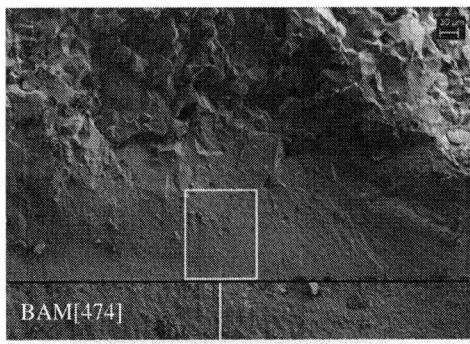

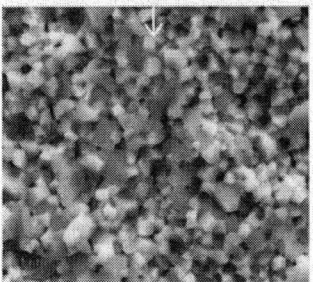

Figure 6: SEM micrograph of fracture surface of the sandwiched laminate NiCuZn/BAM[474] with 1540 μm thick ferrite layer (top) and the fine-grained microstructure of ferrite near the interface with BAM[474] (bottom).

The permeability results of embedded NiCuZn-ferrite with different thickness compared to monolithic ferrite are shown in figure 7. As expected, the measured permeability for embedded ferrite is lower than that for monolithic ferrite which amounts to about 800. According to the thickness of the embedded ferrite, permeabilities between 230 and 570 were measured. With decreasing thickness of the embedded ferrite, the permeability decreases. One reason for this reduction of permeability is the decrement of volume fraction of large ferrite grains with decreasing ferrite thickness [1, 6].

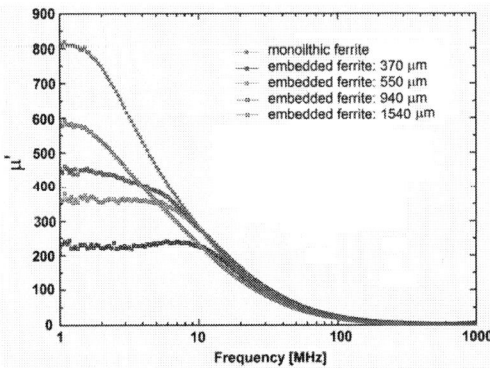

Figure 7: Permeability of monolithic NiCuZn-ferrite and embedded NiCuZn-ferrites with different thicknesses.

On the other hand, there is a big difference between the permeability of embedded ferrites with thickness of 370 μm and 550 μm, although both samples exhibit similar microstructure as shown in figure 5. It is assumed that the residual stresses caused by constrained sintering, also has an influence on the permeability [4], which needs to be confirmed in future investigations.

Co-firing of MnZn-ferrite in LTCC

Linear sintering shrinkage of low fired MnZn-ferrite is shown in figure 8. More than half of the total linear shrinkage of MnZn-ferrite proceeds MnZn-ferrite with dielectric tapes a good adaptation of sintering shrinkage of both materials is required. However, processing of a dielectric material with such a high sintering shrinkage during dwell time as MnZn-ferrite is very challenging because of high volume fraction of glass frit in standard dielectric materials. By using a suitable glass and filler components in BAM[562] dielectric material a high sintering shrinkage of about 4 % during dwell time could be achieved.

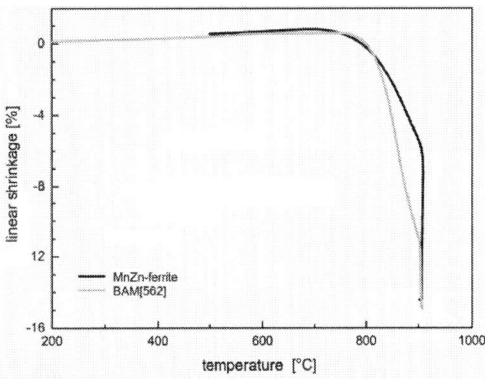

Figure 8: Shrinkage curves of MnZn-ferrite and BAM[562] dielectric material

The CTE curves of MnZn-ferrite and BAM[562] dielectric materials are shown in figure 9. The CTE curve of BAM[562] is similar to that of MnZn-ferrite which prevents cracking of sandwiched laminate of these materials during cooling phase of co-firing.

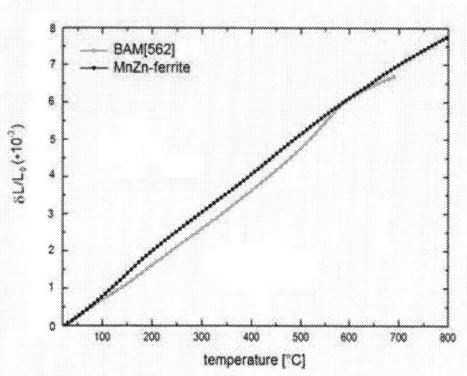

Figure 9: CTE curves of MnZn-ferrite and BAM[562] dielectric material

Cross section of MnZn-ferrite co-fired with BAM[474] dielectric tapes on top and bottom is shown in figure 10. No delamination or cracks are visible. The measured open porosity was less than 0.5 %. The microstructure of embedded MnZn-ferrite in figure 11a shows a two phase microstructure consisting of a dark grey MnZn-rich phase into which a few light grey Fe-rich grains are embedded [4]. The volume porosity of embedded ferrite is around 2 %. The volume fraction of light grey Fe-rich grains is much higher in embedded MnZn-ferrite compared to monolithic MnZn-ferrite as shown in figure 11. The Bi_2O_3–B_2O_3–SiO_2–ZnO glass used as sintering aid segregates at the grain boundaries.

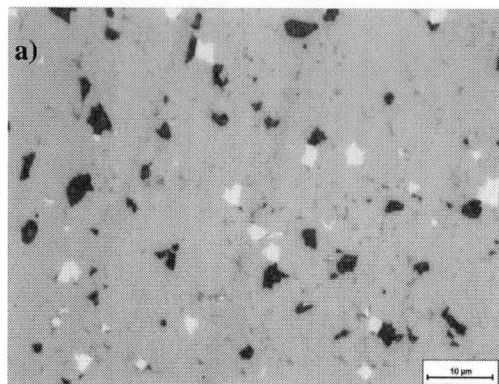

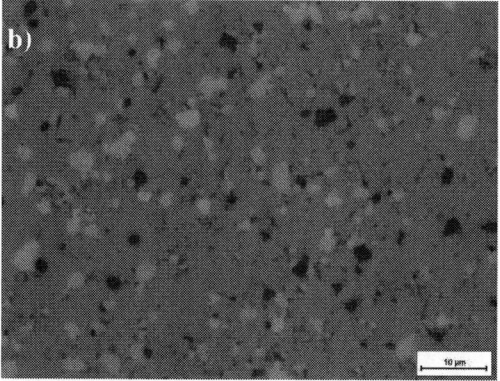

Figure 11: LM-micrograph of a) monolithic MnZn-ferrite and b) embedded MnZn-ferrite, revealing dark grey MnZn rich and light grey Fe rich phases.

The XRD analysis of embedded MnZn-ferrite compared to monolithic ferrite in figure 12 shows a higher ferric trioxide formation in embedded MnZn-ferrite. This result agrees well with microscopic results.

Figure 10: Cross section of the sandwiched laminate MnZn-ferrite/BAM[562].

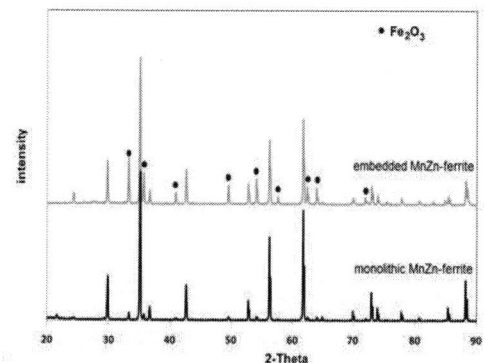

Figure 12: XRD patterns for monolithic and embedded MnZn-ferrite.

The permeability of sandwiched laminate ring cores with ferrite thickness of 200 μm was around 300 (up to 6 MHz), i.e. a factor of two below that of monolithic MnZn-ferrite (μ'=600). One important reason for this permeability reduction in embedded MnZn-ferrite is the partial ferrite decomposition as shown in microscopic (figure 11) and XRD results (figure 12). Prevention of this ferrite decomposition is subject of future investigations.

Conclusions

To integrate ferrite tapes into the LTCC substrate low fired MnZn- and NiCuZn-ferrites were developed which can be sintered at 900 °C with permeabilities up to 800. The ferrite tapes embedded into new dielectric tapes (BAM[474] or BAM[562]) can be sufficiently densified during co-firing at 900 °C without any cracking and open porosity.

For NiCuZn-ferrite embedded into BAM[474] dielectric tapes permeabilities between 230 and 570 (at 2 MHz) according to the thickness of ferrite layer were measured which is much lower than the permeability of monolithic NiCuZn-ferrite (μ' = 800). An important reason for this permeability reduction is the modified microstructure of embedded ferrite especially near the interface with dielectric tapes. Other criteria such as residual stresses could also cause the permeability reduction.

The permeability of MnZn-ferrite tapes embedded into BAM[562] dielectric tapes is also reduced compared to monolithic MnZn-ferrite (μ' = 600) and amounts to 300. It is assumed that decomposition of embedded MnZn-ferrite during co-firing and formation of ferric trioxide caused this permeability reduction.

Acknowledgement

This work was supported by the Bundesministerium für Bildung und Forschung (Germany) within the framework of research project "ALFERMO" (FKZ 13N10666). The authors are very grateful to the BAM colleagues S. Benemann for the SEM micrographs, R. Schadrack for the CTE measurements. We thank B. Pawlowski and S. Barth (Fraunhofer, IKTS) for supplying the NiCuZn-ferrite tapes.

References

[1] T. Rabe, H. Naghib-zadeh, C. Glitzky, J. Töpfer, „Integration of Ni-Cu-Zn Ferrite in LTCC-Modules", Proceedings of CICMT Confrence, San Diego, 2011.

[2] J. Mürbe, J. Töpfer, "Ni-Cu-Zn Ferrites for Low Temperature Firing: I. Ferrite Composition and its Effect on Sintering Behavior and Permeability", J. Electroceramics, Vol. 15, pp. 215-221, 2005.

[3] J. Mürbe, J. Töpfer, "Ni-Cu-Zn Ferrites for Low Temperature Firing: II. Effects of Powder Morphology and Bi2O3 Addition of Microstructure and Permeability", J. Electroceramics, Vol. 16, pp. 199-205, 2006.

[4] R. Matz, D. Götsch, R. Karmazin, R. Männer, B. Siessegger, „Low temperature cofirable MnZn ferrite for power electronic applications", J. Electroceram", Vol. 22, No. 1-3, pp. 209–215, 2009.

[5] C. Glitzky, T. Rabe, M. Eberstein, W.A. Schiller, J. Töpfer, S. Barth, A. Kipka, „LTCC-Modules with Ferritic Layers – Strategies for Material Development and Co-Sintering", J. of Microelectronics and Electronic Packaging, Vol. 6, No.1, pp. 1-5, 2009.

[6] J. Töpfer, J. Mürbe, E. Müller, F. Bechthold, "Sintering behavior of Ni-Cu-Zn ferrites for multilayer inductors", Ceramic Transactions, Vol. 174, pp. 225-235, 2006.

Embedded RFID TAG inside PCB board to improve supply chain management

Julien Viret, Axel Bindel, Paul Conway, Laura Justham, Heinz Lugo, Andrew West.

Wolfson school of Mechanical and Manufacturing Engineering, Loughborough University, LE11 3TU, UK

Tel :+44 (0)1509 227676, Fax :+44 (0)1509 227648 and E-mail Address : j.g.e.viret@lboro.co.uk

Abstract

Radio Frequency Identification (RFID) is a powerful tool used in several application areas. In this paper a method to embed a passive Ultra High Frequecy (UHF) RFID chip within Printed Circuit Boards (PCB) is presented, along with how this can be used to improve the quality and tracking of boards along the electronics supply chain. Embedding of the RFID tag at the very beginning of the supply chain aims to widen the visibility of the product from a very early stage. The embedding of the tag is performed during the bonding process when the multi-layer board is assembled. From this stage onwards it is possible to visualize, record and trace every movement and operation the board goes through along the supply chain.

Key words: RFID, UHF, Embedded, Supply Chain.

Introduction

Globalisation and international commerce have caused a migration from competition between single isolated companies to competition between supply chains [ref]. Companies are forced to closely integrate business processes across their supply chain. [1]. Furthermore, customer demands for high quality, low cost and short time to market has made product visibility thoroughout the supply chain an important requirement for businesses [2].

However, data transfer between supply chain partners and product visibility along the supply chain is often incomplete and inaccurate, leading to inappropiate supply chain decision making (e.g. when tracing back errors, inventory management), which has economic consequences in areas such as inventory inaccuracies, misplaced stock, wrong information links to products etc., all of which lead to substantial decreases in profit and a long term negative effect on the company's image. Being capable of tracking and tracing products at any point across the supply chain and recording product related information using RFID technology is a possible solution to such problems.

In this paper, the electronic manufacturing supply chain is used as an example of our current research. Section 1 summarizes the basics of the RFID theory. The reasons that have led to the research are presented in Section 2. The proposed embedding method along with test results is provided in Section 3. An example of how the functionality offered by the embedded technology can be used within the electronics manufacturing supply chain is shown in Section 4. Finally conclusions are presented in Section 5.

Basic RFID theory

RFID is a wireless system that uses a reader (emitter) and a tag (transponder) to store and transmit secure data via an internationally standardized protocol [3]. Different types of tags exist and they are classified according to a class number. The relation between the different classes of RFID tags and the functionality offered are shown in Figure 1, where bar code and 2-D Matrix are used as reference technologies, which provide read-only identification.

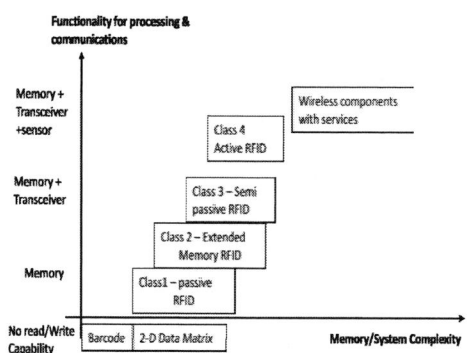

Figure 1: Overview of the different tagging technologies. Shown are the dependency between system complexity and functionality with respect to the power consumption.

Class 1 and 2 are passive tags, meaning they require no external power supply (e.g. batteries) to work. Harvested RF energy is useed to power the tag to respond to the reader commands. Class 3 and above need an external power supply to work and can offer more functionality (e.g. sensors and greater read range distance [4]) than just a memory. In this paper a class 1 passive tag is used.

An RFID tag is an IC with an attached antenna structure that stores a unique ID and / or user data and is powered via the received RF field (See Figure1).

Figure 2: Design of Alien squiggle antenna

The antenna is tuned to the chip and needs to be matched for the specific frequency of operation. A standard RFID tag, which uses an Alien squiggle antenna with a Higgs 3 chip mounted on it, is shown in Figure 2. The operating frequency ranges are between 840MHz and 960MHz [5].

Frequency regulations vary in different regions of the world and have to be considered when choosing the reader frequency *[6]*.

Why embed a tag inside a PCB?

Building an electronic product requires at the minimum two different manufacturing stages, which typically involve two companies:

- A PCB manufacturer: This company is in charge of building the core, following a specific layout (generally a multi-layer core in FR4) defined by the electronic assembly company.
- A printed circuit board assembler (Component Equipment Manufacturer CEM): a company in charge of populating and testing the Printed Circuit Assembly (PCA).

The process to create an electronic product PCB board is as follows:

1. Electronic CAD data (e.g. Gerber file, netlists, drill files) with the manufacturing data for the PCB layer is sent to the PCB manufacturer.
2. PCB manufacturer builds each layer of PCB board and assembles them in a stack to create a rigid multi-layer board structure after bonding
3. The electronic manufacturer populates the boards and tests them according to customer specifications.

Within each step of the production process several tests e.g. electrical tests, optical inspections and functionality tests are conducted to prove the quality of the product. The general information flow can be classified as inter (see Figure 3) and intra-company communication, i.e. information flow between companies (typically paper based or electronically passed on via email) and within each company (controlled by internal MES, ERP and PPS systems). Altough information is transfered with the PCB it is not physically attached to it. At this point we can

easily understand what kind of problem we could experience: information loss, data corruption

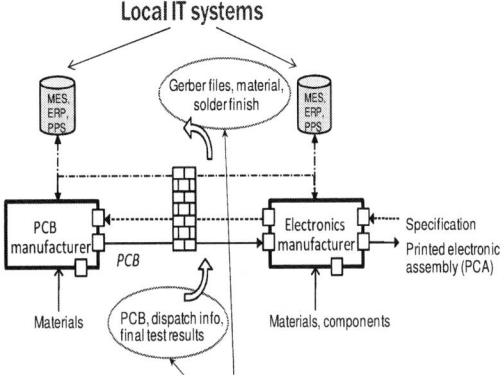

Figure 3: Fabrication of an electronic product

Another issue emerges when trying to find this information when the board leaves production and arrives at the final customer. Generally a simple bar code would allow identification of the board enabling specific queries of board data from distributed databases. However, what happens when the electronic product is not accessible or when the bar code is not readable?

The idea pursued in this research is to associate an RFID tag with internal memory to each board so that critical information can be stored inside the memory. Because of the wireless communication nature, access to the memory does not require a direct line of sight. In the current case, the RFID user memory has been separated into two principal sections, with each one used for a different task:

- A fixed memory part to store information to be exchanged between companies (date of board creation, materials used, test results, etc...).
- A dynamic part to hold internal information exchanged within each manufacturing step. (record of specific test, etc...). This part will be erased and used by each company internally.

Due to the functionality offered by RFID, tracking of the board in real time within manufacture is also possible.

If we follow the information flow used (see Figure 3) to build a PCB board we can easily understand why we need to have an RFID chip embedded inside the PCB board. The embedding of RFID tags at the very beginning of the supply chain aims to widen the visibility of the product from a very early stage. Considering the manufacturing process the earliest process where the tag can be embedded is during the bonding process. From this stage on it is possible to visualize, record and trace every movement and operation that the board goes trough along the supply chain.

Embedded chip procedure and optimisation

The chip chosen was the the Higgs 3 chip. This IC (Figure 4) is composed of an Electronic Product Code (EPC 96 bits) and a user memory of 512 bits. There are four connection pins, but only two of them (The bigger in size: RF and GND in Figure 4) are used for connection with the antenna.

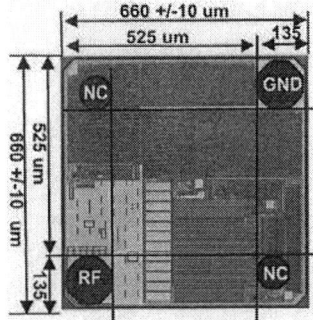

Figure 4: Alien Higgs-3 IC [7].

As shown by previous research [8] the RFID tag is embedded using a flip chip assembly of a bare die chip (e.g.Alien, NXP) which is then soldered to an antenna structure on the core layer. This core posteroiously bonded with an external layer as shown in Figure 5.

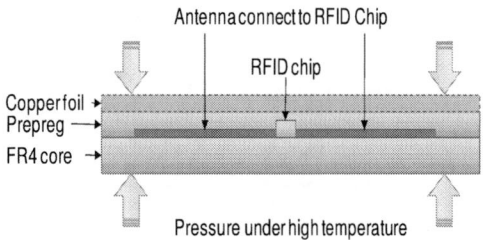

Figure 5: Alien Higgs-3 IC.

The assembly of the multilayer board for the bonding process used is shown in Figure 6. A cavity is made on pre-preg layer 2 to avoid pressure on the chip during the bonding process.

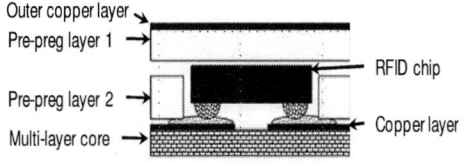

Figure 6: Assembly of different layer for bonding process.

The assembly process for the bare die was optimised to ensure good electrical contact both before and after the embedding of the chip during bonding. Furthermore, different process parameters for the bonding process were optimised to ensure the functionality of the chip after the bonding cycle. As shown in figure 7 (left picture), incorrect process parameters led to cracking of the chip or misalignment.

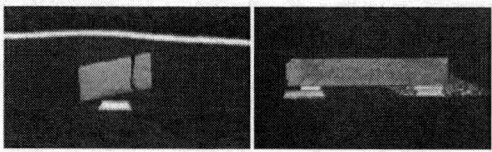

Figure 7: Cross section of an embedded RFID chip. On left failed-embedded on right correctly-embeded.

If, on the other hand, all parameters are optimised, good electrical connectivity and alignment of bonding pads is realized, this is illustrated in figure 7 (right picture) and Figure 8.

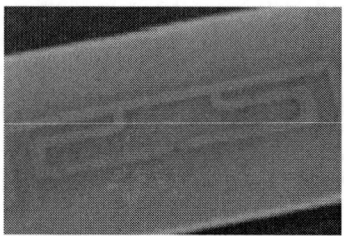

Figure 8: An RFID Higgs3 chip and antenna assembled tag after bonding process.

After conducting several tests the following parameters have been identified as the most relevant for the success of the embedding process:

- Pick and place of chip (from wafer, from waffle pack).
- Interconnection type (Isotropic or anisotropic adhesive, solder paste).
- Recess hole (shape and size).
- Bonding profile (pressure and temperature).

For each test phase a panel (24'' x 18'') with 63 antenna structures was used (Figure 9).
Due to the limitations of the pick and place machine used, the blue antenna structures shown in Figure 9 have no RFID chip connected.

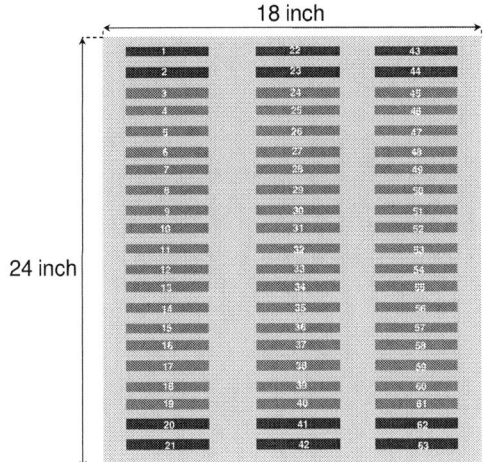

Figure 9: Placement of our Tag on a panel

An increase of yield up to 100% during the bonding process was observed after variation of the process parameters. The yield result of 100% was reproduced without trouble. It was found that the most significant parameter is the form and size of the cavity in the prepreg layer.

As shown in figure 10, an investigation of different test phases on the size and form of the cavity inside the first prepreg layer was made.

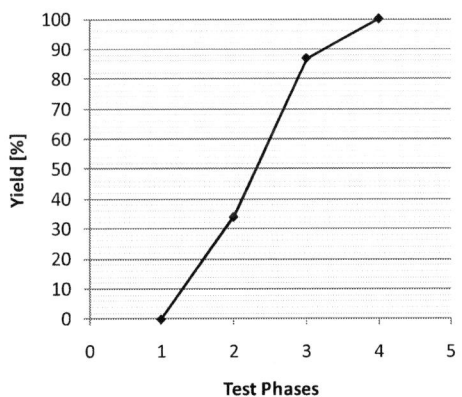

Figure 10: Different phase step of bonding.

Before and after the bonding process the tag is tested for successful functionality (read and writes on user memory).

Test and utilization of RFID tags inside manufacturing supply chain

The second part of our study focused on the test and utilisation of RFID tags along the electronic manufacturing supply chain to improve the product visibility.

We have divided our test into 8 steps, summarised in Table 1:

Table 1: Test made in laboratory to see the feasibility of supply chain tracking

	Process	Readability of RFID tag
1	Made a main core	No
2	Pick and place RFID chip and cured it	yes after cure
3	Bonding	No, due to copper foil
4	Insolate, revelate, etching	Yes after
5	Baking	Yes
6	Put solder and pick and place component	No
7	Solder with reflow oven	yes before and after
8	Rework (reflow oven X time)	yes before and after

We first created an FR4 sample (A4 size) with an RFID tag (Higgs 3 with squiggle antenna, Figure 11). This board is made in 2 steps: the core is bonded with pre preg on top, as shown in the embedded process (step 3). A copper foil on the top then used to make the top layer of our board.

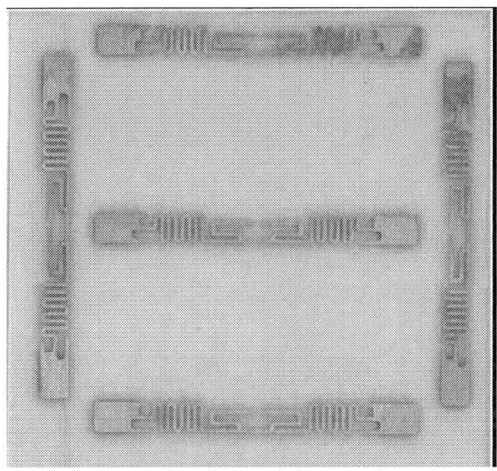

Figure 11: FR4 used for bonding

The Tag was programmed with specific EPC and user data. Due to a lowmemory capacity on the existing RFID Chip (512 bit), we coded data and used our own software (Figure 12) to recover information. In the future, these memory capacities shall increase, as exemplified by recent Tego proposals [9]. With such a type of RFID chip, we could record documents directly onto the memory. In our case, EPC is used to store date and time of fabrication with the name of manufacturer. The user data contains simple information, such as test results passed or not passed.

As explained in Table 1, we could read and write to our Tag only after a sequence of steps:
1. We read our Tag only after the second step because no chip is in place before this. This is our earliest stage of supply chain management.
2. It is impossible to have access to our Tag directly after bonding, due to copper foil on top and bottom of the PCB (picture). This copper is used to create Top and bottom tracks on a

multilayer board. After step 4 we can read and write to our Tag.

3. The baking step is to remove all moisture from the board. We can read and write after baking (125ºC in oven for 20 minutes)

4. Step 6 is usally made automatically by two different devices using a conveyor to move the PCB. An AOI is also made after this step. In our test we decide to read and write before and after reflow oven (step 7).

5. Another AOI is performed after the first reflow. We exposed the PCB to three more reflow cycles. Before and after each cycle we read and wrote to the Tag.

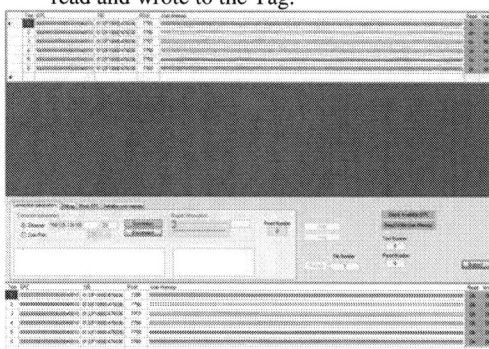

Figure 12: front end software

Figure 11: Front end of read and write software developed at Loughborough.

Specific software developed in VB2005 (see Figure 11) was created to automatically read and write to the Tag memory. A summary of this test is show. in Table 2.

Table 2: Result of read and write after each test determine in Table 1.

	Process	Test	Work
1	Made a main core	No	
2	Pick and place RFID chip and cured it	yes after cure	Yes
3	Bonding	No, due to copper foil	
4	Insolate, revelate, etching	Yes after	Yes
5	Baking	Yes	Yes
6	Put solder and pick and place component	No	
7	Solder with reflow oven	yes before and after	Yes
8	Rework (reflow oven X time)	yes before and after	Yes

In all cases we could read and write to our Tag. The temperatures experienced by the board did not appear to be a problem, even after the reflow oven.

Conclusion

We have demonstrated how to implement, with high yields, an RFID Tag inside a PCB board in a real PCB manufacture. This technology has potential to offer real improvements in supply chain management. We have also demonstrated in Laboratory where we can implement an UHF reader to track a PCB. This demonstration proved the feasibility of this supply chain management. The next step is to implement RFID readers before and after each step of electronics manufacturing. This will allow the tracking and recording of information after each manufacturing process. It may be shown, that real time information about the board location can be realised.

Acknowledgements

The authors wish to acknowledge the financial support of both the Engineering and Physical Sciences Research Council (EPSRC) and Technology Strategy Board under grant references EP/E002323/1 and TP11/HVM/6/I/AB138G respectively.

References

[1] A Summary of RFID Standards, *RFID Journal*, 2008, Available at:http://www.rfidjournal.com/article/view/1335/1/129.

[2] Coltman, T.; Gadh, R. & Michael, K. (2008), RFID and Supply Chain Management: Introduction to the Special Issue, *Journal of Theoretical and Applied Electronic Commerce Research*, (3)1, pp. iii-vi, Available at: :http://works.bepress.com/kmichael/27

[3] EPCGen2, Standart for UHF RFID http://www.gs1.org/epcglobal/standards

[4] Engels, D. & Sarma, S., 2005. Standardization Requirments within the RFID Class Structure Framework.

[5] Alien squiggle Tag with higgs3 chip properties : http://www.alientechnology.com/docs/products/DS_ALN_9640.pdf

[6] Resume of frequency regulation in different country, http://www.gs1.org/docs/epcglobal/UHF_Regulations.pdf

[7] Alien EPC Class 1 Gen 2 RFID Tag IC Higgs 3 data sheet, http://www.alientechnology.com/docs/products/DS_H3.pdf

[8] Bindel, A., Conway, P.P., Justham, L.M. and West, A.A., "New lifecycle monitoring system for electronic manufacturing with embedded wireless components", Circuit World, 36(2), 2010, pp 33-39. DOI: 10.1108/03056121011041681

[9] Tego Electronic UHF RFID chip, http://www.tegoinc.com/products/products_tego chip.php

Manufacture of a 3-D Package Using Low Temperature Co-fired Ceramic Technology

Yves Lacrotte, Robert W. Kay and Marc P.Y. Desmulliez

MIcroSystems Engineering Centre (MISEC), Institute of Integrated Systems,
Heriot-Watt University, Edinburgh EH14 4AS, Scotland, UK

Phone: +44 (0)131 451 3783 Fax: +44 (0)131 451 4155, email: yl270@hw.ac.uk

Abstract

This paper presents the feasibility of sculpting 3D microstructures in a single layer of unfired green tape ceramic. To achieve this result, a powder blasting machine was used to pattern the soft layer. This process is based on the abrasive power of micro particles propelled at high speed towards a substrate. To create 3D structures, the angle of incidence of those particles was varied. Using this method, we have realised sloped wall, sharp edges and transverse channels inside the green tape ceramic. Sub-150µm oblique via walls were also produced with angles ranging from 40° to 60°. These vias can also be filled with conductive paste to accommodate microelectronic needs. The assembly of these 3-D layers into LTCC package can be envisaged for the fabrication of innovative packages for microfluidic and electronic applications.

Key words: LTCC, powder blasting, micro-machining, tapered vias.

Introduction

Low Temperature Co-fired Ceramic (LTCC) has become over the years an increasingly important material for micro-system packaging. It is now being used extensively in electronic, RF systems [1,2]. Its properties make it a very attractive alternative to the well established PCB in the microelectronic world [3]. It has a Coefficient of Thermal Expansion (CTE) similar to silicon, a higher dielectric constant and a better thermal conductivity. Moreover, due to its specific fabrication process -each layer of LTCC is independently patterned and/or printed, then stacked and laminated, and then fired together- it is suitable for the high integration of passive components such as inductors and capacitors, microchannels and cavities. This opened up new fields ranging from micro-fluidics and optics to biology.

The green tape ceramic is conventionally patterned using laser-based systems such as the diode laser-pumped and CO_2 system, and mechanical puncher. They are both very well controlled processes, fast and easy to use. The equipment needed however is expensive and requires a high maintenance. Both processes generate also defects inside the tape, such as remelts or slugs [4, 5], that deteriorate the quality of the structures. An alternative to these methods is the powder blasting technique [6]. The technique is based on the casting of fine particles or powder ejected at high velocity from a nozzle directly onto a substrate. The velocity of the particles is controlled by the gas flow pressure exerted inside the powder reservoir. The high energy acquired confers the small particles (10 to 30 µm) enough power to erode the exposed areas of the substrate. Metallic and polymer are the material of choice for the masks used to protect the substrate from unwanted machining [7, 8].

The machining of LTCC using powder blasting has previously been demonstrated by us to be feasible [6]. It is a cheaper alternative able to etch the substrate quickly and precisely. Furthermore, unlike laser and punching machines, the angle of incidence at which the substrate is etched can be changed, offering new structure configurations. On brittle substrates such as glass and silicon, slanted structures have been achieved [9, 10, 11]. This article presents the further progress achieved in the sculpting of 3D structures using powder blasting on LTCC substrates, especially in the view of manufacturing 3D packages.

Experimental Work

The setup used for the experiments is described in figure 1. The powder blaster is commercialised by the company Texas Airsonic and is located at Heriot-Watt University. The nozzle and stage are enclosed in a plexiglas box which allows the particles and chips stripped off the substrate to be aspirated by a dust collector unit. Pressure and intensity, which characterise respectively the speed and the quantity of particles being forced out from the reservoir, are controlled from a wall mounted panel. The maximum pressure supported in the reservoir is 8 bars. The particles used in this paper are aluminium oxide powder, Al_2O_3, bought from the company ETC limited in Kilmarnock, Scotland.

Figure 1: Experimental setup with x-y translation stage (striped platform) and nozzles.

The particle average size is 9μm, but the size variation is large as evidenced from Figure 2 with particles being as large as 30μm and as small as 3μm. The particles are dried for 3 hours prior to utilisation to eliminate moisture contained inside the powder and stored in the reservoir at a constant temperature of 60C°. This simple preparation avoids clogging of the powder inside the reservoir chamber itself. The specific shape of the nozzle holder seen in figure 1 allows an easy change of the angle of incidence of the jet of particles. With this setup, the maximum angle of incidence is 70° with respect to the vertical direction.

The green tape ceramic used for this paper is manufactured by DuPont and is labelled DuPont 951 P2. The tape is 164-/+11μm thick and does not undergo any treatment prior to test.

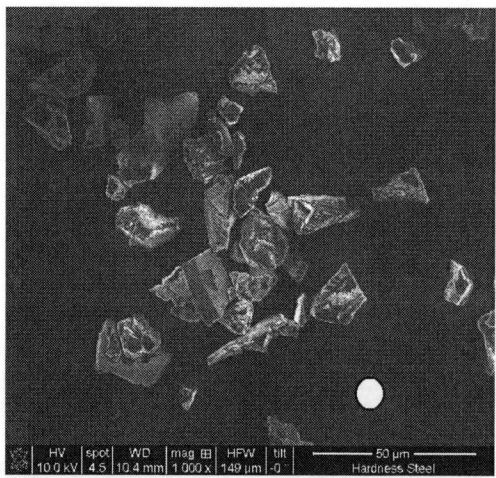

Figure 2: SEM picture of the Al_2O_3 particles. The white circle represents a particle with diameter 9μm.

A few sheets of paper are placed between the ceramic tape and the metallic stage. This intermediate layer prevents the particles from bouncing off the hard metallic stage and from etching the bottom side of the ceramic tape as seen in Figure 3.

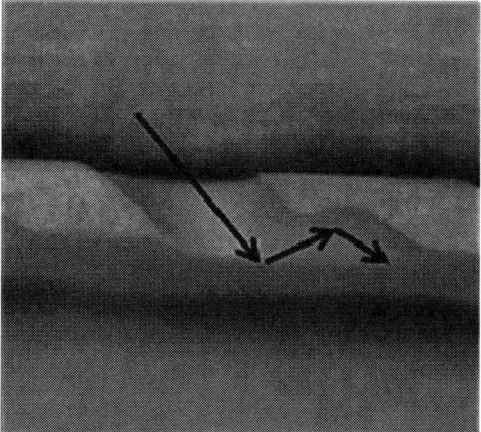

Figure 3: Side view of a transverse via machined at an angle of 45°. The particles underetched the bottom side of the tape after bouncing off the stage.

During our experiments, operational parameters were fixed with the pressure set at 2.75 bars (40 psi). The mass of particles expelled is 210mg per second. The separation distance between the tip of the nozzle and the green tape ceramic has been fixed at 2 cm. As the particles start to scatter after exiting the nozzle, the diameter of the particles beam waist changes from 0.8mm at the nozzle exit to 2.5mm at the surface of the substrate, producing a machinable area of about 4.9mm^2. The full angle at which the jet of particles is scattered is 2.43°. With these parameters, the etching rate in weight at normal incidence, i.e. perpendicular to the substrate, of the green tape ceramic is about 0.57mg per second.

The mask is rastered at a speed of 3mm/s and the rastering interval is 2mm, in accordance with the size of the machinable area. As the ceramic green tape is etched in a few seconds if the substrate is not translated during machining, we have chosen not to utilise the time used to etch the substrate to characterise our process, but rather use the more meaningful parameter of number of passes or scans.

To transfer a pattern to the green tape ceramic, we use an in-house electroformed nickel mask coated with a layer of a silicone based polymer. Nickel electroplating offers the possibility of manufacturing small features size with well defined and smooth edges. The detailed description of the steps of the production of these masks is displayed in figure 4. The Alpha 940 negative dry film resist from Ordyl is laminated at 105°C onto a titanium coated wafer. This negative resist is used as

46

a mould. The film is exposed to UV light (360-380nm) and is developed in Na_2O_3 0.9% concentration. The wafer is then electroplated to the desire metal thickness and knife coated with PDMS or silicone gel, giving the mask its final thickness. The mask is stripped off the wafer and placed onto the LTCC tape. During powder blasting, the dry film left inside the apertures is etched away rapidly before the silicone sustains substantial damage during the abrasive micromaching process. Figure 5 shows an example of the silicon gel coated nickel mask removed from its wafer.

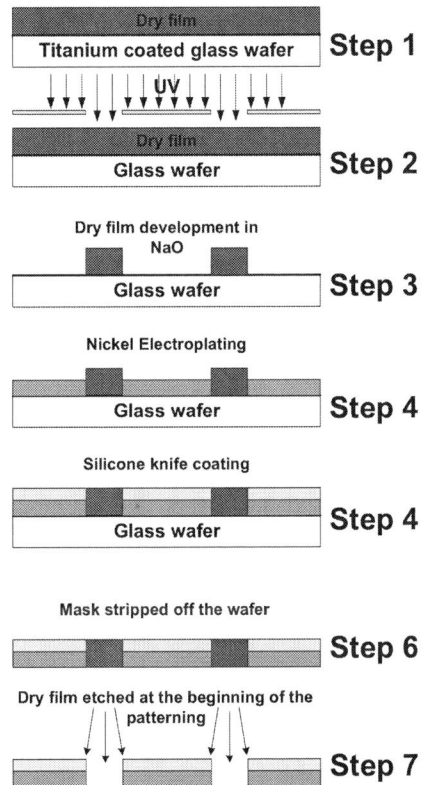

Figure 4: Schematic representing the different steps involved in the fabrication of the mask.

Figure 5: Nickel electroformed mask.

The final thickness of the mask can be changed by varying the number of layers of the dry film resist. Each layer has a thickness of 40μm, but can be stacked and laminated one on top of another, allowing thicker masks to be fabricated. Too thick a mask would reduce its ability to produce small features. Two mask thicknesses were only considered therefore: 40μm and 80μm which correspond to 35 and 75 μm, respectively, after development and electroplating. The mask is held in place by a strong magnetic force generated by magnets clamped onto the substrate, which provides an excellent contact of the mask with the substrate, minimising thereby the amount of underetching. The glueing of the mask onto the LTCC substrate has proven difficult especially in the removal phase. With our apparatus, a thickness of around 20μm of nickel provides magnetic attraction to ensure a good contact during powder blasting.

Results and Discussion

Formation of the tapered vias

The evolution of the formation of oblique vias has been characterised as a function of the number of passes. The particles incident angle was set at 60°. The mask used had a thickness of 40μm. The circular aperture used had a diameter of 300μm. Figure 6 shows 5 cross sections of the ceramic tape as the number of passes increases from 1 to 5. The

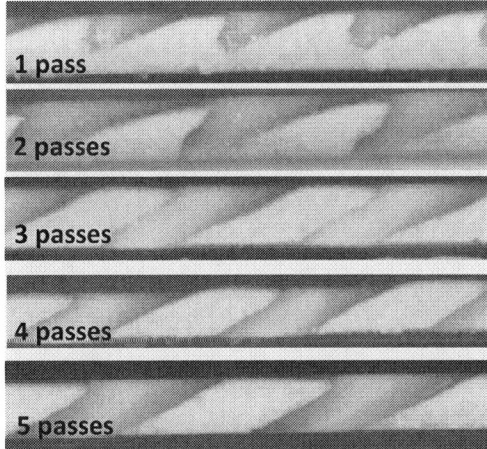

Figure 6: Transverse vias micromachine at 60° angle after 1, 2, 3, 4 and 5 passes (top to bottom).

tape was pierced quickly after 2 passes, but the channel created does not possess flat features and the incline walls are not parallel as the opposite wall is clearly tapered. Also the dimensions of the exits of the vias are too small to be used as evidenced by the curves shown in Graph 1. Transverse vias with near parallel side walls are obtained with 4 passes. Such vias have a diameter of around 420μm as measured from the top of the substrate and 320μm from the bottom. The diameter of the channel obtained is

150μm as shown in figure 7. As the vias have an oval shape, the Graph 1 gives the dimensions of the large (x-direction) and small (y-direction) axes of the oval. Not surprisingly the long axis is parallel to the direction of the passes. Thoese measurements confirmed first that, even with the excellent contact provided by the magnet between the metallic mask and LTCC, under etching was not prevented.

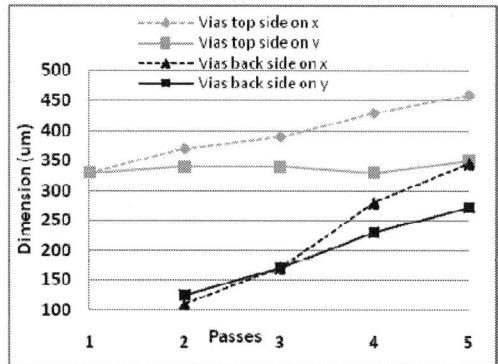

Graph 1: Vias dimensions as a function of the number of passes.

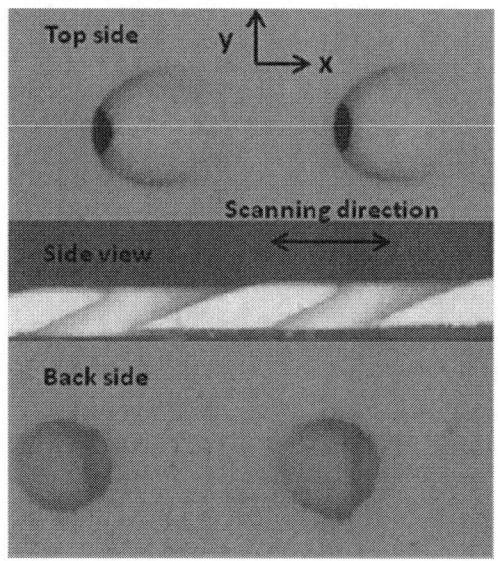

Figure 7: Top, cross section and back side view of transverse vias machined after 4 passes at 60°.

Moreover, due to the inclination of the jet of particles, underetching is more present along the x-direction and increases with the number of passes. Underetching along the y-direction remains constant after 2 passes at around 340μm with is 40μm larger than the aperture size of the hole present in the mask. Graph 2 shows the angles of the opposite and adjacent walls of the machined vias. No large variation of the angle is noticed after 5 passes and both walls become increasingly parallel after 3 passes. Based on these results and in order to minimise the wear of the mask and the scanning

time, the subsequent tests were performed after 4 passes.

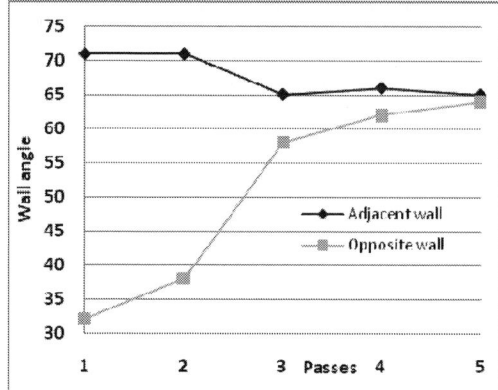

Graph 2: Wall angle of the oblique vias as a function of the number of passes.

Furthermore, the vias obtained after four passes can be easily filled with conductive paste. The paste used to fill the vias is a silver based paste also from DuPont named 6141. Figure 8 shows a cross sections of the transverse vias with a width of 150μm screen printed with silver based paste.

Figure 8: 150 um transverse vias (60°) filled with silver conductive paste.

Limitations of the tapered vias

We have studied the limitations in the fabrication of vias due to the angle of incidence, aperture size and thickness of the mask. These parameters put a constraint on the maximum particle beam waist that can penetrate through the mask aperture. A simple example, with a mask thickness of 75μm, a circular aperture of 300μm and an angle of incidence of 60°, shows the maximum waist of the beam of particles penetrating the mask is 81μm. As we consider that a transverse channel should have at least a waist of 100μm, this configuration would not recommended. But these calculations do not take into account the underetching effect and assume that the beam stays collimated.

To determine the limitations in the manufacture of the vias, 2 masks of 35μm and 75μm thickness, respectively about 15μm and 35μm

48

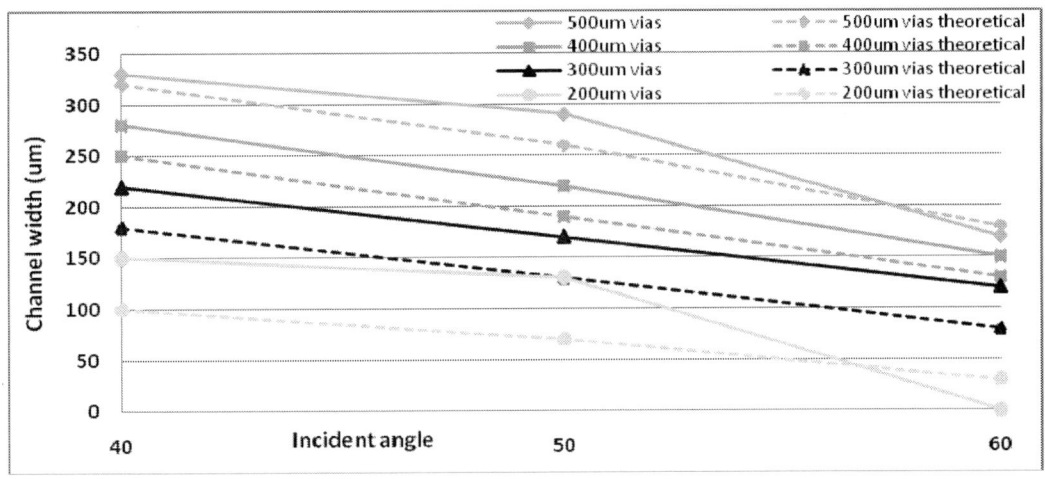

Graph 3: Channel diameter of oblique vias measured with different aperture size and angle of incidence. The mask thickness is 75um.

thickness of electroplated nickel, were used with vias aperture ranging from 200µm to 500µm. The angles of incidences were set at 40°, 50°, 60° and 70°. 4 passes were used.

First the angle of incidence of the particles was transferred accurately to the LTCC channels as evidenced by figure 9. The figure shows oblique vias made with a 400µm aperture size having the same wall angles than the incident beam. This is also true for other the apertures and mask thickness.

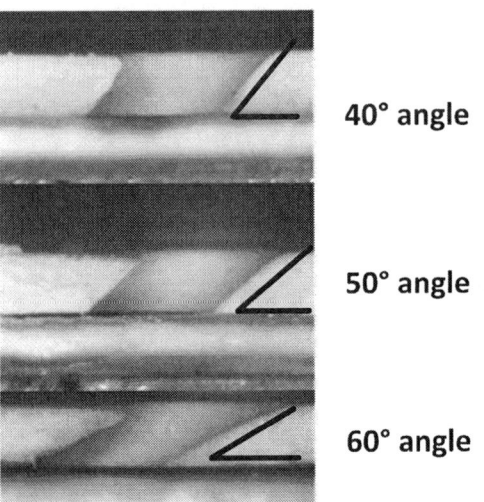

Figure 9: Side view of transverse vias patterned at 3 different angles.

Graph 3 compares the theoretical width of the vias (or channel) obtained for different angles of incidence and aperture sizes and the experimental results obtained in the green tape ceramic with the same parameters. The mask used is 75µm thick. Vias have successfully been pierced at 40°, 50° and

60° with any aperture size, except at 60° using the 200µm aperture. The channel walls are nearly parallel in all cases and are larger than their theoretical counterpart as expected. Figure 10 show a transverse vias patterned at 40° using a 200µm aperture. The width of this channel is 150µm.

Figure 10: Top view of transverse vias machined after 4 passes.

Graph 4 shows comparable results to graph 3, but obtained with a 35µm thick mask. As expected, oblique vias can be more easily machined with a thinner mask. The only aperture that did not produce a through via was the 200µm one at 70°. However, such an angle of incidence introduces high underetching that increases the size of the vias. With an aperture of 300µm, the dimension along the x-direction of the via at the top surface is 630µm as shown in figure 11.

The smallest transverse via produced was for an incident angle of 50°, a 200µm aperture size and a 75µm thick mask. The entry and exit sides are almost circular with dimensions respectively of

49

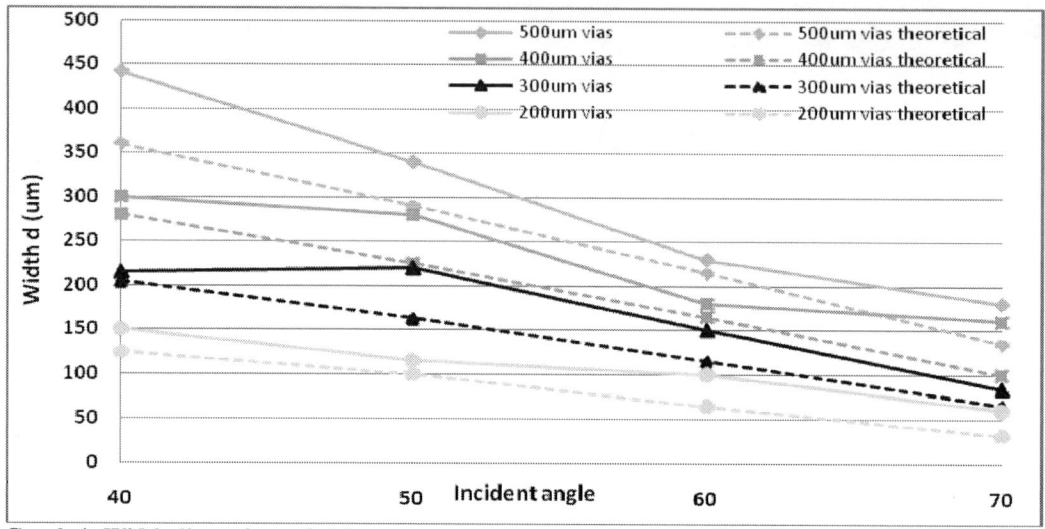

Graph 4: Width dimensions of oblique vias measured with different aperture size and angle of incidence. The mask thickness is 75μm.

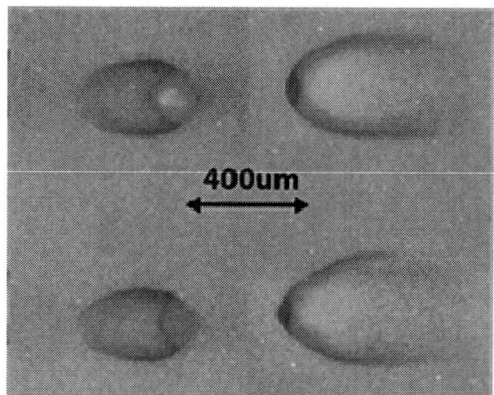

Figure 11: Back side and top side view (respectively on the left and right) of transverse vias machined after 4 passes.

240μm in x-direction, 210μm in y-direction. The width of the via was 130μm.

Manufacturing of blades

The same process was used to carve blades in a single layer of green tape ceramic. Figure 12 shows 2 cross section of blades carved with an angle of incidence of 45°. The features of the mask included 100μm beams spaced by 200μm. The tape was etched through after 5 passes, but the process can be stop earlier to create indentation patterns

Figure 12: Micromachining of tilted blades onto a 164μm thick Dupont 951ceramic green tape.

Conclusion

We have shown the possibility of using a powder blasting machine to pattern vias with an angle of incidence set between 40 to 70°. Below 150μm channels can be made with vias dimension not exceeding 250μm on the top surface and 200μm on the exit side. A thick electroplated nickel mask coated with silicone can be used to produce very fine structures, ensuring an easier handling and better contact. First 3D prototypes were made using slanted microstructures that could find applications for the manufacture of embedded mirrors in integration optics for example.

Acknowledgments

The authors would like ot acknowledge the financial support of the UK Engineering and Physical Sciences Council (EPSRC) through the award of an Engineering Doctorate Scholarship to Mr Yves Lacrotte. The authors would also like to thank the Company Renishaw plc for their continuing support.

References

[1] R.Kulke, C.Gunner, "24 GHz Radar Sensor integrates Patch Antenna and Frontend Module in single Multilayer LTCC Substrate", Proceedings of the 2005 European Microelectronics Packaging Conference (EMPC), June 12-15

[2] S. Marksteiner, "A Miniature BAW Duplexer Using Flip-Chip on LTCC", IEEE Ultrasonics symposium, 2003.

[3] R.Kulke et al, "LTCC- Multilayer Ceramic for Sensor and Wireless Applications", Produktion von Leiterplatten und Systemen (PLUS), Eugen G. Leuze Verlag, Seite 2131-2136, Ausgabe Dezember 2001.

[4] Gangqiang Wang, "Fabrication of Microvias for Multilayer LTCC Substrates", IEEE Transactions on Electronics Packaging Manufacturing, vol. 29, No. 1, January 2006.

[5] S. H. Rhim . S. Y. Shin, "Burr formation during micro via-hole punching process of ceramic and PET double layer sheet", Int J Adv Manuf Technol (2006) 30: 227–232.

[6]: Y. Lacrotte, F. Amalou, W. Yu and M. Desmulliez, "Micro-patterning of green tape ceramic using powder-blasting for LTCC manufacturing", IMAPS CICMT 2009, Denver.

[7] Henk Wensink, Henri V Jansen, "Mask materials for powder blasting", J. Micromech. Microeng. 10 (2000) 175–180.

[8] Henk Wensink, J.W. Berenschot, "High resolution powder blasting micromachining", IEEE 2000.

[9] E. Belloy, A. Sayah, M.A.M Gijs, "Powder blasting for three-dimensional microstructuring of glass", Sensors and Actuators 2000. p231–237.

[10]: E. Belloy, A. Sayah, M.A.M Gijs, "Powder blasting for three-dimensional microstructuring of glass", Sensors and Actuators 86 (2000) 231–237.

[11]: E. Belloy, A. Sayah, "Oblique powder blasting for 3-D micromachining of brittle materials", Sensors and Actuators (2001) 358-363.

TuA – Display Presentations

Non-Destructive X-Ray Mapping of Strain & Warpage of Die in Packaged Chips

Jennifer Stopford[a], Arthur Henry[b], Dionysios Manessis[c], Nick Bennett[a], Ken Horan[a], David Allen[a], Jochen Wittge[d], Lars Boettcher[c], Aidan Cowley[a], and Patrick J. McNally[a].

(a) Dublin City University, The RINCE Institute, Dublin 9, Ireland.
(b) Department of Manufacturing Engineering, School of Manufacturing and Design Engineering, Dublin Institute of Technology, Ireland.
(c) Fraunhofer IZM Berlin, Microperipheric Research Center, Berlin Center of Advanced Packaging, Gustav-Meyer-Allee 25, Berlin, Germany.
(d) Albert-Ludwigs Universität Freiburg, Kristallographie, Institut für Geowissenschaften, Freiburg, Germany.
JStopford.dcu@gmail.com

Abstract

3D through silicon via (TSV) and vertical stacking technologies have become key enablers in the move towards high density, high functionality 3D integrated circuits. The need to address strain, warpage, delamination, etc. in Systems on Chip or Systems in Package (SoC/SiP) is recognised in the International Technology Roadmap for Semiconductors (IRTS 2009). This paper describes the implementation of a novel technique called 3-dimensional surface mapping (3DSM) for the non-destructive measurement of in situ strain and wafer die warpage in thin, chip embedded Quad Flat Nonlead (QFN) packages.

X-Ray Diffraction Imaging (XRDI) is used, in conjunction with 3DSM, to obtain high resolution (~3 μm) strain/warpage maps and qualitative information on the nature and extent of the strain fields in completely packaged chips. QFN packages were examined after 3 different stages of the manufacturing process: chip attach, via electroplating, and post-production as a means of demonstrating the potential for this technique to provide quick feedback towards process improvement and thus reduce major sources of die warpage.

Key words: XRDI, 3DSM, SoC, SiP, strain, stress.

Introduction

Embedding of chips into organic substrates using printed circuit board (pcb) technology has emerged as a promising technology route for 3D component integration [1, 2, 3, 4]. However, IC manufacturing and chip embedding processes can induce stresses in the chip which have the potential to affect device functionality and reliability and ultimately lead to device failure. The management of mechanical stresses is therefore one of the key enablers for the successful implementation of 3D IC technology [3]. The need to address stress and strain in Systems on Chip or Systems in Package (SoC/SiP) is recognised in the International Technology Roadmap for Semiconductors (IRTS 2009). There is [5]: "… a need to develop metrologies that can be used to efficiently measure either stress or strain under both thermal and mechanical loading conditions in thin films (for example in layers within Silicon) in packaged form."

To date test methodologies for embedded chips have consisted of finite element modelling (FEM), micro-Raman spectroscopy, optical inspection, electrical testing and destructive testing such as shear testing [6, 7, 8, 9, 10]. Significant causes of strain, and hence failure of embedded chips are known to come from 3 major sources; materials induced stresses caused by differences in the coefficient of thermal expansion (CTE), wafer / chip warpage and stress from the embedding / lamination [6, 1, 7, 10, 11]. In addition, wafer / die warpage temds to increase greatly at thicknesses less than 100 μm. With so many diverse challenges to the realisation of high volume 3D IC integration, advanced, non-destructive testing methodologies are fundamental to the development of embedded chip design and manufacture.

In this paper X-Ray Diffraction Imaging (XRDI) is used in conjunction with 3D surface modelling (3DSM) to non-destructively obtain qualitative information on the nature and extent of the strain and induced die warpage in thin, chip embedded Quad Flat Nonlead (QFN) packages at different stages in the manufacture, test processes. Micro-Raman spectroscopy complements 3DSM data by quantifying the strain in unpackaged test chips.

Quad Flat Nonlead (QFN) packages

The packages examined in this study are QFN-A packages [1, 6], measuring 160 μm thick and 10 mm × 10 mm in size. Packages were produced at IZM Fraunhofer, Berlin, under the

HERMES project. The package consists of an active die bonded Si chip, 5 mm × 5 mm in size and 50 µm thick, with a peripheral bond pad pitch of 100 µm, embedded face up in a substrate. The chip is covered from the top side with a RCC (resin-coated-copper) dielectric layer 90 – 100 µm thick. A schematic of the chip assembly, processing and testing steps is shown in Figure 1 [1, 6].

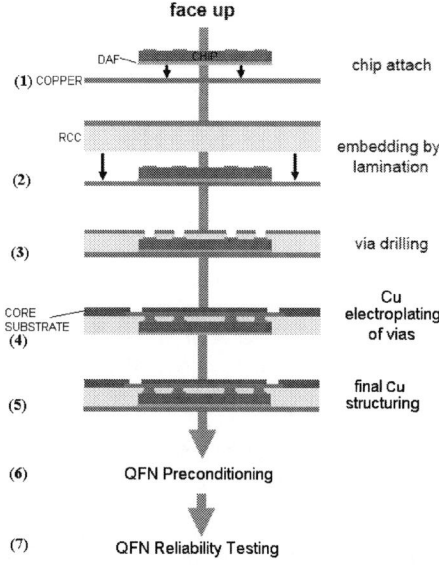

Figure 1: Schematic of 5 major chip embedding steps for QFN-A package. QFN-A chips underwent preconditioning and reliability testing after assembly.

Prior to embedding, chips were prepared by electrolytic deposition of 6-8 µm of Cu to the bond pads. Using a high precision (± 10 µm) die attach machine, double layered die attach film (DAF), an adhesive layer 20 µm thick was used to bond the chip face up, to the Cu substrate (Figure 1 (1)). A RCC layer 90 – 100 µm thick was applied to the chip from the top side using a standard pcb multilayer vacuum lamination process [9], and the epoxy was then cured at ~185 °C for 60 min (Figure 1, (2)). A pulsed 355 nm UV laser was used to drill the microvias through to the chip pads (Figure 1, (3)). The vias were then cleaned to remove excess epoxy resin and improve Cu adhesion, and palladium was deposited on the epoxy surface prior to Cu metallization via electroplating (Figure 1, (4)). Structured Cu conductor lines are required to connect the bond pads and capture pads on the chip. This was undertaken by means of laser direct imaging (LDI), which was used to expose a negative resist, which was subsequently acid etched to reveal the Cu line structure. The final step in the packaging process is Cu structuring on the bottom side of the package (Figure 1, (5). Packages were processed in large panel format, and separated by cutting or

sawing [1, 6] before undergoing pre-conditioning and qualification tests.

Experimental

3-Dimensional Surface Modelling (3DSM) is a modelling technique which uses the high resolution (~3 µm) strain imaging capabilities of XRDI, combined with the advanced freeform modelling capabilities of a 3D CAD design software (*Solidworks*) to build 3D profiles of crystalline misorientation across completely packaged chips. The XRDI/CAD based 3DSM process includes four main phases: XRDI image capture, importing XRDI into Solidworks, creating B-spline curves and constructing the 3D surface model.

The X-Ray Difraction Imaging (XRDI) measurements were performed at ANKA, Karlsruhe, Germany using the TopoTomo beamline [12, 13]. The topographs were recorded in section transmission (ST) geometry as per [14]. In order to obtain section topographs the X-rays are collimated into a narrow ribbon only 15 µm high and the CCD camera was positioned to record the 220 reflection from the Laue diffraction pattern of topographs [14]. Using a precision XY stage ST topographs were taken at either 0.5 or 1.0 mm steps from the pads at the top of the chip stepping sequentially until the bottom of the chip, was reached. The chip was then rotated at 90° and the process repeated; this enabled a grid of ST topographs, as illustrated in Figure 2, to be obtained. ST topographs represent reflections from the {2 2 0} planes of Si.

Horizontal ST topographs were imported into *Solidworks* and positioned so that the distance between adjacent sketch planes was proportional to the distance between two corresponding ST topographs. In order to model the shape of the imported ST topographs the contours of the topographs were followed by positioning points along the exterior contours using the point option of the spline mode in *Solidworks*. B-spline curves were created for each of the horizontal and vertical ST topographs. The imported topographs were then hidden, enabling a set of splines to be displayed as in Figure **2**.

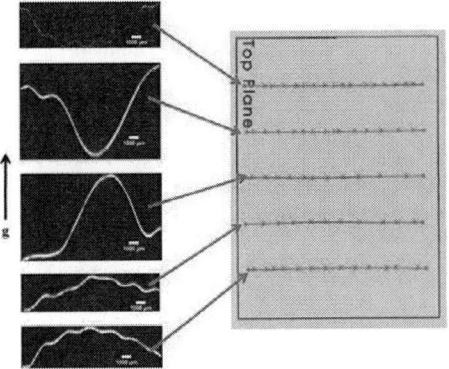

Figure 2: Horizontal topographs and corresponding horizontal B-Spline curves created in *Solidworks*.

The 3D surface model was created using the boundary surface feature in *Solidworks*. Figure 3 shows the boundary surface mesh across 5 horizontal splines; no tangency constraint, i.e. zero curvature is applied in order to give the closest possible fit of the surface to the splines. With the exception of stage 3 samples 3DSM were shaded in order to enable the top and underside surfaces to be identified, with the top surface shaded in green, and the underside shaded in grey.

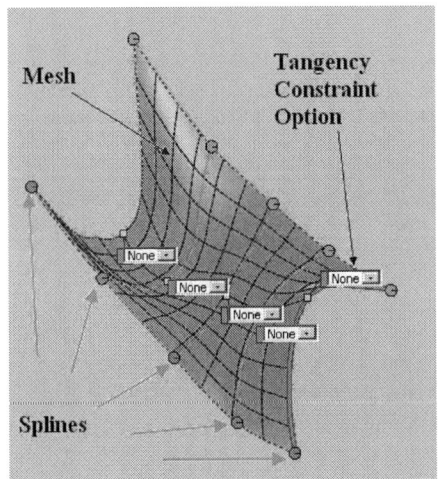

Figure 3: Boundary surface formation showing splines and mesh detail.

Micro-Raman measurements were used to quantify strain levels in the stage 1 samples [15.16]. Using the 488 nm Ar⁺ laser line on a JY Horiba LabRAM HR800 high resolution Raman microscope scans were taken at 3 different positions on the chip, as described in Figure 4 (Positions 1, 2 & 3 to scale).

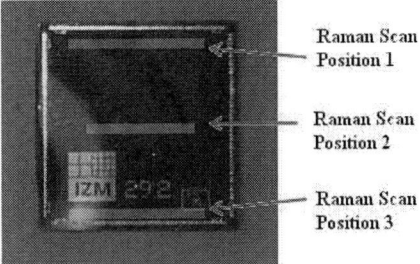

Figure 4: Positions of Raman matrix (2D) scans on Sample 4.

As the total thickness of the chip, die attach film and Cu substrate measured less than 200 µm, flatness of the stage 1 chips was an issue. To compensate for any small deviations in flatness the chip-substrate structure was taped to a microscope slide and the optical auto-focus functionality on the micro-Raman system was used. This enabled signal intensity levels to be maintained across the width of the chip while using a short accumulation time (1s), to prevent heating of the sample.

Results
Stage 1: Chip Attach

After the first processing step, samples consisted of 5 mm × 5 mm squares of 50 µm thick Si embedded face up on a Cu substrate. A single chip, sample 4, was examined using XRDI, 3DSM, and the Ar⁺ laser of the micro-Raman system. Initial tests on stage 1 chips showed them to be highly sensitive to the x-ray beam heat load from the synchrotron, which was estimated to raise the local chip temperature to ~40°C. As a result of this the acquisition time for ST and LAT topographs was set to 0.5 s, though the contrast of ST and LAT topographs was compromised only to a minor extent. Elongation of LAT topographs occurs parallel to projection of the diffraction vector, **g,** and is observed on several samples in the study. This phenomenon is most likely a result of the high degree of lattice curvature present in the chips / packages. We know that the greatest sensitivity to lattice displacement occurs when **g** is parallel to, or antiparallel to the reciprocal lattice vector **h**, XRDI and 3DSM in this paper have therefore been normalised to reflect the square shape of the chip, and thus compensate for any geometric distortion observed.

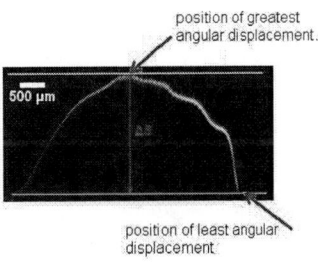

Figure 5: 2 2 0 ST topograph of 1st in series of 10 ST topographs which combine to form 3DSM images of Figs. 6 &7. Yellow lines show maximum angular deviation in ST of ΔS.

The 3DSM are created from a series of up to 10 ST topographs as described previously. Distortion in the ST topographs is formed by the orientational contrast mechanism, where diffracted x-rays from highly strained and hence misorientated, regions in the Si lattice possess an altered Bragg angle. Deviations in the recorded position of the diffracted Si in the ST images (Figure 5) are seen as a result of the lattice planes being distorted due to lattice misorientation throughout the entire chip. For orientational contrast to occur, the lattice misorientation must exceed the x-ray beam divergence, which is a maximum of 2 mrad in the case of the ANKA synchrotron [12].

The magnitude of shift due to lattice misorientation, or warpage is given by:

$$\frac{\Delta S}{2L} \cong \Delta\alpha \qquad (1)$$

where ΔS is the measured angular misorientation across the measured length of the sample, as shown in Figure 5, L is the sample the sample to film distance, and Δα is the maximum apparent shift in the Bragg angle due to the strain-induced tilt of the diffracting planes [17, 18, 19]. $\Delta\alpha_{MAX}$, the maximum misorientation of the Si planes for the reflection imaged (using XRDI), is therefore measured relative to the points of greatest and least angular displacement in the relevant ST topograph, as shown in Figure 5.

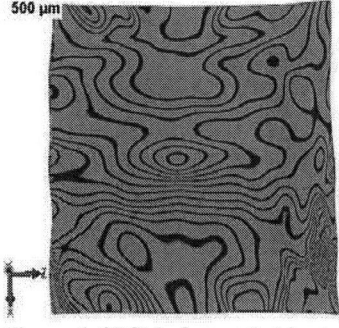

Figure 6: 3DSM of stage 1 chip showing misorientation of (2 2 0) Si planes of sample 4. Contours are shown to indicate degree of warpage of chip.

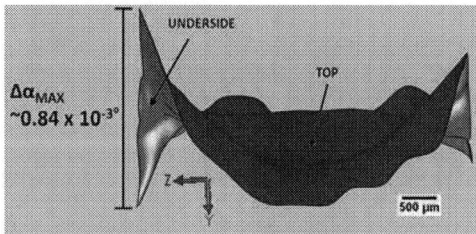

Figure 7: 3DSM showing Δα, the relative misorientation of (2 2 0) Si planes of sample 4 viewed end on, with ST position corresponding to the upper edge of the chip facing the page.

Figures 6 and 7 are 3DSM of the stage 1 sample. The greatest lattice distortion is observed in towards the top of the chip, as seen in Figure 9. $\Delta\alpha_{MAX}$, the maximum {220} Si plane lattice tilt for the stage 1 sample is 0.84×10^{-3} (± 5%) degrees, considerably less than that seen at later processing / reliability testing stages. Micro-Raman spectra were taken using the Ar+ laser, which has a penetration depth of ~556 nm in Si, thus the data are indicative of strain levels close to the surface of the chip. Micro-Raman scans were taken across 3 regions of the sample as per Figure 4. Scans 1 and 3, across the upper and lower regions of the chip respectively, show a single narrow Si peak, indicative of crystalline Si, centred on ~521.7 cm⁻¹. If a uniform biaxial stress model is assumed the stress corresponding to the Raman peak shift can be calculated from [20]:

$$\sigma_{xx} = \sigma_{yy} = \frac{\Delta\omega}{-4} \qquad GPa \qquad (2)$$

where Δω is the change in Raman frequency from the calibration position at 520.07 cm⁻¹, and σ_{xx} and σ_{yy} are the stress components for biaxial strain in the x - y plane.

Strain levels across the scans 1 and 3 range from 345 – 470 MPa, with both regions in tensile strain, and the highest strain levels closest to the corners of the chip. The central micro-Raman, scan 2, shows uniform strain levels of up to ~360 MPa from left to right across the chip, with changes in the strain occurring from the top of the scan downwards. As with scans 1 and 2 tensile stress is largely present, although low levels of compressive stress are also observed. Only a single phase of crystalline Si is present in scan 2, signified by a single narrow peak, with an average FWHM of 4.1 cm⁻¹. Multiple scans on the same region prove that the regular strain pattern is a genuine strain phenomenon rather than a temporal or thermal effect.

Micro-Raman spectroscopy is a well established and widely used characterisation technique for analysis of strain in Si die and packaged chips [16, 20, 21, 22]. As such it provides an appropriate platform from which to evaluate the efficacy of novel, qualitative techniques such as XRDI and 3DSM. The Raman results confirm strain levels close to the surface of the chip of the order of 100's of MPa, with the greatest levels of tensile strain apparent at the edges and corners of the chip. These compare well with the 3DSM (Figures 6 & 7), which exhibit a uniform, relatively undistorted (warped) region at the centre of the chip, and warpage of the {2 2 0} lattice planes along all 4 edges of the chip. Although the magnitude of the induced lattice warpage, as seen in the 3DSM, and measured from the ST topographs is relatively small at this early processing stage, micro-Raman results prove that close to the surface, the induced strain levels are high and have the potential to affect device functionality and/or reliability [1, 11, 23]. Large area topographs (not shown here) cannot image effectively the strain field induced warpage in the chip, demonstrating that detailed information on lattice warpage is not always apparent from conventional XRDI, and that 3DSM is a valuable tool in characterisation of strain in packaged chips.

Stage 3: Via Electroplating

Two samples were anaylsed after Cu electroplating of the vias, as both exhibited similar results only a single sample is discussed here. After stage 3 of processing die are completely enclosed in Cu and no part of the die is visible.

The LAT of this sample (Figure 8) displays a distinctive wave pattern, with a peak to peak pitch of 1050 – 1150 µm and showing typical image elongation parallel to **g**.

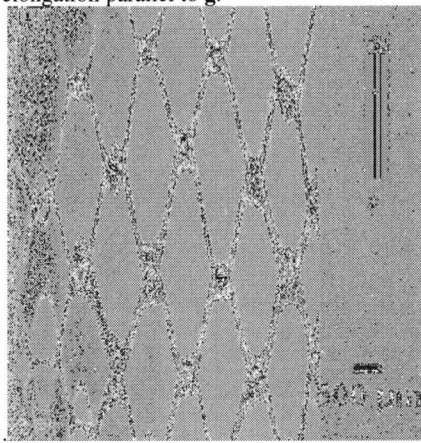

Figure 8: 2 2 0 LAT topographs reflection from central section of Stage 3 sample. Sample was oriented at 90⁰ when LAT was recorded.

The wave pattern is most prominent in the central region of the chip, with peaks smoothing out and becoming less pronounced towards the edges of the chip. Initial examination of the above LAT topograph gives little information on the nature of the lattice distortion which would give rise to such a pattern, however the 3DSM clearly shows the pattern of strain induced die warpage inside the package (Figures 9 & 10). As with the previous sample the shape of the 3DSM have been normalised to reflect the square shape of the chip. Figure 11 illustrates how the bumps on the 3DSM align to the wave pattern on the LAT topograph, and further demonstrates the efficacy of the surface mapping process for packaged chips. $\Delta\alpha_{MAX}$, the relative angular lattice misorientations across the die is $\sim 0.65\text{-}0.9 \times 10^{-3}$ ($\pm 5\%$) degress for these samples.

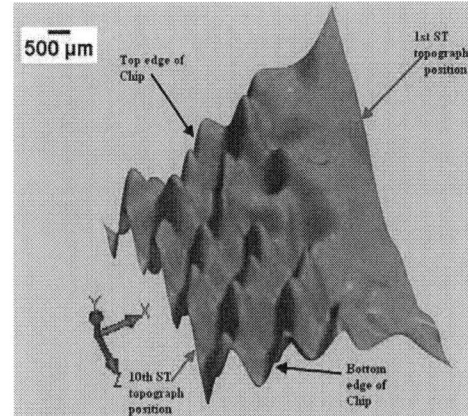

Figure 9: 3DSM of misorientation of (2 2 0) Si planes inside Atiotech 3a sample. Sample was oriented at 90⁰ when ST were recorded.

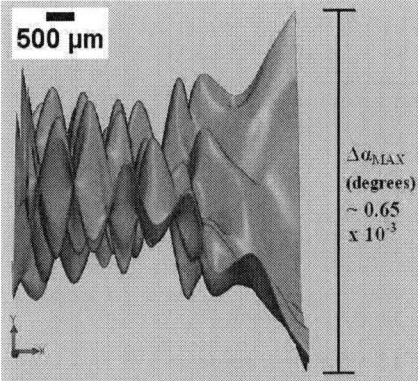

Figure 10: 3DSM of misorientation of (2 2 0) Si planes inside Atiotech 3a sample, showing $\Delta\alpha MAX$. Sample was oriented at 90⁰ when ST were recorded.

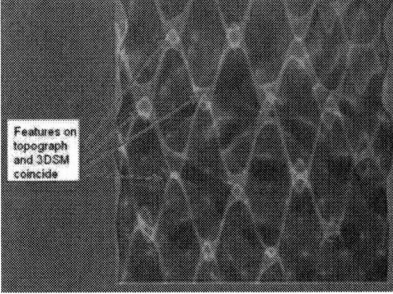

Figure 11: 3DSM of misorientation of (2 2 0) Si planes inside Atiotech 3a sample with LAT overlaid. Scale is same at that for Figure 5.11.

The development of the regular pattern of bumps/peaks at this stage in the process, and its absence at stage 1 suggests the phenomenon is linked to either the chip embedding by vacuum lamination or via drilling processes. Through silicon vias (TSVs) are positioned along the edges of the chip, with 44 vias on each side placed 100 µm apart.

The density and position of the bumps is therefore inconsistent with TSV drilling and placement. The lamination process for QFN-A packages has undergone comprehensive development and optimisation over the course of several projects [1, 6, 9, 24, 25]. Heating rate and pressure are critical parameters with respect to lamination integrity. High heating rates at low pressures (5 – 10 bar) can lead to the occurrence of voids, and excessive epoxy fluid viscosity has the potential to cause insufficient epoxy coverage around the chip and influence package flatness [1, 6]. The distinctive pattern of lattice warpage seen after stage 3 processing is therefore most likely linked to vacuum lamination process, where the presence of voids or variations in epoxy coverage may lead to localised variations in strain in the crystal lattice, resulting in the formation of the bump structures of Figs. 9-11. This is the subject of ongoing investigations.

Stage 5: Post Production

In the case of the post production (qualification) samples an example of a 2 2 0 LAT topograph (sample 09-096 no. 158) is shown in Figure 12) this having been corrected to compensate for geometric distortion, as with Stages 1 & 3. The region circled in red, in Figure 12 corresponds to a highly strained area on the chip. Orientational contrast occurs when diffracted beams overlap or diverge due to lattice misorientation. In the case of topographs recorded using a ccd camera the image contrast is inverted, and regions of low intensity, where loss occurs appear black. The greatest lattice distortion is therefore located towards the centre of the circled region.

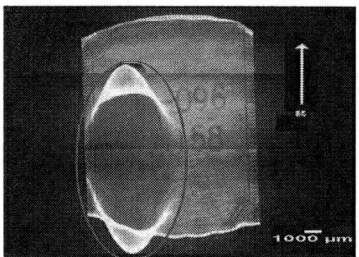

Figure 12: LAT topographs of sample 09-096, no 158, with image corrected to compensate for geometric distortion parallel to g.

Chip support and clamping in die bonding and chip placement processes are critical for die placement accuracy [6]. When support is insufficient warpage can occur [6, 25]. The LAT topographs for two samples examined at Stage 5 both show a similar strain field pattern, with a corresponding elliptical region of high distortion at the same location. The uniform nature of the strain fields, and their presence on 2 packages, suggests they are linked to the manufacturing process. They may be due to the supports or clamps in die bonding and chip placement processes, perhaps where a die or

substrate is supported on two sides, stress from processing can manifest itself in unsupported planes. The presence of vias can also lead to localised stresses in the package. Via drilling was undertaken using a pulsed 355 nm UV laser. We know that localised heating of the Cu surrounding very small diameter (15 μm) vias can lead to warpage [26], and can induce significant strain, of the order of MPa [27]. Die warpage may therefore be partially attributable to vias. TSV can also induce stress by squeezing/stretching the adjacent material leading to material deformation, debonding and delamination.

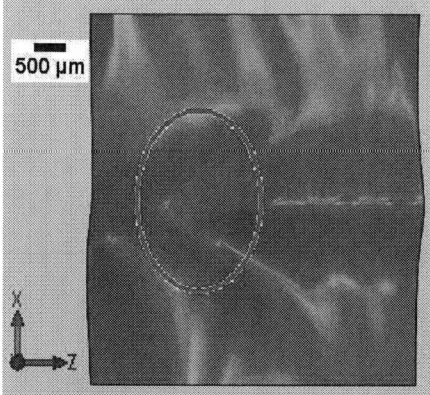

Figures 13: 3DSM maps of misorientation of (2 2 0) Si planes inside a completely sealed package 09-096 no. 158. ST topographs were taken with the package orientated at 0°, i.e. in the horizontal position. Red circles correspond to the elliptical region in Fig. 12.

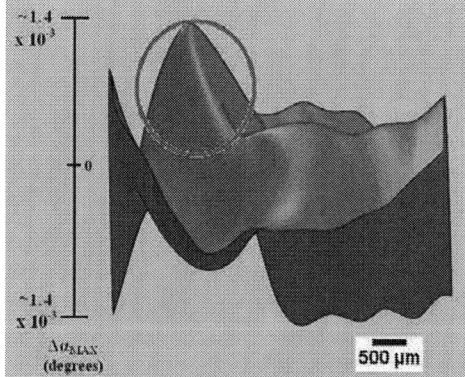

Figures 14: 3DSM maps of misorientation of (2 2 0) Si planes inside a completely sealed package 09-096 no. 158. ST topographs were taken with the package orientated at 0°, i.e. in the horizontal position. Red circles correspond to the elliptical region in Fig. 12.

The maximum lattice warpage for the stage 5 chip ranges from ~ 0.15-1.69×10^{-3} ($\pm$ 5%) degrees, and is dependent on the location on the chip. The 3DSM for this sample, is comprised of 5 ST topographs placed 1 mm apart, and is therefore of lower resolution than the previous 3DSM from stages 1 and 3.

Conclusions

This paper describes the use of 3DSM, a novel technique for the non-destructive measurement of *in situ* strain and wafer die warpage in thin, chip embedded Quad Flat Nonlead (QFN) packages. 3DSM is used in conjunction with XRDI to obtain high resolution (~3 μm) strain/warpage maps and qualitative information on the nature and extent of the strain fields in completely packaged chips. Micro-Raman spectroscopy confirms and complements the 3DSM data by providing quantitative information on the level of strain within the chips prior to the embedding process.

Chips/packages were examined at 3 different stages in the embedding process. Results show that the magnitude of strain induced lattice misorientation/warpage varies from chip to chip within the same process step; however the location of strain maxima and the strain pattern is characteristic of the particular process or reliability test being studied. 3DSM can be generated at each stage of production or reliability testing, and complemented by LAT topographs and angular displacement data.

Lowest levels of strain are apparent before chip packaging/embedding, with strain levels increasing throughout the packaging process. LAT topographs and 3DSM show that all chips have lowest strain levels at corners, irrespective of the processing stage. Analysis of strain patterns using XRDI and 3DSM agree well with the outcomes of FEM, shear testing, SEM and acoustic tomography on similar packages [1, 6, 7, 8, 9]. However, crucially, all these techniques are destructive (and of course FEM modelling is only as good as the assumed input data). The close relation between features observed on the x-ray topographs, and those reproduced in the 3DSM, particularly at via electroplating and post production stages prove that 3DSM is an effective technique for producing accurate maps of the location and magnitude of strain induced die/wafer warpage in side packaged chips.

The need to address non-invasive and non-destructive stress and strain metrology in Systems on Chip or System in Package (SoC/SiP) is recognised in the International Technology Roadmap for Semiconductors (ITRS 2009). 3DSM mapping provides the unique ability to non-destructively assess strain inside sealed packages post production, and after reliability testing, and complements other established characterization techniques. Results prove the potential for this technique to examine non-destructively major sources of die warpage in SoC/SiP, and to provide quick feedback towards process improvement.

Acknowledgements

PMN & JS acknowledge the financial support by EU-FP7 Project No. 216382 SIDAM. This work was conducted under the framework of the INSPIRE programme, funded by the Irish Government's Programme for Research in Third Level Institutions and Science Foundation Ireland's "Precision" Strategic Research Cluster. We would like to thank P. Vagovic˘ and H. Shade at ANKA for their technical support. This work was also supported by the European Community—Research Infrastructure Action under the FP7 "Structuring the European Research Area" Programme. The financial support of HERMES project (FP7-ICT-224611) by the European Commission is highly appreciated.

References

[1] Dionysios Manessis, Shiu-Fang Yen, Andreas Ostmann, Rolf Aschenbrenner, and Herbert Reichl, "Chip embedding technology developments leading to the emergence of miniaturized system-in-packages", Proc. Electronics Components & Technology Conference (ECTC) 2007, Reno, NV, May 29-June 1, pp. 278-285, 2007.

[2] ITRS "More than Moore White Paper", http://www.itrs.net/Links/2010ITRS/IRC-ITRS-MtM-v2%203.pdf

[3] ITRS "SiP White Paper V9.0. The next Step in Assembly and Packaging: System Level Integration in the package (SiP)", http://www.itrs.net/Links/2007ITRS/LinkedFiles/AP/AP_Paper.pdf

[4] M. Töpper et al., "Wafer-level chip size package (WL-CSP) ", IEEE Transactions on Advanced Packaging, Vol.23, No.2, May 2000, pp 233-238.

[5] ITRS 2009, Chapter 2, Assembly and Packaging. http://www.itrs.net/links/2009ITRS/2009Chapters_2009Tables/2009_Assembly.pdf

[6] Dionysios Manessis et al., "Chip embedding technology developments leading to the emergence of miniaturized system-in-packages ", Electronic Components and Technology Conference (ECTC) 2010, pp. 803 – 810.

[7] Fritzsch, T. et al., "3-D thin chip integration technology - from technology development to application", IEEE International Conference on 3D System Integration, 2009, pp. 1-8.

[8] Ligang Niu, Yang, D., Mingjun Zhao, "Study on thermo-mechanical reliability of embedded chip during thermal cycle loading", International Conference on Electronic Packaging Technology & High Density Packaging, (ICEPT-HDP), 2009, pp. 1229 – 1232.

[9] Dionysios Manessis et al., "Breakthoughs in chip embedding technologies leading to the emergence of further miniaturised system-in-packages", Microsystems, Packaging, Assembly

and Circuits Technology Conference, 2009, IMPACT 2009, pp. 174 – 177.

[10] Thenappan Chidambaram, Colin McDonough, Robert Geer, and Wei Wang, " TSV Stress Testing and Modeling for 3D IC Applications", IEEE Proceedings of 16th IPFA - 2009, China, pp.727 – 730.

[11] Daniel N. Bentz, et al., "Modelling thermal stresses of copper interconnects in 3D IC structures", Proceedings of the COMSOL Multiphysics User's Conference, 2005 Boston.

[12] Jacqueline Heinrich, ANKA Instrumentation Book, October 1st, 2007.

[13] Danilewsky A. N. et al., "White beam synchrotron topography using a high resolution digital X-ray imaging detector", Nuc. Instrum. Meth. Phys. Res. B, 266(9), pp.2035-2040, (2008).

[14] J. Stopford, D. Allen, O. Aldrian, M. Morshed, J. Wittge, A.N. Danilewsky, P.J. McNally, "Combined use of three-dimensional X-ray diffraction imaging and micro-Raman spectroscopy for the non-destructive evaluation of plasma arc induced damage on silicon wafers", Microelectronic Engineering 88 (2011) pp.64–71.

[15] R. J. Needs and A. Mujica, "First-principles pseudopotential study of the structural phases of silicon", Phys. Rev. B 51, (1995), pp. 9652–9660.

[16] I. De Wolf, "Stress Measurements in Si Microelectronics Devices Using Raman. Spectroscopy", J. Raman Spectrosc., 30, 877–83 (1999).

[17] D. Noonan, P.J. McNally, W.-M. Chen, A. Lankinen, L. Knuuttila, T.O. Tuomi, A.N. Danilewsky, R. Simon, "The evaluation of mechanical stresses developed in underlying silicon substrates due to electroless nickel under bump metallization using synchrotron x-ray topography", Microelectronics Journal, 37, 11, pp1372-1378 (2006).

[18] J. Kanatharana, J.J. Perez-Camacho, T. Buckley, P.J. McNally, T. Tuomi, A.N. Danilewsky, M. O'Hare, D. Lowney, W. Chen, R. Rantamaki, L. Knuuttila and J. Riikonen, "Evaluation of mechanical stresses in silicon substrates due to lead-tin solder bumps via synchrotron X-ray topography and finite element modeling", Microelectron. Eng. **65** (2003), pp. 209–221.

[19] J. Kanatharana et al., "Mapping of mechanical stresses in silicon substrates due to lead-tin solder bump reflow process via synchrotron x-ray topography and finite element modelling", J. Phys. D: Appl. Phys. 36 pp. A60-A64 (2003).

[20] I. De Wolf, "Micro-Raman spectroscopy to study local mechanical stress in silicon integrated circuits", Semicond. Sci. Tech. **11**, pp.139 (1996).

[21] Jian Chen, De Wolf, I., "Raman spectroscopy as a stress sensor in packaging: correct formulae for different sample surfaces", Electronic Components and Technology Conference, Proceedings. 52nd, 2002, pp. 1310 – 1317.

[22] Jian Chen and Ingrid De Wolf," Theoretical and experimental Raman spectroscopy study of mechanical stress induced by electronic packaging ", IEEE Transactions on Components and Packaging Technologies, Vol 28, No. 3, pp.484 – 492, (2005).

[23] J. Zhang, M. O. Bloomfield, J-Q Lu, R. J. Gutmann, and T. S. Cale, "Thermally Induced Stress in 3D Wafer Stacks", Microelectronic Engineering, 82(3-4), pp 534-547, 2005.

[24] Andreas Ostmann et al., "Industrial and Technical Aspects of Chip Embedding Technology", 2nd Electronics System Integration Technology Conference, Greenwich, UK, pp. 315 – 320, 2008.

[25] Dionysios Manessis et al., "Innovative Approaches for Realisation of Embedded Chip Packages – Technological Challenges and Achievements". 2009 Electronic Components and Technology Conference, (ECTC) 2009, pp 475 - 481

[26] Chien-Wei Chien, et al., "Chip Embedded Wafer Level Packaging Technology for Stacked RF-SiP Application." 2007 Electronic Components and Technology Conference.

[27] J. Stopford, Masters Research Project, School of Electronic Engineering, DCU, 2006.

Design, modeling and fabrication of novel MEMS structure utilizing carbon nanotubes

J. Haze, J. Pekarek, M. Magat, M. Pavlik, R. Vrba

Brno University of Technology, Faculty of Electrical Engineering and Communication,
Dept. of Microelectronics, Technicka 10, Brno, Czech Republic

Phone: +420541146102, Fax: +420541146298, haze@feec.vutbr.cz

Abstract

The paper deals with novel Micro-Electro-Mechanical System (MEMS) structure. The structure was designed to be utilized as capacitance pressure sensor. The most applicable topology of the MEMS structure was selected by means of electrostatic model analysis made by MatLAB software. There were proposed five structures and the best solution was selected as the "chessboard" structure, since it provides the suitable capacity for the sensor measurement range. The overall capacity and also sensitivity of the MEMS sensor was increased by means of carbon nanotubes (CNTs). Therefore the pressure measurement range increased as well. The comparison between conventional capacitance sensor structures and the presented proposal shown, that the improvement depends on the area and number of the growing fields. However this improvement is from tens to hundreds of percents. The article depicts the design process including simulations and analysis of the chosen structure too.

Key words: MEMS, capacitive sensor, carbon nanotubes

Introduction

The MEMS technology and its applications are one of the most growing areas in microelectronic world [1], [2], [3], [4]. Fig. 1 shows typical representative of MEMS sensor for industrial application.

Figure 1: MEMS resonator [5]

The very innovative automotive industry introduced the MEMS in cars in last years to make new cars safer (accelerometers for airbags), more energy efficient (fuel pressure sensors), and more environmentally friendly. There are many other automotive applications such as air flow sensors, tire pressure sensors with automatic built-in tire pumps, "smart" sensors for collision avoidance and skid detection, smart suspension for sport utility vehicles to reduce rollover risk, automatic seatbelt restraint and door locking, vehicle security, headlight leveling, and navigation.

The next progressive application is in variety of disposable medical devices. The most succesfull application in medicine is pressure sensor. It can measure blood pressure, intrauterine pressure during birth, patient's vital signs. Moreover the pressure sensors are used in surgery devices, hospital beds and many other medical devices. The future in this area belongs to home personal medical care "in vitro".

The very promising area of MEMS and nanotechnology usage are biotechnology sciences, ie. the polymerase chain reaction microsystems for DNA amplification and identification, enzyme linked immunosorbent assay, capillary electrophoresis, electroporation, micromachined scanning tunneling microscopes, biochips for detection of hazardous chemical and biological agents, and microsystems for high-throughput drug screening and selection.

It is possible to find various pressure sensors, microactuators, accelecometers etc. in ordinary life devices too.

Therefore many researchers are focusing onto MEMS and nanotechnologies development since the MEMS market is still growing and more application possibilities wait for full exploitation [6], [7], [8], [9].

The paper presents a novel capacitive MEMS pressure sensor, its design, development and fabrication using the CNTs [10], [11], [12]. The paper is structured as follows. First the possible structures and selection of appropriate sensor structure are presented. Simulation results with selected ideal and non-ideal structure are shown afterwards. The simulations of non-ideal structure were mainly focused on mismatching of the sensor plates. The fabrication process and CNTs growth are depicted later on.

Proposed MEMS structure

The most applicable topology of the MEMS structure was chosen by means of electrostatic models analysis. The analyses proceed using of MatLAB software. There were proposed several structures and the best solution was selected as the "chessboard" structure as shown in Fig. 2, since it provides the suitable capacity for the sensor. The other analysed structures were comb structure, pyramid structure and chimney structure. The last two structures were discarded firstly, because of the possible fabrication problems with substrate preparation or non-ability to control of CNTs growth (pyramid structure).

The proposed "chessboard" structure consists of the top and bottom plates forming together the designed pressure sensor. Since the capacity of the sensor depends on the area of the plates, the CNTs were added to increase the overall capacity. Each patch of the plates serves as substrate for CNTs growth. The plates with nanotubes change their capacity according to applied pressure. It is due to the columns of nanotubes, which are more or less fitted together depending on applied pressure.

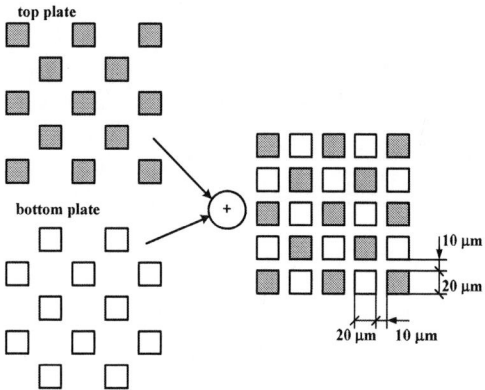

Figure 2: The principle of the selected structure - „chessboard"

The most important dimensions of the presented structure are dimensions of the top and bottom plates (the complete structure, not only single square-shaped part) and distance between both plates. The optimal dimensions were also observed, because the number of growing CNTs depends on the area of each single patch. The less patch area the less number of the CNTs and vice versa. However small patch dimensions (about

hundereds of nm) lead to higher number of patches on the same capacitor area. Therefore the overall area and thus sensor capacity will increase. Consequently this is trade-off between request on sensor capacity and fabrication possibilities, which were taken into account.

Simulation results

The simulation of final dimensions based upon electrostatic model was done by utilization of the MatLAB software as well. The model simulations also had shown the influence of non-symetrical fitting of the plates on overall pressure sensor capacity.

The ideal state, where the symetrical structure is assumed, was analysed at first. The input parameters of the simulated block were patch dimensions and the distance between the plates ($L_{elektrod}$) as depicted in Fig. 3.a). Fig. 3b) shows the electrostatic model of the ideal "chessboard" structure. The area of the plates was 5 x 5 mm at the beginning, but final dimensions were adjusted according to specification of certain application.

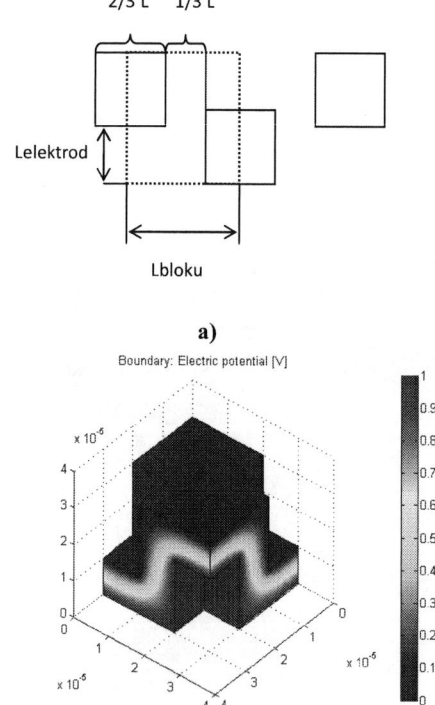

Figure 3: The ideal symetrical „chessboard" structure a) its electrostatic model b)

There has been observed the typical capacity dependence on the distance between top and bottom plate. It is illustrated in Fig. 4. The distance was changed from 4 to 22 µm, which causes the capacity change about 4 pF.

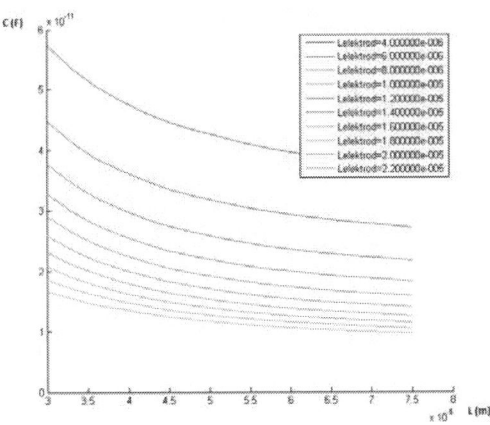

Figure 4: The dependence of the sensor capacity on distance between plates

Since the fabrication process is not perfect, it was also simulated the mismatching of the both sensor plates.

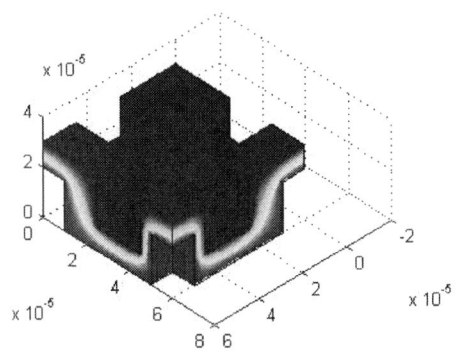

a)

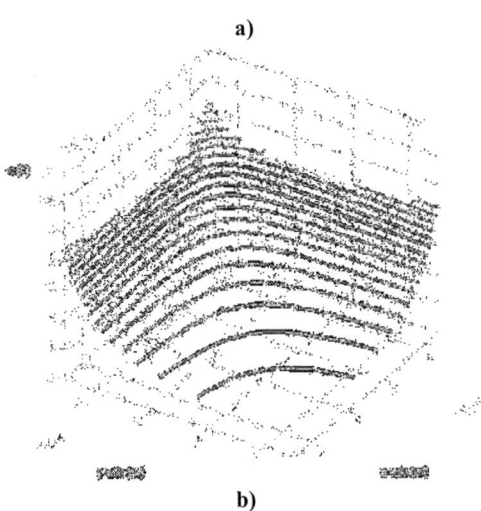

b)

Figure 5: The electrostatic model of the plates mismatching a) its influence on the overall capacity b)

The overall error of the mismatching was calculated for spacing between patches of 10 µm and shift in X and Y axes from 1 to 9 µm. Fig. 5a) shows the electrostatic model of this case, capacity changes are depicted in Fig. 5b). The simulated mismatching causes capacity changes from 4 to 9.2 pF. Thus the accurate patch fitting is very crucial condition to be satisfied.

Sensor manufacturing

The used fabrication process allows us to control the length of each nanotube. Therefore it is possible to fabricate sensor with exact specified capacity. Fig. 6 depicts the fabricated plates. The ferrum catalytic layers with defined shapes and dimensions were prepared for CNTs growth. The various dimensions of these shapes were fabricated to verify the simulation results. There were produced patches with 1 µm, 5 µm, 10 µm and 15 µm dimensions in the first experiment fabrication stage. There was used the negative photoresist to satisfy appropriate layer thickness, homogenity and adhesion.

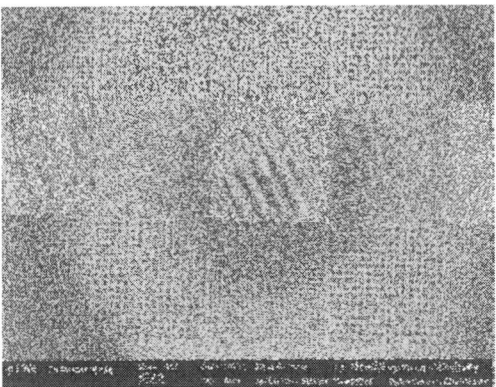

Figure 6: The fabricated „chessboard" prepared for CNTs growth

Patches with 100 nm, 250 nm, 500 nm, 800 nm, 1 µm and 5 µm dimensions were produced in the second fabrication stage. The overall area of the plates for one sensor was 400 µm x 400 µm. The negative photoresist was diluted for patches with dimensions smaller than 500 nm.

There were done tens of experiments, because the original diluent for photoresist diluting was not compatible with ferrum catalytic layer. Therefore the homogenity and mainly adhesion were not acceptable for the fabrication purposes. The optimal diluent "N1" was deposited with rotation speed of 5000 rpm. The promoter HMDS (hexamethyldisilazane) was deposited to improve the adhesion. The final thickness of the photoresist was 93 nm, which had very good homogenity and adhesion properties.

The exposed ferrum catalytic layer was sputtered out, the photoresist was removed and carbon nanotubes growth started.

Deposition of carbon nanotubes

The CNTs were prepared by plasma enhanced chemical vapor deposition (PECVD) on the silicon wafer with a patterned iron catalytic layer.

This unique technique has many advantages, such as absence of the vacuum system and time undemanding deposition. Microwave power from a 2.45 GHz generator is supplied via a rectangular waveguide, matching unit and coaxial line to an iron nozzle electrode.

The CNTs were grown on silicon substrates coated with a thin iron catalytic layer (5 nm thick) which was vacuum-evaporated. The substrates with the catalytic layer were directly used for the deposition of nanotubes. The typical deposition conditions were: flow rates of argon, methane and hydrogen QAr = 1000 sccm, QCH4 = 50 sccm and QH2 = 200 to 300 sccm, respectively, microwave (mw) power of 400 W, substrate temperature 900 to 1100 K, deposition time 1 minute. Thin CNTs with a diameter of about 100 nm were standing vertically perpendicular to the substrate due to a crowding effect. A detailed study of the microwave torch for deposition of CNTs and their characterization were published in [10].

Fig. 7 depicts the CNTs made with proposed "chessboard" structure. The patch dimensions were 10 μm x 10 μm.

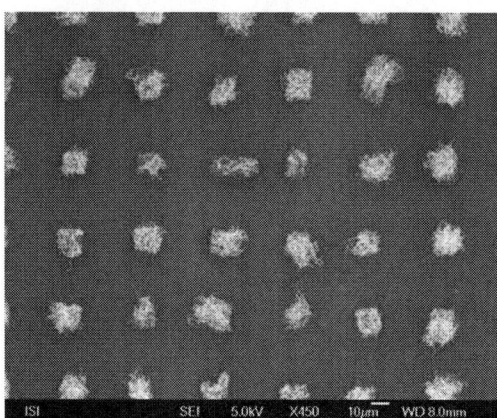

Figure 7: The fabricated „chessboard" structure with grown CNTs

The first sample of the manufactured pressure sensor is shown in Fig. 8.

Figure 8: The fabricated capacity MEMS pressure sensor

Very first laboratory results measured on fabricated structures show very accurate values which correspond with simulated results. The sensors have capacity from pF to hundereds of pF, so it is wide range of pressures, which sensors are able to measure.

Conclusions

The paper describes design, simulation and fabrication of novel MEMS capacity sensor. There were proposed several structures of the sensor. The "chessboard" structure was selected as optimal to requirements defined at the beginning of the research. The chosen structure was analysed utilizing MatLAB software. There were simulated ideal and mismatched sensor structures. The simulation results helped to define real dimensions of the sensor plates for sensor fabrication. Carbon nanotubes were used as well. The overall capacity increased, because the area of the final sensor construction increased too.

Acknowledgements

The research is supported by Czech Ministry of Education and Sports in the frame of Research Program MSM0021630503, by the Ministry of Industry and Trade of the Czech Republic under the MPO 2A-1TP1/143 and by the Czech Science Foundation under 102/09/1601 project IMINAS.

References

[1] CH. Liu, "Foundations of MEMS", Pearson Education, Inc., New Jersey, USA, 2006, ISBN 0-13-147286-0.

[2] T., R., Hsu, "MEMS and Microsystems – Design, Manufacture, and Nanoscale Engineering", John Wiley and Sons, Inc., second edition, New Jersey, USA, 2008, ISBN 978-0-470-08301-7.

[3] O., Brand, G., K., Fedder, "CMOS – MEMS, Advanced Micro and Nanosystems", Wiley-Vch Verlag GmbH and Co. KGaA, volume 2, 2005, Weinheim, Germany, 2005, ISBN 978-3-527-31080-7.

[4] Web pages dedicated to MEMS and Nanotechnology Therory and Applications – "MEMSnet", http://www.memsnet.org/.

[5] Project "Long-term stability of vacuum-encapsulated MEMS devices using eutectic wafer bonding", Proposal/Contract no.: IST-2001-34224.

[6] Web pages dedicated to MEMS - http://www.allaboutmems.com/.

[7] Web pages dedicated to MEMS - http://www.memx.com/.

[8] K., D., Wise, J., M., Giachino, H. Guckel, G., B., Hocker, S., C., Jacobsen, R., S., Muller, "Microelectromechanical systems in Japan", JTEC Panel report, 1994, ISBN 1-883712-35-1.

[9] RTO Educational Notes, "MEMS Aerospace Applications", North Atlantic Treaty Organisation, 2004, ISBN 92-837-1113-0.

[10] R., Ficek, R., Vrba, L., Zajíčková, O., Jašek, F., Matějka, "Carbon nanotubes pressure sensors modified for chemical analysis", Proceedings of the 12th International Conference on Miniaturized Systems for Chemistry and Life Science. San Diego, USA, pp. 245 – 247, 2008, ISBN 978-0-9798064-1-4.

[11] J., Pekárek, R., Vrba, R., Ficek, M., Magát, "Carbon nanostructures used in capacitive sensors as the surface increase element", Proceedings of the 15th International Symposium for Design and Technology of Electronics Packages, Budapest, Hungary, pp. 353 – 356, 2009, ISBN 978-1-4244-5133-3.

[12] M., Magát, R., Vrba, J., Pekárek, R., Ficek, "Capacitive pressure sensor modelling", Proceedings of the The Second International Conference on Advances in Circuits, Electronics and Microelectronics, 11-16 October, Sliema, Malta, pp. 81 – 85, 2009, ISBN 978-0-7695-3832-7.

[13] J., Pekárek, R., Vrba, R., Ficek, M., Magát, "Electrodes Modified by Carbon Nanotubes for Pressure Measuring", Proceedings of the 32nd International Spring Seminar on Electronics Technology, Brno, Czech republic, pp. 1-5, 2009, ISBN: 978-1-4244-4260- 7.

[14] J., Háze, R., Vrba, M., Pavlík, "The integrated unit for MEMS based pressure measurement", Proceedings of ICONS 2009 Fourth International Conference on Systems, France, pp. 100-103, 2009, ISBN: 978-1-4244-3469- 5.

Inverse Modeling of the Button Shear Test

Torsten Hauck, Min Ding, Ilko Schmadlak

Freescale Halbleiter Deutschland GmbH

Phone +49-89-92103685, Ilko.Schmadlak@Freescale.com

Abstract

This paper presents computational fracture analysis of the button shear test. The specimen under consideration is commonly used for the assessment of adhesion strength between encapsulating mold compounds and metal leadframes. The finite element method is applied for stress strain analysis of the shear test and for the extraction of fracture parameters for various crack configurations in the material interface. Impacts of geometry and loading conditions of the button shear setup onto load and the crack driving forces at the interface crack tip will be discussed. The simulation results will be correlated with existing experimental data for different material combinations. Means to an estimation of adhesion strength values from the button shear test will be provided.

Key words: Delamination, Interface Fracture, Adhesion Strength

Motivation, Adhesion Strength

Mechanical testing of material and interface strength properties is a fundamental part of electronic package development and qualification. Delamination between the metal leadframe and the mold compound, which encapsulates die and interconnects, is a challenging failure mode at temperature cycling tests during product qualification. The button shear test is an established way to pre-screen a variety of mold compounds and leadframe surface finishes, finding the material pairing with best adhesion strength, even before first prototypes are available. The test is usually of comparative nature. The force, required to shear the button from the leadframe is recorded and used to rate the different material combinations.

The presented work is focusing on ways to enhance the button shear test by means of finite element simulation. A combined effort of test and simulation could increase the value of test results significantly, if it would be possible to conclude the strain energy release rate directly from measured shear forces.

Button Shear Test

The test method for the button shear test is described by the SEMI G69-0996 standard [1]. A frustum shaped button of molding compound on the surface of a lead frame sample is sheared off the leadframe surface by a chisel tool (Fig. 1 and 2). The variation of the shear height can be part of the test in order to account for different shear / tool mode mixity values at the material interface.

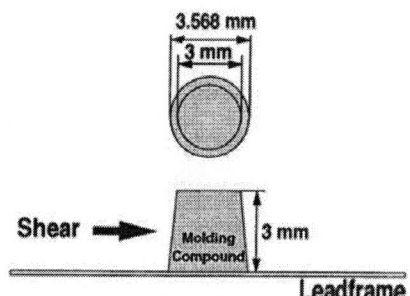

Fig. 1: Sample Button Shear Test (SEMI G69-0996)

Fig. 2: Button Shear Test Specimen

During the process of shearing the required force is measured in order to assess the adhesion that is present in the interface. This is the key result parameter in interface pre-screening with the button shear test (Fig.3).

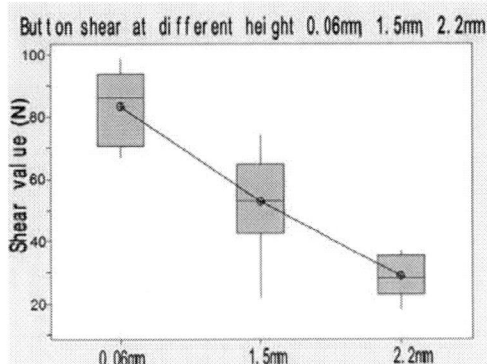

Fig. 3: Example of shear test results (shear force over chisel height

Crack Advance Model

In order to predict delamination by means of modeling and simulation, a critical crack driving force or interface adhesion strength needs to be known. This failure criterion is compared to Energy Release Rate (ERR) values, calculated from the Finite Element simulation results [2]. Unlike stress based simulation methods, the energy based approach avoids the singularity problems at the modeled crack tip, returning always finite results.

In case of the presented button shear simulation, the specimen on a copper leadframe is modeled, considering half symmetry. The chisel is represented by a rigid contact target with variable height (Fig. 3). The force is applied to the leadframe part of the model. The simulations consider the thermo-mechanical pre-stress due to cool down from assembly temperature.

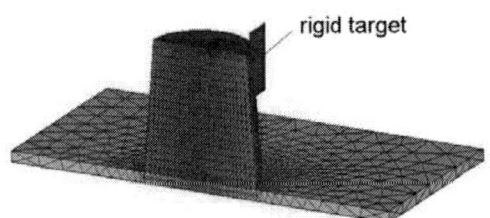

Fig. 3: Button Shear Finite Element model

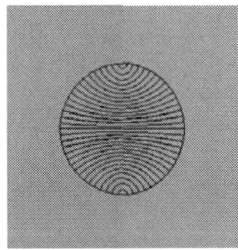

Fig. 4: Button to lead frame interface with predefined crack openings modeled

A simulation model was generated that accounts for 31 predefined crack opening stages as Figure 4 shows. The ERR was calculated for a crack growth, defined by the crack opening stages, until half of the material face is opened.

Global Energy Release Rate

Different approaches to assess the delamination risk with energy based simulation methods have been published [2]. This paper uses a procedure that calculates the fracture energy of predefined cracks of different size and determines the ERR over crack size function in sub sequential steps [3]. In order to identify the fracture energy, two quantities need to be known. These are the adhesion forces in the uncracked interface between mold compound and copper leadframe and crack opening displacements of the crack in the same interface (ref. Fig.5).

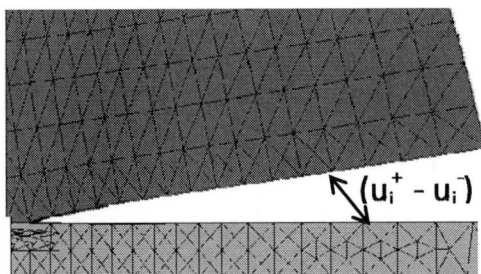

Fig. 5: FE model with in-build delamination

The total fracture energy that is released from the elastic body by an interface crack of a given size is then determined by the adhesion forces and the associated crack opening displacements at each interface node i as follows

$$W = \frac{1}{2} \sum_i (u_i^+ - u_i^-) f_i \qquad (1)$$

By performing a polynomial expansion of the fracture energy as a function of crack surface

$$W = C_2 A^2 + C_3 A^3 + \ldots + C_n A^n, \qquad (2)$$

and differentiating it with respect to the crack surface area, the ERR is obtained:

$$ERR = \frac{\partial W}{\partial A} = 2C_2 A + 3C_3 A^2 + \ldots + nC_n A^{n-1} \quad (3)$$

Figures 6 and 7 show the fracture energy and the ERR for a given chisel height and shear force, plotted over crack face area and crack length, respectively.

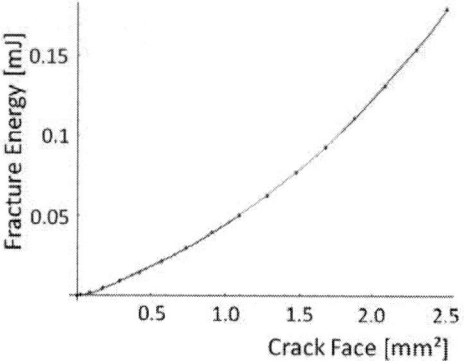

Fig. 6: Fracture energy over crack surface area

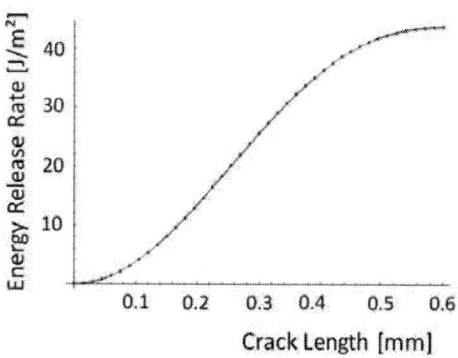

Fig.7: ERR over crack length

Inverse Modeling and Estimation of Adhesion Strength

The aim of this work is a procedure to link the shear force directly with a strain energy release rate, knowing chisel height and initial crack size. This ERR would then be equal to the adhesion strength of the tested interface. The authors propose a set of iso-ERR curves that are plotted in a force over crack length diagram for different chisel heights.

The chart in Figure 8 represents such a family of curves for a chisel height of 2 mm. In the below example, an initial delamination length of 0.2 mm is assumed which, together with a measured shear force of 30N, results in an ERR of about 6 J/m².

In order to plot a sufficient number of these curves, a simulation study, with varying shear force and delamination size needs to be carried out for different shear tool heights. The results are then valid for the mold compound that was considered in the model. Different surface finishes on the lead frame side, such as plating or plasma treatment are not considered in the model and only have effects on the test results. The same iso-ERR diagram would be used in that case.

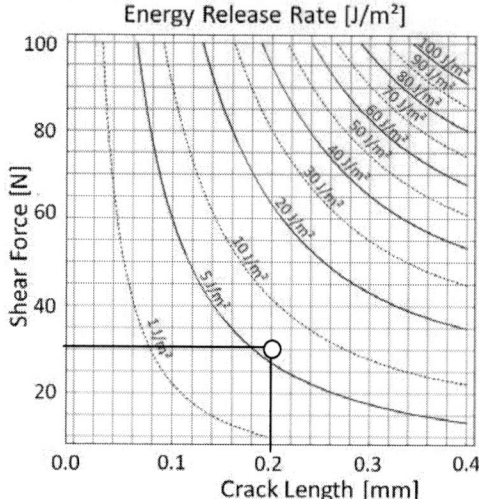

Fig. 8: Iso-ERR curves for shear force over crack length for a chisel height of 2mm

Influence of Shear Tool Height

The measurements (Fig.3), considering three different shear tool heights, revealed a decrease in the required shear force with an increase in shear tool height. This is explained by the different mode mixity that is present at the crack tip. If the shear tool contacts the test specimen at a higher point, the peel component will become more dominant in the stress state at the crack tip. The adhesion strength for the interface would be reduced as a consequence.

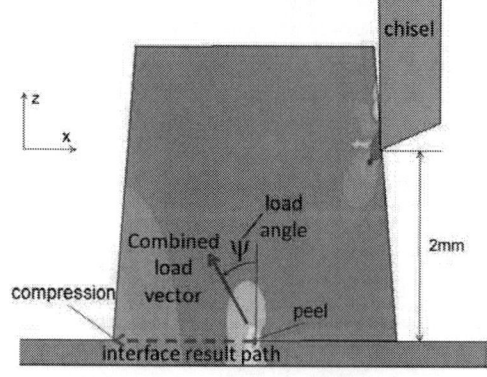

Fig. 10: σ_z stress contour showing result path for load angle

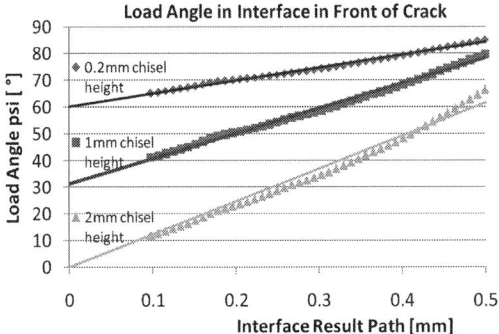

Fig. 11: Load phase angle in the material interface (ligament) for three different chisel heights

This expected behavior is verified by the simulation results. Figure 10 shows the result path along which the shear/peel mode mixity was extracted from the model. The curves in Figure 11 show the load phase angle ψ, a measure for the mode mixity, over the path from the crack tip towards the edge of the button at the 3 different chisel heights of 0.2mm, 1mm and 2mm. An angle of 90° represents a pure shear mode and maximum interface adhesion strength, while 0° equals pure peel mode and lower adhesion strength values.

For the 2mm chisel height scenario, the load at the crack tip is pure peel stress. Moving along the plotted path, away from the crack tip, into the interface, the shear component is increasing until a point of pure shear is passed. At a larger distance to the crack tip, the compressive component is increasing, showing a maximum compression at the button edge (ref. Fig. 10).

If the shear tool height is decreased the load phase angle near the crack tip increases, hence the interface load vector contains a larger shear component and the interface adhesion strength increases.

In order to use the described procedure, an initial delamination or crack would need to be present and its size known. This is not the case for the current button shear test, but could be realized by applying appropriate changes to the sample preparation.

Summary

The paper describes the button shear test as a valuable method for pre-screening different mold compound to copper leadframe interfaces. Differences in the adhesion strength result from different materials, plating, surface finishes or other process parameters. The authors further describe the method of simulating this test with finite element modeling and extract fundamental fracture parameters in order to calculate the strain energy release rate during crack advance. Further, the paper discusses how the inverse modeling of a set of button shear scenarios

could be used to extract the adhesion strength for the interface directly by measuring the shear forces present during the experiment. The proposed method requires knowledge about the initial crack size which is not foreseen in the current test specification. A predefined, initial delamination present in the interface as part of the test is therefore proposed.

References

[1] SEMI G69-0996, "Test Method for Measurement of Adhesion Strength between Leadframes and Molding Compounds", SEMI 1996

[2] R. Dudek, R. Pufall, B. Seidler, B. Michel, "Studies on the Reliability of Power Packages Based on Strength and Fracture Criteria", Conference on Thermal, Mechanical and Multi-Physics Simulation and Experiments in Microelectronics and Microsystems, EuroSimE, Linz 2011

[3] T.Hauck, W.H. Mueller, I. Schmadlak, "Fracture Risk Assessment for Interface Flaws in the 3D-Interconnect Systems of a CMOS Chip", Conference on Thermal and Thermomechanical Phenomena in Electronic Systems, ITHERM, Orlando 2008

Changes in Water Absorption and Modulus of Elasticity of Flexible Printed Circuit Board Materials in High Humidity Testing

Sanna Lahokallio, Kirsi Saarinen, and Laura Frisk

Department of Electronics, Tampere University of Technology
P.O.Box 692, 33101 Tampere, Finland

Phone +358 40 849 0168
Fax +358 3 3115 3394
Email address: sanna.lahokallio@tut.fi

Abstract

Polymers are frequently used as the insulating material in printed circuit boards (PCBs). However, many polymers tend to absorb moisture and this may impair their electrical and mechanical properties. In this study three flexible PCB materials; polyimide (PI), fluorinated ethylene-propylene (FEP) and polyethylene terephthalate (PET) were aged in high humidity test in which a temperature of 85°C and relative humidity of 85 percent were used. The aim was to study whether the high humidity test has an effect on the water absorption and mechanical strength of the parameters examined. Water absorption was measured using two different methods; precision scale and moisture analyzer. The modulus of elasticity was determined with a thermomechanical analyzer (TMA) in order to gain information about changes in mechanical strength. High humidity testing seemed to decrease the water absorption of PI but it had no effect on the modulus of elasticity. On the other hand, water absorption of FEP remained very low after high humidity testing but the modulus of elasticity decreased. For PET the high humidity test was very harsh and it became fragile during aging. Testing did not affect water absorption of PET before it became brittle but after that the water absorption increased markedly. The modulus of elasticity of PET seemed to decrease during high humidity testing but as the standard deviation was very high, it is not certain whether or not the mechanical strength was already impaired before the embrittlement.

Key words: High humidity testing, water absorption, modulus of elasticity, flexible printed circuit board, thermomechanical analysis

1. Introduction

Printed circuit boards (PCB) are essential parts of electronic products. In addition to providing electrical interconnections, the functions of PCBs include mechanically supporting components mounted on them and providing a safe working environment for the whole device, such as offering a path for thermal conduction away from components [1; 2]. In general there are two basic groups of PCBs. These are rigid and flexible PCBs. Flexible PCBs are chosen instead of traditional rigid PCBs when lower weight or thinner package sizes are needed [3; 4]. Additionally, since flexible boards can be bent, they can tolerate considerable vibration, and their heat dissipation capability is better than with rigid PCBs [3; 4]. However, fabrication and processing costs of flexible PCBs are higher, dimensional stability is poorer and fragile behaviour is possible [5; 6].

Polymers are mainly used as the insulating material in PCBs. Many polymers tend to absorb moisture and this may impair important properties. Typical changes resulting from moisture are the impairment of mechanical and electrical properties,

particularly insulating properties [3]. Moisture absorption may also cause hygroscopic swelling, which may increase the package stresses [7; 8] when polymers expand significantly more than the metal wirings. Additionally, at high temperatures the water uptake and swelling frequently increases [9].

Another important function of polymers in PCBs is to provide mechanical support. Therefore mechanical strength is an important parameter for PCB materials. Mechanical strength is typically measured with the modulus of elasticity or tensile strength. Water absorption may impair mechanical properties of polymers and decrease tensile strength.

For PCB materials decreases in insulating properties and mechanical strength can have a great effect on their reliability. In this study flexible PCB materials were aged in high humidity test in order to study the effects of humidity and temperature. The effects of aging are important information in modelling and in reliability engineering. In modelling material parameters are needed in order to create accurate models. This is because material parameters and all changes in them for all the different materials used in the model need to be known so as to predict behaviour in the required environment. In addition, in reliability engineering it is important to know the behaviour of different parts of the system so as to be able to analyze for possible failures.

The effect of high humidity testing on the water absorption and the mechanical strength of flexible PCB materials has not been widely studied. However, the effect of high humidity test on the water absorption of PI has been studied previously by Denton et al. It was reported that the water absorption of PI increases after aging [10]. Additionally Tsukiji and Bitoh have studied the effect of high temperature aging on the tensile strength of PI. It was found that aging at 400 ºC decreases the tensile strength according to the aging time [11]. Oreski and Wallner have studied the effect of high humidity test on the tensile strength of PET. The tensile strength of PET was reported to decrease during aging due to the embrittlement caused by hydrolysis and PET samples aged for 2000 hours fracture before their yield stress was reached [12].

In this paper three PCB materials, polyimide (PI), fluorinated ethylene-propylene (FEP) and polyethylene terephthalate (PET), were aged in high humidity test in which high temperature and high humidity were combined. Additionally the effect of aging on water absorption was measured with two different techniques and changes in mechanical strength were determined by measuring the modulus of elasticity.

2. Experimental

In this study three pure polymer films without metal wirings were used; polyimide (PI), fluorinated ethylene-propylene (FEP) and polyethylene terephthalate (PET). The films were first aged in high humidity test in which temperature of 85ºC and relative humidity of 85 percent were used. The testing time varied from 500 hours to 6000 hours. In addition, non-aged reference samples were also analyzed. After aging, water absorption and modulus of elasticity were measured in order to study the possible changes in chemical and mechanical properties.

Water absorption was measured with precision scale. Additionally, another method, moisture analyzer was used to verify the results. Both of these methods are based on the determination of mass and its changes due to absorption. Water absorption can be calculated from equation [9]

$$\frac{Weight(wet) - Weight(dry)}{Weight(dry)} \cdot 100\%$$

The size of the samples was approximately 5 cm x 5 cm. Thickness of PI was 50 µm, of FEP 127 µm and of PET 125 µm. Three samples from each different aging group were analyzed for each of the materials.

In the precision scale measurements the samples were first dried in a laboratory oven for 2 hours at a temperature of 110ºC in order to ensure that moisture absorbed from indoor air would be removed. After drying, the samples were cooled in desiccators. PET samples were excluded from the oven drying as their glass transition temperature T_g is lower than 110ºC according to the TMA measurement done on the PET samples. The risk of altering the structure of PET during the drying was too high and thus the drying was achieved by keeping the samples in desiccators until their weight remained constant. Dry weights of the samples were measured with precision scale and subsequently the samples were immersed in deionized water. Wet weights were measured after different immersion times; after 20 and 40 minutes, 1 and 2 hours and 1, 2 or 5 days until the absorption reached saturation.

In the moisture analyzer measurements wet, saturated samples were dried with a halogen dryer. The heating temperature was chosen to be the lowest possible, 50ºC, to ensure low vaporization because the test samples were thin and saturated quickly. Samples were assumed to be totally dry when no changes in the weight were observed after ten minutes.

Prior to measuring the modulus of elasticity, the thermal history was removed by heating the samples above their T_g in order to relax thermal stresses [13]. For PI the temperature used was 420ºC, for FEP 290ºC and for PET 85ºC was used. The length of the samples was 23 mm and width was between 2.6 mm and 4.7 mm depending on the sample. The thicknesses were those of the films.

The modulus of elasticity was measured with a thermomechanical analyzer (TMA) which enables the measurement of dimensional changes in length and volume as a function of time or temperature [13]. Stress ramp tests in which the sample is exposed to a ramped stress and the resultant strain is monitored were used. Stress was increased at 1 MPa per minute up to 10 MPa. A constant temperature of 25ºC was used across the whole stress ramp. From the measured stress-strain –curve the modulus of elasticity was determined as an angular coefficient of the curve.

3. Results and Discussion

3.1 High Humidity Testing

The high humidity test had no visual effect on PI or FEP films but it appeared to be very harsh for PET films. The PET samples aged for 4000 and 6000 hours were so fragile that they broke into pieces when handled and controlled cutting was not possible. The behaviour of PET films is caused by slow hydrolysis of ester groups in amorphous parts of the material [14; 15; 16]. This leads to chain scission, which results in embrittlement of the material [14; 15; 16; 17]. As a consequence, measurement of water absorption with precision scale or measurement of the modulus of elasticity with TMA was not possible as the PET samples could not survive drying or moving of the sample nor the sample preparation. The only possible measurement was to measure water absorption of PET with moisture analyzer. Furthermore, the mechanical strength of the PET samples had decreased so much that their use as flexible PCBs would not be practicable or even possible, as the embrittled PET would not be able to provide mechanical support for any electrical device.

3.2 Water Absorption with Precision Scale

Water absorption of PI is shown in Figure 1. The water absorption rate of PI is very high during the first 20 minutes of immersion. During this time the water absorption increases from zero to over one percent. After 20 minutes the absorption rate decelerates and levels off after one and two days. At this point the saturation is reached and film samples cannot absorb more moisture. In Figure 1 the water absorption of some samples seems to decrease between one and two days; this is due to scale inaccuracy. Additionally, as visible water drops were dried manually, the samples were not all necessarily equally dry. The remains of water can easily cause inaccurary in the weights of light samples.

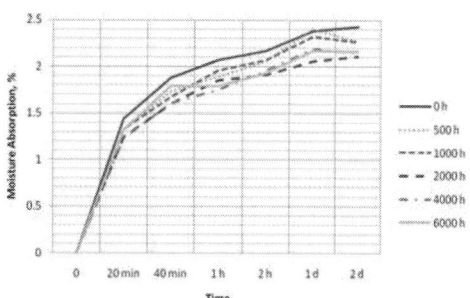

Figure 1: Water absorption of PI measured with precision scale.

In general, water absorption of PI seems to decrease as the aging time increases. According to Turnhout et al., amorphous polymers tend to undergo a vitrification process during aging [18]. This will have an effect on properties which are sensitive to the changes at T_g [18]. Consequently, this would result in a decrease in free volume and water absorption would also decrease as there is less space for water in the polymer structure [18]. As PI is an amorphous material, this could explain the behaviour of PI in this study. On the other hand, T_g for PI is much higher than the aging temperature. However, as humidity can furthermore decrease T_g [19; 20], a decrease of the free volume may also

have occured at the temperature used in high humidity test. However, Denton et al. report the reverse of this behaviour for aged PI [10].

High humidity testing as well as immersion time did not affect the water absorption of FEP as can be see in Figure 2. All the changes in weigh are a maximum ± 0.003 grams which is equal to the linearity of precision scale. Water absorption of FEP fluctuates around zero. If the total average water absorption of all measurements is calculated, the result is zero percent with a standard deviation of 0.02 percent. Even though the FEP films may have absorbed some water, the precisions scale used was not accurate enough to be able to measure it. Furthermore, the water absorption of FEP was very easy to measure as FEP is resistant to water and the surface of the material was very dry. Consequently, drying was easier than with PI and measurements were easier to perform accurately under similar conditions.

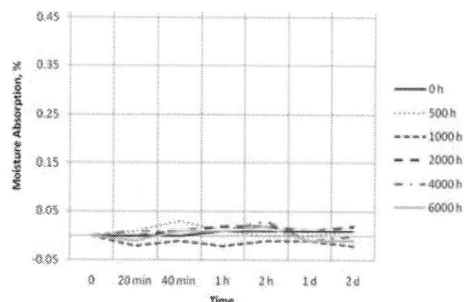

Figure 2: Water absorption of FEP measured with precision scale.

As Figure 3 shows, PET absorps water slowly and the absorption is quite low. PET samples absorbed water gradually and some of the samples did not show clear saturation even after five days of immersion. As the growth of water uptake had, however, levelled off, saturation was assumed to have been reached after five days. Samples aged for 1000 and 2000 hours seem to have already reached saturation after only two days.

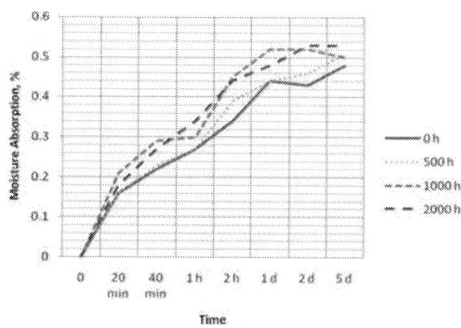

Figure 3: Water absorption of PET measured with precision scale.

72

With precision scale PET samples up to a maximum of 2000 hours aging were analysed because other PET samples were too brittle to handle. Water absorption of non-aged PET samples is lowest but the differences between other PET series are very small, only few tenths of a percentage unit.

3.3 Water Absorption with Moisture Analyzer

Only one measurement from each of the aging groups for each material could be performed with moisture analyzer because the weight of the sample must exceed 0.5 grams. The films were so lightweight that all three samples from each material and aging group had to be combined to exceed this limit.

For PI the water absorption altered randomly between 2.89 and 3.23 percents with repeatability of ± 0.6 percent. If the aging decreased the water absorption of PI, the effect was obscured by the inaccuracy of the device. The water absorption of FEP also changed randomly between 0.05 and 0.14 percent with repeatability of ± 0.15 percent. As the change in weight was a maximum of 0.004 grams, any apparent differences in the water absorption values were caused by the inaccuracy of the moisture analyzer.

However, the moisture analyzer allowed water absorption measurements of brittle PET samples. The water absorption of PET samples aged up to 2000 hours changed randomly between 0.67 and 0.78 percent with repeatability of ± 0.3 percent. However, the water absorption of the PET sample aged for 4000 hours was 0.86 ± 0.3 percent and the sample aged for 6000 hours had water absorption as high as 1.64 ± 0.3 percent. This marked increase in water absorption is probably caused by the cracks and micro cracks which are created in the structure due to the embrittlement. Through them moisture can penetrate the material more easily [21].

For each material the absorption results were markedly higher compared to the precision scale measurements. Additionally the repeatability of the results was low as the samples were so light. The moisture analyzer is not well-suited to measuring the water absorption of lightweigt film samples.

3.4 Modulus of Elasticity

The average values of the elastic modulus determined from stress-strain –curves for PI are shown graphically in Figure 4. The standard deviation is shown with vertical error bars. It can be clearly seen that aging does not seem to affect the modulus of elasticity of PI. The average values vary randomly from 2.5 to 2.9 GPa. Standard deviation is varying quite much but is still quite low, a maximum of few hundred Pascals. It can reasonably be concluded that the mechanical strength of PI does not change after aging for up to 6000 hours.

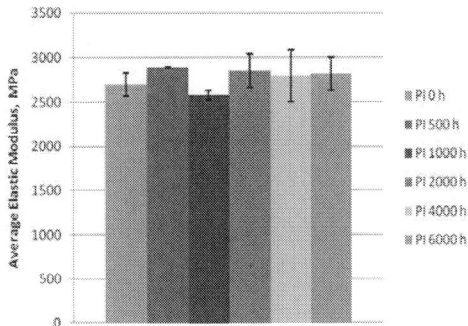

Figure 4: Modulus of elasticity of PI.

The average elastic modulus values determined from stress-strain –curves for FEP are shown graphically in Figure 5. Standard deviations are again represented by vertical error bars. When the values of the elastic moduli are studied, it seems that aging decreases the elastic modulus of FEP. The average moduli values decrease from 448 MPa to 313 MPa fairly systematically. Standard deviation is a maximal of 49 MPa but in most cases it is much less. The results indicate that aging has an effect on the elastic modulus. It seems that this level of aging impairs the mechanical properties of FEP.

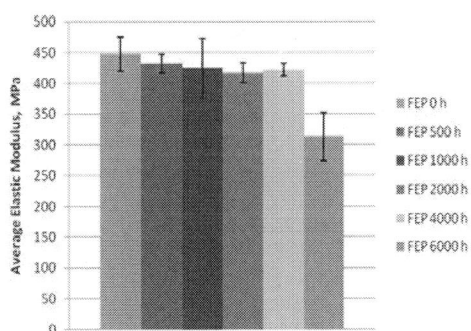

Figure 5: Modulus of elasticity of FEP.

The average elastic modulus values determined from stress-strain –curves for PET are shown graphically in Figure 6.

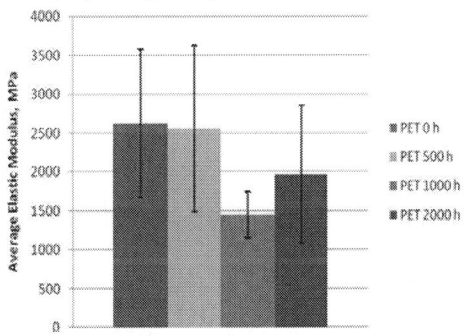

Figure 6: Modulus of elasticity of PET.

As earlier, the standard deviation is represented by the vertical error bars. It seems that the modulus of elasticity decreases during aging. However, because the standard deviation is almost as high as the moduli values themselves, this cannot be concluded with certainty.

4. Conclusions

PI justified its reputation as being a reliable PCB material. Aging did not have an effect on the modulus of elasticity which means that PI maintains its mechanical properties even after 6000 hours of aging. The problematic water absorption seems to decrease during aging. Lower absorption results most likely from the decrease in free volume which limits the space where the moisture can penetrate.

Aging did not affect the water absorption of FEP and its absorption of moisture was very small even after 6000 hours of aging. However, its modulus of elasticity decreased when aging time increased and thus, the mechanical strength reduced although no external, visible changes occured.

For PET the clear result was that between 2000 and 4000 hours of aging the film lost its capability to serve as PCB material because high temperature and high moisture together caused severe embrittlement of the material. PI and FEP films maintain their flexibility even after 6000 hours of aging. As PET is known to be used in low-cost applications whereas PI and FEP are typically used in more challenging environments and in high-tech products, the result is not surprising. The embrittlement of PET under high humidity conditions has also been reported before.

The embrittlement of PET seems to be the major problem of this material. Its water absorption is much lower and slower compared to PI. Aging did not influence the water absorption before the material became totally brittle. Higher water absorption after embrittlement is probably caused by the cracks and micro cracks resulting from the embrittlement, through which the moisture can penetrate the material. On the other hand, the increased water absorption at this point is not important since the PCB is, in any case, unusable at this stage of its life cycle.

References

[1] J.A. Scarlett, "The Multilayer Printed Circuit Board Handbook", Electrochemical Publications Ltd, Great Britain, 1995.

[2] Clyde F. Cooms, Jr, "Printed Circuit Handbook", McGraw Hill, fifth edition, USA, 2001.

[3] Martin Goosey, "Plastics for Electronics", Kluwer Academic Publishers, second edition, Netherland, 1999.

[4] Joseph Fjelstad, "An Engineer's Guide to Flexible Circuit Technology – Materials, Design, Applications, Manufacturing",

Electromechanical Publications Ltd, England, 1997.

[5] Clyde F. Coombs Jr, "Printed Circuit Handbook", McGraw Hill, fifth edition, USA, 2001.

[6] Thomas H. Stearns, "Flexible Printed Circuitry", McGraw Hill, USA, 1996.

[7] Xuejun Fan, "Mechanics of Moisture for Polymers: Fundamental Concepts and Model Study", Proceedings of the 2008 9th International Conference on Thermal, Mechanical and Multiphysics Simulation and Experiments in Micro-Electronics and Micro-Systems (EuroSimE), Freiburg-im-Breisgau, Germany, April 20-23, pp. 1-14, 2008.

[8] H. Walter, E. Dermitzaki, B. Wunderle, B. Michel, "Influence of Moisture on Humidity Sensitive Material Parameters of Polymers used in Microelectronics Applications", Proceedings of the 2010 3rd Electronic System-Integration Technology Conference (ESTC), Berlin, Germany, September 13-16, pp. 1-5 , 2010.

[9] Vishu Shah, "Handbook of Plastics Testing Technology", John Wiley & Sons, second edition, 1998.

[10] Denice D. Denton, Milan C. Buncick, Hartono Pranjoto, "Effects of Process History and Aging on the Properties of Polyimide Films", Journal of Material Research, Vol. 6, No. 12, pp. 2747-2754, 1991.

[11] G. Oreski, G.M. Wallner, "Aging Mechanism of Polymeric Films for PV Encapsulation", Solar Energy, Vol. 79, No. 6, pp. 612-617, 2005.

[12] B. Fayolle, L. Audouin, J. Verdu, "Radiation Induced Embrittlement of PTFE", Polymer, Vol. 44, No. 9, pp. 2773-2780, 2003.

[13] Joseph D. Menczel, R. Bruce Prime, "Thermal Analysis of Polymers – Fundamentals and Applications", John Wiley & Sons, USA, Chapter 4, p. 320, 338-339, 2009.

[14] V. Bellenger, M. Ganem, B. Mortaigne, J. Verdu, "Lifetime Prediction in the Hydrolytic Ageing of Polyesters", Polymer Degradation and Stability, Vol. 49, No.1, pp. 91- 97, 1995

[15] B. Fayolle, X. Colin, L.Audouin, J. Verdu, "Mechanism of Degradation Induced Embrittlement in Polyethylene", Polymer Degradation and Stability, Vol. 92, No. 2, pp. 231- 238, 2007.

[16] A. Launay, F.Thominette, J.Verdu, "Hydrolysis of poly (ethylene terephtalate). A Steric Exclusion Chromatography Study", Polymer Degradation and Stability, Vol. 63, No. 3, pp. 385-389, 1999.

[17] G. Oreski, G. M. Wallner, "Aging Mechanism of Polymeric Films for PV Encapsulation", Solar Energy, Vol. 79, No. 6, pp. 612-617, 2005.

[18] J.Van Turnhout, P.TH.A.Klaase, P.H. Ong, L.C.E. Struik, "Physical Aging and Electrical Properties of Polymers", Journal of Electrostatic, Vol.3, pp. 171-179, 1977.

[19] S. Luo, J. Leisen, C. Wong, "Study on Mobility of Water and Polymer Chain in Epoxy and Its Influence on Adhesion", Journal of Applied Polymer Science, Vol. 85, No. 1, pp. 1-8, 2002.

[20] Gottfried W. Ehrenstein, Gabriella Riedel, Pia Trawiel, "Thermal Analysis of Plastics – Theory and Practise", Carl Hanser Verlag, Germany, 2004.

[21] Pradeep Lall, Michael G. Pecht, Edward B. Hakim, "Influence of Temperature on Microelectronics and System Reliability" CRC Press LCC, New York, 1997.

LTCC-based capacitive pressure sensor in a harsh environment

Darko Belavič[1,4], Marina Santo Zarnik[1,4], Marko Hrovat[2,4], Kostja Makarovič[2,4], Sandi Kocjan[3], Marjan Hodnik[3,4], Borut Grošičar[4], Marko Pavlin[3], Janez Holc[2,4]

[1] HIPOT-RR d.o.o., Šentpeter 18, SI-8222 Otočec, Slovenia
[2] Jožef Stefan Institute, Jamova 39, SI-1000 Ljubljana, Slovenia
[3] HYB, Levičnikova 34, SI-8310 Šentjernej, Slovenia
[4] Centre of Excellence NAMASTE, Jamova 39, SI-1000 Ljubljana, Slovenia

Phone: +386 1 4773479, Fax: +386 1 4473887 and E-mail: darko.belavic@ijs.si

Abstract

In this paper the initial capacitances of pressure sensors consisting primarily of the capacitance of an air-gap capacitor and the parasitic capacitance of sensors realized with LTCC (low-temperature cofired ceramics) materials are described. The capacitances at different temperatures (from 10°C to 75°C) and different levels of humidity (from 20% to 80% relative humidity) were measured with an LCR meter at two frequencies, i.e., 10 and 100 kHz. At the lower humidity the temperature coefficient of the capacitance was around 200×10^{-6}/K, while at the higher humidity the temperature dependence of the capacitance was much higher, up to 800×10^{-6}/K. In all cases the dependences of the capacitance versus temperature were almost linear. The humidity has a significant and nonlinear influence on the capacitance at lower frequencies, while at higher frequencies the influence is smaller.

Key words: capacitive pressure sensor, ceramic pressure sensor, LTCC material, harsh environment

Introduction

In most cases the sensing elements of ceramic pressure sensors are based on the piezoresistive principle, but in some applications the capacitive principle is a very useful alternative. Capacitive pressure sensors have a relatively high sensitivity and therefore they are suitable for low-pressure ranges. They have an extremely low power consumption, so they are appropriate for low-energy applications. On the other hand, the main disadvantage lies in the challenge of measuring a very small fractional change of capacitances (down to 10^{-15} F) induced by the applied pressure. Because of this they are very sensitive to any disturbances, like electromagnetic interference, parasitic capacitances, the influence of temperature and humidity, etc [1-7].

This paper is focused on an investigation of the temperature and humidity dependence of the initial capacitance of capacitive pressure sensors made with low-temperature cofired ceramic (LTCC). The LTCC technology is a three-dimensional ceramic technology utilizing all three dimensions for the interconnect lines, the electronic components, and the different three-dimensional structures, such as cantilevers, bridges, diaphragms, channels and cavities. This technology was chosen because of its mechanical properties, compatibility with thick-film technology, and flexibility in designing, which enable a fast and easy fabrication of electronic devices. In the past ten years the use of LTCC technology for pressure sensors and other microsystems has been presented in many articles and papers [8-17].

Design and fabrication of the capacitive pressure sensor [18]

The basic element of a capacitive pressure sensor is a flat, parallel-plate capacitor, as shown in Figure 1. The principle of the capacitive pressure sensor is based on changes to the capacitance values between two electrodes (plates).

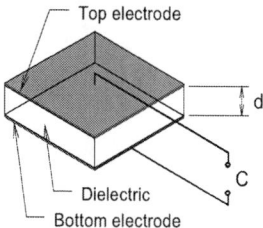

Figure 1: The parallel-plate capacitor

The value of the capacitance (C) of a flat parallel-plate capacitor is expressed by the following equation (1).

$$C = \varepsilon_0 \cdot \varepsilon_r \cdot \frac{A}{d} \qquad (1)$$

where C is the capacitance between the two plates (bottom and top electrodes), ε_0 is the permittivity of free space, ε_r is the relative permittivity of the dielectric material, A is the area of overlap between the two electrodes, and d is the distance between the two plates. Varying any of them will change the capacitor's value, which can be measured quite accurately with an appropriate electronic circuit.

Most ceramic pressure sensors are made with deformable diaphragms, which deflect with the applied pressure. In the case of a capacitive pressure sensor the applied pressure changes the distance between two electrodes, which is inversely proportional to the capacitance (Equation (1)).

The LTCC-based ceramic pressure sensor consists of a circular, edge-clamped, deformable diaphragm that is bonded to a rigid ring and a base substrate. These elements form the cavity of the pressure sensor. The basic element is an equivalent, parallel-plate, air-gap capacitor, as shown in Figure 2. A specific feature of the capacitive pressure sensor is that the bottom electrode is deposited on a rigid substrate and the upper electrode of the capacitor is deposited on a deformable diaphragm. A second specific feature is that the distance between the deformable diaphragm and the rigid base substrate is smaller and must be very well defined.

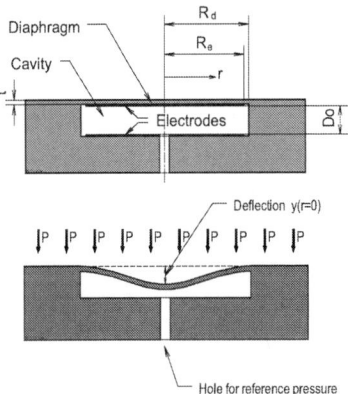

Figure 2: The cross-section of an LTCC-based capacitive pressure sensor with and without the applied measuring pressure (schematic, not to scale).

For the evaluation of the ceramic capacitive pressure sensors the test samples for the pressure range from 0 to 10 kPa were fabricated with the LTCC material DuPont 951. The thickness of the diaphragm is 100 μm and the diameter of the cavity is 8.4 mm. The diameter of both electrodes is 7.7 mm and the distance between the electrodes is about 80 μm. The dimensions of the LTCC structure are 17.5 × 12.6 × 1.0 mm. The distance between the electrodes and the area of the electrodes define the initial capacitance of the air-gap capacitive pressure sensor, which is about 5 pF. The applied pressure generates a relatively small deflection of the ceramic diaphragm, which slightly changes the capacitance of the capacitive pressure sensor.

The LTCC structure includes the diaphragm, the cavity, the two electrodes and the channel. The channel links the cavity with the inlet tube for the reference pressure. The electrodes are connected with two contact pads. One of the fabricated samples of the ceramic capacitive pressure sensors is shown in Figure 3.

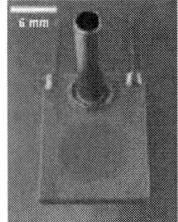

Figure 3: The fabricated test sample of a capacitive pressure sensor.

Results and discussion

The test samples of the pressure sensors were tested at different temperatures (from 10°C to 75°C) and at different relative humidities (from 20% to 80%). The initial capacitances (C) of the capacitive pressure sensors were measured with an LC meter HP4263B at a voltage of 1 V and at two frequencies, i.e., 10 and 100 kHz. The relative changes (ΔC/C) in the initial capacitance of the pressure sensor versus the temperatures at different relative humidities are shown in Figures 4 and 5.

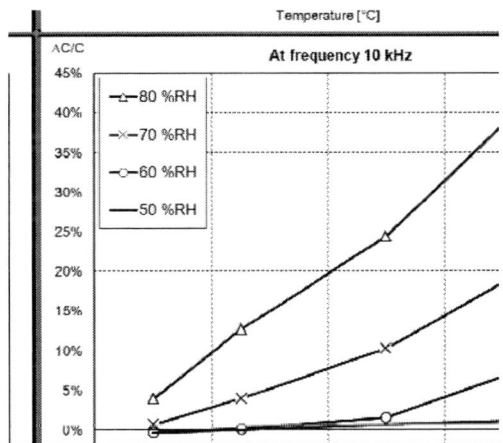

Figure 4: The relative change of initial capacitance versus temperature at different relative humidities for the LTCC-based capacitive pressure sensor. Measuring frequency is 10 kHz.

The temperature dependence of the initial capacitance is almost linear, but very dependent on the relative humidity. Up to a relative humidity of about 50%, the temperature dependence of the initial capacitance is relatively low (the temperature coefficients of capacitance are between 200×10^{-6}/K and 300×10^{-6}/K) and similar for both the measuring frequencies, i.e., 10 and 100 kHz. However, at a relative humidity over 50% the temperature dependence of the initial capacitance increases rapidly. The temperature coefficients of the capacitance are relatively high: up to 5200×10^{-6}/K at a measuring frequency of 10 kHz and up to 1200×10^{-6}/K at a measuring frequency of 100 kHz.

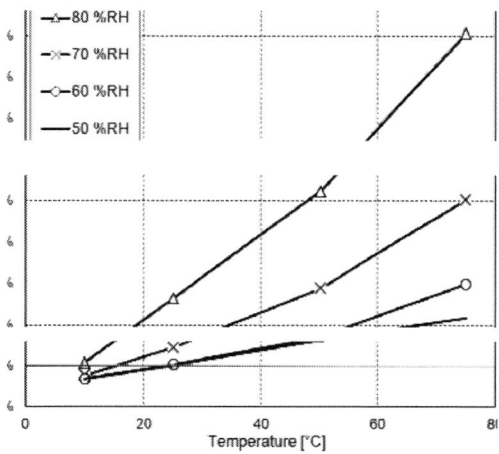

Figure 5: The relative change of initial capacitance versus temperature at different relative humidities for the LTCC-based capacitive pressure sensor. Measuring frequency is 100 kHz.

The relative changes ($\Delta C/C$) in the initial capacitance of the pressure sensor versus the relative humidity at different temperatures for the two different measuring frequencies, i.e., 10 and 100 kHz, are shown in Figures 6 and 7, respectively. The humidity dependence of the initial capacitance is very low and temperature-independent at lower humidities, but increases significantly at higher humidities. The relative changes are increasing at higher temperatures and decreasing at higher measuring frequencies.

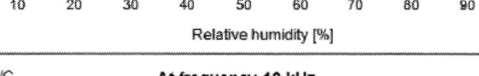

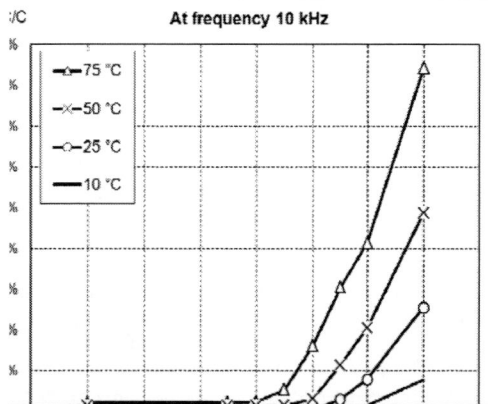

Figure 6: The relative change of initial capacitance versus relative humidity at different temperatures for the LTCC-based capacitive pressure sensor. Measuring frequency is 10 kHz.

The relative changes in the initial capacitance because of the combined temperature and humidity effect are up to 43% at a measuring frequency of 10 kHz and decrease by about 8% at a measuring frequency of 100 kHz.

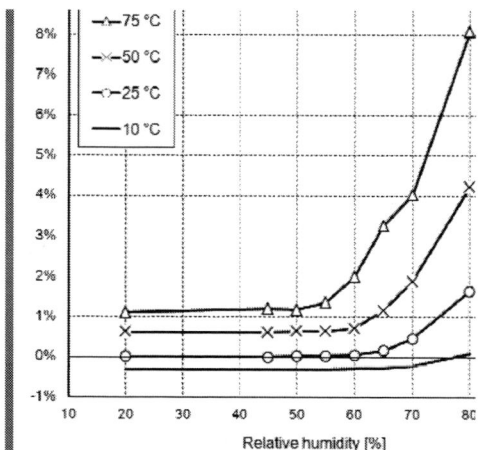

Figure 7: The relative change of initial capacitance versus relative humidity at different temperatures for the LTCC-based capacitive pressure sensor. Measuring frequency is 100 kHz.

Conclusions

The fabrication of capacitive pressure sensors using thick-film and LTCC materials and technology is a challenging opportunity for the pressure-sensors market. The applied pressure generates a relatively small deflection of the ceramic diaphragm, which slightly changes the capacitance of the capacitive pressure sensor. These small fractional changes in the capacitances are of the order of a few 10^{-15} F and are very sensitive to any disturbances.

The ambient temperature and humidity have a significant influence on the value of the initial capacitance. Therefore, special attention must be paid to both the temperature and humidity dependence of the capacitance. The humidity dependence is negligible up to a relative humidity of 50%. However, at higher relative humidities the temperature dependence of the initial capacitance is high and less well defined. However, a higher relative humidity could be avoided by using humidity protection and/or operating at a higher frequency.

The temperature dependence of the initial capacitance is almost linear and could be easily compensated by one of the conventional methods [19].

Acknowledgements

The work is carried out in the frame of the Centre of Excellence NAMASTE (financed by the European Commission and the Slovenian Ministry of Higher Education, Science and Technology), and it is partly supported through the projects L2-0186 and L2-2343 (financed by the Slovenian Research Agency and the company HYB).

The authors wish to thank Mr. Mitja Jerlah for fabricating the test samples.

References

[1] R. Puers: "Capacitive sensors; when and how to use them", Sensors and Actuators A, Vol. 37, 1993, pp. 93-105.

[2] C. B. Sippola, C. H. Ahn: "A thick film screen-printed ceramic capacitive pressure microsensor for high temperature applications", Journal of Micromechanics and Microengineering, Vol.16, 2006, pp. 1086-1091.

[3] D. Belavič, M. Santo Zarnik, M. Jerlah, M. Pavlin, M. Hrovat, S. Maček: "Capacitive thick-film pressure sensor: material and construction investigation", Proceedings of the XXXI International Conference of IMAPS Poland 2007, Rzeszóv, Krasiczyn, Poland, 23 - 26 September 2007, pp. 249-253.

[4] D. Crescini, V. Ferrari, D. Marioli and A. Taroni: "A thick-film capacitive pressure sensor with improved linearity due to electrode-shaping and frequency conversion", Meas. Sci. Technol. Vol. 8, 1997, pp. 71-77.

[5] M. Santo Zarnik, D. Belavič, S. Maček: "Investigation of a thick-film capacitive pressure sensor in a three-dimensional LTCC structure. Proceedings of the 33nd International IMAPS-IEEE CPMT Poland Conference, Pszczyna, Poland, 21 – 24 September 2009, pp. 371-374.

[6] G. Radosavljević, L. Živanov, A. Marić, L. Nađ, W. Smetana, M. Unger: "Performance Improvement of a Resonant Pressure Sensor by Means of Model Based Design Optimisation", Proceedings of the 7th Conference on Sensors (IEEE Sensors 2008), Lecce, Italy, 25 – 29 October 2008, pp. 1008-1011

[7] M. Santo Zarnik, M. Možek, S. Maček, D. Belavič: "An LTCC-based capacitive pressure sensor with a digital output", Informacije MIDEM, 2010, Vol. 40, no. 1, pp. 74-81.

[8] M. R. Gongora-Rubio, P. Espinoza-Vallejos, L. Sola-Laguna, J. J. Santiago-Aviles: "Overview of Low Temperature Cofired Ceramics tape technology for meso-system technology (MsST)", Sensors & Actuators A, Vol. 89, 2001, pp. 222–241.

[9] L. J. Golonka, A. Dziedzic, J. Kita, T. Zawada: "LTCC in microsystem application", Informacije MIDEM, Vol. 32, No.4, December 2002, pp. 272-279.

[10] T. Thelemann, H. Thust, M. Hintz: "Using LTCC for Microsystems", Microelectronics International, Vol. 19, 2002, pp. 19-23.

[11] J.Kita, R.Moos: "Development of LTCC-materials and Their Applications: an Overview", Informacije MIDEM, Vol. 38, No. 4, 2003, pp. 219-224.

[12] H. Birol, T. Maeder, C. Jacq, S. Straessler, P. Ryser: "Fabrication of Low-Temperature Co-fired Ceramics micro-fluidic devices using sacrificial carbon layers". Int. J. Appl. Ceram. Technol., Vol. 2, 2005, pp. 364–373

[13] L. J. Golonka, T. Zawada, J. Radojewski, H. Roguszczak, M. Stefanow: "LTCC microfluidic system". Int. J. Appl. Ceram. Technol., Vol. 3, No. 2, 2006, pp. 150–156

[14] U. Partsch, D. Arndt, H. Georgi: "A new concept for LTCC-based pressure sensors", Proceedings of the IMAPS/ACerS 2007, 3rd International Conference and Exhibition on Ceramic Interconnect and Ceramic Microsystems Technologies (CICMT 2007), Denver, Colorado, USA, 23-26 April 2007, pp. 367-372.

[15] K. A. Peterson et al: "Macro - Meso - Microsystems Integration in LTCC: LDRD Report", Sandia National Laboratories, Albuquerque, New Mexico, 2007, 90 pages.

[16] U. Partsch, S. Gebhardt, D. Arndt, H. Georgi, H. Neubert, D. Fleischer, M. Gruchow: "LTCC-based sensors for mechanical quantities", Proceedings of the 16th European Microelectronics and Packaging Conference & Exhibition (EMPC 2007), Oulu, Finland, 17 - 20 June 2007, pp. 381-388

[17] X. Wang, B. Balluch, W. Smetana, G. Stang: "Optimization of Cavity Fabrication for Micro - Fluidic Systems"; Proceedings of the 31st International Spring Seminar on Electronics Technology (ISSE 2008), Budapest, Hungary, 07 - 11 May 2008, pp. 271-274

[18] D. Belavič, M. Santo Zarnik, S. Maček, M. Jerlah, M. Hrovat, M. Pavlin: "Capacitive pressure sensors realized with LTCC technology", Proceedings of the 31st International Spring Seminar on Electronics Technology (ISSE 2008), Budapest, Hungary, 7 - 11 May 2008. pp. 271-274.

[19] M. Možek, D. Vrtačnik, D. Resnik, B. Pečar, S. Amon: "Digital temperature compensation of capacitive pressure sensors", Informacije MIDEM, Vol. 40, No. 1, 2010, pp. 38-44.

A Comprehensive Packaging Solution For Next Generation IC Substrates

Stephen Kenny, Dave Baron, Bernd Roelfs and Frank Bruening

Atotech Deutschland GmbH

+49 3034985921, Fax +49 3034985 618
Stephen.Kenny@atotech.com
Dave.Baron@atotech.com
Bernd.Roelfs@atotech.com
Frank.Bruening@atotech.com

Abstract

The requirements for reduction of line and space dimensions in IC substrates are driving developments to improve both production yield and capability. In particular the production of substrates with line and space dimension at or below 10 μm is required for the next level of integration. However traditional production techniques using dry film image transfer are already reaching capability limits and are unlikely to achieve satisfactory future production requirements especially considering the production process yield. This paper presents the latest developments in a system for the production of structures based on the copper filling of trenches on a dielectric substrate. The system is being targeted for the manufacture of IC package substrates with capability for sub 8 μm lines and spaces. The resulting package may be characterized by padless vias and shows significant electrical performance improvements in comparison to substrates produced using standard production methods. The trenches are produced by laser ablation of the dielectric which is subsequently metalized, this method of track embedding gives an improved circuit adhesion due to the three point contact to the substrate in comparison to the standard method of production, typically using the semi-additives process (SAP) where tracks are produced with contact only at the base. Recent optimization of the electrolytic copper plating process has resulted in an improved metal distribution and more uniform copper filling of the ablated trenches. The latest results in circuitisation are shown together with data on substrate capability using the trench filling technology.

Key words: Substrate packaging, embedded trace, laser ablation, Via².

Introduction

The production of conductive traces on IC substrates using laser structuring and laser ablation of a suitable base material was introduced in a joint paper from Amkor, Unimicron and Atotech in a paper "Unveiling the Next Generation in Substrate Technology" [1]. Further development of the process was described in [2]. The process is targeted at the production of IC substrates with feature size in the range 10μm to 8μm and lower. For this type of substrate limitations in the conventional production process become critical and yields are under pressure. Also the constant demand for reduction in trace dimension is pushing the conventional photolithography process to its limits; the trace resolution is dependant on the photolithography in the SAP process and cannot meet future requirements. Use of laser ablation for production eliminates a number of process steps in photolithography and gives a method to meet future requirements. There are various names to describe the process such as Via² or LEC (Laser Embedded Conductors). The process uses laser ablation of a suitable substrate to create a trench which is subsequently metalized using a thin electroless copper and then filled with electrolytic plated copper. The copper plating is over the whole surface of the substrate but with preferential deposition into the trenches, the target being to deposit copper as thin as possible on the surface whilst filling the trenches. After copper plating removal of excess surface copper gives the required circuitisation. Initial tests with the trench filling process have identified the base material as a critical factor and the plated uniformity of the electrolytic copper process. Development has been made to investigate various base materials with varying particle size to identify the optimum material. Parallel to this, optimization of the copper plating process in horizontal continuous plating equipment is being made in a dedicated production line.

The substrate production process starts with the lamination of a SBU material onto a core. The core may be treated with an adhesion promoter to enhance the physical properties of the final package. Then the laser is used to ablate the SBU material and form trenches, pads and blind micro vias (BMV) the latter having a connection to the core copper layer. The so formed embedded structures need to be desmeared before further treatment to ensure good adhesion and are then made conductive by an electroless copper step. After this the complete filling of all structures by electro plating with copper has to be achieved. The subsequent etching step generates the circuitization, this may be a purely chemical etching or by use of the so called chemical mechanical polishing, CMP process. Following circuitisation of the layer the process may be continued with lamination of a new SBU layer and the aforementioned steps to generate a multilayer. This sequence may be repeated to give the numbers of layers required and then completed with the deposition of a suitable final finish for bonding or soldering.

Figure 1 shows a schematic process sequence as follows,

1. SBU material lamination on core substrate
2. Laser structuring into SBU material
3. Desmear and e'less copper plating
4. Electrolytic copper Plating
5. Copper etching
6. Restart at 1. (SBU lamination) to produce a new layer.

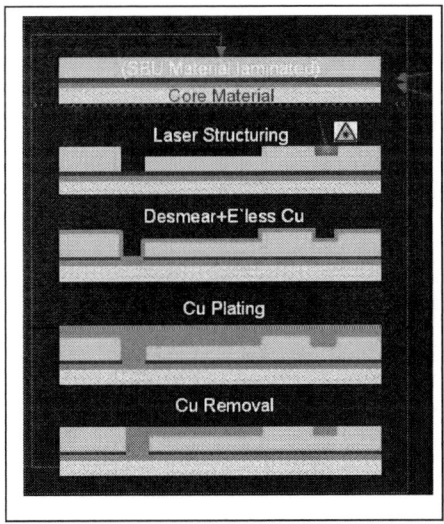

Figure 1: Schematic process sequence of laser embedded track production

This sequence deviates substantially from the traditional circuitization process based on a photo resist application by eliminating the latter completely (passivation, lamination, imaging, developing, stripping), this gives obvious savings and potential yield gains. However to achieve this a number of process obstacles had to be overcome.

The laser performance was the first item to be considered. In the past ablation speed of common lasers was not competitive to meet the through put demand but latest developments have made substantial progress. Today there are pulsed ns or even ps laser systems available which can ablate a complete IC substrate panel (500 x 500 mm² or 19,7 x 19,7 inch²) in less than 3 minutes with lines and spaces below 10μm -15μm. Fig. 2 shows an SEM picture of pads and conductor lines after ablation in a common SBU material.

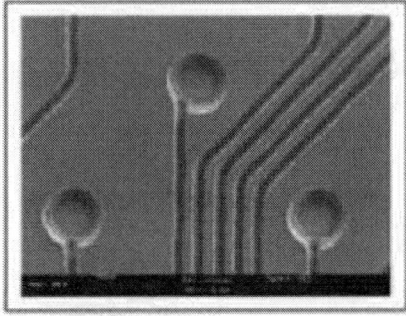

Figure 2: SEM of laser ablated structures in common SBU material. Pads 90μm and 250μm conductor lines 15μm. Structures are 25μmdeep.

In figure 3 the same structure is shown after the trench filling process has been completed and the excess surface copper has been removed so completing the circuitisation of this layer.

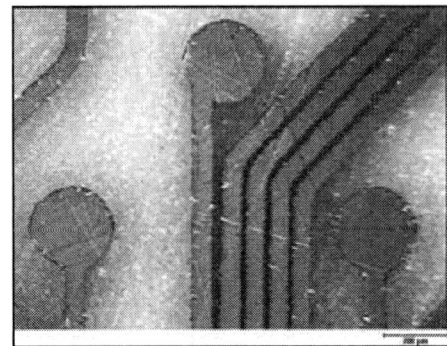

Figure 3: SEM circuitised ablated trenches after removal of excess surface copper.

Material Requirements for Embedded Traces

The base material for the ablation process was the next issue to be considered. The roughness of the conducting track is determined by the surface roughness of the material after laser ablation of the trench. This again is dependant on the material characteristics such as filler particle size and uniformity. As shown in the schematic of the process sequence an integral part of the metallization process is the use of a "desmear" process to remove any debris form the surface after laser ablation and also to give a defined surface roughness. Track surface roughness is important for high frequency applications and has been investigated with varying base materials and also processing parameters. Figure 4 shows a comparison of base materials with different filler quality

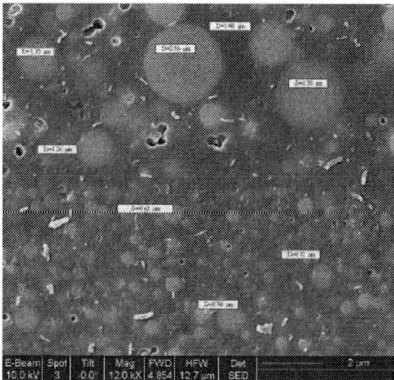

Figure 4: Microsection of two different base materials laminated to show the varying particle size.

In the microsection the material in the upper half has particle sizes in the range from 1.2μm up to 2.5μm, this characteristic is not suitable for the ablation of fine structures due to the wide filler particle variation. The material in the lower half of the microsection has particle size in the range 0.6μm to 0.9μm and is therefore more suited for use in this technology. After ablation the correct process conditions for the desmear before metallization are essential to prevent any excessive roughening of the surface whilst ensuring an optimum adhesion of the plated copper. The impact of a too rough material can be seen in figure 5; this shows the surface of the ablated layer after circuitisation.

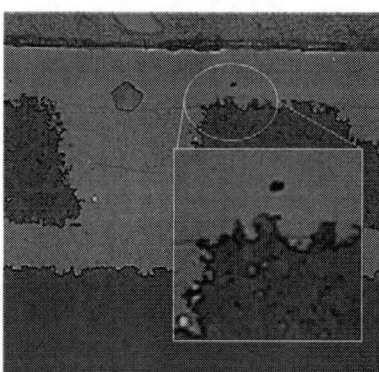

Figure 5: Copper retention on the surface of rough base material after circuitisation

Optimization of Copper Deposition

The targets in the electrolytic copper deposition are to fill the ablated structures whilst keeping the overall plated copper to a minimum. This is to keep the copper overburden low so that subsequent circuitisation may be easy and also cost effective. The copper plating process used is based on reverse pulse plating in a horizontal continuous plating line using dimensionally stable anodes. A key feature of the anode installation is the use of segmentation to enable optimum copper plated surface distribution. The segmentation can be seen in figure 6; each segment of the anode package is supplied from a separate rectifier.

Figure 6: Top view of single anode unit showing four segments.

The electrolyte is a conventional copper sulphate and sulphuric acid base but with the added component of iron sulphate which enables the redox system copper replenishment process. This process has been introduced for many applications in the production of HDI and in particular for the filling of through holes with copper for IC substrate cores as described in [3]. An added advantage of the horizontal plating module is the use of force flood electrolyte delivery to the cathode chamber. This is done using frequency controlled pumps which may be varied depending on the application and the substrate characteristics. For the filling of the ablated trenches a three step plating process was found to give best results. The first step is for initial reinforcement of the electroless copper metallised surface followed by a half filling of the structures. A deposit thickness of 3μm to 5μm on the surface using strong reverse pulse plating conditions is made. Following the reinforcement step a filling step is used to fill the structures as far as possible with a maximum of 5μm plated onto the surface again with strong pulse plating parameters. The strong parameters give a good filling result but there is the disadvantage of surface roughness after this step. To remove the surface roughness and to complete the copper plating process a leveling step is used. This is with medium pulse plating parameters and serves to remove any remaining dimple as well as to give an improved surface appearance. The leveling step is targeted to deposit as thin a layer of copper as possible to give a smooth surface. Figure 7 shows a microsection of a filled structure which has been carefully etched to show the demarcation between the filling steps and the final leveling step.

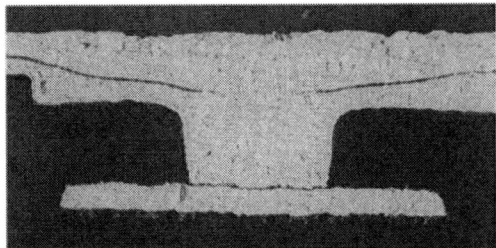

Figure 7: Copper filled structure showing demarcation between filling steps and levelling steps

Production feasibility test result.

Figure 8 shows a microsection through a completed substrate incorporating a conventional "core" made with glass reinforced dielectric together with a layer incorporating the embedded traces. The core is labeled as layer 6 followed by the associated copper foil labeled 5. The embedded traces are in layer 4 and the SAP layer is 3 with the pattern plated track 2 followed by the soldermask 1. The contrast in the grain size of layer 4 with that of layer 3 can be clearly seen. The grain size for the embedded traces is in the nm range whilst that for SAP processing is in the µm range.

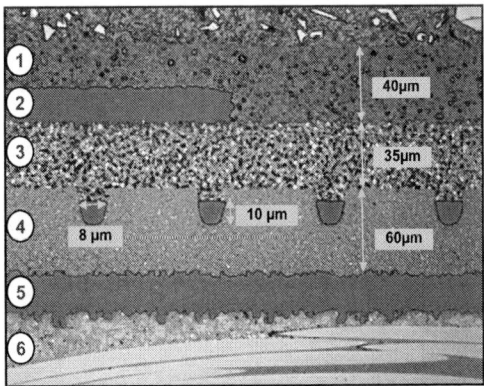

Figure 8: Microsection of embedded 8µm wide and 10µm deep traces in substrate

Summary

Production feasibility tests are continuing with target to further improve the plated copper surface distribution. This factor directly determines the process capability as a high copper overburden requires mechanical or chemical removal and impacts the quality of the plated surface, the removal

of copper has itself a process variation causing a subsequent thickness distribution on the circuitised panel. Keeping the surface copper thickness to a minimum has a direct impact on process cost and overall feasibility. The process based on laser embedded structures offers the possibility to produce smaller features with lines and spaces in and below the range 10µm to 8µm. The resulting substrate offers an improved performance due to the higher degree of miniaturization possible and also the potential to reduce overall production cost. This is due to a reduction in the number of layers necessary for a particular application and also in the elimination of traditional photolithography with its associated process steps and increasing costs.

References

[1] R. Huemoeller, S. Rusli, S. Chiang, T. Chen, D. Baron, L. Brandt and B. Roelfs, "Unveiling the Next Generation in Substrate Technology", Proceedings of the 2007 Pan Pacific Conference, 2007.

[2] T. Fujiwara, B. Roelfs, "Increased Miniaturization with Trench Filling Technology", Proceedings of the JPCA 2007.

[3] S. Kenny, B. Roelfs, "Copper Electroplating Process for Next Generation Core Through-Via Filling", Proceedings of the Pan Pacific Conference, 2009.

Modular Microelectronics by System-in-Packages with Embedded Components

A. Ostmann[1], B. Bruehl[2], D. Manessis[2], M. Seckel[2] and K.-D. Lang[2]

[1]Fraunhofer Institute for Reliability and Microintegration (IZM)
[2]Microperipheric Research Center, Technical University of Berlin (TUB)
Gustav-Meyer-Allee 25, 13355 Berlin, Germany

E-Mail: ostmann@izm.fhg.de, Tel: +49 30 46403 187

Abstract

In coming applications a further increase of diversity and higher complexity of electronic systems is expected. As a consequence it will be an increasingly challenging and time consuming task to design these systems. Although various new components are coming up, the problem of increasing cost for system design, debugging and verification is generally not addressed by conventional technology developments. A new concept to face these challenges will be presented. It is based on the constitution of a system by miniaturised modules, each representing a complete functionality like system control, data storage or sensing. The modules are planar System-in-Packages with embedded active and passive components, having typically 4 contacts on bottom and top. Besides the hardware an integral part of each module is its firmware, a small piece of software which controls a basic functionality, e. g. sensor data filtering and the communication with other modules. The realisation of a Smart Pixel SiP with I²C bus interface and embedded components as well as the design of a modular sensor system will be described. The data communication between modules is also realised by an I²C bus. By combining these modules in different ways, various systems can be realised. The capabilities and challenges of such systems will be discussed.

Key words: chip embedding, PCB technology, System-in-Package, package stacking, modular microelectronics

Introduction

The embedding of electronic components into printed circuit board (PCB) structures has a long history. Already 1968 a first patent was applied by Philips [1]. Since the 1990s experimental work was presented from several groups [2]-[5]. In the first years these activities gained mainly academic interest and it took significant effort to overcome technological problems. Since last year chip embedding is known to be in high volume production [6][7]. Today advanced PCB materials, equipment and processes are ready for the realisation of packages and System-in-Packages (SiPs) for consumer, medical and power products [8]:

The chip embedding starts to replace established technologies for the realisation of QFN, BGA and other package types. They can be used as conventional packages, even when the semiconductor inside is connected in a different way. But packages or SiP with embedded chips can offer much more:

- Capability for complex SiPs with multiple stacked functional layers.
- PCB technology easily provides top to bottom connections inside a SiP by through vias or (stacked) micro vias.
- Conventional planar PCB surfaces on top and bottom provide a good capability for SiP stacking.

Today a typical system is manufactured by the assembly of several active and passive components on a PCB. But these features open the way to the realisation of microelectronics systems by the interconnection of few modules, each representing a complete function.

Several approaches for modular or 3D systems have been presented already. The Match-X system of Fraunhofer IZM [9] is based on a 3D stacking of small SMD board using PCB frames with through vias. 3D Plus [10] and Irvine Sensors [1] offer technologies for aerospace and avionic applications using 3D SiPs with sidewall connections. These approaches offer well-adapted solutions for their specific application fields. But they did not reach high volume manufacturing for a wide range of products. A problem e. g. of the Match-X technology is the difficulty to achieve a stack of several components with 80 and more vertical interconnects. Warpage of PCB material is challenging for a satisfying yield. Technologies like structured side-wall metallisation are a challenge for low-cost manufacturing.

Modular System Concept

SiPs realised in PCB embedding technology can serve as a basis for modular microsystems. A concept for such technology should use the advantages offered by embedding but also regard limitations of the underlying PCB technology:

- Systems are formed by hardware blocks (modules) which can be stacked in arbitrary sequence.
- Each module represents a complete function like sensing, power management, wireless interface or system control.
- The number of interconnects between modules should be low for robust assembly with high yield and high reliability, reducing the yield risk due to warpage.
- The communication in the system between modules should be based on existing serial busses like I²C (Inter-Integrated Circuit), CAN (Controller Area Network) or USB (Universal Serial Bus), only requiring 4-6 physical connections.
- Each module has the capability to communicate over this bus with other modules, at least with the system controller.
- The module communication is driven either by an internal component's hardware or by an internal microcontroller's software.
- A microcontroller in a module can take care for additional tasks e. g. sensor data filtering or storage.

The hierarchy levels of such a modular system are shown in Figure 1.

1st level- Component

The first level is represented by active semiconductors and passive components, as they form today's system. The components can be bare Si dies or packaged chips. In case of packages, they should not exceed reasonable thickness e. g. 1- 2 mm. The embedding of bulky components like high-value capacitors is typically not economic.

2nd level - Module

Next is the level of modules. They consist of one or more embedded components and a communication capability, realised by a small piece of software in a microcontroller or as hardware in one of the embedded components.

3rd level - Sub-System

The following level can be called the sub-system level. Here different modules are stacked together, forming a new block with capability for internal and external signal communication. Such block consists e. g. of a sensor module, a power management module and a controller module.

4th level - System

The last level is the system level. Here one or more sub-systems or single modules can be connected to system board, which may contain a connector for power supply or for another type of bus connection to the outside world, e. g. Ethernet.

In future the design cycle time for such systems could be much faster than today. After system design in a software tool the designer could simply put modules together for first tests, rather than waiting weeks for PCB manufacturing and component assembly. Also the modules represent proven functions with defined mechanical and software interfaces. They can be handled as "black boxes" without knowing their constituting circuits and components. Adding of further features e. g. an additional sensing function can be evaluated in short time, just by adding a further module and extending the controller's signal flow software.

Of course such modular concept can not be successfully used for all types of applications at lower cost than conventional circuits. With increasing complexity and reduced time for the design cycle however, modular systems will gain an increasing importance.

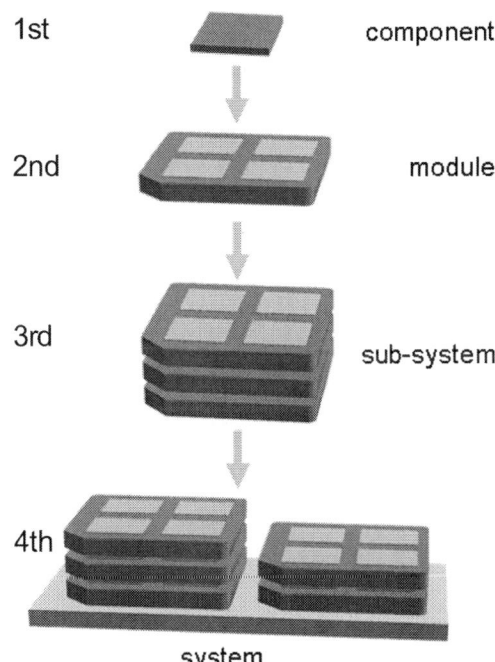

Figure 1: Hierarchy levels of a modular system with embedded components.

In the following two sections examples on the way towards modular microsystems are presented.

Modular LED System-in-Package – Smart Pixel

At Fraunhofer IZM a technology for integrating electronics in textiles has been developed. It is based on stretchable interposers made of Polyurethane in a PCB-like process. They serve as an interface between conventional SMD components and textiles [12]. A demonstrator realised is a fashion dress with 32 warm white LEDs which via a driver were controlled by a microcontroller's analogue outputs. For future versions of textile lighting systems colour LEDs in much higher numbers are desirable. Even by using RGB LEDs with 3 colours integrated in one package the number of required lines would quickly exceed the available I/Os of every microcontroller, in case each LED is directly connected to it. Therefore a SiP called "Smart Pixel" was developed. Smart Pixels can be powered and controlled by an I²C bus with 4 lines (power, ground, clock and data). Each bus can have 64 Pixels. So a microcontroller with 8 I/Os and 2 bus lines can control colour and brightness of 128 Smart Pixel by calling their individual addresses. A Pixel consists of a RGB LED, a LED controller with I²C interface, 2 capacitors and 2 resistors. The controller and passives are embedded in a PCB layer, the LED is assembled on top using SMD technology (Figure 2).

Figure 2: LED SiP and its constituting components.

The process flow for the Smart Pixel realisation is shown in Figure 3). The manufacturing was done using a panel format of ´320x225 mm². First a 2-layer FR4 panel was made containing positions for several Smart Pixels. Then lead-free solder paste was printed and the OFN package of the LED controller and the 0402 passives were placed and reflowed. Next some laser-cut prepreg layers were placed on top, having cavities at the positions of the mounted components. The amount of prepregs was calculated to compensate for the thickness of the QFN (about 1 mm). On top and bottom of this PCB sandwich complete prepreg layers and Cu foils were positioned.

Then this stack was cured in a vacuum lamination press. During this process the SMDs were completely embedded in the PCB structure. Next through vias were drilled and metallised. Then the Cu on bottom and top was structured by subtractive etching, a solder mask was applied and the Cu was coated by an immersion Ag surface finish. Solder paste was printed again on top of the panel. The LEDs were placed and reflowed. Finally solder caps were applied to the bottom side and the individuals Smart Pixels separated by mechanical milling.

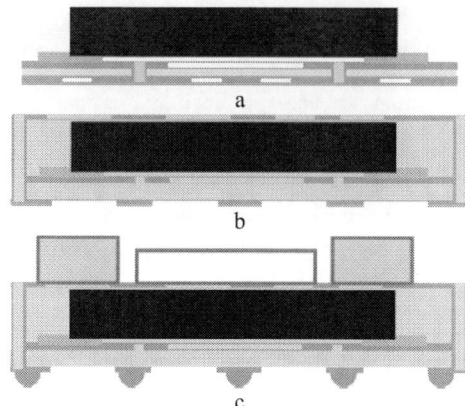

Figure 3: Manufacturing flow of LED SiP: (a) SMD assembly on inner PCB layer, (b) embedding of SMD components, drilling and metallisation of through vias, (c) assembly of SMD components on tops side and solder ball application on bottom side.

Figure 4 shows an x-ray tomography of a ready Smart Pixel. First tests on a demo board have successfully demonstrated their functionality.

These colour LED SiPs represent a first step on the way to modular microsystems. They can not be stacked, since stacking of light emitting devices makes no sense. But they allow a modular thinking and more degrees of freedom in the design process. Instead of thinking about several components, wiring density and circuit layout, textile designers can place colour effects.

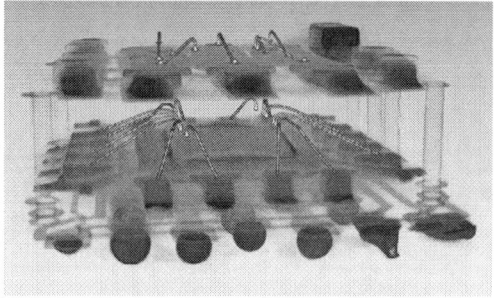

Figure 4: X-ray tomography of LED SiP.

Modular Sensor System

In a German R&D project, together with the partners Hofmann Leiterplatten and SysCom Electronic, the aim is to develop a modular sensor system for technology evaluations. Each module will have a size of 12x12 mm² and a thickness of 1-2 mm, depending on the embedded components. The embedding process is similar to that described in the previous section. But no components will be assembled on top, keeping the surface free for module stacking. All modules are based on the 8-Bit microcontroller ATMega168 from Atmel. Four types of modules were designed:

- a temperature sensing module
- a light sensing module
- an acceleration sensing module
- a controller module with USB connection

Each of the modules contains the ATMega, a packaged sensor chip with I²C interface, several passives like blocking capacitors and 2 LEDs for indication of the current operation mode. In case of the controller module the ATMega is connected to an UART/USB bridge for the outside communication. The sensors are connected to the microcontroller's hardware I²C interface. The I²C communication to the other modules via the top and bottom contacts is realised by software emulation.

As an additional feature for fast and easy testing the modules contain 4 embedded permanent magnets of 1 mm size in their corners. In the evaluation phase the magnets allow sufficiently good contacts without the need of soldering or gluing or the use of mechanical fixtures. Of course the magnets would be too space consuming and expensive for real products.

a

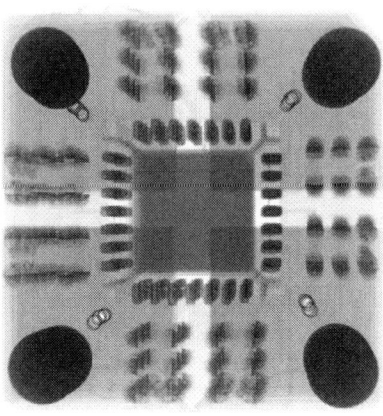

b

Figure 5: X-ray image of 3 interconnection test modules, stacked by isotropic conductive adhesive: (a) side view, (b) top view.

Design and layout of the 4 modules has been finished and the manufacturing and software coding is still ongoing. In order to start first evaluations of module stacking and the reliability assessment of modules and module stacks, an additional interconnection test vehicle has designed and manufactured. It has the same geometries than the functional modules but instead of a microcontroller is contains a QFN dummy package surrounded by 8 LEDs and 8 resistors. The module was realised on a panel format of 320x225 mm² with a manufacturing flow as described in the previous section. Modules were stacked and successfully tested. Intact daisy chain connection could be easily detected by lighting of the LEDs. For these tests the force of the magnets was sufficient to achieve contacts. For the reliability tests modules were stacked using solder or isotropic conductive adhesive. Figure 5 shows x-ray images of stacked modules connected by adhesive. Reliability test of the module stacks are ongoing.

Summary

The chip embedding technology has made its way from research to an established high-volume manufacturing technology. Besides the replacement of conventional package types it offers the capability to realise stackable System-in-Packages for future modular microsystems.

A concept for modular microsystems has been described. Each module should represent a complete function with defined mechanical and software interfaces. This allows designers a much faster realisation and test of prototypes and a fast start of production. Modules can communicate via serial data busses which are widely available in many components.

A first realised example towards modular systems for textile applications is a Smart Pixel. Colour and brightness can be controlled via I²C bus. The assembly of a RGB LED on a small PCB with embedded LED controller and 4 passive components leads to a very compact package with small footprint. The functionality of the Smart Pixel modules was successfully tested.

Finally a first design for a modular 3D sensor system has been described. It consists of 4 modules with different functions, which can be combined in any order. First module stacking tests were performed using an interconnection test vehicle with embedded LEDs.

References

[1] T. TeVelde, "Semiconductor Circuit having Active Devices Embedded in Flexible Sheet", U.S. Patent 3,579,056, filed October 10, 1968.

[2] R. Filion, R. Wojnarowski, T. Gorcyzca, B. Wildi, H. Cole, "Development of a Plastic Encapsulated Multichip Technology for High Volume, Low Cost Commercial Electronics",

Proc. 10th Applied Power Electronics Conf, IEEE, 1994, pp. 805-809.

[3] T. Waris, R. Tuominen, J. Kivilahti, "Panel-sized Integrated Module Board Manu-facturing", Proc. Polytronic Conference, Oct. 21. - 24. 2001, Potsdam, Germany.

[4] H. Braunisch, S. Towle, R. Emery, C. Hu and G. Vandentop, "Electrical Performance of Bumpless Build-Up Layer Packaging", Proc. ECTC 2002, May 28. - 31. 2002, San Diego, USA.

[5] A. Ostmann, A. Neumann, S. Weser, E. Jung, L. Boettcher and H. Reichl , "Realization of a Stackable Package Using Chip in Polymer Technology", Proc. Polytronic Conference, June 23.-26. 2002, Zalaegerszeg, Hungary

[6] J. Stahr, "Embedding Technology Status of Development Applications, Business Fields", Proc. SMT Conference, June 9. 2010, Nuremberg, Germany.

[7] H. Utsunomiya, "Emerging Embedded Devices into Substrate for More than Moore Era", Proc. IMPACT Conference, October 20. – 22. 2010, Taipei, Taiwan.

[8] L. Boettcher, S. Karszkiewicz, D. Manessis and A. Ostmann, "Realization of Power Modules by Chip Embedding Technology", Device Packaging Conference, March 7. - 10. 2011, Scottsdale, USA.

[9] K. Amiri Jam, V. Großer, K.-D. Lang, H. Reichl, "Application of 3D-stacking technology for sensor integration and miniaturization", Proc. Micro System Technologies Conference, October 5. – 6. 2005, Munich, Germany.

[10] Ch. Val, J. M. Benedetto, "High-density Packaging for Spaceborn Electronics", COTS Journal, Issue August 2002.

[11] K. D. Gann, "Neo-Stacking Technology", HDI Magazine, December 1999.

[12] A. Ostmann, R. Vieroth, M. Seckel, Th. Loeher and H. Reichl, "Stretchable Circuit Board Technology in Textile Applications", Proc. IMPACT Conference, October 21. – 23. 2009, Taipei, Taiwan.

Packaging Challenges of Producing Smaller Power MOSFETs & Schottky Diodes for the Automotive Sector

Kevin J Keller

ON Semiconductor, 5005 E McDowell RD, Phoenix AZ, USA

602-244-4983, Kevin.keller@onsemi.com

Abstract

The global automotive electronics market is expected to be worth more the $240 billion by 2017 (according to research conducted by Strategy Analytics). Innovative IC (Integrated Circuit) and power devices for powertrain, lighting and body electronics are helping car manufacturers to reduce the emission levels, improve the fuel economy, and increase the safety of the vehicles they produce.

The total electronics content inside the average vehicle has risen considerably over recent years and will continue to grow at an almost exponential rate. As a result there is an ever greater need for more compact and high performance semiconductor devices for powering all this complex electronics hardware – offering greater efficiency levels and using up less battery life, while simultaneously saving space. Though semiconductor process technologies continue to adhere closely to Moore's Law, allowing smaller and smaller dimensions for IC dies, far greater technological challenges must be overcome if the packages in which they will be enclosed are to keep up. Major advances are called for if the packaging technology is going to be able to deal with major concerns such as ensuring adequate heat dissipation and robustness, as well maintaining overall cost effectiveness.

In order to decrease the amount of board real estate being taken up by power devices, a great deal of engineering effort must now be devoted to issues such as minimizing the interconnect resistance and curbing the lead inductance, while still delivering sufficient thermal management to ensure the long term reliability.

Key words: power density, thermal efficiency

Trends in Automotive Electronics

The automobile has benefited greatly from advances in electronics design, fabrication and packaging. The incorporation of advanced electronics into automotive designs has resulted in greater efficiency, improved reliability, increased safety and an all-round better driving experience for the consumer. These benefits have resulted in a rapid increase in per vehicle electronics content.

Figure 1 provides the historical and projected growth of vehicle electronics content in Europe from 1950 through 2020. The adoption starts with basic power management and entertainment applications; space constraints are at a minimum and operating temperature requirements are moderate. The development of IC technology, improved power components and new regulatory requirements governing emissions and safety generated rapid adoption.

The increase in per vehicle electronics content has combined with other trends in automotive design and manufacturing to place additional requirements on electronic components. The shrinking form factor of automotive electronics modules has created space constraints. Electronics packaging with higher thermal performance is required to accommodate the increasing power density. With the increasing number of infotainment and convenience systems available, package height has also become an issue.

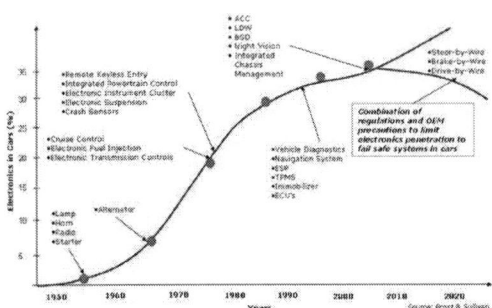

Figure 1: Growth of Vehicle Electronics Content in Europe between 1950 and 2020 (source Frost and Sullivan)

While consumer product lifecycle expectations for portable and computing applications may be measured in months to a few years, the automotive product lifecycle may be 10 to 15 years. The level of customer inconvenience caused by an automotive product failure can be significantly greater than that caused portable and computing applications. Applications such as supplemental restraint system (airbag and pretensioning seatbelts), antilock braking and electronic stability control directly affect occupant safety. Looking forward, applications such as brake-by-wire and steer-by-wire offer greater manufacturing versatility and generate elevated demands on system reliability. These considerations place greater reliability requirements on the electronic content used in automotive designs.

Early automotive electronic control modules (engine, transmission) were typically located in-cabin. The maximum device operating temperature requirements generally did not exceed 125°C. It is increasingly common to locate modules in the under-hood environment. Under-hood electronic device maximum operating temperature requirements are typically 150°C to 175°C. Designers of alternator and in-transmission module applications are requesting maximum operating temperatures of 200°C and higher. This is beyond the capability of most molded epoxy electronic packaging.

The in-cabin module environment is also trending toward higher device operating temperatures. With the number of in-cabin applications increasing there is a growing need to reduce the footprint of each application. Increasing power density and increasing ambient temperature within an electronic module often results. It is not uncommon to see an in-cabin module maximum ambient temperature specified at 125°C. The electronic devices within the module must be capable of operating at a minimum of 150°C with 175°C capability preferred.

The SO-8 Flat Lead Package

The SO-8 flat lead package is part of a new generation of low profile thermally efficient surface mount packages. It is available in two versions, a single flag version and a dual flag version. Top and bottom outlines of both versions of the package including pin numbering are shown in Figure 2.

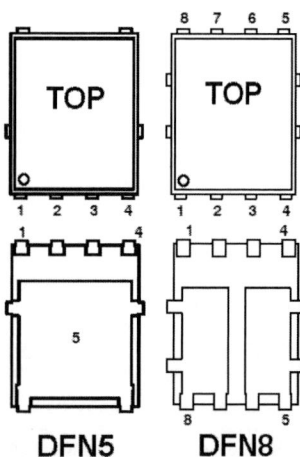

Figure 2: Top and Bottom outline drawings of the DFN5 and DFN8

The single flag version is commonly referred to as the DFN5 (Dual Flat No-lead 5 terminal) or DFN5 5x6 referring to the 5 mm by 6 mm package footprint. The dual flag version is often referred to as the DFN8 or DFN8 5x6.

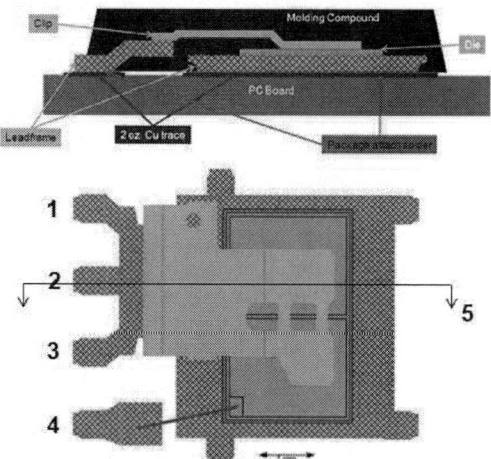

Figure 3: Cross Section and un-Molded top view of SO-8 flat lead Package

A detailed cross section and un-molded top view of the DFN5 is shown in Figure 3. In the case of MOSFETs terminal 4 is the gate connection. This connection is made with a 0.05 mm diameter bond wire. Terminals 1, 2 and 3 are connected by a clip to the source of a MOSFET. Terminal 5 is both an electrical connection and the die bond flag. This configuration provides a superior thermal path to ambient compared to J bend leaded packages.

For rectifier devices, terminals 1, 2 and 3 are connected to a single anode. Terminal 5 is both the die bond flag and cathode. Terminal 4 is a no connect.

For both rectifiers and MOSFETs the clip is soldered to the top surface of the die. Under high current operating conditions this provides lower Rdson or forward voltage drop. It also provides a thermal advantage as the clip provides a better thermal path out of the package compared to multiple small diameter bond wires (large diameter bond wires cannot be used as they would violate loop height rules in such a low profile package).

The internal construction of the DFN8 dual flag package is similar to the DFN5. For MOSFETs the gate connection is made with bond wires and the top side source connections are made using clips. For rectifier devices the bond wires are omitted. A significant advantage of the DFN8 is that the two devices in the package are electrically isolated from each other and may be operated independently. For rectifiers this allows the application designer to use the devices independently, in a common cathode configuration, in a common anode configuration or in a series configuration.

Advantages of the SO-8 flat lead Package

One of the most complex issues automotive electronics module designers must contend with is thermal management. Space constraints are forcing form factors to shrink. Ambient operating temperatures are increasing.

Figure 4 shows a comparison of steady state thermal resistance from junction to ambient versus mounting board copper area for a variety of common rectifier packages (obtained by thermal modeling). The semiconductor die mounted in each package is the same. The minimum copper area is set at 125 square millimeters to allow for a meaningful comparison (a copper area below the minimum footprint of a part will produce erroneous results). The SOD-123 flat lead, Powermite, SMA, SMB, SMC and SO-8 flat lead all use a clip for the top side electrical connection. The DPAK uses bond wires to make the top side electrical connection.

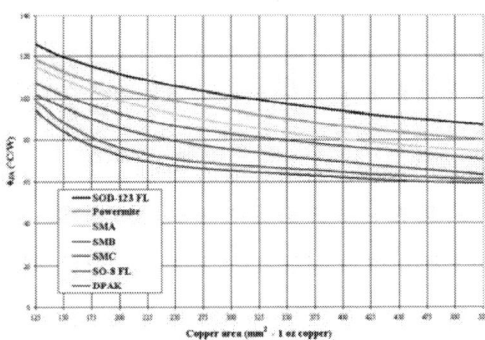

Figure 4: Comparison of Steady State thermal Performance vs. Copper area

Referring to Figure 4, the SO-8 flat lead and DPAK provide the best thermal performance to ambient. This shows the advantage of having a large exposed electrical and thermal pad connected directly to the board. Heat generated by the device is efficiently conducted down through the die structure, the die attach solder joint and into the exposed pad which acts as a heat spreader. The full area of the exposed pad is soldered to the application board, providing a large surface area for heat transfer.

The J bend configuration of the SMA, SMB and SMC limits the thermal performance of these packages. Heat generated by the device must pass through the cross sectional area of the J bend leads to get to the PCB and then to ambient. This cross sectional area is far smaller than that of the exposed pad of the SO-8 flat lead and DPAK. The clip connection to the top terminal of the SMx packages does improve thermal performance, but this cannot compensate for the limited cross sectional area of the J bent leads.

The Powermite and SOD-123 flat lead are significantly smaller than the other packages shown

in Figure 4. This limits the value of a direct comparison. Figure 4 clearly shows that there is a diminishing return on increasing copper area. Eventually, the copper area becomes so large that heat is transmitted to ambient before migrating farther out into the copper pad. A point to be made here is that improving package thermal performance has limits. As device power dissipation increases, there will be a point where a larger package is required.

Figure 5 shows a comparison of the single pulse transient thermal response from junction to ambient of a variety of common rectifier packages (obtained by thermal modeling). For this comparison the board copper area has been fixed at 140 square millimeters of 1 ounce copper for all packages. The die (semiconductor device) is also constant.

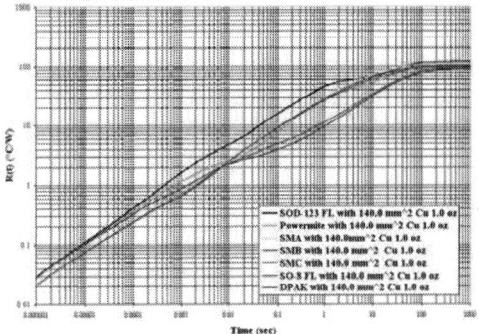

Figure 5: Comparison of Single Pulse Transient Thermal Response to Ambient

The SO-8 flat lead and DPAK again produce similar results. Both packages also out-perform the SMx, Powermite and SOD-123 flat lead packages. The most significant difference in performance occurs between approximately 0.1 second and 100 seconds. During this time, heat generated by the device in the SO-8 flat lead and DPAK packages has migrated through the package structure and is interacting with the ambient environment. The large contact area of the exposed pad improves heat transfer during this time. Compared to the SMx J bend lead packages, the thermal path from junction to ambient is also shorter.

For short duration pulses, all packages produce similar results. In this time frame, the transient thermal response is dominated by the semiconductor die characteristics. Heat generated in the junction of the semiconductor device has not had time to migrate through the package structure and interact with the ambient environment. If better short duration transient thermal response is desired, it is necessary to increase the die size of the device or change the electrical characteristics of the device.

Another advantage of the SO-8 flat lead can be described as packaging efficiency. This is a measure of the maximum die size the package can accommodate relative to its footprint on the PCB. Figure 6 shows a plot of maximum die size versus package footprint for a variety of common MOSFET packages. The SO-8 flat lead is capable of accommodating a die almost as large as the DPAK while occupying half the board space and 22% of the volume.

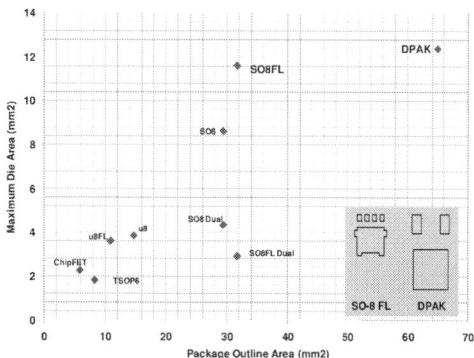

Figure 6: Plot of maximum die size vs. Package footprint

The DPAK is a wire bonded package. Electrical connections to the top surface of the die are made using relatively large diameter bond wires. To ensure bond wire integrity, a certain amount of wire loop is required. This limits the minimum package height and reduces the flag area available for die mounting. The SO-8 flat lead uses a copper clip to make the electrical connection to the top surface of the die. In the case of MOSFETs, the gate connection is made with a small diameter bond wire requiring minimum wire loop. This configuration allows for larger die bond flag area and a lower package height. With respect to both area and volume, the SO-8 flat lead is a very efficient packaging solution.

Design Considerations for the SO-8 flat lead

A common complaint of system designers using newer low profile high performance packages is that there is no standard outline. Often, designers or procurement groups do not want to be in a sole-source relationship for a power discrete device. The older generation of power packages and most IC packages conform to a standard outline, often JEDEC. This is no longer true of power packages.

The new generation of power packages often have design variations from vendor to vendor. This does not imply that the packages cannot be used interchangeably. Figure 7 provides a comparison of the mounting pad attributes of SO-8 flat lead packages from different vendors. Two vendors single pad versions and three vendors dual pad versions are provided. A possible board level solder footprint overlay is placed on each image in light gray. While the mounting pad details differ, it is possible to create one solder footprint that will accommodate packages from different vendors. Sole sourced designs can be avoided with proper board layout.

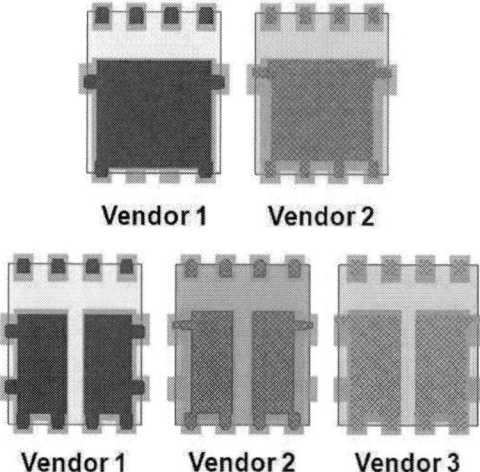

Figure 7: Backside Comparison of Single and Dual SO-8 flat lead Packages from Different Vendors

Another board level design consideration encountered using compact power packages is the ability to verify solder joint integrity after reflow. This problem is common to saw singulated packages without exposed lead ends. Saw singulated packages often require X-ray to ensure proper formation of solder joints. Figure 8 shows the external lead construction details of the SO-8 flat lead. This is a punch singulated package with exposed leads. All leads protrude from the molded edge of the package body. All metallic surfaces except the lead ends are tin plated to provide a solderable surface

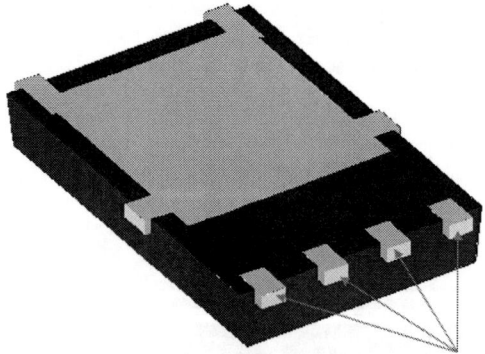

Leads Protrude from mold edge

Figure 8: Lead Construction Detail of the SO-8 Flat Lead Package

Having the leads protrude from the package body allows for visual inspection of the solder joint after reflow. Figure 9 provides a pictorial image of this.

Figure 9: Image of Solder Joint between SO-8 flat lead and PCB

Applications

Having superior transient thermal performance in the 0.1 second to 100 second time frame gives the SO-8 flat lead a distinct advantage in applications such as antilock braking systems (ABS). Such systems operate in a kind of "burst" mode, turning on for some number of seconds and then shutting down. The ability to quickly remove heat from the active semiconductors is a must.

Designers of ABS systems typically use either the DPAK or D2PAK for power MOSFETs and rectifiers. With the shrinking form factor of this application, there is a need for a more compact package. The SO-8 flat lead provides very similar thermal performance while using half the board space and 22% of the volume a DPAK.

Direct injection engine control is another application that can benefit from the superior thermal performance and compact size of the SO-8 flat lead. Such modules are often placed under hood where ambient temperature ratings are typically 125°C. A package with good thermal performance is required to prevent devices from exceeding their rated maximum operating temperatures.

Most automotive LED lighting is low power. LED headlights are the exception. LED headlights benefit from the use of a switching power supply to regulate voltage and current. The SO-8 flat lead provides compact solution for the power devices in this switching power supply.

The increasing number of infotainment and convenience applications are consuming more power and occupying more space. These applications benefit from both the superior thermal performance of the SO-8 flat lead and low package profile.

Conclusion

Automotive electronics is one of the most demanding environments designers must contend with. Applications must operate over a wide temperature range and maintain long term reliability.

Systems are subject to the same shrinking form factors and cost pressures faced at the consumer level. New higher performance packaging solutions are required to keep pace with new designs. The SO-8 flat lead offers automotive designers a new and much needed alternative to some to the older power packaging options.

Application of an Angular Exposure System to Fabricate True-Chip-Size Packages for SAW Devices

Barbara L'huillier[a], Michael Hornung[a], Dietrich Tönnies[a], Michael Jacobs[b], Christian Bauer[c], Frank Hammer[c], Thorsten Heuser[c] and Gregor Feiertag[d]

[a]SUSS MicroTec Lithography GmbH, Schleissheimer Str. 90, 85748 Garching, Germany
[b]SUSS MicroTec Lithography GmbH, Ferdinand-von-Steinbeis-Ring 10, 75447 Sternenfels, Germany
[c]EPCOS AG A member of TDK-EPC Corporation , Anzinger Strasse 13, 81671 Munich, Germany
[d]Munich University of Applied Sciences, Lothstrasse 64; 80335 Munich, Germany

+49 89 32007 366, +49 89 32007 390, barbara.lhuillier@suss.com

Abstract

Surface Acoustic Wave (SAW) filters are key components of mobile phones. In SAW components mechanical waves propagate on the surface of the chip, so the package must provide a cavity. Solid materials on the chip would inhibit the propagation of the surface waves. Reduction of size and cost, improved reliability and electrical performance are the main trends of SAW component evolution. Therefore, advanced cavity package technologies are required for SAW components. The Die Size SAW Package (DSSP) developed by EPCOS is a true chip size wafer level package that meets these requirements. The cavity package of the DSSP causes a significant topography. As a consequence the following rerouting process has to deal with the photo-lithographical challenge in terms of conformal resist coating and resist exposure. While conformal coating can be achieved by a spray process the exposure of patterns at sidewalls of topographic structures needs a customized solution. Exposure at a selectable angle instead of standard perpendicular exposure will be discussed in this paper based on EPCOS' DSSP technology.

Key words: 3D Lithography, SAW Filter, Surface Acoustic Wave, Wafer-Level-Package, MEMS Package

Introduction

Surface Acoustic Wave (SAW) filters have conquered one application after another over the past 30 years, first in TV sets and later in mobile phones [1]. Since SAW filters have a superior filter functionality and a small size, they play an important role in the evolution of mobile phones. The reduction in size of SAW filters has made a major contribution first to the miniaturisation of mobile phones and later to extending their functionality [2-4]. Several billion SAW filters are now produced per year, more than any other Micro-Electromechanical (MEMS) component.

The size of SAW filters can be reduced by reduction of the package size. With the Die Size SAW Package (DSSP) the package is as small as the chip. So DSSP is a true chip size wafer level Package. For electrical interconnect CuNi traces, defined by photolithography have to be routed across the vertical sidewall of the package.

Standard photo-lithography exposes the wafer with UV irradiation directed perpendicular to the wafer plane. As long as 3-dimensional structures have shallow sidewall profiles such topographic structures can be patterned with perpendicular illumination as well.

But if sidewalls are steep or even perpendicular standard wafer exposure from the top will not properly expose the resist film anymore. In this case the effective thickness of the resist that needs to be exposed at the sidewalls is equal to the height of the topographic structure. This height can easily represent multiples of the nominal film thickness. Even if the photo resist is suited for such thicknesses the exposure dose will be a multiple of the dose required for the resist layer deposited on plane structures. Under these conditions a uniform exposure of the entire conformal resist film is impossible.

The best compromise to expose conformal resist layers on severe topography will often be to expose the wafer from an angle of approximately 45°. Inclined exposure was used in the past to make on a laboratory scale 3-D-photo resist structures [5, 6].

Examples for exposure of vertical sidewalls

SAW filters are key components for mobile communication. In a SAW device mechanical waves propagate on the surface of a piezoelectric bulk material like $LiNbO_3$ and $LiTaO_3$. Therefore the package must not directly connect to this surface but has to provide a cavity between die and package. Thus SAW components require dedicated packaging solutions.

Size reduction of SAW filters allows for further miniaturisation of mobile phones and an extension of their functionality. The mobile communication industry, therefore, demands advanced cavity package technologies to be developed for SAW components. These packages have to offer high performance, high reliability, low cost while maintaining a form factor as small as possible. To address these requirements EPCOS has developed the Die Sized SAW Package (DSSP), a true chip-size wafer level package for SAW filters, see Figure 1. The key for the realisation of this package are the 3D-interconnects.

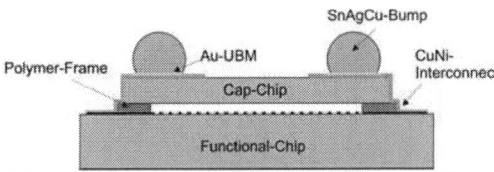

Figure 1: Left: Schematic layout of the DSSP package.

Application of the Angular Exposure System for SAW Wafer-Level-Packages

The Angular Exposure System was integrated in the process flow for the wafer-level-package for DSSP components at EPCOS. For high frequency applications the metallization of the SAW-transducers has a thickness of approx. 100nm and sub-micron critical dimensions. Therefore the transducers are very sensitive to any mechanical stress and moisture. The cavity-package has to protect the transducers against moisture and mechanical stress and must provide the electrical connection between the transducers and the PCB.

The cavity is fabricated by adhesive wafer to wafer bonding, where the polymer frame on the functional-chip and the cap-chip creates the cavity. The inter-connects are fabricated by electroplating in combination with spray coating and the angular exposure. On the interconnects a Under Bump Metallisation (UBM) is patterned and after bumping with SnAgCu solder balls the wafer is singulated.

Coating solution: Spray coating

For high topography wafers, a good resist thickness uniformity is necessary for the subsequent process steps, which can not be achieved with spin coating. Lamination of dry film resist is also not suitable because the topography of the packages is too high and the trenches between the caps are too narrow to allow a good conformal lamination. Spray coating is developed for high topography coatings

[7-9] and this method is applied for the customized lithography step, which is described in this paper. Negative tone resists and lift-off resists have been evaluated but for these resists the coverage of the upper edges is not sufficient. With positive tone photo resist, the coated film uniformity meets the required coating uniformity.

Technical Solution with Mask Aligner

The spray coated resist provides a nearly conformal layer that has to be exposed. As the interconnects are running across the sidewalls the photo resist needs to be exposed and developed completely in these areas. For the standard perpendicular exposure the effective resist film thickness corresponds to the topography depth (Figure 2), that can be easily 10 times the film thickness on the even planes.

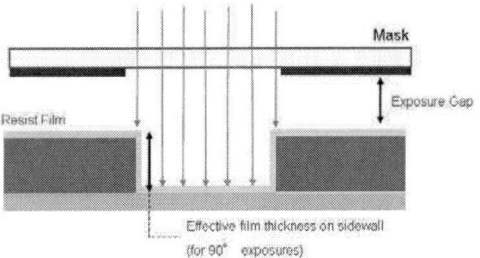

Figure 2: Standard perpendicular exposure of deep trenches and vertical sidewalls.

For negative tone resist this is less critical because the photo resist is cross linked by exposure and it might not be necessary to expose the resist down to the resist-wafer-interface. Especially, when a post exposure bake is required or at least optional, this can allow such a difference in film thickness. But for spray coating of negative resist the required conformity could not be achieved, therefore positive tone resist must be used.

For positive tone resist it is essential, to expose the complete film layer from top to the wafer surface to make sure, that the exposed areas can be developed completely. Even if two exposures are performed, one for the thin film areas and a second one for the sidewall exposure, exposing such deep sidewalls is still challenging for positive tone resist. Chemically amplified resists seem to solve this problem, but even for these resists, the thickness (corresponding to the sidewall depth) might be too high.

Due to resist bleaching during the exposure the absorption of the UV light depends on the resist thickness. Further by refraction of the incident light at the rounded upper edge of the resist layer a part of the side walls is not exposed. This is partly compensated by refraction at the lower edge.

95

The perfect solution would be to expose the sidewall with light perpendicular to its surface, this can not be realized with a mask aligner.

Therefore, SUSS developed the angular exposure solution.

If the wafer is exposed with light of an inclination angle of 45° the effective film thickness is nearly uniform across the resist surface. The film thickness on the vertical sidewalls is only 1.4x of the nominal film thickness (Figure 3).

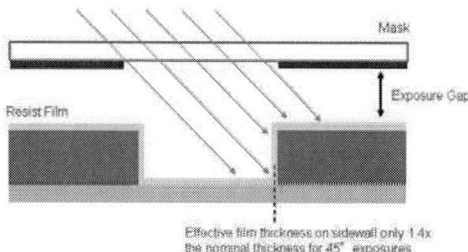

Figure 3: Angular exposure of deep trenches and vertical sidewalls.

The possibility of angular exposure with inclination angles of 45° and 60°, respectively, is an additional option to the standard perpendicular exposure of wafers. As the optical system itself needed some changes, no upgrade is possible for mask aligners in the field.

Figure 4 shows the light path for the two extreme angles, 45° and 90°, respectively. Further some changes of the optical system will be described according to Figure 4.

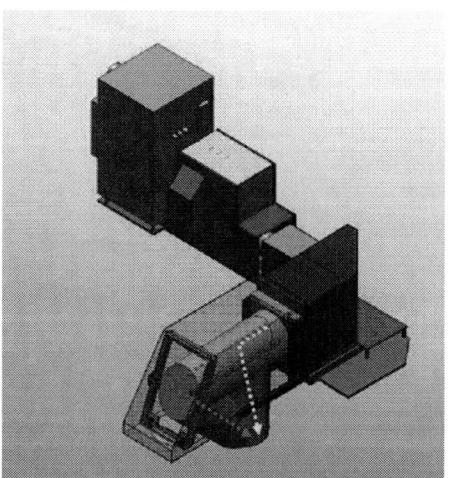

Figure 4: Schematic design for light path in standard perpendicular (yellow line) and 45° (red line) exposure direction, respectively.

In figure 4 it can be seen that the mirror house needs to be moved to the front compared to the 90° exposure position. This is necessary to achieve that the light beam with a 45° and 60°

inclination angle, respectively, exposes the same wafer area on the chuck like the perpendicular exposure. The position for the 60° exposure is between the 90° position and the 45° position.

As the pre-aligner and the linear transport system from the pre-aligner to the exposure chuck are placed behind the chuck moving the mirror house is the only possibility to arrange the proper wafer exposure for all exposure angles.

Besides the different mirror house positions the deflecting mirror has to be tilted in different angles, too, to deflect the light in a 45° and 60° angle onto the wafer. Both positions of mirror house and of deflecting mirror have to be adjusted manually. In order to help avoid operator errors sensors monitor the position of mirror house and deflecting mirror. The exposure program will start only if the actual positions match the settings in the process recipe.

The possibility to adjust the exposure angle made it essential to change the position of the front lens compared to a standard mask aligner. If the front lens would be at its standard position directly above the wafer it would be necessary to tilt the front lens when exposing the wafer at an angle different from 90°. This would require too much space in the machine. Therefore, the front lens is now located inside the optical tube approximately at the starting positions of the yellow and red line in Figure 4. As this changes the optical path, a reconfiguration of the field lens and a longer optical path was necessary. Also maintenance is slightly more extensive as the front lens needs to be accessed from the top now.

Due to the inclination angle of the exposure light the shape of the exposed area changes. For a 90° exposure the exposed area is circular while it changes to ellipsoidal shape for angular exposure. The diameter in y-direction gets larger while the diameter in x direction decreases. This effect is even stronger for the 45° exposure compared to the 60° exposure. Thus, the angular exposure is possible only for wafers up to 150mm, while the 90° exposure is possible for up to 200mm wafers. To increase the exposure area for 200mm wafers for angular exposure, extensive changes are necessary for the optics configuration and more space would be needed.

As for angular exposure the mirror house needs to be moved to the front, the housing of the machine needs to be changed. More space is needed in the front. Due to the new position of the front lens, the optical path is longer and the housing is extended in the width, too. Both changes make the footprint of the angular exposure system slightly larger then the footprint of the standard system.

Exposure effects: Shadows and Reflections

As already explained the angular exposure system is used for exposing wafers with high topography structures. The high topography structures are leading to shadow effects due to the inclined light flow. Further, reflections from the pattern of the opposite side change the exposure dose in areas close the sidewalls of the structures.

As can be seen in Figure 5, the exposed trench can be separated into three areas: Depending on the inclination angle, one area is exposed by direct light and additionally by reflected light from the sidewall (area I). The higher exposure dose in area I has a severe influence on the CD control. The design rules for the mask layout therefore have to compensate this effect. Another area is exposed by direct light only (area II) and one area is not exposed, because it is in the shadow of the neighboured sidewall (area III). In case the structured material is transparent to UV-light, like some semiconductor materials or epoxy resins, this area is slightly exposed by transmitted light, but of course the exposure dose here is not enough to expose the photo resist completely.

For this reason, exposure of all sidewalls in one shot is not possible with angular exposure. Up to 4 exposures will be necessary to structure the interconnects at all 4 sidewalls. Typically, each exposure requires a separate mask. For proper exposure of all interconnects accurate alignment is essential.

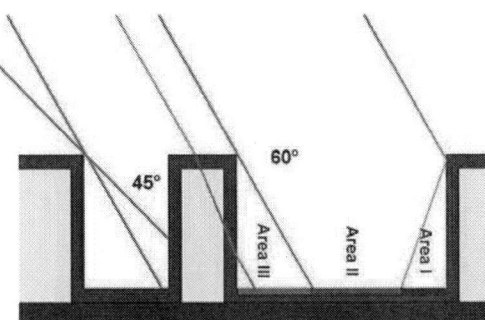

Figure 5: Schematic illustration of exposure effects.

Changes in Alignment Process

The alignment process is slightly different from standard proximity printing. With standard perpendicular exposure the position of the exposed structures does not change compared to the mask positions.

For angular exposure, the printed image in the resist is shifted compared to the original position. The shift depends on the exposure gap (Figure 6). It could reliably be predicted by geometrical considerations from the exposure gap and the inclination angle. This has to be noted and can be corrected either by adjusting the mask layout or by using the off-set function in the alignment software.

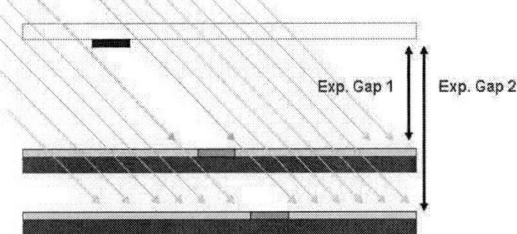

Figure 6: Schematic illustration of pattern printing shift caused by angular exposure.

The possibility of correcting the pattern shift with the mask layout requires the exposure gap to be constant. The advantage of correcting the shift with the mask layout makes it also possible to allow the alignment check with the alignment key before exposure, if only the functional structures are corrected for the shift. Whereas using the off-set function allows additional corrections in the exposure gap when this is required for any reason. The main disadvantage of using only the offset function is the limitation of the offset correction. However both possibilities can be combined also, even though calculation for the shift correction becomes more extensive.

The dependence of the alignment accuracy on the exposure gap requires a precise control of the exposure gap. Due to the strong gap dependency the alignment accuracy is approx. +/-5µm for the angular exposure while it is +/-0.5µm (direct alignment) and +/-1µm (standard alignment), respectively, for perpendicular exposure. Any unevenness or particles between chuck and wafer changes the exposure gap and thus influences the alignment accuracy.

Performance of the Angular Exposure System

The light intensity and the light uniformity across the exposure area are comparable to the standard exposure system. For perpendicular exposure the light uniformity is below 3% ((max-min)/(max + min)) for a 200mm wafer area and for 60° inclination angle the light uniformity is below 2.5% for a 150mm wafer area. The light intensity is around 30mW/cm^2 for both exposure angles (i-line). For measurement of the light intensity a special probe holder is used that tilts the probe surface towards the inclination angle to keep the measurement error small.

The resolution results are in the same range for the angular exposure system as for the standard perpendicular exposure. The Figures 7 and 8 demonstrate successfully processed interconnects after lithography and after metal patterning for the DSSP package., respectively

Figure 7: SEM image of spray coated, exposed and developed resist pattern for the metal interconnect.

Figure 8: Successfully processed metal connects.

Reliability of the SAW Wafer-Level-Packages

The thermo mechanical properties of the package were simulated by Finite Element Modeling (FEM), to analyze the mechanical stress on the 3D-interconnects and thus the reliability of the package (Figure 9). The simulations show only moderate stress at the interconnects and the bumps and therefore a very reliable package is expected.

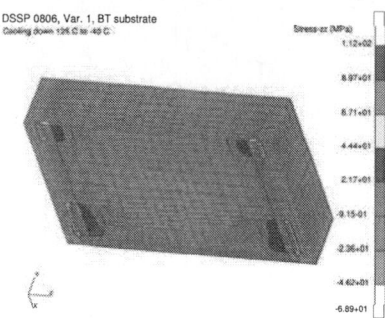

Figure 9: Example for FEM of stress during thermal cycling of a DSSP mounted on BT-substrate.

The typical requirements for mobile communication components include more than 500 cycles for the Temperature Cycling Test (TCT) from -40°C to 125°C. According to the simulation approx. 1200 cycles are expected before the first parts would fail in this test.

For the experimental verification Daisy Chains of the DSSPs with different sizes and pinnings were fabricated and mounted on test boards (Figure 10).

Figure 10: DSSP Daisy-Chains mounted on BT test board.

After preconditioning with moisture sensitivity level 2a according to JEDEC standard 020D (with 6 reflows) the test bords were cycled 2000 times from -40°C to 125°C. Figure 11 shows the weibull chart of this test. The failure rate is less than 5% at 1500 cycles with a confidence level of 90%. The experimental results are correlating very well to the simulation. The requirements of components for mobile communication are fulfilled and the mass production of the Die Sized SAW Package was already started successfully.

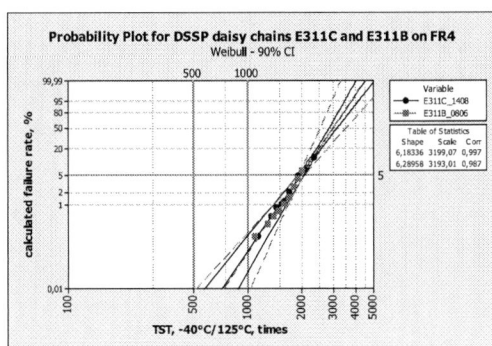

Figure 11: Weibull plot of TCT.

Summary

SAW components require low cost, small size, high reliability and good electrical performance. These requirements do not only relate to the SAW component itself but also to its packaging. True chip size packaging as DSSP meets these reqirements but poses a challenge to the photo-lithographical process due to its packaging caused topography.

For high topography structures like true-chip-size packaged SAW devices 3D-lithography is required.

Conformal coating of such high topography wafers is solved by spray coating of positive tone resist.

For the exposure of the interconnect structures on the vertical sidewalls of the package, SUSS and EPCOS developed a solution for this challenge with the angular exposure system.

The inclination angle gap exposure causes a shift in the print that can be corrected by the mask layout or by the offset parameters of the alignment software to achieve accurate alignment. As the correction by off-set parameters is limited, a combination of both methods is recommended. As the shift depends on the exposure gap, the correction factor needs to be calculated by geometrical considerations. For exposure of all sidewalls of the structures, up to 4 exposures are necessary.

Realization of the metal connects for high topography structures on the wafers is done successfully with SUSS lithography equipment.

Simulations and experimental resuslts proofed the fulfillment of the requirements of the DSSP and the processing technology successfully.

References

[1] C. Ruppel, L. Reindl, R. Weigel, "SAW Devices and their Wireless Communications Applications", IEEE microwave magazine, June 2002, pp 65-71

[2] P. Selmeier, R. Grünwald, A. Przadka, H. Krüger, G. Feiertag and C. Ruppel, "Recent Advances in SAW Packaging", Proc. 2001 IEEE Ultrasonic Symp., pp 283-291

[3] Robert D. Koch, Martin Schwab, F. Maximilian Pitschi, J¨urgen E. Kiwitt, and Robert Weigel, "Ultra Low-Profile Self-Matched SAW Duplexer with a Flip-Chip HTCC Package for W-CDMA 2100 Mobile Applications" Proc. IEEE Int. Mirow. Symp., 2009

[4] G. Feiertag, H. Kruger, and C. Bauer, "Surface Acoustic Wave Component Packaging," in European Microelectronics and Packaging Conference, 16th, June 2007, Oulu, Finland

[5] C. Beuret, G. –A. Racine, J. Gobet, R. Luthier and N. F. de Rooi, "Microfabrication of 3D Multidirectional Inclined Structures by UV Lithograpy and Electroplating" Proc. IEEE Workshop on MEMS, 1994

[6] Y.-K. Yoon, J.-H. Park, andM. G. Allen, "Multidirectional UV Lithography for Complex 3D MEMS structures" in J. Micromech. Systems, October 2006, pp 1121-1130

[7] K. Fischer and R. Süss, "Spray Coating – a Solution for Resist Film Deposition across Severe Topography" Proc. Semicon West, July 2004

[8] K. A. Cooper, C. Hamel, B. Whitney, K. Weilermann, K. J. Kramer, Y. Zhao and H. Gentile, "Conformal Photoresist Coatings for High Aspect Ratio Features" Proc. IWLP Sept. 2007

[9] K. Cooper, K. Cook, B. Whitney, D. Toennies, R. Zoberbier, K. J. Kramer, K. Weilermann and M. Jacobs, "Lithographic Challenges and Solutions for 3D Interconnect" Proc. IWLPC 2008

Silver/Palladium Pastes for Aluminium Nitride Applications

Kathrin Reinhardt, Christel Kretzschmar, Richard Schmidt, Markus Eberstein

Fraunhofer Institute for Ceramic Technologies and Systems

Winterbergstraße 28, 01277 Dresden Germany

Phone: +4935125537837 Fax: +493512554338

Email: Kathrin.Reinhardt@ikts.fraunhofer.de

Abstract

Silver/palladium resistor pastes on aluminium nitride (AlN) are presented. These resistor pastes are very interesting for power electronic and heater applications. Currently realised paste systems are a resistor paste based on AgPd as conductive phase, which has sheet resistivities up to 1 ohm/sq and a RuO_2 resistor paste system with a sheet resistance higher than 6 ohm/sq. However both pastes are not mixable. These results make it necessary to develop AgPd pastes in this resistance gap. Pastes with sheet resistivities between 100 mohm/sq and 10 ohm/sq and low TCR values (thermal coefficient of the resistance) are under development. By addition of glass and an inert additive it was possible to achieve with a AgPd ratio of 1/1 different surface resistances. All formulated pastes show low TCR values between 0 and $100*10^{-6}$ K^{-1}. Resistance measurements were done, to characterise the different formulated pastes. To perform comparing measurements of the particular resistances under electrical load different layouts are designed, which allows for each substrate the same surface heating. Electrical load measurements at 200 °C and 300 °C and cycled tests for 3000 pulses were done. The influences of long-term electrical load and ageing at higher temperature were investigated. Film structures were analysed by optical microscopic measurements and FESEM.

Key words: aluminium nitride, heater, Ag/Pd pastes, thick film

Introduction

In the recent years aluminium nitride (AlN) became more important as substrate material in microelectronic components and power electronics. Compared to other materials like aluminium oxide (Al_2O_3) and Low Temperature Cofired Ceramics (LTCC), AlN has an excellent thermal conductivity up to 180 W/mK and very good mechanical properties as well as a good thermal shock resistance. Those properties make AlN an interesting material for heater applications [1]. Currently realised paste systems are a resistor paste based on AgPd as conductive phase, which has sheet resistivities up to 1 ohm/sq and a RuO_2 resistor paste system with a sheet resistance higher than 6 ohm/sq [2, 3]. The disadvantage is that both systems are not blendable. At present new resistor pastes with AgPd as conductive components are under development with sheet resistivities up to 10 ohm/sq. This study shows the influence of the content of glass and an inert additive in AgPd resistor pastes on the film properties as well as on the heater characteristics.

On the basis on silver, palladium, an inert filler and a lead oxide free glass, which has a thermal expansion coefficient (TEC) near that of AlN, a resistor paste system is developed with sheet resistivities between 0.1 and 10 ohm/sq and low TCR values (thermal coefficient of the resistance).

Therefore the influence of the AgPd ratio on the sintering behaviour and the thermal coefficient of the resistance (TCR) were investigated. The resistor films were characterised under electrical and thermal load.

Experimental Procedure

Materials

All AgPd pastes were done at the IKTS. Also for the contacting an IKTS silver-based ink was used. The inks were printed on 1 x 1 inch² aluminium nitride substrates (thickness 625 μm).

Paste formulations

The powders were mixed together in a tumbling mixer for one hour. The mixture was ground together in a dissolver with a solution of ethyl cellulose in texanol to prepare a printable paste. Afterwards the pastes were passed through a three roll mill to ensure that there are no large particles or agglomerations.

Printing test structure and measurements

A semi-automatic screen printing machine, Microtronic II, from EKRA Automatisierungs-systeme GmbH (Bönnigheim, Germany) was used to print the test structure for the AgPd pastes. The layout of the test structure is shown in Figure 1. The test layout consists of four meander structures, where each meander has a line width of 500 μm and 200 squares.

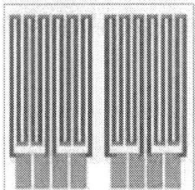

Figure 1: Test layout for resistance measurements

All prints were done with stainless steel screens from Koenen GmbH (Ottobrunn, Germany). The contact paste (Fig. 1: blue structure) was printed with a 200 mesh and the resistor paste (Fig. 1: red structure) with a 325 mesh screen. After each printing step the pastes levelled 10 minutes at room temperature and were heated to 150 °C for 15 minutes to remove volatile organics. Both structures were fired separately in a belt furnace at 850 °C for 10 minutes in air atmosphere. Using these structures the resistances were measured by a quasi-4-point probes method. The measured resistances were converted into sheet resistivities and normalised at a 21 μm dried thickness. The relative resistance (resistance at the measured temperature divided by the resistance at 25 °C – R/R25) is evaluated in dependence of the temperature. The temperature behaviour of the resistances was measured between -55 °C and 150 °C.

To perform comparing measurements of the resistances under electrical load different layouts were designed, which allows for each substrate the same surface heating (Fig. 2, Tab. 1). To realise these structures four prints have to be done. At first the resistor paste (Fig. 2: red structure; 325 mesh), second the contact pads (Fig. 2: green structure; 200 mesh) and finally two overglaze layers (200 mesh). All inks were dried and fired like described above, with exception of the overglaze layers, which are fired at 650 °C in air atmosphere.

Table 1: Test layout for STOL measurements

Paste	Resistivity [Ω/□]	Layout	Number of squares
A	0.1	Heater 0.1	145.4
B	1	Heater 1	16.6
C	10	Heater 10	2.11

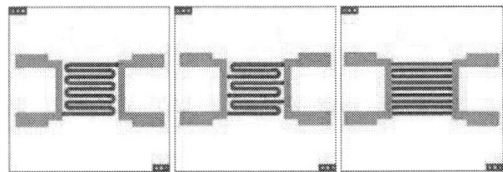

Figure 2: Layouts for STOL measurements (from left to right: heater 0.1, heater 1 and heater 10)

For the investigations of the resistances under electrical load STOL measurements (short term overload) were done. For this purpose the substrates were thermal connected to an Al cooling block with a Ag-containing heat-conducting paste. Initial resistance was measured at room temperature. The electric load was applied as 5 second pulses. The power increases stepwise and the resistance was measured 60 seconds after every pulse. This procedure is shown in Figure 3.

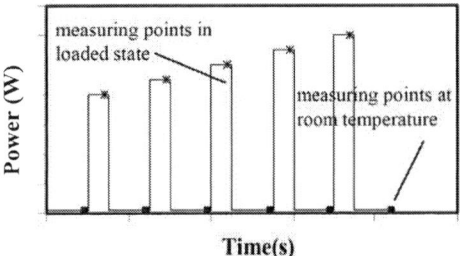

Figure 3: Loading step test with electric pulses

During the measurement the maximum temperature is determined with a pyrometer. The change of the resistance at room temperature for the overload tests is observed. In addition to this long term measurements at 200 °C and 300 °C with and without electrical load were carry out. The temperature distribution of the resistor and substrate during and after the load was measured with an infra-red camera.

The surface was characterised by FESEM, using a dual-beam focused ion beam and scanning electron microscope NVision 40 (company Carl Zeiss AG) at 8 kV.

Results and Discussion

Resistivity and TCR

To achieve heater pastes with resistivities between 0.1 and 10 ohm/sq, pastes with different Ag/Pd ratios were tested: 1/1; 10.3/1 and 23.1/1. Figure 4 shows the resistivity and its temperature coefficient as a function of the glass and inert additive concentration for the three Ag/Pd ratios. The resistivity diagrams show, that a resistances range between 0.1 and 10 ohm/sq could only achieve with a Ag/Pd ratio of 1/1. To set the right resistance the glass ratio was varied. With the inert additive precise adjustment of the resistance could be done. Furthermore the inert additive was inserted to avoid the formation of blowholes, which are formed due to high glass content. Looking at the TCR diagrams, it can be seen that the different AgPd ratios determined the temperature coefficient of resistance in a high degree. The variation of glass and inert additive show an insignificant part at the TCR adjustment. The TCR varies between 0 and $1500*10^{-6}$ K^{-1} depending on the AgPd ratio. This means that resistor pastes with a resistivity range between 0.1 and 10 ohm/sq results in low TCR values. So for our applications a Ag/Pd ratio of 1/1 is used, where TCR's between 0 and $100*10^{-6}$ K^{-1} were achieved.

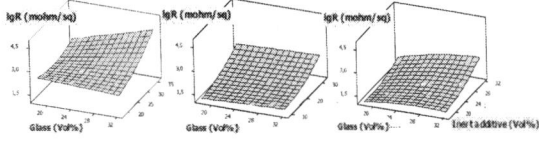

AgPd = 1/1   AgPd = 10.3/1   AgPd = 23.1/1

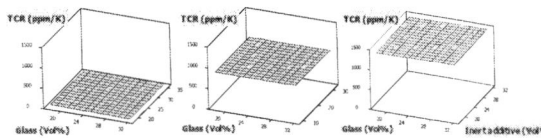

Figure 4: Resistivity and thermal coefficient of resistance depending of the solid composition

After these characterisations three pastes with resistances of 0.1 ohm/sq, 1 ohm/sq and 10 ohm/sq with low TCRs were formulated.

For all three pastes the temperature dependencies of the resistors were characterized. The results are shown in Figure 5. The resistances show a maximum at 175 °C, 110 °C and 90 °C. The cold-TCR values (-55 °C … 25 °C) are between 50 and $77*10^{-6}$ K^{-1} and the hot-TCR values (25 °C…150 °C) between -5 and $35*10^{-6}$ K^{-1}.

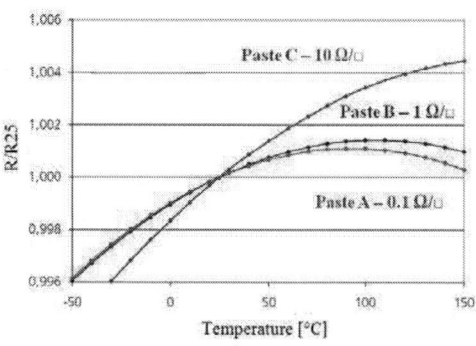

Figure 5: Relative resistance as a function of the temperature

Heater characterizations

The heaters were investigated by electrothermic pulse load measurements. Different layouts were developed for a uniform surface heating, which allows a comparison between the resistor pastes (0.1 Ω/□, 1 Ω/□ and 10 Ω/□). Figure 6 shows the infrared images of the three heater pastes. Thereby Figure 6a) presents the 0.1 Ω/□, 6b) the 1 Ω/□ and 6c) the 10 Ω/□ paste. It can be seen, that a uniform heating of each film for the respective layout could be achieved. Furthermore presents Figure 7 the electrical power of the films in correlation to the measured temperature. For each

film the same electrical heating was determined. The substrate temperatures increase directly with the electrical power. By applying about 250 W all samples heat to 180-200 °C

To compare and characterise the curves of Figure 7 and to evaluate the three STOL layouts, the thermal resistance was estimated. The thermal resistance is the specific factor of a device and conduce to the characterisation of cooling elements and heat-conducting pads. This means for our applications, that a same thermal resistance results in comparable STOL layouts for the particular resistor pastes. The slope of the curves in Figure 7 between 0 and 300 W relates to the thermal resistance. It can be seen that the thermal resistances of all three resistors are very similar in this power range and are between 0.62 and 0.63 K/W. The results show, that the layouts for the different resistor pastes could be used and are comparable to each other.

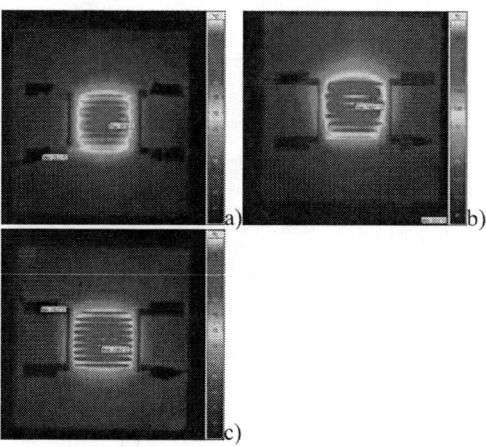

Figure 6: IR-measurements of the films for testing low ohmic resistances: a) heater 0.1, b) heater 1 and c) heater 10

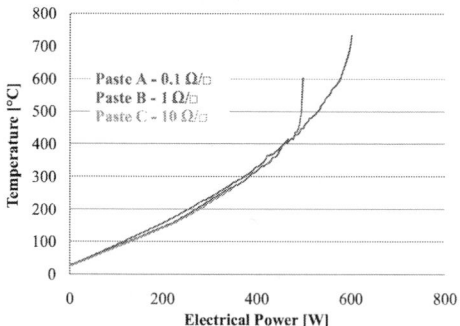

Figure 7: Heater temperature measurements by pulse electrical load

Deterioration due to electrical pulse load

Figure 8 shows the resistance change during the pulse electrothermic loading of the films. The resistance measurements were carry out at room temperature after every electrical pulse, when the substrates were cooling down at 25 °C. It can be seen, that the resistance change increase with the increasing content of glass and inert additive. Paste C shows the highest resistance change, which is at 1% by an electrical power of 350 W. In contrast to this paste A shows at 350 W only a resistance change of 0.1%, which is one tenths lower than this of paste C.

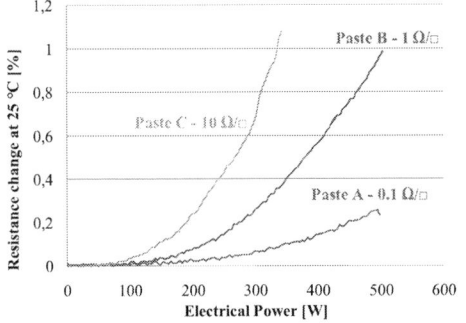

Figure 8: Resistance change at 25 °C in dependence of the electrical power

In addition to these investigations long term measurements at 200 °C and 300 °C were done. Thereby at 200 °C all three pastes were tested, meanwhile at 300 °C only the behaviour of pastes A and B were analysed. Each paste was loaded with the particular electrical power to 200 °C or 300 °C for 3000 pulses, which relates to a complete loading time of 4 hours for each temperature. Between the pulses the resistance was measured at room temperature. The results are presented in Figure 9 (200 °C) and Figure 10 (300 °C). Paste C shows already a resistance change of 5% after 1000 cycles and conducting paths were broken. In comparison, paste B show a 2.8% change and paste A just a 0.6% change of the resistance after 3000 cycles and no damages of the heater lines could be detected. Also at 300 °C the resistance changes of paste A only for 2.3% after 3000 cycles, whereas paste B exhibits a change of 25.4%. The overglaze layer of resistor B shows cracks in the middle of the structure.

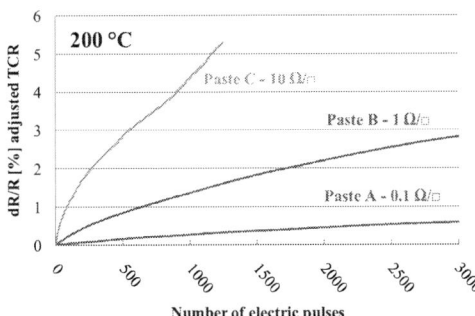

Figure 9: Resistance change for pulsed electrical load measurements at 200 °C – 3000 pulses

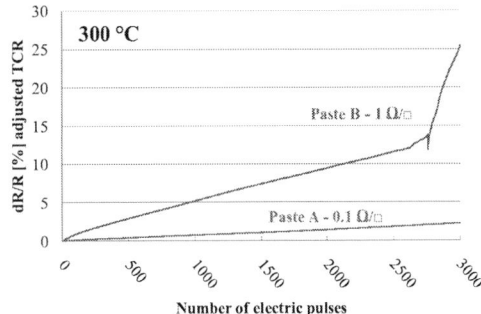

Figure 10: Resistance change for pulsed electrical load measurements at 300 °C – 3000 pulses

Deterioration due to continuous electrical load

The correlation between the continuous electrical load and the paste properties were investigated. The fired pastes were electrical loaded for 50 and 100 hours at 200 °C and 300 °C without any interruption. Table 2 shows the measurement values for all pastes and loading times. The resistance change was measured for each condition. Paste A with a resistivity of 0.1 $\Omega/\square$ shows no significant resistance change for 50 hours as well as 100 hours continuous electrical load. Also paste B (R_{sq} 1 $\Omega/\square$) shows for the long-term electrical loading at 200 °C only a small resistivity change of 1.05 % for 100 hours loading time. The measurement values of paste B at 300 °C are not available yet and will be done next. For paste C, which has the highest sheet resistance of 10 $\Omega/\square$, no more than the resistance after 50 hours / 200 °C continuous electrical load could be determined. Thereby the resistance change was over 125 %. All other values were not measureable because of defects in the conductive paths, which were broken at 200 °C.

Table 2: Measurement values of the continuous electrical load for 50 and 100 hours

Paste	Temp. [°C]	Change of resistance [%]	
		50 hours	100 hours
A	200	0.05	0.24
0.1 $\Omega/\square$	300	0.31	0.33
B	200	0.04	1.05
1 $\Omega/\square$	300	Values are not available yet	
C	200	125.58	-
10 $\Omega/\square$	300	Values are not available yet	

Deterioration due to temperature

The ageing influence of the temperature is considered in more detail. The fired pastes were storage for 50 hours at 200 °C and 300 °C, which is nearly twelfth times more than the loading time at 3000 pulses (4 hours). Figure 11 shows the results for all pastes and ageing temperatures. No significant resistance change could be determined for any paste and all structures were free of defects in the conductive paths as well as in the overglaze layers.

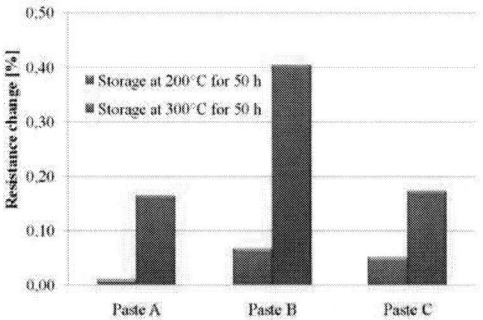

Figure 11: Resistance change for long term storage at 200 °C and 300 °C for 50 hours

Correlations between electrical and structural properties

The results of the resistance deterioration under different electrical and thermal conditions show that defects occur in the formed layers. Thereby possible failures could be:

- Temperature
- Electrical damage or
- Thermo-shock impact

The influence of temperature was done by resistance measurements of substrates which were storage at 200 °C and 300 °C for 50 hours. No significant resistance change could be determined for any paste. For this reason an ageing influence only in case of temperature could be excluded.

The examinations of the long term electrical load measurements show, that the higher ohmic pastes show defects in the conductive paths (see Fig. 12) after loading the substrates pulsed with electrical power for a longer time.

Figure 12: Microscopy image of a fused conductive path of paste C (10 Ω/□) after pulsed electrical load measurements

A possible cause could be an electrical damage of the layers, which has lower amounts of conductive paths inside the sintered layer. Figure 13 to 15 present FESEM images of the surface of the AgPd films A, B and C. For the higher resistance pastes, like paste C, the conductive paths (AgPd – light grey regions) in the filmd are clearly in a minor amount than in paste A. In case of the higher glass and inert additive content (dark grey regions), the formed conductive paths are not so continuous arranged like in paste A. Therefore it is possible that a breaking path shows a higher resistance change.

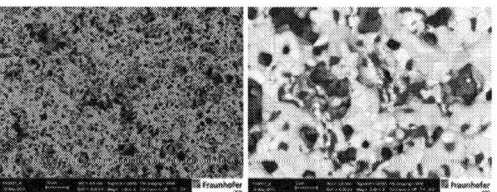

Figure 13: FESEM images of the surface of AgPd-film A – 0.1 Ω/□

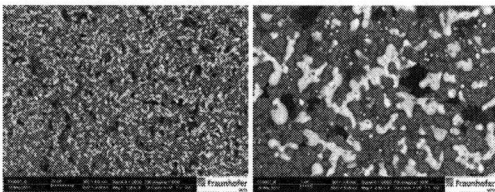

Figure 14: FESEM images of the surface of AgPd-film B – 1 Ω/□

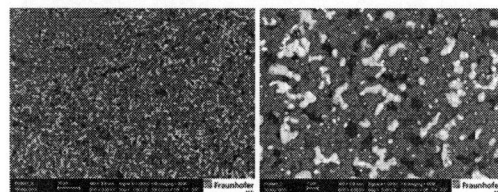

Figure 15: FESEM images of the surface of AgPd-film C – 10 Ω/□

Furthermore long term measurements under permanent electrical load for 50 and 100 hours were done. These investigations show that higher resistor pastes like paste B and C show broken conductive paths after the continuous electrical load over 50 and 100 hours at 200 °C or 300 °C. These results exhibit that the electrical load has a direct influence to an electrical damage of the conductive paths.

Electrical load measurements with increasing power or at constant power but for many cycles exhibit cracks in the overglaze layer like it is shown in Figure 16 b. The figures presents microscopic images of a defect-free (Fig. 16 a) and a damaged (Fig. 16 b) overglaze layer with subjacent conductive paths of a 1 Ω/□ resistor paste.

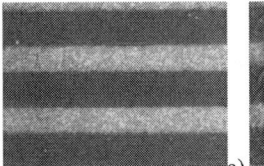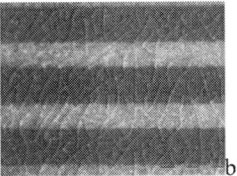

Figure 16: Microscopy images of encapsulated conductive paths of paste B (1 Ω/□) after long term electrical pulse loads: a) 200 °C / b) 300 °C

For both kinds of measurements the temperature of the substrates changed continuously in short periods of time between 25 °C and 200 °C / 300 °C. Thereby a large temperature difference effects on the substrate which can cause a TEC mismatch between the overglaze layer, the AgPd film and the AlN substrate. Tensile stress inside the glass can cause cracks, which results in higher resistance change and a damage of the heater structure. Figure 16 shows that a temperature of 300 °C cause a higher tensile stress to the encapsulating layer, which affects more cracks on the overglaze surface. In this case a glass with an adapted TEC value has to be applied.

Table 3 gives a summarizing overview of the three AgPd paste systems after the long-term temperature measurement and the continuous and pulsed electrical load measurements. With these tests the influences of temperature, electrical load and thermo-shock were determined.

Table 3: Summary of the AgPd pastes and their stabilities after long-term temperature measurements and continuous / pulsed electrical load measurements

Paste		Influence of temperature over 50 h	Continuous electrical load	Pulsed electrical load (3000 cycles)
0.1 Ω/□	200 °C	stable	stable	stable
	300 °C	stable	stable	stable
1 Ω/□	200 °C	stable	stable	stable
	300 °C	stable	Values are not available yet	Cracks in overglaze, Not fused
10 Ω/□	200 °C	stable	Fused, Without cracks in overglaze	Fused, Without cracks in overglaze
	300 °C	stable	Values are not available yet	Values are not measured

Conclusion

The development of new resistor pastes for AlN allows the production of coatings with high resistance (10 ohm/sq) and low temperature coefficient. Such pastes are very interesting for power electronics even more after closing the resistor gap between 1 ohm/sq and 10 ohm/sq, which is realized at the IKTS.

The AgPd paste with a sheet resistivity of 0.1 ohm/sq show very good results and stable resistances for temperature ranges up to 300 °C. Also the 1 ohm/sq AgPd pastes could achieve such results for temperatures up to 200 °C. However for higher temperatures (300 °C) the overglaze layer shows cracks. This means the TEC mismatch between the pastes and the substrate should be optimized. The 10 ohm/sq is not stable for temperatures higher than 200 °C. Further investigations relating to lower temperatures will be done.

References

[1] M. Hentsche et al., "Thick film heater for aluminium nitride ceramics", Proceedings of the 2008 41st International Symposium on Microelectronics (IMAPS), Providence, Rhode Island, November 2-6, pp. 1224-1228 , 2008.

[2] C. Kretzschmar et al., "A new paste system for AlN", Proceedings of the 2001 7th International Symposium on Advanced Packaging Materials, Braselton, Georgia, March 11 - 14, pp. 672-675, 2001.

[3] C. Kretzschmar et al., "Thick Film alloy resistor system on AlN", Proceedings of the 1998 43rd International Scientific Colloquium, Ilmenau, Germany, September 21-24, pp. 92-97, 1998.

Miniaturization - Solder Paste Attributes for Maximizing the Print & Reflow Manufacturing Process Window

Karthik Vijay

Indium Corporation of Europe, Milton Keynes, UK

kvijay@indium.com ph.+44 (0) 7584 643 677

Abstract

Current technology drivers span a broad spectrum that include miniaturized hand-held devices where real estate is at a minimum as well as high density server assemblies that have high-reliability requirements. SMT (Surface Mount Technology) assembly consequently involves overcoming challenges that include- complex components such as sub-0.4mm CSPs, 01005s, LGAs; Area Ratios of 0.6 and below; Stencil apertures sub-250 microns dia, Stencil thicknesses between 75-110 microns; Solder-Mask vs Non-Solder-Mask defined pads; varied surface finishes such as OSP, ImAg; and long reflow profiles with reflow in air.

With miniaturization and smaller stencil apertures, some of the key manufacturing challenges include (a) achieving a good print brick deposit, maximizing paste transfer, minimizing print-to-print deviation; (b) eliminating HiP (head-in-pillow) which is the biggest issue facing the industry; (c) achieve complete solder coalescence and prevent clumpy/grainy solder joints for the small paste deposits that see longer reflow profiles.

These conditions impose heavy demands on the solder paste flux chemistry. To add to the complexity, halogen-free flux requirements in a Pb-free process are pretty much becoming the norm. Pb-free solders mean higher temperatures and therefore the flux needs to do more work to reduce oxide. This is more difficult for the smaller paste deposits because of the higher surface area to flux volume ratio. Halogen-free fluxes mean that the fluxes have lesser "juice" when compared to normal halide-containing fluxes, but still need to address higher temperatures associated with a Pb-free process.

This study details the development of a flux technology platform to (a) achieve print consistency (maximize paste transfer, minimize deviation) for small apertures (sub-300 microns diameter, area ratios of 0.5 – 0.6); (b) eliminate head-in-pillow with an enhanced oxidation barrier approach as opposed to just making a flux more active; (c) achieve complete solder coalescence and prevent clumpy solder joints; (d) achieve very low voiding. The flux in the solder paste is a complex optimized chemistry and is truly the "secret sauce" that helps maximize the process window (print, reflow) towards achieving high yields.

Defining the Print Process Window

80% of all SMT defects can be attributed to the print process. So achieving a good print process is crucial to improving yields. The 2 metrics that were used to measure the print process were (a) Print Transfer Efficiency- transfer efficiency is the ratio of the actual paste volume deposited on the PCB pad (through the stencil aperture) over the theoretical maximum paste volume that can be deposited on the pad, and is expressed as a percentage. The actual paste volume deposited on the PCB pad is measured using a Koh Young optical inspection machine with a high Gage R&R. The theoretical maximum paste volume is calculated using the stencil aperture dimensions. The key is to maximize the print transfer efficiency thereby minimizing the paste trapped in the stencil aperture. **Figure 1** shows the fish-bone diagram of the factors influencing transfer efficiency; (b) Deviation between print-to-print. The deviation is calculated using the normal statistical formulae based on the measured paste volumes. The key here is to minimize the print-to-print deviation.

The variables to measure print transfer efficiency and print-to-print deviation included (*all testing was done with Type 4 powder on a standard laser cut stainless steel stencil*):

a. **Area Ratio, Stencil Thickness, Aperture Dimensions**

Area Ratio is the ratio of the cross-sectional area over the aperture wall area. Area ratio is therefore a function of the stencil thickness and the aperture dimensions in the x-y plane. **Figure 2** shows how the area ratio is calculated. The smaller the area ratio, the more difficult it is for solder paste to consistently get released from the stencil aperture onto the board.

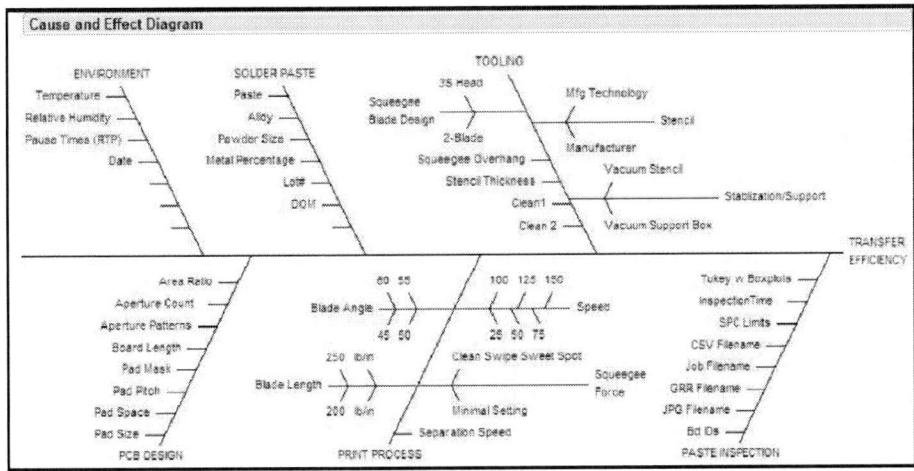

Figure 1. Factors influencing Paste Transfer Efficiency from the stencil onto the board

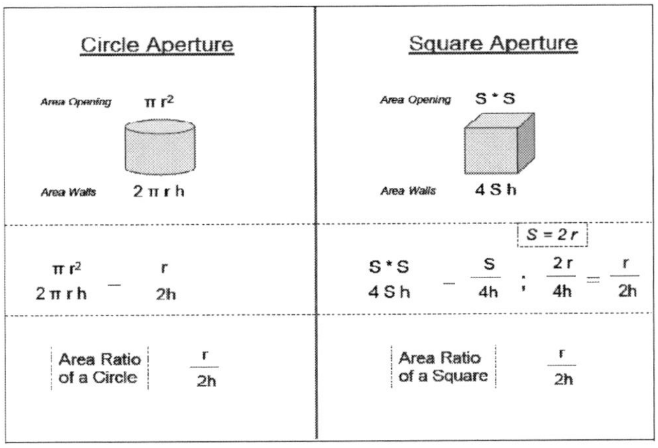

Circle Aperture		Square Aperture	
Area Opening $\quad \pi r^2$		Area Opening $\quad S * S$	
Area Walls $\quad 2\pi r h$		Area Walls $\quad 4 S h$	
		$\boxed{S = 2r}$	
$\dfrac{\pi r^2}{2\pi r h} - \dfrac{r}{2h}$		$\dfrac{S*S}{4Sh} - \dfrac{S}{4h}$; $\dfrac{2r}{4h} = \dfrac{r}{2h}$	
Area Ratio of a Circle $\quad \dfrac{r}{2h}$		Area Ratio of a Square $\quad \dfrac{r}{2h}$	

Figure 2. Area Ratio Calculations

To maximize the paste volume, larger apertures could help, but this is a limit of the pad pitch which is pretty small (sub 0.4mm). To compensate for the small apertures, stencils could be made thinner, but thinner stencils mean lesser paste volume and this could have a direct impact on manufacturing yield (HiP, poor solder coalescence, opens…). Also very thin stencils wear out quickly and extra costs need to be considered. It is therefore crucial to optimize the flux chemistry to maximize paste transferred volume especially for the smaller area ratios. **Figure 3** shows a contour plot of actual measured paste transfer efficiencies for different ratios at different print speeds. The X-axis shows the print speeds from

40mm/sec to 130mm/sec. The inner Y-axis shows the area ratios from 0.5 to 1.0. The outer Y-axis shows the corresponding circular aperture dimensions. So C12 would mean a circular aperture 0.012" diameter. The different colours refer to the measured paste transfer efficiencies. As the area ratio decreases and the print speeds increases, paste volume transferred decreases.

b. Response-to-Pause Testing

This measures paste performance over time when kept idle on the stencil. This mimics production scenarios where equipment is down

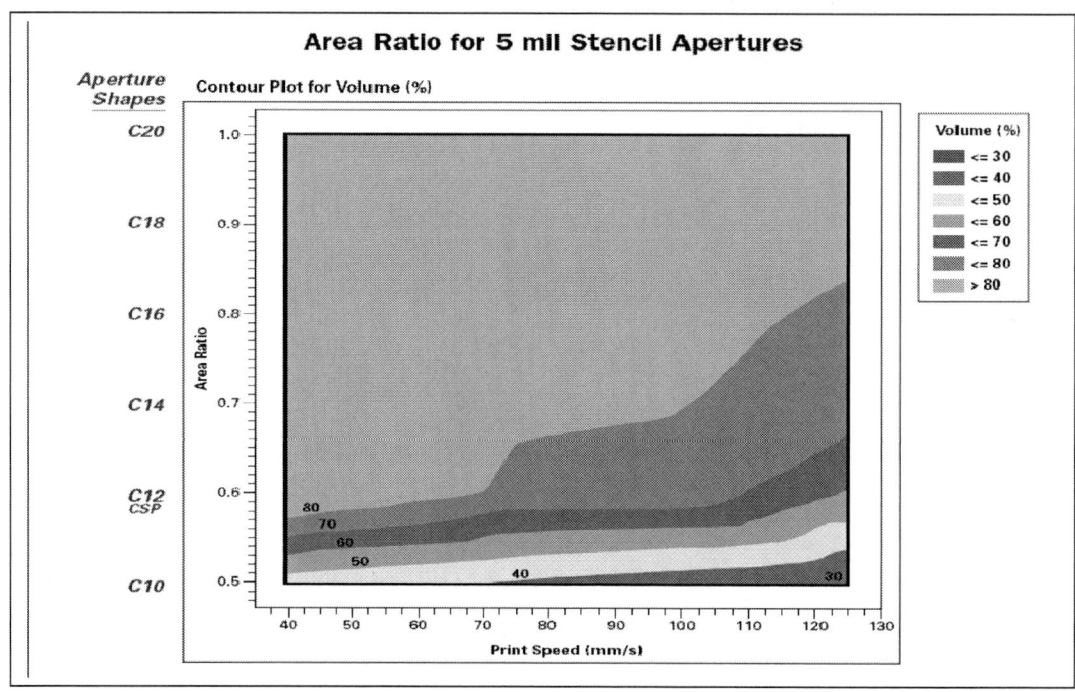

Figure 3. Plot showing Paste Transfer Efficiency for Different Area Ratios and Print Speeds

and the paste is idle. When the paste starts being sheared again by the squeegee, it is important to achieve the same print volumes as in t=0 (at the beginning). This is even more significant for smaller apertures (smaller area ratios) where if the flux formulation was not optimized, there could be huge swings in printed paste volumes and catastrophic failures (opens, HiP) on critical components such as 0.4mm CSPs.

In this test, a certain number of boards are printed at time 0. The actual print volumes deposited for pre-defined pads and the corresponding deviations are measured. The paste is then allowed to sit idle on the stencil for 1 hour. At the end of the hour, 5 more boards are printed and the print volumes, deviations are measured and compared to the boards printed at time 0. The same process is repeated for paste idle times of 2 and 4 hours on the stencil; and the paste volumes and corresponding deviations are compared to t=0. There is no dummy printing or kneading of the paste. **Figure 4** shows the transfer efficiency for response to pause over 4 hours for an area ratio of 0.63 (circular aperture= 350μ dia), where the transfer efficiency is uniform even after the idle times. **Figure 5**

compares the transfer efficiency of 2 different pastes for response to pause over 4 hours for an area ratio of 0.58 (circular aperture = 250μ dia), and at different print speeds of 25, 50 & 100 mm/sec. Paste A showed lesser print deviation than Paste B. During paste evaluations, response-to-pause print characteristics is awarded a high ranking because of its strong correlation with manufacturing yields.

c. **Solder Mask Defined (SMD) Vs Non Solder Mask Defined (NSMD) Pads**
Figure 6 shows the difference between SMD and NSMD pads. As aperture sizes become smaller (350μ diameter and lesser), SMD pads consistently give better results with lesser print deviation, when compared to NSMD pads. This effect is not completely understood as in the normal scheme of things, NSMD pads would provide for better gasketing with the stencil. What could be happening is that with SMD pads and smaller apretures, the mask creates a well and acts as an anchor to 'grab' the paste during the print stroke. **Figure 7** shows the print transfer efficiency for SMD Vs NSMD pads; Circular stencil aperture- 200μ dia; (Area Ratio= 0.5). SMD pads had lower print deviation.

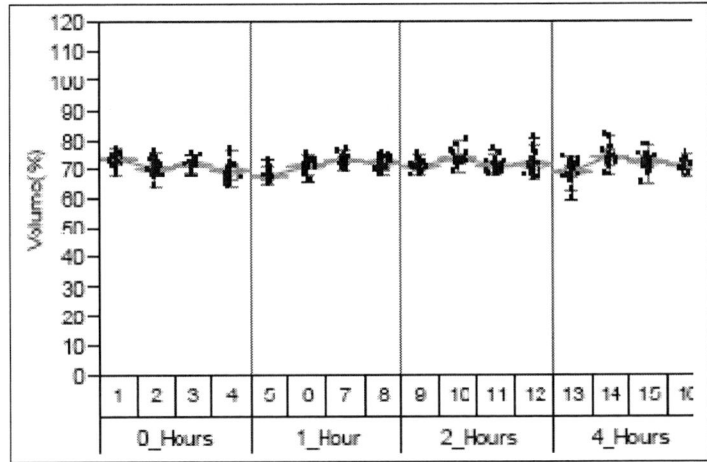

Figure 4. Response to Pause Data- Area Ratio= 0.63, 350µ dia circular aperture

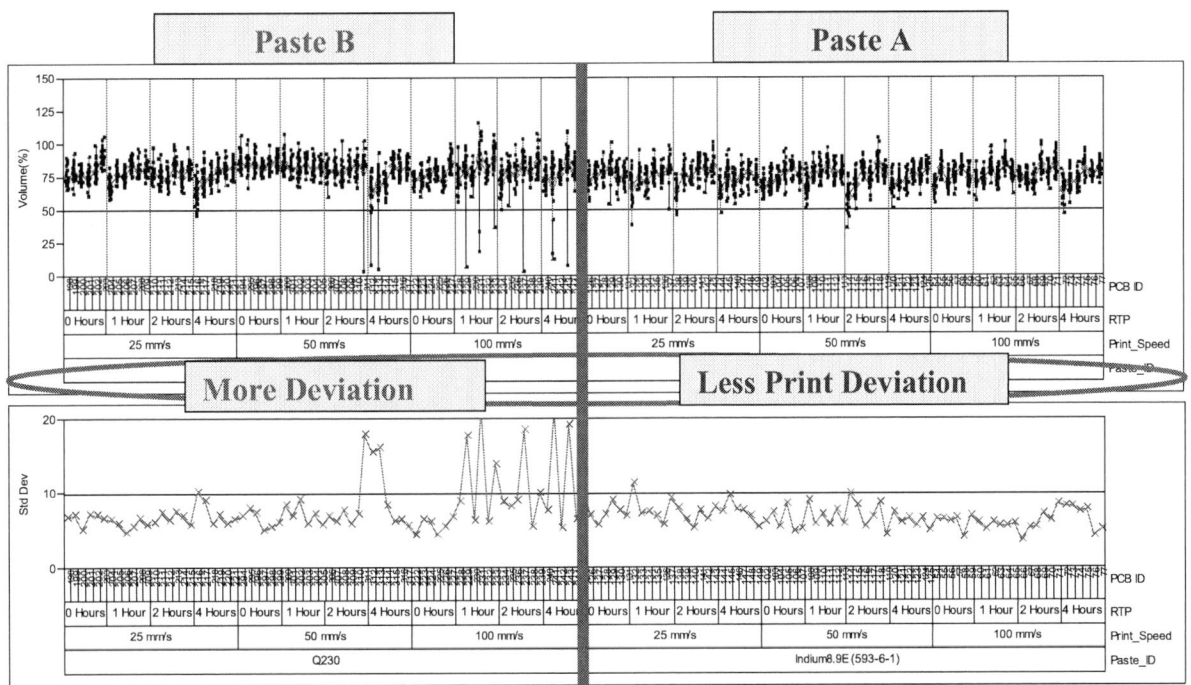

Figure 5. Response to Pause Data Comparing 2 Pastes; Area Ratio= 0.58, 250µ dia circular aperture

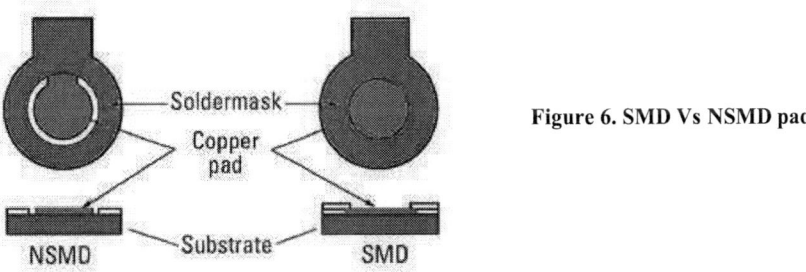

Figure 6. SMD Vs NSMD pads

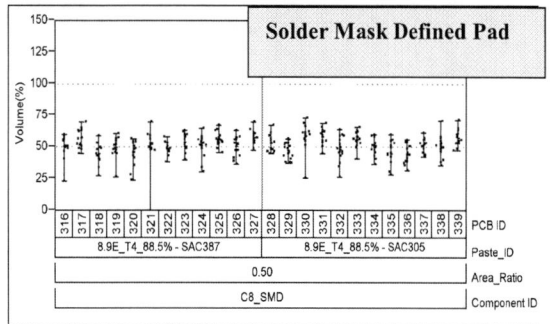
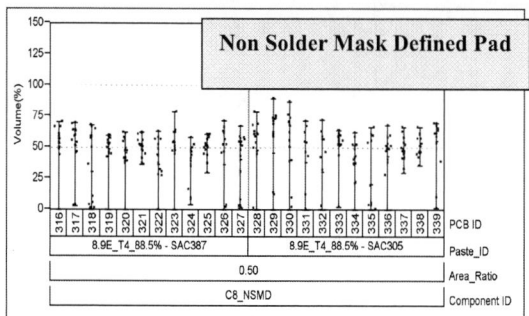

Figure 7. Pint Transfer Efficiency for Solder Mask Defined Vs Non Solder Mask Defined Pads

d. Separation Speed and Squeegee Blade Angle
Separation speed is the speed with which the board drops from the stencil underside at the end of the print stroke. Squeegee blade angle is the angle between the blade and the stencil. In this test, print transfer efficiency and deviation was measured for different separation speeds and blade angles.

Variables:
2 different Area Ratios = 0.7 (250 μ aperture); 0.56 (200 μ aperture)
2 different Separation Speeds = Fast Drop; Slower Drop at 2 mm/sec
2 different blade angles = 45 deg; 60 deg

Fixed Conditions: Print Speed= 50 mm/sec; Squeegee Force= 40N

Figure 8 shows the print transfer efficiency and deviation for the larger area ratio of 0.7. There was not a significant difference between the separation speeds and squeegee angles. For smaller area ratios (below 0.66), the effect of the separation speed and blade angle becomes very pronounced as shown as **Figure 9**. As the aperture sizes become smaller, previously ignored factors become very important and need to be optimized to maximize yields.

Defining the Reflow Process Window
The key purpose of a flux during reflow is to reduce oxide on the surface of the substrates as well as that on the surface of the solder particles, thereby promoting good spreading and most importantly forming reliable bonds with the substrates. This needs to be achieved inspite of challenges including OSP substrates, high thermal mass boards, small

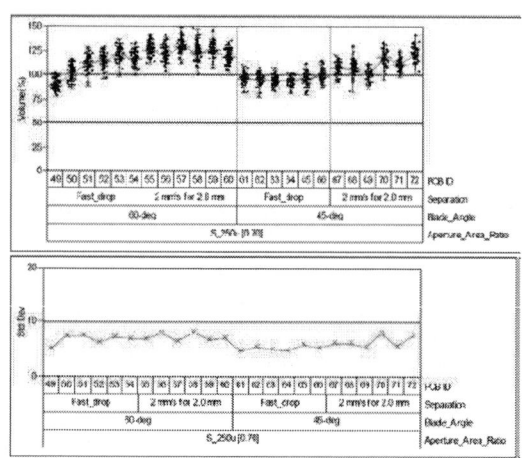

Figure 8. Area Ratio = 0.7; Transfer Efficiency & Deviation for Different Separation Speeds & Blade Angles

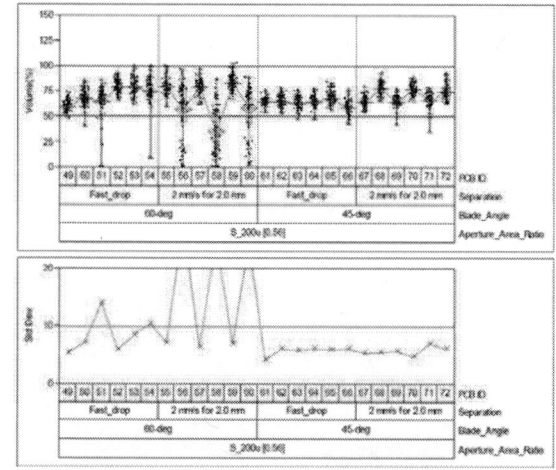

Figure 9. Area Ratio = 0.56; Transfer Efficiency & Deviation for Different Separation Speeds & Blade Angles

components (01005, 0.4mm CSP) harsh profiles (long profile, high peak temperature, long time above liquidus) in air reflow.

Due to miniaturization, stencil apertures become smaller. The paste volume for a smaller aperture is obviously smaller than that of a larger aperture. What is not realized is that the corresponding flux volume to unit surface area (of the solder particles in the paste) is **exponentially lower** for the smaller apertures when compared to the larger apertures. It is not a linear relationship. **Figure 10** illustrates the lower flux volume per unit surface area for different apertures. It can clearly be seen that as the apertures become smaller, the flux per unit surface area reduces dramatically. Therefore as apertures become smaller, it is even more important to maximize paste transfer to maximize the flux content available during reflow, as well as minimize print deviation.

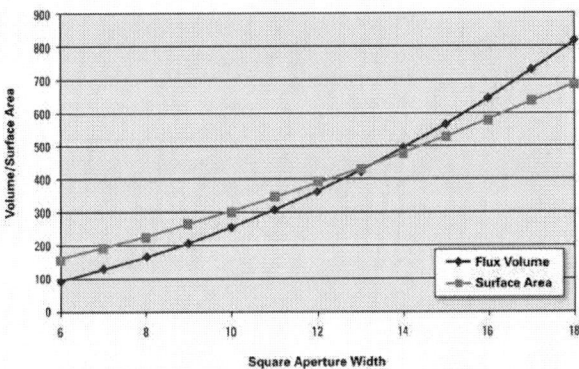

Figure 10. Flux Volume per unit surface area reduces significantly as stencil aperture size decreases

The effects of an optimized flux in addressing the challenges of Head-in-Pillow, Graping (incomplete coalescing of small paste deposits), and Voiding for micro via-in-pad, were studied for small paste deposits reflowed in air.

a. Head-in-Pillow (HiP)

HiP is the single largest issue plaguing the industry today. This is associated with BGA devices. HiP causes incomplete coalescence between the BGA ball and the reflowed paste deposit. A homogeneous joint is therefore not formed. This is a very dangerous scenario as short of destructive testing, it is impossible to visually confirm if 100% of the balls have completely coalesced and formed homogenous joints. Even the best x-ray algorithms are not advanced enough to 100% identify non-homogeneous joints. Physical contact between the individually reflowed ball and solder deposit allows for an electrical connection and this could allow a faulty product to go out through the door. It is impossible to predict when the non-homogeneous joint causes intermittent contact, and when this happens in the field, the results are truly catastrophic that could involve the calling-in of millions of units. **Figure 11** shows a HiP non-homogeneous joint versus a completely coalesced BGA joint.

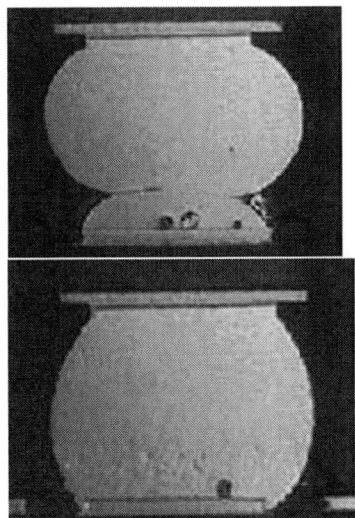

Figure 11. HiP joint (top) versus a completely coalesced joint (bottom)

(i)   HiP due to Excessive BGA Oxide or Contamination
When the flux in the solder paste is not able to break the oxide film on the BGA ball, then there is incomplete coalescence between the BGA ball and reflowed solder deposit. *Flux activity which is the ability of flux to reduce oxides is very important.*

(ii)   HiP due to Excessive Warpage
As more functionality is packed into devices, the number of I/Os increases. Also due to miniaturization, the package has to be very thin. Package thinness increases the possibility of warpage during reflow. If this warpage is excessive during reflow, the BGA ball loses contact with the solder paste and both the ball and solder paste melt and solidify individually without

forming a homogeneous solder joint. *Using an optimized flux to increase printed paste volume reduces the chances of the BGA ball losing contact with solder paste despite warpage.*

(iii) **HiP due to a Combination of Warpage and Oxidation**
This is the most common case. During the initial part of reflow, the BGA balls are initially in contact with the solder paste deposits. As the temperature increases, because of warpage, the BGA balls lose contact with the solder paste. The BGA ball and solder paste now become molten and the package unwarps. The molten BGA ball wants to make contact with the molten paste deposit. However an oxide layer has been formed on top of the molten solder and this oxide prevents the molten solders from coalescing and forming a homogeneous joint. **Figure 12** shows the mechanism of HiP.

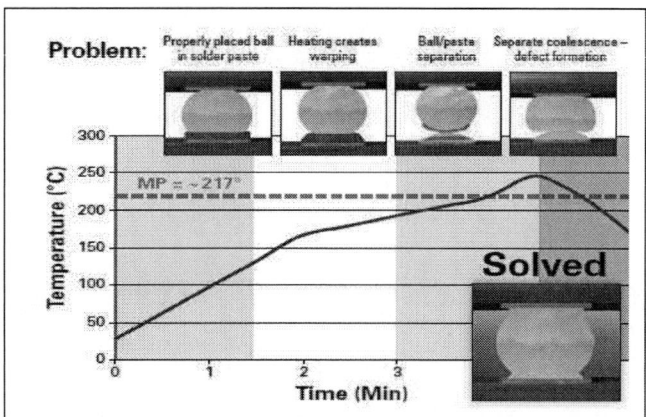

Figure 12. Mechanism of Head-in-Pillow

The role of the flux here is crucial. One approach is to make the flux really active by beefing it with organic acids to increase its ability to reduce oxide. This however is a reactive approach as oxide has been allowed to form in the first place. **A proactive approach involves engineering the flux with an oxidation barrier that protects the molten solder from surface oxidation even when reflowed in air. By eliminating / significantly reducing the surface oxide, coalescence of the** molten BGA ball and molten paste deposit happens and a homogenous joint is formed.

In addition to the flux's oxidation barrier, some of the other attributes engineered in the flux to prevent HiP are: (a) Maximize printed paste volume – this means more flux to address oxide and more paste means lesser chances of the ball losing contact with the paste; (b) Low Slump- prevents the BGA ball from losing contact with the paste; (c) Higher Tackiness- helps the paste to hold onto the BGA ball better and minimize chances for loss of contact.

(iv) **Test 1 – Verifying Flux Oxidation Barrier to Eliminate HiP – Ball Onto Paste Method**
Pb-free Solder paste is printed on a copper coupon and placed on a hot plate set at 200 deg C. The paste is subject to different soak times of 1, 2, 3, 4 minutes. The hot plate temperature is now set to 260 deg C and the solder paste becomes molten. At this point, a pre-baked solder ball (subject to 200 deg C for 25 min) is dropped into molten solder and the ball is now molten as well. The solder is allowed to remain molten for different time intervals of 0, 40, 60, 80 sec and coalescence of the molten ball with the molten paste is observed for the different molten time intervals of 0, 40, 60, 80 sec. **Figure 13** describes the test.

The combination of long soak times and molten times (time above liquidus) simulates a worst-case scenario to check for the flux's oxidation barrier. Oxide formation increases with temperature and time. *An optimized flux with an excellent oxidation barrier is to key to allow for coalescence between the molten ball and molten paste despite the long soak times and times above liquidus.*

Both halogen-free and halide-containing fluxes were tested for ball coalescence. **Figure 14** shows the coalescence results. As the soak times and molten times increase, some pastes allowed for coalescence between the ball and solder

112

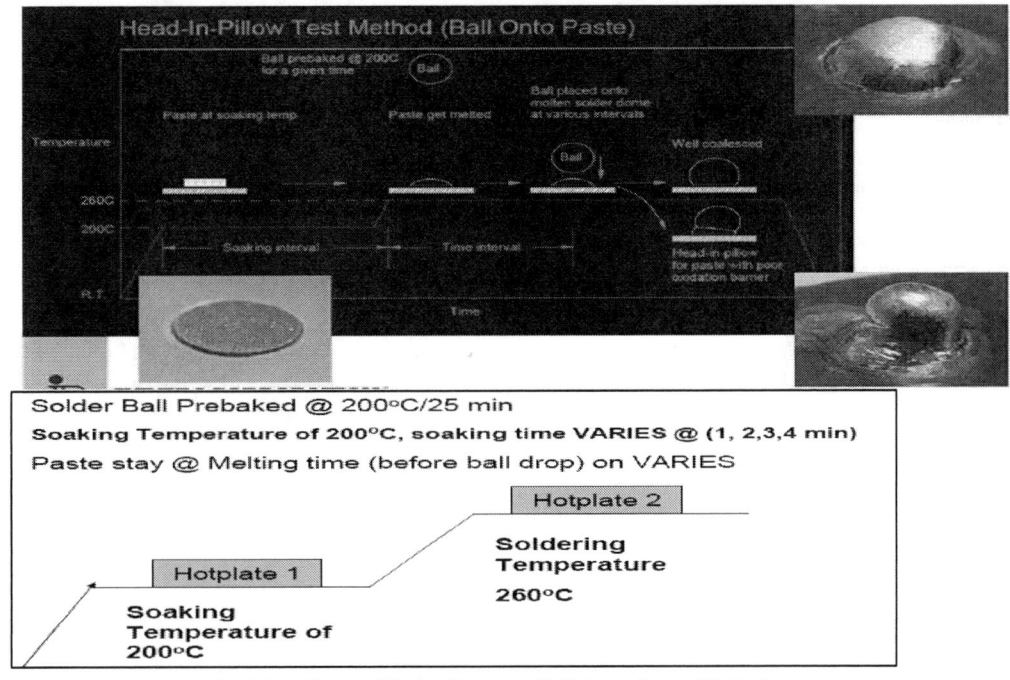

Figure 13. Verifying Flux Oxidation Barrier- Ball Onto Paste Method

Soaking @ 200C for 1 min

Drop Sphere @	0 s After melting	40 s After melting	60 s After melting	80 s After melting
Halo-cont. Paste 1			coalesced	
Halo-cont. Paste 2				
HF Paste 1				
HF Paste 2				
HF Paste 3				

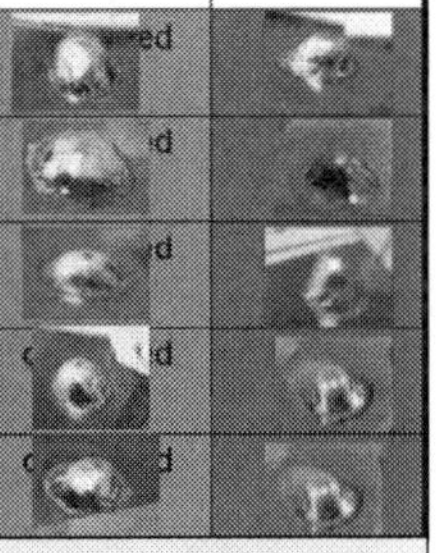

Soaking @ 200C for 2 min

	0 s After melting	15s After melting	20 s After melting	30 s After melting	80 s After Melting	100 s After Melting
Halo-cont. Paste 1					coalesced	HIP
Halo-cont. Paste 2					coalesced	HIP
HF Paste 1	coalesced	HIP				
HF Paste 2			coalesced	HIP		
HF Paste 3		coalesced	HIP			

Soaking @ 200C for 3 min

	0 s After melting	5 s After melting	20 s After melting	60 s After melting	80 s After Melting	100 s After Melting
Halo-cont. Paste 1					coalesced	HIP
Halo-cont. Paste 2				coalesced	HIP	
HF Paste 1	coalesced	HIP				
HF Paste 2	coalesced	HIP				
HF Paste 3	coalesced	HIP				

Soaking @ 200C for 4 min

Drop sphere	0 s After Melting	60 s After Melting	80 s After melting	100 s After Melting
Halo-cont. Paste 1			coalesced	HIP
Halo-cont. Paste 2		coalesced	HIP	
HF Paste 1	X died			
HF Paste 2	X died			
HF Paste 3	X died			

Figure 14. Verifying Flux Oxidation Barrier- Ball Onto Paste Method- Coalescence Results

paste, while others caused HiP. The green boxes denote the combinations that had complete coalescence and the red boxes show the combinations where HiP was observed. In some cases, halogen-free pastes were better than halide-containing pastes in preventing HiP. **Flux Oxidation barrier is the key.**

(v)   **Test 2 – Verifying Flux Oxidation Barrier to Eliminate HiP – Tiny Paste Dot Method**

This is an easy test that can be used by Contract Manufacturers to check the paste's oxidation barrier. The test vehicle involves small pads – sub-300 μ diameter. The stencil apertures for these pads are consequently small and therefore the paste volume deposited is small. As explained earlier, smaller paste deposits have a smaller flux volume per unit surface area (refer to **Figure 10**). The flux for a smaller paste deposit therefore has a tougher job and has to do more work in addressing the oxide as opposed to larger paste deposits. This when combined with a harsh reflow profile (air reflow, long soak, high peak temperature) simulates a worst case scenario for oxide formation. **Figure 15** shows the reflow profile, pad and stencil details.

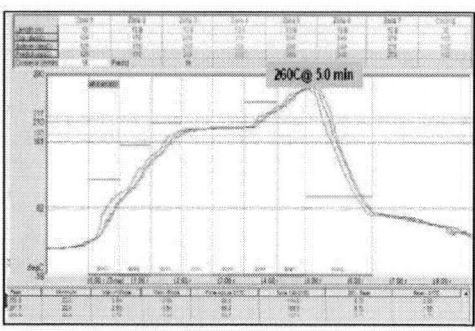

Circular OSP NSMD pads; 245μ pad dia; 127μ stencil thick; 1-to-1 opening; T4 powder Long Harsh profile; **Total of 9 pastes evaluated**

Figure 15. Verifying Flux Oxidation Barrier- Tiny Paste Dot Method

Fluxes with an optimized oxidation barrier will allow for smooth, non-grainy reflowed solder deposits. On the

other hand, fluxes that are not able to prevent oxide formation will result in clumpy, grainy reflowed solder deposits. **Figure 16** shows coalescence results for the Tiny Paste Dot Method. **Flux Oxidation barrier is once again the key.**

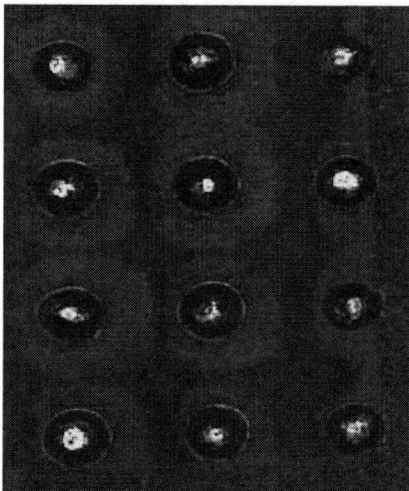

Paste A – Good Solder Coalescence, Smooth reflowed deposits, Good Oxidation Barrier

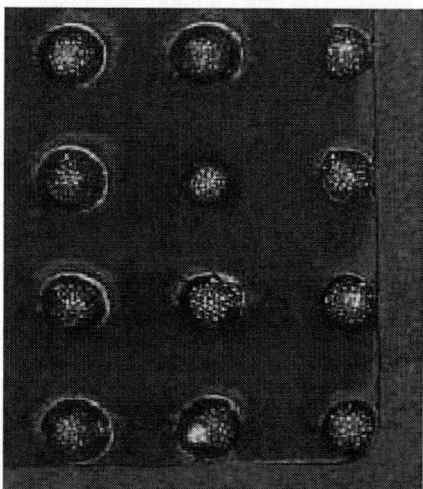

Paste B – Poor Solder Coalescence, Grainy reflowed deposits, Poor Oxidation Barrier

Figure 16. Verifying Flux Oxidation Barrier- Tiny Paste Dot Method- Coalescence Results

b. Graping / Incomplete Coalescence of Small Solder Deposits

Graping or incomplete coalescence of solder deposits is mostly observed on small components such as 0201s. **Figure 17** shows pictures of incomplete coalescence for 0201s and 0402s. Poor solder coalescence may happen even with a short reflow profile. The reasons for graping are the same as mentioned above for HiP – poor flux oxidation barrier coupled with lesser flux volume per unit surface area for smaller paste deposits.

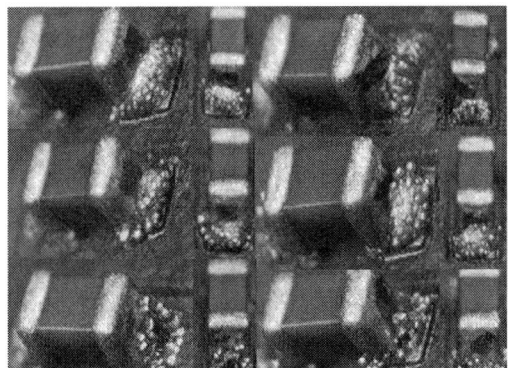

Figure 17. Incomplete Solder Coalescence because of the poor oxidation barrier of the flux in the paste

Therefore using a flux with the properties of an excellent oxidation barrier as well as maximizing paste transfer efficiency, is key to achieving small paste deposit coalescence.

Conclusions

Based on the print, an advanced optimized flux platform in the solder paste based on its rheology could visibly demonstrate (a) maximizing the paste transfer efficiency (> 80%) for small apertures (sub 350µ apertures; (b) minimize standard deviation. The flux enhanced the print performance during response-to-pause testing which is a key test that simulates production downtimes. During the reflow process, it was found that the flux's oxidation barrier was key to eliminating major defects such as Head-in-Pillow and incomplete coalescence / graping.

The rate of miniaturization is only going to increase and this poses major challenges for PCB Assembly. With smaller apertures, the older rules of thumb may no longer be applicable. Achieving an optimized print and reflow process is the key to maximizing manufacturing yields. From a material standpoint, the flux in the paste is what makes this possible.

References

Anglin, C., Babka, G., Sbiroli, D., and Brooks, R., "Sustaining a Robust Fine Feature Printing Process", SMTA International (SMTAI), San Diego, CA; October 4 – 8, 2009.

Anglin, C., Briggs, E., Lasky, R., and Connell, D., "Fine Feature Stencil Printing 0.3mm Pitch Components", International Conference on Soldering and Reliability, Toronto, Canada; May 20 – 22, 2009.

Lee, N C., " Optimizing Reflow Profile Via Defect Mechanisms Analysis", IPC Technical Conference, Las Vegas,NV; March 31 – April 2, 2009

Inkjet Printing of Electrical Vias

Ingo Reinhold,[1,2] Moritz Thielen,[1] Wolfgang Voit,[1,2] Werner Zapka,[1,2] Reiner Götzen,[3] Helge Bohlmann[3]

[1]XaarJet AB, Elektronikhöjden 10, 17526 Järfälla, SE.
[2]KTH Royal Institute of Technology, SE-100 44 Stockholm, SE
[3]microTEC Gesellschaft für Mikrotechnologie mbH, Bismarckstrasse 142, 47057 Duisburg, DE.

+46 8 58088728; ingo.reinhold@xaar.se

Abstract

Inkjet printing of planar and via (through-hole) electrical interconnections is developed to be incorporated into a roll-to-roll manufacturing line. The specific roll-to-roll machine uses rotary RMPD® technology , self-alignment of bare LED dies, and inkjet printing of the electrical connections to the LEDs dies. The key problem to be solved is the inkjet printing of electrical connections through via holes with vertical walls

Xaar126-50pL industrial inkjet printheads were used to print silver nano particle ink at 0.1 m/s to connect through LED vias of 90 μm diameter and up to 50 μm depth. While high throughput sintering techniques are desirable for the specific roll-to-roll machine standard convection oven sintering was applied for the proof of principle described here. Sintering at temperatures as low as 135 °C for 30 min prevented damage to the substrate and LED dies and yielded electrical connections that allowed to drive LEDs with 20 mA at 3V under emission of bright green light.

Key words: inkjet printing, silver nano particle ink, electrical via

Introduction

With the advent of the new generation of industrial inkjet printheads inkjet printing of functional fluids has recently been the topic of numerous development projects. In order to introduce such 'digital fabrication' processes into the manufacturing lines the inkjet printheads, the fluids, the substrates and the necessary pre- and post processes must be adapted to each other in a system approach.

The present paper presents the experiences with the integration of inkjet printing into a roll-to-roll machine for the manufacturing of LED packages. The specific task was to produce the electrical contacts to the LED dies by way of electrical vias.

Within the 'Light Rolls' EU FP7 project a novel roll-to-roll manufacturing process is developed for production of interior lighting devices based on the integration of multiple LED dies into flexible substrates. The substrate is produced in-line by way of a novel rotary RMPD® technology, self-alignment of bare LED dies, and by inkjet printing of the electrical connections to the LEDs dies [1].

In the following we focus on the generation of electrical connections using inkjet printing of silver nanoparticle inks and the evaluation of one potential sintering technology. The LEDs of choice were to be driven with a current of 20 mA at 3V. The project was thus designed to evaluate the printing and sintering processes as well as to assess the operational window of the electrical interconnects.

Apart from the actual inkjet printing process the pre-and post-processes are of importance, and need to be developed to control the spreading of the ink on the substrate, to optimize the adhesion on the substrate, and to provide high electrical conductivity without damage of the substrates and the devices already in place.

Strong spreading of the ink can result in electrical shorts between adjacent electrical connections but may also inhibit the desired electrical characteristics, since very shallow features increase the influence of boundary scattering of electrons during conduction and, therefore, increase the resistance of the track. Several techniques can be employed to control the pattern formation on the substrate and prevent excessive spreading by either modification of surface energies [2], by mechanical constraints [3], by usage of phase change materials [4] or by applying of elevated substrate temperatures in order to rapidly vaporize solvents and thereby reduce spreading.

Post-processing of the inkjet printed tracks of capped silver nano particle inks is necessary to enable metallic contact between the individual nano particles and thus to provide the high electrical conductivity. Potential post-processes are convection oven sintering, IR sintering [5], microwave sintering [6], plasma sintering [7], electric sintering [8], photonic sintering [9] as well as chemical sintering [10]. The candidate method for the specific roll-to-roll application in Light Rolls is required to produce reasonable electrical conductivities on various substrates without damage to substrate and LED dies, and has to comply with short processing times to meet roll-to-roll feed rates. While tests are ongoing with photonic sintering [7] as method of choice for the Light Rolls project, the proof of principle tests of inkjet printed electrical vias were conducted with sintering in a conventional oven.

Experimental Setup

The ink used throughout the experiments was U5603 (SunJet, UK), a solvent-based silver nano particle ink with a solid content of 20 wt%.

Substrates were selected to represent the final application in terms of via dimensions, wetting and thermal characteristics. Test samples were obtained from partner MicroTEC. These samples from the RMPD® process contained various combinations of via dimensions and locations as well as possible routing paths and were fabricated onto gold coated glass substrates, which acted as a common bottom electrode. The circular via opening in the RMPD® layer had a nominal diameter of 90 μm. The thickness of the RMPD® layer varied to provide vias of depth from 25 μm to 65 μm.

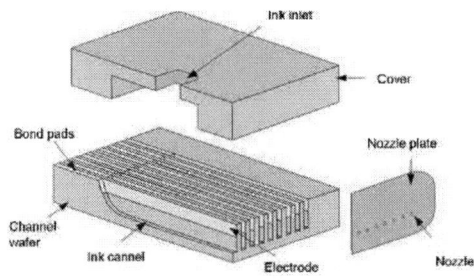

Figure 1: Schematic Xaar-type 'end-shooter' printhead actuator

Printing was carried using a binary Xaar126 printhead with a nominal drop volume of 50 pL and a maximum frequency of 7.5 kHz (see figure 1). Drop formation was evaluated (see figure 2) and a driving voltage waveform was developed that matched the acoustic response of the printhead to the fluidics. The meniscus pressure was held constant at -10 mbar. Single pass print experiments were carried out at 360 dpi print resolution and therefore, required tilting of the printhead by 59.04°.

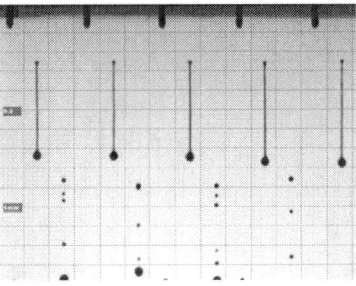

Figure 2: Typical droplet ejection from an 'end-shooter' printhead in 3 cycle operation

A three axis printing system with a heatable and rotatable stage was employed in the print tests. The print rig comprised two linear Anorad tables, the linear x-stage for the printhead, and the linear y-stage for the substrate. A manual z-drive allowed vertical adjustment of the distance between the printhead and the substrate to nominally 1 mm. Alignment was carried out using a digital microscope (Dino-Lite, NL) with a precision of alignment better than 35 μm. Sintering was carried out with a convection oven FD 53 (Binder, USA).

Sintering of Inkjet Printed Tracks

Sintering in convection oven was studied to evaluate the onset of saturation in electrical conductivity of the printed silver nano material. For the electrical measurements, square patterns of size 20 mm x 20 mm were printed onto polyimide foil with 3 layers of 360 x 360 dpi. In this experiment the printhead was mounted perpendicular to the direction of substrate motion and lateral resolution was emulated using multi-passing. The temperature of the substrate was kept at 70 °C, which controlled spreading of the ink on the substrate and facilitated drying of the silver ink.

The inkjet printed samples were exposed to 135°C, 160 °C, 175 °C, 200 °C, and 230 °C for 30 minutes and measured using a four-point probe. The results of the sheet resistance measurements are shown in figure 3, which indicate an exponential decrease in resistivity up to a temperature of about 185 °C beyond which no significant change with additional thermal energy can be observed.

It is well-known that inkjet printed patterns do not form solid bulk structures but a porous network morphology after sintering as result of the insufficient energy to trigger bulk diffusion as well as the remaining organic components of the stabilizing shell. The layer thickness was, therefore, not measured but calculated as a hypothetical bulk layer thickness using the known silver solid content of the ink, the drop volume, the print resolution and the number of layers, and assuming bulk non-porous morphology. Comparison with the sheet resistance of bulk silver indicated that the inkjet printed pattern could achieve a factor of 3 of bulk resistivity of silver at temperatures above 185 °C for 30 minutes. Since, however, the LED dies to be used in the project restrict the applied temperatures to 150 °C the resistivity of the printed electrical tracks was limited to a resistivity a factor 15 higher than bulk silver. See figure 3.

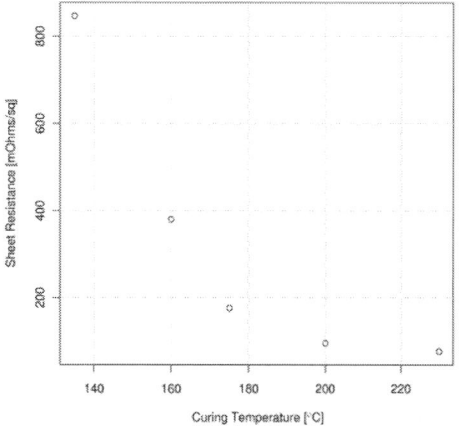

Figure 3: Sheet resistance of inkjet printed SunJet U5603 silver ink versus sintering temperature; sintering time kept constant at 30 minutes

Multi-layer printing of several layers of silver ink on top of each other was shown to decrease the overall ohmic resistance due to the increase in cross-sectional area of the track. Each layer was dried at the substrate temperature of 70 °C between consecutive prints, and a single sintering step was conducted with the complete stack.

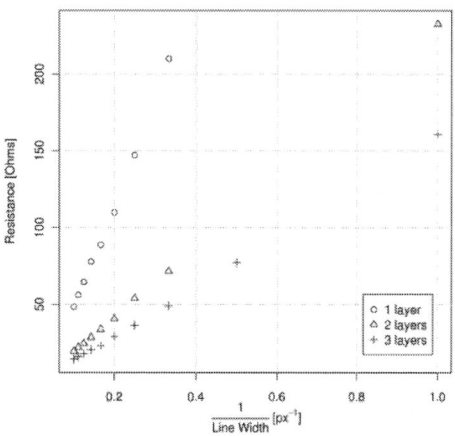

Figure 4: Ohmic resistance over 100 mm silver tracks versus the inverse line width

In a set of print tests, tracks of 100 mm length and a width in the range of 210 μm to 706 μm were printed with 50 pL drop volume at 360 dpi. Single, double and fivefold layer printing was conducted at a substrate temperature of 70 °C with a subsequent curing step of 200 °C for 30 minutes upon completion of all layers.

Assuming a close-to rectangular cross-section of the printed tracks, the dependency of the resistance should be inversely proportional to the width of the track. Figure 4 clearly supports this assumption, while slight deviations from the optimum straight line indicate some deviation in line morphology.

Inkjet Printing of Electrical Vias

The RMPD® process produces vias with almost vertical side walls as shown in figure 5. The resulting sharp convex edges posed a serious challenge to the overall process, as slip was assumed to generate discontinuous metal layers or result in very thin coatings, which in turn could lead to locally high resistances and the tendency to burn and break the electrical connections during usage.

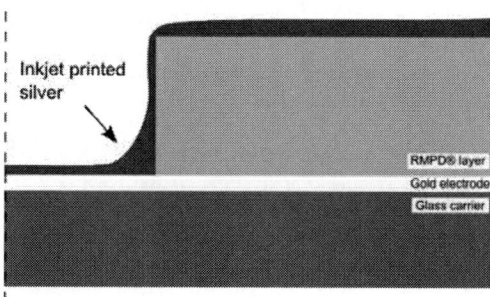

Figure 5: Schematic of an RMPD® layer indicating the problem to cover the sharp convex edge of the via with inkjet printing

The studies of connecting through the circular openings in the RMPD®-layer were conducted using samples with irregularly distributed vias. Print patterns were designed that contained both, several lines of different pixel width on the top surface of the RMPD® layer, and 14 contact pads on the top surface with connection tracks to and into individual vias. The former were meant to allow for the evaluation of ohmic resistance of a connection track on the top RMPD® surface, while the latter resulted in measurement data of the ohmic resistance of the connection track, the ohmic resistance of the thin layer at the rim of the via as well as the contact resistance to the gold bottom electrode. In order to reduce effects from the top layer resistance, printed contact pads were 1.4 x 1.4 mm^2 and connection tracks of 423 μm width, which is much larger than the via diameter of 90 μm. Sintering was carried out at 150 ºC for 30 minutes due to the substrate restrictions in thermal durability.

Samples prepared with single layers of silver nanoparticle ink showed an average resistance of

4.64 Ω with a standard deviation 2.7 Ω, while triple layers improved the resistance to an average of 1.93 Ω with a standard deviation of 0.99 Ω. The value of the standard deviation provides a measure for the reproducibility of the process and includes variations of 10 μm in the thickness of the underlying RMPD® layer, as well as the variation in thickness of the silver deposit at the convex edge of the via.

While the variation in resistance can be partially compensated for by using appropriate circuit layouts, the long-term stability of the electrical characteristics are more critical with respect to the final application. Therefore, samples were tested with defined levels of current from 10 to 800 mA, while monitoring their electrical performance. Visual analysis was performed after the tests to identify the potential failure mechanism.

The results indicate that the generated conductive features could conduct 80 mA at 3 V driving voltage for time scales up to 60 s without changes in the ohmic resistance or any visual damage. Applying the target current values, namely 20 mA for extended time and 40 mA for 600 s, showed no change in resistance and were thereby validated for the anticipated application. Currents above 100 mA resulted in varying values for ohmic resistance thus indicating excessive Joule heating of the features followed by an improvement of conductance, but also eventual disintegration. Applying currents above 400 mA lead to lift-off of the electrical tracks around the rim of the via, likely due to increased current density in the thin layers on the convex edge of the via.

In a following experiment samples with the RMPD®-layer and integrated LEDs dies were used and electrical connectors to the LED pads were inkjet printed and sintered as described. The LEDs could be operated with 20 mA at 3 V under emission of bright green light as shown in figure 6.

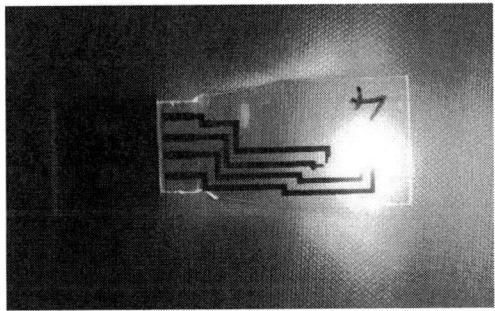

Figure 6: Bright green light emitted from an LED when driven with 20 mA at 3V by way of inkjet printed connections

Conclusions

Electrical connections through vias in RMPD®-layers were inkjet printed and shown to provide the required low ohmic resistance as specified for driving LEDs in RMPD® packages. The inkjet printed tracks were tested positively to conduct the required 20 mA for extended time and 40 mA for a minimum of 10 minutes without degradation. Degradation of the very thin layers at the convex edges of the vias occurred at current levels of 100 mA.

Successful operation could be demonstrated using samples with RMPD®-layer and integrated LED dies, and with inkjet printed connections and sintering in convection oven. Driving such LEDs at 20 mA and 3 V resulted in emission of bight green light.

Acknowledgements

This work was funded by the European Community's Seventh Framework Programme under Grant Agreement n° CP-TP 228686.

References

[1] www.light-rolls.eu, last accessed: May 2011.
[2] T. H. J. v. Osch, J. Perelaer, A.W. M. de Laat, U. S. Schubert, Advanced Materials 20 (2008).
[3] C. E. Hendriks, P. J. Smith, J. Perelaer, A. M. J. v. d. Berg, U. S. Schubert, Advanced Functional Materials 18 (2008).
[4] W. Voit, W. Zapka, A. Menzel, F. Mezger, (Proc. Digital Fabrication Conference 24, Pittsburgh, PA, 2008), 678-683.
[5] R. Bollström, A. Määttänen, D. Tobjörk, P. Ihalainen, N. Kaihovirta, R. Österbacka, J. Peltonen, and M. Toivakka, Organic Electronics 10 (2010).
[6] J. Perelaer, M. Klokkenburg and U. S. Schubert, EU Patent 2009/08675, 2009.
[7] I. Reinhold, C. E. Hendriks, R. Eckardt, J. M. Kranenburg, J. Perelaer, R. R. Baumann, U. S. Schubert, Journal of Materials Chemistry 19 (2009).
[8] M. L. Allen, M. Aronniemi, T. Mattila, A. Alastalo, K. Ojanperä, M. Suhonen and H. Seppä, Nanotechnology 19 (2008).
[9] J. West, M. Carter, S. Smith and J. Sears, NSTI-Nanotech 2 (2010).
[10] Werner Zapka, Wolfgang Voit, Christian Loderer, and Philipp Lang, (Proc. Digital Fabrication Conference 24, Pittsburgh, PA, 2008), 906-911.

Copper Wire Bonding Experiences from a Manufacturing Perspective

Bernd K Appelt, Andy Tseng and Yi-Shao Lai*

ASE Group, 1255E Arques Ave, Sunnyvale, CA, USA
*ASE Group, Nantze Export Zone, Kaohsiung, Taiwan

1.408.768.8533 p; 1.408.636.9485 f; bernd.appelt@aseus.com

Abstract

The rapid ramping of fine pitch copper wire bonding production is a perfect example of a market driven technology implementation. The explosive increase in gold commodity pricing has paved the road for this conversion. Several challenges known from thick copper wire bonding had to be overcome to make this technology viable technically and economically. Some fundamental studies of copper wire bonding to the full spectrum of die surfaces and wafer nodes was available at the beginning of manufacturing implementation but over the last year, a large number of research papers have been published greatly enhancing the basic understanding of mechanism, metallurgy and chemistry of copper wire bonding. Equipment manufacturers have developed dedicated copper wire bonders and tool kits, wire manufacturers have developed softer copper wires as well as oxidation resistant wires, mold compound suppliers have developed low corrosion compounds and wire bond engineers have learned to optimize the bonding process to yield cost effective and reliable copper wire bond die packages. Here, high volume manufacturing experiences of copper wire bond packaging and long term reliability data will be presented.

Key words: copper wire bonding, fine wire and fine pitch, long term reliability

Introduction

Wire bonding is still by far the most widely used method of die to substrate interconnection method and is likely to retain that dominant position for years to come. The explosive growth of the gold commodity price (see figure 1) is however forcing a change in materials from gold wire to copper wire. Copper wire bonding is not new per say but it had not been used for fine pitch wire bonding e.g. wire diameters below 1.2 μ because of a number of challenges like copper oxidation, hardness, corrosion and slow intermetallic compound growth. Additional challenges arise from the advancing wafer nodes and the concomitant, ever more fragile materials. Above recent events have lead to many new fundamental studies on the copper wire bonding mechanism and bond reliability. From a manufacturing perspective the focus has been to develop a repeatable and reliable process. Here, we will describe the methodology for developing a highly reliable process and we will present long term reliability test results. The later data is collected as part of reliability process/product monitoring and shows that copper wire bonds can exceed the typical JEDEC requirements by four to six times. These results apply to all commonly used package types as well as to 40/45 nm wafer node. The initial focus was on package conversion but now the confidence level in this new technology has increased to the

point that users are qualifying devices from the start with copper wire and without a gold wire back-up plan. The latest process development activities are focusing on stacked die packages with copper wires. It will be demonstrated that very low wire loops can be formed in copper wire without breaking the wire neck and with long overhangs without cracking the die. Lastly, a process for reverse bonding has been developed as well for copper wire enabling die to die bonding.

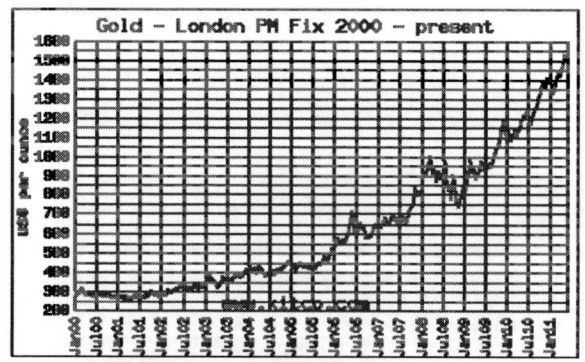

Figure1: Gold commodity price development reached a high of US$ 1543.90 at closing on June 6, 2011.

Manufacturing Process Development

For the purpose of this discussion, fine diameter Cu wires refers to wire diameters below 1.2 mils. In fact, the majority of the experiences described here is based on 0.8 mils diameter wire, either 4N Cu wire Maxsoft from Heraeus or 1X Palladium coated Cu wire from Nippon Micrometal. The wire bonders were KnS models Maxum Ultra and Maxum Plus, Iconn as well as lately Iconn ProCu. All wire bonders are equipped with inert gas EFO kits for Cu or CuPd wires. A proprietary capillary design was used and substrates and dice are customer specified with nodes ranging from 180 to 40 nm.

The first step in wire bonding is the free air ball (FAB) formation. For 4N Cu wire forming gas (95% nitrogen and 5% hydrogen) were used as a shroud and for PdCu wire nitrogen was used. The FAB geometry was tuned to yield a spherical FAB without any surface blemishes. This is achieved as usual by optimizing sparc current and duration as well as the gas flow. The spherical ball shape is a good indicator that an 'oxide-free' ball has been formed. This first step is actually one of the easier steps in the process.

| Good FAB | Void | Asymmetry | Unstable FAB |

Figure 2: Examples of good, spherical and unoptimized free air balls are depicted here.

The second step is the actual bond formation on the die pad. The process of tuning the bonding parameters is essentially the same as for Au wires. Albeit considerably more adjustments are required to ensure that no pad cratering or die cracking occurs due to the more aggressive conditions required to achieve a strong bond. The bond parameter optimization typically follows the standard procedure design of experiment (DOE) as for Au albeit the process window turns out to be considerably smaller than for Au. The boundary conditions are that Al splash, which is usually quite pronounced, must be contained within the bond pad opening (BPO) as shown in Fig. 3. Further, the residual Al thickness has been selected to be 100 nm minimum. It has been shown separately, that this thickness typically survives JEDEC TCT of more than 1000 hrs.

With the proper bond parameter optimization, dice of any node up to and including 40/45 nm can be bonded successfully without pad cratering or cracking. One great analytical tool employed here is focused ion beam (FIB) microscopy which can provide excellent resolution of pysical structures and grain structures. The time and effort for FIB afford only selective analysis rather than line monitoring.

Cross-section examples of dice from different nodes are given in Figure 4. Typically Cu wafers with an Al cap are more robust in bonding than pure Al wafers. Care must be taken to ensure that the via structures under the pad are not disturbed by excessive force during bonding. The via structures under the pad are quite sensitive to the bonding conditions and design rules have been developed to improve the successful process optimization

Figure 3: Al splash is controlled to be within the bond pad opening for ease of inspection.

While the exact bonding parameters are dependent on the particular devices and considered proprietary, the bond attributes are open and specified. The wire pull and ball shear strength at time zero are considerable higher than for corresponding Au wires although the AlCu intermetallic compounds are extremely thin. Part of the process optimization is to obtain adequate IMC coverage. During process development this IMC growth is tracked throughout the entire assembly process and at times through reliability testing. In general, the observations in above literature have been confirmed: initial IMC is very thin and difficult to detect and IMC growth is very slow. Actually, Cu IMC formation after bonding is spotty and grows at elevated temperatures to become a continuous layer of increasing thickness. The growth rate is more than an order of magnitude slower than Au [x].

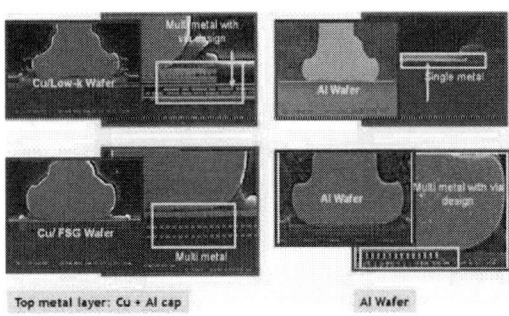

Figure 4 shows cross sections and under pad structures for Al and Cu wafers.

Looping wires from ball to the stitch bond has not presented any difficulties albeit angulation is lagging slightly behind gold wires and loop heights are still slightly higher than the lowest gold wire loops. Especially with latest generation bonders and the additional control parameters available, the challenges of very low loop heights and particular shapes could be minimized. The second bond or stitch bond has not been any challenge to date. No changes in substrate finish has been required either to obtain strong bonds as reflected in the wire pull values. Bond shapes are equivalent to Au. The second bond certainly benefits from an inert gas shroud by minimizing oxidation of the wire even though the temperature is between ambient and substrate temperature. In a high volume manufacturing environment, the shelf / floor life of the wire is not really an issue as the rate of consumption is a fraction thereof. But manufacturing floor management is simpler with a wire of long shelf life like PdCu wire.

The latest advances in bond process development has enabled die stacking with die gaps of less than 50 um, comparable to Au wires and die over hangs of as much as 1 mm. A sample is shown in Figure 5a & b. Like wise reverse bonding for die to die bonding has also been implemented successfully as demonstrated in Fig. 5c.

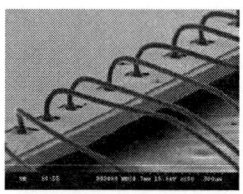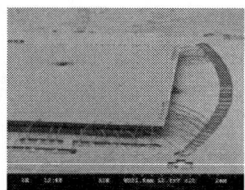

Figure 5a (left) and 5b (right) show a two die stack with 1 mm overhang with 20 µ copper wire and loop height of 50 µ. Die thickness was reduced to 40 µ.

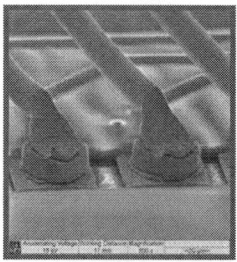

Figure 5c shows a reverse copper wire bond i.e. a stitch bond on a ball bond.

Finally, the die and wire bonds are encapsulated via molding. The mold process and pre-mold plasmas do not require any change other than the usual optimizations of plasma. Concerns have been raised about the reliability of standard mold compounds as do to the propensity of oxidation and corrosion of Cu. Corrosion is a known phenomenon also for Au and the mechanisms are essentially the same for both metals. In both cases several intermetallic compounds (IMC) are formed during and/or during thermal aging. Only

one of the IMCs is subject to attack by chlorine or bromine ions. The end result is the formation of alumina. Because the halogen ions are reformed, the reaction will continue until all IMC is consumed which typically leads to an electrical open. The open is caused by the alumina forming an insulator or initiating a crack because of the concomitant volume expansion. Extensive efforts have therefore been undertaken to reduce the amount of halogens in the mold compounds. This is an effort that has started long ago for Au and is being continued vigorously for Cu. The mold compound suppliers have an extensive repertoire of actions to minimize the effective amount of halogen ions in the mold compound. Apart from screening resins for intial low halogen content, additives act as ion trappers and buffer the pH as well as mdoify the glass transition temperature. The chemistries involved are of course proprietary.

Lastly, as a general practice for wire bonding and especially for Cu wire bonding, strict manufacturing floor management is key for successful high volume manufacturing. This entails strict clean room management, decidcated tools and operators, as well as strict adherence to hold times between operations to minimize oxidation and surface contamination. Cu wire bonds are not reworkable and therefore first pass yields are final yields for bonding. With such a rigorous methodology it is possible to achieve yields equivalent to Au and machine stop rates are actually less than the traditional Au floor. A reflection of the quality of the Cu wire bond process is a device with 1,478 wires which is running in manufacturing now. Of course the learning can be reapplied to the Au floor and improve its perfromance.

Based on above learning experiences on package types: SO, QFP, QFN and BGA as well as dice from virtually all wafer factories, a rigorous methodology for the evaluation of and qualification of new devices has been devised. It is a three step procedure where bondability of a die is established with a base set of parameters. In phase two the bond parameters are optimized and the bond attributes are characterized. In some cases, short loop reliability tests are being performed. Finally in phase three, the actual qualification hardware is being built and tested with the usual JEDEC tests as specified by the application environment. If the new die has some novel attributes like pad stack, metal thickness, etc., additional characterization loops and tests may be performed to ensure the reliability of the package.

As part of the line quality and reliability monitoring, extended reliability testing is being performed for selected packages and wafer nodes. As shown in Table 1, the standard JEDEC tests like HAST, THT and TCT can achieve at least 2x the usual life times. Depending on the application conditions, these JEDEC data represent multiple life times, especially for many mobile and consumer

122

applications. Above testing is continuing to determine the long term reliability as required by networking and automotive applications.

Table 1: Extended EDEC testing shows that typical test durations can be exceeded to pass multiples of standard test requirements.

Package Type	Body Size	PCT hrs	TCT cycles	HAST hrs	HTS hrs	THT hrs
QFN	6 x 6		2,000	200	2,000	2,000
QFN	8 x 8		2,000	200	2,000	2,000
QFN	9 x 9	936*	6500*	178*	4500*	
aQFN	11.5 x 11.5	168	500*		500*	
QFP	14 x 20	2016	6000		3500	
QFP	28 x 28	1272	6,000		4,000	
LQP	10 x 10	336	2,000	192	2,000	
LQFP	14 x 14	264	1,500			
LQFP	20 x 20	336	2,000	192	2,000	
LQFP	24 x 24	264*	1,500*	144*	1,500*	
TQFP	14 x 14		2,000	400	2,000	2,000
HQFP	14 x 20	336	1,000*	548*	3,500*	
TFBGA	9 x 9		3,500	144		
TFBGA	12 x 12		6,000*	864	2,000	
LFBGA	16 x 16		1,500	168	1,500	
HSBGA	27 x 27	336	4,000	336	4,000	
* test in progress						

Conclusions

Fine pitch or fine diameter Cu wire bonding has been introduced successfully in high volume manufacturing. Quality and yield has been advanced to levels equal to Au wire bonding. Reliability has been demonstrated to exceed 2x standard JEDEC testing and is continuing. At present, more than three billion devices have been shipped from six different factories. More than 4,000 wire bonders are running with Cu wire and it is expected that by the end of the year this number will increase to 6,000.

Acknowledgments

The authors would like the engineering teams under Mike Hung, Louie Huang, Scott Chen and Mike Zhao for their discussions and making the data available for publication.

References

[1] B. K. Appelt, "Fine Pitch Copper Wire Bonding – Why Now?", *Proc. 11th Electron. Package. Technol. Conf.*, Singapore, 2009

[2] R. Chylak, "Developments in Fine Pitch Copper Wire Bonding Production", Proc. 11th Electron. Packag. Technol. Conf., Singapore, 2009

[3] C.R. Kao, "Growth of CuAl Intermetallics in Cu and Cu(Pd) Wire Bonding", Proc. 61th Electron. Comp. Technol. Conf., Lake Buena Vista, 2011

[4] I. Qin, "Wire Bonding of Copper and Palladium Coated Copper Wire: Bondability, Reliability and IMC Formation", Proc. 61th Electron. Comp. Technol. Conf., Lake Buena Vista, 2011

[5] S. Flynn, "Die to Die Cu Wire Bonding Enabling Low Cost 3D Packaging, Proc. 61th Electron. Comp. Technol. Conf., Lake Buena Vista, 2011

[6] T. A. Tran et.al,"Copper Wire Bonding on lowK Wafers with Bond over Active (BOA) Structures for Automotive Customers", Proc. 61th Electron. Comp. Technol. Conf., Lake Buena Vista, 2011

[7] A. Rezani et al.,"Free Air Ball Formation and Deformability with Palladium Coated Cu Wire", Proc. 61th Electron. Comp. Technol. Conf., Lake Buena Vista, 2011

[8] D. Lu, "Effect of Heat Affected Zone on the Mechanical Properties of Copper Bonding Wire", Proc. 61th Electron. Comp. Technol. Conf., Lake Buena Vista, 2011

[9] S. Qu et al., "Over Pad Metallization for High temperature Interconnections", Proc. 61th Electron. Comp. Technol. Conf., Lake Buena Vista, 2011

High speed bending of 2nd level interconnects on printed circuit boards for automotive electronics

M.H.M. Kouters[a*], R. Ubachs[b], H.J. van de Wiel[a], A. van der Waal[a], J. van der Veer[a]

[a] TNO Technical Sciences, Materials Performance group, P.O. Box 6235, 5600 HE Eindhoven, The Netherlands
[b] NXP Semiconductors, RQC Europe, 6534 AE Nijmegen, The Netherlands
* M.H.M. Kouters, marcel.kouters@tno.nl, +31(0)88 86 65461

Abstract

Standard drop tests for portable electronics are not representative for the qualification of automotive electronics. High-frequency vibrations are more dominant than abrupt shocks during normal operation. In this work a high speed board bending (HSB) method is developed to mimic the constant cyclic solder joint loading (sinus wave load, 10-200 Hz, <2 mm peak-to-peak). A series of test printed circuit boards with wafer level chip scale packages (WLCSP) and Micropearl SOL lead-free solder balls arrays are daisy-chain interconnected and in-situ monitored to detect failure during loading. After failure defect interconnects are cross sectioned for fractography to determine the corresponding failure mechanism. To determine the maximum stress and strain levels finite element modeling (FEM) is used and compared with the results from HSB testing. Finally, a proof of concept is done for the high speed bending test. Further verification is necessary to use this test as qualification for 2nd level interconnect qualification of automotive electronics.

Key words: PCB, 2nd level interconnect, bending, mechanical, vibration, FE modeling

1. Introduction

Standard drop tests for portable electronics are not representative for the qualification of automotive electronics [1,2,3,4]. For automotive applications, high-frequency vibrations are more typical mechanical loads and dominant over abrupt shocks during operation (e.g. due to engine revs or driving dynamics). This is crucial for the prediction of long term behaviour of 2nd level solder interconnects. Cyclic bending and deformation of the printed circuit board cause repetitive stress and strain in the solder joints. In order to qualify electrical components within a short period of time (hours or weeks) a high speed bending (HSB) test set-up is proposed. A four point bending set-up is connected to an industrial vibration exciter and closed loop controlled with an accelerometer. The daisy-chained interconnects are electrically monitored to detect failure during bending, where a threshold resistance is chosen as failure criterion. Finite element modeling (FEM) is applied to determine and locate maximum stress and strain in solder joints during bending. Solder joint damage analysis and fractography are applied to examine failure initation, mechanisms and locations of damage and cracks in solder joints. Finally, a proof of concept is done for the HSB test, where further verification is necessary to use this test as qualification for 2nd level interconnect qualification of automotive electronics.

2. Experimental procedure

The four-point HSB set-up is designed to apply a high frequent cyclic homogeneous bending moment on the inner section of a clamped PCB. The vibration load can be varied in the range of 10-200 Hz with a maximum amplitude of 1-1.5 mm, which is payload (and thus board stiffness) dependend. The set-up is universally designed in such way that different board dimensions and thicknesses can be tested. The HSB supports are gradually adjustable for board lengths of 40-120 mm and maximum board widths of 80 mm. The maximum board thickness is limited by the vibration exciter capacity. The inner supports are adjustable from 40-60-80 mm; the outer supports from 60-80-100-120 mm. The inner supports are attached to the innerspan, which is connected with a flexible M8 tip to an industrial vibration exciter (Brüel & Kjaer type 4809, shown in Figure 1). The vibration exciter is closed loop controlled with a LDS Comet USB vibration controller (Brüel & Kjaer, Figure 3) with swept sine function, which generates a constant sinus wave load. The board vibration is measured by an accelerometer (Kistler, type 8614A100M1), which is attached on the PCB backside (Figure 2). An 8-digit counter (Kubler) is used for cycle counting. For this test PCBs are clamped symmetrically between the HSB supports with 60-80mm inner/outer support spacing. The HSB is grounded to an airmount suspended table to minimize external vibrations.

Figure 1: HSB set-up on airmount suspended table. Right side device power supply (above) and 8-digit counter (below)

Figure 2: Side view of PCB in HSB set-up with piezo-accelerometer on top. Below M8 flexible tip for vibration feedthrough.

Figure 3: Bruell & Kjaer Comet USB vibration controller (middle), pc for data storage and Tektronix oscilloscope (left side).

Before testing, first, the board electrical contacts are interconnected to signal cables by a conventional soldering process. Secondly the initial contact resistance of the connected devices is measured. Teflon tape is used as spacer between the supports and PCB to avoid clamping damage and short circuits. Finally, the vibration exciter is positioned in the HSB frame and connected to the innerspan by using a flexible M8 tip. This tip is provided with electro-discharged machined cuts to avoid bending or torsion loading on the HSB innerspan due to misalignment of the vibration exciter.

For HSB testing NXP PCAs (printed circuit assemblies) with board dimensions of 101x48x1mm and 8 wafer level chip scale packages (WLCSP) are used. These WLCSP are placed asymmetrically on the board (see Figure 4) and interconnected by 7x7 lead-free solder ball grid arrays (BGA). Full device and PCB specifications are given in Table 1. Micropearl SOL lead-free solder joints (Sn/Ag = 96.5/3.5) are used. These solder balls contain a co-polymer inner core for spacing and shock absorption. The solder balls were reflowed in a dual stage thermal process for lead free solder types with pre-heating of 140-160°C for 50-70s and peak heating of >225°C for 20-40s.

During HSB four daisy-chained WLCSP devices (chip 5, 6, 7, and 8) are empowered continuously by an external power device and each is coupled in series with a 1000Ω resistance. The initial electrical resistances of each device is determined by measuring the daisy-chain resistance before testing with a megger. At the start of the HSB test the sinus wave load is gradually increased until the maximum peak-to-peak displacement is reached. During testing, at frequencies of 80-100Hz, only one single device (chip 5, indicated in Figure 4) is selected for triggering. In case the total resistance of the device exceeds the threshold resistance of 100Ω during HSB, the oscilloscope (Tektronix DPO5034) is triggered and saves the measured signals of all four devices (chip 5, 6, 7 and 8). If the failure criterion is met and intermittent fails occucring the HSB test is stopped, the test board removed and detached from signal cables and failed devices are marked for solder damage analysis.

Table 1: Device and PCB suppliers and specifications

Device	WLCSP49DC-SOL	PCB	
Component supplier	NXP	PWB supplier	Ibiden
Die dimensions (mm3)	2.8x2.8x0.360	Core material (type)	FR4 Matsushita R-1766
Pin count	49 (7x7)	Prepreg	FR4 Matsushita R-1661
Solder pad diameter (μm)	230	PWB thickness and layer count	1mm 8 layers
Solder ball diameter (μm)	250	Build up layer material (type)	RCCu Mitsui R-0880
pitch (μm)	400	Surface finish (type)	OSP
Solder ball type	SOL pearl	Pad structure (viap/noviap)	noviap
Shipment media	tape and reel	Pad diameter, actual (μm)	250
UBM	TiNi	Solder mask aperture (NSMD/SMD)	NSMD
Solder material	Plastic core/Cu/SnAg solder	Solder mask aperture diameter (μm)	325
Solderball supplier	Sekisu		

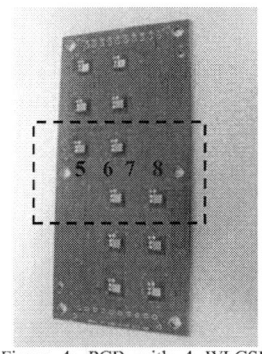

Figure 4: PCB with 4 WLCSP devices (5-8) for measument. Outer devices (4x) are removed for HSB support contact

125

3. Failure criterion

In general the JEDEC standard [5], which is the standard for board level cyclic bend test method for interconnect reliability, is considered as failure criterion, which is a standard for board level cyclic bend test methods for interconnect reliability. It is defined that a threshold resistance of 1000Ω or 5 times the initial resistance, whichever is greater, should be set for failure determination within a period of 1-5 microseconds. However, one aim of this study is to detect initial interconnect damage and its apparent failure mechanism, not to qualify series of test boards. Therefore the trigger resistance in this work is set much lower at 100Ω.

For the series of NXP PCAs the initial resistance of one daisy chained WLCSP deviates between 1.7-2Ω. In the HSB test the soldered board contact resistance varies and therefore significant deviations in initial resistance values are measured. In comparison to the JEDEC criterion, a trigger resistance of 1000Ω appeared to be too large, where a trigger resistance of 10Ω (5x initial resistance) is too sensitive for the trigger signal. Therefore the HSB trigger resistance is set at 100Ω.

Furthermore it appeared that continuous data logging of e.g. 10 times the cyclic frequency gave an enormous amount of data. Therefore the data logger is programmed to only record the resistance when the threshold resistance of 100Ω is exceeded.

4. Finite element modeling

To predict and localize the maximum stress and strain in solder joints during HSB testing two different finite element models were made in Marc & Mentat 2007r2: one global model of the PCB (Figure 5) and one local, refined model of a single solder joint (Figure 6). From the global model absolute displacements for critical (corner) solder joints are determined and used as boundary conditions for the local model. The contact option is used to simulate the forced bending by the inner and outer supports, which are clampled on both sides to the PCB in the HSB test. All model and element properties as used in the global and local model are given in Table 2.

In Figure 7 the total equivalent of plastic strain in the SnAg solder section is shown, where the narrow section around the inner core contains the largest plastic strain (4.66×10^{-2}). The maximum Von Mises stress is 47-51 MPa, which is above the SnAg yield strength of 47 MPa (Figure 8). The contact region between WLCSP and solder contains the highest shear stress (26 MPa). To investigate if damage occurs in the plastic deformed regions solder joint analysis is done on failed devices and explained in next paragraph.

Figure 5: Global model: Von Mises stress distribution in test PCB given for bending maximum at +0.9 mm amplitude.

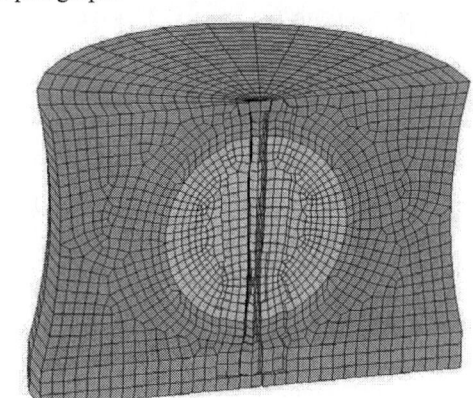

Figure 6: Local model: model half section with polymer inner core, Cu core layer, Cu metallization path and SnAg solder.

Table 2: Finite element and material properties for global and local model

Set	Element	Contact	Material	-	Yield strength [MPa]	E-modulus [GPa]	Poisson ratio [-]
Global model							
PCB	3D solid shell	deformable	PCB	anisotropic	-	$16.0_{11\text{-}22}$, 3.2_{33}	$0.15_{12\text{-}23\text{-}31}$
Solder	2D beam	rigid	SAC	isotropic	47	42	0.4
Product	3D shell	rigid	Si	isotropic	-	169	0.23
Local model							
Cu	3D solid	-	Cu	isotropic	70	124	0.3
Solder	3D solid	-	SnAg	isotropic	47	42	0.4
Micropearl SOL	3D solid	-	Polymer	isotropic	10	5	0.3

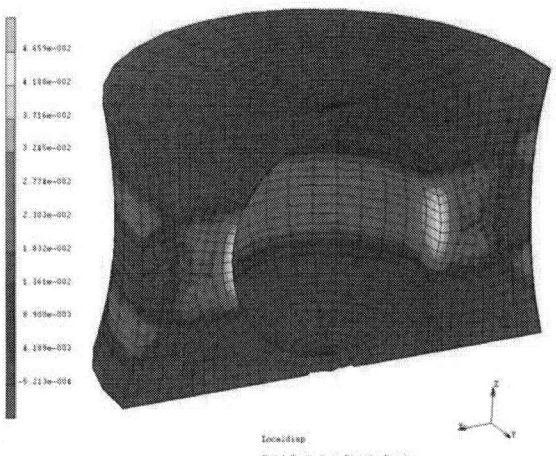

Figure 7: Local model of SnAg solder cross section: Total equivalent plastic strain for +0.9 mm amplitude.

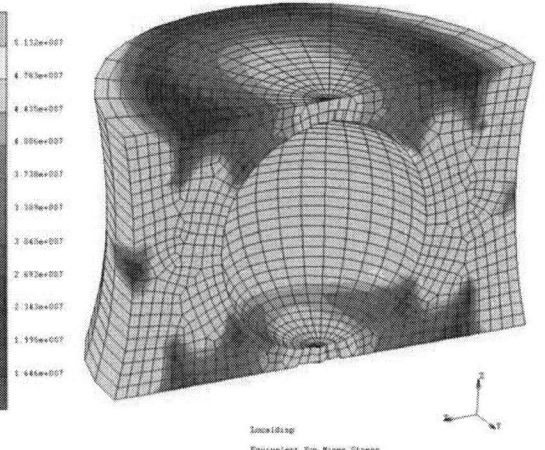

Figure 8: Local model of SnAg solder cross section: Equivalent Von Mises stress for +0.9 mm amplitude.

5. Solder joint damage analysis

Several devices that met the HSB failure criterion were marked and cut out the PCB, embedded in epoxy molding and cross-sectioned for damage analysis. After several grinding and polishing steps they were examined with light microscopy. For crack detection WLCSP are cross sectioned vertically as shown in Figure 9. Solder damage in these devices was not always found during cross sectioning, but other solder abnormalities such as inclusions, voids or micro-structural irregularities were found.

Between solder joints significant differences in geometries were found; solder balls are not perfectly round. In Figure 10 it can be observed that the Cu-coating layer around the inner core is not fully dissolved in the SnAg solder region. In the solder intermetallic phases are observed (dark grey regions), which mainly consist out of the elements Sn, Ag and less Cu.

Figure 11 shows a cross-section image which is made by horizontal (parallel to the PCB) grinding and polishing of the WLCSP in order to expose the daisy-chained BGA solder structure. It can be observed that the co-polymer inner cores are not precisely aligned in the solder ball centers. In Figure 12 a continuous crack in the solder region near the WLCSP metallization is found. In the solder region on the board side also cracks were observed, but for most solder balls damage is found in the solder region on the WLCSP side. This is in agreement with the location of maximum shear stress as found in the solder edge during FE modeling.

With regard to HSB, not all devices which met the failure criterion were examined by cross sectioning. Unfortunately no clear failure mechanism can be attributed to the cracks as found in the analyzed solder joints.

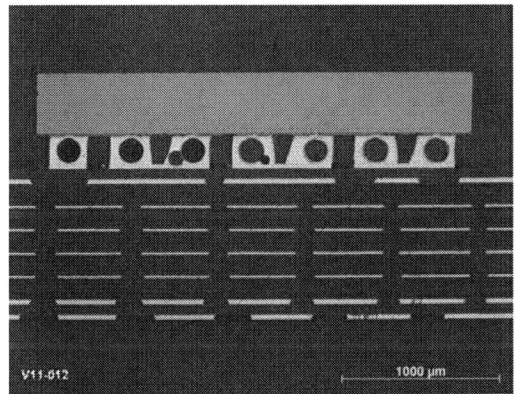

Figure 9: Microscopy image of cross section of WLCSP on PCB (side view). One row of 7 solder joints is visible including voids in solder joints 3 and 4 (from left to right).

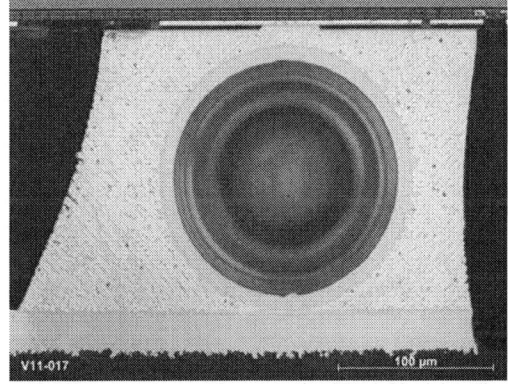

Figure 10: Microscopy image of cross section SOL solder joint. The inner core is a co-polymer coated with copper, solder is SnAg, the under bump metallization is a Cu-path.

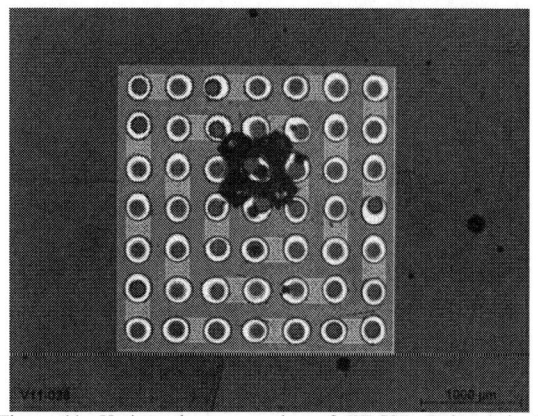

Figure 11: Horizontal cross section of WLCSP. Daisy-chained interconnect structure is visible. In the middle an inclusion is embedded due to cross sectioning. Not all inner cores are aligned perfectly in solder center.

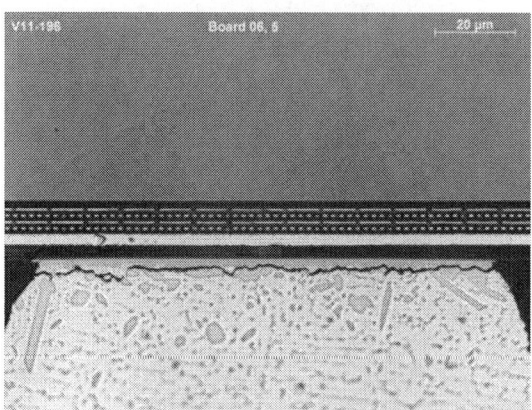

Figure 12: Microscopy image of crack in solder region between solder and device. Cracks propagate along the intermetallic region. The cross section is made outside the solder center; no inner core is visible.

Table 3: Results for HSB test of PCB 6-14, all loaded at 80-100 Hz with 1.8mm peak-to-peak displacement. Values for defect devices deviated from 1×10^4 to 4.8×10^6 cycles. Grey colored values are censored data.

PCB	Test run	Cycles		Duration	Frequency	Peak-to-peak	Chip 5	Chip 6	Chip 7	Chip 8
-	-	Begin	End	hours	Hz	mm	cycles	cycles	cycles	cycles
6	1	0	295.858	1,0	80	1,8	295.858	295.858	295.858	295.858
	2	295.858	6.498.254	21,5	80	1,8	2.200.000	2.200.000	2.200.000	6.498.254
7	1	0	279.629	1,0	80	1,8	279.629	279.629	279.629	279.629
	2	279.629	376.336	0,3	80	1,8	376.336	376.336	376.336	376.336
8	1	0	362.422	1,0	100	1,8	362.422	362.422	362.422	362.422
	2	362.422	2.049.628	4,7	100	1,8	2.049.628	2.049.628	2.049.628	2.049.628
	3	2.049.628	2.888.989	2,3	100	1,8	2.888.989	2.888.989	2.888.989	2.888.989
	4	2.888.989	4.480.642	4,7	95	1,8	4.480.642	4.480.642	4.480.642	4.480.642
9	1	0	1.729.505	5,1	95	1,8	710.000	1.729.505	1.729.505	1.729.505
10	1	0	38.972	0,1	95	1,8	38.972	10.000	38.972	38.972
	2	38.972	43.716	0,0	95	1,8	43.716		-	43.716
11	1	0	1.274.320	3,7	95	1,8	1.274.320	1.274.320	132.000	1.274.320
	2	1.274.320	1.308.933	0,1	95	1,8	1.308.933	1.308.933	-	1.308.933
12	1	0	247.153	0,7	95	1,8	247.153	247.153	247.153	247.153
	2	247.153	548.432	0,9	95	1,8	548.432	548.432	548.432	548.432
	3	548.432	1.052.280	1,5	95	1,8	680.000	1.052.280	1.052.280	1.052.280
13	1	0	3.029.520	8,9	95	1,8	2.880.000	3.029.520	3.029.520	3.029.520
14	1	0	3.021.313	8,8	95	1,8	3.021.313	3.021.313	3.021.313	3.021.313
	2	3.021.313	6.050.135	8,9	95	1,8	6.050.135	6.050.135	6.050.135	6.050.135
	3	6.050.135	8.662.676	7,6	95	1,8	8.662.676	8.662.676	8.662.676	8.662.676
	4	8.662.676	10.690.469	5,9	95	1,8	10.690.469	10.690.469	10.690.469	10.690.469

6. Results and discussion

In Figure 13 an example of signal measurement for test 10 at $3.8*10^4$ cycles is given. Two signals (chip 5 and 8) are still intact, while chip 6 and 7 show continuous resistance increase. Because the resistance trigger is set at chip 5 no trigger signal and data output is given. Besides, the resistance threshold (100Ω at 1V) is not exceeded. Although it appears that in both devices fracture has initiated.

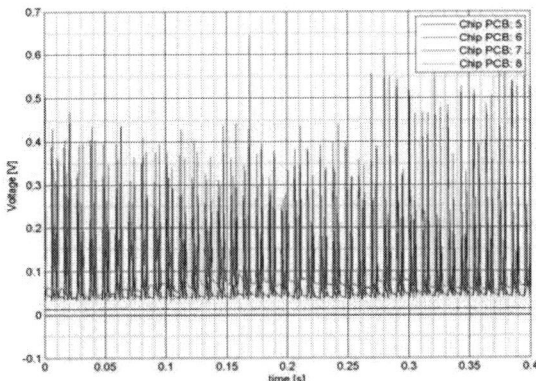

Figure 13: HSB test 10 example of signal measurement: Chip 6 and 7 show intermittent increases in resistance, chip 5 and 8 are still intact.

In table 4 all HSB results from testing of PCB no 6-14 are collected. During pre-testing of PCB 1-5 it appeared that PCB's structurally failed at a HSB amplitudes of 0.9mm, after using various sinus amplitudes from 0.5 to 0.9mm. Therefore test boards 6-14 were loaded at 0.9mm amplitude, which gives a 1.8mm peak-to-peak displacement.

Test frequencies in the range of 80-100Hz are used during testing of PCB 6-14. Finally, 95Hz is chosen as optimum for testing of PCB 9-14, because it was experienced that long-term vibration is limited by the vibration exciter capacity (80Hz) and sensor detachment (100Hz, which is near an Eigen frequency of the system).

For each PCB multiple test runs were done. This is mainly due to practical concerns, but also due to the absence of devices which met the HSB failure criterion, transducer detachments or error messages from the closed loop controller. Values for defect devices deviated from $1x10^4$ to $4.8x10^6$ cycles and were sometimes not determined precisely due to the single-channel trigger signal. A Weibull probability plot is given in Figure 14.

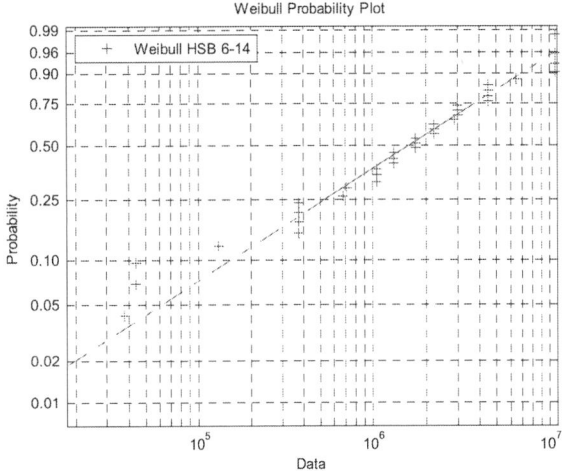

Figure 14: Weibull probability plot for results HSB 6-14

For further verification of the HSB method solder strain rates, which are common for 2nd level interconnects in automotive components, should be taken into account. For HSB this requires the right combination of amplitude and frequency. Furtermore, the detection of events, which are electrical discontinuities of resistance greater then the threshold resistance, should be measured to meet with the JEDEC standard.

Conclusions

- A high speed bending (HSB) method, based on high frequent four point bending moment is optimized for 2nd level interconnect testing of automotive electronics. A first proof of concept is given and verified with FE modeling and fractography. However, further verification of this method is necessary to use it for component qualification according to the JEDEC standard. Multi-channels data loggers or event detector instruments are necessary.

- FE modeling was used to predict the maximum stress and strain in the solder joints as loaded during HSB. This is supported by fractographic analysis of the solder joints. Continuous cracks were found in the device – solder contact region.

- In total 12 NXP test boards were tested; a large deviation in cycles for interconnect failure were found. Due to set-up limitations only 4 channels were monitored and only one channel was used as trigger. Values for defect devices deviate from $1x10^4$ to $4.8x10^6$ cycles.

Acknowledgments

The research leading to these results has received funding from the ENIAC Joint Undertaking and from Senter-Novem in the Netherlands under Grant Agreement number 120009.

References

[1] Y. Liu, F.J.H.G. Kessels, W.D. van Driel, J.A.S. van Driel, F.L. Sun, G.Q. Zhang, Compact drop impact test method using strain gauge measurements, Microelectronics Reliability 49, 1299-1303, 2009.

[2] E.H. Wong. S.K.W. SEAH, C.S. Selvanayagam, R. Rajoo, W.D. van Driel, J.F.J.M. Caers, X.J. Zhao, N. Owens, M. Leoni, L.C. Tan, High-speed cyclic bend tests and board-level drop tests for evaluating the robustness of solder joints in printed circuit board assemblies, Journal of electronic materials, Vol. 38, No. 6, 2009.

[3] S.K.W. Seah, E.H. Wong, Y.W. Mai, R. Rajoo, C.T. Lim, High-speed bend test method and failure prediction for drop impact reliability, Electronic Components and Technology Conference, 2009.

[4] E.H. Wong, S.K.W. Seah, V.P.W. Shim, A review of board level solder joints for mobile applications, Microelectronics Reliability 48, pp. 1747-1758, 2008.

[5] JEDEC Standard JESD22B113, board level cyclic bend test method for interconnect reliability, 10 Electrical monitoring requirements and failure criteria.

Reliable hermetic MEMS chip-scale packaging

G. Spinola Durante[*], R. Jose James[a], C. Bosshard[a], C. Muller[b], J. Baborowski[b],
A. Pezous[b], F. Cardot[b], M.-A. Dubois[b], A. Neels[b], A. Dommann[b]

[*,a] CSEM Alpnach, Untere Gründlistrasse 1, CH-6055, Alpnach Dorf
[b] CSEM SA, Rue Jaquet-Droz 1, CH-2002, Neuchâtel

* Corresponding author Tel.: +41.41.672.7529, Fax: +41.41.672.7500, gsd@csem.ch

Abstract

Reliable hermetic MEMS chip-scale packaging has been developed and tested with gold-tin soldering on $2x1.5$ mm^2 chips. The test vehicle configuration consists of a glass cap with AuSn solder ring bonded onto a silicon resonator dummy chip. Through careful chip layout, UBM layer improvements and soldering process optimization, the shear tests yielded forces above 5 kg on a bonding ring area of ~0.5 mm^2. With a resulting shear stress larger than 100 MPa it is now possible to address a wider application domain for vacuum-encapsulated MEMS, where highly reliable hermetic solutions are required. Promising applications of hermetic and reliable MEMS are found in the medical field and in the space domain where defect-free sealing is a must. In the consumer electronics domain, where time-to-market and cost reduction are key factors, development time can be shortened for extremely miniaturized packaged MEMS.

Key words: hermeticity, reliability, MEMS, packaging, gold-tin, galvanic

Introduction

During the last few years piezoelectric silicon resonators have been developed at CSEM targeting time refererence applications. In order to guarantee a high performance (high Q value) and longterm stability, vacuum hermetic wafer-level packaging technology is required [1]. At CSEM high-vacuum packaging is required in a variety of projects. The current development objective is focusing on improving the reliability of the AuSn chip-scale bonding technology. The solder eutectic melting takes place at relatively high temperatures (~280 °C) and solidification embeds a large amount of stress in due to thermal mismatch of the package materials. This stress weakens the sealing ring and reduces the overall packaging process yield and reliability.

The chosen methodology to address this issue is to improve the UBM (under bump metallization) technology, to select a suitable AuSn solder deposition method and to optimize the controlled-atmosphere soldering process.

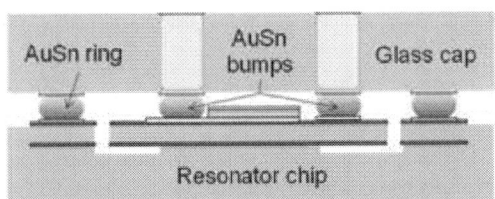

Fig. 1: Silicon resonator packaging configuration

The selected configuration comprises a silicon resonator chip bonded to a glass cap including vias

to electrically interconnect the resonator and to further shrink the footprint of the package (Fig.1). The through-vias technology has a direct benefit on the functionality of the resonator itself, since lower parasitic capacitances can be achieved with direct bonding of the resonator on the driving IC. Vias are part of a parallel development. For the purpose of optimizing the shear strength of the solder joint we will only consider glass caps without vias and silicon resonator dummy chips.

To achieve simultaneous sealing and bump interconnection during bonding it is sufficient to have pads and ring realized with the same UBM metal and solder. In Fig.2 a glass dummy cap is bonded to a silicon dummy chip resonator (*2x1.5 mm^2*) with a sealing ring and four pads.

Fig. 2: Bonded glass cap to silicon dummy chip

The solder is deposited on the glass chip. Challenges in the soldering process come from the required height uniformity of the solder on the sealing ring and on the interconnection pads which must be achieved before the bonding step. In case of a reflow process prior to the bonding step this is not trivial due to morphology changes during the reflow process itself.

UBM technology development

For packaged MEMS typically thermo-cycling, 85/85 and high-temperature storage test survival is expected. The argument for a AuSn solder choice is exactly that it can survive high temperatures and is relatively stable against oxidation and corrosion. This clearly suggests that reliability can only be weakened by a non-optimal UBM selection.

One of the key aspects of soldering is therefore the metallization choice that will form a solid joint by means of a molten solder that wets and creates a solid interconnection during the cooling phase. The molten solder and the under bump metallization will undergo interdiffusion. The interdiffusion process happens simultaneously to intermetallic growth. The thickness and the composition of intermetallics will decide on the strength of the solder joint as well as the reliability and lifetime of the final assembly [2].

Special attention must be given to the consumption of the UBM by the interdiffusion when the solder is in the molten phase, since this is strongly dependent on the solder process temperature-time budget. It is clear that no simple model can describe the diffusion-reaction process of the AuSn alloy soldering in the solid-state. When the solder is in the molten state only semi-empirical formulas can be used to estimate the intermetallic growth and UBM consumption [2].

The experience of CSEM is that several defects can appear if the layer composition is not optimized and the surface is contaminated. As an example in Figs.3-5 it is possible to observe, respectively, voids, dewetting and delamination from the different UBM metal stacks.

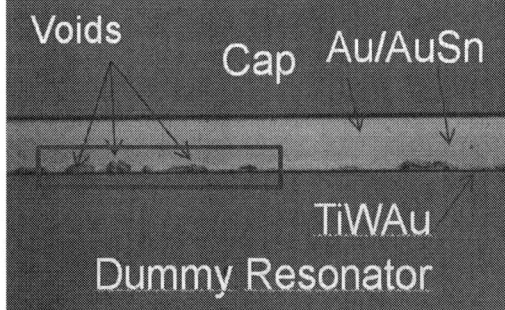

Fig.3: TiWAu 2μm thickness: voiding defects

Careful selection of the UBM layer composition, its deposition method and surface

cleanliness, together with solder process timing optimization has been very important in our investigations. Fig. 6 depicts a TaPt UBM after AuSn soldering showing almost defect-free behavior.

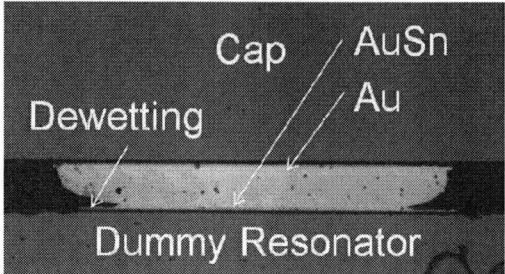

Fig.4: TiWAu 3μm thickness: de-wetting

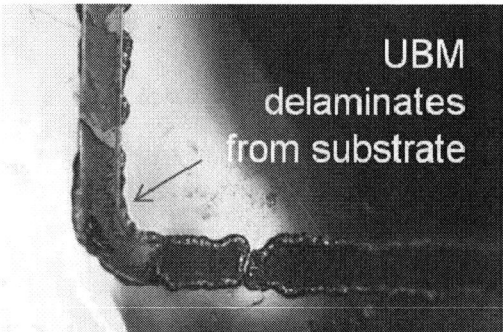

Fig.5: TiNiAu 5μm thickness: delamination

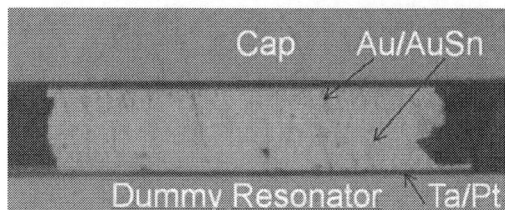

Fig.6: TaPt 0.25μm thickness: no defect observed

This development was further optimized targeting a reliable UBM to package the resonators mentioned above. Trials have been carried out to lower the UBM stress and improve the cleanliness of surfaces. Adhesion tests have been also performed prior to soldering tests. To check the UBM-solder interaction knife tests and shear tests have been performed, together with cross-sections.

Solder deposition method

Two different AuSn solder types have been extensively tested. The first type is a galvanic solder with Au/Sn layers. These layers are very flat and with low roughness (typically R_a 20-40nm). These layers need to be reflowed to get eutectic composition and subsequently soldered. Effort was put to develop a low-stress reflow process and to fine tune the amount of the eutectic AuSn phase

132

which is the solder medium (see Figs.7 and 8). An evaporated multilayer of 4 μm was tested in parallel. The deposition in this case is AuSn near a 75/25 composition deposited on a Au socket. (Fig.9).

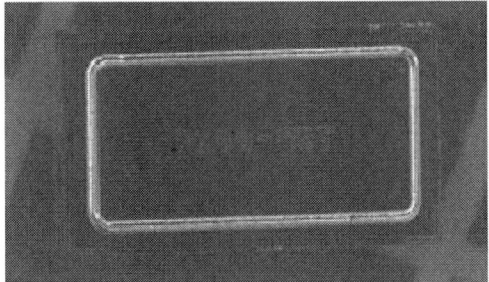

Fig.7: AuSn solder ring reflowed (top view). The ring is 1.65x3.3 mm^2 and 100μm wide

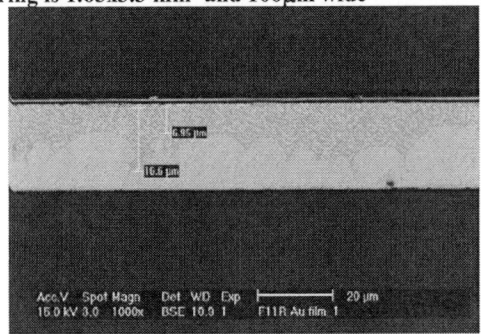

Fig.8: AuSn solder reflowed (cross-section)

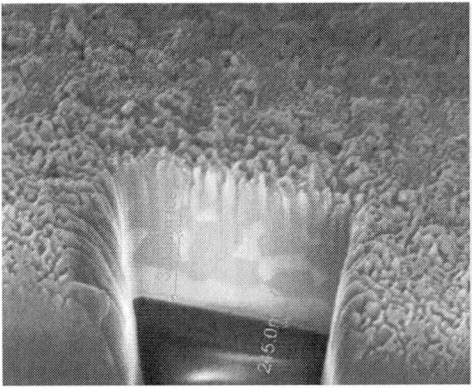

Fig.9: AuSn solder reflowed (cross-section)

Both approaches are suitable for chip-scale hermetic bonding and have been proven to be hermetic.

Controlled-atmosphere soldering process

The AuSn soldering process is very sensitive to the thermal budget of the process. Too short reflow times during the soldering process will lead to non-bonded areas and defects will be embedded in the solder joint and hermeticity can be compromised. Too long reflow-times will lead to dewetting after UBM dissolution, which will also result in non-bonded areas and hermeticity can therefore be degraded.

Vacuum hermetic processes are even more challenging since the oven temperature profiles will be affected by the oven chamber pressure level. These issues have been tested at CSEM and optimized for the resonator UBM yielding reproductible results. Optimized load applied during soldering is also critical.

To some extent solder flowout was observed due to solder composition and loading conditions (see Fig. 10).

Fig.10: AuSn solder flowout on chip

Actions were taken on the process side to strongly reduce the risk of solder flowout and proved to be successful at chip-scale level.

Development of AuSn galvanic deposition

CSEM is also currently involved in the in-house development of AuSn galvanic deposition to better adapt the solder to the packaging requirements.

The main advantages are the following: first to better match the soldering process window, since controlling the solder amount can significantly reduce the issues coming from planarity and non-uniformity when targeting reliable hermetic wafer-level bonding. The second reason is to find the optimum composition to allow for gold diffusion in the solder joint and thus increase the remelting temperature making it possible to activate getter materials. These materials typically need more than 400°C to be activated and the longer the time (>30minutes) at high temperature the more effective the gettering effect will be.

It is also important to keep under control the voids and minimize defects in the deposited solder to avoid any possible small leakage that could compromise high-vacuum sealed cavities.

CSEM develops AuSn galvanic deposition starting from 4" wafers and moving up to 8" wafers. Figs. 11 and 12 show the AuSn solder ring for resonators. On a 4" wafer there are typically >1750 chips.

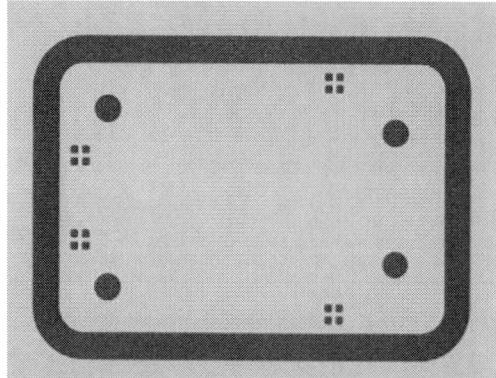

Fig.11: Galvanic AuSn solder on silicon chip. The ring is 100um wide and round pads are 100um in diameter

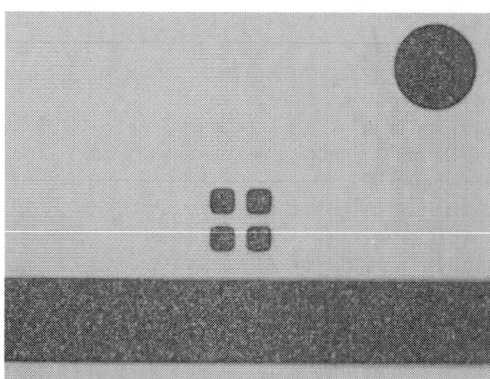

Fig.12: Details of galvanic AuSn solder. . The ring is 100um wide and round pads are 100um in diameter

Alternatives to slow flip-chip alignment process

To further speed-up the chip-scale packaging development, alternatives to a slow flip-chip alignment process have been tested.

On one side a pick&place machine with ~10 μm alignment (Fig. 13) capability was used to tack (Fig. 14) the chips and the glass caps in the range of <1 minute, starting from samples positioned with tweezers in an undefined position within a support of 25x25mm^2. This speed can still be improved. Subsequent soldering is done in a vacuum-capable reflow oven. This approach is faster if compared to a tacking process performed in conventional flip-chip machines that are used for the hermetic sealing and results in more chips available for statistical purposes in process optimization tasks.

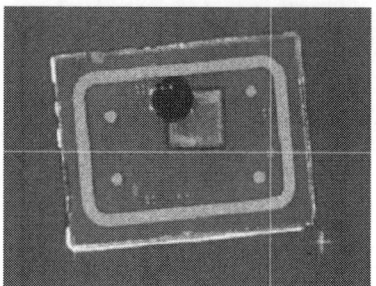

Fig.13: Fast automatic pattern recognition and alignment of a glass cap. White rulers crossing the picture identify align marks on the chip

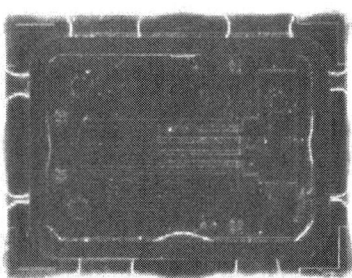

Fig.14: Tacked glass cap on a silicon dummy chip by means of a tacking fluid

A second alternative is shown in Fig. 15 where an array of 7x5 resonators and the corresponding glass array are bonded with AuSn solder.

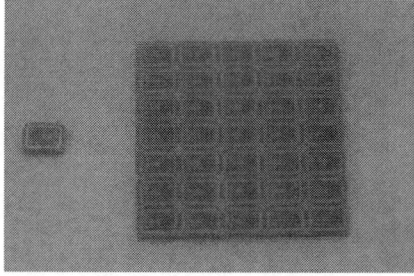

Fig.15: Silicon dummy resonator die 2x1.5 mm^2 in comparison to a 7x5 array

At the moment the bonding yield of 50% (Fig. 16) already leads to a considerable bonding time improvement since we can easily bond 10 of these chip arrays all at once, which corresponds to a 350 sensor assembly. One reflow process lasts around 25 minutes. The oven plate, upon design of a proper assembly tray, makes it possible to assemble more than 100 arrays which corresponds to around 3500 samples and allows small series production. In addition, it allows to demonstrate reliability with the statistics of at least a few hundred samples.

Fig.16: Bonded Silicon dummy resonator 7x5 array showing AuSn solder spreading

Hermeticity testing

The hermeticity of AuSn solder is well established for ceramic packages. At chip-scale level a first simple test that can be performed to check hermeticity is the dye-penetrant test. Since the bonded chip (Fig. 2) is capped with a glass tile, it is possible to use deionized water instead of a fluorescent liquid. Water capillarity will immediately drive water within non-hermetic cavities. In the following pictures a hermetic (Fig. 17) and a non-hermetic (Fig. 18) capped resonator are shown.

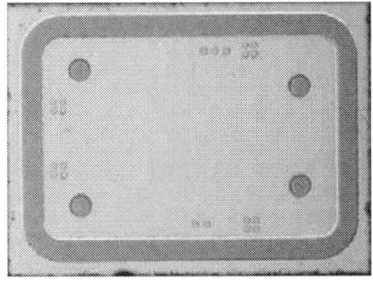

Fig.17: Hermetic silicon dummy resonator

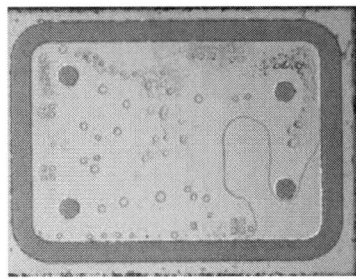

Fig.18: Silicon dummy resonator samples with DI water inside the cavity

As mentioned above, the reason for non-hermeticity is the unplanarity of the solder deposition which results in voids. Particle contamination plays also a role since they can act as spacers across the solder joint. Cleaning of the samples has to be thouroughly executed prior to the alignment during the chip-scale bonding process.

Shear test results

As an outcome of the described steps and efforts, shear tests on AuSn bonded chips of 2x1.5 mm² (Fig. 2) yielded shear forces above 5 kg. This is equivalent to a shear stress >100 MPa if we consider that the bonding ring is 100 microns wide and has a surface of ~0.5 mm². In Fig. 19 the glass cap after performing the shear test is shown.

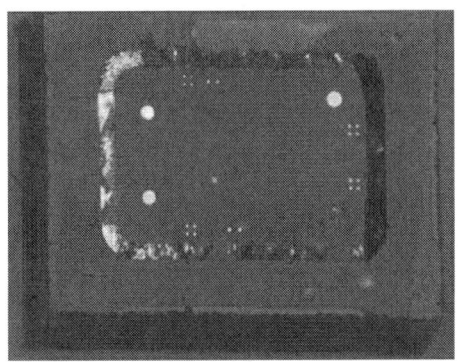

Fig.19: Glass cap after shear test

In this picture the cracks in the solder and in the glass chip itself are clearly visible. Shear test failure modes and shear force values confirm that all key components of the solder joint, including chip UBM, solder deposition, and cap UBM are effectively robust and prone to survive high mechanical and thermo-mechanical stress.

Conclusions & perspectives

The results presented here show that it is now possible to address a wider application domain for vacuum-encapsulated MEMS devices, where highly reliable hermetic solutions are required. Promising applications of hermetic and reliable MEMS are found in the space and medical fields where defect-free sealing is a must. In the consumer electronics domain, where time-to-market and cost reduction are key factors, development time can be shortened for extremely miniaturized packaged MEMS.

Further reliability test will be performed for the packaged resonators to check mechanical robustness and hermeticity through the impact on electrical functionality of the resonator devices.

In conclusion, our results demonstrate reliable hermetic chip-scale packaging and UBM and AuSn solder deposition. The processes can be scaled targeting reliable hermetic wafer-level bonding and small series production.

Acknowledgements

This work was supported by MCCS Micro Center Central Switzerland. CSEM thanks them for their support.

The authors thank Petra Staiger (Injector Solutions AG) for her support in the galvanic AuSn solder deposition developments.

References

[1] J. Baborowski, et al. "Wafer level packaging technology for silicon resonators", Procedia Chemistry 1, pp. 1535-1538, July 2009

[2] S. Anhöck, H. Oppermann, C. Kallmayer, R. Aschenbrenner, L. Thomas, H. Reichl, "Investigations of Au/Sn alloys on different end-metallizations for high temperature applications", IEEE/CPMT Electronics Manufacturing Technology Symposium, pp. 156-165, 1998

High resolution microstructural investigation of leadfree Aluminum-Germanium and Aluminium-Germanium-Copper alloys for high temperature silicon die attach

S.Klengel (Bennemann)[1], B.Böttge[1], M.Petzold[1], W. Schneider[2],

[1]Fraunhofer Institute for Mechanics of Materials, Walter-Hülse-Straße 1, 06120 Halle, Germany

[2]Microelectronic Packaging GmbH, Grenzstrasse 22, 01109 Dresden, Germany

Sandy.Klengel@iwmh.fraunhofer.de; +49 3455589125

Abstract

Due to the RoHS and WEEE legislative requirements the implementation of lead free solder material is of prior significance in electronic manufacturing. As a consequence, many efforts have been started in order to adapt the soldering technology as well as to provide adequate equipment and materials. While European RoHS regulations currently exempt high Pb solders used as component solders and die attaches for automotive and other high temperature applications, there is a strong drive to find Pb-free alternatives for these high temperature electronic applications, as well. The development of lead-free solder materials with improved reliability properties requires a detailed understanding of the microstructure and thermal behavior. Aluminium-Germanium and Aluminum-Germanium-Copper alloys are described in literature as possible alternative solder materials for high temperature silicon die attach. However, no micro structural or process properties are released. In this study, DSC measurements will be performed to analyze the melting point of the different alloys and the crystallization behaviour. High resolution micro structural analyses will be shown at Al-Ge and Al-Ge-Cu alloys in initial condition and after melting processes. Using Focussed Ion Beam techniques (FIB) and Scanning Electron Microscopy (SEM) the alloys will be analyzed in delivery condition and after defined temperatures. Additionally the wetting behaviour of the two alloys will be described for different surfaces. Summarized this study gives a comprehensive knowledge of microstructural phenomena for the specific Aluminium-Germanium and Aluminum-Germanium-Copper alloys.

Key words: AlGe50, AlGe44Cu0.8, die attach, power application

Introduction

Due to the RoHS and WEEE legislative requirements the implementation of lead free solder material is of prior significance in electronic manufacturing. As a consequence, many efforts have been started in order to adapt the soldering technology as well as to provide adequate equipment and materials. While European RoHS regulations currently exempt high Pb solders used as component solders and die attaches for automotive and other high temperature applications, there is a strong drive to find Pb-free alternatives for these high temperature electronic applications, as well.

Promising new alternatives especially for power application are silver sintered components (both pressure and pressure less) [1], [2]. During the last years also Aluminum-Germanium solder materials with different alloyed elements (e.g. Zn) were proposed as possible materials for high temperaterure die attach [3], [4].

This paper focuses on thermal and microstructural investigations of Aluminum-Germanium (AlGe50) and Aluminum-Germanium-Copper (AlGe44Cu0.8) alloys.

The paper is organized as follows. First we provide a summary on previous material investigation of Aluminum-Germanium alloys. After presenting the thermal results achived by DSC analyses this study presents microstructural investigations done by Focussed Ion Beam (FIB) resp. Scanning Electron Microscopy (SEM) and Energy Dispersive X-Ray Analyses (EDS) for AlGe50 and AlGe44Cu0.8 materials before and after reflow as well as bonding behaviour to common silicon dies for power application.

In summary our analyses show the AlGe50 and AlGe44Cu0.8 could be an adequate alternative for high temperature die attach due to excellent bonding behaviour at common silicon dies for power application.

Aluminum-Germanium Solder Material

Aluminum-Germanium alloys are commercial available and were described in several publications before [3], [4], [5]. Figure 1 shows the Aluminum-Germanium equilibrium diagram

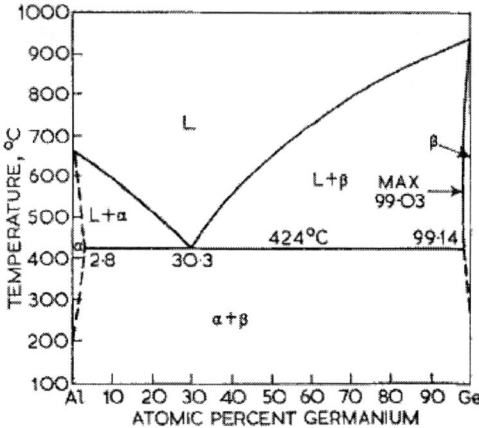

.**Figure 1:** Aluminum-Germanium equilibrium diagram [3].

Surynarayana et al. showed the correlation between cooling rate and solidification behavior of the system [3]. For the Aluminum-Germanium solder material described in this study (AlGe50, AlGe44Cu0.8) two different conditions are able to occure depending on the cooling rate.

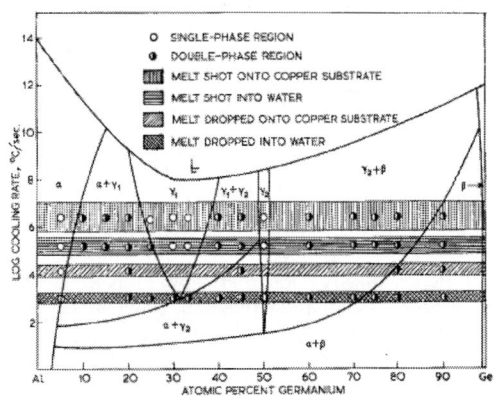

Figure 2: Relation between cooling rate and constitution at room temperature for the system Aluminum-Germanium [3].

DSC Analyses of Aluminum-Germanium Solder Material

DSC analyses with a heating rate of 5K/min were performed and demonstrate the real melting behavior of the AlGe44Cu0.8 and AlGe50 solder material. Figure 3 shows the DSC scan for the system AlGe50. The onset temperature is about T = 433°C and correlates with the in [3] reported melting point. Figure 4 results the melting temperature for the AlGe44Cu0.8 material system with T = 426°C that is slightly lower compared to AlGe50.

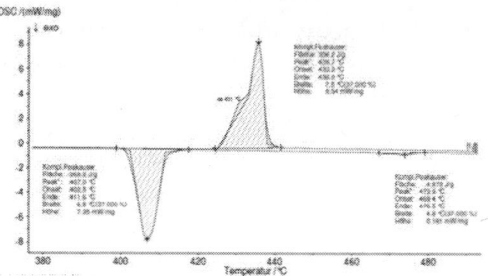

Figure 3: DSC scan with heating rate 5K/min demonstrates the melting behavior of AlGe50 solder material; the blue graph shows the heating behaviour and purple graph shows the cooling behaviour.

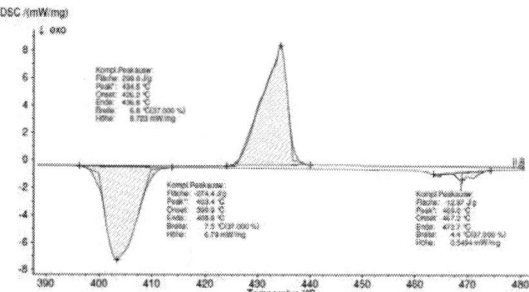

Figure 4: DSC scan with heating rate 5K/min demonstrates the melting behavior of AlGe44Cu0.8 solder material; the blue graph shows the heating behaviour and purple graph shows the cooling behaviour.

Microstructural Analyses of Aluminum-Germanium Solder Material

For microstructure investigations the solder material was investigated in initial condition and after reflow process at T= 480°C.

To show the material behaviour at different silicon dies for power application AlGe50 was used for interconnection without intermediate die metallisation and AlGe44Cu0.8 was used for silicon die interconnection with Ag plating.

The solder/silicon die interconnects were prepared in cross sections using metallographic procedures followed by high quality ion milling [6]. High resolution Focussed Ion Beam (FIB) resp. Scanning Electron Microscopy (SEM) analyses were applied to describe the microstructure phenomena occurring at the material systems both in intial condition and after reflow processes and to investigate particularly the formation of intermetallic compounds (IMC) and potential void- or crack-like defects.

Initial condition

Figures 5 and 6 show the microstructure of the AlGe44Cu0.8 and AlGe50 solder material in initial condition. A fine disperse structure is detectable for both materials consisting of single Aluminum (dark) and Germanium phases (bright).

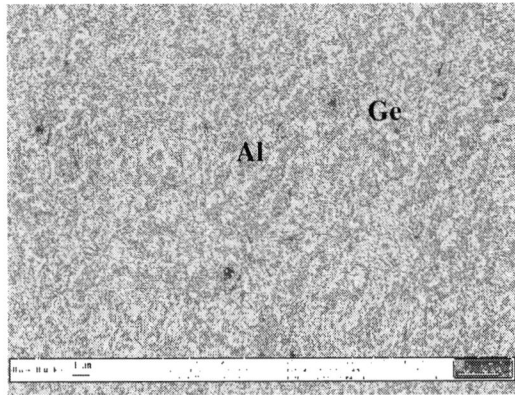

Figure 5: Scanning Electron Microscopy image of AlGe44Cu0.8 in intitial condition shows the fine disperse structure of the material system. Aluminum containing areas appear dark, Germanium containing areas appear light.

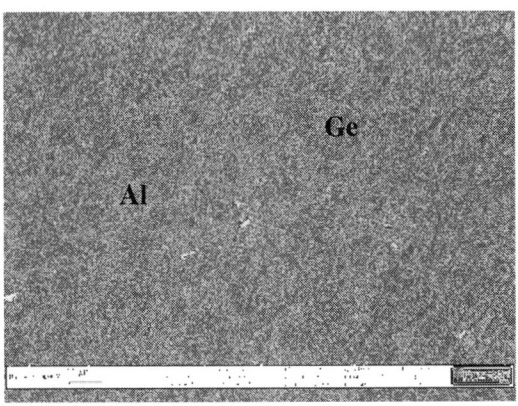

Figure 6: Scanning Electron Microscopy image of AlGe50 in intitial condition shows the fine disperse structure of the material system. Aluminum containing areas appear dark, Germanium containing areas appear light.

Reflow process

After reflow process at T = 480°C the fine disperses structure changes into a more dendritic structure for both alloys (cp. Figure 7 and Figure 9). EDS analyses were done with U = 20kV accelerating voltage and show that there is at least a two phase system containing Aluminum and Germanium (Figures 8 and 10). Copper seems to agglomerate at the Germanium phase (cp. Figure 8) for AlGe44Cu0.8 material system.

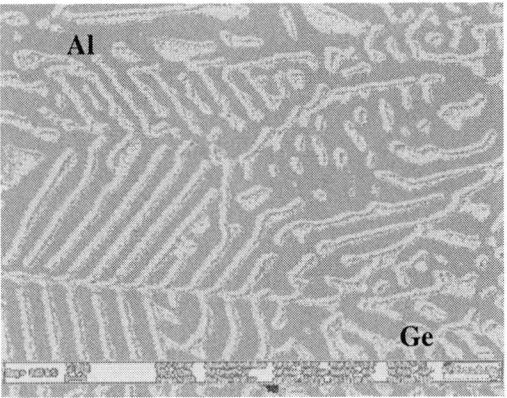

Figure 7: Scanning electron microscopy image of AlGe44Cu0.8 after reflow. The fine disperse structure was transferred into a more dendritic structure after reflow; bright areas contain Germanium, dark areas contain Aluminum

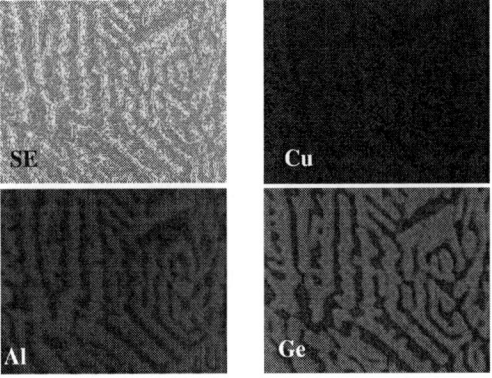

Figure 8: AlGe44Cu0.8 - EDS mapping of the phases formed between aluminum, copper and germanium layer after reflow. Element distribution indicates that germanium exists besides the aluminum and the copper is adapted to the germanium.

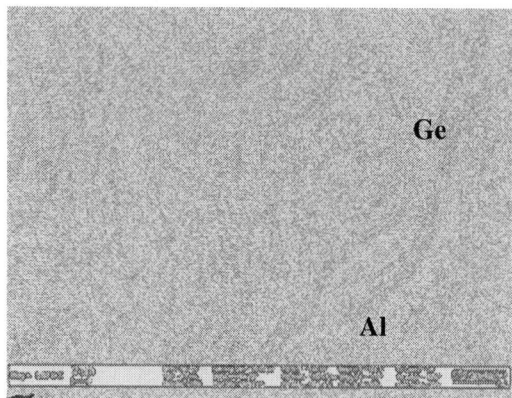

Figure 9: Scanning electron microscopy image of AlGe50 after reflow. The fine disperse structure was transferred into a more dendritic structure after reflow; bright areas contain Germanium, dark areas contain Aluminum

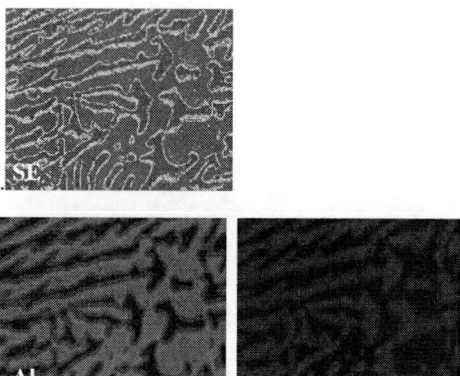

Figure 10: AlGe50 - EDS mapping of the phases formed between aluminum and germanium layer after reflow. Element distribution indicates that germanium exists besides the aluminum.

Soldered components

To show the material behaviour at different silicon dies for power application the AlGe50 solder material was used for interconnection without intermediate die metallisation; AlGe44Cu0.8 was used for silicon die interconnection with Ag plating. Reflow peak temperature was T = 480°C for both material systems.

AlGe44Cu0.8 was soldered to a silver coated silicon die. Aluminum and Germanium are bonded separately in a two phase structure directly to the silicon material. The Ag plating seems to be dissolved completely in the solder material. Additionally there are a lot of dendritic like structures in the solder bulk that could remain from the Ag coating (cp. Figures 11 and 12).

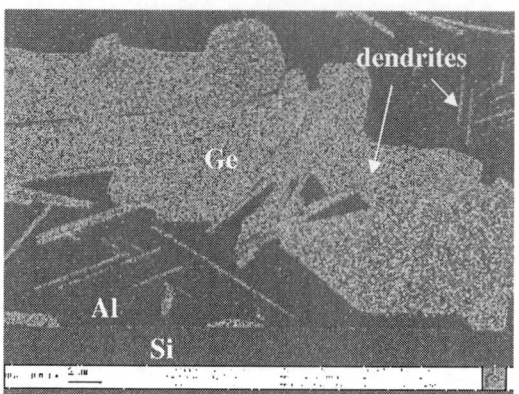

Figure 11: AlGe44Cu0.8 after reflow; SEM image shows bonded interface between silicon die and Aluminum resp. Germanium containing areas. Several dendritic structures are detectable at the solder bulk.

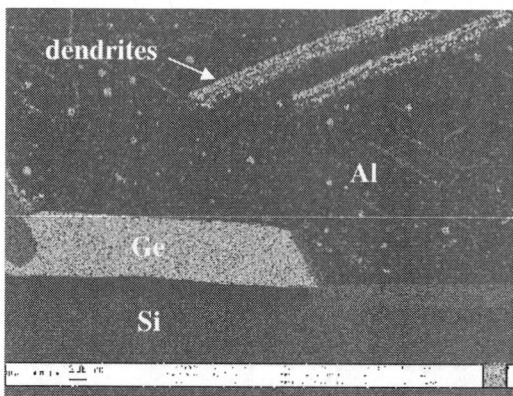

Figure 12: AlGe44Cu0.8 after reflow; SEM image shows detail of the bonded interface between silicon die and Aluminum resp. Germanium containing areas. Ag coating is completely solved in the AlGe44Cu0.8 solder material and is suspected to form dendrites.

AlGe50 was soldered directly to a silicon die without chip metallization. Aluminum and Germanium are also bonded at different areas to the silicon. No dendritic structure could be observed (cp. Figures 13 and 14).

Both material systems showed a good bonding behaviour without the formation of any void-like defects or cracks. No intermediate layers could be observed between solder material and silicon.

However, the formation of the dentritic structure shown in Figure 11 could lead to crack formation and induce reliability problems. Thus, investigations of the behaviour after stress test are absolutely necessary (both microstructural and mechanical).

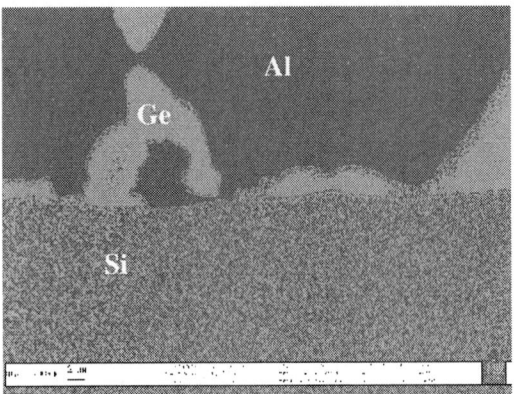

Figure 13: AlGe50 after reflow; SEM image shows bonded interface between silicon die and Aluminum resp. Germanium containing areas. No dendritic structures are detectable at the solder bulk.

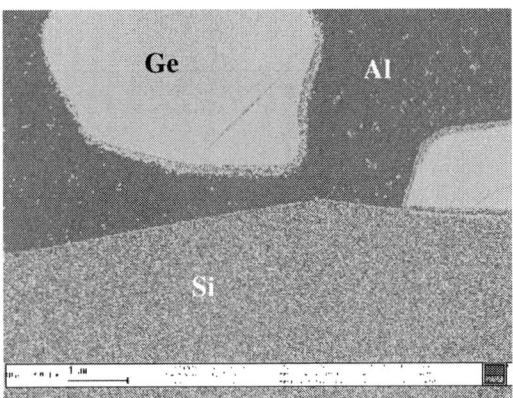

Figure 14: AlGe50 after reflow; SEM image shows detail of the bonded interface between silicon die and Aluminum resp. Germanium containing areas. AlGe50 solder is bonded directly to the silicon without any intermediate layers.

Summary and Conclusion

This study presented thermal and microstructural investigations of Aluminum-Germanium and Aluminum-Germanium-Copper alloys for alternative high temperature silicon die attach.

DSC analyses resulted in the melting point of T = 433°C for Aluminum-Germanium and T = 426°C for the Aluminum-Germanium-Copper alloy. Both alloys are changing the microstructure from fine disperse at intital condition to more dendritic after reflow process.

AlGe50 was soldered directly to an uncoated silicon die. The interface showed no voids or cracks which indicates a very good bonding bahvior.

AlGe44Cu0.8 was soldered to a Silver coated silicon die. The interface between solder material and silicon shows no voids or cracks but the solder

bulk material includes some dendrites which are suspected to be formed by the Ag plating. However, even this material system shows a good bonding behavior directly to the silicon die.

Further investigation will be observed to describe the interface reaction between silicon die and solder material before and after high temperature storage resp. temperature shock as well as mechanical tests to give some information about reliability.

Acknowledgements

This work was partly supported by the German Bundesministerium fur Bildung und Forschung (BMBF) under contract "HiT-Modul" (13N10953). Thanks to Karin Wittich and Jana Fiedler who have done the DSC analyses.

References

[1] J. G. Bai et al., „Thermomechanical Reliability of Low-Temperature Sintered Silver Die Attached SiC Power Device Assembly", IEEE Transaction on device and materials reliability, Vol. 6, No.3, September 2006

[2] U. Scheuermann, "Reliability challenges of automotive power electronics", Microelectronics Reliability, Volume 49, Issues 9-11., September-November 2009, Pages 1319-1325

[3] C. Surynarayana et al., "Solidification of Aluminum-Germanium-Alloys at High Cooling Rates", Journal of Material Sience, p.992-1004, 1970

[4] A. Haque et al.," Investigations on Zn-Al-Ge Alloys as High Temperature Die Attach Material", 3rd Electronic System-Integration Technology Conference (ESTC), 2010

[5] L. Illgen et al.," Preparation of ductile Al-‑Ge soldering foils by PFC technique", Materials Science and Engineering: A Volume 133, 15 March 1991, Pages 738-741

[6] W. Mack, et al., „Ion Beam Polishing as Final Treatment for Embedded Cross Sections and its advantages for FESEM Analysis in Electronic Packaging", *Proc. 30. Int. Sympos. On Testing and Failure Analysis (ISTFA)*, Worcester, MA, Nov. 2004, pp. 338-345

Reliability of ACA joined thinned chips on rigid substrates under humid conditions

Laura Frisk, Kirsi Saarinen, and Kati Kokko

Department of Electronics, Tampere University of Technology
P.O. Box 692, 33101 Tampere, Finland

Phone +358 40 849 0609
Fax +358 3 3115 3394
Email address: laura.frisk@tut.fi

Abstract

Thinned chips are an interesting option for reducing the thickness of an electronics package. In addition to the reduced size, thinned chips are flexible and can dissipate more heat than thicker ones. Joining of the thinned chips can be done using several different techniques. Of these, anisotropic conductive adhesives (ACA) are an interesting option as they have several advantages such as low bonding temperature and capability for high density. The reliability of ACA flip chip joints under thermal cycling conditions has been found to increase when thinned chips are used. However, the effect of humidity has not been fully explored. The effect of humidity on reliability raises specific issues since it causes various types of stress to form in the structure. In this study the reliability of thinned chips under humid conditions was compared to that of thicker chips. Five test lots were assembled with thinned and thick chips using an ACA film and three different FR-4 substrates. The substrates had the same sizes, but their thicknesses and structures varied. A constant humidity test was used to study the reliability of the joints. With one of the substrates the reliability of the thinned chips was noticeably inferior to that of the thicker chips. In order to compare the stresses in the joints during testing, finite element models (FEM) were used. Greater stresses were seen in the thinned chips during humidity testing when compared to those seen in the thicker ones.

Key words: Flip chip technology, anisotropic conductive adhesive, thin chip, humid conditions

Introduction

The trend for ever smaller packaging solutions has increased the use of chip on board (COB) technologies. In these technologies a chip is left unpacked and thereby space is typically saved. One example of the COB technique is flip chip attachment, in which a bare chip is attached to a substrate with the active side towards the substrate. Flip chip technology enables the production of very high density packages. In addition, it has other benefits such as better electrical performance due to the short interconnection path and size reduction [1]. The flip chip technique can be used to attach a chip directly onto a printed circuit board, but also as an attachment method in single-chip packages, such as a ball grid array (BGA) and a chip scale package (CSP), and in multi-chip modules.

Currently mainstream flip chip technology is based on solder bumps [2,3]. These can be produced using both traditional tin-lead and new lead-free solders. However, increased environmental concern, which has led to extensive research in the area of lead-free solders, has also increased interest in environmentally friendly electrically conductive adhesives. In addition to being lead free, they can be used with substrate materials which do not withstand

soldering temperatures. Thus they can be used to solve the problem caused by the high reflow temperature needed by most lead-free solders [4].

Electrically conductive adhesives are polymeric materials containing conductive particles. In isotropic conductive adhesives (ICA) the concentration of the conductive particles is high and they conduct in all directions. On the other hand, in anisotropic conductive adhesives (ACA) the concentration of conductive particles is low and the adhesive conducts in z-direction only after the bonding process. Both of these adhesives can be used in flip chip attachment. Extremely fine pitch can be achieved with ACAs.

Although, the size of a flip chip package is typically very small, its volume can be further reduced by thinning the attached chip. This enables very thin packaging solutions. Another important advantage of the thinned chips is their flexibility. When silicon chips are thinned below 100 um, they became pliant [5] and can be used in solutions where they are bent. Consequently, they can be used in flexible electronics. Thinning also allows the chips to dissipate more heat [6], which is important when the densities of the packages increase. However, thinned chips have certain drawbacks. They are more fragile than thicker chips and this needs to be taken into account when thinned chips are handled. Special tools may be needed since the fragile edges of the thinned chips are easily broken during handling. During the thinning and dicing processes a considerable amount of stress may be induced in the chips [6,7,8]. The dicing process is especially important; it may damage the chip edge. If the chip edge is damaged or there is stress in the chip, it may result in failure during testing and operational life and may even impair reliability compared to that of normal sized chips.

In modern electronics reliability is often the key issue. Under thermal stress, thinning of the chips has been found to enhance the reliability of both solder and ACA flip chip joints [9,10,11,12]. This is caused by the pliability of thin chips, which decreases the CTE mismatch stresses in the package, thereby enhancing the reliability of the joints [9]. However, there are rather few studies of the reliability of thinned chips under humid conditions. Thin and flexible products would have a wide range of applications such as in portable consumer electronics and electronics for medical applications. Such products may be subjected to humidity and may even be immersed in water. Consequently, reliability under humid conditions is often critical for the use of thinned chips in these kinds of applications.

This study compares the reliability of thinned chips under humid conditions with that of thicker chips. The chips used were silicon chips containing a daisy chain structure. Five test lots were assembled with the chips using a commercial ACA film and three different FR-4 substrates. FR-4 was chosen because it is an attractive alternative for making relatively low-cost high-density interconnections. The FR-4 substrates used were of the same size, but their thicknesses varied. In addition, one substrate had a relatively soft resin-coated copper (RCC) layer on it. To study the reliability of the joints a constant humidity test was used. The test temperature was 85 °C and relative humidity was 85 %. The duration

of the test was 2,500 hours. In order, to determine the exact time of failure during testing, the resistance of the test samples was measured using continuous real-time measurement.

Experimental

Three different FR-4 substrates were used in this study. The thickness of these substrates varied and one of them had a resin copper coated foil (RCC) laminated onto it. All substrates used were single-sided. The thickness of substrate S710 was 710 μm. The thickness of the nickel-gold plated copper tracks on the substrate was 19-20 μm. The size of this substrate was 110 mm × 55 mm and it contained 8 sites for 5 mm × 5 mm test chips. Substrate S100 was similar to S710, but its thickness was 100 μm and it contained only one layer of glass fibre cloth. Due to the thinness of this substrate it was relatively flexible. Substrate SRCC was also similar to S710, but it had 60 μm thick RCC laminated onto one side of the substrate. The copper tracks on this substrate were only on the RCC and their thickness was 14-15 μm. Properties of the substrates are shown in Table 1.

Two different test chips were used in this study, both of them silicon chips with a daisy chain interconnection pattern. The size of chip C480 was 5 mm × 5 mm and it had 69 square copper bumps. The size of the bumps was 100 μm × 100 μm and their thickness was 20 μm. The pitch of the peripherally situated bumps was 250 μm. The thickness of the chips was 480 μm. Test chip C80 was similar to chip C480 but it was thinned to a thickness of 80 μm.

Test samples were prepared by attaching the test chips to the substrates using commercially available anisotropic conductive adhesive film (ACF). The thickness the ACF was 40 μm and it contained gold-coated nickel particles of 8 μm in diameter. Some properties of the ACF used are presented in Table 2.

Table 1 Properties of the substrates. The properties given for substrate SRCC are those of the RCC epoxy. The properties of the FR-4 layer of substrate SRCC are similar to those given for substrate S710.

	S710	S100	SRCC
Thickness of FR-4 / μm	710	100	710
Thickness of copper tracks / μm	19-20	19-20	14-15
Thickness of nickel on copper / μm	5	5	1
RCC / μm	No	No	60
T_g / °C	130-140	130-140	150
CTE (below T_g) / ppm/°C (x-y)	12-16	12-16	70
Elastic modulus (Young's modulus) (at 30 C°) / Gpa (x - y)	23.4-24.8	23.4-24.8	3.5
Poisson's ratio (x - y)	0.16-0.14	0.16-0.14	0.38

Table 2 Properties of ACF used.

	ACF
Adhesive type	Epoxy based thermoset
Film thickness / μm	40
Conductive particle, diameter / μm	Au coated nickel, 8
Density of conductive particle / pcs/mm^2	3 000 ± 200
Nonconductive particle, diameter / μm	SiO$_2$, 0.8
Density of nonconductive particle / pcs/mm^2	>10 000
CTE (below T$_g$) / ppm/°C	39
CTE (above T$_g$) / ppm/°C	552
Elastic modulus (at 30 C°)/ Gpa	2.3
Elastic modulus (at 150 C°)/ Gpa	0.047

The chips were assembled using a Toray FC-1000 flip chip bonder. First, the ACF was cut to the correct size to cover the bonding area. The ACF was then aligned to the substrate and prebonded using light pressure and low temperature. The prebonding was performed with the flip chip bonder. After prebonding, the carrier film on the ACF was removed. The final bonding was made with the flip chip bonder. First, a chip was picked by the flip chip bonder. Then the bumps on the chip and the pads on the substrate were aligned. Finally, the chip was pressed onto the substrate and heat was applied to the chip and the substrate. After bonding, the package was cooled to below the glass transition temperature (T$_g$) of the ACF while still under pressure.

Bonding time and temperature were chosen according to the ACF manufacturer's recommendation. The temperature at the bonding tool was 210 °C and 100 °C at the base. The bonding time was 25 s with 15 s additional cooling under pressure. The bonding pressure was selected on the basis of previous studies made with the same ACF and substrates [13]. A bonding pressure of 110 MPa

was used. Six different test lots were assembled using the different substrates and test chips. The test series and numbers of tested samples are presented in Table 3.

The reliability of the test samples was studied using a constant humidity test. The test was performed in a Tabai Espec PL-1KPH test chamber according to the standard EIA/JESD22-A101-B. The test temperature was 85 °C and the relative humidity was 85 %. The duration of the test was 2,500 hours. The test was halted after every 500 hours and the temperature and humidity were lowered to room conditions.

To determine the exact time of a failure during accelerated environmental testing, the resistance of the test samples was measured using continuous real-time measurement with a National Instruments data logger system. A constant, stable current was fed separately through shunt resistors to all channels to be measured. The voltage over the daisy chain structure was measured individually for every channel. Any rise in the measured voltage indicated an increase in the resistance of the measured test sample. As open joints formed on the test samples, the measured voltage rose to the supply voltage. The voltage was measured every 60 seconds during the constant humidity test.

In order to study the structural shear stresses during the constant humidity test, the structures of different test lots were modeled using a finite element (FE) software. The material parameters used in the model for the adhesives and substrates are presented in Table 1 and Table 2. The Poisson's ratio used for both adhesives was 0.3.

Due to the symmetry of the structure, only one quarter of the structure was modeled. The structure of the model was basically the same in the test lots. The model was based on the assumption that the bumps and pads were attached to each other so that sliding and opening between them was not possible. This enabled the calculation of the stresses in the adhesive. The size of the models varied between the test lots and consisted of approximately 14,000 - 17,000 elements and nodes. The contact stresses after bonding in different structures were assumed to be approximately the same and the stress-free condition was chosen to be room temperature, 20°C.

Table 3 Test lots assembled

Test lot	Substrate	Test Chip	Number of samples tested
Test lot 710,480	S710	C480	24
Test lot 710,80	S710	C80	16
Test lot RCC,480	SRCC	C480	16
Test lot RCC,80	SRCC	C80	15
Test lot 100,80	S100	C80	8

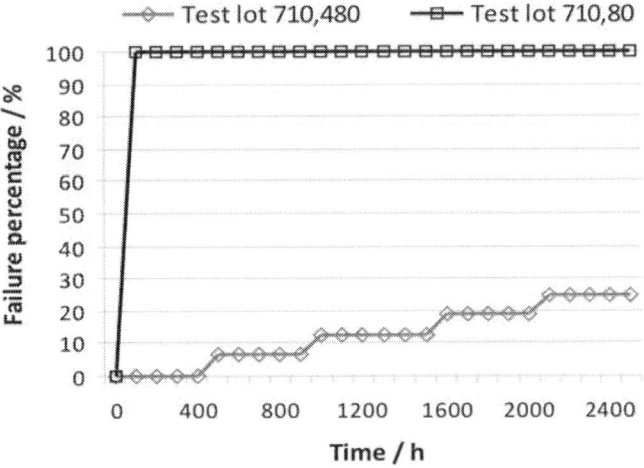

Figure 1Failure percentages for the test lots with thick FR-4 substrate S710.

Results

The reliability of the test samples was studied by measuring their daisy chain resistance in real time during testing. A test sample was considered to have failed when its resistance doubled.

The reliability of the thinned chips when thick substrate S710 was used was very poor. All test samples from this test lot failed within the first 25 hours of testing. On the other hand, only four samples out of sixteen failed from the test lot with the thicker chips. The results of testing for test lot 710,480 and test lot 710,80 are shown in Figure 1. Compared to the reliability of substrate S710, which did not have RCC, the reliability of the test lots with the RCC substrates was markedly better. No failures occurred in either of the test lots assembled with these substrates and their daisy chain resistances remained constant during testing.

The reliability of the test lot with the thinnest substrate and the thinned chips was also good, as only one sample from this test lot failed. This failure occurred within the first 25 hours of testing and, therefore, this failure was may have been caused by problems in the manufacturing process of the chip or in the bonding process. The failure percentages of all test lots after testing are presented in Table 4.

Table 4 Failure percentages for the test lots after 2,500 hours of constant humidity testing.

Test lot	Failure percentage
Test lot 710,480	25 %
Test lot 710,80	100%
Test lot RCC,480	0%
Test lot RCC,80	0%
Test lot 100,80	12,5%

On the basis of the test results, reliability was markedly enhanced by the soft RCC. Similar results were seen when the effect of RCC was studied under thermal cycling conditions [13]. Since the RCC does not contain any fibre reinforcement it is relatively soft. During the bonding process it may deform and thereby decrease the pressure exerted to the chip. The FR-4 used in the substrate also deforms, though this deformation is minor compared to that of the RCC. Additionally, the deformation of the glass fibre-reinforced FR-4 is uneven because it depends on the position of the glass fibres. Uneven deformation may cause stresses to form in the structure and may therefore explain the differences between the test lots. The high bonding pressure needed in the ACA process combined with the uneven distribution may be detrimental for thinned chips since they are fragile and fracture easily. Consequently, for maximum reliability, lower bonding pressure might be needed for the thinned chips than for the thicker chips.

The FE model was used to study the stresses forming into the joints under the test conditions. A separate model was made for each test lot. Additionally, the stresses were calculated for the test conditions separately so that both humidity and temperature were modeled independently. The results for these models are presented in Table 5. The overall stresses in the structures according to the models are also shown.

From the results it can be seen that the stresses change markedly when thinned chips are used. An interesting result is that the stresses are distributed differently in the structures with the thick and the thinned chips. The thinning of the chips seems to increase the stresses, especially between the bump and the chip. This is seen in both test lots with the thicker substrates. In addition, the thickness and the material of the substrate have a considerable effect. The stresses in the structure with

Table 5 Results of the FE model.

	Humidity model		Temperature model		Overall	
Test lot	**Stress between bump and pad/MPa**	**Stress between bump and chip/Mpa**	**Stress between bump and pad/Mpa**	**Stress between bump and chip/MPa**	**Stress between bump and pad/Mpa**	**Stress between bump and chip/MPa**
Test lot 710,480	1,76	2,76	-6,06	3,96	-4,30	6,72
Test lot 710,80	2,38	4,00	-4,00	4,29	-1,62	8,29
Test lot RCC,480	1,21	0,975	5,85	-1,06	7,06	-0,085
Test lot RCC,80	1,12	3,07	4,31	-1,75	5,43	1,32
Test lot 100,80	2,31	3,72	-1,04	2,03	1,27	5,75

RCC are noticeably different from those with the pure FR-4 substrate. The stresses between the bump and the pad are greater with the RCC. However, the stresses between the bump and the chip are considerably smaller. The thinner substrate has a similar stress distribution to the thicker substrate, though the stresses are smaller.

From the results of the constant humidity test, it was seen that the reliability of the test lot 710,80 was very poor. From the models it may be seen that the stresses in this kind of structure between the pad and the bump are the highest of all test lots studied. It has been demonstrated that the shear stress in the structure is critical for the reliability of the ACA joints [14]. Consequently, these high stresses may explain why these test samples failed so early in the test. Differences in the stresses are also seen in the test lots with the RCC layer. However, no failures occurred in these test lots. According to these results it seems that the stress between the chip and the bump is more critical under humid conditions than the stress between the bump and the pad.

The stresses seen in the temperature model are caused by CTE mismatches in the structure. The CTE of the adhesive is considerably greater than that of the chips and the FR-4 material. As the material expands by different amounts, the stresses form to the structure. The CTE of the RCC is relatively near that of the ACF when compared to the CTE of the FR-4. This changes the stresses in the structure and, thereby, causes the changes between the test lots with the different substrates. In addition to the expansion caused by high temperature, under humid conditions water absorption may cause expansion of organic materials. This swelling causes the stresses seen in the humidity model. Furthermore, moisture may change the properties of organic materials and thus cause reliability problems.

In order to determine the kind failure mechanisms these different stresses caused, cross sections were made using test samples from test lot 710,480 and test lot 710,80. The cross sections were studied using SEM. When the test samples from test lot 710,80 were studied, there was no clear indication as to why the test samples failed and the

daisy chain resistance values increased. Delamination between the adhesive and the chip was found in all the test samples studied. However, this delamination did not propagate between the contacts. On the other hand, all of the failed test samples were conducting after testing even though their resistance values had increased. Such failures are often difficult to detect since the changes in the structure may be very small.

The test samples from test lot 710,80 with the thinned chips showed open joints after testing. In these test samples a marked delamination was seen between the pad and the bump. An example of this delamination is shown in Figure 2. The formation of this kind of failure indicates that there are high stresses in the structure. These samples also manifested the highest stresses in the models. In addition to the swelling, moisture absorbed by the adhesive matrix may cause other reliability issues. The moisture may diffuse into the interfaces and thereby decrease adhesion strength [15]. This will facilitate the formation of the delamination. However, no delamination between the pad and the bump was seen in the test samples with the thicker chips. Consequently, even though humidity facilitates the formation of the delamination, the main reasons were most likely the stresses in the structure.

Conclusions

The reliability of thinned chips under humid conditions can be critical in several applications. In this study reliability was investigated by comparing ACA-attached thick and thinned chips in constant humidity testing. In addition to the chip thickness, the effect of the substrate structure and material was studied by using three different substrates.

The reliability of the thinned chips on a thick FR-4 substrate proved to be very poor compared to the thicker chips. Additionally, the reliability of this substrate was poor compared to other substrates. All test samples with thinned chips failed within 50 hours of testing, while 25 % of test samples with thicker chips failed during 2500 hours. Only one failure was seen in the other test lots. Marked

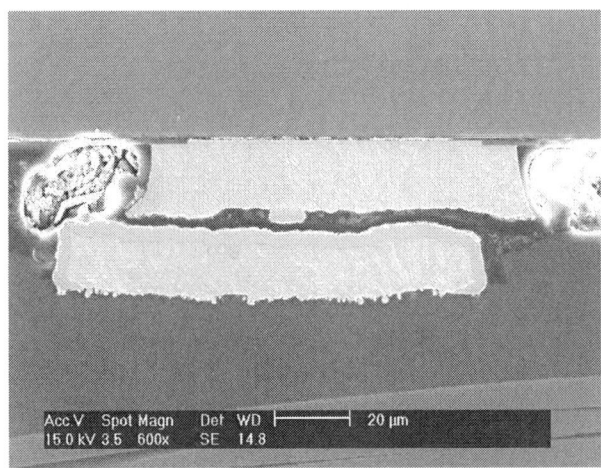

Figure 2 Micrograph showing delamination between pad and bump in test lot 710,80.

delamination was seen in the test samples with thinned chips and thick FR-4 substrate, which indicates high stresses in the structure. These stresses were also seen in an FE model.

According to FE modeling, the stresses in the test lots with a thin FR-4 substrate and a RCC substrate were smaller. Additionally, they were found be in different locations if the substrate had an RCC layer. This different distribution of stress may be a critical factor in reliability. In addition to different pressure distribution, the RCC layer decreases the pressure exerted on the joints and also evens it out. This lower pressure may have been critical to reliability so that the bonding pressure needed for maximum reliability may be lower with the thinned chips than with the thicker chips.

References

[1] J. Lau, "Low Cost Flip Chip Technologies for DCA, WLCSP, and PBGA Assemblies", McGraw-Hill Publishers, New York, 2000.

[2] G. Dou, et al., "The Effect of Co-planarity Variation on Anisotropic Conductive Adhesive Assemblies", Proceedings of the 56th IEEE Electronic Components and Technology Conference, San Diego, CA, USA, May 30 – June 02, pp. 932-8, 2006.

[3] S.-Y. Jang et al., "FCOB (Flip Chip on Board) Reliability Study for Mobile Applications", Proceedings of the 54th IEEE Electronic Components and Technology Conference, Las Vegas, Nevada, USA, June 1-4, pp.62-7, 2004.

[4] Y. Li and C.P. Wong, "Recent Advances of Conductive Adhesives as a Lead-free Alternative in Electronic Packaging: Material, Processing, Reliability and Applications", Materials Science and Engineering R 51, pp.1-35, 2006.

[5] E. Jokinen and E. Ristolainen, "Flip chip joining of thin chips on flexible PEN substrate",

Proceedings of the 34th IMAPS International Symposium on Microelectronics, Baltimore, USA, October 7-11, pp. 600-4, 2001.

[6] H. Jiun et al., "Effect of Wafer Thinning Methods Towards Fracture Strength and Topography of Silicon Die", Microelectronics Reliability, Vol. 46, No. 5-6, pp.836-45, 2006.

[7] M. Feil et al., "The challenge of ultra thin chip assembly", Proceedings of the 54th IEEE Electronic Components & Technology Conference, Las Vegas, Nevada, June 1-4, pp. 35- 40, 2004.

[8] S. Savastiouk et al., "Moore's law, the Next Dimension", Advanced Packaging, September/October, pp.55-8, 1998.

[9] T. Alander et al., "Improving the Fatigue Life of a Bare Die Flip Chip by Thinning", Soldering & Surface Mount Technology, Vol. 15, No. 3, pp. 8-12, 2003.

[10] K. Y. Chen et al., "Ultra-Thin Electronic Device Package", IEEE Transactions on Advanced Packaging, Vol. 23, No. 1, pp. 22-26, 2000.

[11] P. Marjamaki et al., "Modeling Stresses in Ultra-thin Flip Chips", Proceedings of the 4th IEEE Conference on Adhesive Joining and Coating Technology in Electronics Manufacturing, Espoo, June 18-21, pp. 24-7, 2000.

[12] S. Shkarayev et al., "Stress and Reliability Analysis of Electronic Packages With Ultra-Thin Chips", Journal of Electronic Packaging, Vol. 125, No. 1, pp. 98-103, 2003.

[13] L. Frisk and K. Kokko, "Effect of RCC on the reliability of adhesive flip chip joints", ASME Transactions Journal of Electron Packaging Vol.129, No. 3, pp. 260-5, 2007.

[14] W. Kwon, et al., "Thermal Cycling Reliability and Delamination of Anisotropic Conductive

Adhesives Flip Chip on Organic Substrates with Emphasis on the Thermal Deformation", ASME Transactions Journal of Electronic Packaging, Vol. 127, No. 2., pp.86-90, 2005.

[15] S. Luo et al., "Study on Mobility of Water and Polymer Chain in Epoxy and Its Influence on Adhesion", Journal of Applied Polymer Science, Vol. 85, No. 1, pp.1-8, 2002.

The Influence of Copper Trace Routing on the
BGA Lifetime Prediction in JEDEC Drop Tests

Frank Kraemer, Steffen Wiese

Saarland University, Saarbrucken/Germany

f.kraemer@mx.uni-saarland.de

Abstract

Virtual prototyping is able to speed-up the development cycle of new products if based on exact models. In case of dynamic mechanical loads like JEDEC drop tests of BGA modules, broken copper traces at the PCB side are more and more often observed to be the ultimate failure effects. Straightforward FEM simulations showed unrealistic high stress and strain results not matching with the experimentally gained characteristic lifetimes of productive BGA components.

A comprehensive physical investigation of the failure formation during JEDEC drop tests has revealed the complex nature of the dynamic failure formation. In addition to the ultimate fracture of the copper trace, a flaw of the IMC layer on top of the PCB pad is identified by 3-D X-ray tomography. Obviously, both failure mechanism are initiated during the consecutive drop events at different times propagating with different speeds. Both mechanisms are interacting with each other with no single mode alone dominating the failure evolution. Hence, FEM simulations have to account for both failure mechanisms in order to capture the ultimate copper trace stress realistically.

The inclusion of the IMC flaw dramatically increased realism of FEM results. The plastic copper trace strain is reduced by 80 % compared to an ideal interconnection. Now realistic values are calculated by simulations, indicating possible lifetimes as seen in the experiments. The combination of failure modes also affects the copper trace routing effect on the PCB. The new resulting strain map clearly indicates the critical routing direction, which is parallel to the direction of highest PCB deflection and leaving the component body, which is in agreement with experimental observations.

This very realistic 2nd level interconnection model can be applied to assess the lifetime of the investigated packages by combining the routing effect with the interconnection stress. The resulting lifetime model is able to predict the experimental number of cycles-to-failure of all three packages at different PCB positions with less than 25 % deviation. The results presented here set the ground for virtual prototyping, which also includes the BGA drop test endurance.

Key words: JEDEC Drop Tests, FEM Simulations, Lifetime Prediction

Introduction

Virtual lifetime estimations require the precise knowledge of failure mechanism and failure position under all test and service conditions in order to predict reliable numbers of cycles-to-failure. This requires a close interaction between simulation and experiment during the model development and verification.

This paper presents the development of a trustworthy virtual lifetime model based on FEM for mechanical drops of microelectronic components on a PCB. The JEDEC drop test is the basic experimental methodology for this development. The main advantage of this test is its high failure reproducibility. Characteristic cycles-to-failure were obtained by this methodology for each of the 15 components on the well defined JEDEC board. The sequence of failing components at their board position is used to select a proper failure criterion and to evaluate the simulation methodology.

Currently, the accuracy target is set to ±50 % for the prediction based on FEM. Of course this target can only be fulfilled if the dominating failure mode has been identified and transferred adequately to the simulation model.

Experimental setup

The JEDEC standard JESD22-B111 [1] is the best suited drop test experiment, which can be transferred to a FEM-based simulation model. The standard precisely specifies the drop test setup, the specimen and especially the load conditions of a drop test for 2nd level assemblies. This way, highly reproducible results can be achieved with this methodology, which is essential for detailed failure and lifetime analyses by FEM simulations. The schematic of this methodology is shown in fig.1.

In this study, JEDEC drop condition B was used, i.e. the 0.5 ms sinusoidal pulse had a peak acceleration of 1500 G on top of the sledge. This acceleration impulse was continuously measured proofing its conformity during all tests. In contrast to the standard the JEDEC PCB was fixed by 6-screws

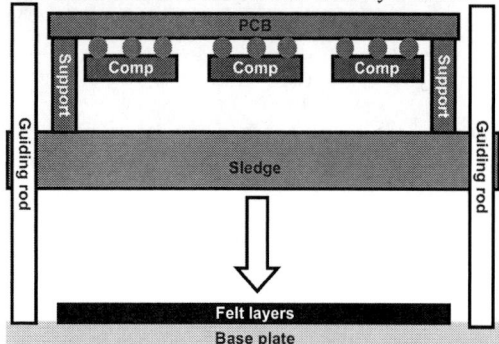

Figure 1: Schematic of the JEDEC drop test setup

with additional screws at the middle of both long board edges. Due to these modified boundary conditions the eigenmodes and eigenfrequencies of the PCB were changed, resulting in different stress and failure distribution of all 15 components on the board. Applying this setup, electrical failures of the daisy-chained packages appeared after some 10 to a few 100 drops depending on the packages investigated.

This investigation included 3 different packages, which are listed in tab. 1. These daisy-chained packages were designated for memory applications, resulting in a ball arrangement of two groups with 3 rows each around a central bond channel. The solder applied here was SnAg1Cu0.5. The SMD copper pads of the components had electro-less nickel/gold finish, whilst the N-SMD PCB pads were covered with Cu-OSP.

Table 1: Geometrical data of the investigated daisy-chain packages

Package type	VFBGA-90	VFBGA-60	TFBGA-60
Height [mm]	0.8	0.8	1.2
Package area [mm]	12.5 x 9.5	10.0 x 9.5	10.5 x 8.0
Number of balls	90	60	60
Ball-out size [mm]	11.2 x 6.4	7.2 x 6.4	8.0 x 6.4
Ball arrangement	15 x 6, full rows	10 x 6, full rows	11 x 6, partly populated

Failure detection was done by single in-situ resistance measurements of all 15 components separately. The measurements were done continuously throughout the test. As defined by the JEDEC standard [1], the failure flag is set when the

event detector captured 1 kΩ or more in three consecutive drop cycles.

Experimental results and failure analysis

Cycles-to-failure were recorded for the 3, or more, most critical package positions on every JEDEC board. These data were analyzed by Weibull distributions, which were generated separately for each critical component position and each package type. The outcomes are characteristic numbers of cycles-to-failure, N_{63}, for all packages at their most critical positions on the JEDEC board. Applying these data, the drop endurance of the investigated packages can be derived for their individual stress level depending on the PCB position.

Qualitatively, the sequence of failing component positions can be derived from the statistical analysis. This sequence of failure is applied to assess the accuracy of different simulation result criteria. Simulation criteria which can not follow the experimental sequence of failure are not suited for quantitative lifetime estimations. Both statistical results are shown in tab. 2.

The results of tab. 2 show that the most critical component positions are situated in the central row of the JEDEC test board see fig. 7. This is caused by the deformation behavior of the PCB under 6-screws boundary condition, which is dominated by a bending along the short edge. The central position #8 is the most critical, followed by neighbor positions. The package size has a strong influence on the characteristic number of cycles-to-failure. The bigger the package the smaller is its drop endurance. Due to this fact the biggest package VFBGA-90 fails earlier and at more component positions than the smaller packages with 60 I/Os.

Table 2: Characteristic lifetime, N63, and sequence of failure of the most critical component positions for all investigated package types

Package type	VFBGA-90	VFBGA-60	TFBGA-60
Comp 06	89 (4)	-	-
Comp 07	94 (5)	291 (3)	161 (3)
Comp 08	43 (1)	90 (1)	68 (1)
Comp 09	58 (2)	141 (2)	113 (2)
Comp 10	70 (3)	-	-

Physical failure analyses of the drop tested samples are done by dye & pry tests. This method is very effective and allows for the analysis of every failed package. As an example, figure 2 shows the result of a TFBGA-60 package. The red colored sections are damaged during the experiment, while the other sections are damaged during the lift-off of the component. The analyses reveal that electrical failures are dominantly caused by broken copper traces at the PCB pads. More detailed analyses show that these cracks do only appear when the copper traces follow the direction of highest PCB bending.

As shown in figure 2, electrical failures do not only appear at the corner joints of a package, which are the most stressed under drop test conditions. In this case the copper trace at the second joint follows the direction of highest PCB deflection and thus causes the failure.

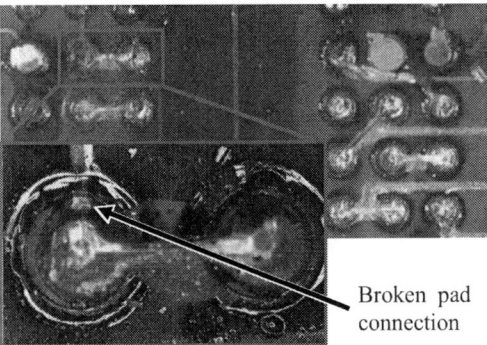

Figure 2: Failure analysis with dye & pry test. Broken copper traces at the PCB pads cause the electrical failures

Further failure analyses are done by 3-D X-ray tomography. A 3-D X-ray scan allows a deeper analysis of multiple failure mechanisms, which can not be found by the dye & pry test. The procedure, however, is more time consuming due to the scanning procedure and thus it is not applicable to inspect every failed component of a batch.

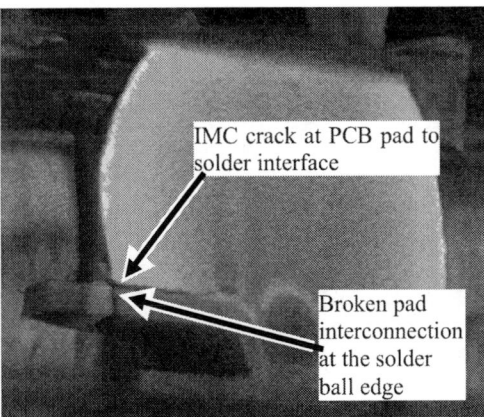

Figure 3: 3-D X-ray tomography analysis of a BGA interconnection

The 3-D X-ray scan surprisingly revealed the complex nature of the failure formation in the investigated samples. In addition to the broken copper trace an IMC flaw at the interface between PCB pad and solder ball was detected, which is shown in fig. 3. Both failures seem to be initiated during the drop event at different times and propagate with different speeds. While the IMC crack is almost stopped probably due to reduced stress at the center of the solder joint, the copper

trace is further damaged with every drop event due to PCB bending and causes the ultimate failure.

The precise knowledge of the failure modes and the failure position under all test conditions is absolutely required to develop precise lifetime models based on FEM simulations. The failure modes detected by the X-ray tomography are mutually interacting with each other with no single mode alone dominating the failure evolution. Hence, FE models need to account for both failure modes in order to capture the copper trace stress realistically.

Global FE simulations of the JEDEC drop test

FE simulations are executed with the explicit code LS-DYNA. In this code the transient shock response of the investigated structure is calculated by small time steps. The size of these time steps is dominantly influenced by the smallest element size of the entire model. In order to achieve a reasonable computing time for a drop event the coarse meshed global model is simulated first followed by a finer meshed sub-model.

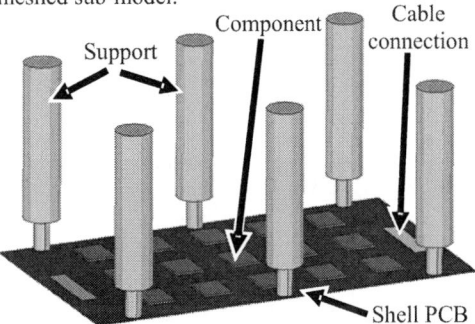

Figure 4: FEM global model of the JEDEC drop test assembly

The global model is a complete 3-D representation of the entire JEDEC test board including all 15 components. As shown in fig. 4, the model also accounts for the cable connection areas on the PCB and the supports. The steal supports transfer the acceleration impulse of the drop event to the PCB. Comparison between experimentally recorded and simulated acceleration curves has shown that the supports do not act absolutely rigid and thus influence the PCB motion [2]. A similar effect is detected for the cable connection areas which have an additional damping effect on the PCB.

The packages are modeled in a rather simple way. The hexahedral component bodies apply effective elastic material properties similar to those of mold compound. The solder joints are modeled as bricks without any copper pads within. The solder bricks apply a rate dependent plastic material model [3]. Only the most critical joints at the outermost rows are modeled as barrels replicating the actual dimensions at both pads as well as at the equator of the joint. The barrels are meshed by 24 elements, while 8 elements are sufficient for the solder bricks.

These components are connected to the shell PCB by offset contacts. Shell elements are required for the PCB in order to apply the material model called composite matrix [4], which is able to capture the resulting PCB properties that are calculated by the laminate theory [5]. This material model is able to capture both tensile and bending stiffness of the PCB, which are strongly differing from each other.

The assumptions presented above are found adequate for the global model requirements. Its main task is to record the displacement data, which is the load input for the sub-model. Basic stress analysis is done with the solder barrels in order to quantify the load levels at the individual component positions.

Sub-model simulations

The sub-models applied here are detailed representations of a single BGA component. The sub-models cover all mechanically relevant parts as shown in fig. 5. The component body includes the substrate, substrate solder mask, die attach film, die and the mold compound. The solder joints are modeled by the substrate pads, the solder ball, the PCB pads and the PCB copper traces. The PCB pads are non-solder-mask defined leaving the pads and a little bit of its trace uncovered. The copper traces may approach the PCB pad in any 45° direction, as shown in fig. 6. The influence of the routing on the copper trace stress can be investigated with this option. The IMC flaw, which has been detected by the X-ray scans, is also represented in the sub-

model. According to the results of the failure analysis twin nodes are generated at the pad-solder-interface in order to create independent elements. The flaw is covering half of the PCB pad starting from the copper trace. Penetration of the adjacent elements is prevented by contact definitions.

The PCB is meshed by shell elements again while all remaining parts of this model are meshed by solid elements. The bottom of the PCB pads and the solder mask are connected to the PCB by offset contacts. Potential sub-model edge effects are minimized by a sufficient PCB size, which is about twice the length and the width of the component. According to the stress distribution, the outermost joints have fine meshes while a more coarse mesh is sufficient for the inner low stressed interconnections.

The material models of the parts within the component body are linear elastic. No large deformations and no or negligible non-linear material behavior is expected in these parts. Again, the PCB material is modeled by the direct composite matrix. Similarly, the solder applies the material behavior presented for the global model. The copper traces and pads use a plastic material model with isotropic hardening [4] following the coefficients provided by [6].

A realistic copper trace strain map

The detailed stress analysis of the copper traces is executed in case of the TFBGA-60 component at the most critical PCB position #8, see fig. 7. This component is chosen due to its distinct experimental failure distribution. The copper trace routing effect is evaluated by eight sub-models with changing routings at each of the outermost PCB pads. It is assumed that the routings of the outermost joints do not affect each other.

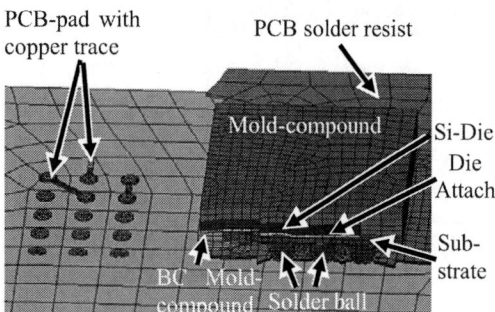

Figure 5: X-section of a TFBGA-60 sub-model with its characteristic constituents

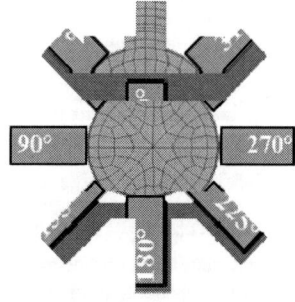

Figure 6: Copper trace orientations at the sub-model copper pads for the investigation of the routing effects

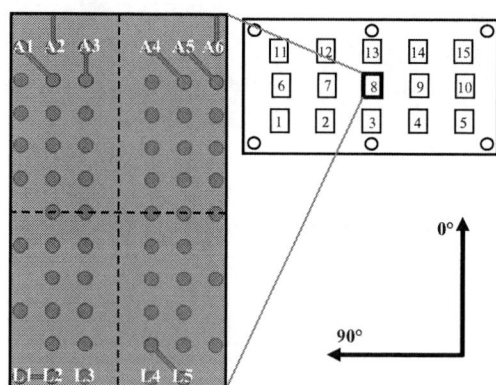

Figure 7: Sub-model orientation on the JEDEC board and experimental/initial PCB pad design of the TFGBA-60

Different result criteria may be applied to assess the mechanical loading of the copper traces. The plastic strain accumulated in the copper trace between the pad edge and the solder mask was chosen due to the results of earlier work [2]. As

152

detected in the failure analysis, maximum copper strain appears at the edge of the solder ball.

The copper trace routing effect is shown for interconnections A1 and L1 in fig. 8. As expected, there is a strong influence of the routing direction on the resulting plastic strain. The maximum plastic strain appears in 0° and 180° direction for joints A1 and L1, respectively. A plastic strain of about 3 % is created in these worst configurations. This is significantly lower than the maximum plastic strain of up to 15%, which was reported in [7] without the consideration of an IMC flaw. Thus the IMC flaw is an important failure mode, which necessarily has to be considered in those simulations in order to yield in a realistic copper trace stress. Consecutive simulations of a single sub-model have shown that this initial amount of plastic strain per drop is even reduced by further 90 % within the first four drops remaining in this stable growth rate. This way the strain increment per drop seems to be reasonably low to pass about 70 drops until ultimate failure.

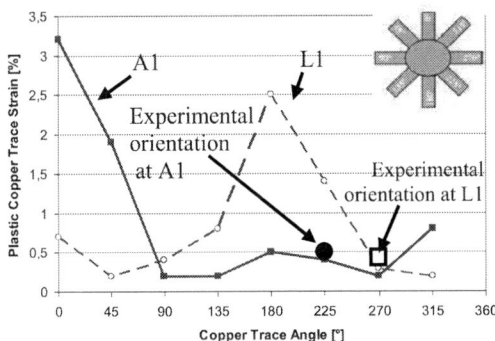

Figure 8: Copper trace routing effect at two interconnections on the component edges

The strain curves of A1 and L1 are very similar. They are simply shifted by 180°. Highest strain appears when the copper traces are routed parallel to the PCB width leaving the component, which fits to the experimental observations. The plastic strain clearly declines the further the traces are routed differently. The most save option is a routing towards the component center.

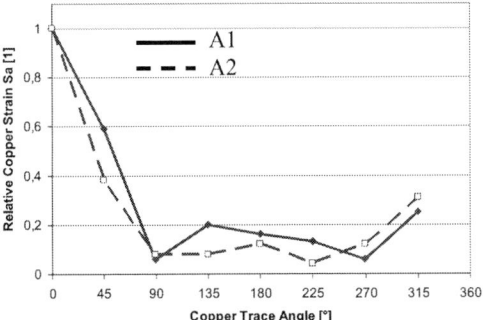

Figure 9: Relative copper trace routing effect of two adjacent interconnections

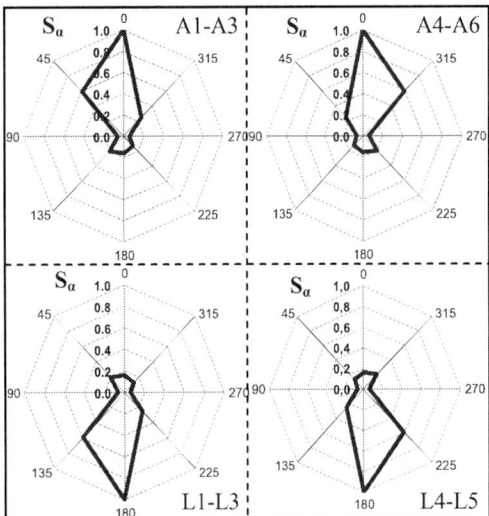

Figure 10: Strain map for 6-screws boundary condition of the Jedec PCB

The relative distribution of the copper trace routing effect is shown in figure 9. This relative strain distribution S_α is simply calculated by setting the plastic strain at any trace orientation in relation to the maximum plastic strain at this interconnection (1),

$$S_\alpha = \frac{\varepsilon_\alpha}{\varepsilon_{max}} \qquad (1).$$

The comparison of the outermost joint A1 with its adjacent joint A2 shows only little differences of S_α between both interconnections. Thus both distributions can be combined into a single group for the upper left component quarter (A1, A2, A3), fig. 7. The relative distribution proofes that any other routings except for the 0°-direction cause less than 40 % of the maximum plastic strain. All other routings are supposed to fail much later than those following the critical 0°-direction.

The combined relative strain distributions for all edges of a package are shown in fig. 10. As stated during the introduction of S_α the relative distribution can be similarly found at every outermost interconnection. This way a symmetrical strain map is derived which is able to show the potential of a failing copper trace routing on the entire ball-out of the component.

The strain map introduced in fig. 10 is applicable to all component positions of the middle row on the JEDEC board (position #6 to #10), since the stress distribution is similar for all these packages if the PCB is supported in a 6-screws configuration. Furthermore, this strain map can be applied for similar package types failing with broken copper traces. Now an appropriate assessment of a failing interconnection can be done with a combination of the global- and the sub-model results.

Lifetime model for JEDEC drop tests

The most important essence of a FEM-based lifetime model is a solid and reliable result criterion. As shown in previous studies [2, 7, 8] SEND (2) has proven to fulfill these requirements completely. High stress components can be identified precisely with this criterion on the entire JEDEC board. However this criterion is not able to identify the failing solder joint, which triggers the point of electrical failure in the experiments.

$$SEND = \int \dot{\varepsilon}_{pl} \cdot \sigma_{Pad} dt \qquad (2)$$

The combination of this basic global failure criterion $SEND_{bas}$ and the sub-model copper strain map S_α should be able to overcome its limitations and identify the failing solder joint correctly, equation (3),

$$SEND_{corr} = S_\alpha \cdot SEND_{bas} \qquad (3).$$

The validity of this equation is proven in case of the TFBGA-60 package. The results of both failure criteria are shown in tab. 3.

The uncorrected energy values of $SEND_{bas}$ predict the first failure to appear at interconnection row L. Based on these values interconnection L5 should fail first and trigger the component failure. But experimental failures never appeared at this interconnection. The consideration of the copper trace routing effect is able to outbalance these experimentally uncritical interconnections and highlight the failing joints A2 and A6. Due to the correction of the energies applying S_α, the values at all other interconnections are much lower. Joints with good-natured copper trace orientations are not suspected to fail at all. The new strain map resulting from the consideration of the IMC flaw highlights the critical interconnections even stronger, since the influence of routing is much higher in these conditions.

Table 3: Corrected and basic values of SEND at critical interconnections of a TFBGA-60 package at board position #8

Intercon-nection (fig. 7)	$SEND_{bas}$ [mJ/mm^3] eq. (2)	$SEND_{corr}$ [mJ/mm^3] (S_α) eq. (3)	Experimental failure analysis
A1	26.6	3.5 (0.13)	Pass
A2	22.6	**22.6 (1.00)**	**Fail**
A3	23.7	3.8 (0.16)	Pass
A4	23.9	4.8 (0.20)	Pass
A5	23.3	4.7 (0.20)	Pass
A6	28.4	**28.4 (1.00)**	**Fail**
L1	36.7	2.2 (0.06)	Pass
L2	30.5	1.8 (0.06)	Pass
L5	42.3	5.5 (0.13)	Pass

The failure energies gained with $SEND_{corr}$ can be applied for lifetime modeling. Typically an inverse power law is the basis for such a lifetime model. All available experimental and simulation

data of the three investigated packages were assembled for this model. The resulting simulated cycles-to-failure are compared with the characteristic N_{63}-values as shown in fig. 11.

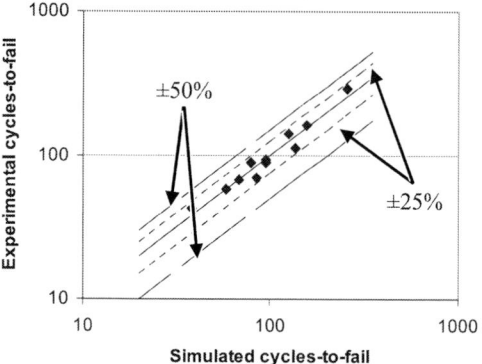

Figure 11: Results of the FEM-based lifetime model compared to experimental results

The comparison shows a close agreement between both curves. The difference between prediction and experimental results is less than ±25 %. This variation is well within the experimental scatter proving the sufficient accuracy of the combined lifetime criterion $SEND_{corr}$.

Conslusion

This paper presents the steps to the development of an appropriate virtual lifetime model for the JEDEC drop test. For this purpose, experiments with industrially fabricated specimens are executed applying the JEDEC drop condition B. Three memory packages are tested with at least six JEDEC boards in order to determine the characteristic number of cycles-to-failure at several component positions on the board for each of these package types.

The detailed failure analysis of all failed components revealed the complex nature of failure development. Copper trace cracks are the dominating failure mode triggering the electrical cuts of the daisy chains. More detailed analysis shows that these copper traces have a high risk to fail if they are routed along the direction of highest PCB deflection. This way, failures do not only occure at the most stressed corner solder balls. Additional analyses by X-ray tomography show that IMC flaws at the solder-pad interface are also generated in these experiments. None of the mechanisms is dominating the failure evolution alone but they are mutually interacting.

The new insights of the failure analyses are transferred to numerical simulations. The IMC flaw is represented at the interconnections of a detailed sub-model. The IMC flaw clearly reduces the stress on the copper traces. The maximum plastic strain reaches up to 3 % per drop if the copper trace is routed in the worst direction. This value is well

154

within the expected amount of plastic strain based on the characteristic number of 68 cycles-to-failure.

A detailed investigation of the copper trace routing effect is done with the sub-model. The resulting strain map highlights interconnections with a bad copper trace routing much stronger than in the case of disregarding the IMC flaw. Copper traces routed in other directions than the critical one see 40% or less of plastic strain as compared to the critical direction.

The final lifetime model applys a combination of the global energy criterion SEND and the copper trace routing. Critical interconnections, triggering the electrical failure during the experiments, are precisely indentified at each package. Having the correct energies of each failing interconnection at hand, the expected lifetime can be assessed very close to the characteristic number of cycles-to-failure. The inaccuracy between experimental and simulated lifetime is less than ± 25 %. This is well within the experimental scatter and thus well within the initial target.

References

[1] JEDEC Standard JESD22-B111, "Board Level Drop Test Method of Components for Handheld Electronic Products", July 2003.

[2] F. Kraemer, et.al. "Lifetime Modeling for Jedec Drop Tests", Proceedings of the 2009 International Conference on Thermal, Mechanical and Multi-Physics Simulation and Experiments in Microelectronics and Microsystems (EuroSimE), Delft, The Netherlands, April 27-29, pp. 309-317, 2009.

[3] K. Meier, et.al. "Mechanical Behaviour of Typical Lead-Free Solders at High Strain Rate Conditions", Proceedings of the 2010 Electronics Packaging Technology Conference (EPTC), Singapore, December 8-10, pp. 825-831, 2010.

[4] Livermore Software Technology Corporation, "LS-DYNA Keyword User's Manual", May 2007, Version 971, www.lstc.com.

[5] F. Kraemer, et.al. "A Multilayer PCB Material Modeling Approach Based on Laminate Theory", Proceedings of the 2008 EuroSimE, Freiburg, Germany, April 21-23, pp. 234-243, 2008.

[6] S. .Wiese, et.al. "Constitutive Behavior of Copper Ribbons used in Solar Cell Assembly Process", Proceedings of the 2009 EuroSimE, Delft, The Netherlands, April 27-29, pp. 44-51, 2009.

[7] F. Kraemer, et.al. "The Effect of Copper Trace Routing on the Drop Test Reliability of BGA Modules", Proceedings of 2010 Electronic Components and Technology Conference (ECTC), Las Vegas, Nevada, June 1-4, pp. 1217-1225, 2010.

[8] F. Kraemer, et.al. "BGA Lifetime Prediction in Jedec Drop Tests Accounting for Copper Trace Routing Effects", Proceedings of the 2010 EuroSimE, Bordeaux, France, April 26-28, p. 1-8, 2010.

Electronic Packaging for Bio-Diagnostic Microfluidics Application

U. Mastromatteo, A. Alzati, R. Brioschi, S. Conoci, M. De Fazio,
P. Magni, A. Maierna, M. Marchi, M. Palmieri, M. Shaw, M. Suardi, F. Ziglioli

ST Microelectronics – Agrate Brianza (MB) – Italy
Amadeo.Maierna@st.com

Abstract

Silicon based semiconductors techniques are creating a solid base for a new category of micro electro mechanical systems (MEMS) with reliable structures for microfluidic funtions, handling either electrical signals or mechanical/physical parameters. A particular category of these early developed devices is constituted by Bio-MEMS, providing electro-microfluidic features in order to integrate in one system all the functional elements needed for a complete bio-assay response. New packaging architectures are then coming much more a reality and in particular some simple concept-structures for DNA analysis on a single silicon device are involved in this bio-tech environment. Some notes on core biological characterizations of PCR and Matrix-Spot detection structures are also present in this paper.

Introduction

The most common definition of Micro-Fluidics physical range is related to the study of structures that do not permit fluid dynamics description with conventional theories of fluids, due to the small dimensions of the involved elements. This limit is conventionally intended for channels of a minimum diameter from 5 to 100 μm, for reservoirs in the range from 10 picolitres to 10 microlitres (1). These structures present dimensional ranges usually involved in silicon-based MEMS dedicated to handle liquids, which flows are controlled by the application of electrical signals to the IC electronic elements, like Power MOS or simple power-resistors. Typically a generic microfluidic chip is characterized by very small dimensions, so an external fluidic system has to be designed with the function of facilitate the "fluidic I/O's" handling, in terms of filling, transfer and washing of liquid samples and reagents. Some different techniques can be used in order to realise the microfluidic package of silicon devices, but only few are considered applicable for a good cost-effective result for an under development product.

Bio-Mems Packaging Architecture

The main purposes of an electro-fluidic packaging architecture is to provide a reliable and viable system for electrical connections of the electronic IC, facilitate the fluidic interconnection with the surrounding lab environment and prevent the functional device from handling damage. In the bio-MEMS field these two features are not sufficient and there is a third key aspect to face: the bio-protection of the analysis micro system from the external environment, in order to prevent cross-contamination by different source of biological samples. For DNA handling this constitutes one of the main challenges for package designers and assembly process engineers. Surfaces of the package elements dedicated to fluid transfer have to face the micromachining from 50 to 200 μm channels depth. Typically, at cartridge level, the filling volumes for samples are in the range of 0.5 – 5.0 microliters, therefore the creation of "dead volumes" is strictly to be avoided. Inlet holes have to match the pipette shapes with a exactly, with the goal to avoid openings and fillets smaller than approx of 1-3 microns, furthermore DNA handling all the materials involved in chambers dedicated to PCR (Polymerase Chain Reaction) or "DNA replication" needs to be made by means of materials with high PCR biocompatibility (4). The basic structure we approached for DNA analysis is shown in the following drawing:

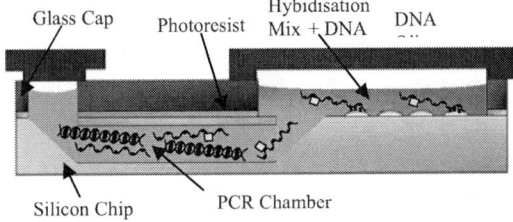

Fig. 1 Schematic Lab on chip

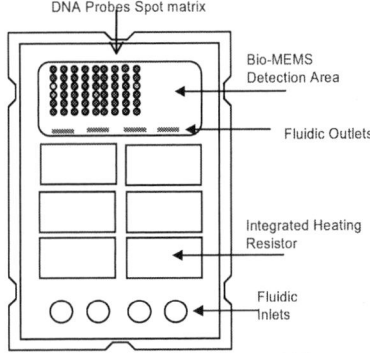

Fig. 2 Lab on chip Silicon Chip

All the operations, starting from samples filling, are usually done manually. Sealing of the inlets, outlets and other processing volumes are provided by clamping elements or by cartridge-level solutions. In the above image is shown the silicon detection area and readings probe layout of the lab-on-chip device involved in these considerations. The silicon device includes an IC with very simple semiconductor technology; this will ensure reliability and low-cost process for the MEMS on high production volumes.

Micro-Fluidic Channel Dimensioning

The overall structure of the fluidic package is strongly driven by the geometrical dimensions of the involved structures, like tanks, reservoirs, channels and fluidic I/O. In the following drawing we can identify the inlet filling pressure P_{in}, and the outlet pressure P_{out} or the input value into the DNA probe detection area, both parameters are important in order to define the fluid dynamics of the system:

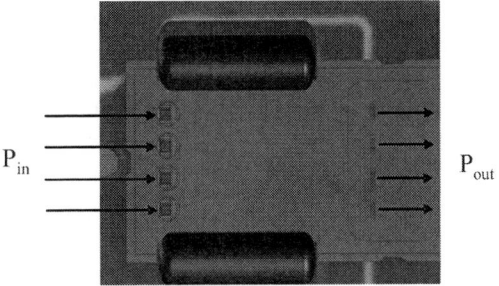

Fig. 3 Flow through Chambers

Inside the microfluidic package four single inlets will be reached by different pressure flow. In order to evaluate P_{out} considering channels with a defined section we shall have to combine only few equations but with a complex structure like the one shown in Fig. 4b, large numbers of equations are involved, so the easiest way for design is the simulation approach with CFD techniques, applied to filling channel system shown.

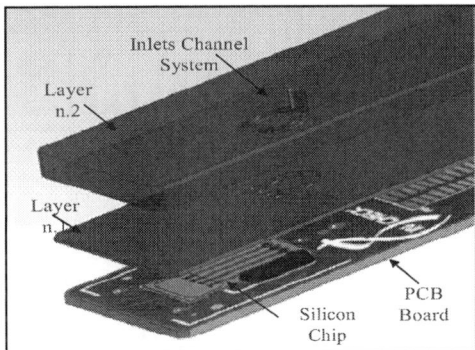

Fig. 4a Cartridge schematic

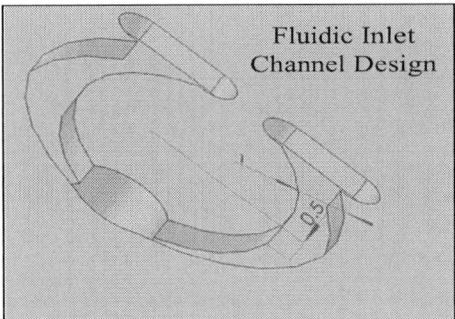

Fig. 4b Fluid inlet design

In order to improve the performance of the package, the fluidic loading flow due to the sample and reagent insertion into the cartridge have to be facilitated and it is important to recognize the key role of the contact angle of the liquid on the walls of the fluidic channel. For lab-on-chip application, involving a procedure for manual insertion of liquids inside the package, we consider only low-velocity models, since the pressures at inlets are substantially controlled by the pipette system, tunable in the range of microliters.

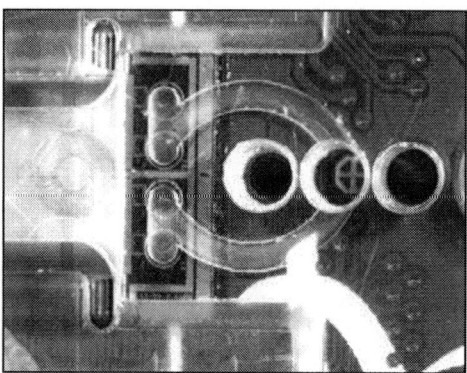

Fig. 5 Fluid Inlet

Also the parameter for contact angle referred to aqueous compounds is an important factor in order to have a good evaluation of filling phenomena. By example, for silicon and glass (4), the reference contact angle can be measured, and considered quite constant from 1 up to 40 PCR cycles (standard samples of 25 µl). In following Fig. 6a is the fluidic simulation for untreated polycarbonate, and in Fig. 6b for the same material after O2 plasma treatment, considering the difference introduced in terms of contact angle.

Table 1 Material Wetting angle

Material	Untreated	With surface treatment
Slicon	$61° \pm 7.7°$	$117° \pm 5.2°$
Glass	$49° \pm 8.1°$	$120° \pm 6.4°$
Polycarbonate Makrolon®	$90° \pm 9.0°$	$25°-30°$
PDMS	$110° \pm 10°$	$28°-30°$ (*)

(*) evaluated with O2 plasma

Inlet Filling Simulations

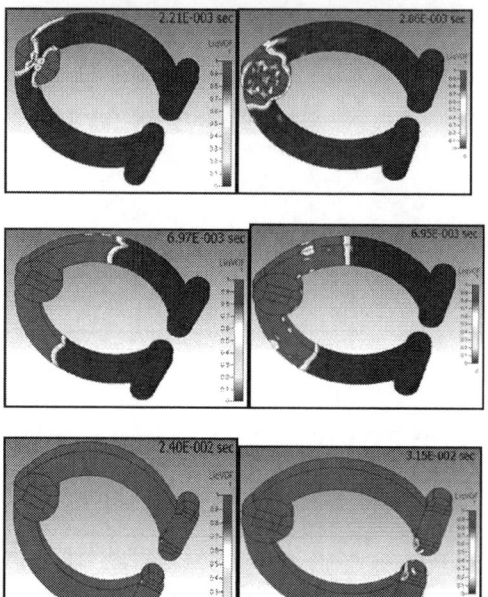

Fig. 6a   Fig. 6b

Contact Angle = 110°   Contact Angle = 30°

Results of fluidic simulations shows that the filling inside inlets channel with plasma-treated surfaces could improve by about 30% the filling time, but with manual handling of the fluids this will not constitute a real advantage, so the plasma cleaning from this point of view should be avoided, simplifying in this way the production flow assembly steps.

From the thermal management point of view the structure of PCR chamber built by means of glass structure (shown in fig. 1) is to be intended in the polycarbonate package (as shown in fig. 4a). The power-on transition for T(set) at 94°C, can be also simulated with different P(diss) on the heaters, obtaining the following transient family. The graphs shows a transient completion in about 6.5 sec to a value of T(set) ± 1°C.

158

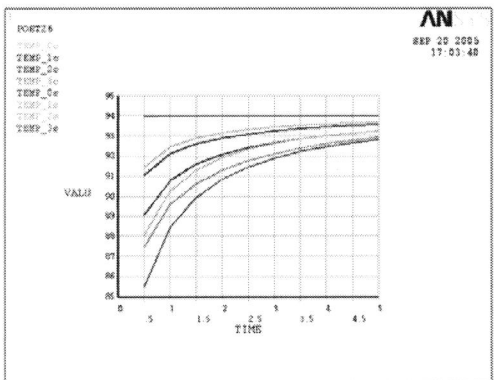

Graph 1 Temperature Transient

Further experimental data confirmed that the temperature performance of this system in the range 60°C – 95°C is effectively operating a control inside T(set) 0.5°C, in line with the performance of a good PCR chamber heating system. This good Temperature range control constitutes one of the main advantage of the silicon based PCR chamber, built starting from MEMS techniques.

Cartridge Structure for DNA-PCR Fluid Handling

The Polymerase Chain Reaction (PCR) is a thermal-induced bioprocess for the DNA denaturation and amplification. Many research groups and MEMS manufacturer worked around the improvements of the performance of PCR chambers in terms of temperature stability and volume uniformity. The approach used for the concept device package under development was designed as in the following draft, showing a concept fluidic package with 2 inlet on PCR chamber and 2 inlets for reagents loading.

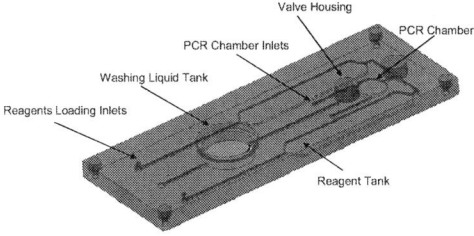

Fig. 7 Cartridge design

For this concept-package the silicon assembly solution is the same as a simple chip-on-board assembly, with the only request of good thermal contact from silicon to the PCR polycarbonate chamber, (fig8), involving the possibility to include a flexible printed circuit.

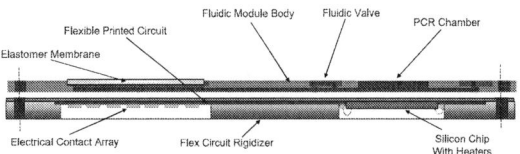

Fig. 8 section cartridges design

As previously mentioned one of the main aspect for fluidic handling inside electronic packages is constituted by the dynamics of the inlet filling. For the system with 2 inlets for reagents we had then to face the 2-liquids mixing. Approaching once more with CFD simulation, some evaluations have been done (Fig 9a,b,c), describing also the dead-volumes evolution during the reagents loading inside the system. The main fluidic handling phases are as follows:

1. Starting T_0 condition: PCR Chamber is fully loaded (Fig 9a light grey volume)

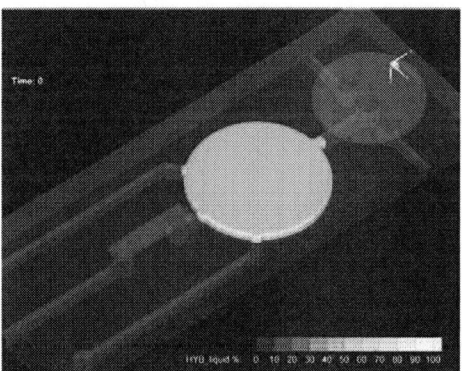

Fig 9a

2. By washing liquid insertion into dedicated inlet, start the liquid flux for PCR Chamber cleaning (Fig 9b dark grey)

159

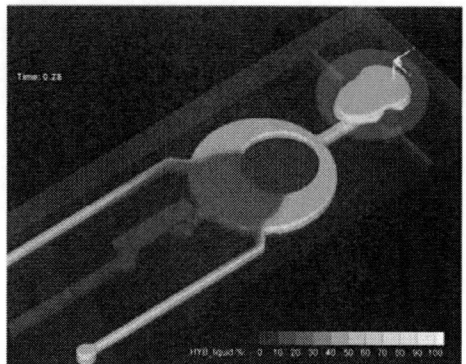

Fig 9b

3. Fluidics transition complete after 600 msec. with standard pipette pressure simulation, residual liquids presence at chamber edge result still present fig 9c

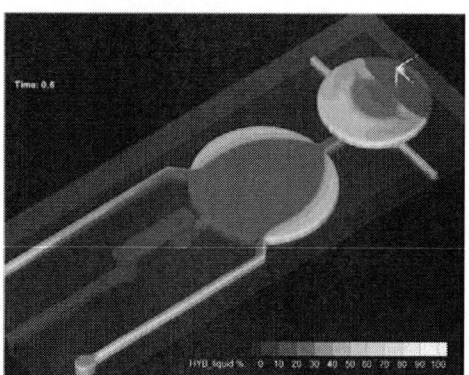

Fig 9c

After the above simulations and design considerations, the dimensioning of the mixing chamber can be considered correct in order to have a good mixing in the time order of about 3 seconds.

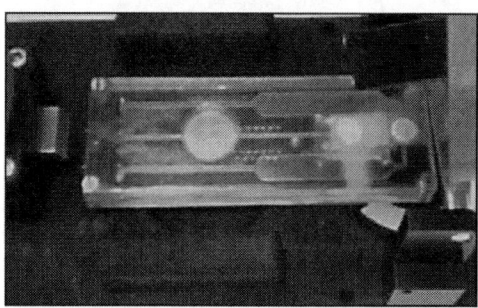

Fig. 9d complete assembly of the fluidic concept-cartridge and housing on thermo-cycles equipment.

Assembly Process

From the manufacturing point of view the introduction of particular surface preparation process and the DNA matrix spotting on the dedicated detection area of the chip constitutes a real difference from the conventional packaging procedure for microelectronics system:

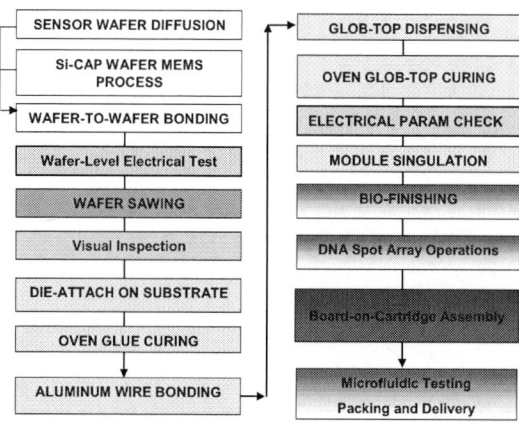

Fig 10 Assembly Flow

Biological Characterisation

The β-globin gene from human genomic DNA was used as target of this model. Both oligonucleotide primers (BgloF and BgloR) and probes were designed with the Oligo primer analysis software, version 6.65, using the β-globin sequence (Accession no. AY260740). The Microarray Layout employed for the present study is reported in fig. 11.

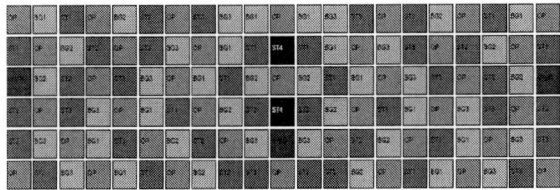

LEGEND

● Empty Position
● Specific Probe
● Hybridization Negative Control Probe
● Orientation Probe
● Hybridization Control Probe

Fig. 11. Microarray Layout on ST's Lab-on-Chip

PCR mixture was prepared in 50ul of volume containing the following concentration of DNA template:

1) Experiment a: 0.25 ng/ul (~$2x10^3$ copies/rx).
2) Experiment b: 0.025 ng/ul (~$2x10^2$ copies/rx).
3) Experiment c: 0.0025 ng/ul (~$2x10^1$ copies/rx).
4) Experiment d: 0.00025 ng/ul (~2 copies/rx).

The **PCR thermal cycling** consisted in:

- Step initial: 5 min 95°C
- Cycle Program (35 cycles): 20 sec 95°C, 45 sec 61°C, 30 sec 72°C

Table 2

Microarray Probe	Function	Results LoC +
AT683	Hyb Control Probe	2/3 probe F635 Median> 50.000 a.u.
AT 730		
AT776		
Bglo1	Specific Probe	(F635 Median$_{Bglo1}$ - F635 Median$_{EMPTY}$) > 1028 (*Cut-off*)
Bglo2		(F635 Median$_{Bglo2}$ - F635 Median$_{EMPTY}$) > 1028 (*Cut-off*)
Bglo3		(F635 Median$_{Bglo3}$ - F635 Median$_{EMPTY}$) > 1028 (*Cut-off*)

The hybridization master mix was prepared in 30 ul of total volume. The Hybridization Step was carried out at 55°C for 3600sec. The washing solution was prepared in 500ml containing 0.1X SSC and 0.1% of SDS..

After the PCR- hyb experiment the chip was read at the optical reader at 250 ms and analysed. For each probe have been calculated the following parameters:

F635 Median: Foreground Median F_{Med} → the median of pixel intensity of fluorescence signal;
B635: Background Mean B_M → the mean of pixel intensity of local background;

Distribution plot and box plot were used for graphical output.
The interpretation of results for a positive test of a qualitative (yes/no) detection of the betaglobine target gene was done according to the rules reported in the table 1.

The results obtained for a Limit of Detection of In-Check system through PCR-Hybridization integrated test by using as DNA target a fragment of Betaglobin gene are below reported

Experiment A: [Input DNA]: 0.25ng/ul (estimated 2000 copies/rx)

Table 3

Probe	Parameter	Value Found (a.u.)	Value Expected (a.u.)	Result
AT683	F635 Median	56384	2/3 probe F635 Median >50.000 a.u.	*passed*
AT730	F635 Median	61524		*passed*
AT776	F635 Median	51579		*passed*
Bglo1	F635 Median$_{Bglo1}$ - F635 Median$_{EMPTY}$	17681	>1028	*passed*
Bglo2	F635 Median$_{Bglo2}$ - F635 Median$_{EMPTY}$	11499	>1028	*passed*
Bglo3	F635 Median$_{Bglo3}$ - F635 Median$_{EMPTY}$	41073	>1028	*passed*

Final Result: *Positive Test*

Experiment B: [Input DNA]: 0.025ng/ul (estimated 200 copies/rx)

Table 4

Probe	Parameter	Value Found (a.u.)	Value Expected (a.u)	Result
AT683	F635 Median	59764	2/3 probe F635 Median >50.000 a.u.	*passed*
AT730	F635 Median	57267		*passed*
AT776	F635 Median	45225		*passed*
Bglo1	F635 Median$_{Bglo1}$ - F635 Median$_{EMPTY}$	8322	>1028	*passed*
Bglo2	F635 Median$_{Bglo2}$ - F635 Median$_{EMPTY}$	5634	>1028	*passed*
Bglo3	F635 Median$_{Bglo3}$ - F635 Median$_{EMPTY}$	15984	>1028	*passed*

Final Result: *Positive Test*

Experiment C: [Input DNA]: 0.0025ng/ul (estimated 20 copies/rx)

Table 5

Probe	Parameter	Value Found (a.u.)	Value Expected (a.u)	Result
AT683	F635 Median	50519	2/3 probe F635 Median >50.000 a.u.	*passed*
AT730	F635 Median	56298		*passed*
AT776	F635 Median	42280		*passed*
Bglo1	F635 Median$_{Bglo1}$ - F635 Median$_{EMPTY}$	4330	>1028	*passed*
Bglo2	F635 Median$_{Bglo2}$ - F635 Median$_{EMPTY}$	2538	>1028	*passed*
Bglo3	F635 Median$_{Bglo3}$ - F635 Median$_{EMPTY}$	11530	>1028	*passed*

Final Result: *Positive Test*

Experiment D: [Input DNA]: 0.00025ng/ul (estimated 2 copies/rx)

Table 6

Probe	Parameter	Value Found (a.u.)	Value Expected (a.u)	Result
AT683	F635 Median	52022	2/3 probe F635 Median >50.000 a.u.	*passed*
AT730	F635 Median	51129		*passed*
AT776	F635 Median	43353		*passed*
Bglo1	F635 Median$_{Bglo1}$ - F635 Median$_{EMPTY}$	2347	>1028	*passed*
Bglo2	F635 Median$_{Bglo2}$ - F635 Median$_{EMPTY}$	2126	>1028	*passed*
Bglo3	F635 Median$_{Bglo3}$ - F635 Median$_{EMPTY}$	7863	>1028	*passed*

Final Result: *Positive Test.*

From the above reported data, it can be noticed that the detection of the betaglobine gene target (Genebank AY260740 from 72 to 311) is achieved up to [Input DNA] of estimated a Limit of Detection attested in 2 copies/rx.

Conclusions

Microfluidic packaging will become in few next years a considerable and actual application for the bio-diagnostic field, taking origin from main concepts peculiar in microelectronics environment. A reliable and cost-effective set of solutions is then mandatory to be faced in order to make the fluidic packaging a reality for device production. As a final consideration we need to underline that the manufacturing steps for MEMS devices and related assembly procedures are compatible with bio-diagnostic concepts and a complete integration between semiconductor assembly technologies and main DNA bio-process will be possible since the near future.

Acknoledgements

We want to tank for micro-fluidic simulations Dr. Gaetano Panvini and Ing. Christophe Canales and Mr. Wolfgang Stoeters (Boehringer Ingelheim) for supporting the basic development. Furthermore the support of AB Engineering Italy was also appreciated in 3D-Flow software simulations. Special thanks to Dr. Patrizia Di Pietro and Floriana San Biagio for the support in the biological characterization.

Bibliography

Articles from conference proceedings

[1] N.T.Nguyen, S.T.Wereley, "Foundamentals and Application of Microfluidics", Artech House, Boston (USA), 2006.

[2] I. D. Johnston, M. C. Tracey, J B Davis, C. K. L. Tan, "Micro Throttle pump employing displacement amplification in an elastomeric substrate", University of Hertfordshire, Hatfield, UK, IOP Publishing LTD, 2005.

[3] Paul Galambos, Gil Benavides, "Electrical and Fluidic Packaging of Surface Micromachined Electro-Microfluidic Devices", Sandia National Labs, Abuquerque, NM, USA.

[4] T.B. Christensen, C.M. Pedersen, K.C. Gröndahl, T.G. Jensen. A. Sekulovic, D.D.Bang and A.Wolf "PCR biocompatibility of lab-on-chip and MEMS materials", Journal of Micromechanics and Microengineering, 17 (2007).

[5] Imants R. Lauks, "Microfabricated Biosensors And Micro-Analytical Systems for blood Analysis", i-STAT Co., Ontario, Canada, 1997.

[6] P.Galambos, G.L.Benavides, M.Okandan, M.W.Jenkins D.Hetherington, "Precision Alignment Packaging for Microsystems With Multiple Fluid Connections", ASME, New York, USA, Proceedings 2001.

[7] K.L.Tan, M. C.Tracey, J. B. Davis, I.D. Johnston, Continuously variable mixing-ratio micromixer with elastomer valves", Science and Technology Research Institute, University of Hertfordshire, Hatfield, UK, 18 Aug. 2005.

[8] Cliff Henke "DNA Chip Technologies" Medical Devices Link, USA 1998.

[9] V. Tesar, R.W.K. Allen, J.R.Tippetts, " Microfluidics – The Challenge of Low Re Flow Control" University of Sheffield, United Kingdom.

[10] A.K.Henning, "Microfluidics MEMS" IEEE Aerospace Conference, NJ, USA, 1998.

[11] A. Han, O. Wang, M. Graff, S. K. Mohanty, T. L. Edwards, KI-Ho Kan and B. Frazier, "Multi-layer plastic/glass microfluidic system containing electrical and mechanical functionality", Georgia Institute of Technology, Atlanta, GA, USA, 2003.

Manufacturing business decisions:
design and manufacture of a new generation "Hybrid" Relay.

Michael A. Lane P.E.

Struthers-Dunn LLC, 407B East Smith Street, Timmonsville, SC 29161, USA

Phone: +1 843-346 4427   Fax: +1 843-346 4465   Email: mlane@struthers-dunn.com

Abstract

Replacement of an MDR (Mercury Displacement Relay) in traffic signal cabinets, by design of a new hybrid relay utilizing a solid-state relay in parallel with an electro-mechanical relay and a control circuit. Subsequent decision paths on prototyping and where to manufacture the final product.

Key words: hybrid   relay   design   manufacturing   costs   India   China

Need for product

All traffic signal control cabinets in the USA have used a mercury displacement relay (MDR) as the main load switch for the last 25+ years. State by state, products containing mercury are now being prohibited and must be replaced. The mercury relay has many performance advantages over traditional electromechanical (EM) relays or solid-state relays (SSRs). It was therefore decided to make a form, fit and function, drop-in replacement for the MDR relay by combining the characteristics of an electromechanical relay and a solid-state relay in parallel in one device. First to replace MDR relays in existing installations (retrofit) and second for new OEM traffic signal cabinets.

Design considerations

The existing traffic cabinet MDR relay works on a 120 VAC input; switching an incandescent tungsten lamp load at 120 VAC, 60 Amps total. The MDR is very good at handling high in-rush currents, voltage spikes, power supply surges and brown-outs, and then recovering and continuing to work. It also has a long life expectancy due to lack of any contacts to wear or weld.

Figure 1: Typical 1-pole MDR

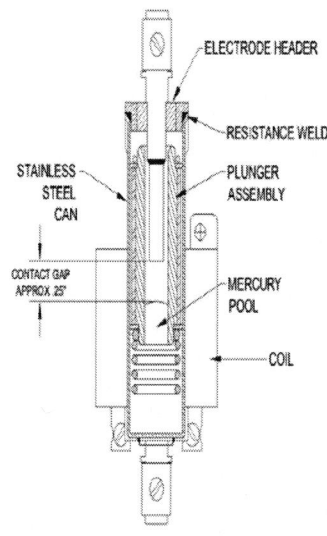

Figure 2: Internal structure of an MDR [1]

A solid-state relay (SSR) is capable of handling such power problems, but to carry 60 Amps continuous current would require a very large heatsink. An electromechanical relay (EM) is capable of handling 60 Amps continuously, but would require large contacts to switch 60A and handle inrush and power surges of tungsten lamp filaments, without the possibility of contacts welding. It would have a shorter life and need frequent replacement.

The design therefore called for the two devices to work in parallel, as a "Hybrid" relay. After input activation, a control circuit would first switch on the SSR to handle the load inrush current and surges for a few cycles at 50/60Hz, then the EM would switch on at almost no load. The control circuit would then switch off the SSR, leaving the EM to carry the load current. The reverse operation would happen at turn-off.

To get the replacement market to easily remove the MDR from the traffic signal cabinet for easy retrofit at roadside intersections, it was decided that the new hybrid would be a drop-in replacement for the MDR. The mounting hole centers, wire clamps (size and position), etc would be as close to the MDR specifications as possible.

Prototyping

A small aluminum base was provided to give the SSR some heatsinking for the Triac during the short SSR operating time, also providing the same mounting hole positions as the MDR and give rigidity to the device. The same copper wire lugs were provided for output terminals and a screw terminal block for the same wire size for input. A plastic housing to cover the working device was also provided, but left the copper wiring lugs exposed as before (not IP20 at this stage).

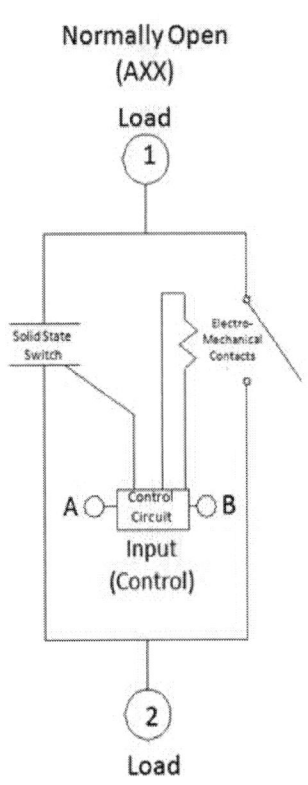

Figure 3: Circuit diagram of a "hybrid" relay

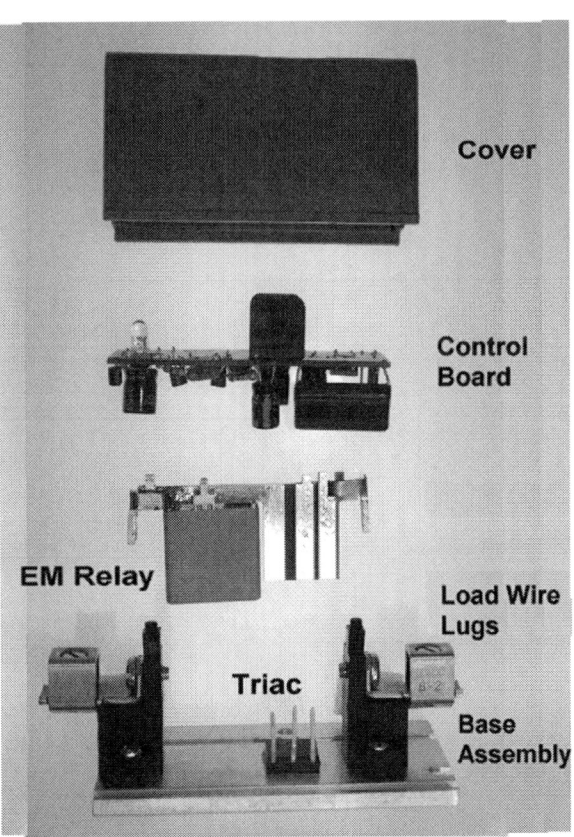

Figure 4: "Hybrid" relay assembly design

Various suppliers in the USA, China and India were then asked to quote tooling and piece part costs.

Figure 5: Quoted tooling costs in USD

	USA	India	China
Cover mold tool	34,000	3,000	2,000
Alu. base extrus.	15,000*	3,000	2,000

* US quote required minimum 100lbs aluminum

It was therefore decided to make the initial 12 pieces of the cover by 3D Stereolith process, at a cost of $400 each. The aluminum Base Extrusion tool was ordered in India, as the extruder agreed to provide a small quantity of extrusions for prototypes to prove tool. However, it was decided to make the 12 prototypes by cutting and milling slots in sheet aluminum, until design was proven. Wiring lug plastic stand-offs were machined from a solid block of plastic for the prototypes. Copper busbars were cut out of copper sheet and hand tooled at the factory.

The prototype design was then tested and the control circuit was optimized for turn-on turn-off performance.

Figure 6: Final prototyped "hybrid" assembly

Volume Manufacture

After proving the design concept with prototype testing, the "where to" manufacturing decisions had to be made. Struthers-Dunn 15 years previously had a factory in Mexico, which has since been closed and has six current vendor partners in China, and three in India.

USA manufacture was out of the question due to high labor rates, social costs and tooling costs. **Mexico** was dismissed due to previous cost and labor experiences. Unfortunately lack of engineering, prototyping and tooling talent, with low productivity, rapidly increasing wages and social costs prohibits manufacture in Mexico, even after the NAFTA agreement. **China and India,** were considered by looking at the "Pros" and "Cons", with particular focus on:
GDP and Exports.
Future Labor rate projections and labor unrest.
Social costs.
Raw material costs.
Currency exchange. Inflation.

China / India comparisons

Looking towards the future, it is first interesting to review the country's GDP (Gross Domestic Product) and then the percentage of its exports with regard to total country GDP. This is a good measure of where the country is today and how the country will grow in the future.

Figure 7: Exports as a percentage of GDP [2]

GDP World Rank	Country	GDP USD (millions)	Exports % GDP
1.	United States	14,657,8002	13%
2.	People's Republic of China	5,878,257	41%
10.	India	1,537,966	15%
14.	Mexico	1,039,121	3%

This is interesting in that it shows the potential for big trading problems for China in the next decade. Almost half of their GDP is currently export based. The domestic economy must rapidly be improved. However, rapid domestic expansion in many areas may lead to a myriad of other problems, (which will be reviewed later).

India on the other hand currently under exports, but has a relatively developed domestic infrastructure, giving the country plenty of room for inward investment and export growth, without extreme growth in costs.

India's GDP growth is a close second to China's, but has a much more domestic bias.

Figure 8: GDP growth Average 2009-Q1 2011 [3]

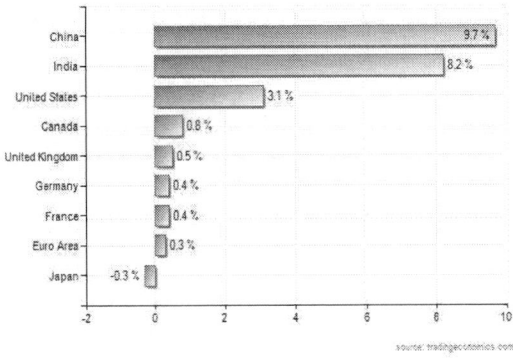

Wages

Of course, blue collar wages are the most important factor when considering any offshore manufacturing. China currently has the edge over India on base wages. However, in a cursory glance, most people do not see the almost 50% government social costs tacked on to the base labor rate for social security and health care type costs in China. India by comparison has a much lower level of "social" costs required of the employer. Escalation of labor rates looking forward are also extremely important if investment is for the next several years, or even decades. China's productivity rates are higher than India's current rates,which offsets the higher social costs, but India's productivity rates are also rising.

In certain coastal areas of China, where most electronic manufacturing currently takes place, hourly wages are now well above the minumum government set rates. At our partner in Guangdong province (previously Canton), China, wages were Renminbi (Yuan) 750 per month in 2010. In 2011, the government minimum wage is now set at Rmb.1100 per month, but our factory is paying Rmb.1200 (approx. USD 200/month) and are planning for it to double again in the next 3 years. [4] India on the other hand has a slightly higher, but much more stable growth in blue collar wage rates. Additionally, China is the one new fast emerging country where the percentage of population under 30 is actually falling. This is due to the one child per family policy, which does not bode well for labor availability in the future.

Social developments

Social unrest in China will increase. The government is trying to contain much of it with restricted and blocked sites on the internet. However, the ever increasingly affluent coastal areas are now spending money on expensive imports. For example BMW and Mercedes have

their fastest growing and soon to be largest export market in China. People want much more freedom to spend their new wealth. It is difficult to buy land and get mortgages in China, so one expression of wealth is the automobile. China is no longer a third world country. Take a look at the new airports in Beijing (for the Olympics) and Shanghai and the 430 km/h Maglev train from Shangahi center to the airport for example. [5] This shows the incredible development of infrastructure in the country. The elevated road system in Shanghai has automated electronic boards just before on-ramps, showing traffic congestion on all nearby roads, so that the least congested route to destination may be chosen .

Fiugure 9: Maglev train and Shanghai airport

On the other hand, the interior farming population which previously came to the coast for work as migrant workers are no longer doing so in such great numbers. A good gauge of migrant worker movement was during 2011's Chinese Lunar New Year. In Guangzhou, the major city in Guangdong province, it was estimated that 1.6 million migrant workers returned home for the holiday[6]. Usually about 90% return to their employer after the holiday. This year it was estimated that only 60-70% returned to their job, an enormous number, considering this area already has a labor shortage.

Ou Zhenzhi, director of Guangdong's department of human resources and social security, estimates that the entire province will have a labor shortage of about 1 million people in labor-intensive industries by 2013. [7] Workers in the poorer interior provinces now have more opportunity for work at home and without reform of the "hukou" law, they will continue to stay there if they can. [8] This rule allows social benefits only in the province where born, encouraging migrant workers to remain home now, if they can find non-farm work locally. [9]

However, the other side of developing the interior provinces for manufacturing industries is the logistical problem of transporting raw materials in and finished goods out, due to the tremendous distances involved, bad roads and lack of rail.

Currency valuation

With respect to the U.S. dollar, both China and India exchange rates have remained relatively flat the last few years, despite pressure from the USA and other developed countries to insist that China allow the Renminbi to float upward with market forces.

This will now happen anyway due to the market forces internal to the country. Although it has not appreciated much yet, the trend is obvious.

Figure 10: U.S. $/Yuan Exchange [10]

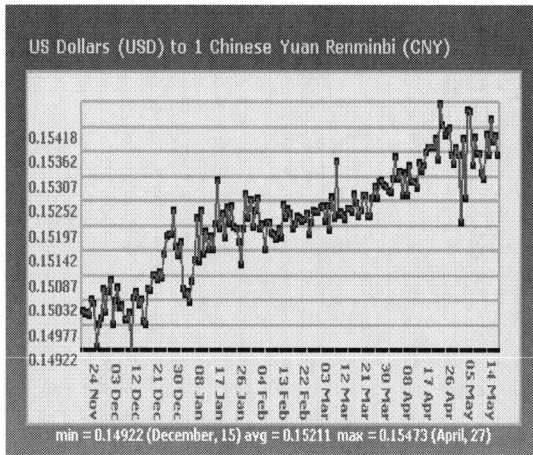

The newly affluent want to spend their money and have access to more and more Western goods. Inflation will increase as a result and become a problem. Imported food and capital goods prices will then soar. Already, from March 2010 to March 2011 the inflation rate is + 5.4% the government goal for all 2011 was 4% max. [11]

For example, although many imports are intentionally blocked to encourage domestic manufacture, and certain industries have been championed with government investment, nearly all automobiles are imported. The Renminbi will be allowed to rise to mitigate some of these soaring import costs, when currency conversion is taken into account. So China's inflation grows, as the currency rises to try to mitigate the cost of imported capital goods.

India on the other hand has had car and truck manufacturing for decades. (Mostly under license). Now Tata Motors has made a play to get entire families, presently traveling on 2-wheeled vehicles, into the 4-wheeled, four-doored "Nano", or the "one lakh" car as it has been called. (One lakh being 100,000 rupees, about $2000). [12] In this way, India will slowly grow internal domestic demand. It is estimated that India will have the

fourth largest market for light vehicles by 2014, ahead of Germany, Russia and Brazil. [13]

Figure 11: The Tata Nano

Wages should grow slowly with moderate increases. Domestic demand will be stimulated, but inflation should be kept under control. Most food is grown locally and with the introduction of mechanization, food production will become even more efficient.

Conclusion

The choice of offshore manufacturing for today, is therefore India, when chosing for the future.

The hybrid relay is now in production in the U.S.A., using parts already manufactured by the Indian partner, with some others temporarily hand-tooled in the U.S. factory. Full manufacturing will be transferred to Bangalore, India shortly, after local sourcing of an EM relay (local supplier found), a 2-pole terminal strip to our specifications, and the pcb components is approved.

Future plans for the product include a normally closed version, needed for a different type of traffic signal cabinet. These traffic cabinet versions are well protected inside the control cabinet with brown-out, surge and other protection provided to the cabinet inputs and outputs. It is therefore intended to make more "ruggedized" versions to replace 1-, 2-, and 3-pole MDRs in many other applications. In particular, street lighting applications for example, where many MDRs are still in use. A larger cover will make the design IP20 on the next revision.

A typical (EM) relay with silver cadmium oxide contacts at 60Amps will give a life expectancy of 100,000 to 200,000 switching cycles. It is expected from extrapolated life tests that a "hybrid" relay may give up to 1,000,000 operations, similar to that of an MDR. This is important in many critical applications, where replacement of relays either requires a shutdown of operations (Electric Power generation, food processing, etc), or a failed device causes

168

significant, or other safety problems (Example: non-working traffic signals)

The partnership of an Indian company, strong in engineering and manufacturing, will allow further development of the product and the supply of a technologically advanced product, with extremely high life expectancy. This product will be provided to the markets demanding it, at a price comparable to that of the old mercury relay.

Acknowledgements

Mike O'Donnell. Engineering Manager, Struthers-Dunn, Timmonsville USA
For finishing the design and taking my concept to reality.

Akash and Ajit Ranka, Paramount Industries, Bangalore, India
For tooling and design advice.

References

Technical manual
[1]   Struthers-Dunn internal QA and Training Manual 2005.
Book
[2]   Central Intelligence Agency. "The World Factbook - China". May 17, 2011
Website
[3]   Trading economics. From NationalBureau of Statistics. May 2011.
      http://www.tradingeconomics.com/china/gdp-growth
Personal Interview
[4]   Beta Electronics. Pan Yu, Guangdong province, China. April 2011.
Photo
[5]   Steve Clemmons. "Shanghai Envy" The Washington Note. July 9, 2008.
Article
[6]   Wen Shidi. "The beginning of the end for Guangzhou?" Guangzhou / China Q&A. April 6, 2011
Quote
[7]   *Ou Zhenzhi*, director of the Guandong provincial department of human resources and social security.
      www.chainadaily.com.cn/china/2011-03/03/content_12106767.htm
      Mar 3, 2011
Article
[8]   Wen Liao. "The Princeling and the Paupers". A public fight over the internal immigration system. Foreign Policy Magazine.
      March 5, 2010

Paper
[9]   Kam Wing Chan and Li Zhang. "The Hukou System and Rural-Urban Migration in China: Processes and Changes". Department of Geography. University of Washington, Seattle, USA.
      Published in China Quarterly, 1999.
Website
[10] www.exchange-rates.org/history/USD/CNY/G/180
      May 17, 2011
Website
[11] www.businessinsider.com/china-inflation-hits-54-in-march-2011-2011-4
      April 14, 2011
Website
[12   Tata Nano – The People's Car from Tata Motors.
      http://tatanano.inservices.tatamotors.com
      May 2011
Carblogspot
[13] Tata Nano: People's Car, fourth largest light vehicle market in the world.
      Wantedcarz.bolgspot.com/2008/05
      May 18, 2008

A Smarter Supply Chain – End to End Quality Management

Kitty Pearsall, B.J. Steele, and Paul Zulpa

IBM, 11400 Burnet Road, Austin, Texas 78758

512-286-7957, kittyp@us.ibm.com

Abstract

For companies in a "Smarter Planet" it is key that they are globally integrated, interconnected and intelligent. For globally integrated supply chains the large enterprises must rely heavily on their suppliers and their suppliers' suppliers for products, services and software. This increased reliance and dependence requires that an extended supply chain have an effective supplier quality management (SQM) system that provides visibility into this complex system. Essential in this SQM system is the ability to provide an end to end (E2E) process that begins with the concept of a new product and manages it through its extended lifecycle. These supply chains of large enterprises are faced with many daunting challenges such as consistent quality, compliance performance, part management (change control, parts obsolescence, counterfeit parts, etc.), continuous improvement, inventory management, product security, on time delivery, and environment regulation compliances. An added complexity to the extended supply chain is that over the past several years there have been disruptions in the supply chain as a result of raw materials, demand changes, commodity volatility and fuel price increases. Per Manufacturing Insights "Worldwide Supply Chain 2010 top 10 Predictions" the #5 Prediction: "intelligent Supply Chains will put Broader Visibility Burden on Supply Chain Organizations, both Owned and Outsourced" further driving the need for a strong E2E Quality Management System. [1] Many major enterprises', like IBM, are implementing or considering implementing the use of business intelligence in their E2E supply management system.

Key words: Supplier Quality, Product Development, Supply Chain

Introduction

It is the intent of this paper to illustrate the E2E quality management process in place at IBM today that is required to effectively detect and prevent these impacts. It will highlight the key processes that are deemed essential for the various stages of a new product introduction into the marketplace and its life cycle. Key stages that will be highlighted are Strategic Sourcing, Supplier and Parts selection, Qualification of both, and continuous monitoring of the SQM process at the various stages. In each of the aforementioned areas, the intent is to highlight the methods and/or processes followed in order to be successful. Associated IT tools will be mentioned in terms of their capability and how they enable the various processes to work seamlessly together. Along with this the interrelationship of each phase with respect to the stage gated process, Integrated Product Design, (IPD) will be noted.

Background

The IBM Integrated Supply Chain, E2E Supplier Quality Management takes a holistic systems view of the hardware engineering design environment, the embedded software development environment, the verification engineering environment and the required IT infrastructure to support the growing complexities of design, development and qualification of new products in today's environment of increasing dependency on Contract Manufacturers, multiple suppliers and their sub-tier suppliers. [2] It is an integrated solution designed to provide a rigorous and disciplined integration of methods, best practices and tools that are used to deliver high quality products to the market quicker with a significant reduction in the number of redesigns. This is all done against a backdrop of the Integrated Product Development (IPD) infrastructure in IBM which is driven via an Integrated Portfolio Management Team (IPMT). Many large enterprise organizations besides IBM, such as HP, utilize a form of IPD, but to varying degrees. [3, 4, 5] Refer to Figure 1 for IBM's IPD Process.

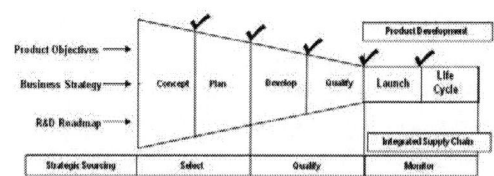

Figure 1: IPD Stages, Concept to End of Product

IPD is based on an integrated product design approach. The innovation stems from designing in from the start of product development those characteristics of the product that apply across a set of functional or system requirements. This definitely includes ease of manufacturing, design for supply chain, and most certainly quality and reliability. Since quality and reliability are the most significant with respect to the E2E Supplier Quality Management System there are four stages that will be discussed with respect to the IPD process: Strategic Sourcing, Supplier and Parts Evaluation and Selection, Qualification of the selected suppliers and products and continuous monitoring of these.

Strategic Sourcing

Through early procurement involvement, the ISC procurement engineer directly interfaces with both the Development engineers and the Global Commodity Managers on the commercial procurement team for any new part introduction. This collaboration is initiated when the procurement engineer receives the first sourcing plan for review. This review is to optimize the source/part selection in the earliest IPD phase. The objective is to secure the part sourcing and minimize business risk through the selection of qualified suppliers, standardization, and common parts. The key is to select the right suppliers and technology up-front, to ensure product quality, to reduce cost at the early stage of new product development, and to decrease Time-to-Market. Strategic sourcing is typically done in a Pre-Concept stage with respect to the IPD process. (Refer back to Figure 1.) It begins with IBM's Commodity Councils as shown in Figure 2 and ends with the production of a Technology Roadmap (TR).

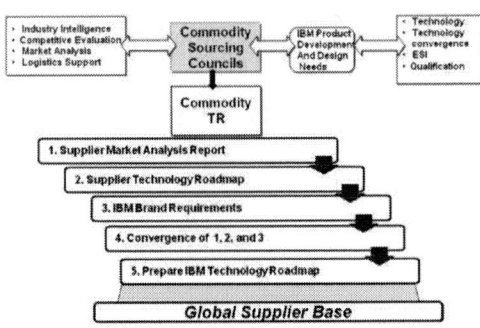

Figure 2: Strategic Sourcing by Commodity TR

Each Commodity Sourcing council initially focuses on the Supplier Market Analysis (SMA) report for their given commodity. In this report key activities & tasks are: first conduct market research, develop market and suppliers' profiles, analyze supply market data, conduct a vulnerability and risk assessment, and provide a definition of value chain cost analysis; second, obtain the relevant supplier technology roadmaps as a result of the SMA. From

here IBM product requirements and supplier technology available are converged via the Strategic Sourcing process which aligns the interests of the company with the interests of all those in the supply network. This results in the definition of an IBM strategic TR by commodity that highlights the parts with the most competitive costs and best performances that have part lifecycles clearly defined that meet the Development requirements.

In today's environment this means that the supply base must be agile and adaptable as it is a well known fact that the ability to influence a product is greatest during its early life design stages, while the cost to change a product increases dramatically as its life progresses. (Refer to Figure 3.) At this point it is best for the procurement engineer to utilize the unique commodity TR developed for the next stage, Supplier Evaluation.

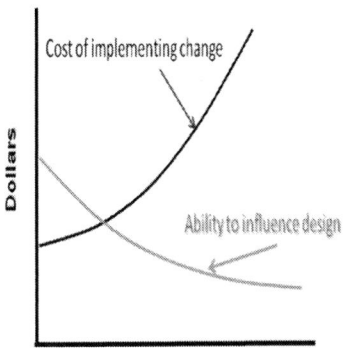

Figure 3: Ability to influence design

Supplier Evaluation and Part Selection

The end deliverable of Strategic Sourcing is a final source plan or an early bill of materials (BOM) that is used for the Supplier Evaluation and subsequent Part Selection. This occurs during the Concept and Plan stages of IPD. Through Early Supplier Involvement (ESI) the IBM Commodity Councils also review both technical (production processes, audits, etc.) and business (D&B, IBM Financial Analysis) standing of the suppliers to assess robustness of their fundamental business. The supplier selections are based on their capabilities, cost, quality, service, and time, to name just a few criteria. IBM then approves and qualifies only those companies considered to be world-class suppliers for their respective commodities. Typically the IBM supplier base for a given commodity is the biggest and most trusted in the industry. These are companies that IBM has a long standing relationship with; and, companies with whom IBM has had a sustained, strong history of quality, reliability, and financial performance. These companies have

171

exhibited strengths in technical depth and problem resolution. They are generally the product manufacturing leaders and/or technology innovators. Only qualified suppliers are put onto IBM's Approved Vendor List (AVL).

At this point the BOM is ready for final part number review. Selected parts on the BOM not already qualified and in use in IBM undergo an extensive qualification process. The goal is to have an end BOM that has a high percentage of preferred parts, a significant amount of part reuse and parts that have more than one supplier. In order for a part to be considered preferred IBM looks at two dimensions: business/technical preferences and supply preferences. The first dimension considers parts that are consistent with platform architectures and TR's, competitive costs, value in the market place with product differentiation, long life cycle, design flexibility, worldwide availability and a proven technology with low risk. The second considers parts that are industry standard and have multiple qualified suppliers, volume flexibility, well established manufacturing processes, short lead times, trusted supplier relationships and low cost adders.

Qualification Process

IBM audits the Supplier Process with a keen focus on their Quality Systems as well as the quality systems of their suppliers and their sub-tier suppliers. IBM, like many other organizations has experienced severe disruptions in its supply chain as a result of a sub-tier's problem. Therefore it is critical to know as much as possible about the Supplier's Quality Management System (SQMS). The Supplier's manufacturing facility is also audited and must itself be qualified to validate its ability to routinely build the part being qualified for use in the product. It is very important to verify that the supplier's non-conforming product process concludes with part destruction.

Suppliers must comply with a very detailed set of IBM Quality specifications by commodity. Suppliers must comply with detailed IBM Technical specifications that may/will vary by commodity. They must also meet all regulatory and environmental regulations both in and outside of IBM. Once this is done IBM releases a unique IBM Part Number (PN) for every part that is procured. Any major follow-on part / process change made by the Supplier requires IBM's expressed approval.

Monitoring of the Suppliers, Parts and Processes

IBM has developed a complete set of methodologies, processes, predictive analytics, and collaboration tools to address todays complex development process. The Methodologies and Processes are listed in Figure 4 with respect to IPD.

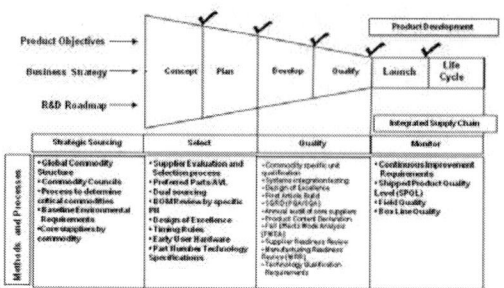

Figure 4: Methodologies/Processes-E2E SQMS

It is critical that continuous monitoring at every stage of the process takes place. Core suppliers receive a minimum of a formal Annual Audit, plus other unscheduled visits. IBM reviews sub-tier management processes initially during the parts qualification process and again during any future audits. The suppliers are expected to perform incoming inspections (including electrical tests) as defined in the statement of work. The suppliers hold monthly "parts quality reviews" with sub-tiers and provide supplier scorecards. Higher assembly parts are clearly marked and labeled according to IBM standards, and bear IBM PNs. Suppliers monitor their parts from their factory to the bonded hubs supporting the fulfillment operations. The Bills of Lading and customs paperwork help authenticate that parts are from the proper source, ensuring that counterfeit parts are minimized entering an IBM end product. Any counterfeit issue, from supplier delivery through end-user location, will be: reported back to quality, the supplier and IBM; investigated to determine the source and its impact to other hardware/customers; and finally subjected to corrective action put into place to prevent reoccurrence. But, in order for these methods and processes to be effective there must be procurement tools to enable this monitoring.

IBM uses its own specific Procurement Tool Set [6] as collaboration tools to communicate with the suppliers. (Refer to Figure 6.) The tools that are noted in bold text in the figure are directly integrated with the suppliers as this improves the efficiency, accuracy and completeness of the knowledge transfer. It has been found that separate, non-integrated communication tools can lead to, and have in the past led to, misunderstandings; delayed findings in either failure analysis or root cause identification, and problems occurring at either the supplier or on the box manufacturing line such that parts with inherent defects continue to be built. This not only results in unnecessary delays in product build,but potential product escapes to the fields, and it can eventually result in increased warranty costs. These tools are discussed next in more detail.

172

Procurement Tool Set – Process Enablers

The Procurement Tool Set consists of four Lotus Notes databases and one dynamic tool integrated to support and fully enable the methodologies and processes discussed in the previous section. A high level description of each will be presented as well as benefits derived from its use.

Part Change Notification (PCN) is a Lotus Notes database that manages any supplier PCN/EOL (end of life) issues through a centralized IBM approach allowing for the integration of Dev/Mfg/Proc changes in a common report format. The cycle time reduction realized and centralized repository equates to savings in component cost and improvements in the supply chain. [7]

Quality Information Network (QIN) is a Lotus Notes database that manages the supplier audits providing a consolidated view of audit data, audit results, and supplier action items. It leads to reduced qualification cycle times and eliminates redundant audits through audit standardization and improved communications, thus increasing the quality of IBM products. [8]

Supplier Problem Log (SPL) is a Lotus Notes database that serves as a collaboration tool that allows the "sharing" of quality problem data among engineers, commodity chairs, and management. This tighter communication between IBM enterprise and the suppliers for quality problem prevention and resolution is a definite benefit. [9]

Technology Qualification Application (TQA) is also a Lotus Notes database that serves as a Checklist qualification tool that insures proper qualification procedures are followed. It provides the mechanism for automatic notification of the product engineers and gives them review capability of a part or process qualification as it progresses. Ultimately there is a reduction in qualification cycle time, elimination of redundant qualification and improvement in communication within IBM and with suppliers. [10]

Complementing the integrated databases IBM uses its Supplier Quality Management System (SQMS). SQMS is a dynamic tool that can alert both IBM and the supplier simultaneously when a problem is occurring, real-time. SQMS contains data received from the suppliers via a Lotus Websphere portal and stored in a DB2 database. It is a proactive supplier quality solution that measures quality as the supplier "makes" and as IBM "uses" the part. The tool has the capability to generate charts and reports for both the supplier and IBM. The major benefit is the improvement in product quality through the management and measurement of supplier and IBM manufacturing processes real-time. IBM plans to apply advanced analytics to this data to detect issues as they emerge. The benefits of

the "early" notification and identification of potential quality issues is that it can lead to the reduction of time to problem resolution and provide problem visibility to supplier, manufacturing, and field data. [11]

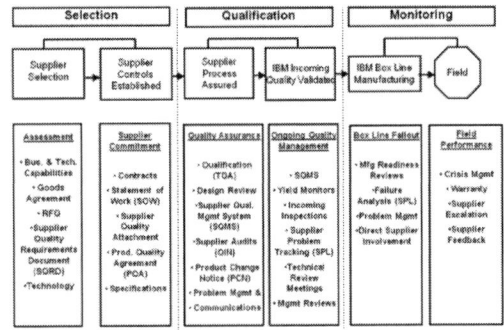

Figure 5: Collaboration Tools used in monitoring the E2E Quality Management

Conclusions

It is essential that large enterprises such as IBM have a supply chain that is globally integrated, interconnected and intelligent in order for it to maintain a common, consistent, traceable and sustainable supply chain in today's fast paced environment. Since the reliance on the supply base (and their suppliers' supply base) is critical to day-to-day operations, one must have visibility on operations without actually being present. An integrated procurement tool set is essential for supplier collaboration to provide that visibility. Dynamic tools such as SQMS are integrating traditional quality techniques with advanced analytics that provide the intelligence to be able to be predictive rather than reactive. All of these enablers add up to a solid E2E Quality Management System.

Acknowledgements

The authors acknowledge the many innovators of not only the Procurement Tool Set but the Dynamic SQMS tool. These include, but are not limited to: John Maresca, Eric Lambert, Eddie Kobeda, Mike Whitney, Ghassan Chidiac, Dan Hayes, IBM Research, Robert Raines, and IBM Business Consulting Services.

References

[1] S. Ellis and K. Knickle, "Worldwide Supply Chain 2010 Top 10 Predictions", Manufacturing Insights: Supply Chain Strategies, December 2009.

[2] S. Zeng, M Cohen, BJ Steele and J, Sairamesh, "A Supplier Performance Evaluation Solution for Proactive Supplier Quality Management", eBusiness Engineering, 2008 ICEBE, IEEE International Conference, December 2, 2008.

[3] "Streamlining the Product Lifecycle - Creating an Integrated Product Development Environment", HP Product Development and Engineering IT Peer Forum, sponsored by HP and Manufacturing Insights, February 17, 2009.

[4] K. Cheng, "Supply Chain Risk Management with Quality Tools", Copyright 5/1/2006.

[5] M. Schaeffer, "DOD Guide to Integrated Product and Process Development (Version 1)", February 1996.

[6] J. Maresca, P. Zulpa, K. Pearsall, BJ Steele, et. al., "Method and System for Providing an End-to-End Business Process for electronic supplier qualification and quality management", US Patent 7,353,497, April 2008.

[7] P. Zulpa, S. Chandra, and J. Maresca, "Tracking System for Monitoring Supplier Product Change Notices from Initial Review until Closure", US Patent 6721746.

[8] M. McKay, S. Chandra, R. Hayes, A. Kostic, J. Maresca, "Method and system for gathering and disseminating quality performance and audit activity data in an extended enterprise environment", US Patent 7,353,497, June 2002.

[9] B. Steele, C. Dishman, R. Dunn, K. Scea, "Method and system for handling production problems in an extended enterprise environment. International Business Machines", US Patent 6804785, October 2004.

[10] E. Lambert, J. Maresca, M. Whitney, "Method and system for electronically qualifying supplier parts", US Patent 7,822,796, July 2002.

[11] E. Blackshear, M. Bolch, C. Bai, G. Hurtis, J. Maresca and P. Zulpa, "Quality/Reliability System and Method in multilevel environment", US Patent 7,147,233, February 2007.

A Cross Supply Chain Collaborative Approach to Addressing the Technical Challenges of Today's Electronics Manufacturing Industry

B.J.Smith, M.Andrews

High Density Packaging User Group Inc.

Abstract

Collaboration across the electronics manufacturing supply chain is now becoming an essential activity to address the future technical challenges of the industry. As research and development budgets contract and outsourcing by the OEMs becomes the norm, engineering expertise is increasingly dispersed throughout the supplier and contract manufacturing base. The need to bring together companies from the full spectrum of the supply chain to resolve pressing manufacturing industry issues is intensifying and driving companies to adopt a new approach to process technology development. This paper discusses the drivers for more open co-operation between producers and suppliers. It looks at the different types of collaborative partnerships currently in place focusing primarily on industry driven R&D consortium based organisations. Key factors that can determine the success of such initiatives are identified.

Key words: Collaboration, Research & Development, Manufacturing Technology, Outsourcing

Introduction

The electronics manufacturing industry has in recent years witnessed significant technical advances and a major change in the business models used to deliver latest generation products to an increasingly demanding consumer market. Aggressive competition across the complete product spectrum has driven major shifts in organisational structures to maintain margins while delivering cheaper goods. These changes have in turn had a direct impact on how products are developed and the manufacturing processes deployed to achieve cost effective solutions. Responsibility for process development has shifted from the original equipment manufacturer (OEM) to the whole of the supply chain. This has driven a new approach whereby multi-company collaborations are now becoming increasingly common as a means to address process related challenges.

Outsourcing and its impact on R&D

The traditional business approach taken by producers up until the mid 1990's was to adopt a vertically integrated structure in which organisations owned their own product development and associated manufacturing capabilities. Major players would install facilities for IC fabrication, PWB manufacture, assembly, and system configuration. Some of the larger organisations would also produce their own PWB laminates and other materials and even design and build their own production equipment. The emphasis was very much on developing customised manufacturing solutions that further enhanced the uniqueness and competiveness of the products offered. Production processes would

be developed by teams of in-house engineers and protected by patents that were every bit as important and plentiful as those associated with product design. Collaboration with other companies, while not uncommon, was carried out under strict confidential arrangements with information shared on a need to know basis. Relationships of this type were typically between the OEMs and specific partners. Very rarely did direct competitors engage in any form of co-operative venture. To minimise the risks of exposure of company secrets, engineers were employed by the OEMs to cover a vast range of expertise which extended across the complete supply chain, thus enabling critical technical decisions to be made in house. The results of this approach, while highly successful in its time, saw an industry flooded with a proliferation of different manufacturing solutions with little standardization and burdened with massive engineering and production overheads.

The outsourcing revolution of the 1990's brought about major changes in how companies organized their manufacturing and R&D strategies. Newly formed companies emerged with little or no in house manufacturing capabilities. Established organisations realized that their overheads associated with owning fabrication facilities that were not efficiently utilized were negatively impacting their competitiveness. This prompted a new sub-contract approach to production. Organisational restructuring saw OEMs concentrate on their core competencies and sell off their manufacturing plants, often under strategic agreements with emerging contract manufacturers that were looking to rapidly expand their operations.

From an engineering perspective, the transfer of responsibilities was a tentative and nervous affair as initial concerns about the depth of technical competence of the sub contract market persisted. A period followed where OEMs micro-managed production processes at their supplier's sites allowing them to develop confidence in their manufacturing partners. Often contract manufacturer acquired sites that were formerly part of a vertically integrated OEM, continued to service their former owners creating a more seamless handover of responsibilities. In the short term, engineering expertise remained with the system producers to oversee outsourced operations and to maintain focus on new manufacturing technologies. At this point in the outsourcing revolution it was also common for production facilities to be in fairly close proximity (i.e. same country) as the customers they supplied. Production process development and qualification projects continued to be driven and overseen by the OEM resident engineers.

A period of consolidation followed in the Electronics Manufacturing Services (EMS) sector which saw contract manufacturers rapidly expand through multiple acquisitions of production facilities previously owned by OEMs and system integrators. In addition EMS providers embarked on major construction of new state-of-the-art factories, many of which were located in geographical regions where labour was cheaper (typically Eastern Europe, Central America and Asia). As common contractual agreements requiring divested manufacturing plants to continue to supply their former owners for a period of time expired; products were transferred to these newer sites regionally remote from the OEMs that they serviced. As the EMS industry matured, the OEMs became increasingly confident to entrust their suppliers and contractors with developing and optimising the manufacturing processes to deliver their products. EMS providers, board fabricators and component package suppliers enhanced their R&D capabilities as they took on greater responsibility for developing reliable cost effective manufacturing solutions. Consequently, the level of hands-on engineering control and intervention previously practised by the OEMs began to fall off and with it the depth of manufacturing technology skills that they retained.

The outsourcing revolution triggered two major changes within the industry. Manufacturing solutions customised to specific OEM product ranges were replaced with universal processes applied to comparable electronic designs marketed by competing companies. Often the same contract supplier would deliver similar services to multiple producers using almost identical processes. Additionally, the highly competitive nature of the EMS and associated contract manufacturing market

resulted in extremely tight margins being required. This impacted the structure and overheads of supplier organisations and significantly restricted the amount of investment in R & D activities.

Technology Challenges Today

Against a background of shifting responsibilities triggered by outsourcing, the need for new manufacturing technologies to support future generations of electronic products has continued unabated. Consumers possess an insatiable demand for greater processing power, increased bandwidth and enhanced portability and functionality. They expect all of these with increased levels of reliability and at lower cost. The industry is also being influenced by a growing worldwide awareness of environmental issues driving the substitution of toxic materials and the introduction of alternative green production processes.

At a manufacturing level these requirements translate into a broad range of physical demands that will inevitably evolve over time. The manufacturing technology road map has multiple branches but a recurring theme across these is the need for higher signal processing and interconnect capacity per square area. This is achieved by greater IC integration (possibly incorporating through silicon vias), smaller component substrate and PWB level features (tracks and vias), an increasing uptake in optical processing, the introduction of advanced materials with improved dielectric constant and loss tangent properties, and greater use of 3 dimensional systems in package structures. In parallel, the impact of environmentally protective legislation has led to research into a new family of solder alloys and flame retardants all of which drive extensive process and product qualification activity. These demands exert pressure on companies located across the entire electronics manufacturing supply chain to invest extensive funds in developing future fabrication capabilities. At a time when eroding margins and other cost constraints are forcing organisations to rein in spending on research and development, the need to provide customers with advanced capabilities that meet latest product requirements presents suppliers with a major dilemma. Increasingly companies are turning to collaborative opportunities as part of their future technology development strategies. Recognizing that the manufacturing industry faces common problems, there is a realization that many of these issues can be addressed by working together with partners across the supply chain and often with their competition. The resultant solutions deployed universally throughout the sector lead to lower costs not only in development but also on-going production.

Collaborative Approaches

Collaboration within the industry is by no means new, however the type of engagements taking place are evolving and expanding as OEMs and their suppliers leverage co-operative activities to reduce their costs. Several approaches to collaborating are deployed. The following four arrangements represent the primary approaches in operation today.

Bilateral Collaborative Agreements

This traditional approach involves a partnership between two companies (possibly more) whereby the focus is on developing a manufacturing solution that will deliver a distinct competitive advantage to a product or range of products. It quite often involves the creation of intellectual property that is classed as confidential to the partners involved. The relationship is typically formalised with binding agreements that include non-disclosure clauses and may involve patent applications.

Government / Regionally Funded Initiatives

These tend to be multi-party activities in which project partners are funded in part or in full by government or regional agencies to investigate/deliver specific technology solutions. Programs of this type often involve formal applications against themed predetermined focus areas and typically run for a period of several years. Typical examples include the Framework Programs run by the European Commission and the DARPA technology grants initiatives within the USA. By the nature of the funding agreements, participation limitations based on geographical and political considerations typically exist. Often universities and research institutes are involved in the projects. Because of the amount of funding involved a formal approach is required and administrative activities (e.g. progress reporting, financial accounting etc.) are fairly onerous. Consortium agreements and NDA's are commonly a pre-requisite.

Research / Academic Institute Programs

This type of arrangement involves a research institute conducting development activities on behalf of a group of companies who provide the funding to support the work. Often the activity is subsidised by governmental/regional administrations particularly if there is clear evidence that there is clear industrial demand for the technology under development. Projects can be structured in various ways and industrial partners may provide in-kind contributions (material resources and expertise) as well as monetary support. The focus of R&D is identified and influenced by the industrial partners who hold the institute to account to deliver. Information dissemination strategies throughout and at the conclusion of the project are normally agreed by all parties at the outset. Typical research organisations adopting this kind of approach include the National Physical Laboratory in the UK, CALCE and the Universal Instruments consortium in the USA and IMEC and the Fraunhofer Institute in central Europe.

Industry R&D Consortia Organisations

The aim of this approach is to engage companies from across the electronics manufacturing supply chain in addressing common industry problems and challenges for the mutual benefits of all parties and ultimately the industry as a whole. Typically these organisations work on a membership model whereby companies subscribe to the organisation. The direction of the consortium and areas of focus are determined by the members. Activities are coordinated by a small facilitation staff. In contrast to the research institute programs described previously, the emphasis is on securing resources and services from the participating members with minimal financial outlay required by any partner. Projects tend to be run on a more informal basis often without project specific agreements in place. They tend to focus on pre-competitive technology areas or on challenges common across the industry. Often projects will address operational issues or focus on reliability studies. Project structures tend to be more flexible and shorter in duration making them appropriate for addressing more urgent issues. Organisations currently operating under this basis include High Density Packaging (HDP) User Group and the International Electronics Manufacturing Initiative (INEMI).

The industry consortia organisations are rapidly gaining popularity in the industry as companies look to deploy R&D budgets more efficiently. Increasingly companies are realizing that the cost benefits of sharing resources to resolve issues outweigh the competitive advantages of developing in house solutions. Moreover, the opportunities to leverage the knowledge of experts from companies throughout the supply chain are invaluable in planning and executing strong projects with a high probability of a successful outcome. In order for co-operative activities to be successful and deliver value for the companies that take part, several factors have to be addressed. The next section discusses the key aspects that can determine whether projects succeed or fail.

Key Factors for Successful Co-operative Initiatives

A vital consideration in committing to a collaborative initiative is how the partners can leverage the results. Traditionally companies would use the output of such projects to supplement and reaffirm the results of in house studies. As R&D

budgets are squeezed it is becoming increasingly important that collaborative programs negate the need to conduct similar in house studies. It is therefore imperative that the credibility of results delivered stand up to scrutiny not only from the companies involved but also from their current and prospective customers. Project planning and on-going leadership and coordination must therefore be executed in a highly professional manner. Engaging technical experts from end user companies (OEMs) in the project is highly beneficial in guaranteeing its integrity. Additionally, soliciting peer reviews throughout the duration of the project assist greatly in ensuring a sound project approach. Industry organisations are particularly effective in this area as they can leverage the views of their complete membership and not solely those participating in the project. Frequent member meetings to review progress on all active projects provide an excellent vehicle for obtaining valuable feedback.

A common argument against collaborative ventures is that they are prone to extended delays and in particular are dependent on the commitment of all active partners. A project member that decommits on resources can have a deleterious impact on the timely outcome of a project. Although formal project agreements can serve to protect against such risks they can also be perceived as a deterrent to signing up to a project in the first place. Organisations such as HDP User Group adopt policies that steer away from bureaucratic contracts of this type finding them counterproductive and difficult to apply penalties to defaulters. A preferred practice is to ensure that, through sound facilitation, all partners fully understand what and when they are committing resources during the planning stage of the project. An environment of openness encourages players to declare any risks at this time so that contingencies can be put in place prior to the implementation phase of the project. A combination of both peer pressure and fellow member support is deployed throughout the project to guard against potential problems of this type.

Management of intellectual property issues often presents a serious challenge to establishing collaborative initiatives. Establishing foreground and background IP content in a project and agreeing on how both can be used and shared has inevitable time delay implications with non-disclosure agreements needing to be generated, reviewed by each partner's legal departments and ultimately signed by senior management. This can be a major source of frustration for projects that require rapid execution. For many of the projects undertaken by membership based industry R&D organisations, becoming embroiled with IP issues can be very complicated. Determining what can be shared between individual project members, the project team as a whole and the overall organisation membership can lead to severe difficulties and the possibility of disharmony within the group. While certain projects, particularly those that involve product development, cannot avoid such issues, timely launch and completion of most projects can be enhanced by steering clear of IP related activities. For many manufacturing related collaborative projects there is a realization by the participants that result dissemination in some form to the industry is beneficial. While guidelines or rules on the timing and extent of information release allows team members the opportunity to derive early and full advantage of project outputs, strict confidentiality clauses are seen as unnecessary and in many cases unhelpful. Indeed sharing of results by project members with their customers and suppliers can be seen as a distinct benefit. A culture of mutual respect and trust between members that they will treat project data responsibly and in the spirit of the organisation's mission principles has been found to be sufficient and productive.

An industry consortium focused on R&D co-operative activities presents members with benefits not typically available from smaller project specific collaborations. Usually the organisation will oversee a broad range of projects and members can decide the extent of involvement in each project. Within HDP User Group participants can provide services and resources to a project, act in an advisory capacity or merely take a passive role in monitoring outputs. Test vehicles are frequently designed such that they are deployable across multiple projects. Results and findings from one project can influence decisions on the approach of newer initiatives. Mechanisms have also been developed to allow non-member companies to take part in projects where they offer a critical contribution not otherwise available within the organisation. This ensures access to all process and material development options within the industry and not just those offered by members. The open and proactive environment encourages technical specialists from across the supply chain to exchange views and work together to identify common challenges and the opportunities to collectively overcome them. Drawing on the extensive knowledge base available within the organisation ensures that projects are unique and address areas not previously investigated in the industry.

Leveraging Cost Sharing Benefits – A Case Study

Embracing an attitude of sharing resources for mutual reward can deliver significant cost and time savings. Identifying those activities that are traditionally duplicated by similar companies (such as material evaluations) and instead conducting them at a group level, not only reduces the burden on individual companies, but also enables broader and more comprehensive studies to be undertaken. The net benefits are illustrated in a case study undertaken by HDP User Group. Several OEM members within the group had expressed concerns that PWB laminate materials marketed as Pb-Free process compatible had not been fully evaluated for applications where complex designs with high layer counts and dense component layouts were involved. Given that many of these designs were for telecommunication and information technology infrastructure products where high performance and long term reliability were essential, the group felt that a more rigid testing regime was required which should be applied to the latest generation of available laminates.

The project team consisting of in excess of 25 companies representing OEMs, printed circuit fabricators, laminate suppliers and assembly houses, designed a specialised test vehicle consisting of detachable coupons that could be despatched to separate test houses. Laminate providers were invited to submit materials for evaluation. Those that were HDP User Group members were granted access to all the results from the study while non-members would receive coded data that would indicate how their material ranked with other anonymous products. The test program was devised to assess the electrical and physical performance of the candidate materials through simulated solder reflow conditions. Previous projects had indicated that due to the high thermal mass of the products that the OEMs were focused on in this exercise, it was possible for the surface of the PWBs to reach up to 260°C during reflow. Multiple assembly processes and contingencies for rework meant that it was conceivable that typical PWBs could be exposed to up to 6 reflow cycles. The project was therefore designed to evaluate change in performance of an as fabricated set of test vehicles versus a similar set exposed to 6, 260°C peak thermal excursions.

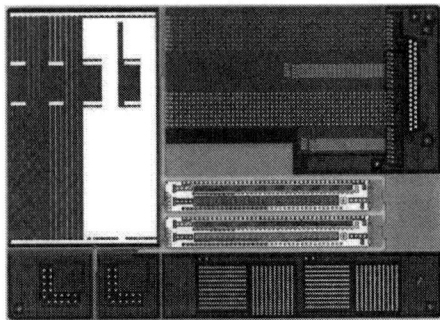

Figure 1: Pb-Free PWB Laminate Evaluation Test Vehicle

The project team secured the required test vehicle fabrication and test resources through in-kind contributions from its members and by leveraging member relationships with additional test service providers. A phased approach was adopted with 23 materials studied in phase 1 and 20 materials in phase 2. Fig 2 details the phase 2 testing schedule and illustrates the estimated cost of producing the test vehicles and conducting the evaluations if the required services were purchased at commercial rates.

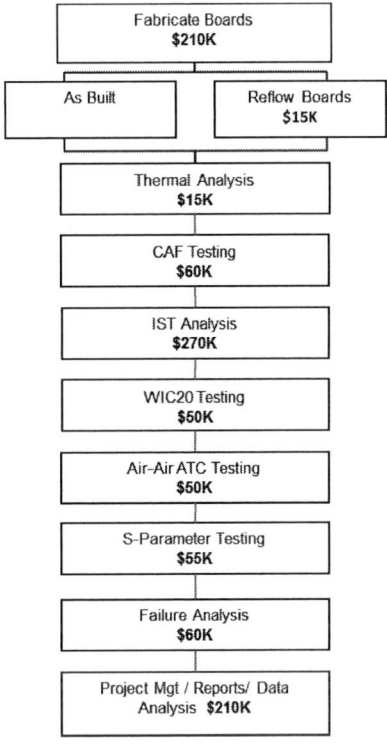

Figure 2: Test program and estimated costs based on commercial rates.

The PWB lead free materials study [1,2,3,4,5,6] clearly demonstrates the benefits of engaging a collaborative approach for this type of activity. While the total value (at commercial rates) of the project was estimated at $995,000 the individual financial outlay of any of the members did not exceed $50,000. Conducting a similar exercise in house would prove prohibitively expensive, even for a fraction of the number of materials evaluated. The combined negotiating power of the team enabled tests to be secured at extremely competitive and often negligible rates while laminate suppliers provided materials free of charge in lieu of the marketing opportunities that they could leverage in being connected with the program. Based on the success of this project the organisation is developing a common test platform approach for future materials evaluations.

179

A common challenge for collaborative programs is how the results are deployed at an operational level. Indeed, in an industrial consortium where competitors work side by side on manufacturing technology initiatives, competitive edge can be gained by individual members in how effectively they apply key findings. Different organisations will interpret and leverage results in different ways depending on the specific challenges they face. It is therefore important that each participant has access to all the results produced so that they can derive their own conclusions from the findings. Raw data rather than general conclusions are therefore regarded as the true value of the project.

Conclusion

The popularity of co-operative activities within the electronics manufacturing industry is growing rapidly as the outsourcing model that OEMs now embrace matures. There is now recognition that common solutions to common manufacturing challenges benefit the whole supply chain. Increasingly companies are turning to collaborative working practises to address urgent manufacturing process needs. By engaging players from all parts of the industry in initiatives of this type, cost effective solutions are more likely to be accepted and deployed across the industry. Risks and expenses are shared minimizing the resources required from each individual contributor. Industry R&D consortia have a major part to play in encouraging and facilitating development projects of this type, providing an environment where ideas can flourish and relationships between competing companies can be established.

References

[1] C. Xu, R. Kopf, J. Smetana, D. Fleming, "HDPUG Pb-Free Board Materials Reliability Project 2 Moisture Sensitivity and Its Effect on Delamination"

[2] Kim Morton, Joe Smetana, Gordon Qin, Thilo Sack, "The Effects of Lead-Free Reflow on Conductive Anodic Filament (CAF) Performance of Materials", APEX2011

[3] Michael Freda, James Frei, Jing Shi, Leoncio Lopez, "Use of Lead-Free Laminate DMA and TMA Data to Develop Stress versus Temperature Relationships for Predicting Plated Hole Reliability", APEX 2011

[4] Bill Birch, Jason Furlong, "High Density Packaging User Group – Pb-free Materials 2 Project Materials Testing of PWB Substrates to Establish Variability of Construction, Estimate Thickness and Determine Survivability through Lead Free Assembly", APEX 2011

[5] Joe Smetana, Bill Birch, Thilo Sack, Celestica, Kim Morton, Marie Yu, Chris Katzko, Erkko Helminen, Laura Luo, "Reliability Testing of PWB Plated through Holes in Air-to-Air Thermal Cycling and Interconnect Stress Testing after Pb-free Reflow Preconditioning", APEX 2011

[6] Joe Smetana, Bill Birch, Wayne Rothschild, "A Standard Multilayer Printed Wiring Board for Material Reliability Evaluations", APEX 2011

[7] Dan Feinberg, "Fein-Lines: Consortia--Worth the Cost and Effort?", Circuitree Magazine, July 2000

Computational Parametric Study on the Strain Hardening Effect of

Lead-free Solder Joints in Board Level Mechanical Drop Tests

Tong Jiang, Xu Zhengjian, S. W. Ricky Lee, Fubin Song, Jeffery C. C. Lo and Chaoran Yang

Electronic Packaging Laboratory

Center for Advanced Microsystems Packaging

Hong Kong University of Science and Technology

Clear Water Bay, Kowloon, Hong Kong

Abstract

Many lead-free solder alloys have been proposed as alternatives to the conventional eutectic Sn-37wt%Pb (SnPb). This replacement induces the drop/impact reliability issues of portable products. Under the drop/impact loading, the main failure mode is interfacial brittle failure between lead-free solder joints and intermetallic compound (IMC) layer. The identification of failure mechanism in the solder joint under drop test becomes a very critical issue. In this paper, a finite element analysis of BGA assembly with lead-free solder joints is conducted to investigate IMC interfacial stresses under repetitive mechanical drops. For the two drop conditions of concern, the dominant strain rates for the solder joints are studied. The results indicate that the stress on the PCB side is larger than that on the package side in both the solder and the IMC. After multiple drops, the stress in the IMC gradually increases due to the strain hardening effect of the solder. During the repetitive drop tests, the increasing stress will eventually exceed the strength of IMC and leads to the brittle fracture of solder joint. The findings in the present study result in a better understanding in the failure mechanism of solder joint under repetitive mechanical drop tests.

Key words: lead-free solder, IMC, drop test, stress

Introduction

With the recent growth in development of consumer electronics, there has come a unique set of reliability issues that are associated with portable products and their use environment. Portable electronics devices are subjected to accidental drop impacts resulting in damage and hence failures on the solder joints. Many lead-free solder alloys have been proposed as alternatives to the conventional eutectic SnPb. However, this replacement induces the drop/impact reliability of mobile electronic devices such as mobile phones and personal digital devices. The lead-free solder joints exhibit much poorer drop/impact performance than the eutectic SnPb solder joint specified by the board level drop test. JEDEC standard about board-level drop test for components of handheld electronic products provides the guideline for conducting drop reliability performance assessment of electronic assembly [1]. Furthermore, under

the drop/impact load，the main failure mode migrates from bulk fracture in SnPb solder joints to interface brittle failure between lead-free solder joints and IMC layer. So the solder joint failure mechanism of board level drop test becomes a very important issue.

Many studies reported that lead-free solders exhibit intermetallic compound (IMC) brittle failure modes during drop/impact reliability test with electroless-nickel immersion (ENIG) or organic solder preservative (OSP) pad finishes [2-8]. While the failure mechanism caused these brittle failure modes is still not well understood. Reiff and Bradley suggested an explanation for the difference in the drop impact reliabilities of PbSn and lead-free solder [2]. SnAgCu was found to deform less and generated greater stress at the pad interfaces, resulting in poorer drop test reliability. D.Y.R. Chong reported that the lead-based solder superseded the lead-free composition regardless of the types of surface finish [4]. The failure modes for the lead-free solder joints were brittle failure at the interfacial layers. S.K.W. Seah and E.H. Wong used high frequency bending test to perform experimental tracking of crack propagation in a single solder interconnection during drop impact [7]. While the crack propagation method under high speed four point bending tests still could not well explain the real failure mechanism of the brittle failure modes during the drop test experiments. F. Song and S.W.R. Lee made a correlation between the high speed ball pull/shear tests [8]. Power law relationships were obtained at different shear/pull testing speeds using the combined results of drops-to-failure. T.Y. Tee studied the drop test reliability and simulation of the TFBGA. The critical solder ball was observed to occur at the outermost corner solder joint, and fails along the solder and printed circuit board (PCB) pad interface [9]. For the correlation cases studied, the maximum normal peeling stresses of critical solder joints correlate with the mean impact lives measured during the drop test. While the simulation only focused on the first drop and stress on the solder joint not on the interface between the IMC and pad. Although there are arguments on the failure mechanism of the solder joints during drop tests, the real reason caused these interfacial brittle failure modes under multiple drop tests are still not very clear. For the drop tests simulation, most of the studies focused on the maximum stress on the solder joint layer under one impact pulse. The stress level on the IMC layer which is the failure location for the drop tests experiments is not well understood.

In this study, drop test finite element simulation for BGA assemblies with Sn3.0wt%Ag0.5wt%Cu (SAC305) lead-free solder joints were conducted to investigate solder joint and IMC drop performance. The input-G level was used as below: 500G acceleration with 1.0 ms duration, 1500G with 0.5 ms duration. For the single drop simulation, the dominant strain rate of the solder joint is investigated. For multiple drops, the simulation was performed considering the strain hardening effect of the SAC solder joint. The layer averaged von-Mises stress on the IMC interface layer is calculated with two different IMC thicknesses. These results provide a better understanding in the failure mechanism of solder 1joint under repetitive mechanical drop tests.

Drop Test Simulation

The 1/8 PCB 3D FE model is built using ANSYS 11.0 as shown in Fig. 1. The dimension of PCB is 142×142×2.2 mm. The BGA package contains a 27×27×0.4 mm

substrate and a 24×24×1.2 mm molding compound. The diameter of the lead-free solder joint is 0.76 mm and the standoff height is 0.52 mm. The pitch between the solder joints is 1.27 mm. The OSP pad is solder mask defined with 0.5 mm opening diameter both on the package side and PCB side. Two Cu_6Sn_5 IMC layer thicknesses 4 μm and 8 μm are chosen for the repetitive drops simulation.

Fig. 1. 3D FE model of board level drop test simulation.

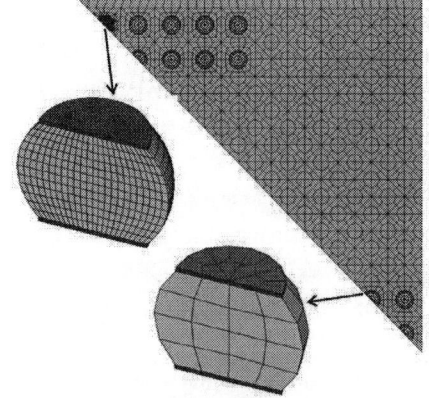

Fig. 2. The solder joint distribution for the board level drop test simulation.

Typically the BGA packaging has many solder joints. During the drop test, most solder joints may be still in the elastic region without plastic deformation. In order to highlight the plastic strain hardening effect under drop test, the solder joint distribution is designed as shown in Fig. 2. The 4×4 solder joints at the center are to hold the package to ensure the stand of height of the solder joints. And there are two lines solder joints and five each at the corner of package. Because the solder joint on the corner of the package is the failure location during the drop tests. The critical solder joint is fine meshed and the other solder joints are rough meshed.

The symmetric boundary condition was used and input-G method was applied at the screw hole on the corner of the PCB. The material properties of the board level drop test are shown in Table 1. The bilinear elasto-plastic material property is used for the lead-free solder joint. The yield stress of SAC305 is used as 55, 60, 65 and 70 MPa to study the strain rate. The tangent modulus of SAC305 lead-free solder is 120 MPa. The Young's modulus is from the previous nanoindentation tests on the Cu_6Sn_5 IMC [10].

Table 1. Material properties for the 3D FE model.

Material	Density (kg/m^3)	E (GPa)	v
SAC305	5770	56	0.35
PCB	2130	21.6	0.39
Substrate	1800	15.2	0.3
EMC	2100	25	0.25
Cu pad	8900	130	0.3
IMC	8520	117	0.3

Table 2. The drop test conditions.

JEDEC Drop Condition	Peak Acceleration (G)	Pulse Duration (ms)
A	500	1.0
B	1500	0.5

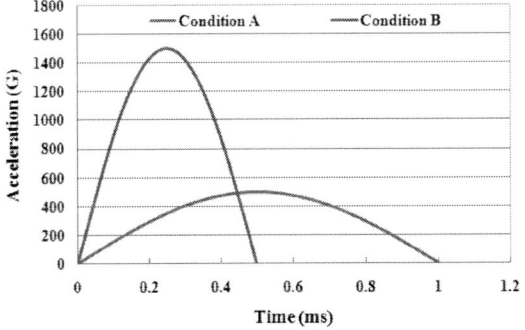

Fig. 3. The impact pulses of the drop test simulation.

The Input-G method for 4-screw fixation with impact pulse is applied in this model. With this method, the drop table, fixture, contact surface, and friction of guiding rods are not simulated, but their complex effects are considered indirectly by using the same impact pulse as drop test experiment [11]. This approach is very convenient to study the board level drop test. With this method, any type of impact pulse can be input to the FE model, and the dynamic responses of the solder joints and IMCs can be calculated. The input-G level was used as shown in Table 2. The drop condition is as specified in JEDEC-B110A. The condition-A is 500G acceleration with 1.0 ms time duration and condition-B is 1500G with 0.5 ms time duration.

Results and Discussion

Fig. 4 shows the dynamic responses of PCB center calculated by numerical modeling using Input-G method. After one acceleration pulse, the PCB vibrates freely for 9ms. The deflection of PCB reaches the largest value at the first peak and the damped gradually with the free vibration. The deflection of condition B is much larger than condition A because of higher peak acceleration. This means more violent impact will induce larger deformation for the board level drop test. For the first peak, the maximum deflection of PCB happens at 1.2 ms for condition A and 0.95 ms for condition B.

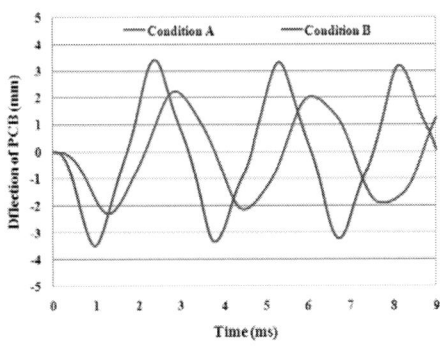

Fig. 4. The response of PCB under single impact pulse.

Fig. 5 shows the von-Mises stress distribution of solder joint with IMC. It indicates that the corner solder joint is critical solder joint with highest stress of IMC. At the same time, the stress of IMC on PCB side is larger than that on the package side. Therefore, the interface on the PCB side is susceptible to higher stress and much easier to fail.

183

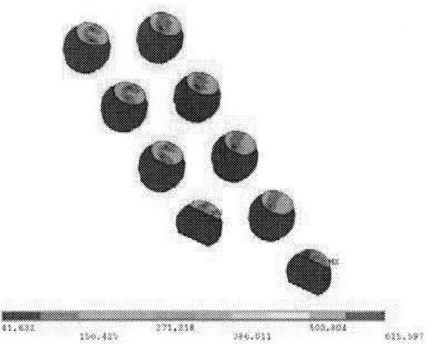

Fig. 5. von-Mises stress contour of solder joint with IMC.

Because the lead-free solder is strongly dependent on the strain rate, the chosen of the material property becomes an important issue. For high strain rate of SAC solder, all the data are from the Hopkins bar tests which are strongly dependent on the sample size and testing machines. In this study, the material property at high strain rate is estimated from previous study on the SAC305 lead-free solder alloy [12]. The Fig. 6 shows the comparison of the simulation strain rate and the material property strain rate under two drop conditions. The simulation strain rate is calculated from the first peak of the impact pulse on the solder joint layer connecting the IMC. From the results, the higher impact pulse generates higher strain rate on the SAC solder joint. From the Fig. 6, it can be found that the strain rate matches well with the experimental data at the yield stress of 60 MPa under condition-A and 70 Mpa for condition-B. So the 60 MPa yield stress is chosen for the condition-A and 70 MPa for the condition-B as the material properties of SAC solder joint for the multiple drop tests simulation. The dominated strain rate for condition-A is $30s^{-1}$ and $112s^{-1}$ for condition-B.

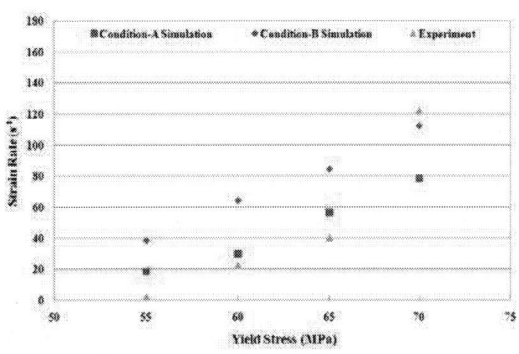

Fig. 6. The strain rate of the solder joint for the first peak under single impact pulse.

From the layer averaged stress history of IMC on the PCB side under single drop response, the first peak generates the highest stress. Then the response of the IMC gradually damped out as shown in Fig. 7. So the first peak is chosen for the multiple drop tests simulation.

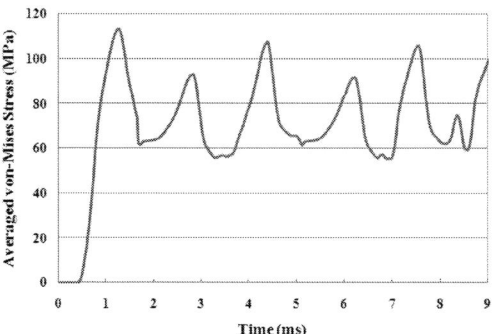

Fig. 7. The layer averaged von-Mises stress history of IMC under single impact pulse.

The Fig. 8 shows the nodal averaged von-Mises stress comparison of SAC solder layer and IMC layer on the package side and PCB side. For the same drop loading condition, the stress on the PCB side is larger than the package side. This can also correlate with the brittle failure location during the drop test experiment. Also the stress on the IMC layer is much larger than that on the SAC solder layer. So under multiple drop loadings, the stress on the IMC layer will be studied. For two drop conditions, the condition-B generates higher stress level than condition-A. This means under higher impact loading, the IMC will endure higher stress and much easier to fail.

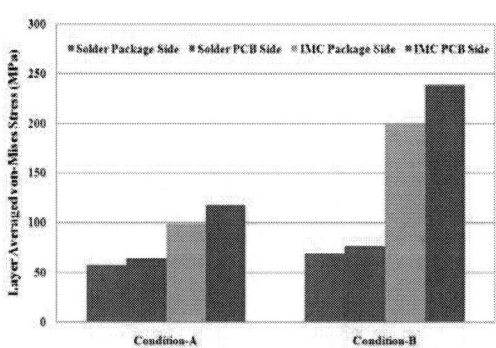

Fig. 8. The comparison of layer averaged stress on the package side and PCB side.

To simulate the drop test under repetitive loadings, the time integration solver is turned off to relax the stress in the solder joint after the first peak of PCB deflections [13]. For drop condition-A the deflection stopped at 2.1 ms and 1.7 ms for condition-B. Then, the time integration solver is turned off for the relaxation of stress. One drop loading is finished at the end of the relaxation. At the same time, the time integration solver resumes for the substantial drop loadings. Fig. 9 shows the PCB deflection time history of 5 drops under two drop conditions.

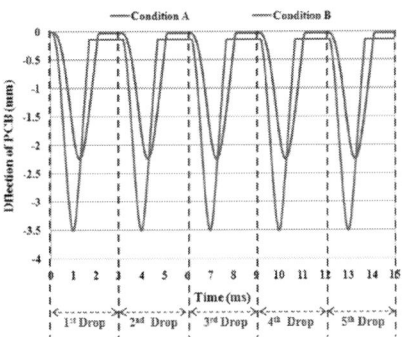

Fig. 9. Deflection of PCB under repetitive drop loadings.

For the board level drop tests with SAC lead-free solder joints, the failure mode is always brittle failure at the interface between the IMC and Cu pad. Since the stress is the root cause for the brittle crack of critical solder joints, the von-Mises stress is chosen for studies of the solder joints and IMC under repetitive drop impact loadings.

For the multiple drops, the stress on the IMC layer is calculated by the averaged nodal von-Mises stress. The Fig. 10 and Fig. 11 show the layer averaged von-Mises stress on the two IMC thickness layers. For these two drop conditions, the stress on the IMC layer gradually increases under repetitive drop loadings. The stress of 8 μm IMC thickness is larger than the 4 μm IMC. That means the thicker IMC layer will endure higher stress under the same impact condition. The stress of condition-B is higher than the condition-A with the same IMC thickness.

Based on the results from the simulation, the possible failure mechanism for solder joint drop failure is proposed. The stress on the IMC interface layer will increase after each drop due to the irreversible strain hardening effect of SAC solder joint. The failure mode is usually IMC interfacial brittle failure in drop test experiments. The increased stress on IMC is applied to the brittle IMC layer, and will eventually exceed the fracture strength of the IMC. Therefore, brittle failure will occur after a number of repetitive drops. In summary, the failure mechanism of solder joints subject to repetitive drops is a competition between the stress on IMC layer and the brittle fracture strength of IMC.

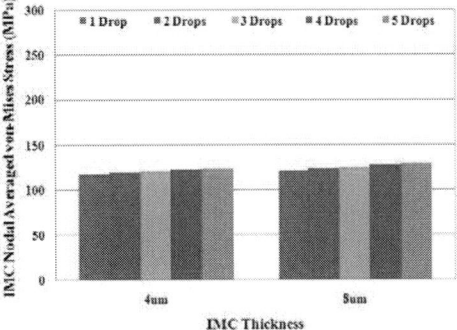

Fig. 10. Nodal averaged von-Mises stress of different IMC thickness under drop condition-A.

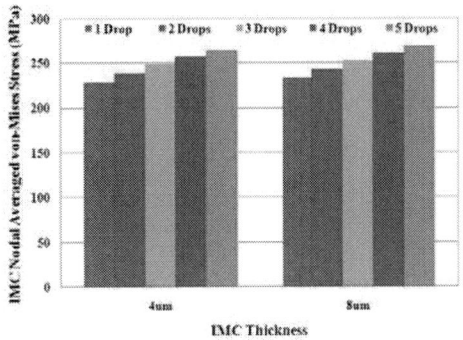

Fig. 11. Nodal averaged von-Mises stress of different IMC thickness under drop condition-B.

Conclusions

In the present study, 3D FE modeling of board level drop test was performed considering the strain hardening effect under repetitive mechanical drops. Based on the simulation, the plastic strain hardening effect of lead-free solder was proven as one important effect on the solder joint reliability under repetitive drops. The stress in IMC on the PCB side is larger than that on the package side. For a single drop simulation, the dominant strain rate of the drop condition-A is 30 s^{-1} and 112 s^{-1} for condition-B. Compared with the 4 μm thick IMC, the stress in the 8 μm thick IMC became higher and more fatal. For the same IMC thickness, the higher acceleration of the drop test generated higher stress in IMC. The interfacial layer averaged von-Mises stress in the IMC layer was found increasing with drop numbers. During the drop test, the interfacial stress would gradually increase through repetitive mechanical shock loadings. Eventually the stress would exceed the strength of IMC layer and brittle failure of the solder joint will occur.

Acknowledgments

This study was supported by Research Grants Council of Hong Kong through a General Research Fund project #615208 to the Hong Kong University of Science and Technology. The authors wish to acknowledge this support.

References

[1] M. Abtew, G. Selvaduray, "Lead-free Solders in Microelectronics," *Material Science & Engineering*, 27(2000), pp. 95-141.

[2] D. Reiff, E. Bradley, "A Novel Mechanical Shock Test Method to Evaluate Lead-free BGA Solder Joint Reliability," *Proc. 55th Electronic Components & Technology Conference*, Orlando, FL, (2005), pp. 1519-1525.

[3] A. Syed *et al.*, "Effect of Pb-free Alloy Composition on Drop/Impact Reliability of 0.4, 0.5 & 0.8mm Pitch Chip Scale Packages with NiAu Pad Finish," *Proc. 57th Electronic Components & Technology Conference*, Reno, NV, (2007), pp. 951-956.

[4] D.Y.R. Chong, F.X. Che, J.H.L. Pang *et al.*, "Evaluation on Influencing Factors of Board-level Drop Reliability for Chip Scale Packages (Fine-pitch Ball Grid Array)," *IEEE Transactions on Advanced Packaging*, Vol. 31(1), (2008), pp. 66-75.

[5] W.H. Zhu, L. Xu, J.H.L. Pang *et al.*, "Drop Reliability Study of PBGA Assemblies with SAC305, SAC105 and SAC105-Ni Solder Ball on Cu-OSP and ENIG Surface Finish," *58th Electronic Components & Technology Conference*, Lake Buena Vista, FL, (2008), pp. 1667-1672.

[6] D.Y.R. Chong *et al.*, "Drop Impact Reliability Testing for Lead-free and Lead-based Soldered IC Packages," *Microelectronics Reliability*, Vol. 46(7), (2006), pp. 1160-1171.

[7] S.K.W. Seah *et al.*, "High-speed Bend Test Method and Failure Prediction for Drop Impact Reliability," *Proc. 56th Electronic Components & Technology Conference*, San Diego, CA, (2006), pp. 1003-1008.

[8] F. Song, S.W.R. Lee *et al.*, "High-speed Solder Ball Shear and Pull Tests vs. Board Level Mechanical Drop Tests: Correlation of Failure Mode and Loading Speed," *Proc. 57th Electronic Components & Technology Conference*, Reno, NV, (2007), pp. 1504-1513.

[9] T.Y. Tee, H.S. Ng, C.T. Lim, E. Pek, Z. Zhong, "Impact Life Prediction Modeling of TFBGA Packages under Board Level Drop Test," *Microelectronics Reliability*, Vol. 44, (2004), pp. 1131-1142.

[10] T. Jiang, F. Song, C. Yang, S.W.R. Lee, "Nanoindentation Characterization of Lead-free Solders and Intermetallic Compounds under Thermal Aging," *Proc. 43rd International Symposium on Microelectronics (IMAPS)*, Raleigh, NC, (2010), pp. 314-318.

[11] J. Luan, T. Y. Tee, "Effect of Impact Pulse Parameters on Consistency of Board Level Drop Test and Dynamic Responses," *Proc. 55th Electronic Components & Technology Conference*, Orlando, FL, (2005), pp. 665-673.

[12] T. Jiang, F. Song, S.W.R. Lee, "Microstructure and Mechanical Properties of Sn-3.0wt%Ag-0.5wt%Cu Solder Alloy under Different Strain Rates," *Proc. 11th EMAP*, Penang, Malaysia, (2009), EP09-23-08. (on CD-ROM)

[13] C.L. Yeh, Y.S. Lai, C.L. Kao, "Evaluation of Board-level Reliability of Electronic Packages under Consecutive Drops," *Microelectronics Reliability*, Vol. 46(7), (2006), pp. 1172-1182.

RELIABILITY MODEL FOR ASSESSMENT OF LIFETIME OF LEAD-FREE SOLDER JOINTS

O. Švecová, P. Kosina, J. Šandera, I. Szendiuch

Brno University of Technology, Faculty of Electrical Engineering and Communication,
Department of Microelectronics, Technická 10, 616 00, Brno, Czech Republic

Phone: +420 604076151, fax: +420 541146298, e-mail: xrussk00@stud.feec.vutbr.cz

Abstract

Legal limitation on dangerous substance resulted into ban on use of lead soldering alloy in electrical industry. The era of lead-free alloys based on tin is coming. Lead-free solders are "young" and their characteristics haven't been surveyed sufficiently yet. Many research teams try to answer an easy question: "How reliable the lead-free solders are?" The given study deals with possible ways of prediction of solder joint made by lead-free alloy SAC 305 lifetime. By combination of several predictive methods we can check up on solder joint lifetime data acquired by different ways and efficiency of methods itself. Experimental, computerized and analytical methods were treated in the study and they proved coincident results.

Key words: lead-free solder, solder joint, ANSYS, reliability.

Introduction

The reliability of lead-free solders is a long discussed theme world-wide. It is possible to predict their lifetime by various methods. Theoretically by using statistical methods and/or fatigue models, experimentally by various tests, for example strength tests, accelerated reliability tests, climatic tests etc. and also mathematical modeling using simulation programs such as ANSYS, COMSOL et.

These software programs use for calculation the finite element method (FEM) which is the only precise method of the day enabling us to find out, already in the product design phase, the state and features of analyzed parts or units under the load limits they were projected for. Upon ascertained results, e. g. the state of deformation and tension (heat distribution, construction stability etc.), it is possible to determine how much mechanical characteristics of material are utilized. If the computation shows, the part is overequipped or alternatively doesn't meet requirements, we can use optimalization analysis furthermore. On the basis of FEM computations results it is possible to carry out a lifetime evaluation, possibly a probability analysis. Computerized methods need quality input data which are acquired by experimental ways. Thanks to acquired data we can describe operating environment of a device in a simulation program exactly and define border specifications for realization of a simulation.

Some of predictive methods do not take into consideration all factors which enter a solder joint creation process, e.g. intermetallic compounds which arise in a solder joint within soldering process. That is the reason why variances in results

of different methods come into existence. Of course, discussion on the topic "which of predictive methods is more precise when calculating solder joints lifetime" is still open.

Experimental Details

In the given study a temperature cycle test which belongs to accelerated reliability tests was used to determine solder joints lifetime experimentally. The aim of accelerated tests is to make failures or accumulate damages by the same damage mechanisms as in real operation, but in a shorter time.

Accelerated tests may be carried out under constant or step-by-step (gradual) load. Under constant load a time-dependant spread of damages is observed. When applying step-by-step (gradual) accelerated tests a load of a surveyed sample is being increased successively. The step which results in damage is observed. In the following case the firstly mentioned method based on constant load by outdoor temperature is used.

Temperature range was chosen under the IPC-SM 785 standard, i.e. from 0°C to 100°C. Technical parameters the experiment ran under follow:

• Components being soldered – chip resistors YAGEO 2512/0R in the ceramic package (the package size 6.325 mm x 3.036 mm), leads Ag/Ni/Sn;

• Methods of soldering – vapour reflow soldering;

• Solder paste for the vapor reflow soldering – SAC 305 (Sn96.5 Ag3 Cu0.5), flux Kester 979;

• The material of the printed circuit board – FR4, thickness of the copper – 35 μm;

• The surfacing of the printed circuit board – immersion tin, two layer PCB;

• Temperature cycle from 0 °C to +100 °C, dwell time on the maximum temperature 15 minutes, cycle's duration about 45 minutes;

•Metal stencil for print – thickness of the stencil 150 μm.

Test samples underwent aging under constant temperature before temperature loading. Test samples were aged during 300 hours and under temperature of 100°C according to recommendation of [6].

When the experiment had been brought to an end, the evaluation based on Weibull distribution was carried out in the statistical program STATISTICA and a probability curve of the solder joints lifetime was obtained (see. Fig. 1).

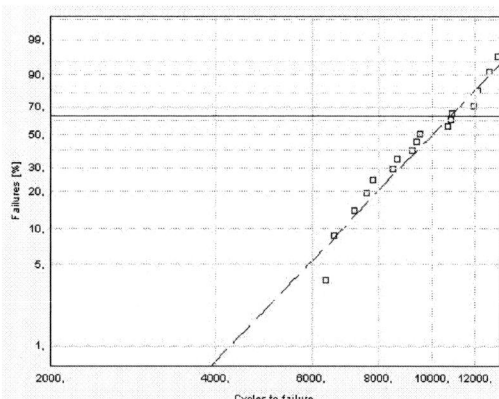

Figure 1: Probability chart

The chart shows the statistic of solder joints lifetime N_f (time at which 63.2 % of solder joints become damaged) that is equal to 11 thousand cycles.

Computing Life Prediction Model

Basic hypothesis of a used fatigue model says that damages arising in a solder joint within temperature cycling are caused by accumulation of the steady state creep strain. Such processes as a primary creep or time independent plastic deformation can influence damage mechanism. However we assume that these mechanisms are smaller and we can include them into rate model of deformation [1].

Instead of traditional definition of fatigue, cycling loading is considered a special case when the creep appears as a result of inoculated cycling loading. To describe an aforementioned process we can use two equations. The first one is the Monkman-Grant relation (see. Eq. 1) of ductility and rapture where t_r is time needed to reach the break.

$$t_r = \frac{\varepsilon_f}{\varepsilon_{cr}} \qquad (1)$$

This value is reversely proportionate to the steady state creep strain rate - $\dot{\varepsilon}_{cr}$ during testing and directly proportionate to creep ductility- ε_f.

In case of varying stressed repeated a special form of a time traction rule (see. Eq. 2) can be used for estimation of number of cycles to failure.

$$N_f \left(\sum_{i=1}^{n} \frac{\Delta t_i}{t_{ri}} \right) = 1 \qquad (2)$$

where :
N_f – number of cycles to failure,
n – number of steps within a cycle,
Δt_i – time spent at stress level σ_i within a cycle,
t_{ri} – rupture time for stress level σ_i.

When applying the Monkman-Grant relation, the relation no. 2 gets the following form (see Eq. 3):

$$N_f \left(\frac{\sum_{i=1}^{n} \Delta t_i \dot{\varepsilon}_{cri}}{\varepsilon_f} \right) = 1 \qquad (3)$$

where: $\dot{\varepsilon}_{cri}$ – rate of steady creep for stress level σ_i. Numerator of the equation no. 3 is accumulation of creep strain during time Δt_i. Summary of all steps n during a cycle sets the accumulated creep strain ε_{acc} during the whole cycle. The relation no. 3 can be easily described as (see Eg. 4):

$$N_f = \left(C' \varepsilon_{acc} \right)^{-1} \qquad (4)$$

where:
N_f – number of cycles to failure,
ε_{acc} – accumulated creep strain per cycle,
$C' - 1/\varepsilon_f$ inverse of creep ductility.

The strain exponent of the fatigue model is equal -1. [2] The Hyperbolic Sine Constitutive Equation (see Eq. 5) is used for calculation of accumulated creep strain:

$$\dot{\varepsilon}_{cr} = A_1 [\sinh(\alpha\sigma)]^n exp \left(\frac{-H_1}{kT} \right) \qquad (5)$$

Material constants used in calculations are given in the table no. 1.

Table 1: Material constants [3]

A_1, s^{-1}	a, MPa^{-1}	n	H_1/k
441000	0.0015	4.2	5412

The Hyperbolic Sine Constitutive Equation used to computation of accumulated creep strain was implemented into environment of a simulation program by standard implicit creep equation (TBOPT=8).

Analytical Life Prediction Model

The Engelmaier fatigue model was used for analytical computation. The model was developed at the beginning of the year 1980. The analytical model is based on a general fatigue model for metals. Used formulation is based on shear strain assumption resulting from easily measured parameters, such as solder joint height or distance from a neutral point. The model is half-empirical. Therefore it needs a factor counting with imperfection. The Engelmaier model uses the F factor which takes into consideration effects of imperfection as cyclic deformation of components or PCB, imperfect geometry of a solder joint, intermetallic compounds, inaccuracy in parameters values.

The Engelmair model (see Eq. 6) consists of two parts. The first one is physical. It gives number of cycles to 50 % of total number of compounds which fail. The second part is statistic. It presents total variation within representative sampling according to the two-parametric Weibull distribution.

$$N(x\%) = \frac{1}{2}\left(\frac{2\varepsilon'_f}{\Delta D}\right)^{-\frac{1}{c}}\left(\frac{\ln(1-0.01x)}{\ln 0.5}\right)^{\frac{1}{\beta}} \quad (6)$$

where:

ε'_f – fatigue ductility coefficient;

ΔD – the solder cyclic creep-fatigue damage (see Eq. 8);

c or m – creep-fatigue ductility exponent (see Eq 9).

β – the shape parameter of the Weibull distribution.

In the given computation a simplified form of the Engelmaier model is used. It consists just from the physical part.

$$N(50\%) = \frac{1}{2}\left(\frac{2\varepsilon'_f}{\Delta D}\right)^{-\frac{1}{m}} \quad (7)$$

Hereinafter there are formulas used for computation of other values entering the computation.

$$\Delta D = \left[\frac{F \cdot DNP \cdot \Delta CTE \cdot \Delta T}{h}\right] \quad (8)$$

where:

F – engineering factor, ~1.2 to 0.7 for SJS with fillets, ~1.5 to 1.0 for SJS without fillets;

DNP – distance from the neutral point/plane;

ΔCTE – CTE-mismatch;

ΔT – cyclic temperature excursion;

h – solder joint height.

$$\frac{1}{m} = c_0 + c_1\overline{T}_{SJ} - c_2\ln\left(1 + \frac{t_0}{t_D}\right) \quad (9)$$

where:

$\overline{T}_{SJ}$ – mean cyclic solder joint temperature (see Eq. 10);

t_D – half-cycle dwell time in minutes;

t_0, c_0, c_1, c_2 – specific coefficients.

$$\overline{T}_{SJ} = \frac{1}{4}\left(T_C + T_{C,0} + T_S + T_{S,0}\right) \quad (10)$$

where:

T_C, T_S – high cycle temperatures of component and substrate, respectively;

T_{C0}, T_{S0} – low cycle temperatures of component and substrate, respectively.

In the table no. 2 there are specific coefficients for computation of cycles to failure based on the aforementioned fatigue model for lead and lead-free solder.

Table 2: Specific coefficients for the Engelmaier fatigue model [4]

Specific coefficients	Values
ε'_f	0.425
c_0	0.480
c_1	$9.3 \cdot 10^{-4}$
c_2	$-1.92 \cdot 10^{-2}$
t_0	500

Results and Disscussion

Surfacing of printed circuit boards is not taken into account at calculation according to the Engelmaier relation. It is necessary to measure the height of a solder joint to calculate relative stress in a solder joint. The metallographic cut was realized for this purpose.

Figure 2: Height of a solder joint for analytical calculation

On recommendation [5] the height of a joint is measured at the level of the point of intersection of angel axis which comes from the corner formed by a solder and a compound (see Fig. 1), and the solder meniscus. The measured value of height is equal to 273 μm.

For simplicity the key values were put in the table. The tables no. 3 and 4 show operational conditions and general physical characters of single components entering the experiment. Results of calculation are shown in the table no. 5.

Table 3: Physical parameters of components in the experiment

Physical parameters	2512 RC	PCB (FR4)
F	1	-
DNP (mm)	3.1	-
CTE(ppm/°C⁻¹)	7	14
h (mm)	0.271	-

Table 4: Operating conditions of the experiment

Parameters	Values
T_{min} (°C)	0
T_{max} (°C)	100
T_{sj} (°C)	50
ΔT (°C)	100
t_D (min)	7.5

Table 5: Results of calculation

Parameters	Values
m	2.244
ΔD	0.008
N_f (50%) (cycles)	17607

Geometry for solder joints of a test formation was created in the ANSYS simulation program to verify the analytically acquired result (see Fig. 3).

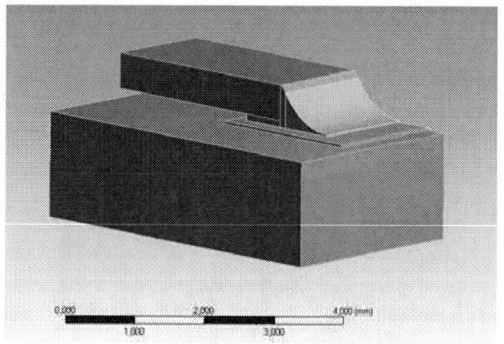

Figure 3: ¼ of test formation

A symmetric model (1/4 of a solder joint) was designed in DesignModeler to ease and accelerate calculation. It means that under the given way of defining border conditions deformation results acquired within simulation will be in compliance with symmetry of a body.

Within the whole experiment the temperature was measured on a solder joint, a component package and a printed circuit board to define border conditions more precisely.

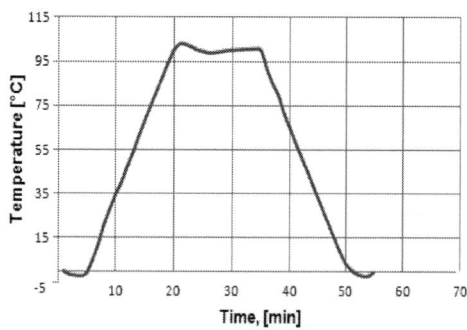

Figure 1: Temperature profile of the cycling

The designed structure will be explored throughout four full temperature cycles which correspond to the temperature cycle measured in environment of a test temperature chamber (see Fig. 4).

Four temperature cycles were used for reasons of creep stabilization and slight changes $\Delta\varepsilon_{cr}$ among cycles (see tab. 6)

Table 6: Change of creep per cycle

Temperature cycle	Change $\Delta\varepsilon_{cr}$
first	0.475 e^{-2}
second	0.191 e^{-2}
third	0.151 e^{-2}
fourth	0.126 e^{-2}

Figure no. 5 shows the point of creep concentration which arises in the solder joint during four test temperature cycles.

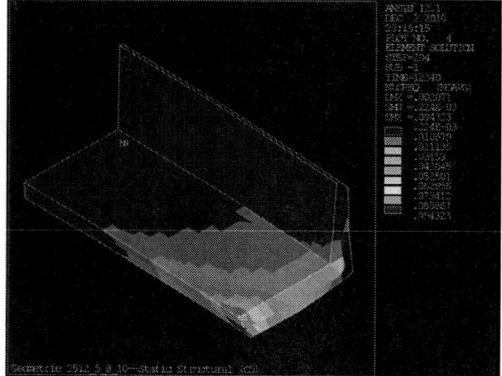

Figure 2: Creep of solder joint – fourth cycle

Regarding the fact a heat transfer coefficient of every material is different the coefficient for each of test formation materials was determined experimentally before starting calculation.

The simulation verified the solder joint lifetime calculated according to the chosen fatigue model is equal to 15443 cycles which agrees approximately to analytical calculation.

Also other simulation results are presented: elastic strain of the formation (see Fig. 6), stress arising in the solder joint (see Fig. 7) and total deformation of the solder joint (see Fig 8).

Note: for hereinafter mentioned outputs the shown results are several times multiplied for better illustration of forces and actions which act in the solder joint.

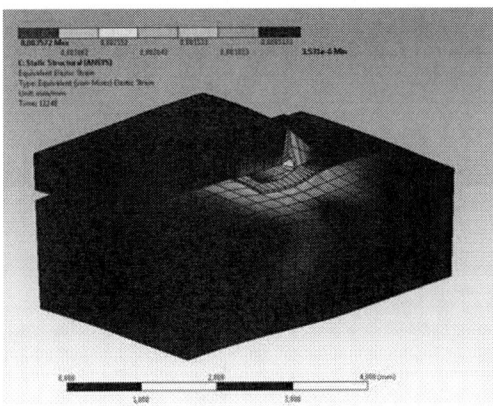

Figure 3: Equivalent elastic strain

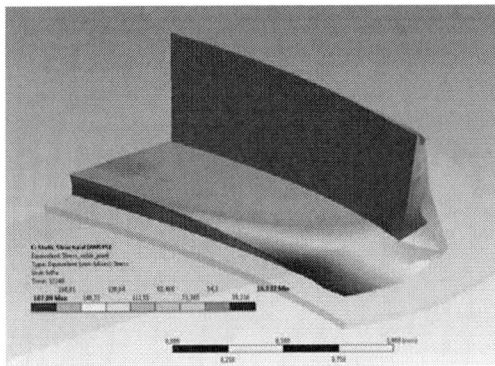

Figure 4: Equivalent stress of solder joint

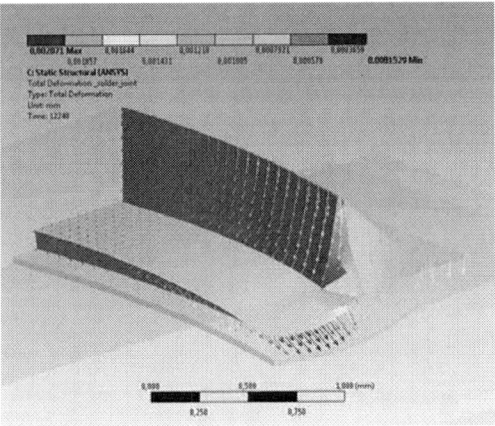

Figure 5: Total deformation of solder joint

Conclusions

The given study dealt with prediction of solder joints made by lead-free alloy SAC 305 lifetime. By means of temperature cycling numbers of cycles to failure were acquired and statistically utilized. It was proved that Weibull characteristic lifetime of a given test formation is equal to 11 thousand cycles.

With the help of the ANSYS simulation program the Hyperbolic Sine Constitutive Equation was implemented in the accumulated creep deformation was determined. During four temperature cycles the number of cycles to failure was determined as well. It equals to 15443 cycles.

The Engelmaier fatigue model was used for analytical calculation. It sets the number of cycles to failure at 50 % of representative population. The result of this analytical method is equal approximately to 17 thousand cycles.

Three different methods of solder joints lifetime prediction were considered in the study. They proved agreeing results. However at analytical calculations it is necessary to take into account that all inlet parameters are not considered.

Acknowledgements

Funding for this research was obtained through grant project from the Czech Ministry of Education MSM 0021630503 MIKROSYN „New Trends in Microelectronic Systems and Nanotechnologies", the Czech Ministry of Industry and Trade, project named TIP FR-TI1/049 "Use of Modern Assembly Techniques and Materials in Electronic Industry" and by Grant project FRVS 2738/2011 "Workplace for solder joint strength testing".

References

[1] J. .H. Lau, "Ball Grid Array Technology", McGraw-Hill Publishers, New York, Chapter 13, pp. 379-439, 1995.

[2] A. Syed, "Accumulated Creep Strain and Energy Density Based Thermal Fatigue Life Prediction Models for SnAgCu Solder Joint" Proceedings Paper ECTC 2004, pp. 737-746. ISBN 0-7803-8365-6

[3] Dongkai Shangguan, "Lead-free solder interconnect reliability", ASM International, USA, Chapter 8, pp. 181-196, 2005.

[4] W. Engelmaier, "Pb-free solder creep-tatigue reliability models updated and extended". Global SMT and Packaging, Vol. 9, No.9. September, 2009. ISSN 1474-0893

[5] W. Engelmaier, "Solder joint reliability prediction for chip components, MELFs, TSOPs, SOTs, ets", Global SMT and Packaging, Vol. 9. No.6, June 2009. ISSN 1474-0893

[6] Standard IPC-SM-785 "Guidelines for Accelerated Reliability Testing of Surface Mount Solder Attachments".

[7] M. Novotný, I. Szendiuch "Lead-Free Solderinf Quality Investigation", Vol. 6, No. 4, pp. 18-27, October, 2006.

Modelling and Experimental Measurement of Multiple Joint Lead Free Solder Interconnects Subjected to Low Cycle Mechanical Fatigue

E. Kamara[1*], H. Lu[1], O. Thomas[2], D. Di Maio[2], C. Hunt[2], I. Fulton[2]

1. University of Greenwich, Park Row, Greenwich, London, SE10 9LSUK
2. National Physical Laboratory,
Queens Road, Teddington, Middlesex TW11 0LW, UK
*E.Kamara@gre.ac.uk

Abstract

Ball Grid Arrays (BGAs) are one of the dominant technologies for high-end packaging solutions of electronic circuits, especially where a high interconnect density is required. This paper aims at studying the solder isothermal fatigue failure characteristics in test structures that resembles BGAs. Test structures comprising between 1 and 8 small solder joints have been made using 96.5Sn3.0Ag0.5Cu lead-free solder and subjected to cyclic mechanical loading that produced a shear force across the joints. The same structures were modelled using finite element analysis. The loading response distribution profile through the joints is analysed and the regions of likely failure identified to be along the shear band and at the stress concentration areas in the corners of the joints. Volume weighted average accumulated creep strain per cycle is used to analyse the failure of the individual joints where it is shown that greater damage occurs in the outer joints. Solder joint models of three different shapes are investigated: rectangular, convex and concave shapes. Results have shown that less damage occurs in the concave joints..

Key words: Solder joint, BGA, low cycle fatigue, lead-free

Introduction

Ball Grid Array (BGA) is a surface mount packaging technique that is popular in electronics manufacturing industries. It offers high interconnection density, easy assembly, and low thermal resistance. Typically, a square array of solder balls connect a laminate substrate and a semiconductor die. Because of their geometry, the ball-shaped solder joints in BGAs are not able to easily deform to accommodate thermal-mechanical and other stresses [1]. Due to the Coefficient of Thermal Expansion (CTE) mismatch between Printed Circuit Board (PCB) substrate and the package, solder joints experience a cyclic strain in temperature varying conditions that induce fatigue in the solder joints and other parts of the assembly, leading to failure of the solder resist, delamination at interfaces of materials, or in the bulk solder joints [1,2]. In this paper, we shall focus on isothermal fatigue testing of specimens containing multiple solder joints. The loading conditions on solder joints are similar to those found in BGAs that are subject to thermal-mechanical loading. The aim of the work is to establish an effective low cost test system that can be used for the evaluation of solder alloys that are used for BGAs, and to provide material data that can be used in physics-of-failure based reliability prediction for BGA or other similar devices with solder interconnects.

Experimental Procedure

Solder interconnect samples with multiple joints were fabricated from copper coupons of size 51×6×1 mm. A low speed diamond saw was first used to produce a 0.3 mm cut running centrally along almost the entire length of the coupon. This slit was placed over a jig comprising an oxidised stainless steel shim with a specified number of slots corresponding to the size and number of joints required in the sample. Before soldering, this space was filled in excess with the desired solder paste. In this study, 96.5Sn3.0Ag0.5Cu lead free solder was used. The jig was then placed onto a hot plate in order for to allow the solder to reflow. After about 5 seconds into the reflow process, the jig was removed from the hot plate and placed on a cold metallic surface to cool such that the time above the reflow temperature was between 10 to 30 seconds. Abrasive paper was used to remove excess solder and slots were cut in the sample (Figure 1) such that when pushed/pulled, a shear force is applied to the solder joints. Samples with 1 to 8 joints were prepared using this procedure and these are reffered to as M1-M8.

The samples were tested by applying isothermal cyclic displacement loading using the Interconnect Properties Testing Machine (IPTM), an instrument described in detail elsewhere [3, 4]. The ends of the samples were clamped and pulled/pushed along the length to produce cyclic deformation in solder joints. A load cell was used to monitor the supported force and laser displacement sensors were used to measure the distance between the fixed and moving stages of the instrument. The displacement was controlled through the thermal expansion of a stainless steel tube connected to the moving stage. As such, the machine can accurately control the strain applied to the sample, whilst measuring the supported load. In addition, a camera with microscope attachment was placed over the sample to take time-lapse photos of the sample during the test.

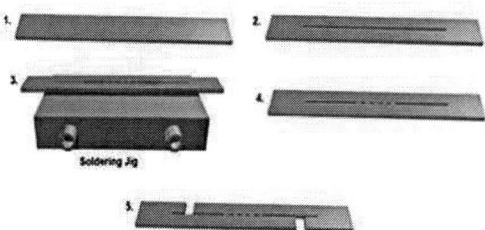

Figure 1: Illustration of sample preparation.

The range of displacement that is applied to the samples is -30 to 30 μm. The ramp rate is 10^{-7} m/s and the dwell time is 5 minutes, producing a cycle time of 30 minutes (see Figure 2). With increasing number of solder joints, the supported force load increases and the strain in the copper also increases, whilst the strain in the solder joints is reduced as the result. To produce a controlled strain in samples with different number of solder joints, the compliance of the copper needs to be taken into account when applying the displacement. To apply this correction, a copper coupon with the same geometry but without solder joints was first tested in the IPTM. Since the supported force is in the elastic region of the copper, it is possible to correct the solder joints tests in real time with the maximum applied displacement adjusted to take into account the extra displacement in the copper. The applied displacement is obtained from Eqn 1.

$$d(t) = w(t) + kF(t) \qquad (1)$$

where $d(t)$ is the displacement applied at time t, $w(t)$ is the waveform displacement required, $F(t)$ is the force and k is the experimentally found compliance. With this correction, the same displacement waveform could be applied to each of the different number of joint samples.

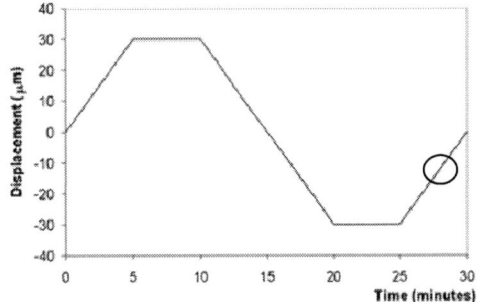

Figure 2: The required displacement waveform.

Modelling Procedure

The tests have been simulated using Finite Element Analysis(FEA) software ANSYS Mechanical APDL [6]. Figure 3 shows the model of a sample with 6 solder joints.

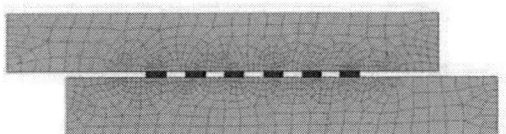

Figure 3: FEM meshing of a 6 joints sample.

The elastic material properties used for the simulations are as shown in Table 1. The loading profile is the same as the one in the experiment (Eqn. 1 and Fig. 2). Since we are dealing with low cycle fatigue, the effect of creep is significant and should be taken into account for in the simulation. In this study the Garofalo's creep model [5] is used to simulate the effect of creep in solder specimens and the creep strain rate equation is shown in Eqn. 2.

$$\dot{\varepsilon}_{cr} = c_1 \left[\sinh(c_2 \sigma) \right]^{c_3} \exp\left(-\frac{c_4}{T} \right) \qquad (2)$$

where c_1, c_2, c_3 and c_4 are material constants, T is the temperature, σ is the effective stress and ε_{cr} is the creep strain.

Table 1: Material properties. E and ν are Young's modulus and Poisson's ratio respectively.

Material	E (GPa)	ν
Sn 3.0Ag 0.5Cu	50	0.4
Copper	120	0.33

In the current experiments, solder joints are approximately cuboid in shape. In order to understand the effects of geometry on the stress state in solder joints, three different solder interconnect shapes are modelled: rectangular, convex and concave shapes (illustrated in Fig. 4).

193

Each solder joint is modelled using 400 ANSYS SOLID185 elements.

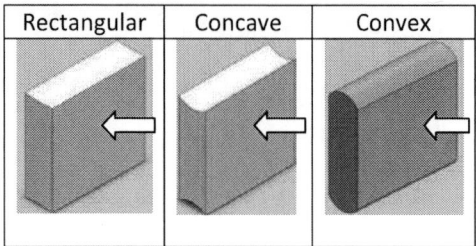

Figure 4: Solder joint shapes investigated. The arrows indicated where the attachment will be between the solder and copper

To predict the damage caused by cyclic loading on the solder joints and hence their lifetime, the volume weighted accumulated equivalent creep strain (ANSYS output variable NLCREP) per cycle ε_{acc} is obtain. A typical accumulated equivalent creep strain value is shown in Fig. 5 for the first three cycles of loading. The value of ε_{acc} is not a constant but from the third cycle the change becomes very small.

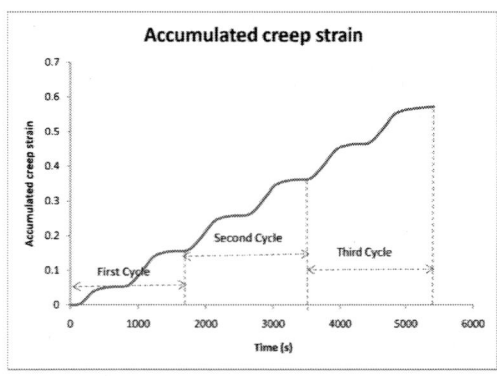

Figure 5: Accumulated creep strain over three cycles for five sample solder joint

ε_{acc} is a measure of the damage in solder joint and it can be linked to the solder joint lifetime through an empirical Coffin-Manson type lifetime model. In this work, the lifetime model used to calculate the lifetime of solder joint is as given by Syed [7]:

$$N_f = \left(0.0513\varepsilon_{acc}\right)^{-1} \qquad (3)$$

Results and Discussions

Each sample was mechanically cycled until the load dropped to 50% of the maximum supported load. Figure shows how the load drops of for 3 M5 samples. In a few instances the experiment was

stopped earlier than the 50% load drop criteria due to various reasons (i.e. computer crash). The variability in the rate at which the load drops for samples of the same type is mainly caused by the statistical nature of the fatigue process [8], but also by unavoidable voids and defects in the samples, and small geometrical differences.

Optical micrographs of the fatigue damage produced in some of the samples are shown in Fig. 7.

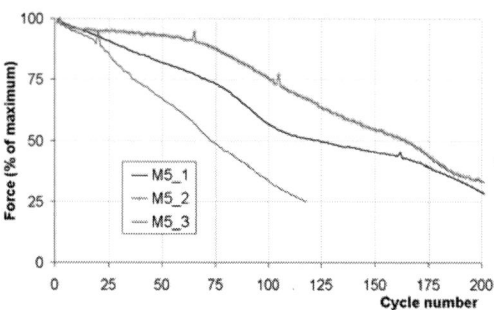

Figure 6: Force drop (as % of maximum) vs cycle number for the sample with five joints.

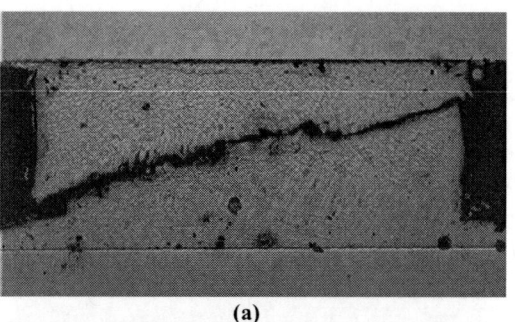

(a)

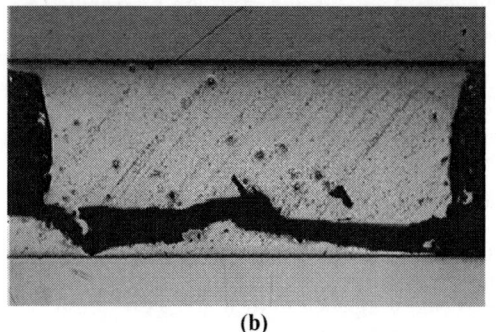

(b)

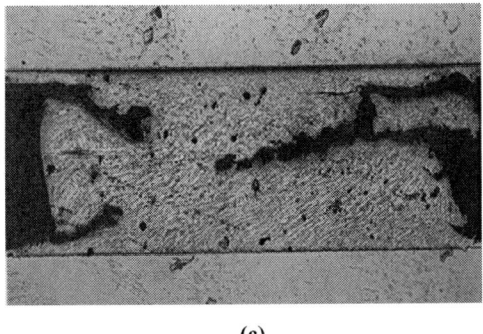

(c)

Figure 7: (a) Crack in the central joint of a M5 sample; (b) crack in the third joint of a M8 sample (c) multiple cracks in an M5 sample and crack along the interface.

In the simulation, stress and strains are obtained. Fig. 8 shows the effective stress as a function of time at the corner of one of the outermost solder joint in a M5 sample. It shows that the stress is highest just before the onset of dwell times. Figure 9 shows a snapshot of the Von Mises and shear stress distribution in the M5 sample at 2088 s. The maximum stress is located at the solder-copper interface of the corner solder joints. Because of the stress relaxation, the stress level in solder joints exhibits much less inhomogeneity than in copper. The shear stress is concentrated in solder joints but it is not uniformly distributed. As the figure shows, the copper parts bend slightly under the loading and deformation in solder joint is therefore not pure shear and stress states are expected to vary from one joint to another. At the outermost solder joints, deformation occurs in the direction perpendicular to the loading because of the bending of copper. This situation is similar to what happens in BGA solder joints under thermal-mechanical loading. Another interesting phenomenon that can be observed is diagonal shear band, in the edge solder joints in particular.

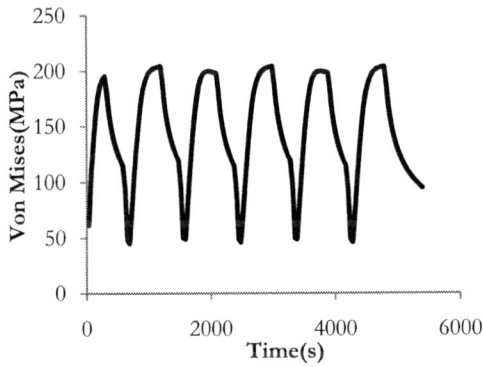

Figure 8: Von Mises stress at the corner of the edge solder joint of a 5-joint model (M5).

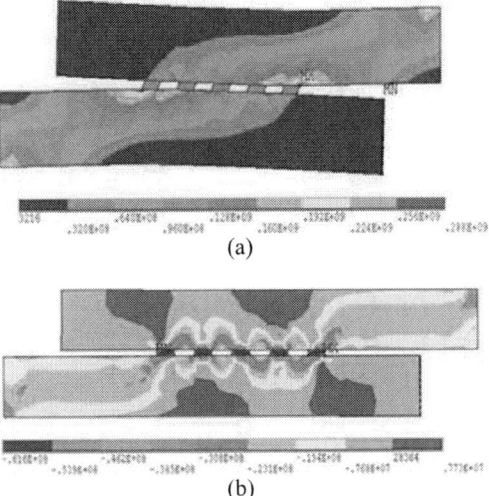

(a)

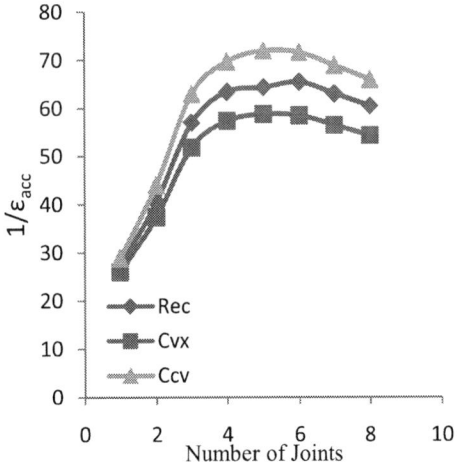

(b)

Figure 9: (a) Von Mises Stress and (b) Shear Stress XY in a modeled 5 joint sample at the start of the first dwell at 30 μm displacement. Displacement in image (a) has been magnified 20 times.

Fig. 10 shows how the three different shapes investigated affect the calculated lifetime for the different number of joints investigated. The results suggest that the convex shapes will have less damage and hence longer lifetime than that of the concave and rectangular shapes.

Figure 10: $1/\varepsilon_{acc}$ vs. number of joints in the sample for the rectangular (Rec), Convex (Cvx) and Concave (Ccv) solder joint shapes.

In Fig. 11, the values of the inverse damage, i.e. $1/\varepsilon_{acc}$, for each individual solder joints are plotted. The results show that outermost solder joints suffer more damage than internal ones and will therefore fail first.

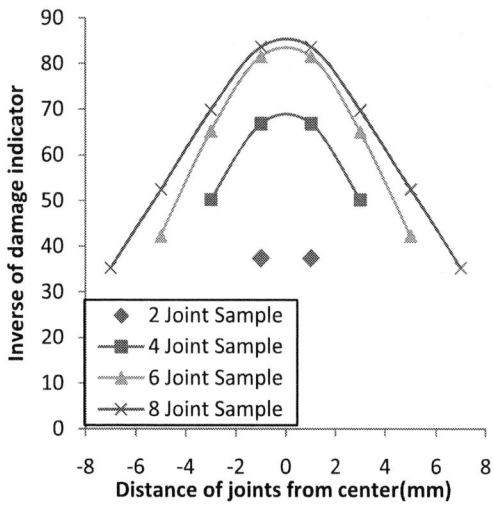

(a)

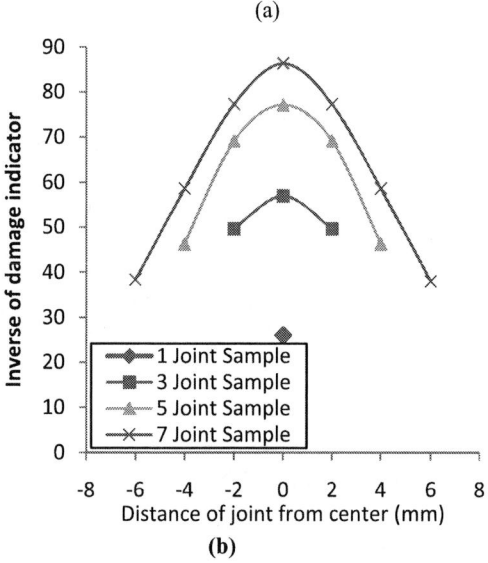

(b)

Figure 11: Modelled joint inverse damage for (a) even and (b) odd number of joints samples.

Fig. 11 also shows that in general, samples with different number of joints have different average damage. This is confirmed by Fig. 12 where the average damge is shown for samples with 1 to 8 joints. The result indicates that the single joint sample will be the most damaged of all the multi-joint models and that the least damaged will be the M5 sample, and M4 and M6 showing similar results.

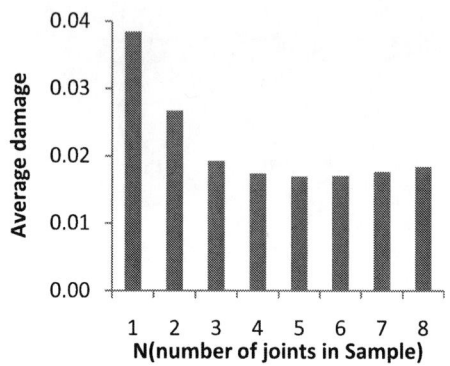

Figure 12: Average damage in samples with N number of joints.

Discussion

The experimental results for the number of cycles required to reach 10%, 20% and 50% of load drop from the maximum load, is plotted in Figure 13 against the number of joints per sample. The relationship between the number of joints and the cycles to reach a pre-defined failure is not that clear but it show that the sample with the longest lifetime is the M5, which is cnsistent with the simulation result.

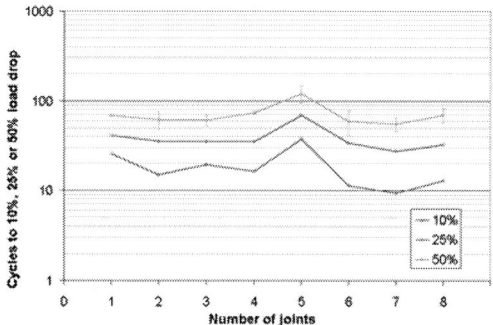

Figure 13: Cycles to 10%, 20% and 50% vs number of joints in sample. Error bars on the 50% load drop data represent the repeatability observed from testing on average 3 samples for each number of joints.

Using the time-lapse photography technique, 4 main different types of failure/crack propagation mechanisms were identified. Type 1 consists of a crack that runs diagonally across the sample as in Fig. 7(a). Type 2 consists of a crack that runs parallel to the interface, as in Fig. 7(b). Type 3 consists in a crack along or very close to the interface (Fig. 7(c)). This is similar to what Lai *et. al* have found in their study [9]. Finally, type 4 is a surface damage that causes the solder to appear darker but without any visible crack.

The initiation of the crack was also not univocal; sometimes cracks initiated in one place, sometimes in several places, with multiple cracks growing and combining as the sample gets more damaged (see Fig. 6(c)).

From these experiments, it was observed that generally, voids in the solder bulk aided the propagation of cracks confirming the findings of Yunus *et al.* [10]. Cracks propagated in all joints almost simultaneously, with minimal changes in crack propagation rate between different joints. However, a tendency for the outer joints to fail first compared to the inner joints was also observed, which is consistent with simulation results.

The computer simulation has shown that solder joints experiences both shear and normal stress of appreciacle magnitude. This proves that the stress states in the multi-joint test samples are similar to those in BGA solder joints and solder is exposed to different stress states, both direct and shear. For the samples that have 5 solder joints are predicted to have the longest lifetime. The most significant difference between the modelling results and the experimental results is that in experiments, samples with one or eight solder joints are not the ones that have the shortest lifetime. This needs more investigation. In future work, lifetime models will be derived based on experiments on the solder joint specimens.

Acknowledgements

The authors acknowledge the financial support of the IeMRC, EPSRC, and the National Physical Laboratory.

References

[1] Moore, T.D., and J.L. Jarvis. "Failure analysis and stress simulation in small multichip BGAs." *IEEE Transactions on Advanced Packaging* 24, no. 2 (May 2001): pp216-223.

[2] Zhang, Bo, Han Ding, and Xinjun Sheng. "Reliability study of board-level lead-free interconnections under sequential thermal cycling and drop impact." *Microelectronics Reliability* 49, no. 5 (May 2009): pp530-536.

[3] D. Di Maio, O. Thomas, M. Dusek & C. Hunt, " Novel Testing Instrument for Lead-Free Solder Characterisation", IEEE Transactions on Instrumentation and Measurements, In Press.

[4] D. Di Maio, C. Murdoch, O. Thomas, C. Hunt, "The Degradation of Solder Joints under High Current Density and Low-Cycle Fatigue", 2010 11th International Thermal, Mechanical & Multi-Physics Simulation, and Experiments in Microelectronics and Microsystems (EuroSimE), pp1-6

[5] Hongtao Ma. "Constitutive models of creep for lead-free solders." *J Mater Sci* (May 2009) 44:pp3841–3851

[6] ANSYS Mechanical APDL 12.0.1, ANSYS Inc. 2009

[7] A. Syed, "Accumulated Creep Strain and Energy Density Based Thermal Fatigue Life Prediction Models for SnAgCu Solder Joints", Corrected versions of the paper presented in the 54th Electronic Components and Technology Conference, Las Vegas, USA, 1-4 June, 2004, pp. pp737-746

[8] Palmer, M A. "Thermomechanical Fatigue Testing and Analysis of Solder Alloys." *Stress: The International Journal on the Biology of Stress* 122, no. March (2000): pp48-54.

[9] Lai, J.K.L., K.C. Hung, Y.C. Chan, and P.L. Tu. "Comparative study of micro-BGA reliability under bending stress." *IEEE Transactions on Advanced Packaging* 23, no. 4 (2000): pp750-756.

[10] Yunus, M, K Srihari, J Pitarresi, and a Primavera. "Effect of voids on the reliability of BGA/CSP solder joints." *Microelectronics Reliability* 43, no. 12 (December 2003): pp2077-2086.

Effect of heating method on microstructure of Sn-3.0Ag-0.5Cu solder on Cu substrate

H. Nishikawa, N. Iwata, T. Takemoto

Joining and Welding Research Institute, Osaka University, Osaka Japan

+81-6-6879-8691, nisikawa@jwri.osaka-u.ac.jp

Abstract

With the miniaturization of electronic productions and the use of heat sensitive electronic components, the traditional reflow soldering process often has difficulties. As an alternative soldering process, the laser soldering process has been recently proposed. The laser soldering process brings several advantages in terms of localized heating, rapid rise and fall in temperature, non-contact and easily automated process. In this study, we investigated the effects of heating method such as reflow soldering and laser soldering on the microstructure of Sn-3.0Ag-0.5Cu solder on Cu substrate and the impact properties of the solder joint. In the as-soldered condition, the intermetallic compound layer at the joint interface formed during the laser soldering process was very thin. Therefore, the impact properties of joints formed with the laser process were superior to those of joints formed with the reflow process.

Key words: lead-free solder, laser soldering process, microstructure, impact properties, joint interface

Introduction

Lead-free soldering in the electronic industry is showing a global trend. Among the various lead-free solders, Sn-Ag-Cu solders are considered the most promising for both conventional wave and reflow soldering processes. In lead-free soldering process, the operation temperature is higher than that in conventional Sn-Pb soldering process due to high melting temperature of Sn-Ag-Cu solders. Then, as one of basic characteristics of lead-free solders, the high reaction rate of metals in molten lead-free solders; that is, the high dissolution rate of metals, is pointed out compared with Sn-Pb eutectic solder. Therefore, during heating, a thick intermetallic compound (IMC) layer is easily formed at the interface between lead-free solder and Cu pad. The IMC thickness at the interface for Sn-Ag-Cu solders drastically increases compared with that for Sn-Pb eutectic solder. It is well known that a IMC thickness of the interface between the solder and substrate is very important for the reliability of the solder joint. Therefore, Cu dissolution from the Cu pad must be strictly controlled in industrial production to form the desired IMC layer at the interface. For example, nickel in lead-free finishes is usually used as a solderable diffusion barrier to prevent the rapid reaction between a solder and a Cu pad. On the other hand, there is limited information of the effectiveness of heating method to form the IMC layer at the interface and to concern the mechanical properties of the solder joint.

Laser soldering process has been introduced in industry. There is limited discussion of the laser soldering process and the performance of the resultant joints. Compared to reflow soldering process, the laser soldering process has several advantages, such as are a non-contact technique, localized heating, and rapid heating and cooling cycles, which minimize distortions and heating damage [1]. In addition, the laser process enables mounting of individual components and customized modifications of PCB assemblies. The purpose of this study is to investigate the effects of heating method such as reflow soldering and laser soldering on the microstructure of the IMC layer at the interface and the joint reliability of the solder joint.

Experimental

Sn-3.0 mass%Ag-0.5 mass%Cu solder ball with a 1 mm diameter was used in this study. Cu pad with 800 μm diameter and 35μm thickness on an FR-4 printed circuit board was utilized as a substrate. The commercial rosin mildly activated (RMA) flux was used in this study. Before soldering, the Cu pads were immersed in a 4 % HCl solution for 120 s and then the Cu pads and the solder balls were ultrasonically rinsed in an ethanol solution. After the solder balls were dipped in flux, they were placed on the Cu pads. The solder balls on the Cu pads were heated by laser soldering process in air using a laser soldering system (UNIX-413L2, JAPAN UNIX Co., Ltd.). Figure 1 shows the schematic experimental setup of the laser soldering. The laser used in this laser soldering system is a high-power diode laser (λ = 940 nm) with a 1.2 mm beam diameter on a focal plane. Laser irradiation conditions were controlled by laser power and heating time listed in Table 1. As a reference, reflow soldering was performed at 523 K for 60 s in a nitrogen atmosphere. After soldering,

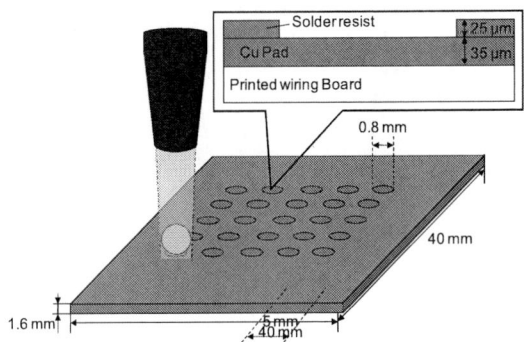

Figure 1: Schematic experimental setup of laser soldering.

Table 1: Laser irradiation conditions used in this study.

Laser power (W)	Heating time (s)
20	20, 30, 40
30	0.5, 1, 5
40	0.2, 0.5, 1

the specimens were ultrasonically cleaned in an ethanol solution to remove residual flux from the solder joints. Some solder joints were then subjected to isothermal aging at 423 K for 168 h and 504 h.

To evaluate the solder joints, an impact test [2] was performed using a micro-impact tester (MI-S, YONEKURA Mfg. Co., Ltd.). The test conditions were as follows: an impact hight of 100 μm and an impact speed of 1 m/s. In this study, the maximum load and total energy of the load-displacement curve obtained from the impact tester were investigated. The maximum load corresponds to the IMC strength, and the total energy represents the toughness of the solder joint. Impact tests were conducted on 10 solder balls for each soldering condition to obtain an average value. After the impact tests, the fracture surfaces of the specimens were observed using optical microscope (OM).

Some specimens before the impact test were cut and their cross-sections were polished to observe the solder/Cu pad interface and the solder microstructure by scanning electron microscopy (SEM).

Results and discussion

The influence of the heating parameters of the diode laser on the maximum load obtained from the impact test was investigated. The relationship between the maximum load and heating parameters is shown in Fig. 2. If the laser input energy was too low, the wetting behavior of the solder by the laser soldering process was not occurred on a Cu pad. Under the conditions within this study, maximum loads for each laser power increase with the increase

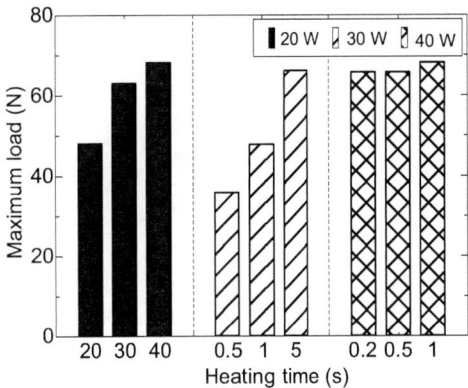

Figure 2: Effect of heating parameters of diode laser on maximum load obtained by impact test.

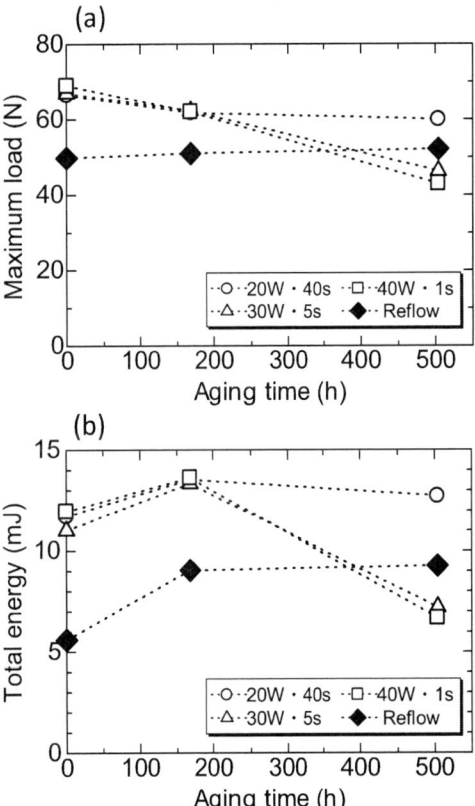

Figure 3: Effect of aging time on impact test for each soldering condition. (a) Maximum load, (b) Total energy

of heating time. In the case of laser power at 20 W, the maximum load is the largest for 40 s. In the case of laser power at 30W, the maximum load is the largest for 5 s. In the case of laser power at 40 W, the maximum load is the largest for 1 s. So, three laser irradiation conditions to examine in detail were

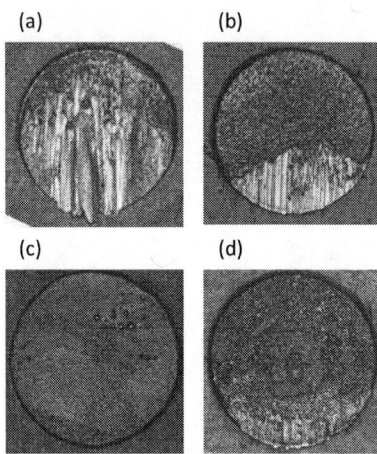

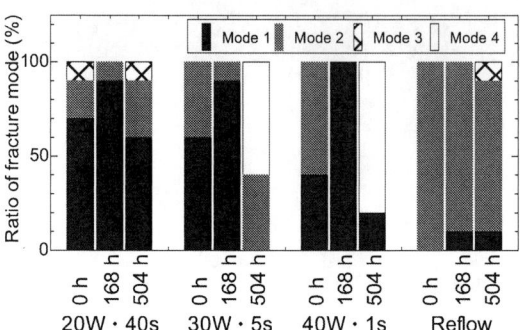

Figure 5: Effect of aging time on fracture mode for each solder bump.

Figure 4: Optical microscope images of possible failure modes subjected to the impact test.
(a) Mode 1, (b) Mode 2, (c) Mode 3, (d) Mode 4

chosen: at 20 W for 40 s, at 30 W for 5 s, and at 40 W for 1 s.

Fig. 3 shows the effect of aging time on the impact test for maximum load and for total energy. Fig. 3(a) shows the effect for maximum load, and Fig. 3(b) shows the effect for total energy. In the as-soldered condition, both the maximum load and total energy of joints soldered by the laser process were clearly higher than those soldered by the reflow process. After aging for 168 h, the maximum load gradually decreased. In contrast, the total energy increased in all solder joints. After aging for 504 h, the maximum load and total energy of the joints soldered by the laser processes at 40 W for 1 s and at 30 W for 5 s were lower than those of the joints soldered by the reflow process. However, in the cases of soldering at 20 W for 40 s by the laser process and soldering by the reflow process, the maximum load and total energy were maintained. So the impact property of the joints soldered by the laser process at 40 W for 1 s and at 30 W for 5 s is very sensitive to the aging treatment, and the maximum load and total energy drastically degrade during aging process from 168 h to 504 h.

To clarify the cause of the degradation of the impact property of the joints soldered by the laser processes at 40 W for 1 s and at 30 W for 5 s, the fracture surface after impact test and the interface between the solder and the Cu pad were investigated.

Optical microscope images of possible failure modes of the solder bump subjected to the impact test are shown in Fig. 4. Four failure modes were classified: a failure through the bulk solder (Mode 1), a fracturing at the IMC and solder (Mode 2), a fracturing at the IMC and Cu pad (Mode 3), and a fracture at the IMC and Cu pad (Mode 4).

Fig. 5 shows the effect of aging time on fracture mode for each solder bump. In the case of

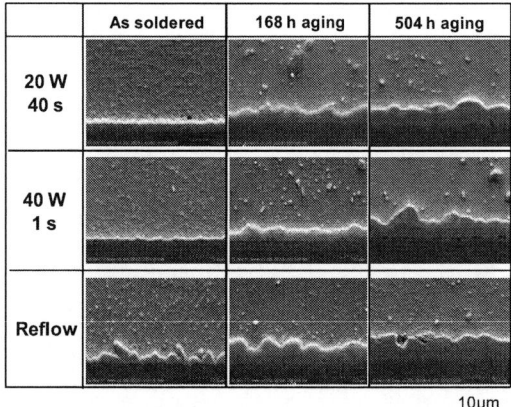

Figure 6: SEM images of joint interfaces for reflow process and laser process in as-soldered condition and after aging for 168 h and 504 h.

reflow soldering, the main fracture mode was observed at the solder and the IMC (Mode 2), and the fracture through the bulk solder (Mode 1) was observed in aged joints. In the case of laser soldering, the fracture through the bulk solder and the fracture at the solder and the IMC were found in the as-soldered condition. The ratio of fracture through the bulk solder increased after aging for 168 h. In the case of soldering at 40 W for 1 s and at 30 W for 5 s, the fracture at the IMC and Cu pad was dominant when the aging time was extend from 168 to 504 h.

Fig. 6 shows SEM images of the interface between the solder and the Cu pad after soldering and after aging for 168 h and 504 h. In the as-soldered condition, the grain size of β-Sn phase soldered by the laser process at 20 W for 40 s and at 40 W for 1 s became small and the width of the eutectic network band decreased became small compared to the reflow process. The IMC layer was formed at the interface between the solder and the Cu pad. As can be seen in this figure, the IMC thicknesses of the joints soldered by the laser processes were less than 1 μm. On the other hand,

200

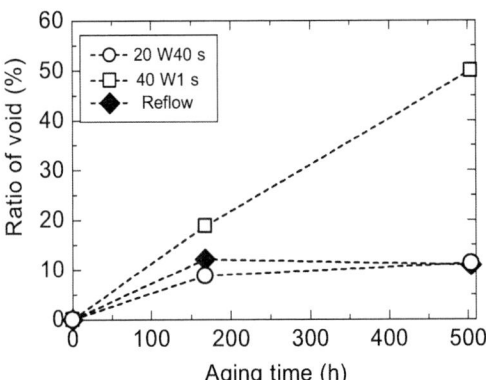

Figure 7: Ratio of voids to length of joint interface as a function of aging time for each soldering condition.

with reflow process, a scallop-like IMC layer with a 3.5 μm thickness was formed at the solder/Cu interface. After aging for 168 h and 504 h, a second layer was clearly observed at the interface between the Cu pad and the first layer and the total thickness of the IMC laser increased. Then, after aging for 504 h, the IMC thicknesses of the joints soldered by the laser processes were almost similar with that of the joint soldered by the reflow process. The growth of the IMC layer was faster by laser soldering than by reflow soldering. Number of small voids can be observed both at the second layer/Cu pad and within the second layer at the IMC laser after aging for 168 h and 504 h.

In general, it is known that a void in the IMC layer provides initiation sites and propagation paths for cracks, which degrades the reliability of solder joints [3,4]. So, the ratio of the length of the voids at the joint interface was measured and was summarized in Fig. 7. The percentage of voids at the interface soldered by the laser process at 20 W for 40 s and by the reflow process was approximately 10 % after aging for 168 h and 504 h. The voids of the joint soldered by the laser process at 40W for 1 s accounted for 50 % after aging 504 h. These voids may be responsible for the degradation of the impact property of the joints soldered by the laser processes at 40 W for 1 s as shown in Fig. 3.

Conclusion

The effects of heating method on the microstructure of the IMC layer at the interface and impact properties of solder joints were investigated. The results obtained are summarized as follows.
1. In the as-soldered condition, the grain size of β-Sn phase soldered by the laser process became small and the width of the eutectic network band decreased became small compared to the reflow process. The impact properties of joints formed with the laser process were superior to those of joints formed with the reflow process due to the thin IMC layer at the joint interface formed during the laser process.
2. After aging for 504 h, the maximum load and total energy of the joints soldered by the laser processes at 40 W for 1 s and at 30 W for 5 s were lower than those of the joints soldered by the reflow process. However, in the cases of soldering at 20 W for 40 s by the laser process and soldering by the reflow process, the maximum load and total energy were maintained. The impact properties of joints formed with the laser process at 20 W for 40s were superior to those of joints formed with the reflow process.

References

[1] C. Chaminade, E. Fogarassy, D. Boisselier, "Diode laser soldering using a lead-free filler materials for electronic packaging structures" Applied Surface Science, vol. 252, pp.4406-4410, 2006.

[2] Y.-S. Lai, H.-C. Chang and C.-L. Yeh, "Evaluation of solder joint strengths under ball impact tes" Microelectronics Reliability, vol. 47, pp. 2179-2187, 2007

[3] K. Zeng, R. Stierman, T-C. Chiu, D. Edwards, K. Ano and K. N. Tu, "Kirkendall void formation in eutectic SnPb solder joints on bare Cu and its effect on joint reliability" Journal of Applied Physics, vol. 97, 024580, 2004.

[4] T. –C. Chiu, K. Zeng, R. Stierman, D. Edwards and K.Ano, "Effect of Thermal Aging on Board Level Drop Reliability for Pb-free BGA Packages" Proceeding of the 54th Electronic Components and Technology Conference, pp. 1256-1262, 2004.

Electro-migration Study of Nano Al Doped Lead- Free Sn-58Bi on Cu and Au/Ni/Cu Ball Grid Array (BGA) packages

I. Shafiq, Y.C. Chan[*], S. Xu, Q.Q. Li

[1] *Department of Electronic Engineering, City University of Hong Kong, Kowloon Tong, Hong Kong SAR*
* Corresponding author: Tel. - +852 27887130; Fax - +852 27887579.
E-mail address: eeycchan@cityu.edu.hk (Prof Y.C. Chan)

Abstract

In order to study the effect of nano particles addition to solders on the electro-migration properties, Al doped-Sn58Bi solders were prepared by mechanically dispersing Al particles additive in Sn-58Bi solders. The interfacial morphologies of the plain solder and doped Sn-58Bi solders under a direct current (DC) of 2.5 A at 75°C temperature with Cu pads and Au/Ni/Cu pads on daisy chain type ball grid array (BGA) substrates were analyzed. Unlike the plain solder, the doped solder does not have obvious formation of Bi-rich or Sn-rich IMC extrusions on the anode or cathode side, respectively. The Sn-Al-Cu ternary phase particles in Cu BGA and Sn-Al binary phase particles in Au BGA are formed in the solder matrix in addition to the Cu-Sn and Ni-Sn phases which were formed near the cathode interface after the first-reflow, blocked the movement of metal atoms/ions, but then induced current crowding. In addition, fracture occurs at the IMC interfacial region during shear testing with a ductile fracture mode. In the solder ball region β-Sn matrix of Sn-58Bi solder joints with a refined microstructure and inter-metallic compound particles Ag were observed, which resulted in an increase in the shear strength, due to a second phase dispersion strengthening mechanism.

Keywords: Nano doping; Flip Chip solder joints; Microstructure; Electro-migration; Shear strength

Introduction

Continuous downscaling and increasing current density in the electronic packages has led to extensive research on one of the main reliability issue in electronics interconnects namely, the Electro-migration (EM) [1,2]. Investigation of solder joints under high current stressing and various under bump metallization (UBM) have been extensively carried out for the past few years [3-8]. EM happens due to combination of thermal and electrical effects on mass transport. It can occur in any material carrying electric current, but in many applications, it is negligible due to the low atomic mobility by diffusion. EM is an enhanced diffusion process of metal atoms in the direction of the current flow. Joule heating-induced degradation, current crowding-enhanced EM, Voids and failure of solder joints are some reliability related failure mechanisms due to EM.

Lead (Pb) has been widely used in the electronics industry for more than fifty years. The most notable example is the use of the eutectic tin-lead (Sn37Pb) solders. However, due to increasing awareness of the toxicity of Pb, a significant shift to Pb-free has taken place in the commercial electronics industry recently [9]. The European Union (EU) legislation, "Restriction of the Hazardous Substances (ROHS)'' (Directive 2002/95/EC) effectively bans the use of Pb and several other substances in electrical products. The eutectic Sn-58Bi alloy is a promising candidate among other lead-free solder for low temperature applications due to its low melting point (139°C) [10, 11]. Few studies on EM have been reported that interfacial segregation of bismuth could occur in the Sn-58Bi/Cu solder joint under high DC current stressing [10]. A continuous Bi layer was found to be formed near the anode side, while a Sn-rich band was formed at the cathode side during the experiments carried out by various researchers. Microstructural evolution in the Sn-58Bi flip-chip solder joint against Au/Ni metallization with and without current stressing was also demonstrated [5]. It has been argued that EM affects IMC formation, phase coarsening and mass accumulation of Bi in the solder joints. Current crowding at the UBM - solder interface for the flip-chip structure, will lead to non-uniform interfaces after the EM test making it hard to estimate the average thickness of the IMC layers and other layers for a flip-chip structure [12].

Being widely investigated in electronic packaging, the nano-particle based additives to solders can enhance the mechanical property and refine the microstructure through second phase strengthening mechanism. Nano-sized metallic particle (Ni, Al, and Ag) reinforced composite solders were developed, and the results showed that the mechanical properties, such as micro-hardness and creep resistance were increased [13-15]. It has been reported that nano-sized Ag reinforced Sn-9Zn composite solders and their joints showed better wetting ability and mechanical properties, as well as longer creep-rupture life than plain solder [15]. While much of the focus is on the mechanical and material characteristics, little attention has been given to the electro-migration; thermo-migration; and current stressing of solder joints and interconnects.

In this study, doped-Sn58Bi solders were prepared by mechanically dispersing Al nano particles additive to Sn-58Bi solders. Based on the ongoing research trend, nano-sized Al particles were selected as reinforcements in composite solders The interfacial morphologies of the plain solder and doped Sn-58Bi solders under a direct current (DC) of 2.5 A at 75°C temperature with Cu pads and Au/Ni/Cu on daisy chain type ball grid array (BGA) substrates were investigated.

Experimental

Composite solders were prepared by mechanically dispersing Al nano-particles (0 and 3wt %) into a eutectic Sn–58Bi (Showa Denko JUFFIT-E) solder paste. The mixture was blended manually for at least 30 min to achieve a uniform distribution of Al nano-particles with a water-soluble flux (Qualitek Singapore (PTE) Ltd.). Then, the paste mixture was printed on to alumina substrates using a stainless steel stencil with a thickness of 0.15mm and reflowed in a reflow oven (BTU PYRAMAX 100N) at 215°C to prepare approximately 320 μm diameter solder balls.

The substrates (both large and small) used to fabricate the BGA solder joints were made of flame retardant 4 (FR4) materials. The thickness of each substrate was 1.5 mm. The BGA pads were a solder-mask-defined type, and the opening diameters of the pads on the small substrate and large substrate were 280 μm and 250 μm, respectively. The thickness of the Cu pads was 35 μm with organic solder-ability preservative (OSP) surface finishing. The pads were linked with designed Cu traces with a thickness of 35 μm for electrical continuity. For the Au/Ni/Cu pad, a triple layer structure of immersion Au (0.3 μm)/electroless Ni-P (5 μm)/Cu (35μm) was used. The pads were linked with designated copper traces (35μm × 500μm in cross section) for electrical continuity.

The small substrate was attached to the large substrate using flip chip bonding technology by a Flip-Chip bonder (SUSS, FCM). The assembly process was similar to that of a conventional surface mount technology (SMT) procedure, including flux printing, ball placement, reflow soldering and cleaning. The aligned samples were reflowed using a hot air convection reflow. The peak temperature in the temperature profile was 176°C and the duration above the melting point of Sn58Bi alloy was about 85 s.

The melting temperature of the solder alloys were measured using a differential scanning calorimeter (DSC Q 10). For DSC analysis, a piece of about 10 mg solder was placed into an Al pan. For melting properties data, the sample was initially

scanned from 25°C to 100°C at a rate of 25 °C min⁻¹ and then at a rate of 5°C min⁻¹ up to 175°C under nitrogen gas atmosphere. To measure the shear strength, die shear tests were performed on the reflow samples using a shear testing machine (Dage Series 4000 Bond Tester) with a 50μm shear tool height and 500μm/s shear speed. The test load for shear was 5000g for every test. After shear testing, the fracture surfaces and compositions were investigated thoroughly using SEM and EDX techniques.

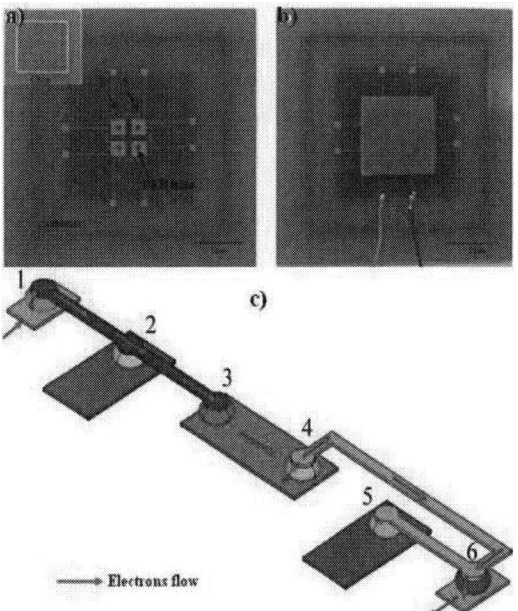

Fig 1 (a) Big and Small Substrate (inset) used for the Electro-migration test.(b) Flip chip substrate after the bonding process (c) Schematic of the Circuit.

Figure 1a shows the images of a large substrate and small substrate (inset) before the assembly. Figure 1b shows the image of a prepared structure, and daisy chains formed at the four edges of the small substrate. Solder joints 1–6 were prepared for each chain. The prepared structure with one chain supplied with DC current and was put into the furnace set at 75°C. Figure 1c shows the electron flow through the solder joint chain. For the solder joints 1, 3, 4 and 6, the electric current passes through the solder bumps. While the electric current passes by the top of the solder bump of joint 2 instead of entering it. No current passes by or passes through the solder bump of joint 5. After predetermined periods of time, samples were taken out of the furnaces and quenched in air.

Every sample was mounted and cross sectioned carefully towards the center of the solder joint. To characterize the microstructures, the reflowed samples were cross sectioned and mounted in resin, then ground by different grit sized emery papers and polished with 0.5μm alumina powder. Finally, the interfacial morphology at the solder alloy/BGA substrate interface was observed using a scanning electron microscope (SEM, Philips XL 40 FEG) using the back-scattered electron (BSE) imaging mode. Before SEM observation, the samples were sputter coated with Au to avoid the effects due to charging.

Results and Discussion

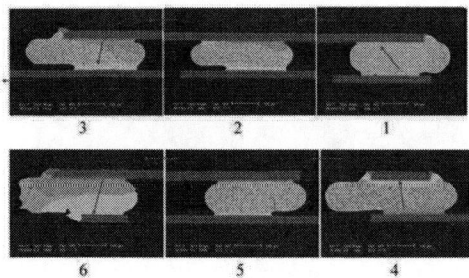

Fig 2 SEM image of one daisy chain Sn58Bi solder joint on Cu pad under 2.5 A DC current stressing at 75°C

Figure 2 shows an actual SEM image of one daisy chain Sn58Bi solder joint on Cu pad under 2.5 A DC current stressing at 75°C. The electron flow path through the solder joint chain is indicated in red arrow. For the solder joints 1, 3, 4 and 6, the electric current passes through the solder bumps. While the electric current passes by the top of the solder bump of joint 2 instead of entering it. No current passes by or passes through the solder bump of joint 5. The solder joint structure was designed to study both the electro-migration and thermo-migration under current stressing. But this study will be confined to the effects of electro-migration only. The Sn-rich areas are shown darker and the Bi-rich areas are shown brighter in the BSE image of the flip chip structure.

DSC analysis was carried out in order to investigate the fundamental thermal reactions that occur on heating and to estimate the melting point of the solder alloys. The following figures 3a and 3b shows the typical DSC curves obtained for Sn–58Bi and Sn-58Bi-3wt%Al alloys on heating, respectively. On heating, the prominent endothermic peak for composite solders appeared at around 142°C, which is closer to the eutectic temperature of the Sn–58Bi binary system (138°C). Addition of nano Al particles shifted the endothermic peak from 138°C to this slightly higher temperature. It has already been proven in other lead free eutectic solders like Sn-Ag-Cu system that the liquidus line of the alloy raises with an increase in the Al content [16]. It is noteworthy that the height of the first peak decreases with an increase in the Al content and a shoulder was observed at around 142°C for 3% Al containing

solders.

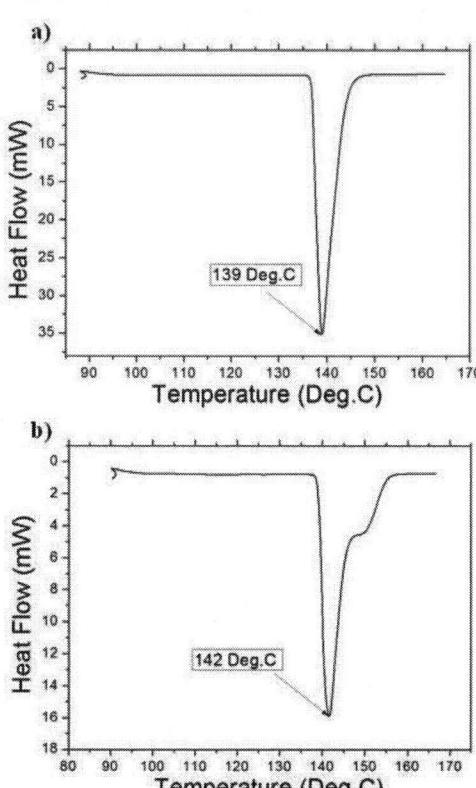

Fig 3 (a) DSC curve of Sn-58Bi solder alloys on heating. (b) DSC curve of Sn-58Bi-3Al solder alloys on heating.

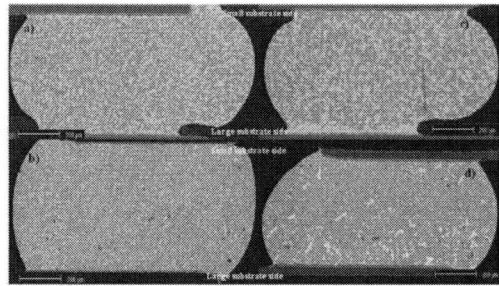

Fig 4 (a) Sn-58Bi Joint 1 on Cu pad BGA substrate after one reflow cycle. (b) Sn-58Bi-3wt% Al Joint 1 on Cu pad BGA substrate after one reflow cycle.(c) Sn-58Bi Joint 1 on Au/Ni/Cu pad BGA substrate after one reflow cycle (d) Sn-58Bi-3wt% Al Joint 1 on Au/Ni/Cu pad BGA substrate after one reflow cycle

The backscattered electron imaging mode of the SEM was utilized to help identify distinguishable interfacial phases between plain Sn-58Bi solder and solder joints containing Al nano-particles with Cu pads and Au/Ni metallized Cu pads on ball grid array (BGA) substrates. Figure 4(a)

204

and (b) shows BSE image of a typical cross section of an as-reflowed solder joints without and with Al nano particles on a BGA substrate with Cu pad reflowed at 215°C for one cycle. Figure 4(c) and (d) shows a typical cross section of an as-reflowed solder joints without and with Al nano particles on a BGA substrate with Au/Ni/Cu pad reflowed at 215°C for one cycle. It was found that the microstructures at both solder to Cu and solder to Au/Ni/Cu interfaces were quite similar. But in case of the nano Al doped solders the matrix seems to be consisting of refined phases of Sn and Bi phases compared to the plain solder joints. Also the Cu_6Sn_5 IMC is formed at the Cu-pad and Ni_3Sn_4 IMC is formed at the Au/Ni/Cu interfaces. The entire Au layer has entirely disappeared and dissolved in the solder. The IMC formed are very thin in order of 1.6 μm.

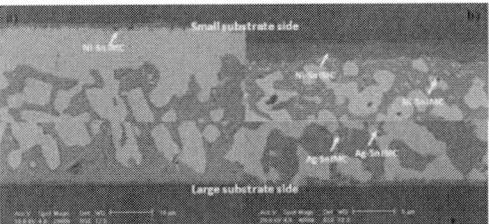

Fig 6 (a) Sn-58Bi Joint 1 stressed in Au/Ni/Cu BGA for 100 hours at 75 °C (b) Doped Sn-58Bi Joint 1 stressed in Au/Ni/Cu BGA for 100 hours at 75 °C

Figure 6(a) shows the BSE images of joint 1 of the plain solder joint interfaces with Au/Ni/Cu pad after being stressed with a current density of 5×10^3 A/cm^2 for 100 hours at 75°C. There is a large migration of Sn and Bi regions towards the electrodes IMC very similar to the Cu BGA structures. Figure 6(b) shows the BSE images of joint 1 of Al doped solder joints Sn-58Bi with u/Ni/Cu pad stressed with a current density of 5×10^3 A/cm^2 for 100 hours at 75°C and reveals the not so severe EM phenomenon and the Bi-rich and Sn-rich regions are not much severely migrated due to the presence of Al-Sn binary IMC particles.

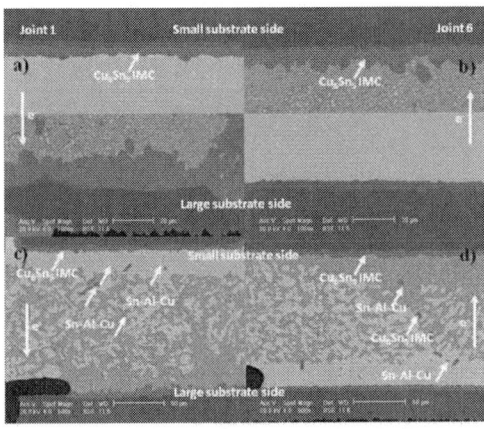

Fig 5 (a) and (b) Sn-58Bi Joint 1and 6 stressed in Cu BGA for 360 hours at 75°C, (c) and (d) Al doped Sn-58Bi Joint 1 and 6 in Cu BGA stressed for 360 hours at 75°C, respectively

Figure 5(a) and 5(b) shows the BSE images of joint 1 and joint 6 of the plain solder after being stressed with a current density of 5×10^3A/cm^2 at 75°C for 360 hours. The light areas are Bi-rich phases and the darker areas are Sn-rich phases. It is obvious that a Bi-rich layer is formed at the anode side, while a Sn-rich band is formed at the cathode side, which suggests a significant mass transport due to EM. Also, the large Bi-rich grain formed in the bulk solder is due to the phase coarsening under the high current density and temperature. Figure 5(c) and 5(d) gives the BSE images of joint 1 and joint 6 of the Al doped solder joint interfaces with Cu pad after being stressed with a current density of 5×10^3 A/cm^2 for 360 hours at 75°C. It can be seen that the Bi-rich and Sn-rich regions are not much severely migrated and the presence of Al-Sn-Cu IMC particles due to the addition of nano Al particles.

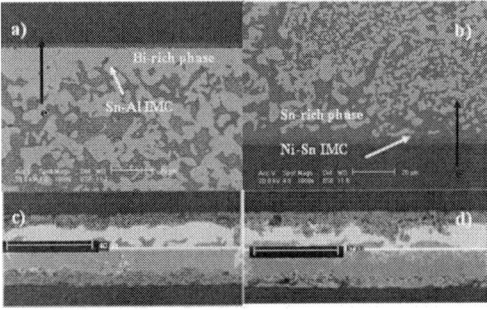

Fig 7 (a) Sn-58Bi Joint 1 stressed in Au/Ni/Cu BGA for 360 hours at 75°C (b) Doped Sn-58Bi Joint 1 stressed in Au/Ni/Cu BGA for 360 hours at 75°C (c) Sn-58Bi Joint 6 stressed in Au/Ni/Cu BGA for 360 hours at 75°C (d) Doped Sn-58Bi Joint 6 stressed in Au/Ni/Cu BGA for 360 hours at 75°C

Figure 7(a) and (b) shows the BSE images of joint 1 and joint 6 ,respectively, with the plain solder and Al doped solder joint interfaces and Au/Ni/Cu pad stressed with a current density of 5×10^3 A/cm^2 for 360 hours at 75°C. There is a medium scale migration of Sn and Bi regions towards the electrodes in the doped sample and the presence of IMC binary phase particles on Sn-Al due to doping is a major factor for retardation of EM. Figure 6(c) and (d) shows the joint 6 of Au BGA package and reveals that the migration is not so severe similar to the joint 1.

A comprehensive study on the effect of Zn-doped SnAg solders on microstructure and EM was previously reported. Minor Zn doping, about 0.6 wt %, in a SnAg solder alloy is found to be effective in

205

stabilizing solder microstructure and improving EM reliability. Early EM failure modes in Zn-doped solders are significantly suppressed, resulting in a longer EM lifetime with tight distributions [18].

Similarly effects of silver doping on EM behavior of the eutectic Sn-58Bi solder were also reported. EM can induce the Bi migration along with the direction of electron flow in the eutectic Sn-58Bi solder, thus Bi depletes at the cathode side but accumulates at the anode side. Plate-like Ag_3Sn compound is formed as 0.5 wt% silver is doped into the solder. The Ag_3Sn plates behave like roadblocks which can intercept the Bi migration from the cathode side to the anode side. Consequently, a great inconsistency is found between the Bi depletion at the cathode side and the Bi accumulation at the anode side [19].

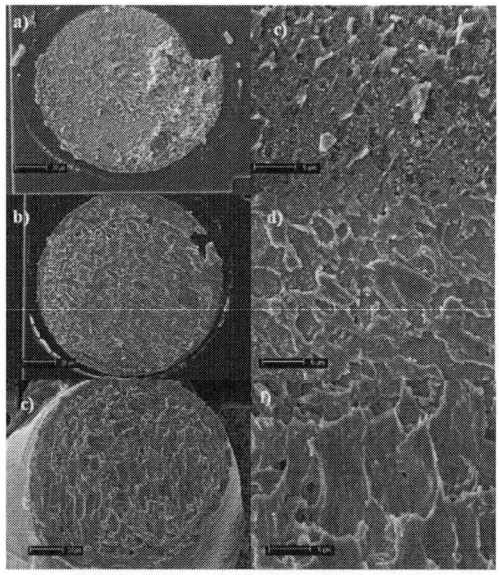

Fig 8 (a) Fracture surface of the substrate side with Sn-58Bi solder on Cu BGA stressed for 100 hours at 75°C (b) Fracture surface of the substrate side with doped Sn-58Bi solder on Cu BGA stressed for 100 hours at 75°C (c) Fracture surface of the chip side with doped Sn-58Bi solder on Cu BGA stressed for 100 hours at 75°C

Consequently effects of Cu doping on the microstructural evolution in the eutectic Sn-58Bi solder stripes under annealing and current stressing reported that the coarsening of the Bi grains was observed in the eutectic Sn-58Bi solder upon annealing at 85°C. Doping of 1wt. % Cu could significantly reduce the grain coarsening rate. In addition to grain coarsening, mass accumulation of Bi at the anode and solder depletion at the cathode of the eutectic Sn-58Bi solder stripe stressed by a current at 85°C. Doping of 1wt. % Cu could also reduce the grain coarsening of the solder under current stressing; however, it resulted in an enhancement of the EM effect. Accumulation of Bi

at the anode and the solder depletion at the cathode became more severe in the Cu-doped solder stripe [20].

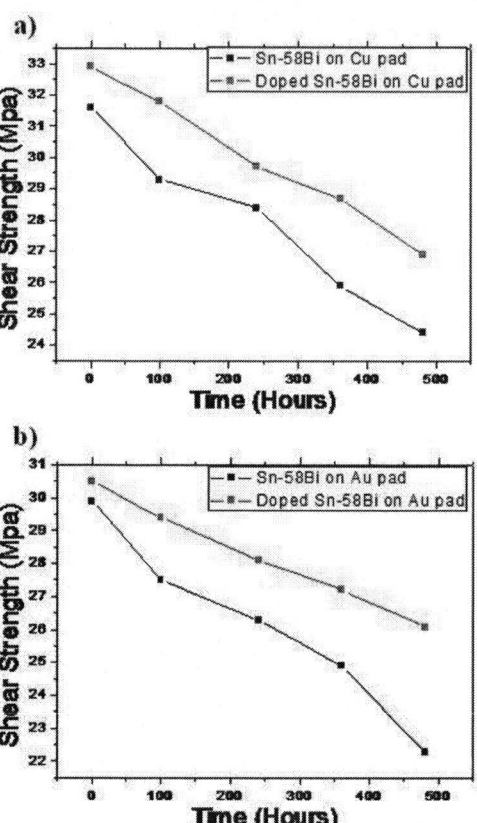

Fig 9 (a) Shear strength of un-doped and doped Sn-58Bi solder joints on Cu pad (b) Shear strength of un-doped and doped Sn-58Bi solder joints on Au/Ni/Cu pad

Figure 9(a) and (b) shows the shear strength of the solder joint without and with Al nano particles doping, respectively. It can be seen that the Al doped Sn-58Bi solder joint have consistently higher shear strength due to a second phase dispersion strengthening mechanism and refined β-Sn microstructure. But the shear strength of Doped solder in Cu pad was higher than the Au/Ni/Cu pad due to the presence of the Cu_6Sn_5 IMC. Ni_3Sn_4 is formed in the case of the Au/Ni/Cu pad joint is more prone to failures due its brittle nature.

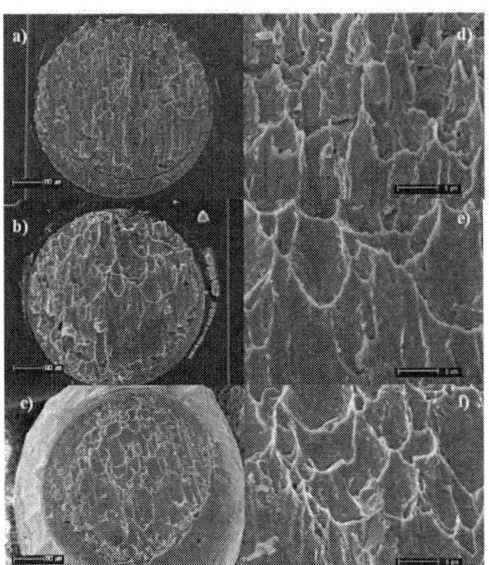

Fig 10 (a) Fracture surface of the substrate side with Sn-58Bi solder on Au/Ni/Cu BGA stressed for 100 hours at 75°C (b) Fracture surface of the substrate side with doped Sn-58Bi solder on Au/Ni/Cu BGA stressed for 100 hours at 75°C (c) Fracture surface of the chip side with doped Sn-58Bi solder on Au/Ni/Cu BGA stressed for 100 hours at 75°C

Figure 10(a) illustrates the SE image of a typical shearing fracture of the un-doped solder joint stressed for 100 h at 75 °C with Au/Ni/Cu on the chip side, which was primarily characterized as a brittle behaviour. As shown in Figure 10(b) and (c), the fracture of the solder joint on the chip side (cathode side) exhibited ductile transition morphology, with big dimples visible on the ductile plastic fracture. According to the equation of atomic fluxes, atoms migrated from the high temperature side to low temperature side by the thermal gradient force, and hence a reversed flux of vacancies moved to the high temperature side. Consequently, it caused vacancy accumulation and void formation at the interface of the solder and UBM. Meanwhile, large stress was easily built and accumulated around the voids due to thermal mismatch. But the doped solder joints showed more ductile fracture indicating a better mechanical strength because the nano-particles forming IMC particles can act as a roadblock and reduce Bi migration.

Conclusion

The solid-state interfacial reactions between plain Sn-58Bi solder joint and Sn-58Bi containing Al nano particles with Cu pad and Au/Ni/Cu pads ball grid array (BGA) substrates were investigated with relation to electro-migration and mechanical strength. In the Cu BGA substrate Cu-Sn inter-metallic compound layer was formed and there is a large migration of Bi atoms towards anode which was

relatively less and prevented due to the Sn-Al-Cu ternary IMC particles formed in the doped Sn-58Bi composite solder. Also the mechanical strength of the solder joints is consistently reliable in cased of the doped Sn-58Bi solder.

After the reflow process, the topmost Au layer dissolved very quickly into the molten solder in case of the Au/Ni/Cu BGA solder joint and Ni_3Sn_4 inter-metallic compound layer was formed at their interfaces. Also the Bi segregation on the anode was relatively controlled and the microstructure was refined due to the addition of the Al nano particles. In addition, the fracture surfaces of doped Sn-58Bi solder joints containing nano particles appeared to be ductile with very rough dimpled surfaces.

Acknowledgements

The authors acknowledge the financial support provided by EPA Centre, City University of Hong Kong and Research Grants Council (RGC) of Hong Kong through the project "Electro-migration and Thermo-migration Studies in Nano-scaled Solder Joints and Interfaces under High Electric Current Stressing and Cumulative Service Loading for Nano-electronics Applications" (Project No.: 9041486; CERG grant of RGC Hong Kong, RGC ref. no. 111309).

References

1. Y.H. Lin, C.M. Tsai, Y.C. Hu, Y.L. Lin, and C.R. Kao, "Electro migration induced failure in flip chip solder joints," J. Electron. Mater., Vol.34, No.1, pp.27-33, 2005.
2. C. Chen and S. W. Liang, "Electromigration issues in lead-free solder joints," J. Mater. Sci: Mater. Electron., Vol. 18, No.1-3, pp. 259–268, 2007.
3. K. N. Tu, "Recent advances on electromigration in very-large-scale integration of interconnects," J. Appl. Phys.,Vol 94, No. 9, pp. 5451-5473, 2003.
4. Q. L. Yang and J.K. Shang, "Interfacial Segregation of Bi during Current Stressing of Sn-Bi/Cu Solder Interconnect," J. Electon. Mater. Vol., 34, No.11, pp.1363-1367, 2005.
5. L.-T. Chen and C.-M. Chen, "Electromigration study in the eutectic SnBi solder joint on the Ni/Au metallization," J. Mater. Res., Vol. 21, No. 4, pp.962-969, 2006.
6. Y.S. Lai, K.M. Chen, C.L. Kaoa, C.W. Lee, Y.T. Chiu, "Electromigration of Sn–37Pb and Sn–3Ag–1.5Cu/Sn–3Ag–0.5Cu composite flip–chip solder bumps with Ti/Ni(V)/Cu under bump metallurgy," Microelectron. Reliab., Vol. 47, No.8, pp. 1273-1279, 2007.
7. T.Y. Lee, K.N. Tu, D.R. Frear, "Electromigration of eutectic SnPb and SnAg3.8Cu0.7 flip chip solder bumps and under-bump metallization" J. Appl. Phys., Vol. 90, Iss. 9, pp. 4502 – 4508, 2009.
8. S.H. Chae, X.F. Zhang, K.H. Lu, H.L. Chao,P. Ho, M. Ding, P. Su, T. Uehling, L. Ramanathan, "Electromigration statistics and damage evolution

for Pb-free solder joints with Cu and Ni UBM in plastic flip-chip packages" J. Mater. Sci: Mater. Electron., Vol. 18, No.1-3, pp. 247–258, 2007

9. M. McCormack, S. Jin, "Progress in the design of new lead-free solder alloys" The Journal of the Minerals, Metals, and Materials Society., Vol. 45, Iss. 7, pp. 36-40, 1993.

10. C.-M. Chen, L.-T. Chen, and Y.-S. Lin, "Electromigration-Induced Bi Segregation in Eutectic SnBi Solder Joint," J. Electron. Mater., Vol.36, No.2, pp. 168-172, 2007.

11. Z. Mei, J. W. Morris, "Characterization of eutectic Sn-Bi solder joints" J. Electron. Mater., Vol.21, No.6, pp. 599-607, 1993.

12. H.-W. Miao and J.-G. Duh, "Microstructure evolution in Sn–Bi and Sn–Bi–Cu solder joints under thermal aging," Mater. Chem. and Phy., Vol.71, No. 3, pp.255-271, 2001.

13. "A.K. Gain, T. Fouzder, Y.C. Chan, A. Sharif, N.B. Wong, W.K.C. Yung, "The influence of addition of Al nano-particles on the microstructure and shear strength of eutectic Sn-Ag-Cu solder on Au/Ni metallized Cu pads", J. Alloys and Compd, Vol. 506, Iss. 1, pp. 216-223, 2010.

14. A.K. Gain, Y.C. Chan, W.K.C. Yung, "Effect of nano Ni additions on the structure and properties of Sn–9Zn and Sn–Zn–3Bi solders in Au/Ni/Cu ball grid array packages", Mater. Sci. Eng., B, Vol. 162, Iss. 2, pp. 92-98, 2009.

15. A.K. Gain, Y.C. Chan, A. Sharif, N.B. Wong, W.K.C. Yung, "Interfacial microstructure and shear strength of Ag nano particle doped Sn–9Zn solder in ball grid array packages", Microelectron. Reliab., Vol. 49, pp. 223–234, 2009.

16. K.L. Lin, C.C. Hsiao, K.I. Chen, "Microstructural evolution of Sn–Ag–Cu–Al solder with respect to Al content and heat treatment", J. Mater. Res., Vol. 17: pp. 2386-2393, 2002.

17. Zeng and K. N. Tu, "Six cases of reliability study of Pb-free solder joints in electronic packaging technology", Mater. Sci. Eng., R, Vol. 38, Issue 2, pp 55-105, 2002.

18. M.H. Lu, D.Y. Shih, S. K. Kang, C. Goldsmith, P. Flaitz, "Effect of Zn doping on SnAg solder microstructure and electromigration stability", J. Appl. Phys., Vol. 106, pp. 053509 (1-5), 2009.

19. C.M. Chen, C.C. Huang, "Effects of silver doping on electromigration of eutectic SnBi solder" J. Alloys and Compd., Vol. 461, Iss. 1-2, pp. 235-241, 2008.

20. C.M. Chen, C.C. Huang, C.N. Liao, K.M. Liou, "Effects of Copper Doping on Microstructural Evolution in Eutectic SnBi Solder Stripes under Annealing and Current Stressing", J. Electon. Mater., Vol. 36, No. 7, pp. 760-765, 2007.

Characterisation of ion transportation during electroplating of high aspect ratio microvias using megasonic agitation

S.Costello[1], N.Strusevich[2], M.K.Patel[2], C.Bailey[2], D.Flynn[1], R.W. Kay[1], D. Price[3], M. Bennet[4], A.C.Jones[4], R. Habeshaw[5], C. Demore[5], S. Cochran[5] and M.P.Y. Desmulliez[1]

[1]MicroSystems Engineering Centre (MISEC), School of Engineering and Physical Science, Heriot-Watt University, Edinburgh EH14 4AP, UK
[2]Computational Mechanics and Reliability Group, School of Computing and Mathematical Science, University of Greenwich. London, SE10 9LS, UK
[3]Merlin Circuit Technology Ltd, Hawarden Industrial Park, Manor Lane, Deeside, CH5 3QZ, UK
[4]School of Chemistry, Joseph Black Building, The University of Edinburgh, Edinburgh EH9 3JF, UK
[5]IMSaT, Dundee University, Wilson House, Wurzburg Loan, Dundee Medipark, DD2 1FD, UK

Phone: +44 (0) 131 451 3774, Fax: +44 (0) 131 451 4155 and E-mail Address: S.Costello@hw.ac.uk

Abstract

This paper presents the characterisation of the flow streams of an electrolyte solution during the electroplating of high aspect ratio microvias using megasonic agitation. The electrolyte fluid flow without acoustic streaming is firstly examined using micro-particle imaging velocimetry (micro-PIV) to establish the velocity of the flow at the mouth and within the microvias. Secondly, the average acoustic pressure developed by the transducer used for the megasonic agitation of the solution is measured using a fibre optic hydrophone. The fibre optic hydrophone is also used to determine the pressure in the mouth of microvia in a test Printed Circuit Board (PCB). The experimental results are then used to validate the flow and acoustic models created to optimise the set of experimental parameters which ensure the successful electroplating of high aspect ratio microvias in a PCB. The lack of fluid flow and, hence, transport of ions is demonstrated in blind microvias. It is also proved that acoustic pressure directed into the via promotes ion transport.

Key words: High aspect ratio microvia, electroplating, megasonic agitation, acoustic streaming, micro-PIV.

Introduction

The continuing drive for enhanced performance, functionality and integration density is the stimulus for the three dimensional (3D) integration of passive and active elements in the Printed Circuit Board (PCB) and Semiconductor industries. A key enabler to 3D integration is Through Silicon Vias (TSVs) processes which involve the fabrication of high aspect ratio microvias and their filling through electrodeposition of copper. Such processes can be encountered not only in the microelectronic but also the PCB industries.

Currently PCB manufacturers are limited by the aspect ratio (AR) of via that can be successfully and reproducibly filled, not often more than 2:1 (height: diameter). For this reason, Sequential Build Up (SBU) techniques are used to achieve high density interconnects (HDI). Figure 1 displays a schematic representation of the various types of microvias incorporated with SBU for HDIs.

Filled vias are typically produced by solder printing, also called plugging, or by electrodeposition [1]. The latter technique offers several advantages over via plugging. Solid copper filled vias are inherently reliable and have a higher

electrical conductivity than organic or conductive pastes and copper electroplating is a technology well known and extensively used in PCB manufacturing [2].

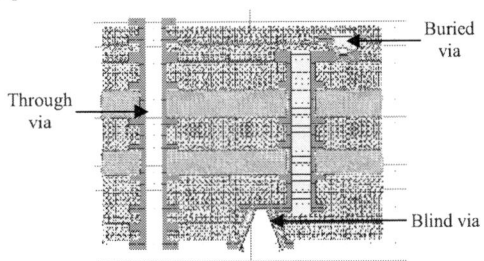

Figure 1: Schematic representation of different types of PCB vias

However, the electrodeposition process usually produces unreliable and inconsistent results for PCBs or silicon wafers of aspect ratios exceeding 2:1. These unsatisfactory results either stem from the difficulty in depositing the initial seed conductive layer within the via or from the filling of the vias itself. As the aspect ratio increases, the standard electrodeposition technique tends to produce voids or streams within the via, as shown in figure 2, affecting the electrical and thermal performance of the interconnect.

Figure 2: Voids in electroplated copper through via, diameter 200μm and length 1.6mm, 8:1 AR

The excessive deposition of copper on the surface of the substrate during via filling also incurs expensive post processing stages such as mechanical polishing. Beside the use of additives (suppressors, levelers, etc.), a possible approach to improve the quality of electrodeposition by eliminating voids, is to enhance the ion transport within the vias. This can be achieved, for example, by submerging a megasonic transducer into the electrolyte solution, which leads to a phenomenon known as acoustic streaming.

In this paper, micro Particle Imaging Velocimetry (micro-PIV) experiments have been conducted to determine the velocity of fluid flow in a microfluidic chip designed to replicate the cross section of microvias in a PCB or silicon wafer. The Sonosys™ submersible transducer array operating at 1MHz has been characterized using a Precision Acoustics fibre optic hydrophone in a water tank. The hydrophone was also used to determine the acoustic pressure exerted by the transducer at the mouth of a test PCB through via. The results of experiments are used to validate flow and acoustic models detailed in this paper.

Experimental Method

A. Micro-PIV

Micro-PIV has been used in our experiments to determine the flow rate of the fluid passing through a microchannel which replicates a microvia in a PCB. A cross-section of blind vias that could be found either within a PCB or a silicon wafer is replicated in transparent materials to allow optical access to the structure as shown in figure 3. The large microfluidic channel aims to replicate the flow conditions happening at the top substrate of the PCB and at the entrance to the holes.

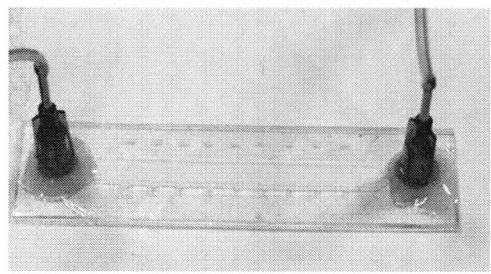

Figure 3: Example of blind via test structure.

The test structures are made of a triple stack layer of 0.5 mm thick PMMA patterned using an Epilog™ CO_2 engraving laser. The lower layer contains two ports which have an etched surface to ensure that the fluid is held on the surface.

The centre layer replicates the cross section of the via. The top layer is a lid with holes for the fluidic input and output ports. The three layers are carefully stacked, pressurized and heated until the PMMA sheets are bonded. To allow fluidic access to the chip, two needle caps are bonded using epoxy glue over the connection ports. The test structure is then positioned in the micro-PIV set-up above the light source and the fluidic channels are illuminated from below, as shown in figure 4.

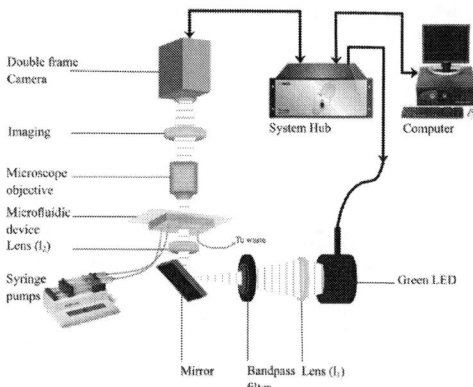

Figure 4: Schematic of the micro-PIV experimental set-up.

A syringe pump is used to pump through the test chip a solution containing polystyrene microspheres of 0.4 μm diameter. These spheres are used to follow the flow streams and indicate the velocity of the fluid. The test fluid contains enough microspheres to ensure that at least 20 of them can be seen in the filed of view of the camera. The objective lens of the PIV set-up is focused into the fluidic channel onto the moving microspheres. Measurements are taken using a Dantec system that coordinates the illumination set-up, the capturing of the images and the processing of the data.

The experimental set-up consists of three parts: the Dantec hub, the camera and the illumination source. When conducting the experiment, the interrogation area is determined to allow the software to correctly map the flow field. The particle flow rate and timings of the pulses are set to ensure that the particles do not move more than 20% of the interrogation area between two pulses. The two frames are transferred to the PIV processor to produce flow field maps by performing average correlation for each integration area on the pairs of images collected.

B. Characterisation of the Transducer

A Precision Acoustic™ fibre optic hydrophone has been used to characterize the intensity of the transducer submerged in the electroplating tank filled with in degassed water as shown in figure 5.

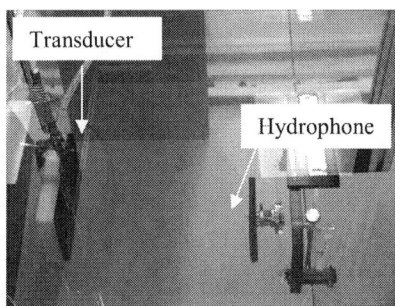

Figure 5: Top-view of set-up inside water tank used to characterisation configuration of the acoustic field originating from the transducer.

The transducer is operated at 1 MHz and 20% power, 100W. A test PCB containing through vias of diameters ranging from 800μm to 200μm in 100μm steps is shown in figure 6. The length of all vias is 1.6mm. The test PCB is used with the fibre optic hydrophone to determine the acoustic pressure at the mouth of the via. The PCB is secured to the hydrophone holder which is moved by a mechanical stage in the tank. The fibre optic hydrophone is inserted into the via from the backside of the PCB upto the mouth of the via facing the transducer.

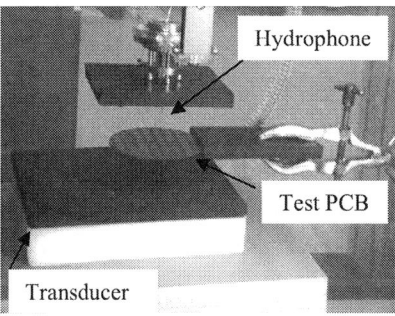

Figure 6: Fibre-optic hydrophone positioned through test PCB via.

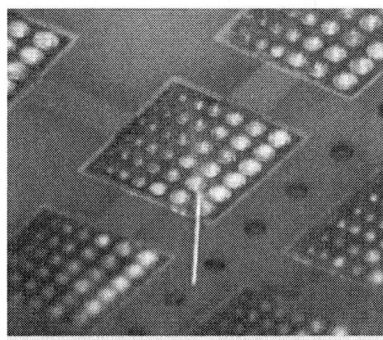

Figure 7: Fibre-optic hydrophone protruding through the via towards transducer.

Figure 7 shows the hydrophone protruding from beneath the PCB. The tip of the hydrophone is held level with the PCB surface to measure the acoustic pressure at the mouth of the via.

Experimental Results

A. Micro PIV

The vias in the microfluidic chip have varying dimensions and aspect ratios ranging from 2:1 to 8:1. The speed of the flow was measured using micro-PIV for a flow rate of 50 μL/min at the inlet of the microfluidic chip. This flow rate translates to a fluid velocity of 333μm/s.

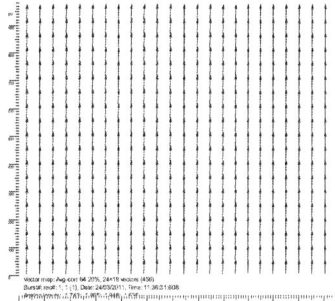

Figure 8: Flow field map of particle velocity in the large channel of the microfluidic chip.

The measured average velocity of the particles in the fluid has been measured at 324μms^{-1} as shown in figure 8. Considering the errors involved in the measurement and manufacturing tolerances of the sample, it was concluded that the particles were moving at the same speed as the fluid. The micro-PIV measurements of particle speed are therefore representative of the velocity of the fluid flow. Previous studies have also shown that particles with a small Stokes number require only a very short time to respond to changes in the surrounding flow field, and therefore are good indicators of the velocity of the fluid streams [3]. With a Stokes number of the order 10^{-10}, it is clear that the polystyrene microspheres used in this experiment move at the same rate as the fluid in which they are immerged.

The velocity of the fluid was measured in two blind microvias, 2:1 AR and 3.2:1 AR. Measurements of particle velocity were taken at five positions in the via, as shown in figure 9, to determine the velocity map of the flow in the mouth of the via.

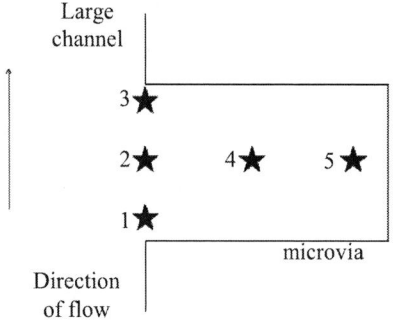

Figure 9: Positions of micro-PIV measurement within microvia.

The velocity profile plots across the 2:1 AR and 3.2:1 AR via mouths measured in positions 1, 2 and 3 are plotted in figures 10(a) and 11(a), respectively. Figures 10(b) and 11(b) show the velocity plot of flow into the via measured in positions 2, 4 and 5 in the 2:1 AR and 3.2:1 AR, respectively.

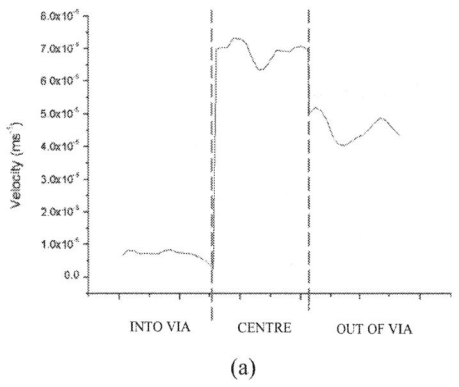

(a)

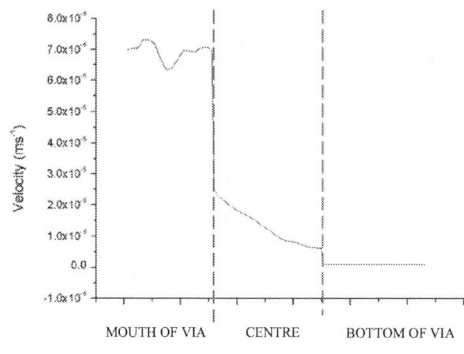

(b)

Figure 10: 2:1 AR via micro-PIV results (a) velocity profile across the mouth of the via (b) velocity profile plot along via length

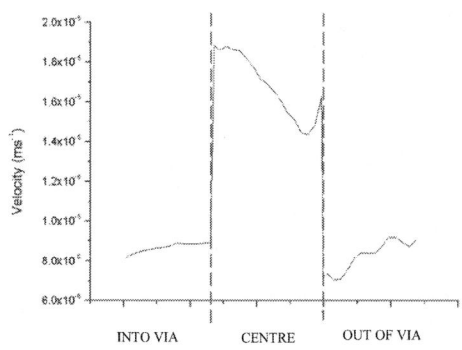

(a)

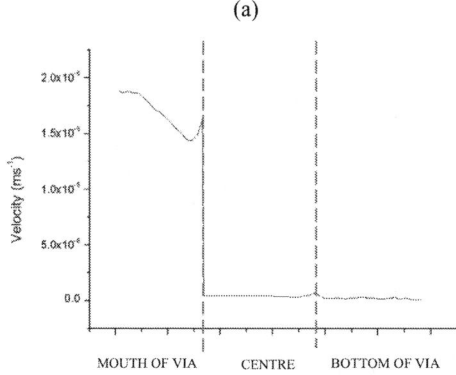

(b)

Figure 11: 3.2:1 AR via micro-PIV results (a) velocity profile across microvia mouth (b) velocity profile plot along via length

The plots are drawn together to indicate the variation in velocity across the via mouth and into the via. The vertical dotted lines in figures 10 and 11 indicate measurements in non-continuous areas. Figures 10(a) and 11(a) show that the fluid velocity is the largest at the centre of the mouth. Figures

10(b) and 11(b) show that the velocity of the flow into the blind microvias has a large magnitude only at the surface. The velocity of flow is reduced in the centre of the 2:1 AR via and Brownian motion can be observed at the bottom. Brownian motion is observed from the centre of the 3.2:1 AR via onwards. Figure 12 shows a flow field plot from the bottom of the 3.2:1 AR motion which is indicative of Brownian motion.

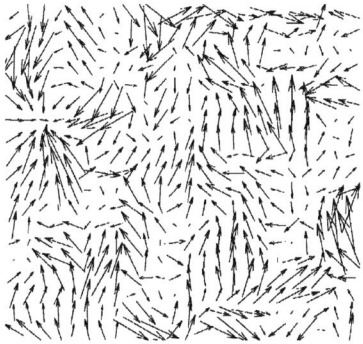

Figure 12: Brownian motion near the bottom of the 3.2:1 AR via. The area considered is 0.04mm².

B. Characterisation of the Transducer

The transducer was firstly X-rayed to determine the position of the sub-elements within the waterproof casing. Figure 13 shows that the transducer is made up of four elements electrically connected.

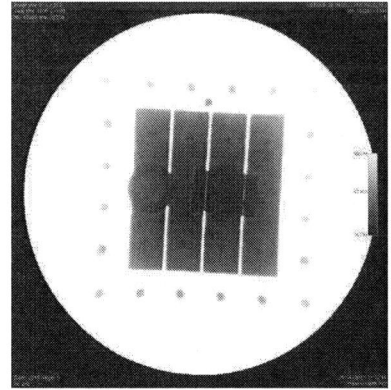

Figure 13: X-ray picture of the Sonosys™ submersible megasonic transducer.

The intensity of the transducer was measured as close to the transducer surface as possible, approximately 25mm and at a large distance of approximately 25cm to determine the behaviour of the acoustic pressure field. Not surprinsigly, the signal from the transducer begins to even out as the distance from the source is increased as shown in figure 14. It is expected that optimum electroplating results will be achieved in the far field, which is

172cm from the source. However, the acoustic pressure would have to to measured at that distance so see whether the drop in acoustic pressure compromise the efficiency of the agitation. Over 25cm from the transducer surface no significant drop in the acoustic pressure was measured and so attenuation through the fluid can be considered low.

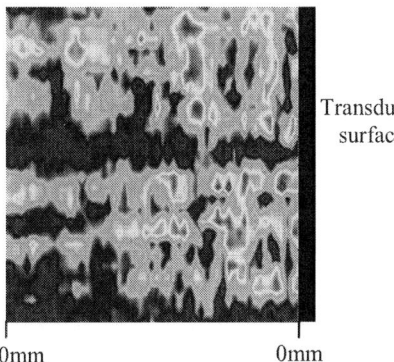

Transducer surface

250mm   0mm

Figure 14: X-Z surface plot of acoustic pressure from transducer operated at 1MHz, 100W.

The hydrophone was then inserted through a via of the test PCB as shown in figure 6. A scan of the acoustic pressure through the microvia when the PCB is held 20mm away from the transducer surface was conducted and the results are shown in figure 15. The sinusoidal behaviour of the signal stems from the time behaviour of the transducer output signal. The scan begun with the fibre optic hydrophone protruding from the bottom of the PCB, through the via hole by a few millimeters. In figure 15 the scan was begun with the hydrophone positioned at 5mm according to the stage coordinates and moved back through the via to the back side of the PCB at 0mm. As the hydrophone was released from the PCB, a sharp peak in the signal was observed as shown in figure 15, at approximately 1mm. This was due to the hydrophone moving suddenly when released from the via hole. This peak can be used to find the exact position of the microvia mouth since the thickness of the test PBC is 1.6 mm. The average pressure measured in the via mouth was found to be 91.5 kPa +/- 14%.

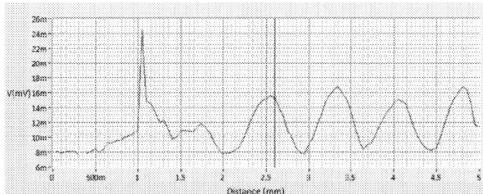

Figure 15: Recorded voltage from the hydrophone through a via held at 20mm from the transducer surface.

Modeling of the Flow in the electrolytic Solution

A. Model of Micro PIV Experiment

A flow model using Navier Stokes equation was created in Comsol to describe the flow in the blind via microfluidic test sample used in the micro-PIV experiment. This model assumed zero pressure at the outlet and the inlet flow velocity was defined as the calculated value, 333μms⁻¹.

B. Model of Acoustic Streaming

An acoustic model was also created in Comsol which used the measured pressure at the transducer surface as the boundary condition. The frequency was set to 1MHz and the distance between the transducer and the PCB was defined as 20mm. Since the acoustic wavelength is 1.5mm, 13 wavelengths should be visible between the transducer and PCB surfaces. To find the pressure at mouth of the via, the pressure on the PCB surface is integrated over the surface area by

$$\overline{P} = \frac{1}{S}\int_S P ds.$$

Modeling Results

A. Micro-PIV

The model of the micro-PIV experiment produced similar results to those achieved in the experiment itself. Figures 16(a) and 17(a) show the flow field maps of the 2:1 and 3.2:1 AR via, respectively. The plots of the velocity profile across the 2:1AR and 3.2:1 AR via mouths are plotted in figures 16(b) and 17(b). Figures 16(c) and 17(c) show the velocity plot at the centre and along the length of the via of the 2:1 AR and 3.2:1 AR via, respectively. These plots demonstrate that the penetration of vertical velocity is 1/4 and 1/3 of the via length.

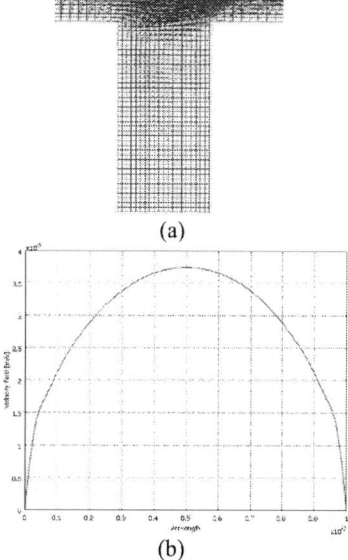

(a)

(b)

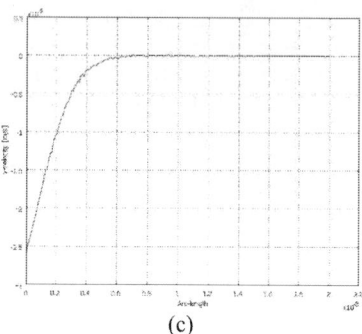

(c)

Figure 16: 2:1 AR via modeling results (a) field flow map (b) velocity profile across microvia mouth (c) velocity profile plot along via length

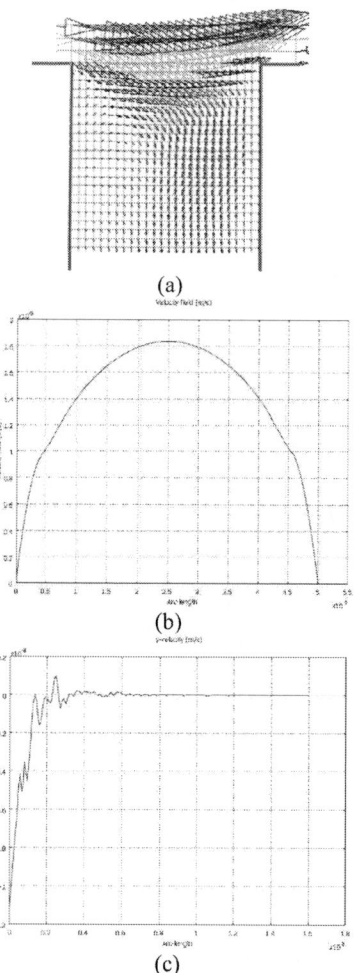

(a)

(b)

(c)

Figure 17: 3.2:1 AR via modeling results (a) Field flow map (b) velocity profile across the mouth of the via mouth (c) velocity profile plot along via length

B. Acoustic Streaming

In the acoustic model, the experimental result of the average pressure close to the transducer surface was used to model the pressure in the mouth of the via. Figure 18 shows the result of the time-harmonic model when the pressure boundary condition for a plane wave is applied on the transducer surface, and the sound-hard boundary condition, $\dfrac{\partial P}{\partial n} = 0$, is used on the PCB surface.

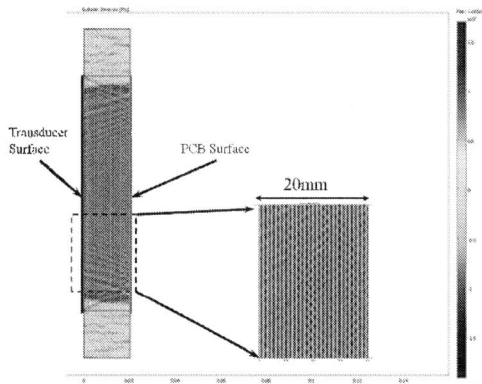

Figure 18: Model of the acoustic pressure from 1 MHz, 100W transducer.

The pressure in the mouth of the microvia was calculated to be 99.8kPa, which is in good agreement with the measured acoustic pressure.

Discussion

The micro-PIV experimental results and modeling results show that the maximum velocity at the centre of the 2:1 AR microvia is 70 μms^{-1} and 38 μms^{-1}, respectively. In the 3.2:1 AR microvia the experiment gives a maximum velocity of 19 μms^{-1} and the model results show 18 μms^{-1}. Both experiment and model show that velocity is highest at the centre of the mouth of the via slowing down towards the edges of the via. Given the errors associated with the precision of the microfluid chip fabrication, it is reasonable to assume that the experimental results are consistent with the model.

The model shows that the speed of flow is negligible one third of the way into the 2:1 AR via and only a quarter of the way into the 3.2:1 AR via. Although it was not possible to measure the exact distance into the via when conducting the micro-PIV experiment, the results have shown that flow slows significantly in the centre of the 2:1 AR via and is insignificant in the centre of the 3:1 AR via. These results are therefore in agreement with the findings of the model.

The average acoustic pressure measured 20mm away from the transducer surface was 91.4 kPa +/- 14%. The model, using the measured average pressure at the surface of the transducer, found the pressure on the surface of the PCB 20mm away from the transducer to be 99.8kPa. Given the error in the calibration factor of the hydrophone

alone, this result is excellent agreement with the experiment.

Conclusions

Micro-PIV experiment and modeling results of the blind via have been demonstrated to be in good agreement. The micro-PIV experiments have validated the flow model without acoustic streaming. The pressure measured experimentally at the mouth of the via was consistent with the result obtained using the acoustic model. The flow and acoustic models have therefore been validated by experiments described in this article.

The micro-PIV experiments have shown that fluid flow, hence ion transport in blind via is poor beyond one third of the total length of the via. The hydrophone experiments have also demonstrated that acoustic pressure can be directed into microvia as shown in figure 15.

Future work will include the modelling of the deposition of copper in microvia using acoustic streaming to promote ion transport. The model can then be used to optimize the experimental parameters in order to achieve successful high aspect ratio microvia filling.

Acknowledgements

The authors would like to thank Dr. Rosario Sanchez-Martin and Juan Manuel Cardenas-Maestre from the University of Edinburgh for supplying the microbeads used in the micro-PIV experiments. The authors would also like to thank the staff from COSMIC in the School of Physics and Astronomy for their assistance with the micro-PIV work.

Funding from the British Engineering and Physical Sciences ResearchCouncil (EPSRC) is gratefully acknowledged through the funding of the Innovative Manufacturing Research Centre (IMRC) at Heriot-Watt University.

References

[1] M. Suppa, "The process and pastes for the via hole plugging of HDI PCBs", pp 25-31, 2008, DOI 10.1108/03056120810848761.

[2] J.H. Lau, C. Chang, "An overview of microvia technology", Circuit World, Vol: 26, pp 22–32, 2000.

[3] Y. Zhang, R.W. Barber and D. Emerson, "Particle Separation in microfluidic devices – SPLITT fractionation and microfluidics", Current Analytical Chemistry, Vol. 1, pp 345-354, 2005.

Electrical Properties of an Isotropic Conductive Adhesive Filled with Silver Coated Polymer Spheres

S. Jain[1], D.C. Whalley[1,3], M. Cottrill[1,2], H. Kristiansen[3], K.Redford[3], C.B. Nilsen[3],

T. Helland[2], C. Liu[1]

[1]Wolfson School of Mechanical and Manufacturing Engineering, Loughborough University, Loughborough, UK

[2]Mosaic Solutions AS, Skjetten, Norway

[3]Conpart AS, Skjetten, Norway

Phone: +44-1509-227682, Fax: +44-1509-227648 and E-mail: s.jain2@lboro.ac.uk

Abstract

In the present study the electrical performance of newly developed epoxy resin based Isotropic Conductive Adhesives (ICAs) filled with silver coated mono sized polymer spheres have been investigated and compared with conventional solid silver particle/flake filled ICAs. The effects of particle size on the volume resistivity and percolation threshold of the new ICA have been studied. Two different diameters, i.e. 30 μm and 4.8 μm, of silver coated mono sized spherical polymer particles have been used in this study. The results show that, for the same volume fraction, the volume resistivity of the adhesive with 4.8 μm particles is lower than that with 30 μm particles. The adhesive formulated with 4.8 μm particles also exhibits a lower percolation threshold than that with 30 μm particles. The resistivity of the adhesive containing 4.8 μm particles was found to be of same order as that of currently commercially available ICAs, but with a significantly reduced silver content.

Key words: isotropic conductive adhesive, percolation threshold, volume resistivity, metal-coated polymer spheres, pathway.

Introduction

Isotropically conductive adhesives (ICAs) have been used in a variety of electronic applications for several decades [1]. Extensive process and reliability data have been generated and published, but there are still many insights to be gained and challenges to be overcome before ICAs may garner widespread acceptance as an alternative to soft solders [2]. No materials and processes that are capable of universal applicability have been identified which are capable of providing high junction/joint stability as a long term solution. Design of a new ICA material requires understanding and optimization of a number of factors including final electrical conductivity, rheological properties during application and the application process window, total metal content (a significant cost issue in the case of noble metal filler materials) and thermo-mechanical properties. These factors will be much more application specific than for solder alloys.

Traditionally ICAs have been manufactured by mixing solid metallic filler particles or flakes, with a non-conductive adhesive matrix, typically an epoxy resin. The current commercially available ICAs typically contain a high volume fraction, i.e., 25% to 40%, of silver particles or flakes in the adhesive matrix. The high filler content is mainly to achieve good electrical conductivity. The filler is usually silver (Ag), because of its high electrical conductivity and highly conductive oxide [3]. However, silver is expensive and such high volume fractions of metal flakes also have other disadvantages. These include a reduction in the impact strength and flexibility of the adhesive, which may affect the reliability in harsh environmental conditions [4].

Our approach to the improvement of ICA properties, whilst potentially also reducing cost, is based on the optimum use of material properties by replacing the solid metal flakes by metal coated polymer cored spheres having a very narrow size distribution. This approach can significantly reduce the required silver content whilst also improving the overall mechanical properties and hence reliability, of the adhesive. The elastic modulus of silver (70 GPa) is roughly 20 times higher than that of typical epoxy resins (below Tg). During mechanical loading of the silver-epoxy composite, this will cause a significant stress concentration and potentially micro-cracking at the interface between the two materials. The first anticipated reliability-benefit of this approach is that the flexible nature of the polymer-based spheres will allow them to better absorb mechanical stresses. The second benefit is the smaller thermal expansion mismatch between the adhesive matrix and conductor particles as compared to the metal particles in traditional isotropic conductive adhesives. Both of these factors are expected to improve the long-term reliability of the new ICA.

Previous studies by Nyugen et al., and Gakkestad et al., have demonstrated the feasibility of using metal coated Mono-sized Polymer Spheres (MPS) in ICAs. Nguyen et al. [5] found that die shear strengths obtained for an ICA filled with 45 volume% (vol%) of Ag-coated polymer spheres (Ag-MPS) and a conventional Ag flake filled ICA were similar over their whole range of test temperatures from 20 °C to 120 °C. Furthermore, neither of the two ICA materials was found to have a drop in the die shear strength near to or above the glass transition temperature.

Gakkestad et al., [6] also reported promising performance in temperature cycling and mechanical tests for an ICA using 30 µm diameter Ag-MPS as the electrically conductive filler. This adhesive was used to connect MEMS test structures directly to a PCB used within a medium calibre ammunition fuse. These test structures, with and without underfill, passed the firing test, in which the structures were exposed to an acceleration of more than 60,000g. After this realistic firing test the adhesive showed an insignificant reduction in die shear strength and only minimal changes in electrical contact resistance.

One concern when loading an ICA with spherical particles is the higher percolation threshold compared to ICAs loaded with traditional flake/particle fillers [7, 8]. This concern is important as a higher volume fraction will increase viscosity and hence affect the printability of such adhesives. Viscosity measurements by Nguyen et al. [9] have shown that a uniform suspension can be obtained up to 55 vol% of spherical polymer particles, which is very close to the theoretical limit. Although their experiments were conducted with non-metalized spherical polymer particles, it should be noted that the effect of metallization on the rheological properties, if any, would be to eliminate possible electrostatic forces. Therefore, these observations can be claimed to be relevant for a real system, where the adhesive matrix is loaded with metal coated spherical polymer particles. This observation has reinforced the potential applicability of the novel ICA, as it indicates the possibility of processing adhesives with high volume fractions (up to 55 vol%) of spherical particles.

The present paper presents results from varying the loading of spherical particles in a resin to determine the electrical percolation threshold. Further, particles of two different diameters, i.e., 30 µm and 4.8 µm, have been used to determine the effect of particle size on the electrical performance and percolation threshold. An adhesive formulated by loading silver flakes into the same adhesive matrix and also a commercially available ICA have been used in this study for comparison.

Materials

Silver coated polymer spheres having a narrow size distribution were produced and supplied by Conpart AS, Norway, while the silver flakes were supplied by Johnson Matthey, UK. The flakes were supplied treated with a surfactant. Table 1 gives the specifications of the conductive fillers used. A commercially available two-component epoxy adhesive was used as the non-conductive adhesive matrix. A widely used commercially available two part silver filled epoxy was used for comparison.

The substrate materials, onto which the adhesives were stencil printed, were both plain glass microscope slides and specially designed printed circuit boards (PCB). These PCBs were manufactured on FR4 laminate material with a standard Cu/Ni/Au, surface finish and have contact stripes for the adhesive which are connected to probe pads for making resistance measurements.

Table 1 Specifications of conductive fillers used

Type of filer	Size	Density (g/cc)
Ag-coated polymer sphere	4.8µm core with 150 nm Ag coating	2.50
Ag-coated polymer sphere	30 µm core with 130 nm Ag coating	1.44
FS34 silver flake	1.4 -11.0 µm with 1 um thickness	9.60[10]

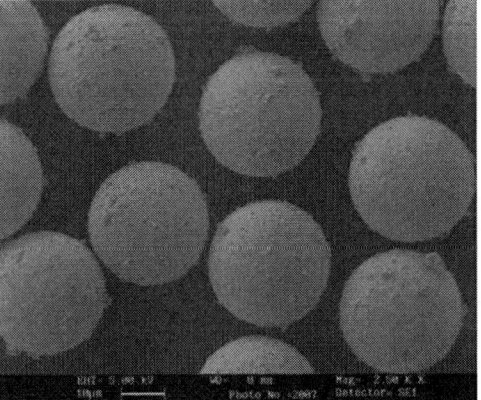

Figure 1 SEM image of 30 µm silver coated polymer spheres with a narrow size distribution

Equipment

A Flack Tek Speed Mixer DAC 150 FVZ-K from Hauschild, Germany was used to mix all of the adhesives. This mixer works on the dual asymmetric centrifuge principle i.e. by spinning a high speed mixing arm at speeds up to 3,500 rpm in one direction and the basket in the opposite direction. A balance with a resolution of 1 mg was used to weigh out the different components of the adhesives. For printing on the glass substrates a 0.02cm thick chemically etched brass stencil with 0.12cm x 4.2cm apertures was used to print stripes of the formulated adhesives, whilst for the PCB boards a 0.009cm thick laser cut plastic stencil with 0.2cm x 5.8cm apertures was used.

A JEM 310 single zone convective reflow oven was used to cure the samples for 15 minutes at 150 °C and a DataPaq data logger was used to continuously monitor the time/temperature profile within the oven. A Keithley 580 micro Ohm-meter was then used to measure all of the sample resistances.

Figure 2 SEM image of FS34 silver flakes used in the study

Adhesive Preparation

In this paper the adhesives loaded with 30 μm and 4.8 μm particles are henceforth denoted as ICA-30A and ICA-4.8A respectively. The ICA filled with FS34 silver flake is referred to as ICA1-Ag, while the commercially available ICA is denoted as ICA2-Ag.

To prepare ICA1-Ag, ICA 30A and ICA 4.8A, the two non-conductive adhesive components, resin and hardener, were first mixed according to the data sheet. A sample of the conventional silver filled ICA was also prepared by mixing the two components according to the data sheet.

Preparation of Adhesive Samples

The percolation threshold for the silver coated polymer spheres was predicted to be higher than that for conventional silver flake/particle filled ICAs [7]. However, a volume fraction of 20%, which is lower than the typical percolation threshold for silver flake filled materials, was used as the initial filler volume fraction to ensure the onset of conduction was not missed. The ICAs were prepared by mixing measured amounts of the pre-mixed non-conductive adhesive with a known weight of conductor particles to obtain the required 20 vol% filler content. The density of the epoxy was found

from the data sheet, whilst the density of the 30 μm Ag-MPS was measured using a Pycnometer. However the average density of the 4.8 μm particles was estimated from the silver coating thickness and polymer core density. To detect the percolation threshold the loading of filler was then increased in intervals of 2-4 vol%. After each particle addition the adhesive was remixed. This process was repeated until the point was reached where printing of the adhesive was difficult. In this way ICA 30A and ICA 4.8A with a range of volume fractions were prepared. The same procedure was followed to prepare samples of ICA1-Ag using a density for the silver flakes calculated from [10].

Thin stripes of the adhesives were manually printed onto the substrates using the stencils and a razor blade. For each batch of the adhesive – particle mixtures, resistance measurements on six stripes for both the glass and the PCB substrates were made.

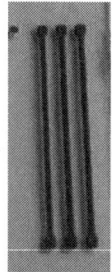

Figure 3 Printed samples on glass substrate (left) and FR4-PCB (right)

Volume Resistivity Measurements

Volume resistivity was measured using ASTM D2739 "Standard Test Method for Volume Resistivity of Conductive Adhesives". The micro-ohmmeter was used to make four wire (Kelvin) resistance measurements, as shown in Figure 4.

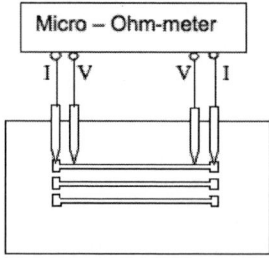

Figure 4 Schematic of four point resistance measurement method

Once the resistance measurements were made the volume resistivity, ρ, of each printed stripe was calculated using:

$$\rho = R \cdot \frac{w \cdot t}{L} \qquad (1)$$

Where R is the stripe resistance, w is the width, t is its thickness, and L is the distance between the voltage probes.

Effect of Shape and Size of Filler on the Percolation Threshold

Percolation theory describes the percolation threshold as a point at which long-range connectivity suddenly appears in a system. For conducting systems it is the point at which a sharp phase transition from non-conducting to conducting occurs. In practical situations with many uncertainties it is defined as a certain range within which the transition is possible [11]. Figure 6 illustrate percolation curves for ICA4.8A, ICA30A and ICA1-Ag on the PCB substrates. In line with the above description the percolation threshold has been found to lie around 24 vol% (40weight %) for the 4.8μm particles, 33 vol% (36 wt%) for 30μm particles and 15 vol% (40 wt%) for silver flakes. The percolation threshold values obtained for the Ag-MPS materials are in close proximity to the empirically derived value of 27% found for random close packed structures made using balls [11]. However, the experimental data also demonstrates, as expected, that the percolation threshold for the Ag-MPS is higher than for silver flakes. The high surface to volume ratio of flakes in comparison to spherical fillers can be expected to result in more particle-particle contacts at lower volume fractions and therefore could provide an explanation for the lower percolation threshold value. This could be explained by considering the flakes as discs of radius r_F and thickness t_F then the surface area S to volume V ratio is given by:

$$\frac{S}{V} = \frac{2}{r_F} + \frac{2}{t_F} \qquad (2)$$

For spheres the surface area to volume ratio is given by:

$$\frac{S}{V} = \frac{3}{r} \qquad (3)$$

where r is the radius of the sphere [7]. Using equations (2) and (3) a S/V value of 2.2 to 4 μm^{-1} is obtained for flakes as compared to 0.2 μm^{-1} for the larger spheres and 1.25 μm^{-1} for the smaller ones. Horizontal alignment of the silver flakes in the commercial adhesive can be seen in figure 6, which also shows the wide flake size distribution. This can also be expected to boost flake connectivity in the "XY" plane, thereby initiating in-plane conduction at a lower volume fraction. This may be an advantage if the material is used to create printed tracks, as here, but when used as an adhesive this orientation of flakes may reduce through thickness, conductivity. The MPS filled materials are not expected to display this anisotropy of resistivity.

Figure 5 SEM image of the top surface of an ICA2-Ag sample

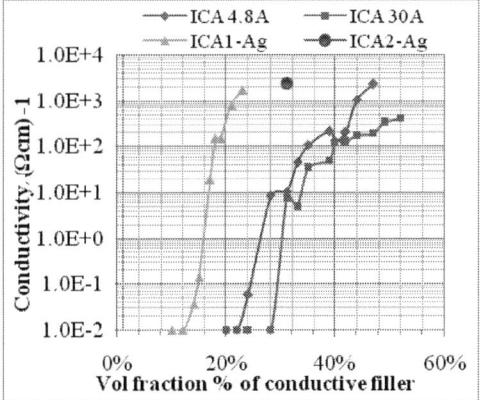

Figure 6 Percolation curves for Ag-MPS and silver flakes with a comparison to the commercially available ICA

The results are also in agreement with the theoretical calculations that if equal mass/weight of large and small mono-sized spheres are considered then:

$$\frac{N_S 4\pi r_S^{\,3}}{3} = \frac{N_L 4\pi r_L^{\,3}}{3} \qquad (4)$$

Solving for N_S gives:

$$N_S = N_L \cdot \frac{r_L^{\,3}}{r_S^{\,3}} \qquad (5)$$

and hence the ratio of surface areas is given by:

$$\frac{A_S}{A_L} = \frac{N_S 4\pi r_L^{\,2}}{N_L 4\pi r_S^{\,2}} = \frac{r_L}{r_S} \qquad (6)$$

where A_S and A_L are the surface areas of the small and large spheres respectively, N_L and N_S are the particle densities of the large and small spheres respectively, and r_L and r_S are the radii of the large and small spheres. This yields a larger surface area for the smaller spheres, accounting for their lower percolation threshold [7].

Small deviations from the trend are seen at certain points on the percolation curves. These variations can largely be attributed to variations in the manual stencil printing process. Another factor which can lead to variations is associated with the pot life of the ICAs, as the mixture that was initially prepared continued to be used by further adding the particles to form ICAs with an increasing volume fraction of particles, rather than fresh resin being used for each batch.

Effect of Sphere Size on Resistance

The resistance measurements show that, for the same filler vol%, ICA 4.8A has a lower resistivity than ICA 30A. This can be explained by assuming that for higher volume fractions of the Ag-MPS particles they form a close packed structure in the adhesive matrix and the resistance of the printed structure is therefore determined by an array of conductive pathways through the silver coating on the spheres. The resistance of each pathway through this network like structure can be determined as the series resistance of the number of particles in the pathway. The overall resistance of the printed structure is then determined by the area of conduction i.e. the thickness of the printed stripe multiplied by its width. This shows that the smaller the particles the greater is the number of parallel pathways and hence the lower the resistance. In addition, the slightly thicker Ag coating on the small spheres may have augmented the pathway area, which can further explain the lower resistance of ICA 4.8A.

Figure 7 shows the averaged results from the resistance measurements (converted to log values) for the ICAs at different volume fractions together with minimum and maximum value error bars. From these resistance measurements it has been found that at low volume fractions of conductive particles the adhesive remains non-conductive, but above a certain point, the percolation threshold, the adhesive becomes conductive and the conductivity increases rapidly as the percolation threshold is exceeded, whilst the spread in resistance reduces.

At the percolation threshold the amount of filler is just sufficient to create at least one path of conduction through the whole system. The large variation in the resistance values around the percolation threshold can be attributed to the variable number and lengths of conduction paths. When the filler volume fraction is well above the percolation threshold there exists a multitude of different similar pathways, hence low and repeatable resistance values are obtained.

Effect of Silver Content and Substrate Type on Volume Resistivity

The effect of the volume fraction of silver on the adhesive resistivity is shown in Figure 8. It is

clearly shown by this graph that with use of the 4.8 μm Ag-MPS as the conductive filler resistivity values similar to the flake filled and commercially available ICAs was obtained using less than 50% of their silver content.

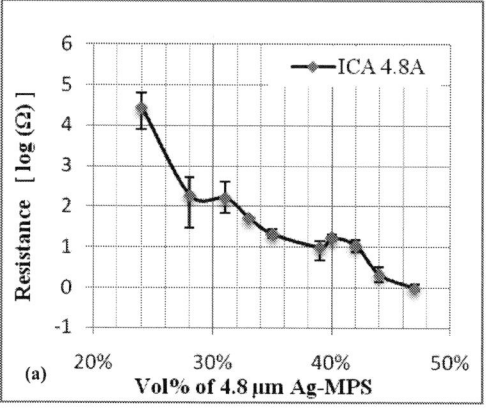

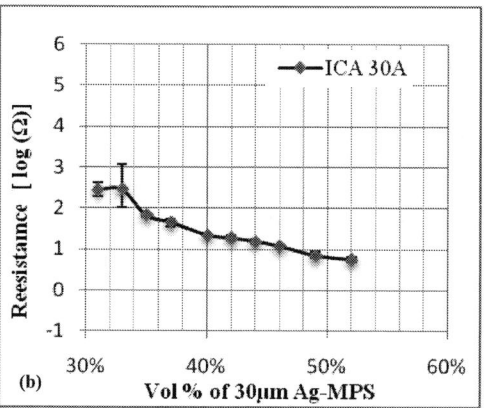

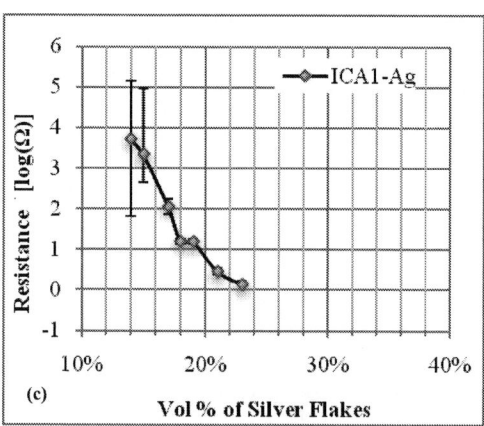

Figure 7 Resistance versus filler volume fraction for (a) ICA 4.8A (b) ICA 30A (c) ICA1-Ag

Furthermore it is been found that reasonably good resistivity values can also be obtained for very low amounts of silver with the 30 μm Ag-MPS filler particles.

The effect of substrate material on the ICA resistivity has been evaluated by comparing the resistance results for the PCB and glass substrates. Figure 9 shows the trends in resistivity on these two substrate materials. The differences in resistance for the two substrate materials are hard to discern at higher volume fractions. These minor differences could be because the PCB material contracts much more than the glass as the temperature is reduced from curing at 150 °C to room temperature.

However, for volume fraction near to the percolation threshold, the samples on the PCB substrate show conduction at lower volume fractions than those on glass. This difference may also be due to the different substrate thermo-mechanical properties, but may also be attributed to the very small contact area with the resistance measurement probes and possibly to settling of the conductor particles to the bottom of the stripe and therefore formation of a thin particle free adhesive layer on the top of the printed stripe. Both of these could hamper contact of any conducting path with the probe when using glass slides, but this is not the case for the PCBs as the adhesive stripe contacts the PCB conductors across its entire width. Further sample analysis is required to confirm this hypothesis.

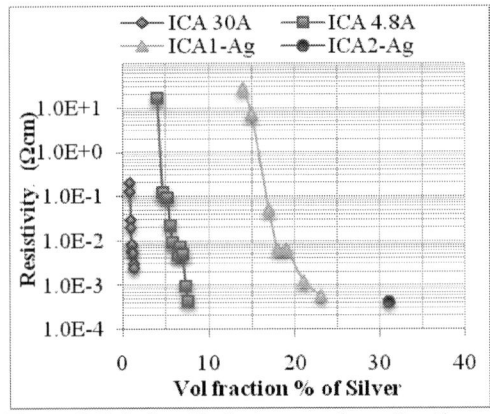

Figure 8 Relationship between volume fraction of silver and resistivity

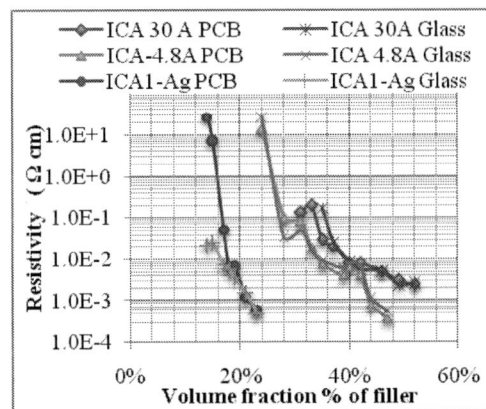

Figure 9 Comparison of resistivity values for PCB and glass substrates

The resistance spread near to the percolation threshold is approximately the same for the two substrate types and further differences in the resistivity values are negligible above percolation threshold for the two substrates, as shown in Figure 9. Therefore only the resistivity values for the PCB substrates were used in the earlier discussion.

Conclusions

The results presented demonstrate that low resistivity values, similar to those for flake filled commercially available ICAs, can be obtained by the use of Ag coated polymer spheres as a conductive filler, but with a greatly reduced Ag content. The results also indicate that, for a similar metallisation thickness, a lower resistivity can be achieved with smaller spheres compared to large ones. However the Ag content for the larger particles is even less than for the smaller particles and for applications where very low resistivity is not required they may offer even greater potential cost benefits.

Further work is still required to deeply understand the causes of the differences observed and the effects of variables such as coating thickness and morphology, the properties of the polymer core material and substrate materials, and the manufacturing process conditions, on electrical resistance. The effect of constriction and tunnelling resistance on the resistivity of this new class of ICAs also needs to be investigated. Long term joint stability and reliability will also require evaluation.

Nonetheless, the excellent electrical behaviour seen in this study, in addition to the high shear strengths obtained previously, indicate that these silver metallised polymer spheres will be suitable for the production of robust and low cost ICAs following their optimization.

Acknowledgements

The authors gratefully thank the Research Council of Norway (RCN), Mosaic Solutions AS, Norway and Loughborough University, UK for funding this project.

References

[1]   O. Boyle, D. C. Whalley, et al., "A study of the Process Parameters Involved in the Manufacture of Conductive Adhesive Joints", Proceedings of the 30th ISHM Nordic Conference, Oslo, pp.138-149, 1992.

[2]   L. Li, J. Morris, "An Introduction to Electrically Conductive adhesives", International Journal of Microelectronics Packaging, Vol.1, No.3, pp.159-175, 1998.

[3]   E. Suhir, Y. C. Lee, et al., "Electrically Conductive Adhesives: A Research Status Review", Micro- and Opto-Electronic Materials and Structures: Physics, Mechanics, Design, Reliability, Packaging, Springer US, pp. B527-B570, 2007.

[4] Y. Li, C. P. Wong, "Recent advances of conductive adhesives as a lead-free alternative in electronic packaging: Materials, processing, reliability and applications", Materials Science and Engineering: R: Reports, Vol. 51, pp. 1-35, 2006.

[5] H.-V. Nguyen, H. Kristiansen, et al., "Temperature Dependence of Mechanical Properties of Isotropic Conductive Adhesive Filled with Metal Coated Polymer Spheres", Proceedings of the 61st IEEE Electronics Components and Technology Conference, Florida, May 31- June 3, pp. 639-644, 2011.

[6] J. Gakkestad, H. Kristiansen, et al., "Use of Conductive Adhesive for MEMS Interconnection in Military Fuze Applications", Journal of Micro/ Nanolithography, MEMS and MOEMS, Vol. 9, p. 04118, 2010.

[7] J. E. Morris, "Conduction Mechanisms and Microstructure Development in Isotropic Electrically Conductive Adhesives", Conductive Adhesives for Electronics Packaging, Electrochemical Publications, Isle of Man, 1999.

[8] M. Mundlein, J. Nicolics, "Modeling of Particle Arrangement in an Isotropically Conductive Adhesive Joint", IEEE Transactions on Components and Packaging Technologies, Vol. 28, No.4, pp. 765-770, 2005.

[9] H.-V. Nguyen, H. Kristiansen, et al., "Spherical Polymer Particles in Isotropic Conductive Adhesives a Study on Rheology and Mechanical Aspects", Proceedings of the 3rd IEEE Electronic System-Integration Technology Conference (ESTC), Berlin, Germany, September 13-16, pp. 1-6, 2010.

[10] G. W. Brassell, D. R. Fancer, "Long-term Strength Characteristics of Conductive Epoxies [Online] http://www.epotek.com/SSCDocs/whitepapers/Tech%20Paper%2022.pdf".

[11] R. Zallen, "The Percolation Model", The Physics of Amorphous Solids, A Wiley-Interscience Publication, New York, pp. 135-204, 1998.

An Investigation of Lead-free Thick-film Resistors on LTCC Substrates – Preliminary Results

Marko Hrovat[1,2], Konrad Kielbasinski[3,4], Kostja Makarovič[1,2], Darko Belavič[5,2], Malgorzata Jakubowska[3,6]

[1] Jožef Stefan Institute, Jamova 39, SI-1000 Ljubljana, Slovenia
[2] CoE NAMASTE, Jamova 39, SI-1000 Ljubljana, Slovenia
[3] Iinstitute of Electronic Materials Technology, Wólczyńska 133, 01-919 Warsaw, Poland
[4] Institute of Microelectronics and Optoelectronics, Warsaw University of Technology, Koszykowa 75, 00-662 Warsaw, Poland
[5] HIPOT-RR, d.o.o., Šentpeter 18, SI-8222 Otočec, Slovenia
[6] Faculty of Mechatronics, Warsaw University of Technology, Św. Andrzeja Boboli 8, 02-525 Warsaw, Poland,

Corresponding author (Marko Hrovat) - e-mail: marko.hrovat@ijs.si

Abstract

The lead-free thick-film resistors with different ratios of conductive and glass phases were investigated. Four resistor materials with nominal sheet resistivities from 50 ohm/sq. to 50 kohm/sq. were prepared with different combinations of two lead-free glasses with reflow temperatures at 940°C and 1240°C, respectively, and two RuO_2 powders (fine grained and coarse grained RuO_2). Thick-film resistors were printed and fired on alumina and LTCC substrates (Du Pont 951) with the aim to study the influence of interactions between resistor layers and rather glassy LTCC substrates on electrical and microstructural characteristics. The resistors were investigated by X-ray powder diffraction and by SEM/EDXS analyses. Sheet resistivities, TCRs, gauge factors and noise indices were measured. X.ray spectra indicated that the RuO_2 do not react during firing with either of two evaluated glasses. The gauge factors were below or around 3, which indicates stable resistance characteristics. The resistivity vs. temperature dependences were remarkably linear with R-squared equal to or better than 0.99. For most resistors sheet resistivities were comparable to the nominal resistivities and noise indices were around or below 0.2 uV/V. However, resistors prepared from the coarse grained RuO_2 and the glass with higher reflow temperature the resistivities were an order of magnitude too high and noise indices were between 7 and 10 uV/V.

Key words: Thick film resistors, lead free, LTCC, characterization

Introduction

Thick-film resistor pastes consist basically of a conducting phase, a silica-rich lead-borosilicate-based glass phase and an organic vehicle. This organic material is burned out between 300°C and 400°C during the high-temperature processing. The ratio between the conductive and glass phases roughly determines the specific resistivity of the resistor. During the firing cycle the conductive phases of the resistor materials interact with the glass phase. In most modern thick-film resistor compositions the conductive phase is either ruthenium oxide or electrically conducting pyrochlores; mainly lead or bismuth ruthenates[1-3].

As mentioned above, the glass phase in commercial thick-film resistors is based on lead borosilicates. The European environmental legislation, i.e., the RoHS Directive (RoHS -

restriction of the use of certain hazardous substances), requires the elimination of lead, or at least a minimising of the lead content in electrical and electronic equipment to below 0.1 wt.% [4]. Thick-film materials are currently an exemption from the directive [5], but the producers of thick-film materials are already developing new material systems, mainly conductors and dielectrics, in accordance with the directive [6]. While there are many lead-free conductor and dielectric compositions available, the development of thick-film resistors with lead-free glasses is still mainly at the experimental stage. At least to the best of the authors' knowledge, no commercially available thick-film lead-free resistor series with characteristics comparable to "conventional" resistors is on the market.

However, in the open literature there are many articles reporting investigations of the characteristics of lead-free thick-film resistor materials. Obviously, the lead ruthenates are not a feasible option for a conductive phase. If the bismuth ruthenates are used as a conductive phase in lead-free thick-film resistors they decompose during the firing due to an interaction with the silica-based glasses ([7,8]). Therefore, the conductive phase in most lead-free thick film resistors is based on RuO_2.

Modern, advanced microelectronic packages in many cases are created by a multilayer ceramic module with integrated electronic components and sub-circuits. The Low Temperature Co-Fired Ceramics (LTCC) technology is considered as one of more advanced technologies for the fabrication of these packages. The LTCC materials are sintered at low temperatures. To sinter to a dense and non-porous structure at these, rather low, temperatures, LTCC materials have to contain some (or a great deal of) low-melting-point glass phase. LTCCs are either based on crystallisable glass or a mixture of crystallisable glass and ceramics; in most cases alumina ([9-15]).

Thick-film resistors are often integrated on the top of multilayer LTCC structures. However, most of the thick-film resistor materials are developed for firing on relatively inert alumina substrates. The compatibility and interactions with the rather glassy LTCC substrates, leading to changes in the electrical characteristics, need to be evaluated. The aim of this paper is to evaluate the characteristics of thick-film lead-free resistors fired on LTCC substrates. The results will be compared with results obtained on alumina substrates.

Lead-free thick-film resistors were developed at the Institute of Electronic Materials Technology (ITME) in Warsaw. Two different lead-free boro-silicate-based glasses were chosen, and RuO_2 was used as a conductive phase. Some characteristic temperatures of the glasses are summarized in Table 1 ([16,17]). For the conductive phase, two RuO_2 powders were chosen: fine grained and coarse grained with specific surface areas of >50 m^2/g and 0.5 m^2/g, respectively. The denotations of the thick-film resistor materials together with other relevant data are shown in Table 2. Denotations are FL, FH, CL and CH for resistors prepared with fine (F) and coarse (C) RuO_2 powders and with low (L) and high (H) sheet resistivities, respectively.

Table 1. Some temperature characteristics of lead-free glasses ([16,17])

Glass	Glass A	Glass-B
Characteristic temperatures		
softening	700°C	710°C
melting	840°C	810°C
reflow	1240°C	940°C

Table 2 Lead-free thick-film resistors – denotations, RuO_2 powders, glass phases and nominal resistivities

Resistor	RuO_2 powder	Glass	Nominal sheet resistivity (ohm/sq.)
FL	fine	glass B	50
FH	fine	glass A	5 k
CL	coarse	glass B	100
CH	coarse	glass A	50 k

Experimental

The resistors were screen-printed and fired for 10 min at 850°C on 96% alumina and on LTCC (951, Du Pont) substrates. The LTCC substrates were made by laminating three layers of LTCC tape at 70°C and at a pressure of 200 bar. The prepared LTCC substrates were fired first for 1 h at 450°C (burn-out of organics) and then fired at maximum temperatures of 850°C. The thick-film resistors were terminated by Pd/Ag electrodes that were pre-fired at 850°C on alumina substrates, and cofired together with the printed resistors on LTCC substrates. The dimensions of the resistors for the microstructural analysis and the X-ray diffraction (XRD) analysis, which were printed and fired without conductor terminations, were 12.5x12.5 mm^2.

For the microstructural investigation the samples were mounted in epoxy in a cross-sectional orientation and then cut and polished using standard metallographic techniques. A JEOL JSM 5800 scanning electron microscope (SEM) equipped with an energy-dispersive X-ray analyser (EDS) was used for the overall microstructural and compositional analysis. The resistors were analysed by X-ray diffraction (XRD) analysis with a Philips PW 1710 X-ray diffractometer using Cu K_α radiation. X-ray spectra were measured from 2 $\Theta=20°$ to 2 $\Theta=70°$ in steps of 0.02°.

Temperature coefficients of resistivity were calculated from resistivity measurements attemperatures between -25°C and 125°C. The current noise was measured in dB on 100-mW loaded resistors using the Quan Tech method (Quan Tech Model 315-C).

The gauge factor (GF) of a resistor is defined as the ratio of the relative change in resistance (dR/R) and the strain (d l/l). Geometrical factors alone result in gauge factors of 2-2.5. Gauge factors higher than these values are due to microstructural changes, i.e., changes of the specific conductivity.

$$GF = (\Delta R/R) / (\Delta l/l) \qquad (1)$$

The changes in resistance as a function of substrate deformation were measured with the simple measurement set-up shown in Fig. 1. The ceramic substrate is supported on both sides. The load that induces a controlled tensile strain in the resistor is applied to the middle of the substrate with a micrometer screw. The magnitude of the strain is given by equation (2) ([18]).

$$\varepsilon = \Delta l/l = (d * t * X * 12)/L^3 \qquad (2)$$

d = deflection (m)
t = substrate thickness (m)
L = distance between the support edges (m)
X = distance between the supported edge and the resistor (m)

For X=L/2, i.e., when the resistor is located in the middle of the substrate, where the strain is greatest, the equation transforms into:

$$\varepsilon = \Delta l/l = (d * t * 6)/L^2 \qquad (3)$$

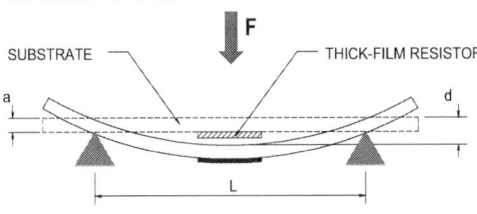

Fig. 1. Method for measuring the changes of resistivity as a function of substrate deformation

Results and Discussion

XRD characterization

X-ray spectra of thick-film resistors prepared on alumina and LTCC substrates are shown in Fig. 2.a for FH resistors – glass A – and in Fig. 2.b for FL resistors – glass B. In both Figs. the spectrum of RuO_2 is added. The results indicate that neither of glasses reacts with the RuO_2 based conductive phase during firing.

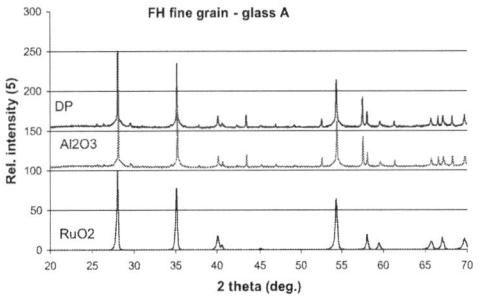

Fig. 2.a X-ray spectra of FH resistors – fine grained RuO_2, glass A – do not react with the glass phase during firing. The spectrum of the RuO_2 is added.

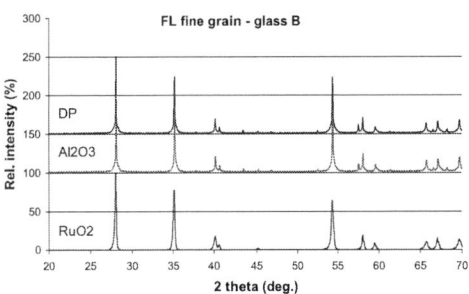

Fig. 2.b X-ray spectra of FL resistors – fine grained RuO_2, glass B – do not react with the glass phase during firing. The spectrum of the RuO_2 is added.

Microstructural characterization

The surfaces and cross-sections of the resistors fired on LTCC substrates were investigated by SEM. The microstructures of the surface and the cross-section of the CL samples are presented in Figs. 3.a and 3.b, respectively and of the HL samples in Figs. 4.a and 4.b, respectively. The EDS analysis identified a light particle in the resistor layers as the ruthenium oxide. In the LTCC substrates at the bottom of Figs. 3.b and 4.b the darker grains are alumina particles and the lighter phase is the glass phase. The possible interactions between the LTCC substrates and active layers were not observed by SEM.

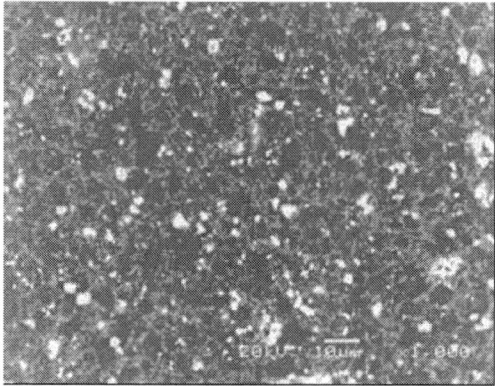

Fig. 3.a. The microstructure of the surface of the lead-free thick-film resistor FL fired on the LTCC substrate (coarse RuO$_2$, glass B)

Fig. 3.b. The microstructure of the cross-section of the lead-free thick-film resistor CL fired on the LTCC substrate (coarse RuO$_2$, glass B)

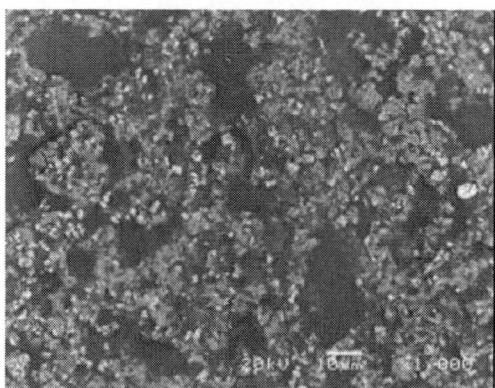

Fig. 4.a. The microstructure of the surface of the lead-free thick-film resistor FL fired on the LTCC substrate (fine RuO$_2$, glass A)

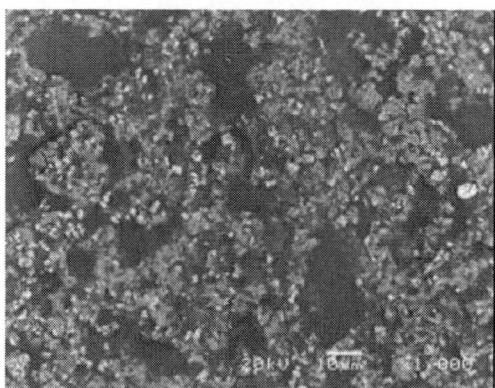

Fig. 4.b. The microstructure of the cross-section of the lead-free thick-film resistor FL fired on the LTCC substrate (fine RuO$_2$, glass A)

Electrical characterization

In Table 3, the sheet resistivities, temperature coefficients of resistivity, noise indices and gauge factors are summarized. Noise indices are given in (dB) and in (uV/V). These two units are related with a simple equation:

$$noise\ (dB) = 20 \times \log noise\ (\Box V/V) \qquad (4)$$

For most resistors sheet resistivities on LTCC substrates are slightly higher than on alumina, which could be tentatively attributed do the diffusion of the glass phase from the LTCC substrates into the resistor films during firing. Noise indices are low - around or below 0.2 uV/V. However, the sheet resistivities of the CH resistors (coarse grained RuO$_2$ and glass A) are an order of magnitude too high and the noise indices are very high – between 17 and 20 dB or 7 and 10 uV/V on alumina and LTCC substrates, respectively, These results indicate that for this thick-film resistor material the firing temperature of 850°C could be too low. Noise indices are graphically shown in Fig 5. Note that the values in the diagram are cut off at 2 uV/V and the values for CH resistors are between 7 and 10 uV/V.

The typical example of resistivities vs. temperature dependences is shown in Fig. 6 for the resistors prepared with coarse grained RuO$_2$. The resistivity vs. temperature curves are remarkably linear with an R^2 more than 0.99. This is presumably due to the high positive linear coefficient of resistivity of the ruthenium oxide.

Gauge factors are between 1.5 and 5.5. However, the interesting and unexpected result is that the gauge factors of resistors with the high concentration of conductive phase and therefore low nominal resistivities are higher than gauge factors of resistors with high resistivity regardless of the grain sizes of RuO_2 powders. Data from the literature indicate that the gauge factors for most if not for all commercial resistor series increase with increasing sheet resistivities and – which is same - with decreasing concentrations of a conductive phase.

Table 3. sheet resistivities, temperature coefficients of resistivity, gauge factors and noise indices of the lead-free thick film resistors fired on alumina and LTCC substrates.

Res.	Sub.	R Ω/sq.	TCR x10^{-6}/K	GF	Noise dB	Noise uV/V
FL	Al_2O_3	75	445	5.0	-18.7	0.05
	DP 951	75	570	5.5	-19.0	0.06
FH	Al_2O_3	6.7k	360	1.5	-14.4	0.19
	DP 951	5.6k	330	1.5	-14.5	0.19
CL	Al_2O_3	185	290	3.5	-25.6	0.12
	DP 951	245	230	4.0	-24.9	0.11
CH	Al_2O_3	41k	500	2.5	17.3	7.3
	DP 951	69k	450	2.5	20.4	10.5

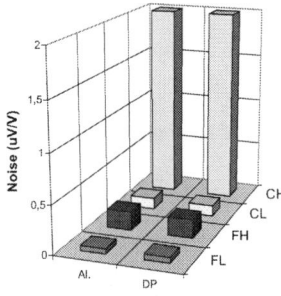

Fig. 5. Noise indices of the resistors fired on alumina – denoted "Al" and on Du Pont 951 – denoted "DP" substrate. Note that the values in the diagram are cut off at 2 uV/V and the values for CH resistors are between 7 and 10 uV/V.

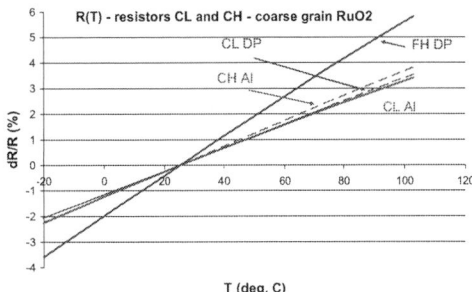

Fig. 6. The resistivities vs. temperature dependences of the thick-film resistors prepared with coarse grained RuO_2. The resistivity vs. temperature curves are remarkably linear with an R^2 more than 0.99.

Summary

Lead –free thick film resistor materials with nominal resistivities from 50 ohm/sq. to 50 kohm/sq. were evaluated on alumina and LTCC substrates. Resistor materials were prepared by mixing either fine or coarse grain RuO_2 powders and two different lead-free glasses with different viscosity vs. temperature characteristics. The aim was to study the influence of possible interactions between resistor layers and rather glassy LTCC substrates on electrical and microstructural characteristics of resistors. .Fired resistors were investigated by X-ray powder diffraction and by SEM/EDXS analyses, and sheet resistivities, TCRs, gauge factors and noise indices were measured.

X-ray spectra of thick-film resistors prepared on both alumina and LTCC substrates indicate that neither of two glasses reacts with the RuO_2 based conductive phase during firing. For most tested resistor materials sheet resistivities on LTCC substrates are slightly higher than on alumina, which could be tentatively attributed do the diffusion of the glass phase from the LTCC substrates into the resistor films during firing. However, these interactions between the LTCC substrates and active layers were not observed by SEM. Noise indices of most resistors are around or below 0.2 uV/V. The exception is the resistor material denoted CH (coarse grained RuO_2 and glass A) for which the sheet resistivities are an order of magnitude too high and the noise indices are very high – between 7 and 10 uV/V on alumina and LTCC substrates, respectively, These results indicate that probably higher firing temperatures are needed for this material.

The resistivity vs. temperature curves for all resistors are remarkably linear with an R^2 more than 0.99. Gauge factors are between 1.5 and 5.5. It is interesting that the gauge factors of resistors with the high concentration of conductive phase and therefore low nominal resistivities are higher than gauge factors of resistors with high resistivity.

Acknowledgments

The authors wish to thank Mr. Mitja Jerlah (HYB) for printing, laminating and firing the LTCC structures. The financial support of the Slovenian Research Agency is gratefully acknowledged.

References

1. J. W. Pierce, D. W. Kuty, J. L. Larry, The chemistry and stability of ruthenium based resistors, Solid State Technol., 25, (10), (1982), 85-93
2. R. W. Vest, "Materials science of thick-film technology", Ceram. Bull., 65, (4), (1986), 631-636
3. M. Hrovat, J. Holc, D. Belavič. J. Bernard, Subsolidus phase equilibria in the PbO poor part of RuO_2-PbO-SiO_2 system, Materials Letters, 60, (20), (2006), 2501-2503
4. On the restriction of the use of certain hazardous substances in electrical and electronic equipment (ROHS), directive 2002/95/EC of the European Parliament and of the Council, 2002
5. Frequently asked questions on directive 2002/95/EC on the restriction of the use of certain hazardous substances in electrical and electronic equipment (RoHS) and directive 2002/96/EC on waste electrical and electronic equipment (WEEE), European Commission, Directorate-General Environment, 2006.
6. A. C. Buckthorpe, J. Cocker, L. Garreau-Iles, D. Greenhill, R. Parr, T. Sweeny, Lead-free solutions for ceramic modules, Proc. 4th European Microelectronics an Packaging Symposium EMPS-2006, Terme Čatež, Eds. D. Belavič, M. Kosec, I. Šorli, 2006, 115-120
7. B. Morten, G. Ruffi, F. Sirotti, A. Tombesi, L. Moro, T. Akomolafe, Lead-free ruthenium-based thick-film resistors: a study of model systems, J. Mater. Sci.: Materials in Electronics, 2, (1), (1991), 46-53
8. M. Hrovat, T. Maeder, J. Holc, D. Belavič, J. Cilenšek, J. Bernard, Subsolidus equilibria in the RuO_2 – Bi_2O_3 – SiO_2 system, J. Eur. Ceram. Soc., 28, (11), (2008), 2221-2224

9. C.-J. Ting, C.-S. Hsi, H.-J. Lu, "Interactions between ruthenium-based resistors and cordierite-glass substrates in low temperature co-fired ceramics", J. Am. Ceram. Soc., vol. 83, no. 12, pp. 23945-2953, 2000.
10. W. K. Jones, Y. Liu, B. Larsen, P. Wang, M. Zampino, "Chemical, structural and mechanical properties of the LTCC tapes", Proc. 2000 Int. Symp. on Microelectronics IMAPS-2000, Boston, pp. 669-674, 2002.
11. M. Hrovat, D. Belavič, J. Kita, J. Holc, J. Cilenšek, L. Golonka, A. Dziedzic, Thick film resistors with low and high TCRs on LTCC substrates, Informacije MIDEM, vol. 35, no. 3, pp. 114-121, 2005.
12. M. Hrovat., D. Belavič, J. Kita, J. Cilenšek, L. Golonka, Dziedzic, A., Thick-film temperature sensors on alumina and LTCC substrates", J. Eur. Ceram. Soc., vol. 25, no. 15, pp. 3443-3450, 2005.
13. Y. Imanaka, Multilayer low temperature cofired ceramic (LTCC) technology, Springer Science + Business Media, Inc., 2005, 1-56
14. C. J. D. Kumar, E. K. Sunny, N. Ranghu, N. Venkataramani, A. R. Kulkarni, Synthesis and characterization of crystallizable amortize-based glass for a low-temperature cofired ceramic applications, J. Am. Ceram. Soc., 91, (2), (2008), 652-65515.
15. R. Muller, R. Meszaros, B. Peplinski, S. Reinch, M. Eberstein, W. A. Schiller, J. Deubener, Dissolution of alumina, sintering, and crystallization in glass ceramic composites fot LTCC, J. Am. Ceram. Soc., 92, (8), (2009), 1703-1708
16.. K. Kiełbasiński, A..Młożniak, M. Jakubowska, "High ohm eco-friendly resistors in thick film technology", 32nd International Microelectronics and Packaging IMAPS-CPMTPoland Conference, Warsaw-Pułtusk, September 2008.
17. K. Kielbasinski, M. Jakubowska, A. Mlozniak, M. Hrovat, J. Holc, D. Belavič, Investigation on electrical and microstructural properties of thick film lead-free resistor series under various firing conditions, J. Mater. Sci.: Mater. Electron., 21, (10), (2010), 1099-1105
18. C. Song, D. V. Kerns, Jr., J. L. Davidson, W. Kang, S. Kerns, Evaluation and design optimization of piezoresistive gauge factor of thick film resistors", IEEE Proc. SoutheastCon 91 Conf. (Vol. 2), Williamsburg, 1991, 1106-1109

Characterization of intermetallic compounds in Cu-Al ball bonds: layer growth, mechanical properties and oxidation

M.H.M. Kouters[a*], G.H.M. Gubbels[a], O. O'Halloran[b], R. Rongen[b]

[a] TNO Technical Sciences, Materials Performance group, P.O. Box 6235, 5600 HE Eindhoven, The Netherlands
[b] NXP Semiconductors, RQC Europe, 6534 AE Nijmegen, The Netherlands
[*] M.H.M. Kouters, marcel.kouters@tno.nl, +31(0)88 86 65461

Abstract

In high power automotive electronics copper wire bonding is regarded as most promising alternative for gold wire bonding in 1st level interconnects and therefore subjected to severe functional requirements. In the Cu-Al ball bond interface the growth of intermetallic compounds may deteriorate the electrical, thermal and mechanical properties. The layer growth and properties of these intermetallic compounds are crucial in the prediction of the long term behavior. To mimic the growth of intermetallic compounds during and after copper ball bonding, diffusion couples of aluminium and copper were annealed at 225-500°C and chemically analyzed by SEM/EDS. Also five separate intermetallic compounds were melted together from the pure elements and aged in evacuated quartz ampoules for 240 hours at 500°C. In this work values for the indentation Young's modulus, load independent hardness, fracture toughness and volumetric densities are given. For four intermetallics in the diffusion couple indentation Young's modulus and hardness were determined using a universal nano-indenter (UNAT). The combination of nano-indentation (10-500mN), micro-Vickers (0.5-2N) and Vickers hardness measurements (10-49N) gave hardness values over a wide range of loads. The reduction of the apparent hardness is explained by the indentation size effect. Fracture toughness was measured with Vickers indentation, volumetric densities with Archimedes method. It can be concluded that the Cu-rich intermetallics Cu_9Al_4 and Cu_3Al_2 are less sensitive to fracture and have lower average densities than the other intermetallic compounds. The volumetric decrease during formation causes internal stress. Finally, the difference in oxidation behaviour of the intermetallic compounds and the Cu solid solution was demonstrated through experiments performed on the diffusion couples.

Key words: Wire bonding, Cu-Al intermetallics, thermal aging, mechanical properties, oxidation

1. Introduction

More than 90% of integrated circuits manufactured today use wire bonding for interconnections [1, 2]. The most commonly used metal in IC wire bonding today is gold, however due to its soaring cost, alternatives are being investigated by IC manufacturers. Copper ball bonding is growing in market share not only due to its lower cost; higher electrical conductivity, increased thermal conductance and resistance to wire sweep during encapsulation make it an attractive alternative to gold. Mechanically copper is stronger and stiffer than gold. The increased stiffness improves looping for long wires especially when subjected to forces from the mould compound [3]. For the new generation high pin-count devices copper is required as gold is unsuitable when the pad pitch is less than 50 μm due to its lack of stiffness [4]. Fine pitch bonding is becoming more popular due to the drive for smaller packages. Higher electrical conductivity means that smaller diameter wire can be used for equivalent current. The increased thermal conductance allows copper to drain more heat from the package than gold. Another advantage of copper is that intermetallics grow a lot slower in the copper-aluminium interconnection compared to gold-

aluminium, which is expected to improve the long-term reliability of products.

The thicknesses of copper-aluminium inter-metallics are a factor 100 times lower than gold-aluminum intermetallics after ball bond formation [5]. However, as the intermetallic layers present in copper wire bonded samples are very thin, it is difficult to determine the chemical constitution and identity of intermetallic compounds present at the Cu-Al interface [6]. This can result in false identification and attribution of properties to the Cu-Al intermetallics.

In our previous study the identification and layer growth rates were determined [7, 8] and compared with data published by Funamizu [9]. The detection of thin intermetallic layers is cumbersome; at lower temperature regimes intermetallic phases are not present in measurable amounts. In particular the phase boundary between Cu_9Al_4 (γ) and Cu_3Al_2 (δ) is not obvious (see Figure 1). In the Cu_3Al_2(δ) layer the original interface (Kirkendall plane) is visible due to presence of contaminations at this interface which are hardly avoidable. These contaminations may mechanically weaken this phase and make this plane most susceptible to fracture.

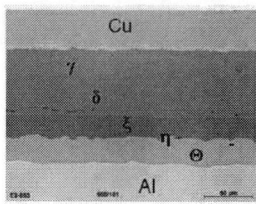 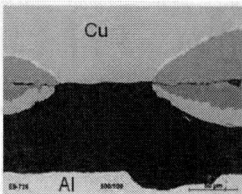

Figure 1: Light microscope images of the diffusion zone developed between Al and Cu (500°C, 100hrs in vacuum). The right picture shows the edge of the diffusion couple and suggests that the original interface (i.e. the Kirkendall plane) is present in the (δ) layer [7].

Some mechanical properties for Cu-Al intermetallic compounds are reported in literature, however Young's moduli and fracture toughness areas yet not documented. Wulff [10] presented the physical parameters as listed in Table 1. With regard to the mechanical properties the Vickers hardness is given; densities for Cu_4Al_3 and Cu_3Al_2 are still missing.

Table 1: Physical parameters of Cu-Al intermetallics [10]

Phase	Cu	HV	Resistivity	CTE	Density
	At%	kgf/mm²	μΩcm	ppm/°C	g/cm³
Al	0-2.84	20-50	2.4	23.5	2.7
$CuAl_2$	31.9-33	324	7 - 8	16.1	4.36
CuAl	49.8-52.3	628	11.4	11.9	2.7
Cu_4Al_3	55.2-56.3	616	12.2	16.1	na
Cu_3Al_2	59.3-61.9	558	13.4	15.1	na
Cu_9Al_4	62.5-69	549	14.2-17.3	17.6	6.85
Cu	80.3-100	60-100	2	17.3	8.93

In order to complete the overview of mechanical properties, hardnesses and indentation Young's moduli were determined by using the combination of UNAT nano-indentor, micro-Vickers and the conventional Vickers method. In addition, fracture toughness data and volumetric densities were determined. The obtained values are crucial for future reliability modeling. We would like to stress that the thermal expansion data as presented in the work of Wulff are presented a factor of ten in error. Unfortunately this is reproduced in subsequent work [11].

2. Experimental procedure

To obtain homogeneous intermetallic compounds as present in Cu ball bonds, diffusion couples were made by galvanic deposition of a 0.25 mm thick Cu layer on an Al sheet (supplier Goodfellow, purity 99.999%, 2 mm thick). The main steps of this process are as follows. After degreasing, a zincate step and a Rochelle copper strike, the galvanic layer was deposited during six hours in an acid copper bath (4 A/dm²) at room temperature. The aluminium sheet with galvanic copper was sealed in evacuated quartz capsules (pressure: ~10^{-5} mbar) and subsequently heated for various times (70-1225 hours) and at various temperatures (225-500°C). All 5 intermetallic

compound layers are visible using a light microscope even without etching, see Figure 1. The chemical composition of the grown intermetallic compound layers is measured with Scanning Electron Microscope equipped with an Energy Dispersive Detector SEM(EDS) using an acceleration voltage of 7 kV while detecting the Al-K and Cu-L characteristic radiation. The different intermetallic compounds are identified based on this measured composition.

Five separate intermetallic compounds were melted together in graphite crucibles under a $N_2/5\%$ H_2 protective atmosphere from the pure elements: Al (purity 99.999 wt%, supplier Good Fellow) and OFHC copper (99.95 wt% supplier Good fellow). The formed intermetallics were further aged in evacuated quartz ampoules for 240 hours at 500°C, see Figure 2a. The weight amounts of Cu and Al (including losses due to melting) are inside the homogeneity ranges of the various intermetallic compounds.

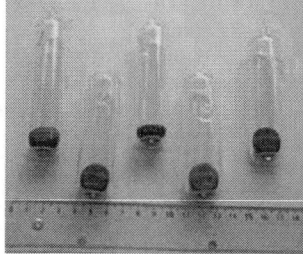 

Figure 2: (a) Intermetallic compounds enclosed in evacuated quartz ampoules and (b) the Universal Nanomechanical Tester (UNAT)

Hardness and indentation Young's modulus of the intermetallic layers formed in the diffusion couples are measured using an Universal Nanomechanical Tester (UNAT, Figure 2b). The UNAT has a lateral indentation precision of <0.5μm and a maximum normal force of 2N. Oblique line trajectories with 120 indentations were made with a Berkovich tip with indentation loads of 10, 20, 50 and 500mN.

The term *indentation Young's modulus* is used because this Young's modulus calculated from the phenomena occurring during and after each indent. After that the maximal load with the Berkovich tip is applied the load is gradually released. The upward tip displacement is a measure for the amount of elastic deformation in the (compressed) indentation region. The calculated indentation Young's modulus is based on an assigned Poisson ratio. In this study a Poisson ratio of 0.30 is assumed for the brittle intermetallics.

To measure hardnesses for a wide range of indentation loads the micro-Vickers method (50-200gr) and a Vickers hardness apparatus (HV1.0-5.0) were used. The indentation fracture toughness was determined for each separate intermetallic compound by using the Vickers indentor with relatively higher loads [12]. Density determination, based on Archimedes method (the buoyancy effect), was measured with a Mettler kit type ME 33340.

3. Results and discussion

3.1 Indentation Young's modulus

Indentation trajectories with the UNAT Berkovich-indentor were made oblique over the diffusion couple with 10, 20, 50 and 500mN indentation loads. A minimum step size of 2μm was chosen (except 3μm for 500mN), which is a pragmatic compromise to distinguish the properties of the thin intermetallic phases. The distance between indentations should be more than 2.5x the indentation diameter apart to avoid interaction between the work-hardened regions. Brittle phases have a smaller work-hardened region and the thinnest intermetallic layer CuAl is very thin (~1μm).

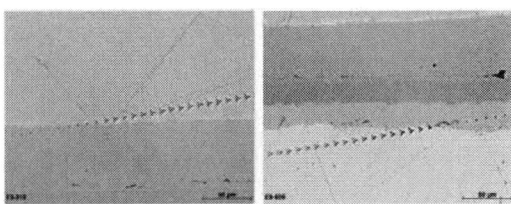

Figure 3: UNAT micro indentation trajectory on (a) Cu-side of the diffusion couple. The increase of the hardness of the solid solution of Al in copper is visible from the smaller indents at the Cu/γ phase boundary. The hard γ Cu_9Al_4 layer gives smaller indents in comparison to the soft Cu and (b) Al-side

In Figure 3 the indentation depths of 20mN in the relatively soft Cu and Al region are adequate for measurement but much deeper than the shallow indentations in the harder intermetallic layer regions. Therefore higher indentation loads of 50 and 500mN were applied for the determination of hardness and indentation Young's moduli of the intermetallic layers.

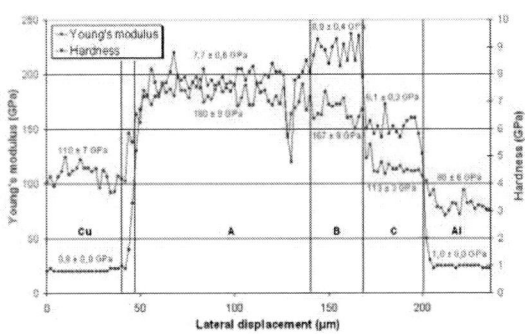

Figure 4: UNAT 240μm line scan with 20mN indentation load to measure indentation Young's moduli and hardnesses of Cu-IMC-Al layer system from left to right: Cu solid solution (ss), Cu_9Al_4 and Cu_3Al_2 (A), Cu_4Al_3 (B) and $CuAl_2$ (C) and Al solid solution. The CuAl layer is too thin to discriminate it from the adjacent layers.

The UNAT trajectory line-scan in Figure 4 shows that the intermetallic layers have significantly higher hardnesses and indentation Young's moduli than the Cu and Al solid solution regions. Due to the relatively fast rate of diffusion of Cu-atoms into the Al-lattice

there is an extensive region (> 30 micron) of solid solution of Al (~2at %) in Cu [7, 9]. Hardness and indentation Young's moduli for the Cu_9Al_4 and Cu_3Al_2 phases are almost similar; in addition, no clear (physical) phase boundary was observed, either in light or scanning electron microscope.

Between Cu_3Al_2 and Cu_4Al_3, at a lateral indenter displacement of 125μm, a decrease in both values, of hardness and Young's modulus, is observed, due to the presence of contaminations at the Kirkendall plane. Furthermore, the CuAl layer thickness is too thin to discriminate it from the adjacent $CuAl_2$ and Cu_4Al_3 layers. In Table 2 the average indentation Young's moduli, determined from multiple line scans with 10, 20, 50 and 500mN indentation load, are presented.

Table 2: UNAT measurements for the indentation Young's moduli: values are averages with standard deviation

	$CuAl_2$	$CuAl$	Cu_4Al_3	Cu_3Al_2	Cu_9Al_4
	Θ	η	ζ	δ	γ
E [GPa]	123.5 ±6.6	180.2 ±12.5		174.4 ±19.5	186.8 ±9.0

3.2 Hardness and indentation size effect

Hardness determination on the intermetallic compounds was done in 3 different regimes: (1) UNAT Berkovich indentor with 10, 20, 50 and 500mN load (2) Micro-Vickers with 0.5, 1 and 2N load and (3) the conventional Vickers apparatus with 9.8, 19.6 and 49N indentation load (1, 2 and 5 kg). The loading time for the Vickers hardness measurements is 15 seconds. In Table 4 all measured hardness values are summarized and discriminated by method. In Figure 5 the apparent, average hardness values are plotted with respect to the corresponding indentation loads.

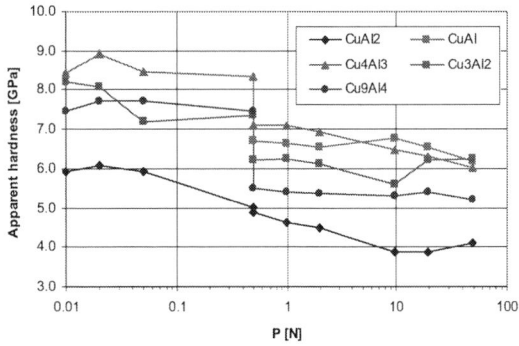

Figure 5: Apparent hardnesses [GPa] vs. indentation load P [N] for UNAT nano-indentation (10mN-500mN), micro-Vickers (0.5N-2N) and Vickers (9.8N-49N).

Significant differences in apparent hardness were found by the measurements in the three different regimes. Apparently indentation size effects are involved due to the large variance of indentation loads [12]. For both Vickers indentation methods the hardness HV is defined as the ratio between the applied load P and the squared length of indentation diagonal d_i multiplied by a geometrical factor for the indentor angle:

$$HV = \frac{1.854P}{d_i^2} \qquad\qquad 1.1$$

This relation does not account for elastic recovery, which implies that the measured hardness is still a function of the load. When the indentor is removed from the surface, the material will partially recover and the indentation size will decrease a little. This is graphically explained in Figure 6, where the indentation diagonal after relaxation is denoted as d_m.

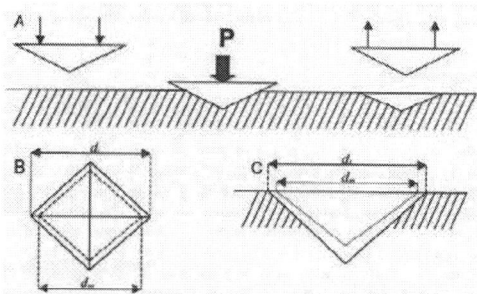

Figure 6: Principle of Vickers indentation (A) top view (B) and cross-section (C) of the indentation, before (d_i) and after relaxation (d_m)

In addition, if the indentor goes deeper into the material, micro-structural properties such as surface condition and lattice irregularities are less of influence (e.g. grain boundaries, impurities, pores or oxidation) on the hardness measurements. Also due to these effects it can be explained that UNAT measurements gave a higher hardness value than measured with the both Vickers methods. In order to evade the indentation load the indentation diagonal after relaxation d_m is measured for both Vickers methods. The Vickers hardnesses versus the corresponding indentation diagonal after relaxation d_m are plotted in Figure 7.

For UNAT measurements with Berkovich indentor, a different approach is needed, no d_m is available. For this reason, in this evaluation, the UNAT results are not taken into account. In order to obtain the load-independent hardness (H_v^0) from the measured indentation size (d_m) equation 1.1 is extended to the linear equation [12]:

$$d_m = \left(\frac{1.854}{H_v^0}\right)^{\frac{1}{2}} P^{\frac{1}{2}} - \delta \qquad\qquad 1.2$$

Thus a plot of d_m versus $P^{1/2}$ should yield a straight line with intercept $-\delta$, the elastic recovery parameter and slope $(1.854/H_v^0)^{1/2}$. In Figure 8 a best linear fit to the raw data results is given. Equations are extracted out of the five plots in Figure 8 and given for each intermetallic in Table 3 below:

Table 3: Equations for determining the load independent hardness H_v^0 and elastic recovery parameter δ of the Cu-Al intermetallic compounds

Phase	Equation	H_v^0 [GPa]	δ [µm]
$CuAl_2$	$d_m = 0.02169\ P^{1/2} - 0.00107$	3.94	1.07
$CuAl$	$d_m = 0.01732\ P^{1/2} - 0.00104$	6.18	1.04
Cu_4Al_3	$d_m = 0.01768\ P^{1/2} - 0.00170$	5.93	1.70
Cu_3Al_2	$d_m = 0.01722\ P^{1/2} + 0.00057$	6.25	-0.57
Cu_9Al_4	$d_m = 0.01889\ P^{1/2} - 0.00055$	5.20	0.55

The elastic recovery parameter δ is calculated from the intercept of the linear trend line. Out of the slope $(1.854/H_v^0)^{1/2}$ the load independent hardnesses are determined and collected in Table 3. These hardnesses are in agreement with results in literature [10]. $CuAl_2$ has the lowest hardness of 3.94 GPa. The intermetallics $CuAl$, Cu_4Al_3 and Cu_3Al_2 are in the same order of hardnesses. As the elastic recovery parameter is independent of the indent diameter its influence is greatest at low load. For that reason the apparent hardness measured with Vickers indentation comes closer to the load-independent hardness than the apparent hardness measured with the UNAT.

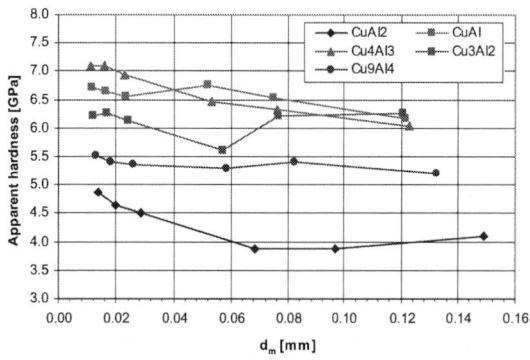

Figure 7: Apparent hardnesses [GPa] vs. indentation diagonal d_m [mm] for micro-Vickers (0.5N-2N) and Vickers (9.8N-49N).

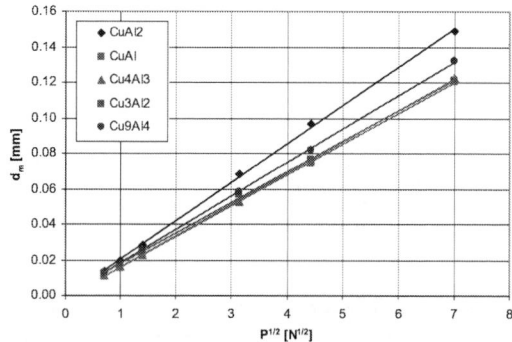

Figure 8: Vickers indentation diagonal d_m vs. $P^{1/2}$ gives straight lines for each intermetallic compounds

Table 4: Hardness measurement overview for UNAT nano-indentation (10mN-500mN), micro-Vickers (0.5N-2N) and Vickers (9.8N-49N). No values available for CuAl with UNAT nano-indentation due too thin layer thickness. All Vickers measurements are measured 3 times; blue coloured cells indicate one or multiple cracks at the indent corners.

Method		UNAT Berkovich								μ-Vickers						Vickers					
		10mN 0.010		20mN 0.020		50mN 0.050		500mN 0.500		50 gram 0.490		100 gram 0.981		200 gram 1.961		1.0 kg 9.807		2.0 kg 19.613		5.0 kg 49.033	
Load N		GPa	±	Gpa	±	GPa	±	GPa	±	GPa	dm	GPa	dm	GPa	dm	GPa	dm	GPa	dm	GPa	dm
CuAl2	1	6.10	0.64	6.08	0.35	5.78	0.32	5.33	0.41	5.12	13.33	4.55	20.00	4.44	28.63	3.82	0.0690	3.70	0.0992	3.84	0.1539
	2	5.78	1.12			6.06	0.34	4.70	0.19	4.69	13.92	4.73	19.61	4.50	28.43	3.95	0.0678	4.04	0.0949	4.26	0.1462
	3									4.83	13.73	4.64	19.80	4.56	28.24	3.88	0.0685	3.88	0.0969	4.20	0.1473
CuAl	1									7.03	11.37	6.71	16.47	6.68	23.33	7.23	0.0502	6.53	0.0747	6.27	0.1204
	2									6.57	11.76	6.55	16.67	6.57	23.53	6.73	0.0520	6.59	0.0744	6.10	0.1221
	3									6.57	11.76	6.71	16.47	6.41	23.82	6.35	0.0536	6.50	0.0748	6.16	0.1216
Cu4Al3	1	8.30	1.02	8.92	0.40	8.65	0.48	8.50	0.33	7.15	11.27	7.21	15.88	6.80	23.14	6.77	0.0519	6.30	0.0760	6.11	0.1220
	2	8.57	0.88			8.28	0.41	8.16	0.57	7.03	11.37	7.03	16.08	7.28	22.35	6.32	0.0537	5.98	0.0780	5.77	0.1255
	3									7.15	11.27	7.03	16.08	6.74	23.24	6.34	0.0536	6.70	0.0737	6.22	0.1210
Cu3Al2	1	8.29	0.70	8.07	0.24	7.07	0.97	7.33	0.42	6.15	12.16	6.55	16.67	5.96	24.71	5.44	0.0579	6.33	0.0759	6.35	0.1198
	2	8.15	0.59			7.31	1.39	7.41	0.27	6.15	12.16	6.25	17.06	6.26	24.12	5.69	0.0566	6.34	0.0758	6.10	0.1222
	3									6.35	11.96	5.97	17.45	6.15	24.31	5.69	0.0566	6.01	0.0778	6.35	0.1197
Cu9Al4	1	7.37	0.74	7.73	0.39	7.63	0.48	7.51	0.27	5.43	12.94	5.41	18.33	5.31	26.18	5.30	0.0586	5.35	0.0825	5.21	0.1322
	2	7.52	0.82			7.82	0.42	7.38	0.16	5.69	12.65	5.41	18.33	5.51	25.69	5.38	0.0582	5.40	0.0821	4.93	0.1359
	3									5.43	12.94	5.41	18.33	5.27	26.27	5.22	0.0590	5.49	0.0814	5.47	0.1290

3.3 Intermetallic volumetric density determination

The metallographic preparation of the Al-rich intermetallic compound, i.e. sawing or polishing, was very difficult due to its crumbling nature. Light microscope images of the intermetallic compounds were made, see Figure 9. On micro-scale it is can be seen that the samples are not homogenous, but contain a small amount of secondary phase (white phase in CuAl, see Figure 9). Some pores are also present.

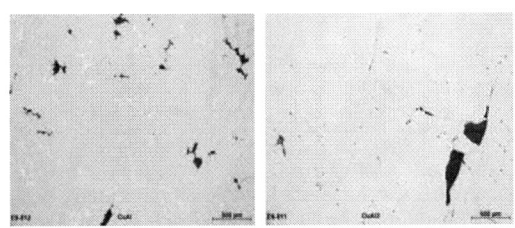

Figure 9: CuAl (left) and CuAl2 (right) intermetallics appear to be inhomogeneous and contain pores, cracks and inhomogeneities.

Nonetheless, the volumetric densities were determined and listed in Table 5. Values are consistent with values presented by Funamizu [9]. In addition, the densities of Cu and Al were measured for verification and comparison.

Table 5: Mass and volumetric density overview of Cu-Al intermetallic compounds

Phase	CuAl2	CuAl	Cu4Al3	Cu3Al2	Cu9Al4
	Θ	η	ζ	δ	γ
d [gr/cm³]	4.27	5.31	5.60	6.25	6.65
Va [cm³/grat]	9.05	8.53	8.44	7.91	7.70

The average atomic volume V_a was determined for all Cu-Al intermetallics and pure Cu and Al. In Figure 10 average atomic volumes V_a are plotted with respect to the corresponding atomic aluminium fraction x_{Al} to evaluate their deviation from the average (dotted line) of pure Cu and Al (Vigard's law).

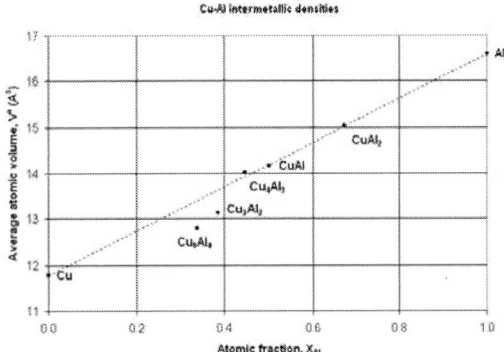

Figure 10: Atomic volume vs. atomic Al-fraction in the Cu-Al system. These results are in agreement with Funamizu [8]. A clear substantial volumetric shrinkage (~4 %) during the formation of Cu-rich intermetallics (Cu9Al4, Cu3Al2) is predicted from these measurements.

3.4 Indentation fracture toughness

Homogenic regions of the intermetallic surface were indented with the Vickers diamond using high load in order to produce cracks in addition to the indents. From the length of the cracks the fracture toughness is determined. If at each indentation corner equal cracks were observed, as shown in Figure 11, indentation fracture toughness was evaluated. Not all indentations in the intermetallics exhibited cracks and often no symmetric crack pattern was found.

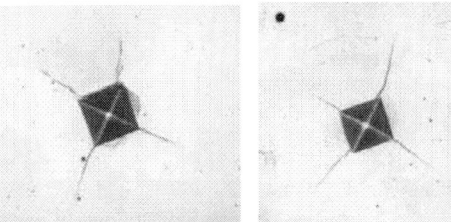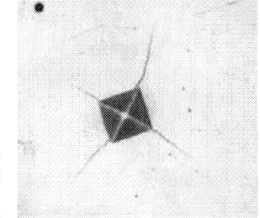

Figure 11: Vickers indentations with symmetric cracks at the corners for the CuAl2 (1.0 kgf) and Cu4Al3 intermetallics (0.3 kgf).

The indentation fracture toughness was estimated from Vickers indentation crack measurement using the following formula [12]:

$$K_{1C} = 0.016 \left(\frac{E}{H} \right) \left(\frac{F}{C^{1.5}} \right) \qquad 1.3$$

where E is the indentation Young's modulus, H the load independent hardness, F the Vickers indentation load (N) and C the crack length mm from centre of indention to crack end. Note that for the indentation fracture toughness determination the UNAT measured indentation Young's moduli and load independent hardness were used. As the indentation Young's modulus of CuAl was not measured, in this case the value of Cu_4Al_3 was taken.

Table 6: Indentation load and apparent fracture toughness with standard deviation for all 5 CuAl intermetallics

Phase	$CuAl_2$	CuAl	Cu_4Al_3	Cu_3Al_2	Cu_9Al_4
	Θ	η	ζ	δ	γ
Load [kgf]	0.5-1.0	0.3-0.5	0.3	5.0	5.0
K_{1c} [MPa√m]	1.54 ±0.34	1.08 ±0.17	1.18 ±0.25	3.57 ±0.81	4.00 ±0.60

No cracks were observed in the relatively soft and ductile Al and Cu bulk material. The Al-rich intermetallics fractured at low indentation loads, where $CuAl_2$ (0.5-1.0 kgf) appeared to be tougher then CuAl and Cu_4Al_3 (0.3-0.5 kgf). The Cu-rich intermetallics Cu_9Al_4 and Cu_3Al_2 only fractured at indentations within the higher loading range (5.0 kgf). Cracks as measured in this study were not deflected and were propagated in diagonal direction from the indentation corners.

3.5 Oxidation

It is well known that aluminium forms a tight dense protecting oxide layer when heated in air. However, copper, notwithstanding its precious properties forms a porous not-protecting oxide layer. In the wire bond interface always some aluminium dissolves in the copper, forming a solid solution of aluminium (range 0 to ~17 at% Al). According to Plascencia [14] aluminium only forms a protecting alumina coating on the copper solid solution if the Al content is higher than 4 wt% (approximately 9 at%). It is expected that in this case the aluminium content is lower; Al operates as a contamination and increases the susceptibility for oxidation of the copper solid solution due to the very high heat of formation of alumina. This increased oxidation attack of the Cu solid solution with a low content of Al is visible in experiments where a cross section of an annealed copper-aluminium diffusion couple is exhibited – for a long time - to ambient air at various temperatures (250 and 500°C).

An image of such an oxidized cross section is shown in Figure 12. It is clearly seen that a very thick oxide layer is present in the copper where the Al rich intermetallics show a much thinner or no oxide layer. At the copper solid solution interface, in all experiments the copper with a small amount of aluminium was attacked more than the pure copper or the yellow Al rich solid solution or the γ-intermetallic phase.

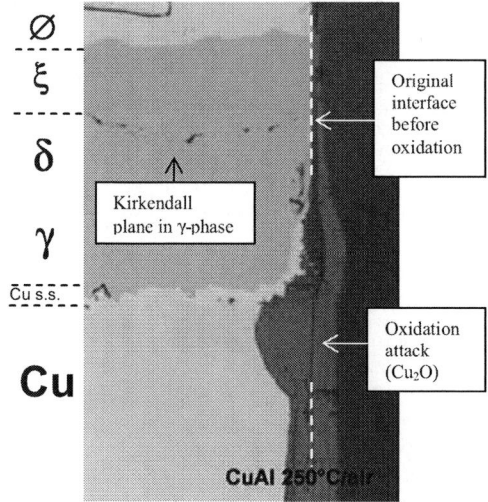

Figure 12: Oxidation attack from the right due to heating at 250°C in air for 150 hr. The yellow region is the Al-rich Cu solid solution.

Hang [15] observed cracks over a length at the interface between the γ-intermetallic phase and the copper wire in annealed Cu-wire bond samples.

4. Conclusions

- The layer growth of all five intermetallic compounds as function of time and temperature is described by Arrhenius plot. At high temperatures (>400°C) all the five intermetallic phases are present, while at temperatures below 300°C only Θ ($CuAl_2$), η (CuAl) and γ (Cu_9Al_4) are detected. The original interface (the Kirkendall plane) is present in the (γ + δ) phase.

- A large variation in apparent hardness was found between the results of the UNAT nano-indentation, micro-Vickers and Vickers hardness measurements. Due to elastic recovery, the indentation size effect causes a large variation in apparent hardness particularly at small loads. The load independent hardnesses of the Cu-Al intermetallics are determined and are in agreement with apparent Vickers hardness measurements at high load published in the literature.

- The Cu-rich intermetallics Cu_9Al_4 and Cu_3Al_2 are less sensitive to fracture than the Al-rich intermetallics. However, these Cu-rich intermetallics have a smaller atomic volume and will show a volumetric shrinkage during formation. This is expected to cause a large internal stress during thermal aging.

- From the experiments it is clear that a solid solution of an intermediate amount of Al is most prone to oxidation. Copper with an Al content between 0.1 (detection limit SEM/EDX) and 5.4 at% is most sensible to oxidation attack.

Finally, all measured mechanical properties for the Cu-Al intermetallics are summarized in Table 7:

Table 7: Mechanical properties overview of Cu-Al intermetallics

Phase	CuAl$_2$	CuAl	Cu$_4$Al$_3$	Cu$_3$Al$_2$	Cu$_9$Al$_4$
	Θ	η	ζ	δ	γ
H$_v^0$ [GPa]	3.94	6.18	5.93	6.25	5.20
E [GPa]	123.5 ±6.6	180.2 ±12.5		174.4 ±19.5	186.8 ±9.0
K$_{1c}$ [MPa√m]	1.54 ±0.34	1.08 ±0.17	1.18 ±0.25	3.57 ±0.81	4.00 ±0.60
d [gr/cm^3]	4.27	5.31	5.60	6.25	6.65
V [cm^3/grat]	9.05	8.53	8.44	7.91	7.70

Acknowledgments

The research leading to these results has received funding from the ENIAC Joint Undertaking and from Senter-Novem in the Netherlands under Grant Agreement number 120009. The authors thank Arthur Eijck for preparing the cross sections and images.

References

[1] H.M. Buschbeck, Euro. Semicond., Dec 2004, pp. 27-30

[2] B. Swiggett, Prismark, Proc. K&S Copper Summit Conference 2008

[3] M. Deley, L. Levine (K&S Publication)

[4] H. Shouyu, H. Chunjing, W. Chunqing, Proc. 6th Int. Conf. Elect. Pack. Technol., IEEE, 2005

[5] C.W. Tan, A-R Daud, J. Elect. Packag. Dec. 2003, Vol 125, pp. 617-620

[6] H. Xu, C. Liu, Growth of intermetallic compounds, J. Electronic Mat. 39 (2010) 124

[7] G.H.M. Gubbels, M.H.M. Kouters, O. O'Halloran, R. Rongen, Growth and properties of intermetallics formed during thermal aging of CuAl ball bonds, Proc. ESTC Berlin, 2010

[8] M.H.M. Kouters, G.H.M. Gubbels, O. O'Halloran, R. Rongen, Mechanical properties of intermetallics formed during thermal aging of Cu-Al ball bonds, Proc. EUROSIME Linz, 2011

[9] Y. Funamizu, K. Watanabe, Interdiffusion in the Al-Cu system, Trans. JIM 12, 1971, 147

[10] F.W. Wulff , C.D. Breach, a.o. IEEE EPT Conference (2004) 348

[11] F.W. Wulff, C.D. Breach, D. Stephan, Saraswati, K.J. Dittmer and M. Garnier, Proc. Semicon, Singapore 2005

[12] D. de Graaf, Hardness and indentation size effect in Ln-Si-Al-O-N glasses, J.Mat.Sci. 39 (2004) 2145

[13] G. Gosh, Elastic properties, hardness, and indentation fracture toughness of intermetallics relevant to electronic packaging, J. Mater. Res., Vol. 19, No. 5, 2004

[14] G. Plascencoia, The oxidation Resistance of Copper-Aluminium alloys, JOM jan.(2005)80

[15] C.J. Hang, C.Q. Wang, M. Mayer, Y.H. Tian, Y. Zhou, H.H. Wang, Growth behavior of Cu/Al intermetallic compounds and cracks in copper ball bonds during isothermal aging, Microelectronics Reliability 48 (2008) 416

Effects of additional Co atoms on microstructural evolution in Sn-Ag-Bi-In solder under current stressing

[1]Kiju Lee, Katsuaki Suganuma, [2]Keun-Soo Kim, [3]Minoru Ueshima

[4]Hans-Juergen Albrecht, Klaus Wilke, Joerg Strogies

[1] ISIR, Osaka University, Osaka, Japan

[2] Fusion Technology Lab., Hoseo University, Asan 336-795, Korea

[3]Senju Metal Industry CO., LTD.

[4]Siemens AG, Corporate Technology, Berlin, Germany

Abstract

Electromigration behavior of Sn-3.0wt%Ag-3.0wt%Bi-4.0wt%In (SABI) and SABI+Co solder joints with respect to the crystallographic orientation of Sn grains was investigated. The test sample had direct contact on Cu substrate with Cu dummy chip and the applied direct current was 20 A at 125 °C solder joint temperature. Before current stressing SABI solder solidified to form large Sn grains. After current stressing, failure of the solder joint was caused by the dissolution of Cu electrode on the cathode side. Fast dissolution of Cu atoms occurred along the c-axis of Sn in the SABI solder joint when the c-axis parallel to the electron flows. On the other hand, SABI solder joint with a small amount of Co shows a different microstructural evolution as compared with SABI, i.e., randomly orientated smaller Sn grains compared to SABI. Also, slight Cu dissolution was observed at the cathode side under same test condition since the Co addition has a substantial effect on the microstructural evolution on SABI solder.

Key words: electromigration, Sn–Ag–Bi–In, crystallographic orientation of Sn, EBSD

1. Introduction

The Important issue in electronic packaging is lowering the melting temperature of lead-free solder to prevent the heat damage of electronic device during reflow process. The Sn-Ag-Cu solder has been widely used as a reliable lead-free solder. Nevertheless, its melting temperature is about 34 °C higher than that of eutectic Sn-Pb solder and the adaptation of lower melting temperature lead-free solders are still indispensable. Among several low temperature solders, Sn-Ag-Bi-In solder is one of the leading candidates with its good mechanical properties and reliability[1].

One of the important reliability issues is the influence of the phenomena caused by the miniaturization trend of electronic devices. As continuously decrease the dimension of the solder joint, the current density in each solder joint increases rapidly. Moreover, the difference in the electrical resistance and the feature size between a solder and an adjacent trace arises a drastic increase of the current density in the solder joint area. The emphasized current density will cause an enhanced massive migration of metal atoms in the direction of electron flow, namely, electromigration (EM)[2]. The directional atomic migration can induce significant microstructural changes in the solder joint, such as void formation, crack propagation, phase segregation, solder extrusion/whiskers, growth of intermetallic compounds (IMC) and dissolution of under bump metallization. Those severe microstructural changes may result in an early failure of the solder joints. Thus, EM becomes one of the serious reliability issues. Recently, a number of investigations have been performed about EM behavior in lead-free solder joints. EM in a flip-chip joint with Sn-Pb, Sn-Ag and Sn-Ag-Cu solders has been investigated in the past researches. Also, EM of Sn-Pb, Sn-Zn and Sn-Bi solders using solder strip model have been reported. From the previous literatures, fast failure of the solder joints mostly occurred due to the dissolution of UBM and failure times are depend on the crystallographic orientation of Sn such as the grain size and direction of c-axis of Sn. Many researchers pointed out the importance of the controlling crystallographic orientation of Sn.

In the present work, EM behavior of the Sn-3.0wt%Ag-3.0wt%Bi-4.0wt%In (SABI) / Cu solder joint and effects of small addition of the Co have been reported.

2. Experimental procedure

Two types of solder alloys, SABI Solder with or without small addition of Co were prepared by Senju Metal Industry. Prior to soldering, dummy Cu chip of height 0.6 mm, width 1.25 mm, and length 2.0 mm was aligned to form a solder joints. Cu chip were aligned and fixed on the Cu substrate. Reflow process was performed in an reflow oven at 230°C peak temperature under nitrogen atmosphere.

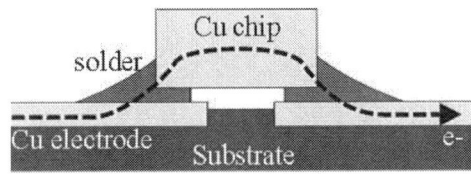

Figure 1: Schematic of EM test sample

The EM test was performed by applying a direct current to both ends of samples by using Cu wire, as shown in Figure 1. The entire test was performed immersed in a silicone oil bath to reduce temperature rise that may occur during current stressing because it may induce thermomigration in the solder joint. The solder joint temperature was monitored using a K-type thermocouple, which was placed close to the solder joint. During current stressing, the monitored temperature of the solder joint was 125 °C. A direct electric current of 20 A was supplied by a power supply. Finite element calculation revealed typical current densities of 8-10 kA/cm² and peak densities up to 18 kA/cm². To investigate microstructural changes in solder joints, joints were mounted in epoxy and polished mechanically followed an argon ion beam polishing by using a cross-section polisher (SM-09010, JEOL). To examine microstructural changes and the crystallographic orientation, back-scattered scanning electron microscopy (BS-SEM; JEOL-5510S, JEOL), an electron backscattered diffraction system (EBSD; TSL crystallography, EDAX) were adopted.

3. Results and discussion

Figure 2 shows cross-sectional BS-SEM images of joint interfaces between the Cu substrate and solder alloys after the reflow process at 230 °C. The analysis area was indicated in the Fig. 2(a) with black square. A scallop type η-Cu_6Sn_5 IMC layer of a thickness of approximately 3 μm was uniformly formed at the interface of Cu and the solder. Sub-micron sized η-Cu_6Sn_5 IMCs were also randomly distributed in the β-Sn matrix. And, a small quantity of additional Co does not significantly affect the microstructure when compared to the Cu/SABI solder interface.

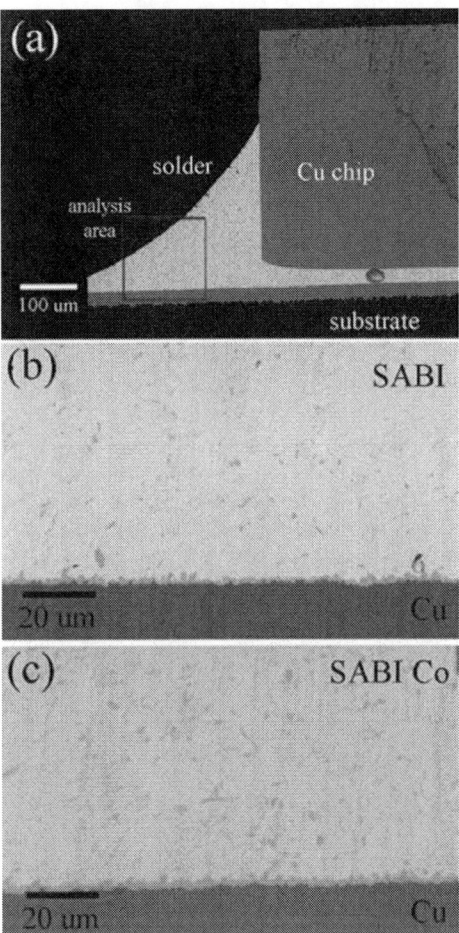

Figure 2: Cross-sectional BS-SEM images, (a) test sample, (b) SABI solder joint, and (c) SABI+Co solder joint

Figure 3(a) shows cross-sectional BS-SEM images of test samples after current stressing for 500 h at 125 °C. As indicated by arrows, electrons flow from the bottom left side (cathode) to the upper right side (anode). Large defect at the cathode interface was observed in SABI alloy. Additionally, large η-Cu_6Sn_5 IMCs were formed at the anode side (dummy Cu chip side) of the SABI solder as well as within the solder matrix. From the enlarge view of the cathode side, the primary failure of the solder joint was caused by dissolution of Cu electrode at the cathode interface. On the contrary, slight microstructural changes were observed in the the SABI+Co solder joint compare to the solder joint without Co addition. Defect between the solder and Cu-Sn IMCs at the cathode side was also observed. However, dissolution of Cu electrode was barely found and small amount of IMCs formed at the anode interface as well as within the solder.

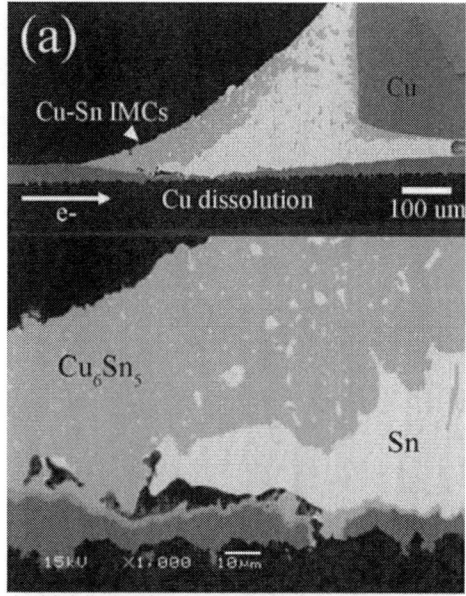

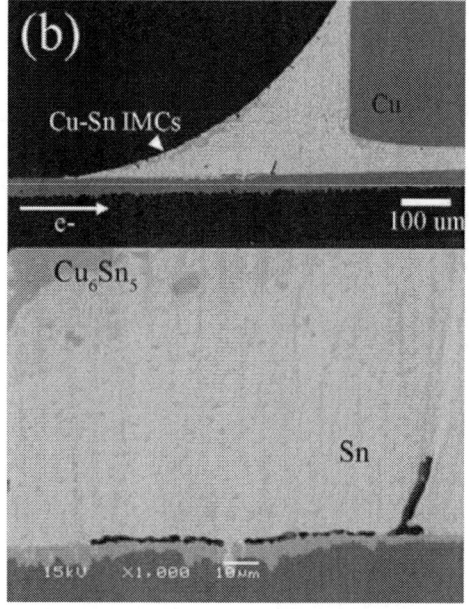

Figure 3: Cross-sectional BS-SEM images of the solder joint after current stressing for 500 h, (a) SABI, (b)SABI+Co

EM is diffusion controlled mass transportation of metal atom. At 125 °C, the diffusivity of Cu in the Sn matrix ranges from 1.1×10^{-7} cm^2/s to 6.4×10^{-6} cm^2/s with regards to the axis direction of Sn while the self-diffusivity of Sn ranges from 6.7×10^{-14} cm^2/s to 1.3×10^{-13} cm^2/s at the same temperature. Therefore, the Cu atom is considered to be the primary mover as it migrates to the anode, while Sn atoms displaced in the opposite direction (cathode side). Therefore, η-Cu$_6$Sn$_5$ IMC layers formed at the cathode were decomposed, and IMCs formed at the anode during current stressing.

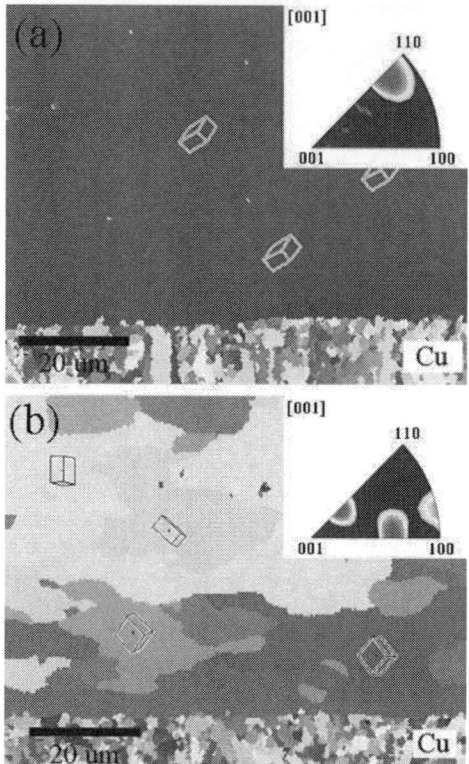

Figure 4: EBSD analysis of the solder joint before current stressing, (a) SABI, (b)SABI+Co

The Sn inverse pole figure before current stressing for SABI/Cu and SABI+Co/Cu was indicated in Figure 4. Unit cells located in each Sn grain indicated the crystallographic orientation of Sn. In Fig. 4(a), SABI solder joint solidified into a large randomly orientated Sn grain as same as pure-Sn and Sn–Ag–Cu solder joints[3, 4]. Large Sn grain is comprised of small sub-grains with a low-angle grain boundary. On the contrary, a reduction in the Sn grain size was clearly observed with regards to the Co addition in solder alloys. In Fig. 4(b), SABI+Co solder joint solidified into randomly orientated small Sn grains with the size of a tens of um, approximately.

Cross-sectional EBSD analysis of the solder joints after current stressing for 500 h is indicated in Figure 5. Same areas are analyzed as indicated in Fig. 3. Electrons flow from the Cu electrode (bottom side) to the solder (upper side) as indicated by arrow. Some Cu-Sn IMCs are encompassed by dotted lines to prevent confusing them with the Sn grains. SABI solder joint which composed a large one grain shows strong dependence to c-axis of Sn. In general, Sn has a highly anisotropic diffusion characteristic. Especially, when the c-axis parallel to the electron flow, fast EM failure occurred due to the fast interstitial diffusion of Cu [5, 6]. In Fig. 4(a), c-axis of Sn in SABI solder joint almost parallel to electron flow, fast dissolution of Cu electrode at the cathode side was occurred and large Cu-Sn IMCs are formed near the cathode side and anode side. However, SABI-Co/Cu solder joint different EM behaviors were observed. In Fig. 4(b), c-axis of Sn also close to the parallel to the electron flow. Nevertheless, dissolution of Cu electrode at the cathode side was barely occurred and formation IMCs are not that serious as SABI solder joint did. This abnormal EM behavior is caused by the size of Sn grain. At high temperature, Cu migration occurred along the interstitial site of Sn grain and lattice diffusion is a predominantly occurred[2]. Owing to highly anisotropic diffusion characteristic of Sn, c-axis provides a high diffusion site for Cu. Therefore, when the c-axis of Sn is parallel to electron flow, migration of Cu atoms occurred easily. In addition, Cu atoms migrated to the anode side without any interference when the grains size of Sn is large. On the contrary, lattice diffusion of Cu was interrupted by grain boundary when the size of Sn grain was smaller and Cu diffusion also suppressed.

Based on results presented above, the EM failure of the SABI and SABI+Co solder joints are caused by the propagation of defect at the cathode interface due to the dissolution of Cu electrode and. During EM, the migration of Cu atoms often exhibits anisotropic behaviors in the Sn matrix. At high temperatures, lattice diffusion is considered to be a primary diffusion mechanism for Cu atoms. Owing to highly anisotropic diffusion characteristics of Sn, the c-axis provides a very fast interstitial diffusion site for Cu atoms as opposed to the a-axis and b-axis. Therefore, during EM tests, the crystallographic orientation of Sn is important factor in Sn-based lead-free solders.

4. Conclusions

The EM behavior of SABI (+Co) solder alloys was investigated and we found remarkable differences depending on the small addition of Co. The primary reason for the EM failure of solder joints was determined to be the propagation of defect between cathode IMCs and solder alloys due to the dissolution of Cu electrode. Upon the addition of Co, the size of the Sn grain is substantially reduced compared to the SABI solder joint. Finer SABI+Co solder can improve EM resistance because the interstitial diffusion of Cu along the Sn lattice was suppressed by uniaxial fine grains.

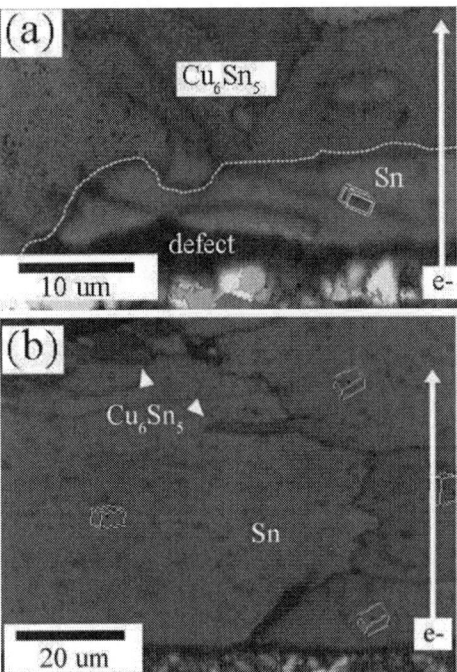

Figure 4: EBSD analysis of the solder joint after current stressing for 500 h, (a) SABI, (b)SABI+Co

References

[1] K.S. Kim, et al., "Properties of low temperature Sn-Ag-Bi-In solder systems", Microelectronics Reliability, Vol. 47, No. 7, pp. 1113-1119, 2007.

[2] K. Tu, "Recent advances on electromigration in very-large-scale-integration of interconnects", Journal of Applied Physics, Vol. 94, No. 9, pp. 5451-5473, 2003.

[3] A. Telang, et al., "Grain-boundary character and grain growth in bulk tin and bulk lead-free solder alloys", Journal of Electronic Materials, Vol. 33, No. 12, pp. 1412-1423, 2004.

[4] L. Lehman, et al., "Growth of Sn and intermetallic compounds in Sn-Ag-Cu solder", Journal of Electronic Materials, Vol. 33, No. 12, pp. 1429-1439, 2004.

[5] K. Lee, et al., "Effects of the crystallographic orientation of Sn on the electromigration of Cu/Sn-Ag-Cu/Cu ball joints", Journal of Materials Research, Vol. 26, No. 3, pp. 467-474, 2011.

[6] M. Lu, et al., "Effect of Sn grain orientation on electromigration degradation mechanism in high Sn-based Pb-free solders", Applied Physics Letters, Vol. 92, No. 21, pp. 211909, 2008.

Influence of Titanium Surface Contamination on the Reliability of Al Wire Bonds

G. Khatibi[1], S. Puchner[2,3], B. Weiss[1], A. Zechmann[2], T. Detzel[4], H. Hutter[3]

University of Vienna, Faculty of Physics, Boltzmanngasse 5, A-1090, Vienna Austria, 2) KAI
Kompetenzzentrum Automobil – Industrieelectronik GmbH, Villach Austria, 3) Institute of Chemical
Technologies and Analytics, TU Vienna, Austria, 4) Infineon Technologies Austria AG, Villach Austria

Phone: +43 1 4277, Fax: +43 1 4277 9513, golta.khatibi@univie.ac.at

Abstract

Organic and inorganic contaminations on the metallization surface of wire bonds are known to reduce the reliability of the devices. In this study the influence of Ti contamination on bondability and quality of Al thick wire bonds was investigated. Ti layers with thicknesses between 2 nm to 15 nm were deposited either directly on AlSiCu metallization pads or after an atmospheric exposure of the wafer. Characterization of the metallization by means of ToF-SIMS revealed two types of surface configurations: An intermixed surface layer composed of Ti- and Al- oxides was formed after a direct deposition of Ti on the pads. Exposure of the wafer to atmospheric environment before Ti deposition resulted in formation of a Ti passivation layer at the top of metallization. Evaluation of the bonding quality by shear tests and an accelerated mechanical fatigue testing method showed that deposition of a 2 nm thin layer of Ti leads to higher shear strength and fatigue resistance. A clear reduction of the bonding strength was observed by increasing the thickness of titanium to 10 nm and 15 nm. Microstructural investigations of these samples showed increased non-bonded areas in the interface due to presence of thick oxide particles which hinder a free metal- metal contact and formation of a defect free interface during the wire bonding process.

Key words: Al wire bonds, Ti surface contamination, accelerated testing, bonding strength

Introduction

Al ultrasonic wire bonds on Al metallization pads are widely used in power semiconductor devices. The performance and reliability of Al wire bonds is highly dependent on the quality and strength of the interface between the Al wire and the metallized surface of the bond pad. One of the main factors affecting the strength of the bonding interface is the quality of metallization which depends on parameters like chemical composition, thickness, hardness, roughness, and surface contamination of the bond pad. Contaminants in electronic devices are foreign (undesired) atoms, molecules or particles which are chemically and physically different from the underlying material [1]. Common contaminations of the bond pad are organic and inorganic substances like fluorine, chlorine, carbon, silicon metals and metal oxides which may occur during the wafer fabrication process, transportation, storage and assembly of the devices. It is generally accepted that contaminants may cause corrosion, reduce or inhibit the bondability and degrade the reliability of the wire bonds. Advanced cleaning techniques like UV-Ozone and Oxygen Plasma methods are used to avoid or remove these foreign particles throughout the various stages of wafer manufacturing and subsequent bonding process [1, 2, and 3].

Still contamination count as a serious problem in electronic industry. Extensive studies have been conducted on effect of surface contaminants of the bond pad on the reliability of the bonding interface with focus on organic materials and halogens [1, 3]. Detailed investigations on the influence of metallic contaminants and oxides on the surface of pad metallization are scarce. As a general rule tarnish or oxide layers on metallization surfaces have been known to cause degradation of the bonding interface. Foreign metallic elements like Cu, Ni, and Ti may lead to alloying and oxide formation on the pad surface causing corrosion and non-bondability of the interface [2].

In a recent investigation on process induced bond pad contaminations traces of Ti were found on the surface of aluminum metallization of Si wafers [4, 5]. Ti contamination was related to the Titanocene which is used as a radical starter in polyimide resins. It was found that even several process steps after introducing the polyimide layer containing this particular contamination, titanium is still present on the wafer surface [4]. The influence of such thin Ti layers on the Al metallization surface and the resulting surface layer composition on the bondability and reliability of Al-Al wire bonds has not been studied so far.

Thus systematic investigations were conducted to study the influence of Ti contamination layer thickness on the quality of Al wire bonded interconnects. For this purpose Ti- layers with thicknesses between 2 nm to 15 nm were deposited on AlSiCu metallization surface of standard MOSFET devices on which Al wires were bonded. Shear tests and a novel accelerated mechanical fatigue testing technique were used to evaluate the bonding strength and to determine the lifetime of the devices. Time-of-Flight Secondary Ion Mass Spectrometry (ToF-SIMS) measurements were performed to characterize the metallization surface composition. Scanning Electron Microscope (SEM) techniques were used to study the microstructure of the interface layer and analyze the fracture surface of the wire bonds after lift-off failure.

Experimental procedure

In the following section the sample preparation procedure, the analytical specimen characterization methods and mechanical testing techniques are described.

Sample preparation

Titanium layers with thicknesses of 2 nm, 10 nm and 15 nm were deposited on silicon wafers with a 5μm AlSi1Cu0.5 metallization by means of physical vapor deposition at 200°C (Fig. 1). The titanium layer was deposited either directly after AlSiCu sputtering or after a short exposure of the wafer to ambient atmosphere (air-break). This was done in order to investigate the effect of surface composition which may result from different contamination sources. Depending on the pattern of the chip the surface topography was about 0.5 μm with an additional aluminum roughness of approximate 100 nm. Subsequently Al wires with a diameter of 400 μm were ultrasonically wedge bonded onto prepared bond pads and reference devices (without Ti- deposition) using standard and identical bonding parameters for all samples.

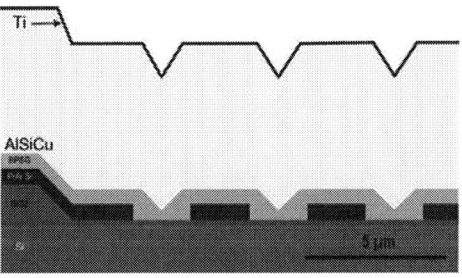

Figure 1: Schematic device structure with Ti contamination layer

Surface characterization by ToF-SIMS

Among the various analytical methods like Auger electron spectroscopy or Energy-dispersive X-ray

spectroscopy which are used for surface characterization of semiconductors, Tof-SIMs is known as an excellent tool for detection of traces of a broad range of substances in this application field [6].

In the present study a ToF-SIMS device (from ION-TOF GmbH located at TU Vienna) was used to characterize the chemical composition and distribution of the surface contaminants as well as depth profiling of the bond pad metallization. The measurements were mostly performed on a sample area of 99.6 x 99.6 μm² directly after titanium deposition, as well as on the fracture surface after life-time tests. The details of the measurements including the setting of the device and applied measurement modes are given in [4, 5].

Static shear test

Shear strength of the wire bonds was determined by using a standard shear testing device (Condor 100 at Infineon). Twenty to forty specimens per condition were glued to a supporting plate and tested with a velocity 250 μm /s. The tip of the shear tool was set at 160 μm and the test distance was 900μm.

Accelerated mechanical shear fatigue testing

Evaluation of long time reliability of wire bonds in high power semiconductors is commonly assessed by thermal and power cycling tests. Accelerated mechanical shear fatigue testing techniques has been suggested as an alternative to these time consuming testing procedures [7]. Using this technique fatigue life curves can be obtained in a significantly shorter period of time in comparison with conventional testing procedures. Furthermore, the predominant failure mechanism of power cycling tests, bond wire lift-off can be reproduced with this method. A subsequent failure analysis of the fracture surface of tested samples may provide important information for improvement of the bonding quality.

The experimental set- up for fatigue testing of wire bonded interconnects consists of a special specimen set-up in combination with an ultrasonic fatigue testing system working at 20 kHz and a laser Doppler vibrometer for velocity measurements. The wire bonded chips (micro-component) is attached to the vibrating testing system in a manner that the coupling of the chip to the system is provided only by the interconnect area. During loading the system is excited to forced longitudinal vibrations (Fig. 2). Due to inertia, shear strain is induced in the interface of the wire bond to the Si-chip resulting in fatigue failure due to wire- bond lift-off. The value of shear stress/strain is related to the mass of the chip, contact area of the joint (wire bond foot print) and the acceleration of the chip (measured by LDV) during the loading. Lifetime curves of wire bonds are obtained in the range of high cycle fatigue representing shear stress ($\Delta\tau$) as a function of loading cycles (N_f). More information on the experimental set-up and the applied method for calculation of the shear stress is given elsewhere [7].

Metallographic investigations of the cross-sections of the wire bonds and fracture surface analysis on the wire bonds after fatigue were conducted by using SEM techniques.

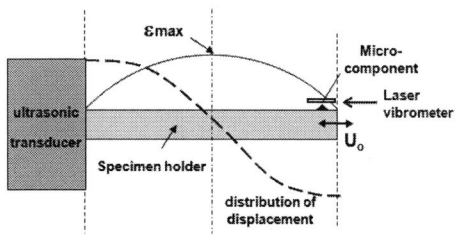

Figure 2: Schematic set-up for accelerated mechanical testing of interconnects

Results and discussion

ToF-SIMS measurements of bond pad

Figures 3a and 3b show the Tof-SIMs profiles of the surface of a sample with 6 nm Ti-layer deposited before and after an air-break. The distribution of AlO$^-$ and TiO$^-$ ions are also representative for other sample configurations, considering the relative change of oxide layer thicknesses with respect to the Ti deposit layer.

A direct deposition of titanium on the aluminum layer allows diffusion of titanium into the AlSiCu metallization and a reaction with the constituent elements. An ambient exposure of the wafer during the further production steps results in the reaction of all four elements Ti, Al, Cu and Si with the atmospheric oxygen depending on their reactivity. Although titanium is the first element which comes into contact with atmosphere, the final uppermost surface layer of the metallization consists of aluminum oxide due to the higher affinity of aluminum to oxygen and the poor barrier effect of Ti for Al.

In case of an air-break (exposure of the wafer to atmosphere before titanium deposition) a native Al-oxide layer with an approximate thickness of 6 nm is readily formed on the surface of the AlSiCu metallization. This Al$_2$O$_3$ layer acts as a diffusion barrier for Ti and prevents the interaction of Ti with under-laying elements. The deposited Ti subsequently forms a TiO$_2$ layer on the top of Al oxide layer. The oxide layer also prevents diffusion of Cu and Si towards the surface. The effective

thickness of Ti deposit layers on samples with air-break was slightly increased due to oxidation of titanium. The measured thicknessess for samples with 2 nm, 6 nm and 10 nm titanium deposition were 3 nm, 8 nm and 12 nm respectively.

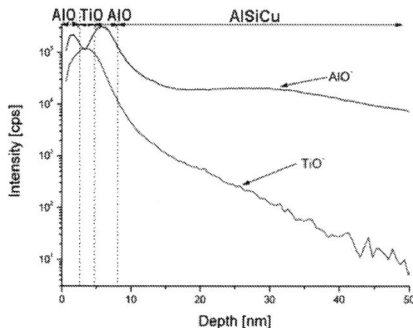

Figure 3a: ToF-SIMS profiles of a sample with 6nm Ti layer without air- break (measured in negative mode). Titanium diffuses into the AlSiCu-metallization

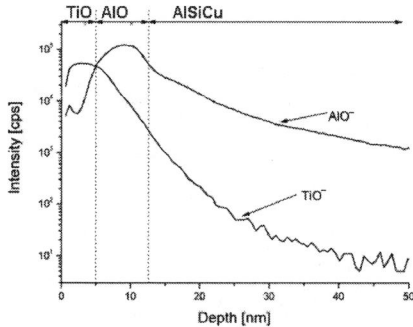

Figure 3b: ToF-SIMS profiles of a sample with 6 nm Ti layer with air-break (measured in negative mode). The aluminium oxide formation suppresses the interaction of the AlSiCu-metallization with the deposited titanium.

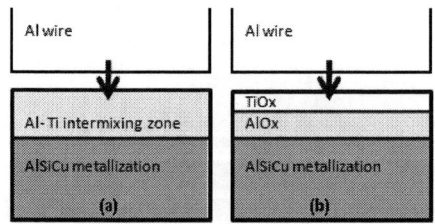

Figure 4: Schematic metallization structure with titanium contamination without air-break (a) and with air-break (b)

The depth profiling measurements show two distinctive surface compositions of the oxide layer on the AlSiCu metallization as represented in the

242

schematic models in figures 4a and 4b. A direct deposition of Ti on AlSiCu results in formation an Al-Ti intermixing zone on top of the metallization surface. In case of an air-break two separate layers of Ti and Al oxides are formed with Ti oxide being the uppermost layer [4]. The thickness of the oxide layer on different samples depends on the thickness the deposited Ti layer and is slightly increased in samples with an air-break.

Further results and experimental details of these investigations are given in [4] from which selected results are presented only.

Evaluation of bonding strength

Shear test results

During the Al wire bonding process, due to the longitudinal vibrations and the pressure of the bonding tool, the brittle native Al-oxide layer is broken and the oxide particles and other contaminants are swept away. Clean surfaces of the wire and the metallization are brought into intimate contact to form the bonding interface. Trapped oxide layers or other contaminants on the bond pad surface act as barriers to material inter-diffusion, prohibit the free metal- metal contact and leave un-bonded areas and in severe cases may lead to non-bonding of the interface [1, 2, 3]. Removal of hard Al oxide layers is achieved by friction forces which are given by the settings of ultrasonic bonding parameters. An increased oxide layer thickness requires modification of ultrasonic power, time and the bonding force which in turn might lead to other reliability problems like high degrees of deformation of the wire or bond pad or cracking of the Si chip [2, 8]. In this study identical standard bonding parameters were applied to all sample configurations in order to compare the influence of thickness and composition of the surface oxide layer on the bonding quality.

The results of shear tests on wire bonded samples with different layers of titanium with and without an air-break are depicted in table 1.

Table 1: Results of static shear test on Al-wire bonds with various Ti- layers

Nr.	Specification	Shear force [g]
1 (Ref.)	no Ti deposition	1606 ± 119
2	2 nm Ti	1832 ± 121
3	2 nm Ti / air-break	1800 ± 160
4	6 nm Ti	1690 ± 141
5	6 nm Ti / air-break	1625 ± 150
6	10 nm Ti	1682 ± 124
7	10 nm Ti / air-break	1522 ± 116
8	15 nm Ti	1395 ± 266

Unexpectedly it can be observed that deposition of a very thin layer of titanium on the metallization pad leads to improved bonding quality for Al-wire bonds. A clear increase of bonding force of about 10-15% is observed for samples with a Ti layer of 2 nm, which is reduced to less than 5% for a 6nm Ti deposit layer. The results show that presence of a total surface oxide layer of above 20 nm results in a significant drop of the bonding force as observed for samples with Ti-layer of 15 nm and 10 nm with air-break. A clear negative effect of deposition of Ti subsequent to an air-break on bonding strength was otherwise not observed. It can be concluded that under given bonding conditions, a considerable amount of thick oxide particles may hardly be broken and removed from the bonding interface. These remnants result in an increased non-bonded area and drop of shear force.

Life-time curves

Fatigue life curves of Al wire bonds with Ti-layer thicknesses of 2 nm, 10 nm and 15 nm in comparison with the reference sample are presented in figure 5. The samples were chosen based on the results of shear test in which a clear difference between the bonding strength of these configurations was observed. The lifetime curves show the shear strength ($\Delta\tau/2$) of the bond interface as a function of number of loading cycles (N_f) in the range of 10^6 and 10^9.

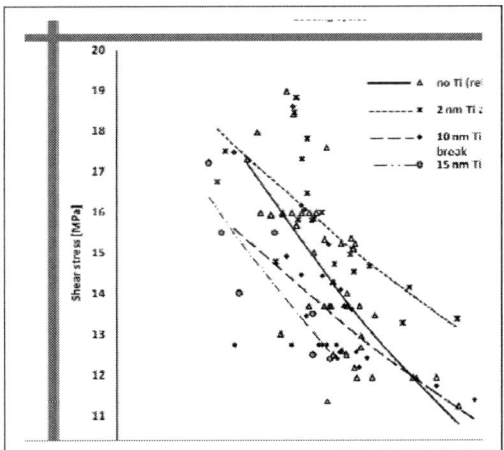

Figure 5: Lifetime curves for Al-wirebonds without Ti-layer, and with 2 nm, 10 nm- and 15 nm Ti-layers.

The cyclic response of the tested samples is in consistency with the results of the shear tests. The best fatigue performance is obtained for the wires bonded on pads with a thin layer of 2 nm Ti deposition and those with a 15 nm Ti layer show the lowest fatigue resistance. Lifetime curves for the reference wire bonds and those with a 10 nm Ti layer show almost a similar behavior. The value of shear stress for all samples in the whole region of high cycle fatigue life is between 11 MPa to 19 MPa with a rather high scatter of stress values. This deviation is higher that that obtained for static shear force of the wire bonds (± 10%) which may be partly related to the statistical nature of fatigue especially in the high cycle regime.

Characterization of the bonding interface

In the following, evaluation and interpretation of the fatigue life data of the investigated sample is discussed with regard to the microstructural features of their bonding interface. Several metallographic cross sections of wire bonded samples with different surface compositions were prepared and their interface region was investigated. Figure 6a and 6b demonstrate the cross sections of reference wire bonds and those bonded on a metallization pad with 10 nm Ti layer (air-break). Typical microstructure of the interface between the Al wire and AlSiCu metallization for the reference sample and a wire bond with 10 nm Ti layer with air-break are shown in figures 7a and 7b. Figure 7a is also representative for other samples with a high quality interface (2 nm and 6 nm Ti) and figure 7b shows the typical non bonded area as observed in sample with inferior quality (10 nm, 15 nm Ti). Due to the extremely high degree of plastic deformation induced by the ultrasonic bonding process, microstructure of the bonded area differs completely from that of the original wire. In the interface region the microstructure of the wire and the metallization consists of highly deformed grains or dislocation cell structure in the range of one to few μm. The inhomogeneous structure of the bonding interface consists of regions of nanometer scaled grains, crystalline and amorphous oxides, voids and non-bonded areas [2]. Independent of the composition of the metallization surface layer, in most of the investigated wire bonds regions of non-bonded areas were observed. It was found that there exists a direct relationship between the abundance and shape of these defects with the thickness of the oxide layer. Increased thickness of oxide layer resulted in accumulation and enlargement of the non bonded areas and presence of clearly separated interface regions with a typical shape as shown in figure 7b. Longitudinal cross sections of highly contaminated samples frequently showed larger non bonded regions in the periphery of the bond and especially at both integration points of the wire to the bond pad resulting in a smaller bond foot print and reduced bonded area (Fig. 6a). An example of a fracture surface of wire side and chip side of a sample with 10 nm Ti layer is shown in figure 8a. The areas with a flat topography indicate a lift-off due weak bonding site and interface separation. The central area with remnants of Al wire on the bond pad after lift-off failure indicates a good adherence of the Al wire to the interface. In this case fatigue failure occurs due to crack propagation inside the Al wire. Figure 8b demonstrates one of the typical interfacial bonding defects which corresponds to the non bonded areas shown in figure 7b. This image

shows that the original pattern of the metallization has been retained due to lack of contact between the wire and the pad.

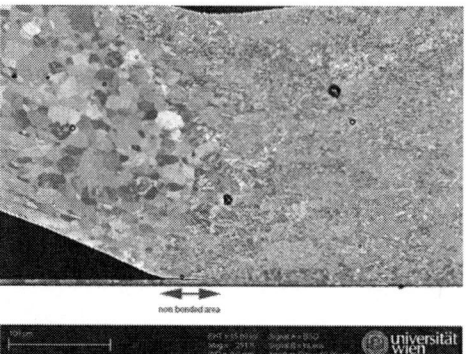

Figure 6a: SEM- ECC image of the cross section of the reference wire bond (without Ti)

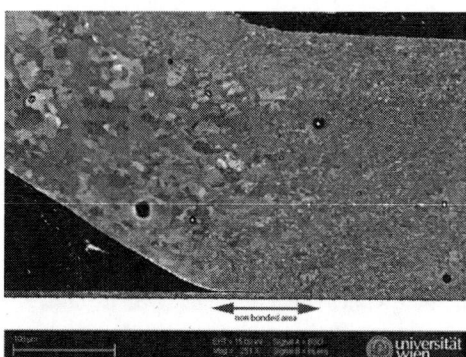

Figure 6b: SEM- ECC image of the cross section of a wire bonded sample with a 10 nm Ti layer with air-break

ToF-SIMS measurements were performed on the foot print of selected fractured wire bonds to analyze the distribution of Ti on the interface region. Figure 9 shows that for a specimen with a Ti layer of 10 nm with air-break depending on the layer thickness, traces of Ti can be found on the bond foot with a higher concentration at the peripheral region (bright areas). Distribution of titanium can be correlated to the fracture surface (Fig. 8a) and the cross section of a wire bond with the same configuration (Fig. 6b). Concentration of Ti at the outer margin of the bond foot corresponds to the regions with a flat topography on the periphery of the bond foot print. Also in comparison with the cross section of a reference specimen (Fig. 6a) rather large non-bonded regions between the bond foot and the pad and especially at the integration points of the wire to the pad can be observed. From figures 6b, 8a and 9 it can be concluded that during the ultrasonic bonding, the thick surface oxide particles are swept from the center of the bonding interface and accumulated in the periphery of the bond foot resulting to a weaker interfacial bond at the periphery and a stronger bonding in the centre of the wire bond. Similar fracture surface features were observed in a study on the influence of oxide layer thickness on bondability of gold ball bonds on Al pads [9]. Otherwise it has been reported that during ultrasonic bonding, a strong bond is mostly formed in the periphery of the contact region and the central parts remain largely non-bonded [2, 8].

Figure 7a: SEM-ECC image of a high quality bonding interface of the reference sample.

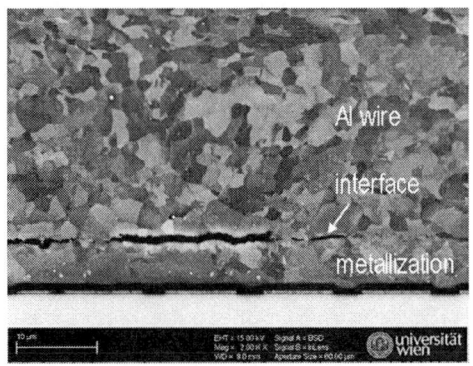

Figure 7b: SEM-ECC image of bonding interface of a sample with 10 nm Ti layer with non-bonded areas.

To our knowledge studies on the effect of Ti contamination on bond wire reliability are scarce. As a general rule metallic contaminates are known to adversely influence the bonding quality due to formation of hardly removable oxide films on the bonding pad. The relationship between the bondability and shear strength of gold ball bonds on AlSi1bond pads was studied by [9]. Thickness of surface oxide layer was varied in the range of 5 nm up to 100 nm. Increasing the film thickness above that of native Al-oxide (about 2- 5 nm) resulted in a continuous drop of shear force from the initial value.

Wire bonding on bond pads with an oxide layer thickness above 20 nm resulted in reject bonds [9].

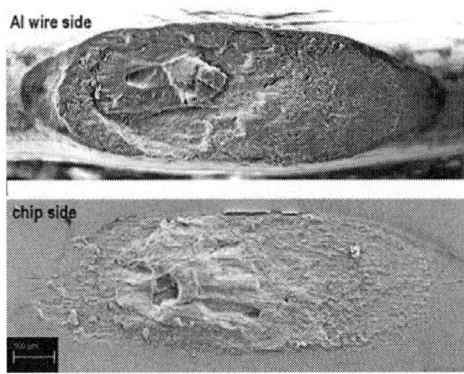

Figure 8a: SEM image of fracture surfaces of a sample with 10 nm Ti with air –break after lift-off failure (wire side and chip side).

Figure 8b: SEM image of a non-bonded area on foot print of a wire bond.

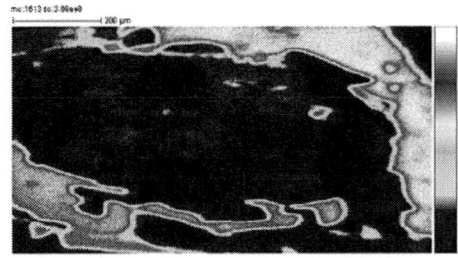

Figure 9: ToF-SIMS element mapping of the Ti distribution of the fracture surface of the bond foot with a Ti layer of 10 nm-air-break.

In one of the few available studies on the effect of thin titanium films on wire bond quality, Ti passivation layers with different thicknesses were deposited onto the copper pads to improve the bondability of thermosonic Au-Cu ball bonds. Ti passivation layer was used to protect the surface of Cu pads from oxidation [10]. Evaluation of the interface quality showed that a titanium passivation layer of 3.7 nm thickness results in improved bondability and an optimum interface quality. Ti-films above this thickness could not be removed by an appropriate range of ultrasonic power during thermosonic bonding resulting to low quality bonds or nonbonding [10].

Though there are crucial differences between Al and Cu metallization especially with regard to the properties of the surface oxide layer, these results resemble strongly to those obtained in our investigations. Deposition of titanium with an optimum layer thickness seems to be beneficial for the quality of the wire bond. Due to the high affinity of Al to oxygen exposure of the metallization film to atmospheric environment results in an immediate formation of a hard passivation layer of Al_2O_3 film of a few nm. A further growth of the film occurs only after an exposure to high temperatures or moisture. In contrary Cu oxide grows continuously in ambient atmosphere resulting to formation of soft Cu oxides and corrosion of the pad [2, 10]. Similar to Al, Ti forms a stable and hard passivation film on the surface of Ti and its alloys which is composed of various titanium oxides. The passivation film is mostly reported to be TiO_2 with a hardness lower that of Al_2O_3 [11]. It has been reported that due to easily removable Ti passivation layers, Ti alloys are suitable candidates for ultrasonic welding [12]. Our results show that deposition of few nm Ti on the bond pad may result in formation of a favorable oxide configuration on the upper layer of AlSiCu metallization. Under the given bonding parameters this layer can be efficiently broken and removed from the pad surface providing the necessary direct contact of the Al wire with the metallization.

An increased thickness of the surface oxide layer on the bond pad results in a degradation of the bonding interface caused by accumulation of non-bonded areas in the interface region. The smearing effect of thicker and softer metallic oxides mitigates the effective friction force between the metallic pair. Remnants of thick oxide particles cannot be removed from the interface under standard settings of the ultrasonic bonding device, impeding a continuous clean metallic surface contact. So far the thickness of the oxide layers remains below a certain value (6 nm Ti layer in addition to 6nm native Al-oxide layer), the structure of the surface passivation layer seems to be rather beneficial to the bonding properties. An excessive oxide contamination layer results to serious reliability problems.

Acknowledgements

The authors thank Austrian Research Promotion Agency (FFG) and ZIT-Comet program, Federal Ministry of Economics and Labour of the Republic of Austria (Contract 98.362/0112-C1/10/2005) and the Carinthian Economic Promotion Fund (KWF) (contract 18 911 | 13 628 | 19 100) for financial support.

References

[1] M.Ohring, Reliability and Failure of Electronic, Materials and Devices. Academic Press Limited, 1998

[2] G. Harman, "Wire Bonding in Microelectronics", 3rd Edition, McGraw Hill, 2010.

[3] Daniel Lu, C. P. Wong, editors "Materials for Advanced Packaging", Springer, 2009.

[4] S. Puchner, "Characterization of Contaminations on Semiconductor Surfaces and Thin Layer Systems with Time of Flight Secondary Ion Mass Spectrometry", Dissertation, January 2011.

[5] S. Puchner, A. Zechmann, Th. Detzel, H. Hutter, "Titanium Layers on Aluminum Bond Pads: Characterization of Thin Layers on Rough Substrates", Surf. Interface Analysis, Vol. 42, p. 779–782, 2010.

[6] B. Pathangey, A. Proctor, Z. Wang, Z. Fu, and R. Tanikella, "Application of ToF-SIMS for Contamination Issues in the Assembly World", IEEE Transactions on device and Materials Reliability, Vol. 7, No. 1, 2007

[7] G. Khatibi, W. Wroczewski, B. Weiss, T. Licht, "A Fast Mechanical Test Technique for Lifetime Estimation of Micro-joints", J. of Microelectronic Reliability Vol. 48, p.1822–30, 2008.

[8] Y. Ding, J. K. Kim, P. Tong "Numerical Analysis of Ultrasonic Wire Bonding: Effects of Bonding Parameters on Contact Pressure and Frictional Energy", Mechanics of Materials, Vol. 38, p.11–24, 2006.

[9] P. Meier K. D. Lang M. Petzold, L. Berthold " Surface Oxide Films on Aluminum Bondpads: Influence on Thermosonic Wire Bonding Behavior and Hardness", Microelectronics Reliability, Vol. 40, p. 1515-1520, 2000.

[10] Jong-Ning Aoh und Cheng-Li Chuang, "Thermosonic Bonding of Gold Wire onto a Copper Pad with Titanium Thin-Film Deposition", Journal of Electronic Materials, Vol. 36, No. 7, 783-797, 2007.

[11] P. Patnaik "Handbook of Inorganic Chemicals" first edition , McGraw-Hill, 2002.

[12] M. Bloss, K. Graff, "Ultrasonic Metal Welding of Advanced Alloys:The Weldability of Stainless Steel, Titanium, and Nickel-Based Superalloys", in Trends in Welding Research, Proc. of the 8th Int. Conference 2008, Pine Mountain, Georgia, June 1-6, p 348-353.

Polymer cored BGA ball reliability optimisation

D.C. Whalley[1,2], H. Kristiansen[2] and L. Halbo[2]

[1]Wolfson School of Mechanical and Manufacturing Engineering, Loughborough University, Loughborough, UK

[2]Conpart AS, Skjetten, Norway

Phone:+44-1509-227661 Fax:+44-1509-227648, E-mail: d.c.whalley@lboro.ac.uk

Abstract

This paper presents the results from simulation studies of the mechanical performance of BGA interconnects based on the use of polymer cored solder balls (PCSBs) as an alternative to the solid solder balls that are typically used. PCSBs have previously been shown to offer improved reliability and this study was performed to improve understanding of the effects of two key PCSB design variables, i.e. metallisation thickness and polymer modulus on their fatigue performance. The results show that minimising the polymer modulus is beneficial in reducing plastic strains in both the solder and copper metallisation on the polymer ball, but that the metallisation thickness has a much less significant effect unless the polymer core modulus is extremely low.

Key words: metal coated polymer spheres, polymer cored solder balls, FEA, ball grid array.

Introduction

Polymer cored solder balls (PCSBs) are an alternative to the solid solder balls commonly used in area array interconnection applications such as ball grid array (BGA) and chip scale packages (CSP). Figure 1 shows cross sectional schematics of such interconnects. The obvious candidate for the metallisation on the ball is copper due to its high electrical and thermal conductivity, good solderability and the availability of suitable plating chemistry. The primary driver for the use of PCSBs is that they have been demonstrated to provide a more reliable alternative to solid solder balls. For example, Okinaga et al. [1] showed a 4 fold increase in fatigue life, Schwanke et al. [2] showed a 2 times increase, Movva and Aguirre [3] showed a 2-3 times improvement, Galloway et al. [4] showed a 60% improvement and Kangasvieri et al. [5] showed a 50% to 75% improvement, as compared to that for conventional solder balls, whilst Sun et al. [6] presented results for a PCSB based assembly which demonstrated a drop impact performance 1.5 to 3 times that for solid Sn3.5Ag solder balls. Other unpublished measurements made at Conpart typically show a four times improvement in fatigue life. This reliability advantage is as a result of the polymer balls increased compliance in comparison with a solid solder ball, thereby reducing the level of stress imposed on the solder joints.

Most published studies of polymer cored balls have, for practical reasons, only examined a very limited range of interconnect geometries and materials properties, however computational modelling techniques such as finite element analysis (FEA) offer a cost effective method to achieving a detailed understanding of the effect of particle properties and dimensions and solder joint geometries on the mechanical performance by performing a series of "virtual experiments" which would be prohibitively expensive if conducted on physical samples. Such modelling techniques therefore offer a route to selecting the appropriate materials for the polymer spheres and their conductive coatings, and for establishing optimum design parameters such as ball diameter, conductive coating thickness, solder pad diameters (for both the component and substrate), and solder fillet volumes. For example Guillén Marin et al. [7] presented results from such simulation based studies which indicated that even greater thermal fatigue lifetime improvements than cited above may be possible through optimization of the interconnection geometry and materials selection. However some of the interconnect dimensions will be constrained by practical considerations and in particular where a "drop in" replacement is required for use with an existing component package style then various features, such as solder pad dimensions, may be fixed.

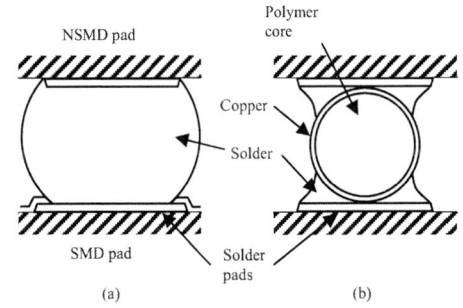

Figure 1. Schematic cross sections of (a) conventional and (b) polymer cored interconnect

This paper will present an analysis of how two critical parameters over which the PCSB manufacturer has control will affect plastic strain levels in the solder fillets and in the copper coating on the polymer ball and therefore the fatigue life. These two parameters are the copper thickness and polymer stiffness.

Modelling Methodology

A similar FEA based approach was used to that presented in [7] and [8] i.e. another computational modelling tool, the Surface Evolver [9], was first used to predict the solder fillet shape for the polymer cored interconnects. Figure 2 shows a typical Evolver model of the solder fillet geometry. A series of co-ordinates taken from the Evolver simulation results were then used to define the fillet geometry for a series of two dimensional (2D) structural models of the interconnects. These models were constructed using the parametric modelling capabilities of the commercially available COMSOL multi-physics FEA software. This parametric modelling feature allows variables such as a dimension or material property to be swept through a range of values in order to assess the sensitivity of performance to that variable.

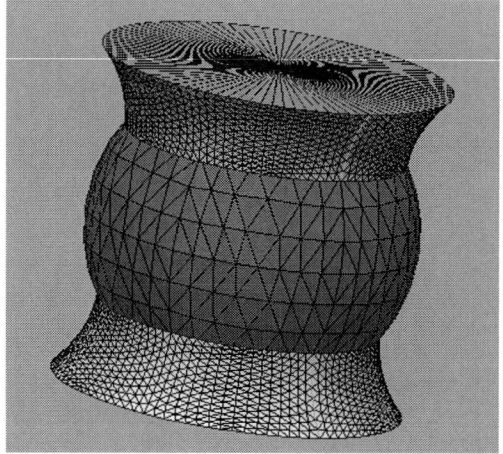

Figure 2. Evolver model of PCSB solder fillets

Model dimensions and material properties

The polymer ball was modelled as a perfect sphere with a diameter of 0.76 mm, including a copper coating which was varied in thickness between 7 and 30μm. The solder pads on PCB and component were both modelled as being copper and with a diameter of 0.86mm. The solder volume per fillet was 43.6×10^{-3} mm^3, equivalent to that printed through a 150μm thick stencil over the entire solder pad area using a 50% by volume metal paste, and the wetting angle of the solder to the copper was 34°. The thickness of the PCB pads was 35μm, whilst that of the components was 10μm. For conventional BGA type interconnects it is well known that whether the solder pads are solder mask defined (SMD) or non-SMD (NSMD) has a significant effect on solder fatigue life, as the main concentration of stress and therefore plastic strain in the solder occurs where the solder meets the edge of the pads. For the polymer cored interconnects the area of stress concentration moves to the solder/ball interface and SMD versus NSMD designs were therefore not considered. However the stress levels in the coating on the ball can exceed the copper yield stress and therefore unlike conventional solder balls fatigue of the copper may present a reliability hazard. Plasticity of both the solder and copper were therefore considered in the FE models.

Table 1 lists the materials properties used for the models. The polymer material was assumed to behave in a linear elastic manner, whilst the copper and solder were given elastic/plastic properties with a simple linear isotropic hardening behaviour represented by a tangent modulus.

	Young's Modulus (GPa)	Poisson's Ratio	Yield Strength (MPa)	Tangent Modulus (MPa)
Copper*	126	0.335	38.0	295
Polymer	0.1 - 0.7	0.300	-	-
Solder*	52.1	0.347	33.3	154

Table 1 Materials properties *[10]

Figure 3 shows an FEA mesh for the structure modelled, however it should be noted that the mesh in this figure is deliberately coarsened to improve legibility – the actual mesh density was considerably finer.

The applied boundary conditions were that the lower face of the PCB pad was restrained from movement and a horizontal, i.e. shearing, displacement of 5μm was applied to the upper face of the component pad to simulate relative thermal expansion between the component and substrate.

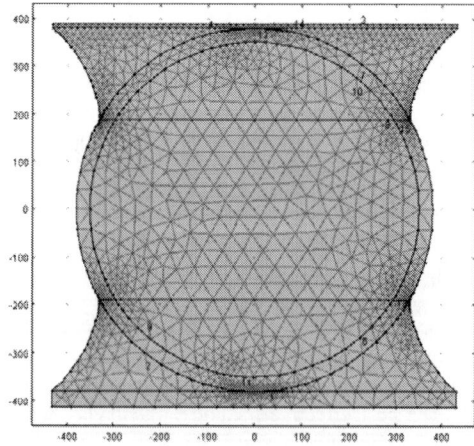

Figure 3 Typical FEA model of PCSB assembly

Results

Figure 4 shows a typical predicted vonMises stress distribution plot for the interconnect, whilst Figure 5 shows the predicted distribution of plastic strain in the metals for one quarter of the model. The stresses can be seen to be concentrated in the areas of solder near to the PCSB and in the copper metallisation of the PCSB with particularly high stresses in the vicinity of the solder/PCSB interfaces. The plastic strains can be seen to be even more localised in these regions.

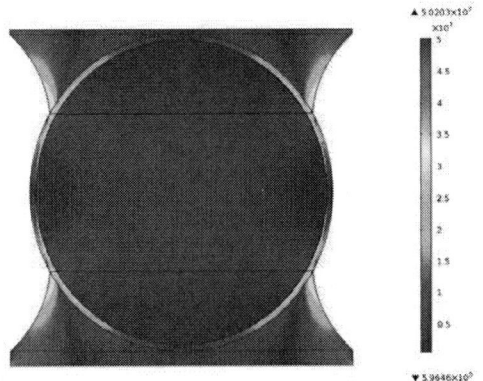

Figure 4 Predicted vonMises stress distribution for 15μm thick copper

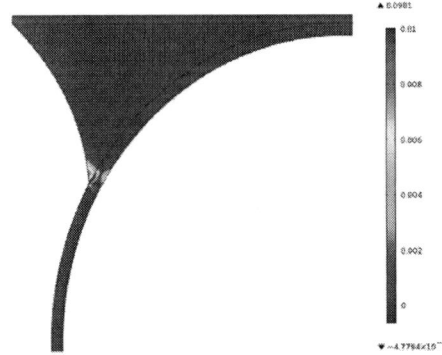

Figure 5 Predicted plastic strain distribution

The location experiencing the greatest cyclic plastic strain will be the location where the material ductility will first become exhausted and therefore where failure will initiate. Although functional failure may not occur until the damage has propagated further through the material the relative maximum cyclic strains will provide a good basis for estimating the effect of different parameters on cyclic life through the use of for example the Coffin-Manson model. Figure 6 shows a cross section of a soldered polymer cored interconnect after thermal

cycling to failure. The failure is in the copper metallisation and solder, and its location is in good agreement with the FE model. In several previously published studies, e.g. [5], the quantity of solder present was much greater, forming an almost continuous column encapsulating the polymer ball. The locus of failure in such studies is therefore different, however the predicted and observed failure locations in figures 5 & 6 do correlate well with those identified in [11] where smaller solder fillets, similar to those presented here, were present.

The raw plastic strain results were post processed to identify the maximum plastic strains in each of the relevant regions of the model and Figure 7 shows the predicted effect of the Young's modulus of the polymer core of the ball on the maximum strains in both the solder and copper. It can be seen that these strains increase with the polymer modulus and in particular that they increase more rapidly when the modulus exceeds 0.5 GPa. A series of simulations were also included where the polymer was completely removed and these are represented in this graph by the point at 0 GPa. This situation was included because some polymer materials will experience shrinkage during exposure to high temperatures, such as those experienced during soldering, and if this led to complete delamination of the metallisation from the core then this might be expected to result in similar stresses to there being no polymer present. The results show a slight reduction in strains compared with a low modulus polymer being present, however this is not seen as being a desirable situation but the results show it may not be disastrous if this situation does occur.

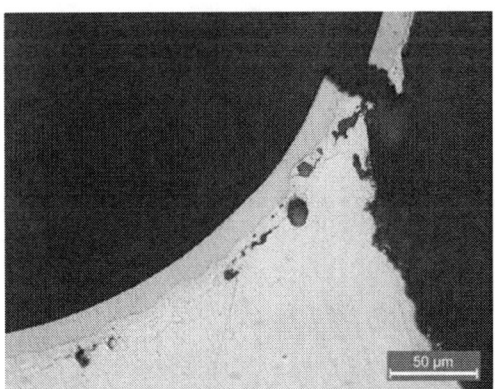

Figure 6 Micrograph of failure after thermal cycling

Figure 8 shows the predicted effect of metallisation thickness on the maximum strains, both for no polymer present and for a 0.5 GPa polymer present. It can be seen that where the polymer is present there is a slight increase in strain with metal thickness, but this may be compensated for by there being a greater thickness of metal for a crack to propagate through before functional failure. Where the polymer is removed the strains are predicted to be more strongly dependent on the metal thickness, but again this cannot be seen as a desirable situation although these results show it may not be a cause for major concern.

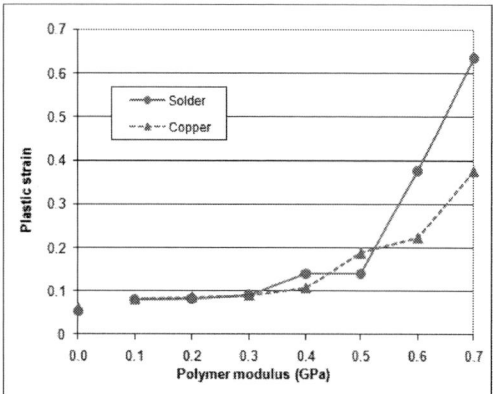

Figure 7 Effect of polymer modulus on predicted maximum strains

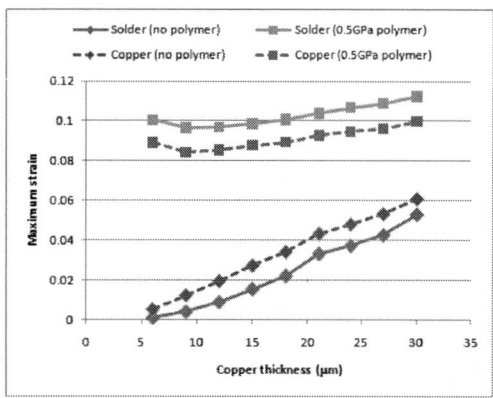

Figure 8 Effect of copper thickness on predicted maximum strains

Conclusions

This study has shown that the modulus of elasticity of the polymer core within PCSBs will have a substantial effect on the non-linear (plastic) strains that develop within the solder and ball metallisation. This is easily understood in the case of solder, as an increased polymer modulus will transfer more strain to the solder. In the case of the copper, it is probably a more local effect. It can therefore be concluded that selection of a low modulus polymer is critical to maximising reliability.

The models have also shown the effect of metallisation thickness on plastic strains is relatively small and it may therefore be concluded that

selection of the metal layer thickness can be made based on the trade-off between the need for high electrical and thermal conductivity, increased cost of a thicker layer and the requirement to be thick enough not to be significantly affected by the amount of copper dissolved during the multiple reflow soldering cycles that will occur during attachment of the ball to the component, of the component to the PCB, and any subsequent rework that may be required.

The results also show that delamination of the metallisation from the polymer core may not necessarily have an adverse effect on reliability, however partial delamination has not been considered and, depending on the specific area of delamination, the effect of this on the stress distribution could be undesirable and further work on this will be required if delamination is found to occur in practice.

A future paper will present experimental results for thermal cycling tests of BGA components of the same dimensions as for these modelling results together with further refinement of the models.

Acknowledgements

The authors gratefully thank the Research Council of Norway and Loughborough University, UK for funding this work.

References

1. Okinaga, N., Kuroda, H. and Nagai, Y., "Excellent reliability of solder ball made of a compliant plastic core", Proceedings of the 51st Electronic Components and Technology Conference, 29 May-1 June, 2001, pp 1345 – 1349

2. Schwanke, D., Haas, T., Zeilmann, C., Halser K. and Klein M., "Reliability Investigations of BGA-Interconnections between RF-LTCC-Modules and PCBs", 14th IMAPS EMPC, Friedrichshafen, June 2003

3. Movva, S. and Aguirre, G., "High reliability second level interconnects using polymer core BGAs" Proceedings of the 54th Electronic Components and Technology Conference, Las Vegas, USA, June 2004, Vol. 2, pp 1443 – 1448

4. Galloway, J., Syed, A., WonJoon Kang, JinYoung Kim, Cannis, J., Ka, Y., SeungMo Kim, TaeSeong Kim, GiSong Lee and SangHyun Ryu, "Mechanical, thermal, and electrical analysis of a compliant interconnect" IEEE Trans. CPT, Vol. 28, No. 2, pp 297-302

5. Kangasvieri, T., Nousiainen, O., Putaala, J., Rautioaho R. and Vähäkangas J.,

"Reliability and RF performance of BGA solder joints with plastic-core solder balls in LTCC/PWB assemblies", Microelectronics Reliability Vol. 46, No. 8, pp 1335-1347

6. Sun, R-D; Okinaga, N; Matsushita, K. and Okuda, M., "Study on improving the drop impact reliability of plastic core solder balls", Proceedings of the ICEP, 2009, pp 1-6

7. Guillén Marin, F., Whalley, D.C., Kristiansen, H. and Zhu, Z., "Mechanical performance of polymer cored BGA interconnects", Proceedings of the 10th IEEE Electronics Packaging Technology Conference, Singapore, December 2008, pp. 316-321

8. Whalley, D.C. and Kristiansen, H. "Current Density Simulations for Polymer Cored CSP Interconnects" Proceedings of the 3rd IEEE CPMT Electronic Systemintegration Technology Conference, Berlin, September 2010, 5 pp

9. Brakke, K., "The Surface Evolver", Experimental Mathematics, Vol. 1, No. 2, pp 141-165

10. COMSOL Multiphysics material library http://www.comsol.com/

11. Loeffler, M. "Polymer-Core Solder Balls: An Alternative to Solid Solder Balls?" CircuiTree Magazine, May 23, 2006, accessed: http://www.circuitree.com/

Power Loss due to Periodic Structures in High-Speed Packages and Printed Circuit Boards

Priya Pathmanathan[i,ii], Christine Madden Jones[i,iii], Steven G. Pytel[i,iv], David L. Edgar[v], and Paul G. Huray[i]

[i]University of South Carolina, Columbia, South Carolina, 29208, USA
[ii]Intel Corporation, DuPont, Washington, 98327, USA
[iii]U. S. Navy, North Charleston, South Carolina, 29419, USA
[iv]ANSYS Inc, Canonsburg, Pennsylvania, 29054, USA
[v]ANSYS UK Ltd, Bracknell, Berkshire, RG12 7BW, UK

Phone: +1 803 777 9520, E-mail: huray@sc.edu

Abstract

In this paper we apply the basic principles of electromagnetic wave propagation in periodic media to explain the high frequency power loss resonances of Printed Circuit Board (PCB) interconnects observed in recent simulations and measurements. For this purpose we consider a simplified PCB trace buried in a one dimensional periodic media consisting of alternating dielectric materials resembling glass weave and resin patterns found in a typical PCB. We formulate distributions of electric and magnetic field components in each homogeneous region and solve them to get the relationship between fields in TE and TM waves in two equivalent layers with the same effective dielectric constant. We apply Floquet's theorem to relate the fields in similar periodic structure media and find the dispersion relationship between frequency (ω), propagation constant (ß) and Bloch wave number (k) separately for TE and TM modes. From this relationship we identify regions where k is purely real and where k becomes complex, which corresponds to propagating and evanescent wave modes respectively. These modes result in resonant power losses, observed in simulations and measurements. To validate this theory we create a parameterized, full-wave simulation model of a typical glass-resin substrate using ANSYS' HFSS™ and HFSS-Transient™ to resemble the theoretical assumptions. Resonant power losses are clearly observed through the simulated s-parameter plots in the frequency domain and the simulation results are well matched to theoretical predictions. Finally we also discuss the importance of considering resonant power loss frequency regimes in the design of high-speed packaging and PCB design.

Keywords: Periodic Wave Propagation, Fiber Weave, High Frequency, Insertion Loss, Power.

Introduction

Man-made materials often consist of periodically placed structures and substances. Starting from late Nineteenth century, scientists [1] analyzed the problem of wave propagation in periodic lattices. Periodic wave propagation theory was exploited in designing electrical filters in the form of periodic networks, optical reflectors, beam splitters and polarizers in the form of stratified optical thin films. During the 1950s periodic structures were heavily used in antenna engineering and Electronic Band Gap (EBG) structures. Frequency dependent propagation loss properties were well studied in the twentieth century with a focus on crystalline structures and Bragg diffraction. Léon Brillouin [2] studied periodic wave propagation in atomic structures and introduced the concept of Brillouin Zones. Since then, several researchers have analyzed electromagnetic and generic wave propagation in greater detail [3]. However there has been no in-depth analysis done so far to understand the periodic structures encountered in digital signal transmission channels and their implications to signal integrity of system busses. The main reason for this lack of attention is possibly due to the fact that the periodic wave propagation effects were well above the predominant high frequency content of the signaling.

Ever increasing signaling rates with faster rise and fall times, in accordance with Moore's Law, demand high speed interconnects with higher bandwidths, often in the sub millimeter wavelength regime. Therefore, understanding propagation properties and controlling losses of interconnects is critical to maintain the Signal Integrity (SI) of busses carrying critical data.

1. Fiber weaves and PCB interconnects.

Printed Circuit boards (PCB), packages, and connectors are the common interconnecting medium for modern computer systems that communicate between various processors and chipsets. Typical PCB manufacturing processes embed layers of woven fiberglass cloth in resins to mechanically strengthen the PCB. Generally the fiberglass and resin have different electrical properties including dielectric strengths and constants. This difference creates a periodically loaded dielectric medium. PCB construction is not the only source of periodic loading in modern high speed signal interconnects. Routing through a pin grid array, placing surface mount discrete components near a signal carrying trace on a PCB or package, and braided shielding on cables are a few examples where periodic loading could potentially affect the signal integrity. For example, Figure 1.a) illustrates an example of periodic loading introduced by routing through a dense pin grid array. However, the scope of this paper is limited to the periodicity effects introduced by fiber weaves of PCBs.

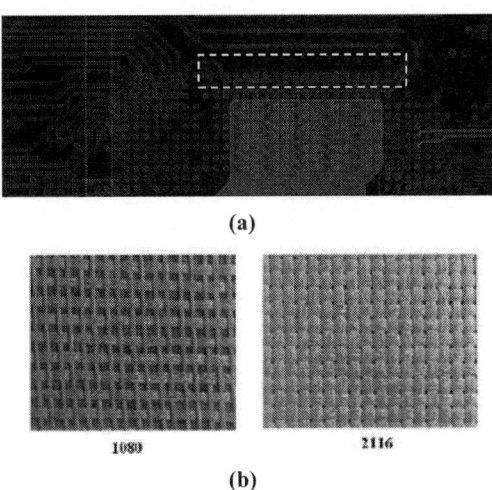

(a)

(b)

Figure 1: Periodic Structures. a) Routing through a Pin Grid Array b) Common Types Woven Fibers used in PCB construction [4]

A number of variables must be taken into consideration when selecting the fiber style for PCB construction. The weave pattern and the yarn thickness are the most critical parameters that will determine the cost and quality of the PCB. These laminates are available in various constructions from PCB material vendors to cater to the specific needs of customers. Two very common types of woven fabric used in PCB construction are shown in Figure 1.b) [4]. The mechanical and electrical properties of the composite material typically depend on the amount of impregnation, and this is affected by the density of the weave and balanced/unbalanced count of the warp and weft yarns. Therefore, good understanding of frequency dependent properties is

necessary to build cost optimized PCB designs with exceptional signal integrity properties to cater to high speed signaling.

Despite the inhomogeneous composition of PCB laminates discussed above, PCBs have been treated as homogeneous electrical medium with averaged dielectric properties for many years. However, as data rates continue to increase, it has become evident [5, 6,7], that the local variations of the material properties can no longer be neglected because they play an important role in the performance of high speed interconnects.

Recently, fiber weave effects on PCBs were studied extensively with emphasis on propagation delay skew within a pair of differential signals and its implications to eye margins [8]. The majority of these studies were analyzed and mitigated through empirical measurements without emphasis on loss characteristics. In [8] it was found that local variations of the dielectrics can lead to unbalanced lines on high speed buses because it is likely that some lines will lie on top of the fiberglass bundles while others will lie in-between the bundles.

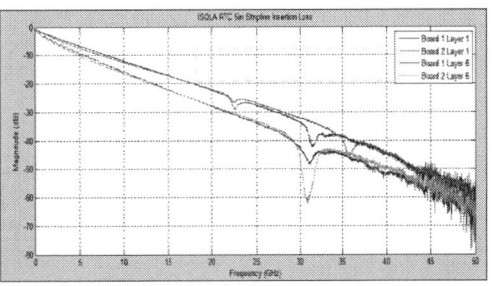

Figure 2: VNA Measurements of Insertion Loss for ISOLA 620 RTC Test Boards and Layers [7]

Recent simulation and measurement based investigations [5],[6] led to the discovery of power losses at high frequencies (above 20GHz) causing signal integrity degradation in PCBs due to periodic loading of the fiber weave structures. Extensive insertion and return loss s-parameter measurements were made in [7] to investigate the resonances. Through controlled experiments several possible factors including humidity effects, structural defects and measurement equipment errors were ruled out as potential causes for these resonances. Figure 2 shows insertion loss for a 6 layer PCB on a 5 inch microstrip with ISOLA 620 prepreg dielectric, 7628 FR4 core and Reverse Treated Copper (RTC) foils. The boards were fabricated with traces on a 10 degree angle to the fiber glass "working panels." These measurements show resonances studied during the remainder of this work. Researchers highlighted periodic fiberglass effect as the most promising cause for these measured resonances. In other independent experiments, full wave simulations were performed [6] and correlated to measurements. First order resonance location in a frequency band was identified by relating the wavelength to periodic pitch of the fiber weave structure. In [6], it was found that the periodic loading of interconnects introduced by the fiberglass bundles can lead to resonances in the insertion and return loss characteristics. In the same work, comprehensive numerical and graphical studies of resonances were performed. In particular, the roles of the pitch between fiberglass bundles and trace angle (with respect to the bundles) were studied.

In this paper we approached this problem from a more theoretical point of view by solving fundamental electromagnetic wave equations. Electromagnetic wave propagation in a periodic medium is thus presented and applied to a simplified PCB fiber weave structure to predict theoretical resonance frequencies. Full wave simulations were performed and matched to theoretical predictions. Effects of periodic loading on system eye margins were studied and high level design guidelines are suggested to mitigate the effects.

2. Wave Propagating in a Periodic Medium

Periodic wave propagation media can generally be classified into two main categories. 1) a waveguide with uniform dielectric medium but with periodic physical boundaries. 2) a medium which is composed of periodically varying dielectric properties. PCBs fall under the second category. In this section we will consider a simple periodic medium resembling fiber weaves of a PCB.

Consider a one dimensional, infinitely long periodic medium consisting of two alternating layers of materials with different dielectric properties as illustrated in Figure 3. Let l_1 and l_2 be the thickness of materials along the periodic direction, x, with period $L = l_1 + l_2$. Due to the periodic nature, the refractive index profile of this one dimensional periodic unit cell can be described by

$$\eta(x) = \begin{cases} \eta_1 = \sqrt{\varepsilon_{r1}\mu_{r1}}, & 0 < x < l_1 \\ \eta_2 = \sqrt{\varepsilon_{r2}\mu_{r2}}, & l_1 < x < l_1 + l_2 \end{cases} \quad (1)$$

where the periodic condition of $\eta(x)$ is given by

$$\eta(x + L) = \eta(x) \quad (2)$$

A typical electric field intensity component of a propagating wave in the z direction can be expressed as

$$E(x,z) = E(x)e^{j\beta z} \quad (3)$$

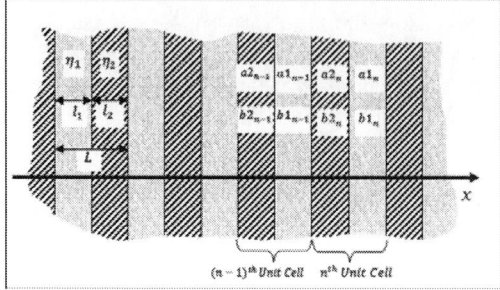

Figure 3: One dimensional, two tone infinite periodic medium.

Plane waves in each homogeneous layer of the nth unit cell can be expressed as combination of incident and reflected waves. Complex amplitudes of the incident and reflected waves in medium 1 and medium 2 can be written in column matrices as,

$$\begin{pmatrix} a1_n \\ b1_n \end{pmatrix} and \begin{pmatrix} a2_n \\ b2_n \end{pmatrix} \quad (4)$$

where $a1_n$ and $a2_n$ represent incident waves and $b1_n$ and $b2_n$ represent reflected waves of unit cell n. Electric field intensity distributions in these regions can be expressed as,

$$E1(x,z) = \left\{ a1_n e^{jk_1(x-nL)} + b1_n e^{-jk_1(x-nL)} \right\} e^{j\beta z}$$
$$E2(x,z) = \left\{ a2_n e^{jk_2(x-nL)} + b2_n e^{-jk_2(x-nL)} \right\} e^{j\beta z} \quad (5)$$

where the complex propagation constants are,

$$k_1 = \sqrt{\left[\left[\left(\frac{\omega}{c} \right) \eta_1 \right]^2 - \beta^2 \right\}}$$

$$k_2 = \sqrt{\left[\left[\left(\frac{\omega}{c} \right) \eta_2 \right]^2 - \beta^2 \right\}} \quad (6)$$

where ω and c represent angular frequency and the speed of light respectively.

Electric and magnetic field intensities are related through continuity conditions at the interface. For Transverse Electric (TE) modes we can impose continuity on the E vector and on its derivative $\partial E/\partial x$ at any arbitrary interface in y-z plane.

On $(n-1)th$ unit cell enforcing continuity on $E(x)$ leads to

$$a1_{n-1} + b1_{n-1} = e^{-jk_2 L}a2_n + e^{jk_2 L}b2_n \quad (7)$$

while enforcing continuity on $\partial E/\partial x$ leads to

$$k_1 \left(a1_{n-1} - b1_{n-1} \right) = k_2 \left(e^{jL}a2_n - e^{jk_2 L}b2_n \right) \quad (8)$$

Similarly considering the nth unit cell and enforcing continuity on $E(x)$ leads to

$$e^{-jk_2 l_1} a2_n + e^{jk_2 l_1} b2_n = e^{-jk_1 l_1} a1_n + e^{jk_1, a} b1_n \qquad (9)$$

Again enforcing continuity of $\partial E/\partial x$ on n^{th} unit cell on $\partial E/\partial x$ leads to

$$k_2 \left(e^{-jk_2 l_1} a2_n - e^{jk_2 l_1} b2_n \right) = k_1 \left(e^{jk_1 l_1} a1_n - e^{jk_1 l_1} b1_n \right) \qquad (10)$$

Equations (9) and (10) can be written in matrix forms as

$$\begin{pmatrix} 1 & 1 \\ 1 & -1 \end{pmatrix} \begin{pmatrix} a1_{n-1} \\ b1_{n-1} \end{pmatrix}$$
$$= \begin{pmatrix} e^{-jk_2 L} & e^{jk_2 L} \\ \dfrac{k_{2x}}{k_{1x}} e^{-jk_2 L} & -\dfrac{k_{2x}}{k_{1x}} e^{jk_2 L} \end{pmatrix} \begin{pmatrix} a2_n \\ b2_n \end{pmatrix} \qquad (11)$$

$$\begin{pmatrix} e^{-jk_2 l_1} & e^{jk_2 l_1} \\ e^{-jk_2 l_1} & -e^{jl_1} \end{pmatrix} \begin{pmatrix} a2_n \\ b2_n \end{pmatrix}$$
$$= \begin{pmatrix} e^{-jk_1 l_1} & e^{jk_1 l_1} \\ \dfrac{k_{1x}}{k_{2x}} e^{-jk_1 l_1} & -\dfrac{k_{1x}}{k_{2x}} e^{jk_1 l_1} \end{pmatrix} \begin{pmatrix} a1_n \\ b1_n \end{pmatrix} \qquad (12)$$

Formulations (11) and (12) can be simplified after eliminating $\begin{pmatrix} a2_n \\ b2_n \end{pmatrix}$ as

$$\begin{pmatrix} a1_{n-1} \\ b1_{n-1} \end{pmatrix} = \begin{pmatrix} A & B \\ C & D \end{pmatrix} \begin{pmatrix} a1_n \\ b1_n \end{pmatrix} \qquad (13)$$

where the *ABCD* matrix, or the chain matrix in (13), relates the complex amplitudes of incident and reflected plane waves in adjacent mediums of a unit cell. A, B, C and D is given by

$$A = e^{-jk_1 l_1} \left[\cos k_2 l_2 - \frac{1}{2} j \left(\frac{k_2}{k_1} + \frac{k_1}{k_2} \right) \sin k_2 l_2 \right]$$
$$B = e^{jk_1 l_1} \left[-\frac{1}{2} j \left(\frac{k_2}{k_1} - \frac{k_1}{k_2} \right) \sin k_2 l_2 \right]$$
$$C = e^{-jk_1 l_1} \left[\frac{1}{2} j \left(\frac{k_2}{k_1} - \frac{k_1}{k_2} \right) \sin k_2 l_2 \right]$$
$$D = e^{jk_1 l_1} \left[\cos k_2 l_2 + \frac{1}{2} j \left(\frac{k_2}{k_1} + \frac{k_1}{k_2} \right) \sin k_2 l_2 \right] \qquad (14)$$

and since this periodic medium is a reciprocal network,

$$AD - BC = 1 \qquad (15)$$

Similarly considering the continuity of H vectors in $y - z$ plane, a similar *ABCD* matrix for TM mode propagation can be derived as

$$A' = e^{-jk_1 l_1} \left[\cos k_2 l_2 \right.$$
$$\left. - \frac{1}{2} j \left(\frac{\eta_2^2 k_1}{\eta_1^2 k_2} + \frac{\eta_1^2 k_2}{\eta_2^2 k_1} \right) \sin k_2 l_2 \right]$$
$$B' = e^{jk_1 l_1} \left[-\frac{1}{2} j \left(\frac{\eta_2^2 k_1}{\eta_1^2 k_2} - \frac{\eta_1^2 k_2}{\eta_2^2 k_1} \right) \sin k_2 l_2 \right]$$
$$C' = e^{-jk_1 l_1} \left[\frac{1}{2} j \left(\frac{\eta_2^2 k_1}{\eta_1^2 k_2} - \frac{\eta_1^2 k_2}{\eta_2^2 k_1} \right) \sin k_2 l_2 \right]$$
$$D' = e^{jk_1 l_1} \left[\cos k_2 l_2 \right.$$
$$\left. + \frac{1}{2} j \left(\frac{\eta_2^2 k_1}{\eta_1^2 k_2} + \frac{\eta_1^2 k_2}{\eta_2^2 k_1} \right) \sin k_2 l_2 \right] \qquad (16)$$

3. Floquet's Theorem

Floquet's theorem is a fundamental theorem explaining the concept of wave propagation in periodic media [9]. Floquet's theorem states that if $F(z)$ represents a wave propagating in positive z direction of a periodic medium with periodicity , then the following condition must be satisfied

$$F(z + p) = e^{-\gamma p} F(z) \qquad (17)$$

where, γ is the periodic propagation constant. In general, the solution for γ may be complex. Real and imaginary parts of γ determines attenuation and propagation properties introduced by the periodic structure.

Electric field intensity components (3) in a periodic medium with a periodic pitch of L satisfies

$$E_k(x + L) = E_k(x) \qquad (18)$$

Floquet's theorem can be applied to electric field intensity components of (3) to obtain

$$E_k(x, z) = E_k(x) e^{jKx} e^{j\beta z} \qquad (19)$$

Thus, it is evident that E depends on K, which is also known as the "Bloch Wave Number"[16].

4. Forbidden Bands or Brillouin Zones

We can apply the Floquet's result in (17) to equation (13) in which incident waves (a_{n-1} and a_n) and reflected waves (b_{n-1} and b_n) are separated by the period L. This forms the following Eigen value relation.

$$\begin{pmatrix} A & B \\ C & D \end{pmatrix} \begin{pmatrix} a_n \\ b_n \end{pmatrix} = e^{-jKL} \begin{pmatrix} a_n \\ b_n \end{pmatrix} \qquad (20)$$

where, e^{-jKL} is the eigen value of $\begin{pmatrix} A & B \\ C & D \end{pmatrix}$.

Eigen value relation (20) can be re-arranged as

$$\begin{pmatrix} A - e^{-jKL} & B \\ C & D - e^{-jKL} \end{pmatrix}\begin{pmatrix} a_n \\ b_n \end{pmatrix} = 0 \ \text{ or}$$

$$e^{-2jKL} - (A+D)e^{-jKL} + AD - BC = 0 \qquad (21)$$

This quadratic equation can be simplified by applying the reciprocal condition in (15). Solutions to this equation are:

$$e^{-jKL} = \frac{1}{2}(A+D) \pm \sqrt{\left\{\left[\frac{1}{2}(A+D)\right]^2 - 1\right\}}$$

This is also known as the Brillouin relation between K, β and ω. $K(\beta, \omega)$ can be expressed as,

$$K(\beta, \omega) = \frac{1}{L}\cos^{-1}\left[\frac{1}{2}(A+D)\right] \qquad (22)$$

Expressing $K(\beta, \omega)$ as an inverse cosine function is useful to understand the characteristics of this propagation constant.

When $\left\|\left[\frac{1}{2}(A+D)\right]\right\| \leq 1$, K is real and the wave is propagating. When $\left\|\left[\frac{1}{2}(A+D)\right]\right\| > 1$, K is imaginary and the wave is evanescent.

Imaginary K can be written as

$$K = \frac{m\pi}{L} + jK_i \qquad (23)$$

The expression $\left\|\left[\frac{1}{2}(A+D)\right]\right\| = 1 \qquad (24)$

thus defines band edges in which wave propagation due to periodic condition is evanescent. Evanescent frequency bands defined by these boundaries are known as Forbidden Bands or Brillouin Zones.

We can express eigen vectors for the first unit cell as

$$\begin{pmatrix} a_0 \\ b_0 \end{pmatrix} = \begin{pmatrix} B \\ e^{-jK\Lambda} - A \end{pmatrix} \qquad (25)$$

so the complete solution for "1" layer of the nth cell is given by

$$E(x,z) = \left[\left\{a_0 e^{jk_1(x-nL)} + b_0 e^{-jk_1(x-nL)}\right\}e^{-jK(x-nL)}\right]e^{jKz} \qquad (26)$$

For a given periodic geometry $y = \frac{1}{2}(A+D)$ can be plotted against frequency using a software application like Matlab®. The evanescent zone boundaries can be located by the intersections of $|y| = 1$. In following sections we will apply this technique to waves propagating on simplified PCB routing and compare it to full wave simulations of a structure resembling periodic placement of fiber weaves.

5. 3D full-wave FEM EM field solver setup

Theoretical evanescent zones discussed in the previous section can be validated by solving a simplified periodic structure using a 3D full-wave finite element method (FEM) field solver. Ansys® HFSS® version 13 was used for all 3D full wave simulations. First we created a parameterized model as shown in Figure 4.a), to match the theoretical predictions to simulations. This model was built with alternating rectangular bars with dielectric properties of typical PCB laminate resin ($\varepsilon_r = 3.5 + 0.002i$) and glass ($\varepsilon_r = 6.0$). All geometrical parameters and material properties were parameterized to enable sweeping and what-if analysis. The total length of the model was 1 inch. Symmetric pitch models with 60, 45, 30 and 15 mil variations were created to demonstrate the dependency on pitch.

Another model was created to analyze more realistic fiber weave effects in a 1080 laminate. In this model a Glass bundle cross-section was approximated by elliptical cylinders (Figure 4.b) with a 1.5 mil offset from the trace. To simplify the model and to simulate a worst case impact, weaves on both top and bottom of the traces were kept symmetrically on top of each other.

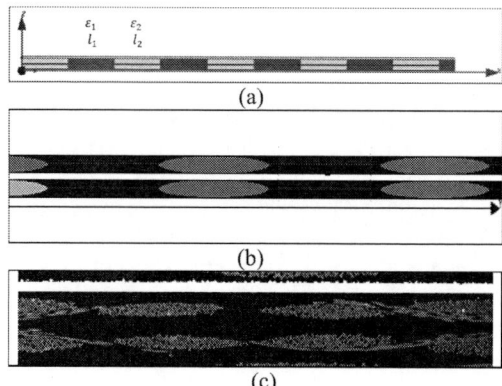

Figure 4: Cross-sections of a PCB and HFSS simulation model. a) An oversimplified HFSS model. b) HFSS model of a 1080 weave approximated to elliptical glass. c) Cross sectional image of a 1080 PCB [17][18].

The HFSS model was configured to produce terminal s-parameters in 500 MHz intervals up to 110 GHz.

6. Simulation results and theoretical Brillouin Zones.

Insertion loss profiles from 3D EM simulation of the simplified periodic medium model are shown in Figure 5.a). Insertion loss versus frequency corresponding to 15, 30, 40 and 60 mil pitch periods are overlaid in this plot. It is evident that the resonant frequency bands are dependent on the pitch of the period. As expected, the 60 mil pitch showed a lower resonance at around 25 GHz while the lowest resonance of the 15mil pitch was at around 95 GHz and much deeper than the larger pitch resonances. Figure 5.b) shows the plot of equation (24), for only values greater than one, for 60 mil pitch geometry. This plot defines Brillouin zone boundaries in which the waves are evanescent. It can be seen that the resonant regimes in the insertion loss profiles from simulation match well with theoretical predictions. Theoretical predictions were also verified for other pitch geometries but not shown in this graph to improve clarity.

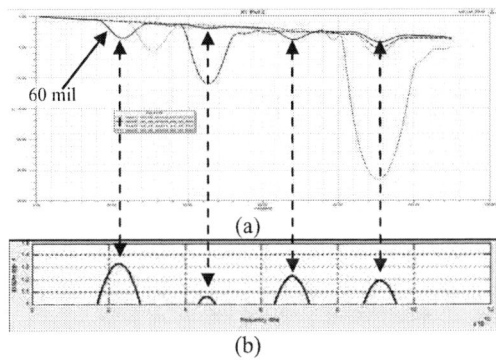

(a)

(b)

Figure 5: Simulated and theoretical Brillouin zones in ideal periodic medium. (a) Insertion loss profiles of 60, 45, 30 and 15 mil periods. (b) Theoretical evanescent zones of the 60 mil period.

Figure 6 compares simulated and theoretical lossy regimes of the more realistic model in Figure 4.b). Theoretical condition (24) is not directly applicable for a curvilinear periodic boundary as in an elliptical fiber bundle so we simplified the geometry with averaged dielectric properties for theoretical calculations. The average dielectric constant ε_{ave_g} of glass regions was calculated based on the content fraction of glass in a cross-sectional area. The average dielectric constant ε_{ave} can be calculated from Glass (ε_G) and Resin (ε_R) as

$$\varepsilon_{ave\_g} = \varepsilon_G \times \frac{\pi}{4} + \varepsilon_R \times \left(1 - \frac{\pi}{4}\right) \tag{27}$$

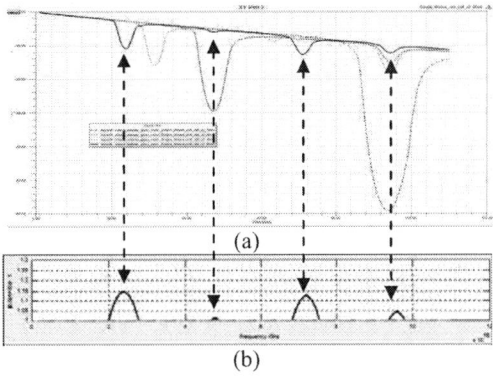

(a)

(b)

Figure 6: Simulated and theoretical dispersion zones elliptical periodic medium. (a) Insertion loss profiles of 60, 45, 30 and 15 mil periods. (b) Theoretical evanescent zones of the 60 mil period.

As seen in Figure 6, simulated and theoretical resonance regimes are closely matched. A theoretical prediction using a weighted average dielectric for the elliptical bundles is a good approximation to identify the resonance locations for practical designs.

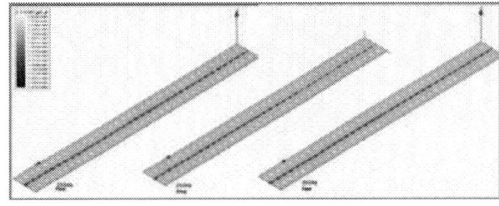

Figure 7: E field propagation at selected frequencies on a trace placed on a 15 mil periodic medium.

Electric field plots generated within the PCB structure at selected frequencies show distinct behavior associated with the pass bands and forbidden gaps calculated theoretically in above sections. As shown in Figure 7, electric field plots at pass band frequencies (15 GHz and 35 GHz) show propagation through the trace, where as plots generated at frequencies near the center of theoretical forbidden gap (25 GHz) decay within a short distance due to the periodic loading.

7. Time dependency of an evanescent propagating wave in a periodic medium.

To better understand and visualize the effects in the time domain a Finite Element Discontinuous Galerkian Time Domain (FEDGTD) simulation [15] was performed to the previously described structures in Figure 4. The FEDGTD method utilized for this work solves Maxwell's equations using a hybrid implicit/explicit solver with tetrahedral mesh elements [16].

$$\frac{\partial \boldsymbol{D}}{\partial t} + \boldsymbol{J} = curl\,(\boldsymbol{H}) \qquad \frac{\partial \boldsymbol{B}}{\partial t} = -curl\,(\boldsymbol{E})$$

$$\tag{28}$$

$$div(\boldsymbol{D}) = \rho \qquad\qquad div(\boldsymbol{B}) = 0$$

FEDGTD methods have a distinct advantage over commonly used Finite Difference Time Domain (FDTD) methods due to the ability to accurately mesh arbitrary geometric objects.

The simulation setup used wave port excitations, referencing ground return paths located 5.1 mils above and below the stripline trace. A 15ps TDR rising edge was used for the excitation source. The FEM mesh from the frequency domain solution was linked (applied) to the FEDGTD solution to ensure a highly accurate time domain solution.

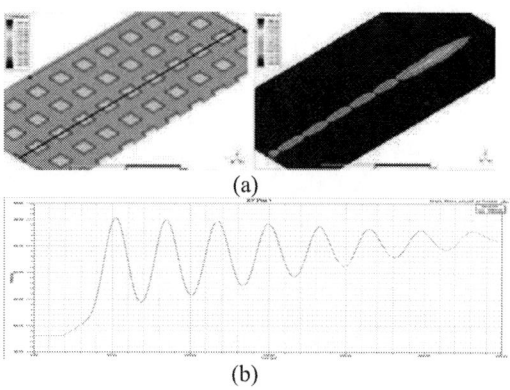

(a)

(b)

Figure 8: (a) E(t) propagation down a stripline trace with uniform, periodic glass weave in an FR4 epoxy. (b) Characteristic impedance as a function of time (propagation delay).

An example of a time domain Electric field propagation is shown in Figure 8.a) in which a reflected component can be seen related to the stop band behaviour. In addition to the time domain field visualization, resonances appear in the TDR results corresponding to the fiberglass weave effect causing an oscillation in the characteristic impedance, seen in figure 8.b). The period of oscillation corresponds to propagation delay of the unit cell.

8. Periodic Fiber Weave impact on system eye margins.

In this section we investigate the effect of periodic loading on system eye margins. For this purpose we created an HFSS model of a stripline trace with a simplified periodic structure of alternating materials. The materials were assigned relative dielectric constants of 3.5 and 6 to resemble resin and fiber weaves respectively. For this

comparison we arbitrarily selected a 60 mil period. ANSYS HFSS and Ansoft DesignerSI™ software was used to extract S-parameters and to compare time domain responses of a 16 Gb/s bit stream, with 15ps rise time, using eye diagrams. Figure 9 compares the impact of periodic loading in an ideal transmitter - channel interconnects - receiver setup. The channel interconnect length was set to six inches, by concatenating six s-parameter blocks of the relevant HFSS model. We observed up to 60mv and 4.25ps eye closure due to periodic loading compared to the non-periodic case and so it's clear that the periodic loading effects cannot be ignored at elevated data rates in the future.

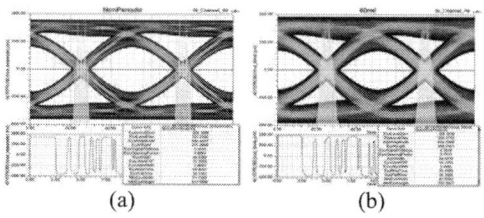

(a)   (b)

Figure 9: Comparing the effect of periodic loading on System Eye margins at 16GB/s for a 6 inch length: a) Eye diagram with homogeneous dielectric. b) Eye diagram with 60 mil pitch periodic loading.

9. Applications to PCB design

As signaling rates for system buses reach multi-Gigabits, homogeneity assumptions of PCB materials break down due to the size of woven fiberglass bundles. Advanced techniques [15], [12] of modeling a transmission line with a non-homogeneous medium comprehends and mitigates potential skew problems in high-speed designs. In addition to skew, as demonstrated in previous sections, deep resonances due to periodic loading at sub-millimeter wavelength frequencies can also degrade voltage and timing eye margins. Ideally, these high frequency resonances can be eliminated by homogenizing the material in exotic ways such as removing the periodic weave structure by layering glass fibers or by embedding randomly oriented glass whiskers in the resin propagating medium. However, these approaches are often cost prohibitive in a High Volume Manufacturing (HVM) setting.

Undesired high frequency resonances can be mitigated by optimizing pitch of the fiber weaves. Techniques such as altering dielectric properties or re-orienting trace routings in PCBs can help to optimize the design for a target signaling rate by pushing the resonance bands away from dominant harmonics of the fundamental frequency. With better understanding of the periodic behaviors and ability to precisely locate the resonances, current

258

PCB materials can be optimized for performance and cost.

We have formulated equation (24) to predict deep resonant frequency bands in terms of dielectric properties and periodic pitch geometry for PCB transmission lines. We have also demonstrated that this theory can also be applied to complex periodic structures by using equivalent dielectric properties of complex periodic structures such as elliptical fiber weave bundles. Once the periodic type is identified, designers can easily find the resonance regimes with reasonable accuracy for practical applications.

10. Conclusion

As the signaling rate and edge rates of system busses increase in concert with Moore's law, critical harmonics of fundamental signaling frequency will be pushed out to sub-millimeter wavelengths where signaling will suffer additional power losses due to the periodic nature of PCB materials. Resonances due to periodic loading can be theoretically predicted and the effects can be eliminated by data driven selection of material properties for target design frequency. Under situations where periodic materials cannot be avoided, time domain solution space must be verified in simulations with models generated from full wave simulations or by theoretically correcting non periodic models to comprehend periodic loss and skew effects.

Acknowledgements

The authors would like to acknowledge Matthew Commnens from ANSYS, Inc. for his invaluable insight into the FEDGTD method.

References

[1] Waves in Active and Passive Periodic Structures: A Review. CHARLES ELACHI, Proceedings of the IEEE, vol. 64, no. 12, December 1976.

[2] Brillouin, L., Wave Propagation in Periodic Structures. New York: Dover, 1953.

[3] Yeh, Pochi et. al., Electromagnetic Propagation in Periodic Stratified Media. I. General Theory. California Institute of Technology, Pasadena, California 91125.

[4] McMorrow, Scott, and Chris Heard. 2005. "The Impact of PCB Laminate Weave on the Electrical Performance of Differential Signaling at Multi-Gigabit Data Rates." Paper presented at the annual DesignCon Conference, Jan 31 – Feb 3, in Santa Clara, CA.

[5] Miller, Jason R. et. al., "Additional Trace Loss Due to Glass-Weave Periodic Loading." Design Con 2010, Santa Clara, CA.

[6] Romo, Gerardo. et. al., "Stack-Up and Routing Optimization by Understanding Micro-Scale PCB Effects." DesignCon 2011

[7] Jones, Christine Madden, "Measurement and Analysis of High Frequency Resonances in Printed Circuit Boards" (2010). Theses and Dissertations. Paper 197.

[8] "Fiber Weave Effect: Practical Impact Analysis and Mitigation Strategies," Jeff Loyer, Richard Kunze and Xiaoning Ye, Proceedings of DesignCon 2007.

[9] Field Theory of Guided Waves, 2nd Edition, Robert E. Collin. December 1990, Wiley-IEEE Press.

[10] H. Songoro, M. Vogel and Z. Cendes, "Keeping Time with Maxwell's Equations," IEEE Microwave Magazine, vol. 11, no. 2, pp. 42-49, April 2010.

[11] M. Commens, "Transient Electromagnetic Field Simulation with Finite Elements", DesignCon 2011, Santa Clara, CA.

[12] Advanced Signal Integrity for High-Speed Digital Designs S. Hall, and H. Heck, John Wiley & Sons, Hoboken, 2009.

[13] Leachy, Charles, "Waves in Active and Passive Periodic Structures: A Review." Proceedings of the IEEE, Vol. 64, No. 12, December 1976.

[14] Chrysostom, J. et. al., "Electromagnetic Wave Propagation in Multilayer Dielectric Periodic Structures." IEEE Transactions on Antennas and Propagation. Vol. 41 , Issue 10

[15] Paul G. Huray, The Foundations of Signal Integrity. John Wiley & Sons, Hoboken, 2010.

[16] Pochi Yeh, Optical Waves in Layered Media. John Wiley & Sons, New York, 1988

[17] Pytel, Steven G. and Paul G. Huray. 2006. "Oak Ridge National Laboratory (ONRL) Measurements HIML Proposal No. 2006-083: Analysis of Printed Circuit Board Copper Conductor and Insulation Material." Presentation given on Oct 24, 2006.

[18] Pytel, Steven G. 2007. Multi-Gigabit Data Signaling for PWBs Including Dielectric Losses and Effects of Surface Roughness. PhD diss., University of South Carolina.

DoE Simulations and Measurements with the microDAC Stress Chip for Material and Package Investigations

F. Schindler-Saefkow[1,3], F. Rost[3], A. Otto[3], S. Rzepka[3], B. Wunderle[2,3], B. Michel[3]

[1] AMIC Angewandte Micro-Messtechnik GmbH, Berlin; [2] Technische Universität Chemnitz;
[3] Fraunhofer ENAS, Micro Materials Center, Chemnitz;

Abstract

The in-situ detection of failures in microelectronic packages in an experiment is still a big challenge. The reliability of most packages will be qualified by measuring the elec-trical resistance of daisy chain structures. The moment of failure in the electrical sig-nals or the changes in the resistance are used for reliability or lifetime estimations. But the correlation of electrical resistance in the metallization and the packages or system reliability is very low. Extremely time-consuming investigation is needed to localize package failure after the experiment. Therefore, a chip, the MicroDAC stress chip, has been developed in a publicly funded project that is able to measure stress induced by thermo-mechanical loads. Different components of the stress tensor can be read out, as e.g. the in-plane stress difference and the in-plane shear stress on the chip surface within a 300 μm grid. This enables in-situ determination of the stress state even when the die is packaged and molded over. Residual stresses induced by processing steps as well as degradation within the materi-als or interfaces can thus be detected and measured. /1; 2/. A further advantage is the simple read out procedure which needs only four wire bond or flip-chip bump connec-tions. With this chip it is possible to get answers about what happened with the package during the temperature cycling tests. How fast is the failure growing from one cycle to the next and when is the failure mechanism changing in the experiment? What is the influence of vibration or moisture on the stress?

Key words: CMOS stress chip measurements, reliability, simulation, parameter identification, microDAC

Motivation

This paper shows some DoE investigations to see the main influences of mechanical loads and stress states in the chip. Geometric design (chip size, chip thickness, substrate and adhesive thickness), but processes (curing temperature) and material parameters (Young's modulus, CTE, glass transition temperature) are also investigated. These simulations are the basis for measurements with this chip. Furthermore, it is planned to combine the stress measurements in the chip with microDAC measurements (micro Deformation Analysis by means of Correlation) at the surface of the package, a method which is often used to very accurately measure deformations contactless and with high accuracy, to compare to the simulations.

The MicroDAC stress chip is assembled on a COB (Chip on Board) setup. A plain glob top surface is necessary for better MicroDAC results. So in the experiment the chip is embedded into the substrate with a cavity for the chip and the glob top.

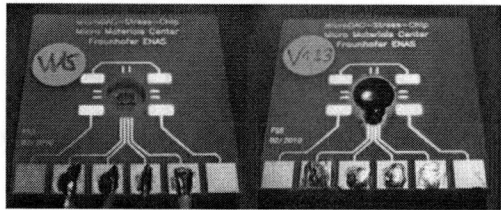

Fig. 1 MicroDAC stress chip setups with and without glob top

Fig. 1 shows the experiment setup with the substrate, the chip with the 4-wire bond, and the glob top that is filled into the chip cavity.

Simulations & boundary conditions

A very simple parameterized FEM model for a COB setup has been made. It consists of a substrate, an adhesive, the chip and a glob top (Fig. 2). The MicroDAC stress chip can measure the in-plane stress at the surface of the chip. So the stress in the chip will be read out in the post-processing of the FEM simulation. A DoE software (Design of Experiments) will calculate the model many times, each time with a different material and geometry setup.

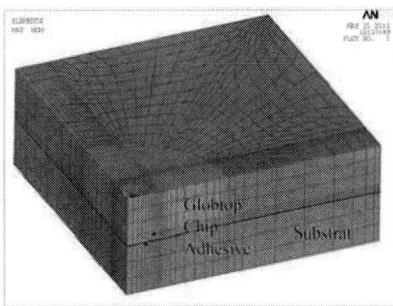

Fig. 2 The 1/4 FEM Model

A thermal mechanical simulation is realized with a heating-up load step from room temperature to a parameterized high temperature as boundary condition for this simulation. Further investigated parameters for the DoE simulation are: glob top young's modulus, glob top transition temperature, glob top depth, glob top radius, chip and adhesive length and width, chip thickness, adhesive thickness, substrate length, and substrate thickness. In this setup, it is possible to verify which parameter changes in the assembly can be detected by the MicroDAC stress chip measurements.

The ANSYS script will automatically read out the stress state of the MicroDAC stress chip and write the data into a file that can be read out by the DoE software. After calculation of a specific minimum of different parameter sets it is possible to build a Response Surface Model (RSM) to see the results' dependency on the input parameters.

DoE results

First investigations should show the influence of different glob top material properties on the stress state of the chip.

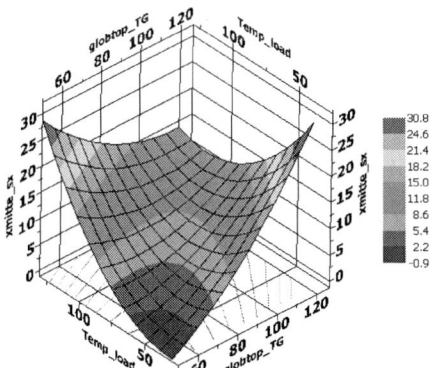

Fig. 3 Normal stress in x-direction in the middle of the chip dependent on the transition temperature and the heating temperature.

Fig. 3 show the stress state in the middle of the chip. This calculation shows that the stress state of the chip has a minimum if the temperature load is rising from a temperature below the transition temperature of the glob top material to a temperature above the transition temperature. This DoE simulation shows that it is possible to detect the glob

top transition temperature by changing the load temperature and find the minimum by measuring the stress state of the chip.

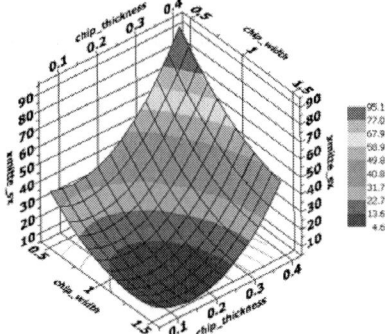

Fig. 4 Normal stress in X direction in the middle of the chip dependent on the chip width and thickness

An often asked question is about the influence of size on the chip. The behavior of the thermal mechanical system chip, adhesive, glob top and substrate should be detectable by the stress chip. Fig. 4 shows the mechanical stress dependent on chip width and thickness. All other described parameters are constant. The measurable mechanical stress varies from 10 to 90 MPa. The chip has an accuracy of about ±1 MPa /1/. A higher stress state will produce results with more accuracy.

microDAC deformation measurements

MicroDAC stress chip setups without glob top (Fig. 1) are investigated by microDAC measurements. The COB setup is heated up in the thermal load device. The plain surface of the glob top has advantages for the focus plan in high resolution microscope pictures. The microDAC measurements /3/ calculate the deformation and strain between two pictures at room temperature and 100 °C.

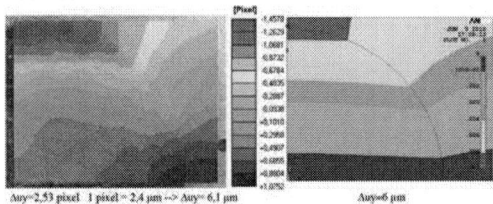

Fig. 5 Measurements and simulation of the x-deformation by comparison (6 μm in simulation and experiment)

Fig. 5 show a top view of a quarter of the glob top encapsulation. The simulation and the microDAC measurements show a deformation in y-direction of about 6 μm.

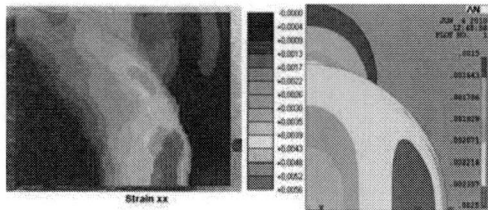

Fig. 6 Strain xx in measurements (max 5.6ppm) and simulation (2.5 ppm) by comparison

The simulation result of the strain in x-direction shows to be half of the value of the measurement. A possible reason for that difference is that the material parameter of the glob top material could not be very accurately verified at this setup. But Fig. 5 and Fig. 6 show these measurements with a more realistic outcome. Next steps are to combine the microDAC deformation measurements with the measurements of the mechanical state of the microDAC stress chip and to identify material parameters.

Stress measurements and simulation

The microDAC stress chip (Fig. 1) is read out under a thermal load. Every stress cell result is written in a text file. The stress results are loaded in Matlab to visualize the measurement results. This way, the results of hundreds of measurements can be visualized in minutes. That is an advantage for time period investigations, for example to visualize the stress development changes at 30 second intervals.

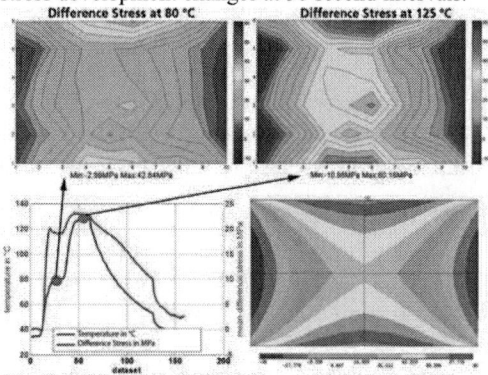

Fig. 7 In-plane difference stress in simulation (-30MPa - 30 MPa @ 135°C) and experiment (-14MPa - 52MPa)

Fig. 7 shows the in-plane difference stress that is developing by heating up the microDAC stress chip from room temperature to 80°C and 125°C. The 60 stress cells on the chip measure a stress field that is in good agreement with that of the simulation.

The stress development of the microDAC stress chip setup is investigated with a heating and cooling down temperature profile. Every stress cell can also measure the temperature. The diagram in Fig. 7 shows the temperature in one of the 60 stress cells on the chip and the difference stress measured in that cell. The diagrams show the result of 125

measurements in a time period of about 2 hours (every 30s). If the temperature is rising, the stress level is rising too.

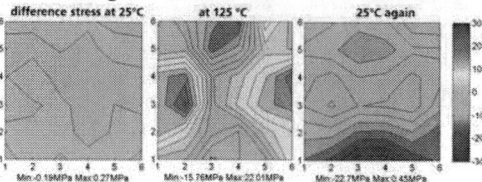

Fig. 8 Difference stress at room temperature - at 125 °C - and at room temperature again (-16 MPa -+22 MPa)

Fig. 8 shows the difference stress of the chip surface at the beginning at room temperature with no stress, at the high temperature load of 125 °C and the cooling down stress at room temperature again. It can clearly be seen that a stress will be induced by this temperature cycle. Further, it is interesting to see that the stress is not going back to the initial situation. Possible reasons for that could be strain hardenings in the glob top and/or in the adhesives above the transition temperature of the glob top. Further investigations have to be done to clarify this effect.

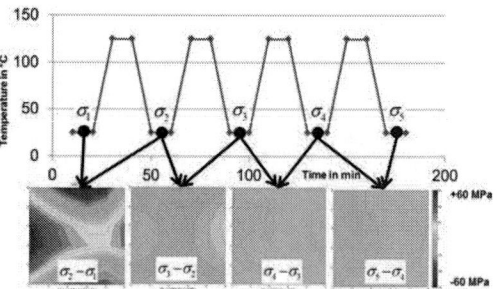

Fig. 9 Stress development between 2 room temperatures within temperature cycles

Fig. 9 shows the measured stress induced by each temperature cycle (TC). The adhesive at the beginning of TC1 is not cured. The first picture (σ1-σ2) shows the difference of the stress after and before TC1. So this picture shows the stress induced by the hardening and the shrinkage of the adhesive and by the CTE mismatch during the cooling down phase. Picture 2 (σ3-σ2) shows further induced stress. That could explain the not fully cured adhesive within TC1. In the additional temperature cycle 2 and 3 nearly no more stress is induced into the microDAC stress chip. The adhesive is fully cured and in two measurements at the same temperature the CTE mismatch has no influence.

Parameter identification

The aim of this verification is to identify certain simulation parameters of materials with the help of measurement results of the microDAC stress chip, which deals with the identification of the mechanical stress on a chip surface of a Chip-on-Board package. Therefore an optimization is needed, which adapts the unknown simulation parameters so long as the difference between measurement and FEM result is small enough.

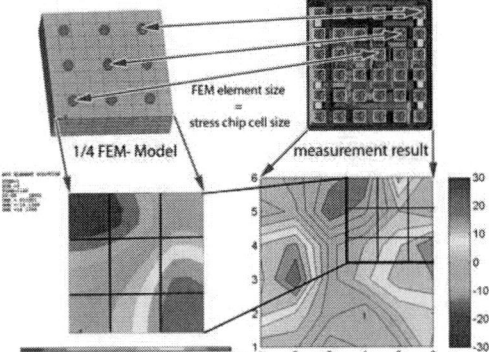

Fig. 10 ¼ Model FEM result in comparison to the microDAC stress chip result

Fig. 10 shows the FEM result of on the chip surface that will be compared with real stress chip measurements. In consideration of the not absolute symmetric result in the measurements, a mean value of all 4 quadrants of the stress chip results will be calculated. This can be compared with the ¼ Model result of the simulation.

Optimization

To present the gap between simulation and experiment result, a failure-factor is needed, which can be minimized later with the FEM tool. Therefore a failure-factor F is defined, which detects the difference of stress of nine certain places of the chip surface and calculates the sum of their single error-square.

$$F = \sum_{i=1}^{9} \left(\sigma_{i,Sim} - \sigma_{i,Exp} \right)^2$$

This F presents now an aberration factor, which converges to zero, if the experiment and simulation are identical.

In the first step, the practical uses of the optimization will beproven. For this, the stress results of the experiment ($\sigma_{i,Exp}$) will define equal to the simulation results. Like this, on the one side the value for F is zero and on the other side all parameters are known.

Now the values of parameters of CTE and Young's Modulus of the adhesive will be defined in ranges, like this they are variable. They are now the input arguments of the optimization.

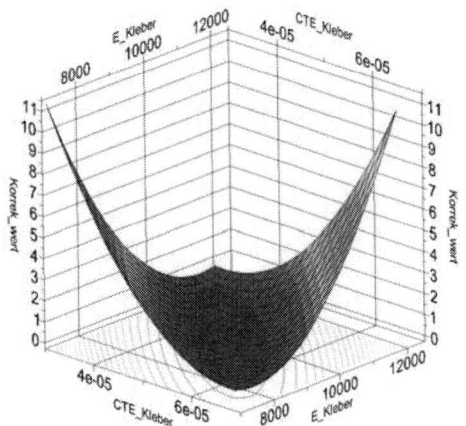

Fig. 11Response surface of the failure factor F, depending on CTE and Young's Modulus of the glue material

Fig. 11 shows the response surface of about 100 FEM simulation results. At the minimum of the response surface, the failure factor F is nearly zero and the adhesive material parameters are identified. With this method it is possible to identify booth parameters with an accuracy of about 5 %. If only one parameter is chosen for identification, the accuracy goes down to about 0.1%.

Additionally, the convergence graph in Fig. 12 shows the working method of the optimization algorithm.

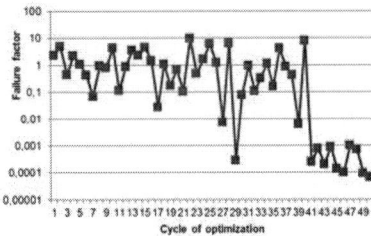

Fig. 12 Convergence graph of the optimization

Therefore this algorithm of optimization reaches a very high accuracy and proves the practicability of this method.

Material parameter identification with real microDAC stress chip results

Once the method is proofed, the identification of parameter can be used for a real application. Values of the experiment are using as the initial boundaries for the optimization and the young's modulus of the adhesive represents the input parameter in a range of 3000 MPa – 10000 MPa, which should get identified.

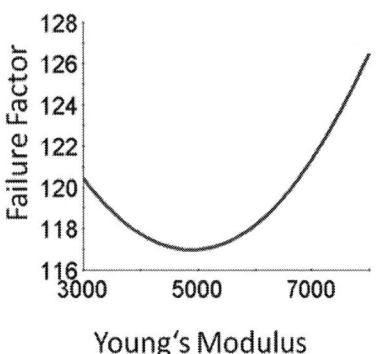

Fig. 5 Model of Young's modulus identification

Fig. 5 shows the first results of the identification. The minimum of the failure factor locates at a Young's modulus of the adhesive about 4900 MPa.

Furthermore, the value of the failure factor reaches a minimum at 116. This represents, how big the gap is between the simulation model and reality. But, this result doesn't contain the temperature dependency behavior of the adhesive material and we had only one measurement result for this setup. Therefore, this result should be used with caution.

But it shows the complexity of the field. Further investigations have to be done to use the stress values of the experiments in a better way, where the identification of the material parameters with a higher accuracy is possible.

Conclusion

It has been shown that the microDAC stress chip package setup is useful to combine stress and microDAC deformation measurements. The microDAC stress chip can read out stress changes induced by the load. This method makes it possible to measure and visualize effects that have never been visible before.

With this chip, it is possible to measure the real stress development in the package within a temperature load. Many other applications are also possible. For example, the stress development within the hardening process of a glob top, molding compound, adhesive or other embedding technology steps.

It then presents a future challenge to combine the stress measurements and the microDAC results with the DoE simulation to identify material parameters (e.g. shrinkage) or to optimize the package setup for more reliable packages.

References

[1] H. Kittel, Novel stress measurement system for evaluation of package induced stress, Robert Bosch GmbH, SMART SYSTEMS INTEGRATION Barcelona, Spain 9 –10 April 2008

[2] F. Schindler-Saefkow, Package Induced Stress Simulation and Experimental Verification,

Fraunhofer IZM, SMART SYSTEMS INTEGRATION Barcelona, Spain 9 –10 April 2008

[3] J. Keller, D. Vogel, A. Schubert, and B. Michel: Displacement and strain field measurements from SPM images. In B. Bhushan, H. Fuchs, and S. Hosaka, editors, *Applied Scanning Probe Methods*, pp. 253–276, Springer, 2004.

Contact:

Florian Schindler-Saefkow,

c/o Micro Materials Center at Fraunhofer ENAS,

Technologie-Campus 3, 09126 Chemnitz, Germany

e-mail: florian.schindler-saefkow@enas.fraunhofer.de

Advancements in Fracture and Failure Simulation for Electronic Package Applications

D. Cadge, D. Reid, and S. Krishna

SIMULIA, 166 Valley Street, Providence, RI 02909, USA

+1-401-276-4400 and david.cadge@3ds.com

Abstract

Advancements in simulation technology over the last several years now make it much easier for electronics engineers to solve the types of fracture and failure problems common in this industry, whilst also enabling the solution of problems that were previously not possible.

This paper will describe the eXtended Finite Element Method (XFEM) and its application to electronic industry workflows. XFEM is an extension of the conventional finite element method which uses special purpose element formulations to capture the presence of discontinuities without the requiring the mesh to match the geometry of those discontinuities. It is therefore a very attractive and effective way to simulate initiation and propagation of a discrete crack along an arbitrary, solution-dependent path without the requirement of remeshing in the bulk materials. Furthermore, XFEM can be combined with surface-based failure mechanisms, which are best suited for modeling interfacial delamination, to simulate cracks that run along interfaces and then may break into the bulk material, and vice versa.

Real world customer examples will be used to highlight the advantages of this new technology, including: ease of use for building curved crack fronts in complex, curved geometries; solution-dependent crack propagation, not requiring apriori crack path definition; cyclic loading for fracture mechanics; the global/local modeling approach; and, co-simulation of implicit and explicit dynamics to enable XFEM for drop test applications.

Key words: Fracture, Failure, XFEM, Abaqus

Introduction

Fracture and failure simulation is an important topic in the electronics industry. Consumers are driving development of smaller devices, with more capabilities and more power, causing larger thermal stresses on ever smaller interconnects. Many modern handheld devices have full-size touch-screens made up of large glass panels. These small interconnects, large glass panels, and other small components are all susceptible to failure when undergoing thermal loading, cyclic loading, drop events, etc.

Modeling stationary a crack with the conventional finite element method requires a mesh that conforms to the geometric discontinuity. Also, considerable mesh refinement is needed in the neighborhood of the crack tip to capture the singular asymptotic fields adequately. Modeling a growing crack is even more cumbersome because the mesh must be updated continuously to match the geometry of the discontinuity as the crack progresses.

The extended finite element method (XFEM) alleviates the shortcomings associated with meshing crack surfaces. In this study, the XFEM implementation in Abaqus [1] is applied to several typical electronics modeling scenarios. The basis of this method is to allow mesh-independent crack initiation and propagation. Study of stationary cracks can also be performed to compare to traditional fracture mechanics methods. The XFEM technique greatly simplifies the mesh generation requirements for models with pre-existing cracks, whilst also allowing for solution-dependent crack initiation and propagation; so the user does not need to pre-define where a crack may occur or where it might propagate before running the simulation.

Extended Finite Element Method (XFEM)

The extended finite element method was first introduced by Belytschko and Black [2]. It is an extension of the conventional finite element method based on the concept of partition of unity by Melenk and Babuska [3], which allows local enrichment functions to be easily incorporated into a finite element approximation. The presence of discontinuities is enabled by the use of special enrichment functions in conjunction with additional degrees of freedom. The finite element framework and its properties, such as sparsity and symmetry, are retained.

XFEM allows the presence of discontinuities in an element by enriching degrees of freedom with special displacement functions,

$$u = \sum_{I=1}^{N} N_I(x)[\boldsymbol{u}_I + H(x)\boldsymbol{a}_I + \sum_{\alpha=1}^{4} F_\alpha(x)\boldsymbol{b}_I^\alpha$$

where $N_I(x)$ are the usual nodal shape functions; the first term on the right-hand side of the above equation, $\boldsymbol{u}_I$, is the usual nodal displacement vector associated with the continuous part of the finite element solution; the second term is the product of the nodal enriched degree of freedom vector, $\boldsymbol{a}_I$, and the associated discontinuous jump function $H(x)$ across the crack surfaces; and the third term is the product of the nodal enriched degree of freedom vector, $\boldsymbol{b}_I^\alpha$, and the associated elastic asymptotic crack-tip functions, $F_\alpha(x)$. The first term on the right-hand side is applicable to all the nodes in the model; the second term is valid for nodes whose shape function support is cut by the crack interior; and the third term is used only for nodes whose shape function support is cut by the crack tip.

The level set method is a numerical technique for describing a crack and tracking the motion of the crack. It couples naturally with XFEM and makes it possible to model 3D arbitrary crack growth without the need for remeshing. The method uses signed distance functions to describe the crack geometry, so that no explicit representation of the crack is needed, and the crack is entirely described by nodal data.

Two level sets are required to represent a single crack, as shown in Figure 1

- The first, Φ (phi), describes the crack surface,
- The second, Ψ (psi), is constructed so that the intersection of two level sets gives the crack front

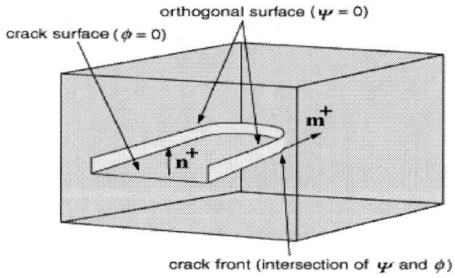

Figure 1: Representation of a non-planar crack

For propagating cracks, elements are allowed to split, and damage initiation and evolution laws control when and where these cracks occur. Damage initiation can be based on the stress or strain state in an element, linear elastic traction-separation, or a user-defined damage initiation criterion. Damage evolution is based on the energy release.

Example 1: J-Integral Calculation

This example follows on from the work done by Wang et al [4] to demonstrate that XFEM is capable of producing similar J-integral results to the traditional fracture mechanics approach. This work investigates a curved crack front in a solder joint, so in curved geometry. Building a traditional finite element mesh for this model is a challenging and time consuming process; a focused crack tip mesh is needed, the elements at the crack tip need to be collapsed and their mid-side nodes moved to the quarter-points. In the work presented by Wang et al, the focused mesh for the crack tip had to be built independently of the rest of the model and then tied to the remaining mesh for the solder joint due to these complexities; this mesh can be seen in Figure 2.

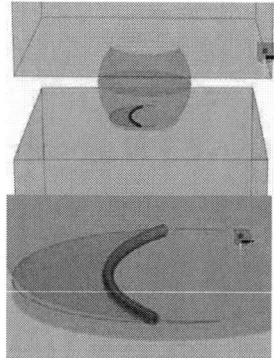

Figure 2: Crack tip mesh for traditional approach

This model is actually a submodel, being automatically driven from a larger, global model of a PCB undergoing a standard mechanical shock test (JEDEC JESD22B111 [5]). The quarter symmetry PCB global model can be seen in Figure 3, along with the location of the submodel. Boundary conditions for the submodel are automatically transferred from the global model, without the need for the meshes or the solution time increments to match between the two simulations.

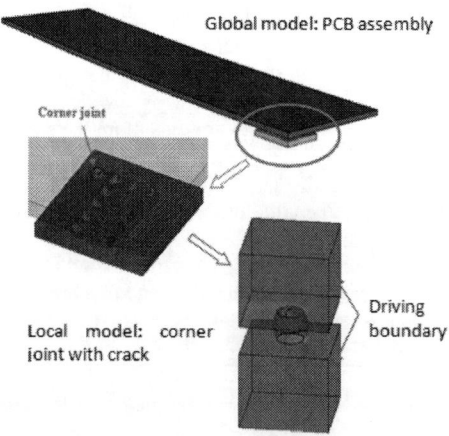

Figure 3: Global-local models of PCB, package, and solder joint

The crack in the equivalent XFEM submodel can be seen in Figure 4. This plot is a contour of the second level set value Ψ on the surface defined by the first level set value Φ. So, the colored region is the crack front, which shows that the mesh does not need to conform to the crack front. Additionally, the crack plane on which this plot is made is part way through a set of elements and so is also independent of the underlying mesh. This mesh independence makes model generation and meshing much quicker with the XFEM approach; no special techniques are needed and any general-purpose meshing tool can be used. The XFEM approach is particularly efficient if several crack lengths need to be simulated. With the traditional approach, each new crack length will require a completely new mesh, with each taking a significant time to generate. With the XFEM approach, the same mesh can be used for all crack lengths, only the initial crack location need to be updated, providing potentially large time savings.

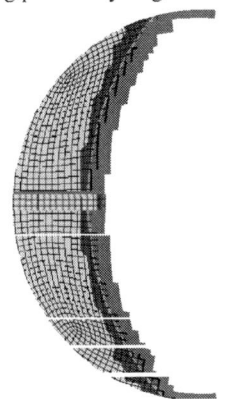

Figure 4: Crack front in the XFEM model

Figure 5 shows a comparison of the J-Integral values for the fifth contour between the traditional approach and the XFEM model.

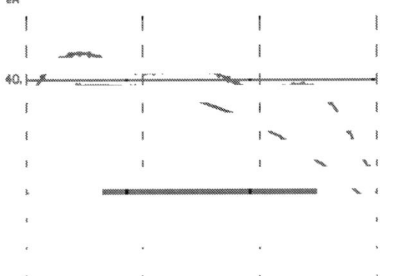

Figure 5: J-Integral results

These results show reasonable agreement and could be used to make engineering judgments on the likelihood of crack propagation. With the given mesh, it was necessary to apply a smoothing to the J-Integral values from the XFEM model. It is believed that using a more refined mesh for the XFEM model would overcome this issue. The major advantage that XFEM offers for this model is in the reduced time taken for mesh generation.

Example 2: Crack Propagation Through Low Cycle Fatigue

This example demonstrates the use of XFEM for crack propagation during solder joint fatigue in a 100-pin plastic quad flat pack (PQFP) [6]. The applied load for this model is a sub-critical temperature cycle. The mismatch of thermal expansion coefficients between the various materials in the package causes stress reversals and thereby fatigue failure eventual occurs in the solder material.

The onset and growth of the fatigue crack are characterized using the Paris law, which relates the relative fracture energy release rate to crack growth rate. XFEM is used to initiate and then propagate a crack in the solder material as the assembly is subjected to 800 temperature cycles. The low cycle fatigue solution procedure, available in Abaqus, provides a very efficient way of simulating this number of load cycles without having to solve for each one; automatic damage extrapolation is included within this procedure.

The model for this example can be seen in Figure 6. The solder joints and leads near the corner of the PQFP experience the highest deformations since they are farthest from the center of the assembly. Because failure is most likely to occur in this area, these joints and leads are more finely meshed so that sufficiently accurate stresses and strains can be calculated. The corner solders and leads are meshed with a total of 7296 first-order brick and wedge elements. The remainder of the model is discretized with 4121 elements. The mesh incompatibility at the refined bond pad/PCB and lead/chip interfaces is accommodated with surface-based tie constraints.

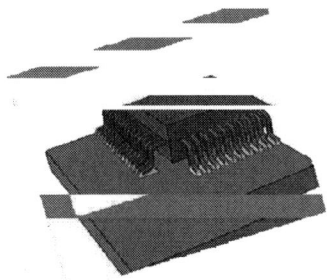

Figure 6: Quarter model of PQFP

Figure 7 shows the location of crack initiation and then the propagated crack after completing the 800 load cycles.

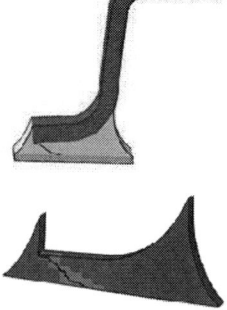

Figure 7: Propagated crack in the solder of the corner lead

Figure 8 shows a photograph from a physical test that was published by Delphi [7] and, at least qualitatively, this shows a similar crack initiation location and propagation path.

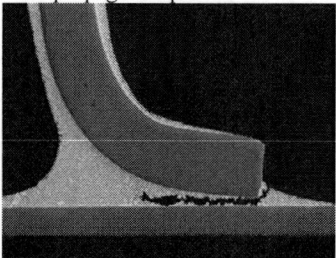

Figure 8: Cross-section of a leaded IC package after thermal cycling

The advantage of XFEM for this type of load scenario and model response is that there does not need to be an initial crack or flaw in the model (XFEM can capture crack initiation) and the propagation of the crack is independent of the mesh (XFEM does not require that the crack propagate along a pre-defined path or along element boundaries).

Example 3: Drop Test With Co-Simulation

This example looks at the potential failure of a solder joint in a BGA on the circuit board inside a cordless mouse undergoing a 1-meter drop test on to a hard floor [8]. Abaqus/Explicit has been used extensively to examine the behavior of electronic devices experiencing mechanical shock and drop loading [9,10,11]. Accurately capturing crack initiation and propagation requires extremely fine meshes, and such small elements would drive down the stable time increment for the explicit dynamic solution method, leading to long simulation runtimes.

Co-simulation can be used to solve part of a model with explicit dynamics and part with implicit dynamics. The implicit dynamic solution method is unconditionally stable and so does not have the same stable time increment limitations as the explicit dynamic method. So, an implicit dynamic solution can be used for the fine meshed, detailed region of the PCB where failure might be expected to occur. Also, XFEM is currently only available with the implicit solvers in Abaqus. An explicit dynamic solution efficiently handles the large amount of contact and large deformations typically present in a drop event. Co-simulation automatically couples the two regions of this model and the two solvers together, facilitating the load transfer across the interface and synchronizing the time incrementation as needed during the solution.

The model for the drop test can be seen in Figure 9, with the PCB showing the chip, to be solved in implicit dynamics with XFEM, in the BGA in Figure 10.

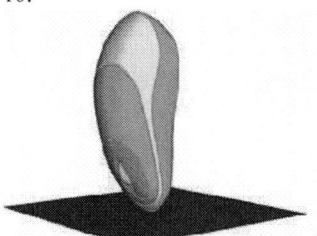

Figure 9: Cordless mouse drop model

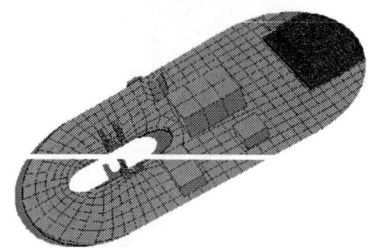

Figure 10: PCB with refined chip for implicit dynamic (in red)

The results of the drop test can be seen in Figure 11, showing that the cases of the mouse unclip and become separated during the drop event.

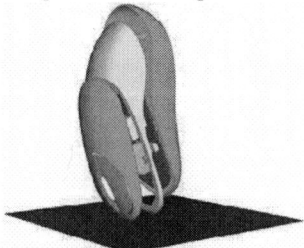

Figure 11: Drop test results showing case separation

The response of the implicit dynamic portion of the model can be seen in Figures 12 and 13, showing the stress in the chip and the locations of failure at the upper and lower interfaces of the corner joints in the BGA.

Figure 12: Mises stress in the chip

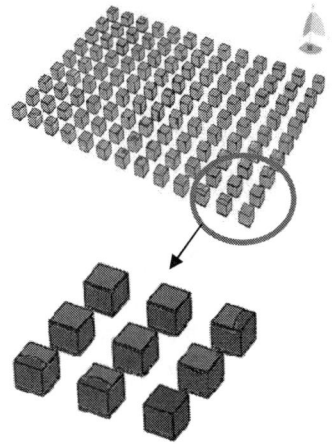

Figure 13: Mises stress in the BGA with the failed elements (red) at the corner joint interfaces

The use of co-simulation in this example enables detailed failure modeling to be performed in a drop test simulation, requiring a much finer mesh that would usually be practical in an explicit dynamic simulation. Previously this type of solution might have been performed using submodeling, but with this approach there is only a one-way coupling as the results of the global model are used to drive the submodel in a separate simulation. With co-simulation there is real-time, two-way coupling between the refined region of interest and the rest of the model; both are run as a single simulation. Co-simulation also enables the use of XFEM, which is currently only implemented in the implicit solver within Abaqus, hence allowing mesh-independent failure initiation and propagation to be investigated.

Conclusions

The XFEM technique implemented in Abaqus can be used to predict failure in electronic components. The use of XFEM provides significant time savings when building models with pre-existing cracks. XFEM also enables solution-dependent crack initiation and propagation, without the need to pre-define where the crack location or path. XFEM can be used with the low cycle fatigue procedure in

Abaqus to investigate crack propagation caused by any cyclic loading conditions. The sub-modeling technique allows for very detailed fracture mechanics simulation in small parts, such as solder joints. A global model can predict the critical location or joint but does not contain the resolution to predict the precise location of crack initiation and then crack propagation. The co-simulation interface in Abaqus enables the use of XFEM in an implicit portion of a model to be included within an explicit dynamic simulation, such a drop test.

References

[1] Abaqus User Manual, version 6.11, 2011.

[2] T. Belytschko and T. Black, "Elastic Crack Growth in Finite Elements with Minimal Remeshing," International Journal for Numerical Methods in Engineering, vol. 45, pp. 601–620, 1999.

[3] J. Melenk and I. Babuska, "The Partition of Unity Finite Element Method: Basic Theory and Applications," Computer Methods in Applied Mechanics and Engineering, vol. 39, pp. 289–314, 1996.

[4] Wang, Seah, Wong, and Cadge "Fracture Mechanics Study of Fatigue Crack Growth in Solder Joints Under Drop Impact," ECTC 2008

[5] JEDEC Standard JESD22-B111, "Board Level Drop Test Method of Components for Handheld Electronic Products", July 2003, JEDEC Solid State Technology Association.

[6] "Low-Cycle Thermal Fatigue of a Surface-Mount Electronics Assembly," Abaqus Technology Brief, Apr-2009

[7] K. Donovan and M. Chengalva "It's All in the Solder Joints" SIMULIA INSIGHTS magazine, Oct-2007

[8] "Drop Test Simulation of a Cordless Mouse," Abaqus Technology Brief, Apr-2007

[9] M. Theman and P. Poskiparta, "Evolution of Mobile Phone Drop Test Simulation at Nokia," SIMULIA SCC-2005

[10] S. Thiruppukuzhi, "Probabilistic Simulation Applications in Product Design at Motorola," SIMULIA SCC-2008

[11] M. Takashi, "Digital camera fall impact analysis at Olympus," SIMULIA Japan RUM-200

New Improvements in Thermal Management: Thick Print Copper Thick Film as a Replacement for Direct Bond Copper

Sarah Groman[1], Tracey Smolinsky[1], Jim Williams[2], Cetin Toy[2]

[1]Heraeus Materials Technology LLC, Thick Film Materials Division, 24 Union Hill Road, West Conshohocken, Pennsylvania 19428 Ph: (610)825-6050, Fax: (610)825-7061, Sarah.Groman@heraeus.com

[2]Marlow Industries Inc., Vista Park Road, Dallas, Texas, 75238, Ph: (214)340-4900, CToy@Marlow.com

Abstract

Direct Bond Copper (DBC) is a widely proven process used in areas such as high power handling applications where thermal management is a priority. The benefits of DBC include excellent thermal conductivity and high current carrying capability, as well as good mechanical strength and adhesion. However, there are several drawbacks to traditional DBC technology. These include thermal cycling performance, minimum copper foil thickness, and pattern limitations due to the fact that the foil must be etched, which is a relatively expensive subtractive process. With new applications such as Concentrated Photovoltaics (CPV), High Brightness LED Packaging, and Insulated Gate Bipolar Transistors (IGBT) a new material solution is needed for high power electronics. Thick film copper can offer improvements over DBC by improving thermal cycling performance, as well as filling a gap where a thickness of less than 150μm is difficult to achieve with the standard processing procedures of DBC. Tighter resolutions are easier to accomplish with the thick film Cu approach enabling the use of a thinner layer of copper for more versatility, and improving line resolution to allow increased design flexibility.

This paper examines a direct comparison of DBC to Thick Print Cu on the substrates BeO and alumina. Not only are there the advantages in thickness, lifetime testing, and pattern guidelines, there is the distinct cost-savings advantage of applying an additive process for thick print Cu. These results are compared and show improvements that have been made in optimizing the performance of Thick Print Cu.

Key Words: Thick Film, DBC, Thermal Management

Introduction

Direct Bond Copper (DBC) technology is found in applications where high power and high electrical current are needed in various power circuits. DBC became such an attractive choice because of the lower electrical resistivity and good thermal conductivity of copper compared to traditional refractory metallization that was developed in the early 1950's. DBC has been well known in the power electronics industry since the early 1970's when Cu foil was first bonded to alumina (Al_2O_3) substrates. This technology continued to advance and in the 1980's the copper foil was able to bond onto other substrates including aluminum nitride (AlN) and beryllium oxide (BeO).

DBC is formed because of the Cu-O eutectic bond which is formed at 1065°C when pre-oxidized copper foil is fired against an alumina based oxide surface in a nitrogen atmosphere where there are tightly controlled heating and cooling cycles. A layer of cuprous oxide forms on the surface of the metal that wets the oxide ceramic surfaces and forms the bond between the copper foil and the ceramic substrate. Tight control of the amount of gaseous

species that is allowed to form at the bond interface is crucial because it directly affects the overall yield and product quality in the form of voids at the bond interface. The area of DBC and voiding has a parallel relationship, as the area of the pattern is increased, the amount of voiding at the bond interface is also increased. There are two world wide recognized approaches for DBC metalized ceramics:

1. Near net shape formation of circuit traces from pre-stamped metalized foils with smaller cross-section. (Process A)
2. Bonding of large foils followed by conventional lithographic methods for conductor trace formation. (Process B)

There are several engineering approaches to overcome void formation; however this is still the dominant factor for yield losses in manufacturing.

As the industry continues to advance manufacturing requirements are made tighter and designs made smaller and thinner to reduce the cost of power circuits. With new applications such as Concentrated Photovoltaics (CPV), High Brightness LED Packaging, and Insulated Gate Bipolar Transistors (IGBT) a new material solution is needed for high power electronics. Thick film copper printing is a way to obtain difficult specifications that traditional DBC cannot accomplish such as, thinner print thickness, increased design flexibility, and an easier additive process. This paper examines a direct comparison of the two methods: DBC and thick print thick film copper.

Metallizations

Three different metallizations were tested for comparison. The first is C7300 a thick print copper paste. C7300 is a screen printable copper thick film paste that can print to a fired film thickness ranging from 25μm- 300μm depending on processing conditions. The second metallization is another thick film copper system that is plateable. The system uses C7401 on the bottom as an adhesion layer with C7300 printed on top; the system is plated using electroless nickel and immersion Au (ENIG Plating). This combination has the same flexibility in thickness parameters. The third and last type of metallization is DBC produced with process B. This paper examines both thick film copper metallization systems at three different thicknesses: 100μm, 200μm and 300μm, and compares to DBC ceramic with 300μm copper film. All materials were processed according to standard operating procedures.

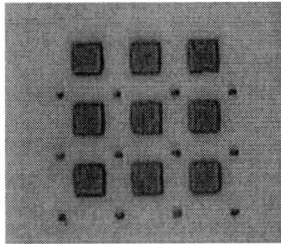

Figure 1: The test samples that were prepared followed the schematic shown above.

Experimental

There were six characteristics evaluated on each of the samples: thickness, solderability, adhesion and aged adhesion, thermal cycling, SEM analysis of the microstructure, and cost. These characteristics were chosen due to their importance in processing and performance of the material.

Thick film printing was completed with three different final fired film thicknesses: 100μm, 200μm and 300μm respectively. The thickness was achieved by obtaining a fired print thickness for each pass of copper of 100 μm. Thus, a 100 μm fired deposit required one print/fire step, while the 300 μm fired copper required three print/fire steps. The C7401/C7300 system required a thinner initial print layer. DBC is typically used for power electronics applications, which usually warrant a much thicker lay-down of material. The thicknesses tested push the limits of the thick print copper up to the 300μm limit to evaluate performance across full

thickness spectrum. The DBC was tested only at 300μm thickness. Thicknesses were measured using a CyberScan laser profilometer as well as the Mitutoyo Digimatic Indicator with a 0.51mm diameter flat tip. Thickness measurement results are shown in Tables I and II. Thickness measurement results on DBC is shown on Table III.

Solderability testing was completed on each of the samples to test the wetting characteristics of the metallurgies. This was completed by printing two different types of solder paste, eutectic Sn/Pb solder with a melting point of 183°C and eutectic Sn/Cu/Ni RoHS compliant solder with a melting point of 227°C. The solder was printed using a stencil with a 2mm diameter opening and then reflowed in a nitrogen atmosphere at the reflow profiles recommended for the specific alloys. Figure 2 shows an example of the solder spread on the C7300 metalized surface. Solderability assessment was done simply by measuring the diameter of the spread of the solder paste and then using the reflowed diameter of the DBC "as fired" surface as the base line diameter with a value of 1. Then this diameter was used in determining the spread on the other surfaces.

In Figure 3 the solderability spread is shown on the different surfaces - "as fired" and "acid cleaned"

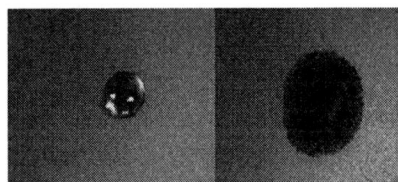

Figure 2: Two different results from the solder paste reflow test are shown. On the left is a reflow test that shows very little spreading of the paste as printed, while on the right is the surface cleaned of oxidation and shows much better spread of the solder.

Adhesion and aged adhesion results were obtained by using the Ametek 500 test stand with the Mark-10 Moded EG10 gage. All wires were attached by dipping the samples in a 63Sn/37Pb solder pot. The flux used was Rosin Flux. The adhesion test samples did require a light sand with sandpaper to remove the peaks of the print. The samples were also mounted on a thicker substrate to reduce breakage during adhesion testing. The wires were attached prior to aging. The parts were aged in air at 150°C for 100 hours and 300 hours. Several pulls were made for each adhesion test condition and averages were reported in following figures. Figures 4A through 4D shows the aged adhesion for each copper/substrate material system. Adhesion was also measured after 100 thermal cycles. Figures 5A through 5D shows the thermal cycling adhesion for each copper/substrate material system.

Thermal cycling was completed using a ESPEC model BTZ-175. The parts were cycled from -40°C to 125°C with a full cycle taking 120 minutes. Sonoscan using the C-Mode Scanning Acoustic Microscope (C-SAM®) at an operating frequency of 75 MHz examined parts before and after the thermal cycling. The purpose of this inspection being to evaluate the bond quality between the metal plate and the ceramic plate in which it was mounted to ensure that any voiding or separation that was found in the part was not there prior to thermal cycling. Acoustic Micrograph pictures were generated for each sample using the Interface Scan Technique. Parts were taken out and examined by Sonoscan at 50 cycles, 100 cycles, 500 cycles and 1000 cycles for Alumina and 50 cycles, 100 cycles, and 500 cycles for BeO. 150μm fired film thickness were used in the Sonoscan evaluation.

Scanning Electron Microscopy (SEM) analysis was performed on each of the parts to examine the cross-sectional microstructure of each type of metallization.

Cost analysis was done comparing the DBC to thick film copper paste systems. This was compared by evaluating the results of this paper as well as the comparative cost of the process and material from each type of metallization.

Results

Thickness

Table I: The thicknesses average and standard deviation results are shown in Table I below. These thicknesses are of the thick film material printed on Alumina substrates in the table on the left, where the table on the right shows the printed and then plated thick film material.

Alumina							
C7300 Thick Copper				C7401/C7300 Thick Copper			
	100μm	200μm	300μm		100μm	200μm	300μm
Average	103μm	187.5μm	302μm	Average	111μm	195.5μm	276μm
Std Dev	10.5	11.2	28.5	Std Dev	7.75	11.5	11

Table II: The thicknesses average and standard deviation results are shown in Table II below. These thicknesses are of the printed thick film material printed on BeO substrates in the table on the left, where the table on the right shows the printed and then plated thick film material.

BeO							
C7300 Thick Copper				C7401/C7300 Thick Copper			
	100μm	200μm	300μm		100μm	200μm	300μm
Average	94μm	207.7μm	307.5μm	Average	112.7μm	186μm	233.5μm
Std Dev	9.75	20.7	41.7	Std Dev	7.5	12.7	11.25

Table III: The thicknesses average and standard deviation results for DBC are shown in Table III below. The thicknesses shown are for DBC Etched Copper and DBC Pre-Stamped Copper.

DBC			
DBC – Etched Copper (Process B)		DBC – Pre-stamped Copper (Process A)	
	300μm		250μm
Average	291.1μm	Average	232.0μm
Std Dev	7.9	Std Dev	7.8

Solderability

The solderability of the copper surfaces when cleaned show the same or better wetting results with thick film as DBC. If C7301 a fritless solderable paste is used as the top layer the solderability is better than DBC in the "as fired" condition.

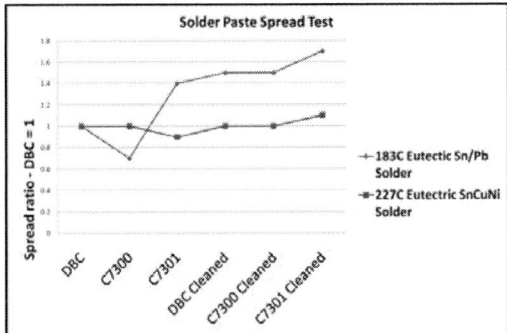

Figure 3: Shows the Solder Paste Spread Test done with both solder types on each of the tested metallurgies.

Adhesion

Direct comparison of adhesion was done only on Al_2O_3. The results show that at 200µm and 300µm thick film print targets using the C7401/C7300 paste system, that the initial and aged adhesion on Al_2O_3 was the same as on DBC. The adhesion of the C7300 system had low aged adhesion on Al_2O_3 and low initial and aged adhesion on BeO.

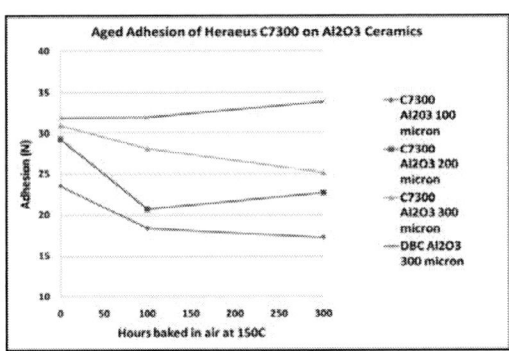

Figure 4A: Shows the aged adhesion of C7300 on Al_2O_3 ceramic at the various thicknesses compared to DBC on Alumina.

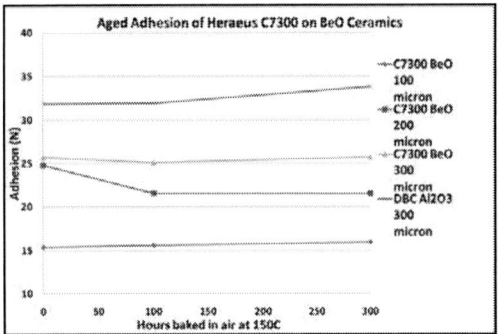

Figure 4B: Shows the aged adhesion of C7300 on BeO ceramic at the various thicknesses compared to DBC on Al_2O_3at a thickness of 300µm.

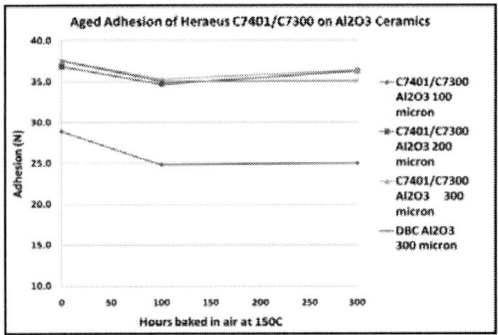

Figure 4C: Shows the aged adhesion of C7401/C7300 on Al_2O_3 ceramic compared to DBC on Al_2O_3

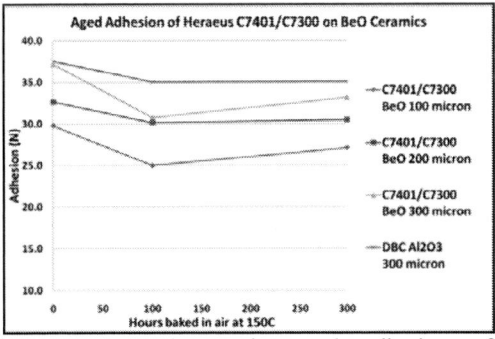

Figure 4D: Shows the aged adhesion of C7401/C7300 on BeO ceramic compared to DBC on Al_2O_3.

Thermal Cycling

The adhesion testing after 100 cycles show a significant difference in the drop in adhesion for the thick film system C7300 on Al_2O_3 as compared to the DBC. In general the thicker the print, which has multiple firings, has higher adhesion.

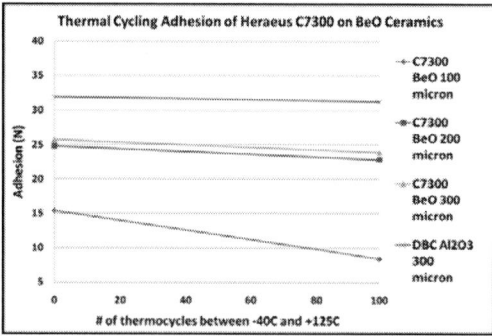

Figure 5A: Shows the adhesion of C7300 on BeO after 100 thermal cycles at the various thicknesses compared to DBC on Al_2O_3 at 300µm after 100 thermal cycles.

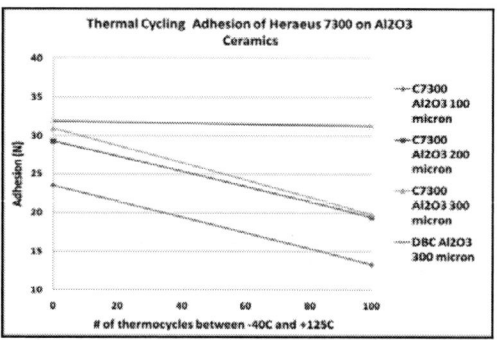

Figure 5B: Shows the adhesion of C7300 on Al_2O_3 after 100 thermal cycles at the various thicknesses compared to DBC after 100 thermal cycles on Al_2O_3.

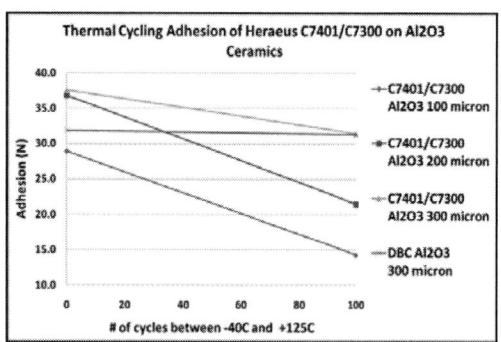

Figure 5C: Shows the adhesion of C7401 and C7300 on Al_2O_3 after 100 thermal cycles at the various thicknesses compared to DBC on Al_2O_3 after 100 thermal cycles at 300µm.

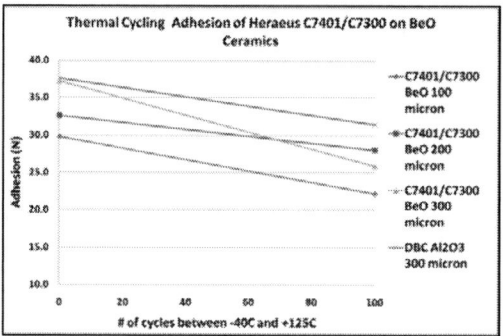

Figure 5D: Shows the adhesion of C7401/C7300 on BeO after 100 thermal cycles at the various thicknesses compared to DBC on Al_2O_3 at 300µm after 100 thermal cycles.

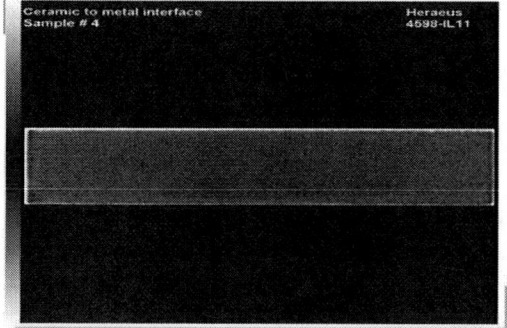

Figure 5E: Shows a C-SAM done on C7300 printed on Alumina before thermal cycling

Figure 5F: Shows a C-SAM done on the same part as Figure 12 after 1000 thermal cycles.

Sonoscan results showed that there was no additional voiding after 1000 thermal cycles when C7300 was printed on Al_2O_3 at 150µm thickness.

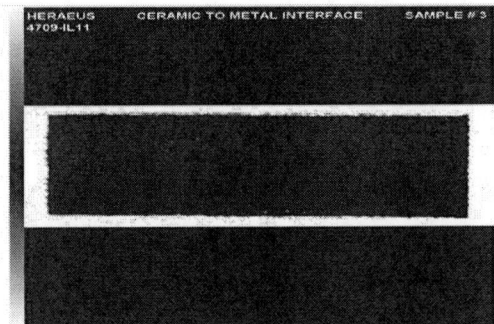

Figure 5F: Shows a C-SAM done on C7300 printed on BeO before thermal cycling.

Figure 5G: Shows a C-SAM done on the same part as Figure 14 after 500 thermal cycles.

Sonoscan results showed a slight increase in voiding after 500 thermal cycles when C7300 was printed on BeO at a fired thickness of 150μm.

SEM
The SEM analysis of both the DBC and the thick film printed copper system show two very different microstructures. DBC is a solid and continuous sheet of material and therefore there is little to no voiding that is seen when examining the cross-sections. Thick film is a combination of copper metal particles, glass along with a vehicle and solvent system. This combination shows voiding in the cross-sections; however, the voiding does not affect the performance of the paste. The SEM results simply show the difference between the DBC and thick film systems characteristic to both production methods.

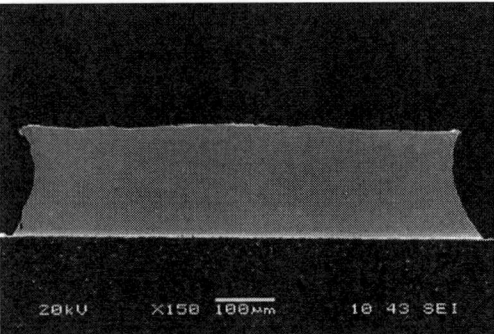

Figure 6A: Shows a cross section of etched copper foil at a magnification of 150x.

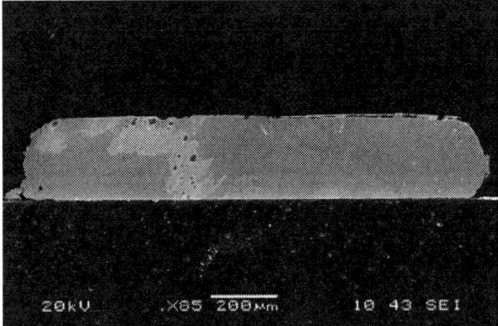

Figure 6B: Shows a cross section of tabbed copper foil at a magnification of 85x.

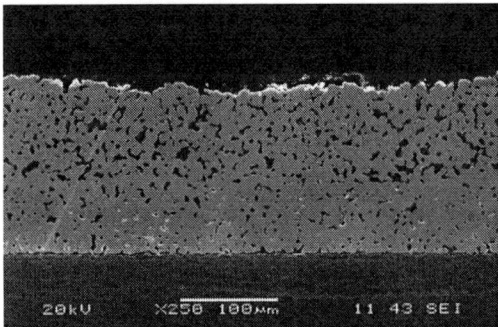

Figure 6C: Shows a cross section of printed C7300 mil shown at a magnification of 250x.

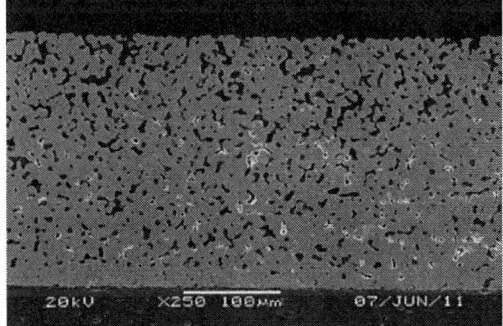

Figure 6D: Shows a cross section of C7401 printed as a base layer and C7300 printed as a top layer. Shown at a magnification of 250x.

Cost

For cost analysis of DBC ceramics and thick film ceramic, the cost of a pre-stamped DBC ceramic was given a cost value of 1X. The part used for this evaluation is shown in Figure 7.

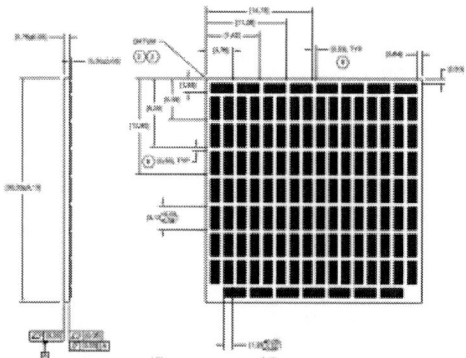

Figure 7: DBC ceramic 30mm square with 300μm thick copper tabs.

The cost comparison of DBC processes is based on printing the parts in manufacturing facilities in Asia.

The estimated cost of each system as follows:

	Process	Cost
1	DBC Tab Placement Process (Process A)	1X
2	DBC Etch Process (Process B)	4.7X
3	One paste system (C7300)	2X
4	Two paste system (C7401/C7300)	2.4X

Design Restrictions

Dimensions	Application dependent **Alumina**	Application dependent **Aluminum Nitride**
Min. width of Cu pattern	0.5 mm / 19.7 mil	0.5 mm / 19.7 mil
Min. spacing between Cu patterns	0.5 mm / 19.7 mil	0.5 mm / 19.7 mil
Minimum spacing between Cu pattern and ceramic edge	0.35 mm / 13.8 mil	0.35 mm / 13.8 mil

Figure 8: Shows the published design restrictions for DBC on Alumina and Aluminum Nitride.

When printing parameters are optimized, the thick film paste system can withstand a wider range of processing parameters than that of DBC due to the additive nature of the printing process.

Discussion

The standard deviation of the print thickness was greater on the thick film pads than on DBC pads. The print thickness target and print thickness variation can be improved with optimization of the screen or stencil. This will be part of a future project

Solderability of the RoHS compliant 227°C melting point solder was the same or better with C7300 as compared with the DBC surface. Cleaning the surface of oxides increased the wetting on all surfaces.

Adhesion improved when fired thick film print thickness was increased; this may be related to the extra firings of the initial pass of copper. This indicates that adjustments to the furnace conditions (belt speed, temperature, and dwell time) may help increase the adhesion of the thinner fired prints. All testing on DBC was completed on the etched DBC (Process B). Previous internal tests on tabbed DBC showed on average that the adhesion was 12N – thus the adhesion on all thick film testing was greater than on the tabbed DBC (Process A) adhesion.

The adhesion of the samples after 100 cycles indicates that additional work is required on the thick print system. There was a significant increase in the percentage of the ceramic fracturing as part of the failure mode after cycling as compared to the initial adhesion state.

Even though the adhesion decreased after thermal cycling, the fact that there were no new voids shown in the C-SAM suggests that the copper film is a close enough thermal match to the substrate that the thermal cycling does not cause delaminations.

Cost analysis demonstrates that the one paste system (C7300) is slightly higher than the lowest cost DBC system (Process A), while both the thick film systems (C7300 and C7401/7300) are significantly lower than the cost of the DBC etched system (Process B).

Conclusion

The C7401/C7300 system is comparable to etched DBC systems of aged adhesion at the 200μm and 300μm thickness. The C7300 paste system is comparable to tabbed DBC system adhesion in the "as bonded" condition.

Based on the C-SAM testing the printed copper paste systems seem to withstand thermal cycling without voiding or lifting, which gives them an advantage to DBC when thermal cycling equals or exceeds 1000 cycles.

Copper thickness has a direct relationship with adhesion. As thickness increases, the adhesion increased on initial, aged and thermal cycling adhesion.

Pricing would be less than the etched DBC processing using Asian based manufacturing facilities for direct labor.

Based on the test results performed for this paper it appears that the Thick Film alternatives to DBC perform equally as well if not better than DBC under optimized conditions. For certain applications thick print copper can offer a cost effective alternative to DBC while maintaining or exceeding the required processing and quality parameters specified by the customer.

Fineline Structuring on LTCC-Substrates for 60 GHz Line Coupled Filters

Jens Müller, Dirk Stöpel, Thomas Mache, Alexander Schulz, Karl-Heiz Drüe, Stefan Humbla, Matthias Hein

Ilmenau University of Technology
Institute of Micro- and Nanotechnologies
P.O. Box 10 05 65
D-98684 Ilmenau, Germany

Phone: +49 3677 69 2606, Fax: +49 3677 69 1204, e-mail: jens.mueller@TU-Ilmenau.de

Abstract

Low Temperature Co-fired Ceramics (LTCC) are used in a wide range of applications. They provide true three-dimensional integration, hermetical sealing, hybrid integration and good microwave properties at moderate costs. In order to take advantage of millimetre wave frequencies a resolution of lines and spaces of less than the screen printable 50 μm is desired. For frequencies in the 60 GHz range integrated coupled line filters require coupling gaps as small as 30 μm. However, the screen printing related accuracy results in dimensional tolerances of line widths or spaces of up to 20%.

This paper describes the technological development of a fineline structuring process based on resinate thin films using electroplating on LTCC.

The initial layer is applied by screen printing of a metal-organic (resinate) paste, a noble metal compound (e.g. gold and silver) dissolved in organic oils. The film thickness after firing is typically less than 1 μm. This layer can be used to define the structures using photolithography and electroplating followed by an etching process. The etching processes show a promising resolution of 20 μm lines and spaces with tolerances defined by the tolerances of the photoresist (typically below ± 1μm) and the substrate flatness. Benefits for the design of microwave and millimetre wave applications are obvious due to the higher resolution and reduced variation of the final structures.

A 60 GHz line coupled band-pass filter is used to demostrate process stability and electrical performance of this new structuring method.

Key words: LTCC, Resinate Technology, Fine Line Structuring, Microwave, Filter

Introduction

At present, the motivation for high density interconnect substrates are component and module miniaturization (related to semiconductor roadmaps [1]) and increased frequencies. The latter, however, does not require a high density of lines, but small lines and spaces in conjunction with low dimensional tolerances in order to allow for the integration of passive elements and components such as inductors or filters.

Low temperature cofired ceramics (LTCC) have been widely used for more than two decades to manufacture microwave components, substrates or modules. Due to the screen printing process to structure conductors made of silver or gold, the pattern resolution is typically worse than 50 μm for lines and spaces (100 μm interconnect pitch). More conservative rules for series production limit the screen printing resolution to 100 μm structures.

Fine line screen printing can be enhanced by screens with trampoline meshes (improved snap-off behavior and screen life time) and hydrophobic coatings at the paste release side. It was reported that 30 μm-structures were achieved in a lab-environment using this method [2].

Alternatively, sintered thick films may be etched or photoimageable pastes such as Fodel® pastes [3] are applied. Both technologies are based on subtractive processes with high material consumption. The minimum line widths and spaces are 30 μm and 40 μm, respectively.

Higher structural resolutions can be achieved by thin films on LTCC. Two thin film systems which have been tested on as-fired LTCC demonstrated lines and spaces of about 10 μm [4]. Electroplating is applied in general to enhance the thickness and conductivity of the thin base conductor layers.

The major drawback of most fine line technologies is that they can only be applied on fired substrates, which limits their use on outer structures. Fine line screen printing and the Fodel® process are the only methods to obtain a higher wiring density on internal layers of LTCC substrates.

Resinate Technology

The excellent thin film resolution is based on the capability to structure thin layers by photolithography. Exposing mask quality, substrate warpage, and surface roughness are often the remaining factors which limit the structural resolution. Since thin film patterns are made up of several layers (e.g. NiCr as adhesion and Cu as plating base), the final arrangement cannot be exposed to temperatures compatible to thick film processing to avoid material diffusion.

An alternative to expensive vacuum processes (sputtering or evaporation) is the resinate technology which allows thin films based on screen printing [5]. Resinate pastes are made of metallo-organic noble metal compounds (e.g. gold, silver), dissolved in an organic suspension. The resinate technology is compatible with a variety of substrates such as alumina, LTCC or glass. It furthermore supports mono-metallic build-up structures which can be fired again at 850°C; an essential step for multilayer circuits [6]. Due to the relatively low metal content in the paste the fired film thickness is typically below 1 μm.

First applications of resinates for sensors on LTCC were published by Kita [7]. The fired resinate film was structured by photolithography and etching for this particular application.

Fig. 1 shows the general manufacturing sequence for the combined resinate/electro-deposition-process.

a) screen printing and firing of the Au-resinate paste on sintered LTCC,
b) spin coating, drying, pre-baking, exposing and developing of the photoresist,
c) electrodeposition of Au (3-4μm),
d) stripping of the photoresist,
e) flash etching of the resinate base layer.

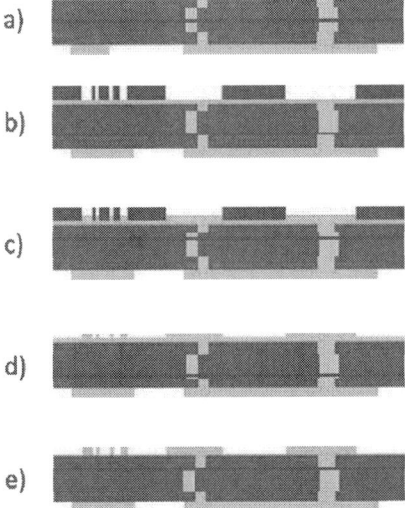

Figure 1: Process flow for resinate on LTCC

Experimental and Test Design

Coupled microstrip band-pass filter designs with a center frequency of 60 GHz are chosen for verification purposes. The test layout is designed on a 6 layer substrate of DuPont 951 PT (single layer fired thickness of about 95 μm) with a substrate size of 58 x 58 mm². The minimum coupling gap is 32 μm. The distance between the ground and the signal layer is about 190 μm (two layers). Fig. 2 shows the schematic layout of a single filter. In order to test process and layout influences (e.g. orientation of the filter structures on the substrate and under etching), multiple structures with different orientation and edge offsets of 4, 6, 8 and 10 μm are implemented on the test design. In addition, several spirals and sheet resistance test pads are placed among the microstrip filters (Fig. 3).

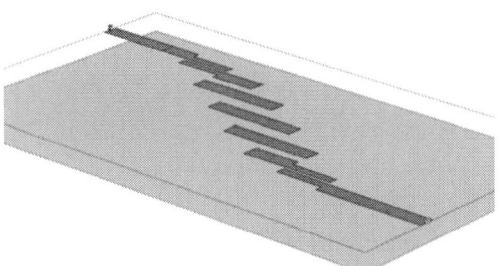

Figure 2: Schematic of the microstrip band pass filter structure with smallest coupling gaps of 32 μm

The inner ground plane metallization and the vias are printed with DP5734 and DP5738 gold paste, respectively. Stacking, lamination, burn out and firing are performed using standard process parameters. After co-firing the resinate paste (Heraeus RP 181208-15%) is printed with a 325 mesh screen with 15 μm emulsion across the entire substrate. The dried resinate layer is sintered at 850°C with a dwell time of 10 minutes, which results in a sheet thickness of about 200 nm. A negative photoresist is applied according to Fig. 1. The polymer mask for contact exposure is plotted using a Gerber photo plotter (MIVATEC) with a resolution of 16.000 dpi.

Electrodeposition (3 to 4 μm) is carried out in a beaker glass with a cyanitic soft gold electrolyte and a current density of about 0,125 A/dm². After stripping the photoresist the base layer is removed by flash etching in KCN.

The test structures are measured after full two port calibration with an Agilent PNA E8361A from 10 MHz to 67 GHz using coplanar GSG microwave probes with 200 μm pitch.

DC-Sheet resistance measurements of the rectangular patches are carried out according to the Van-der-Pauw method and the layer thickness is measured by X-ray fluorescence spectroscopy (XRF) (Helmut Fischer GmbH).

Finally, the galvanic/resinate layers are annealed at 350°C for 2 h and DC- as well as S-parameter measurements are repeated.

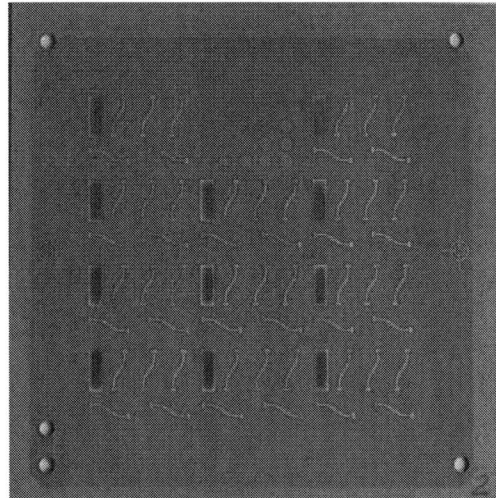

Figure 3: Processed LTCC test substrate

Results

Pattern accuracy

The geometrical differences between the test structures and the design are given in Table 1. These values reveal the influence of the substrate orientation, which was not expected for a photolithographic process. The key reason for this deviation is the polymer photomask (exposure in combination with ageing effects of the photochemistry). All structures on the photomask are undersized by about 10 μm. Therefore, all filters with a 10 μm edge offset meet the specified design values very accurately (± 2 μm). Fig. 4 shows a detailed image of such a filter structure with the most sensitive coupling gap of 32 μm.

Table 1: Differences between design and test substrates

Orientation	Mean value [μm]	Std. deviation [μm]
X-direction	12.13	2.29
Y-direction	10.76	1.4

The sheet thickness varies between 4.1 and 4.9 μm prior to the flash etching step and amounts to 3.1 to 4.2 μm after etching. The wide distribution of thicknesses is assumed to be caused by the beaker glass processing for both the electrodeposition and flash etching.

Figure 4: Coupling structure of an optimized bandpass filter

DC-behaviour

Based on the measured sheet resistance and the thickness of the test pads, the material conductivity is calculated and compared to the bulk conductivity of gold (Table 2). The results demonstrate the strong impact of the annealing step. Nevertheless, the resulting conductivity is still much lower than the bulk conductivity, the tempering step leads to a significant improvement of the specific conductance.

Table 2: Specific conductance of the deposited metal structure (* mean value)

Condition	Specific conductance [10^7 S/m]
Prior to annealing*	1.53
Post annealing*	2.53
Bulk Au	4.52

Microwave properties

Fig. 5 and 6 show the measured transmission and reflection parameters of eight filters with varied edge offsets (oversized Gerber structure). The insertion loss for all filters ranges between 2.8 and 3.3 dB and the return loss is better than -12 dB over the pass-band. The different coupling gaps which are attributed to the mask offsets have an influence on the return loss characteristics and the center frequency of the pass-band. No influence on the insertion loss in the pass-band is observed. A comparison of the insertion losses of filters with different edge offsets (4 μm and 10 μm) is depicted in Fig. 7. The measured response is shifted towards higher frequencies (approx. 680 MHz) for the slightly undersized structures as expected, since the included edge offset has also an impact on the length of the coupling sections. Filters with identical designs have almost identical S-parameters.

Figure 5: Insertion loss versus frequency for a set of band-pass filters (last two digits of the name in the caption denote edge offset)

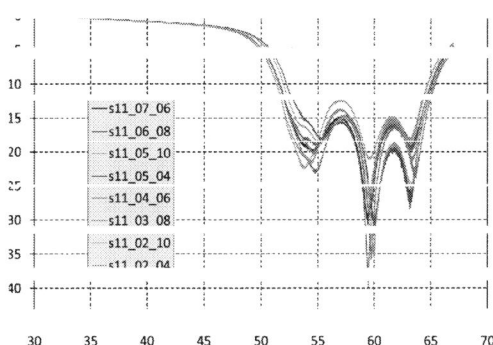

Figure 6: Return loss versus frequency for a set of band-pass filters (last two digits of the name in the caption denote edge offset)

Figure 7: Effect of coupling gap width on the frequency offset of band pass-filters

Even though the annealing process has a large influence on the specific conductance of the gold structures no influence on the microwave behavior is observed in the frequency range around 60 GHz. The slight differences in the return loss curve are most likely due to separate calibrations made before and after annealing (Fig. 8).

Figure 8: Comparison of band-pass filter properties before and after annealing

Conclusion and Outlook

Resinate pastes in combination with electro-deposition enable the use of low-cost screen printing techniques to obtain structural resolutions and tolerances comparable to thin film techniques but at lower costs. A gold resinate paste on LTCC is successfully applied on a test substrate containing microstrip band pass filters for the 60 GHz ISM band. This test verifies as well the electrical interconnect stability between standard LTCC vias filled with thick film paste and the resinate cover layer. Line gaps as small as 32 µm are achieved with excellent reproducibility. Manufacturing tolerances are less than two microns.

Major differences between the designed and the manufactured structures are caused by the inaccuracy of the film-mask photoplotter as well as the developing process. Higher quality mask technologies are mandatory to further enhance the structural consistency between design and manufactured module.

Future work will focus on the application of resinate based structures in multilayer circuits to be able to implement microwave components on inner layers as well.

Acknowledgement

This work has been funded by the German Federal Ministry of Economics and Technology (BMWi) under the project management by the German Aerospace Center (DLR, no. 50YB0622). We appreciate very much helpful technical support from I. Koch, K. Friedel, and B. Hartmann (TU Ilmenau).

References

[1] www.itrs.net
[2] D. Schwanke, J. Pohlner, A. Wonisch, T. Kraft, and J. Geng, "Enhancement of Fine Line Print Resolution due to Coating of Screen Fabrics," presented at CICMT, Munich, 2008.

[3] R. A. Perrone, Erweiterung des Frequenz-bereichs und der Integrationsdichte von LTCC-Modulen mittels Photostrukturierung und Designoptimierung. Ilmenau: ISLE, 2007.

[4] J. Müller, R. Perrone, H. Thust, K.-H. Drüe, C. Kutscher, R. Stephan, J. Trabert, M. Hein, D. Schwanke, J. Pohlner, G. Reppe, R. Kulke, P. Uhlig, A. F. Jacob, T. Baras, and A. Molke, "Technology Benchmarking of High Resolution Structures on LTCC for Microwave Circuits," presented at Electronics Systemintegration Technology Conference, Dresden, Germany, 2006

[5] D. Stöpel, K.-H. Drüe, S. Humbla, M. Mach, T. Mache, A. Rebs, G. Reppe, G. Vogt, M. Hein, and J. Müller, "Fine-Line Structuring of Microwave Components on LTCC Substrates", Proc. ESTC2010, pp. , Berlin, September 2010.

[6] D. Stöpel, K.-H. Drüe, S. Humbla, M. Mach, T. Mache, A. Rebs, G. Reppe, G. Vogt, M. Hein, und J. Müller, „Fineline-Strukturierung auf LTCC-Substraten für HF-Komponenten", Proc. Deutsche IMAPS Konferenz, 12./13. Oct. 2010, Munich.

[7] J. Kita and R. Moos, "Heater for LTCC-Sensors Made of Resinate Pastes," presented at CICMT, Denver, 2007.

Low temperature die-bonding with Ag flakes

S. Sakamoto, K. Suganuma

ISIR, Osaka University

8-1 Mihogaoka, Ibaraki, Osaka 567-0047, Japan

Tel: +81-6-6879-8521

Fax: +81-6-6879-852

E-mail Address: sakamoto@eco.sanken.osaka-u.ac.jp

Abstract

Operating temperature of power semiconductors is expected to rise beyond 150 °C especially such for GaN or SiC. Die-bonding must maintain good reliability in such high temperatures. Currently, the conventional Ag-epoxy conductive adhesives and Pb-free solders have been used for current Si die attachment. They are, however, difficult to satisfy the requirements for such ultra-heat resistant uses. In order to develop of ultra-heat resistant die-attach technology at low temperature, new die-bonding material was proposed, i.e., nano-meter thin Ag flakes. The Ag flakes have large surface area and thickness less than 100 nm. They are mixed with alcohol as a solvent to be formulated into a paste. The other Ag pastes were also made by using Ag micron particles and Ag particles with submicron. These Ag pastes were stencil-printed on a Cu substrate with or without electrolytic plated Au or Ag. Finally, die-bonding was carried out. These samples were bonded in a temperature range from 160 °C to 200 °C for 60 minutes. As the result, the die-bonding with the new Ag nano-meter thin flakes exhibits an about twice times higher bonding strength than the conventional Ag paste die-bonding. By observing cross-sections of the samples after sintering revealed that the nano-meter thin Ag flake die-bonding has denser microstructure than the conventional die-bonding. Shear strength of die-bonding was improved by using the nano-meter thin Ag flakes. It was revealed that the nano-meter thin Ag flakes can form a reliable interconnect below 200 °C.

Key words: Ag flake, die-bonding, Ag paste, low-temperature sintering,

Introduction

Efficient use of the electrical energy and use of the green energy (i.e. Wind Power, Solar, Biomass, Hydroelectricity, Wave, and Biofuels) is recommended to protect global environment. Power semiconductors are emerged as one of the essential devices to conversion of current and control of current, which are used in a wide field of consumer electronic and car. In the past, reduction of power loss and increasing of density of power were improved by development of Si power devices. However, more significant improvement will be not expected since Si power devices almost achieved theoretical efficiency. In recent year, the replacement of Si semiconductors such as SiC, GaN and InP are paid attention to gain breakthrough technology in power semiconductor devices [1-4]. These new power semiconductors possess bigger band gap, density of power and radiation performance and higher operating temperature. Especially, SiC possess about three times as large band gap as Si and about ten times as large breakdown voltage as it. In addition, coefficient of thermal expansion of SiC is 4.9 W/cmK, since the

value is more equivalent the value of DBC substrate than the value of Si, load of substrates and devices decrease. Since SiC power devices which possess large thermal conductivity be able to operate at high temperature, SiC power devices were expected to reduce or remove cooling system. Density of power will be able to improve by efficiently using this newly space caused.

The conventional materials of die-bonding for Si power devices were widely used the solder [5] or the conductive epoxy adhesion [6]. The solder such as the lead solder and the lead-free solder have low melting point of below 300 °C and high reliability of die-bonding. The conductive epoxy adhesive was used when the substrate could not resist die-bonding temperature at about 300 °C. Previously, the solder and the conductive epoxy adhesion enough resisted without degradation and damage since the operating temperature for Si power device was 150 - 175 °C. However, operating temperature of power device is expected to rise beyond 200 °C or up to 300 °C especially such for GaN or SiC power device. Neither the solder nor the conductive epoxy adhesion can provide the performance and reliability beyond 200 °C. The die-

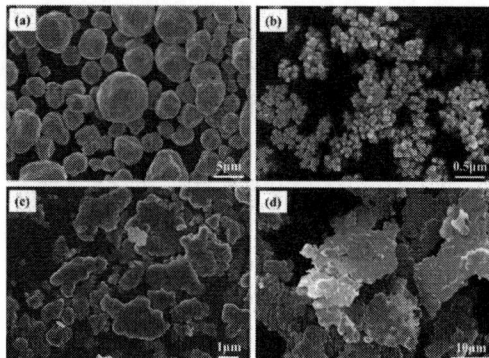

Fig. 1 SEM image of Ag-HWQ, Ag-nano, AgC-239 and XF-301. (a) Ag-HWQ. (b) Ag-nano. (c) AgC-239. (d) XF-301.

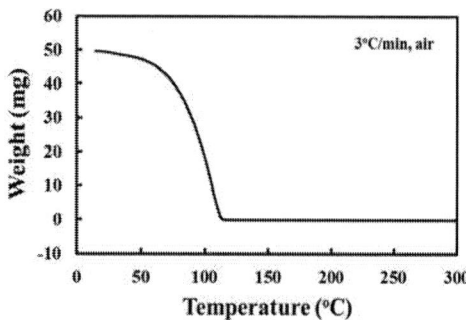

Fig. 2 Thermogravimetry differential thermal analysis (TG-DTA) of the isoamyl alcohol from room-temperature to 300 °C in air with heating rate of 3 °C/min.

attach materials must possess good thermal conductivity and reliability as well as good electric conductivity beyond 200 °C.

There have been several reports on high temperature die-attachment for SiC devices. Au based alloys such as An-Sn (eutectic point: 280 °C) and An-Ge (eutectic point: 356 °C) are an interesting approach for SiC devices assembly [7-8]. These solders are characterized by good wettability to the metallized substrate, high thermal conductivity and very good the strength of die-bonding. Recently, one of the authors has reported SiC die-attachment with pure Zn solder for SiC power devices on a DBC (Direct Bonded Copper) substrate, for which the solder was carried out at 450 °C [9]. It was proved that this die-attach structure can resist severe thermal shock of between -50 °C and 300 °C and is expected to be a low-cost and high-performance candidate for SiC devices. Ag has a great advantage as a high thermal and electric conductive metal. Ag nanoparticle pastes can be applied to bonding metallic materials or metallized components at about 300 °C [10-12]. The die attachment was formed by sintering Ag nanoparticle paste in air at 300 °C under a pressure of 5-50 MPa.

In the present work, the authors try to provide the die-bonding materials that the nano-meter thin Ag flakes can form the interconnect below 200 °C.

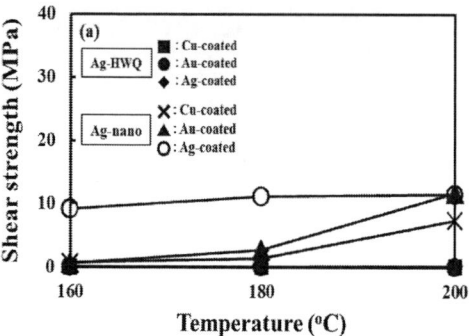

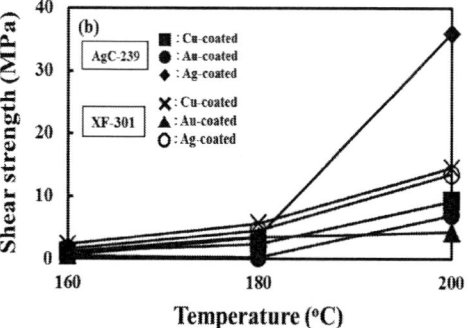

Fig. 3 Die-bonding shear strength of dummy-chips attached on the Cu substrate metallized with or without Au/Ni or Ag as a function of bonding temperature. (a) Test samples were made by using Ag-HWQ paste and Ag-nano paste. (b) Test samples were made by using AgC-239 paste and XF-301 paste.

Experimental

Four kinds of Ag particles were used as the starting materials. Ag-HWQ (Fukuda Metal Foil and Powder Co., Ltd.) particles of an average particle diameter of 5 μm, Ag-C239 (Fukuda Metal Foil and Powder Co., Ltd.) flake particles of an average particle diameter of 8 μm, 576832 (Sigma-Aldrich Co., Ltd.) particles of an average particle diameter of 100 nm and XF-301 (Fukuda Metal Foil and Powder Co., Ltd.) flake particles of an average particle diameter of 6 μm were used. Hereafter, the four Ag particle starting materials are referred to as "Ag-HWQ", "AgC-239", "Ag-nano" and "XF-301", respectively. AgC-239 has a flake shape, of which thickness is 100 nm and possess high specific surface area of 1.2 m^2/g. XF-301 has also a flake shape, but XF-301 has more thinner thickness of 80 nm and higher specific surface area of 1.9 m^2/g. The microstructures of those particles are shown in Fig.1. The solvent was 3-methyl-1-butanol (hereafter, 3-methyl-1-butanol is referred to as "isoamyl alcohol" in this paper), which are typical alcohols vaporizing below the possible lowest sintering temperature, 160 °C (Boiling point of (isoamyl alcohol is 132 °C). Thermogravimetry differential thermal analysis (TG-DTA) was performed by heating the isoamyl

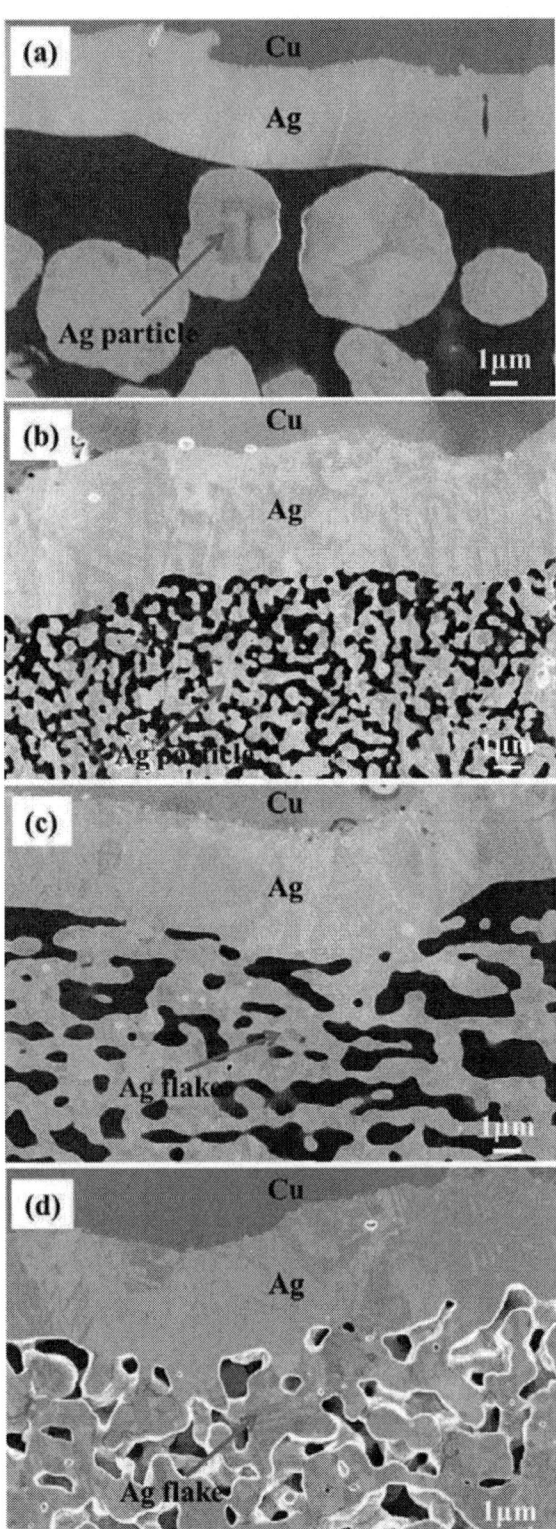

Fig. 4 SEM images of the bonding interfaces of dummy-chips metallized with Ag and sintered layer. (a) Ag-HWQ layer. (b) Ag-nano layer. (c) AgC-239 layer. (d) XF-301layer.

alcohol from room-temperature to 300 ℃. The test was run in air with heating rate of 3 ℃/min. TG analysis of isoamyl alcohol is shown in Fig. 2. It was confirmed that isoamyl alcohol completely vaporized at 114 ℃, and isoamyl alcohol was expected to completely vaporize through sintering.

Four kinds of pastes were prepared by mixing the particles with the solvent. The weight ratio of powders to solvent was 1 : 0.2.

For measuring the bonding strength, test samples were prepared by die attaching metal dummy chips onto metal substrates. Three kinds of metal surface finishes were used as the substrates. The metal substrates were Cu plates of 0.8 mm in thickness. Either Ag or Au was plated on the Cu substrates to enhance bonding strength for Ag die attachment. A Cu substrate without plating was used for comparison. 10μm thick Ag was plated on Cu and 100 nm thick Au/2 μm thick Ni was plated on Cu. The dummy chip was 4 mm square with a thickness of 0.8 mm and was metallized with the same Ag or Au/Ni. The paste was screen printed on a Cu plate and then a dummy chip was place on it. Pressure was slightly applied during die-bonding which was estimated as 0.07 MPa. Bonding was carried out in air at 160 ℃, 180 ℃, 200 ℃, 220 ℃ and 240 ℃ for 60 minutes. The obtained Ag bonding layer was 20 μm to 50 μm in thickness. To measure the bonding strength, the Ag die attached test samples were sheared off by using a die shear test. Scanning electron microscopy (SEM) was used to observe microstructure of interfaces.

Result and Dicussion

Fig. 3 shows the die-bonding strength of samples using Ag-HWQ paste, Ag nano paste, AgC-239 paste, XF-301 paste. In spite of increasing sintering temperature to 200 ℃, the die-bonding strength of samples using Ag-HWQ paste was only 0.2 MPa. Regardless of the substrate and the dummy-chip metallized with or without Au/Ni or Ag, the die-bonding strength of samples using Ag-HWQ paste was the smallest in all the samples. The die-bonding strength of the samples with Ag nano paste increased as increasing sintering temperature. The die-bonding strength of the samples using Ag nano paste was the highest for Ag-coated Cu substrate/dummy-chip at the 200 ℃, which was 11.5 MPa. The die-bonding strength of about 10 MPa was gained by sintering ever at 160 ℃. This strength is the highest in all the samples prepared at 160 ℃. The die-bonding strength of the samples with AgC-239 paste similarly increased as increasing sintering temperature. Especially, the die-bonding strength of samples using AgC-239 paste on the Ag-coated Cu substrate was the highest in all the samples, which reached 36.0 MPa, which is high as the strength for soldering. Finally, the die-bonding strength of the samples using XF-301 paste similarly increased as increasing sintering temperature. The die-bonding strength with XF-301 paste on the Ag coated Cu substrate at the 200 ºC was the highest, which was 13.5 MPa. As metallization of Cu, Ag coating can improve bonding strength better than none-coating or Au coating. Since self-diffusion of Ag is faster than interdiffusion of Ag in Cu or of Cu in Ag of Ag in Au or of Au in Ag, higher shear strength was gained between the sintered Ag and the Ag coated substrate and the dummy-chip.

Fig. 4 (a) ~ (d) shows the microstructure of cross section of samples which were die-attached the dummy chip on the Ag-coated Cu substrate by using Ag-HWQ paste, Ag nano paste, AgC-239 paste and XF-301 paste and sintered 200 ºC for 60 min. In Fig. 4 (a), Micro particles maintained original shape after sintering to 200 ºC for 60 min. Forming of bonded interface was not obtained at all with or without Au or Ag coating. As a result, the die bonding strength of sample using Ag-HWQ paste was very small. SEM image of cross section of sample using Ag nano paste showed clear sintering, neck growth between Ag nano particles and bonded interface between Ag nano particles and the Ag-coated Cu substrates at 200 ºC. As a result, the die bonding strength of sample using Ag nano paste was about 10MPa, which was bigger than that of sample using Ag-HWQ paste. However, since very small of particle size, bonded interface area was small, so that the die bonding strength of only about 10 MPa was obtained. SEM image of cross section of sample using AgC-239 paste showed very clear sintering, large neck growth between AgC-239 particles and bonded interface between AgC-239 particles and the Ag-coated Cu substrate at 200 ºC. Bonded interface area was the largest in the all samples. As a result, the die bonding strength of sample using AgC-239 paste was largest in the all samples. Similarly, SEM image of cross section of sample using XF-301 paste showed clear sintering, neck growth between XF-301 particles and bonded interface between XF-301 particles and the Ag-coated Cu substrate at 200 ºC. Bonded interface area was smaller than that of AgC-239 and, area of pore in the Ag paste after sintering was larger than that of AgC-239. As a result, the die bonding strength of sample using XF-301 paste was smaller than that of AgC-239. For very thin thickness of XF-301, larger pressure was needed at sintering to bring the XF-301 particles close more. However, since pressure was constant while sintering in this paper, bonded interface area was smaller than that of AgC-239 and, area of pore in the Ag paste after sintering was larger than that of AgC-239. From the result of die bonding strength and cross section image, it is found that choice of particles of nano-size or particles of flake shape was able to joint at low-temperature sintering of 200 ºC. However, for nano particle, bonded interface area was small and, area of pore in the Ag paste was large. For Ag particle of flake shape, if that is too thin, more pressure is needed. Choice of thickness of particle flake is important to gain enough the die bonding strength as low temperature sintering.

Conclusions

The material of die-bonding with very good shear strength (36.0 MPa) was successfully developed by using the AgC-239 paste which has shape of flake, thin thickness and high specific surface area. The microstructural morphology of cross-section revealed that AgC-239 paste showed very clear sintering, large neck growth between AgC-239 particles and bonded interface between AgC-239 particles and the Ag-coated Cu substrate at 200 ºC. In addition, bonded interface area was the largest in the all samples. As a result, the die bonding strength of sample using AgC-239 paste was largest in the all samples. On the other hand, Ag-HWQ maintained original shape after sintering to 200 ºC for 60 min. Forming of bonded interface was not obtained at all with or without Au or Ag coating. As a result, the die bonding strength of sample using Ag-HWQ paste was very small. The die bonding strength of samples using Ag-nano paste was obtained only about 10 MPa since bonded interface area was small. The die bonding strength of samples using XF-301 paste was smaller than that of samples using AgC-239 since pressure was lack while sintering. It is found that choice of particles of nano-size or particles of flake shape was able to joint at low-temperature sintering of 200 ºC. However, for nano particle, bonded interface area was small and, area of pore in the Ag paste was large. For Ag particle of flake shape, if that is too thin, more pressure is needed. Choice of thickness of particle flake is important to gain enough the die bonding strength as low temperature sintering.

References

[1] W. Wondrak, "SiC devices for advanced power and high-temperature applications", IEEE Transactions on Industrial Electronics, Vol. 48, No. 2, pp. 307-308, April, 2001.

[2] Y. Sugawara, "SiC devices for high voltage high power applications", Materials Science Forum, Vol. 457-460, pp. 963-968, June, 2004.

[3] M. A. Khan, "GaN based heterostructure for high power devices", Solid- State Electronics, Vol. 41, pp. 1555-1559, October, 1997

[4] A. Katz, "Advanced metallization schemes for bonding of InP-based laser devices to CVD-diamond heatsinks", Materials Chemistry and Physics, Vol. 37, pp. 303-328, May, 1994

[5] J. S. Hwang, "Solder technologies for electronic packaging and assembly", Electronic Packaging and Interconnection Handbook, New York, pp. 6.1-6.83, 2000.

[6] C. P. Wong, "Polymers for electronic packaging", Electronic Packaging and Interconnection Handbook, New York, pp. 1.1-1.21, 2000.

[7] F. P. McCluskey, "Reliability of High Temperature Solder Alternatives", Microelectronics Reliability, Vol. 46, No. 9-11, pp. 1910-1914, August, 2006.

[8] Meyappan K, "Thermomechanical analysis of gold-based SiC die-attach assembly", IEEE Transactions on Device and Materials Reliability, Vol. 3, No. 4, December, 2003.

[9] K. Suganuma, "Ultra Heat-Shock Resistant Die Attachment for Silicon Carbide With Pure Zinc", IEEE Electron Device Letters, Vol. 31 No. 12, pp. 1467-1469, December, 2010.

[10] John G. Bai, "Processing and Characterization of Nanosilver Pastes for Die-Attaching SiC Devices", IEEE Transactions on Electronics Packaging Manufacturing, Vol. 30, No. 4, pp. 241-245, October, 2007.

[11] John G. Bai, " Low-temperature sintered nanosilver as a novel semiconductor device-metallized substrate interconnect material", IEEE Transactions on Components and Packaging Technologies, Vol. 29, No. 3, pp. 589-593, September, 2006.

[12] J. G. Bai, "Thermomechanical reliability of low-temperature sintered silver die attached SiC power device assembly," IEEE Transactions on Device and Materials Reliability, Vol. 6, No. 3, pp. 436-441, September, 2006.

Novel Approaches for Low-Cost Through-Silicon Vias

J.E. Bullema[1], P.M.M.C.Bressers[1], G. Oosterhuis[1], M. Mueller[2],

A.J. Huis in't Veld [1,3] and F. Roozeboom[1,4]

[1]TNO, De Rondom 1, 5612 AP Eindhoven, The Netherlands; [2]ALSI, Platinawerf 20, 6641TL Beuningen, The Netherlands; [3] University of Twente, Postbus 217, 7500 AE Enschede, The Netherlands; [4] Eindhoven University of Technology, Dept. Applied Physics, 5600 MB Eindhoven, The Netherlands

Tel. +31-0)88-8665551; Email jan_eite.bullema@tno.nl

Abstract

3D stacking of integrated circuits is an emerging packaging technology to enable a high degree of functional integration and miniaturization. Footprint reduction in 3D stacking can be achieved by use of Through Silicon Vias (TSV). Creation of TSVs with Deep Reactive Ion Etching (DRIE), laser drilling and pulse reverse plating is established technology. Current TSV technologies are considered as high cost processes due to expensive equipment and long processing times.

In this paper three novel technological approaches to create TSVs are described that potentially lead to a creation of low-cost Through Silicon Vias. The technologies in development discussed here, were identified based upon cost of ownership analysis of current TSV creation processes The paper presents the first results of the different approaches.

Key words: Through Silicon Via, Deep Reactive Ion etching, Laser Induced Forward Transfer, Electrochemical Machining

Introduction

Through Silicon Vias (TSV) are considered as a key enabler for 3D stacking of integrated circuits. An important bottleneck in the use of TSV technology is the relative high Cost-of-Ownership (C-o-O) for the creation and subsequent filling of TSVs [1,2] Figure 1 shows a typical fabrication process flow for creating a TSV based interposer [3]. The EMC3D consortium targeted for both a via-first (*i*TSV™) and via-last (*p*TSV™) process flow a total cost of ownership of under $150 [4].

TSV technology is a relative expensive technology, due to slow processes and relative expensive equipment, in particular for Deep Reactive Ion Etching (DRIE) and plating. In earlier work [5] we have developed a high-speed (3600 chips/hr), high-accuracy (1 μm @ 3σ) pick-and-place solution for chip-to-chip stacking, to overcome an important cost hurdle in application of TSV technology. In this work we explore novel approaches for TSV creation that are applicable in industry and have potential to enable low C-o-O TSV technology.

Besides C-o-O issues for TSV technology there are still a few major challenges in the area of manufacturability, chip design [6] and product reliability.

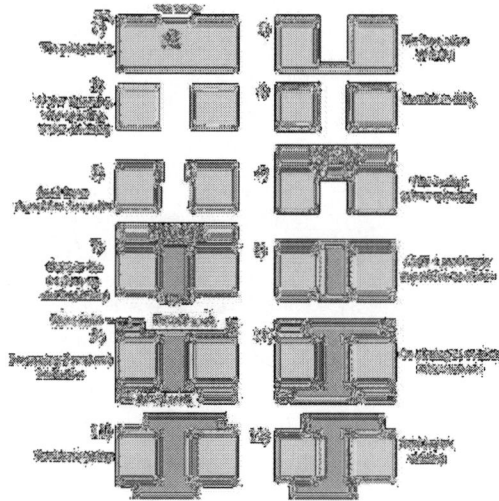

Figure 1. Typical fabrication process flow for a TSV interposer employing bottom-up Cu electroplating; from [3].

Electrochemical Machining

Electrochemical machining of silicon is a well studied subject [7-9]. Many experimental tools for electrochemical machining have been described. For n-type silicon small holes have been drilled using backside illumination. Allongue *et al.* [10] describe an alternative electrochemical method, called electrochemical micromachining (ECM), using ultrashort voltage pulses, to machine silicon substrates in 3D without the application of a mask.

In this work we explore a maskless electrochemical machining (ECM) process that can be used to both drill and fill TSVs. As there are no mask steps involved, the electrochemical machining technology has the potential of becoming a low-cost drilling and filling technology.

Based upon the experimental set up described by Sugita *et al.* [11] we conducted experiments to drill TSV holes with diameters down to 50 µm. In the experimental set-up there is no electrical contact with the silicon substrate (see Figure 2). In an experimental series we have demonstrated relative high drilling rates up to 16.9 µm/min. using aqueous 20 wt % HF solutions. In Figure 3 the experimental relationship between process temperature and drilling speed is given. The experimental set-up leads to a drilling process where the shape of the electrode is transferred into the silicon. Experiments on multicrystalline silicon wafers revealed that the process is isotropic. The quality of the sidewalls can be controlled by the applied potential. No silicon debris or thermal damage is created on the surface of the wafer. It appears to be possible to plate the wafer with metals in the experimental set-up, e.g. when copper is introduced into the bath, a thin copper layer is formed on the surface of the wafer. EDAX analysis shows that the copper layer preferably forms in the vincinity of the electrode tip. The experimental results show that is possible to drill vias and subsequently form a seed layer in the drilled hole.

The next step in this research will be the use of an array of electrodes to simultaneously drill array of holes in a silicon wafer and use the created microvias as microelectrochemical cells that potentially will enable fast plating of the via structures.

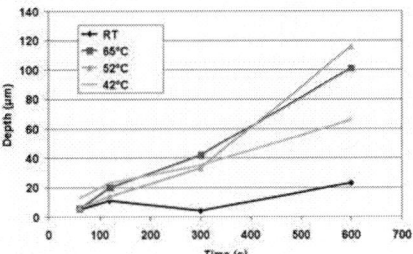

Figure 3. Experimental results of electrochemical drilling at different temperatures in p-doped Si (100) (2 Ω.cm) in HF solution.

Fast Deep Reactive Ion Etching (DRIE)

The conventional DRIE process (or 'Bosch' process) is a room temperature process based on alternating cycles of 1) Si-etching with SF_6 to form gaseous SiF_x etch products, and 2) passivation with C_4F_8 to form a protecting teflon-like polymer deposit on the sidewalls and bottom of the feature. A revolutionary modification of the conventional DRIE process has been conceived that enables fast machining of silicon. Our concept is based on *spatial* rather than temporal separation of the two cycles (SF_6 etching and C_4F_8 wall passivation), combined with gas-bearing technology [12]. This is realized in practice by repeatedly passing substrates horizontally under a gas injection head that delivers the two reactants through inlet channels. These channels are separated by inert gas flow. In addition, the injection head contains outlet channels in the respective half-cycle zones. In Figure 2 the concept for spatial seperated DRIE is schematically shown.

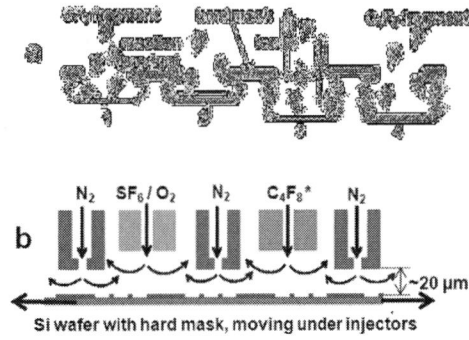

Fig. 2. a) Conventional DRIE (Bosch) process scheme with consecutive etch (odd-numbers) and passivation (even numbers) subcycles. The horizontal bar in grey represents a pre-patterned SiO_2 hard mask; b) alternative spatial process mode where C_4F_8 passivation can be replaced by ALD SiO_2, etc.

Thus the reactor has separate compartments delivering the etch and passivation gas one by one to

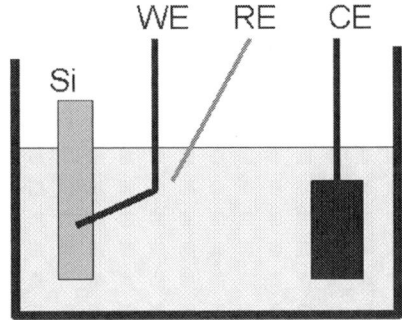

Figure 2. Schematic of the initial electrochemical experiments. Note, that the silicon is not current carrying. A platina wire pressed against the Si substrate was used as working electrode (WE). A second platinum electrode was used as a counter electrode (CE) and a Ag/AgCl electrode was used as reference (RE).

289

the substrate, and simultaneously pumping off the exhaust gases. Between and around the reaction zones, shields of inert gas separate the different gas flows. When operated properly, these gas shields can act as gas bearings, facilitating virtually frictionless movement between reactor and substrate.

Simulations confirm the advantages of this DRIE concept:

• Much smaller reactor chamber dimensions: less intermolecular collisions on the trajectory from plasma to features etched.

• No or little passivation gas interacting during etch step and vice versa; so no or less deposits on the reactor wall that cause conventional DRIE process windows to drift and machine up-time to go down due to regularly reactor cleaning and re-conditioning

• Shorter pulses possible (no flushing), leading to less pronounced scallops/ripples, thus smoother TSV walls.

Our simulations indicate that this concept will enable up to 10 x higher etch rates and thus lower costs, while maintaining the high machining quality of the conventional DRIE process [13, 14].

Another improvement in the spatial approach is the replacement of the CVD-based C_4F_8 passivation cycles by ALD-based SiO_2 deposition cycles. Unlike the C_4F_8 case the ALD-based SiO_2 layer is self-limiting and chemisorptive of nature and easier in its layer thickness control and thus in anisotropy control of the entire DRIE process. Recently Dingemans *et al.* [14] described an efficient plasma-assisted process for low-temperature (50-400 °C) ALD of SiO_2 using $H_2Si[N(C_2H_5)_2]_2$ precursor dosing times as short as ~50 ms, and O_2-plasma. Good conformality over high aspect ratio (30:1) trenches were reported, indicating that recombination of O-radicals in such trenches plays a minor role.

Laser Induced Deposition

For certain TSV dimensions and densities laser drilling is very attractive from a cost point of view [15]. In addition, lasers can be used to deposit metals. Traditionally, Pulsed Laser Deposition (PLD) is known as àn inherently slow process. The laser deposition process demonstrated here, is a fast metal deposition technology with good industrial potential. Laser-induced deposition processes have been widely investigated, as extensively described by Banks [16]. Many varieties of the so-called LIFT (Laser Induced Forward Transfer) process exist [17-19]. In our experiments we used a configuration, where a pure metal donor on a transparent carrier is transferred by ultra-short picosecond (ps) laser pulses on to a moving substrate as illustrated in Figure 4. The donor layer is positioned face-down, close to the substrate surface. The thickness of the resulting air-gap between donor and substrate is typically 10-50 μm.

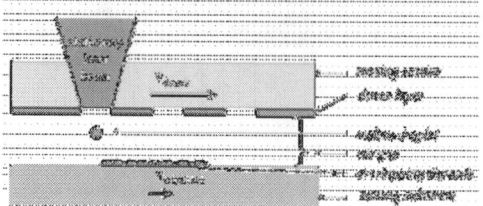

Figure 4. Principle of Lased-Induced Deposition.

In this way, metal deposits in the μm range can be formed using a ps-laser with a spot size of 5-20 μm focused onto the metal layer. To build 2D and 3D structures, overlapping droplets need to be deposited (Figure 4), while scanning the substrate ($v_{substrate}$). The donor layer is fully used for each laser shot. Hence, the donor layer should be refreshed at a higher rate (v_{donor}) than the scanning motion.

Using a straightforward laboratory-scale donor refreshment set-up, experiments have been performed using green and UV wavelengths of the ps-laser machining facility at the Twente University.

First, the deposited feature size was investigated as a function of laser power and air gap size. Subsequently, based upon the identified process parameters high aspect ratio vias were filled with copper.

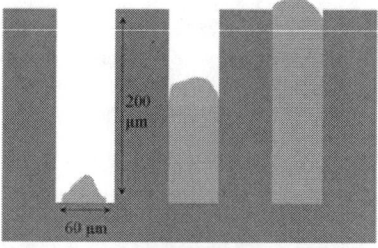

Figure 5: Deposition inside a high AR TSV: a) Schematic overview with the via dimensions used. b) SEM image of the deposited copper droplets on the bottom of the via.

290

In the current stage of the development, industrial feasibility cannot be assessed in an absolute sense. However, the production rate of LIFT can be estimated by extrapolating the current results combined with the properties of state-of-the-art picosecond lasers.

As an example, this has been done for three different processes: TSV filling, conductor lines (like solar busbars) and printed wirebonds. It was shown that, for different types of applications, production rates can be achieved that justify further research [15].

For TSVs, the estimated production rate has been used to perform a preliminary C-o-O analysis, including estimations on tool cost, throughput and consumables.

Based on these estimates, it can be calculated that, by applying LID instead of in-resist plating, the C-o-O for 10 x 50 μm (AR 1:5) via metallization can be reduced from 42 €/wafer to 19 €/w at a via density of 100 TSV/mm^2 and 6 €/w at 10 TSV/mm^2.

This example calculation indicates that for low TSV densities, LID is a promising technology to strongly reduce the costs of complex 3D interconnects.

Cost-of-Ownership (C-o-O) considerations.

Sematech published a C-o-O model [20] that has been used here to evaluate the proposed low cost-of-ownership TSV technologies. Some of the thumb assumptions have to be made concerning C-o-O estimates as the presented technologies are under development. In table 1 an assessment of TSV drilling and filling processes is given. The current TSV processes based upon DRIE and galvanic plating for dilling and filling is expensive, mainly due to low throughput of DRIE processes and slow plating processes combined with high capital expenditures. To come to lower C-o-O for TSV creation faster processes and / or lower capital investments are needed.

Technology	Cost of Ownership	Industrial Maturity
Drilling		
DRIE	High	Production
Laser	Medium	Production
ECM	Low	Research
Fast DRIE	Medium	Research
Filling		
Galvanic	High	Production
ECPR	Medium	Development
ECM	Low	Research
LID	Low	Research

Table 1. Assessment of C-o-O of different TSV technology options.

Discussion

The novel approaches presented here to drill and fill TSVs are still in early development stages. First experimental work and modeling indicates that that the novel approaches have potential. To come to low cost TSV technologies this type of innovative work is necessary.

The ECM technology appears to be limited towards the via diameters that can be achieved while working with electrode arrays. Combination of processes, to simplify logistics end decrease handling probably is an advantage of this technology. ECM technology for TSV technology is especially attractive for TSV-based silicon interposers with medium sized TSV diameters (50 - 100 μm).

Fast DRIE is directed to overcome one of the main disadvantages of the conventional DRIE process, being the low etch rate (now typically 3 μm/min). In combination with gas bearing wafer transport technology, it is expected that throughput numbers of > 100 wafers per hour per tool can be achieved, thus dramatically reducing C-o-O for DRIE.

Laser Induced Deposition is attractive because it has the potential to be a very flexible technology for both writing redistribution layers on wafers as well as filling vias.

Despite the challenges lying ahead for the novel approaches presented here, large scale application of 3D technology requires breakthrough manufacturing technologies to enable lower Cost-of-Ownership. In this work we identified some technology options that can make the desired impact.

Acknowledgements

The authors wish to acknowledge the funding of this work by TNO and by ALSI in the scope of its 3D program.

References

[1] P. Garrou, C. Bower and P. Ramm, eds., Handbook of 3D Integration: *Technology and Applications of 3D Integrated Circuits',* Wiley-VCH Verlag, Weinheim, 2008.

[2] P. Allongue, P. Jiang, V. Kirchner, A. L. Trimmer, and R. Schuster, '*Electrochemical Micromachining of p-Type Silicon'*, J. Phys. Chem. B **2004,** *108,* 14434.

[3] Y.P.R. Lamy, K.B. Jinesh, F. Roozeboom, D.J. Gravesteijn and W.F.A. Besling, '*RF characterization and analytical modelling of through silicon vias and coplanar waveguide for 3D integration'*, IEEE Trans. on Advanced Packaging, **33**, 1072-1079 (2010).

[4] P. Siblerud, EMC3D consortium overview and CoO model, EMC3D Japan/Korea Technical Symposium, April 23, 2007

[5] T. de Hoog, A. Hoogstrate and B. van der Zon, , Mechatronica magazine, Feb. 28, 2011 (in Dutch).

[6] P. Garrou, 3D IC toolset readiness, Cu bonding, interposer failings, Solid State Technology, May 2011, 8.

[7] V. Lehman, Electrochemistry of Silicon: Instrumentation, Science, Materials and Applications. Wiley 2002, ISBN: 3-527-29321-3.

[8] P.M.M.C. Bressers, M. Plakman, and J. J. Kelly, J. Electroanal. Chem. 1996, **406**, 131.

[9] C.-L. Lee, Y. Kanda, S. Ikeda and M. Matsumura, '*Electrochemical method for slicing Si blocks into wafers using platinum wire electrodes*', Solar Energy Materials & Solar Cells **95** (2011) 716.

[10] P. Allongue, P. Jiang, V. Kirchner, A. L. Trimmer, and R. Schuster, Electrochemical Micromachining of p-Type Silicon, J. Phys. Chem. B 2004, **108**, 14434.

[11] T. Sugita, K. Hiramatsu, S. Ikeda, and M. Matsumura, Micromachining of silicon using anodized needle electrodes in HF, Abstract #1151, 219th ECS Meeting, Montreal, May 1-5, 2011.

[12] F. Roozeboom, A.M. Lankhorst, P.W.G. Poodt, N.B. Koster, G.J.J. Winands, and A.J.P.M. Vermeer, '*Ultrafast gas-bearing Reactive Ion Etching*', Patent application EP101554955.8, filed Febr. 26, 2010..

[13] F. Roozeboom, A.M. Lankhorst, G. Winands, N.B. Koster, P. Poodt, A. Vermeer, G. Dingemans and W.M.M. Kessels, '*Concept of spatially divided Deep Reactive Ion Etching with ALD-based passivation*', ALD 2011 Conference, Cambridge (USA) June 23-26, 2011.

[14] G. Dingemans, *et al.*, Electrochem. Soc. Trans. **35** (4) (2011), 191.

[15] G. Oosterhuis, 2D and 3D Interconnect Fabrication by Picosecond Laser Induced Forward Transfer, Proc. 11th EUSPEN Int. Congress, 2011, Lake Como, Italy, May 23-26, 2011.

[16] D.P. Banks, '*Femtosecond Laser Induced Forward Transfer Techniques for the Deposition of Nanoscale, Intact, and Solid-Phase Material*', PhD thesis University of Southampton, Nov. 2008, UK.

[17] S. Bera A.J. Sabbah, J.M. Yarbrough, C.G. Allen, B. Winters, C.G. Durfee, and J.A. Squier, Applied Optics, **46**, (2007). 4650

[18] A. Narazaki et al. 2009, Appl. Surface Science **255** (2009) 9703.

[19] S.H. Ko et al., Nanotechnology **18** (2007) 345202.

[20] R. Doering, Handbook of Semiconductor Manufacturing, CRC Press, 2008, ISBN 1-57444-675-4.

Acceleration factors of combined reliability tests of lead-free SnAgCu BGA interconnections

Przemysław Matkowski

Faculty of Microsystem Electronics and Photonics, Wroclaw University of Technology
ul. Janiszewskiego 11/17, 50-372 Wroclaw, Poland

Phone: +48 713531053, Fax: +48 713531055
E-mail Address: Przemyslaw.Matkowski@pwr.wroc.pl

Abstract

The essence of accelerated reliability test is to accelerate degradation process of tested object and the assessment of its reliability in test and field conditions. Acceleration factor is a unitless parameter that relates a life time of objects at an accelerated stress level to the life time at the use stress level. Acceleration factor of a single loading reliability tests can be calculated on the basis of known models such as Modified Coffin-Manson, Norris-Landsberg or vibration acceleration presented in IPC-SM-785. Acceleration factors of combined tests such as temperature cycle loading combined with vibration loading can not be calculated because such models do not exist. Therefore extensive research are currently carried out. Within the frame of this paper the procedure and the results of the study focused on acceleration factor calculation of combined reliability tests will be presented. Applied method of event detection during accelerated reliability tests and the results of high resolution 3D computed tomography inspections of cracked interconnections will be also presented.

Key words: accelerated reliability test, combined loadings, lead-free, BGA

Introduction

According to 2002/95/EU Directive on Restriction of Hazardous Substances (RoHs) consumer electronics introduced into European markets can not contain hazardous substances such as cadmium, lead, mercury etc. since 1st of July 2006. As the consequence of the restrictions lead-based solder alloys had to be replaced by lead-free alloys. In many research and scientific centers research works on the composition of lead-free alloys have been carried out for many years. Nowadays more than 340 lead-free alloys have already been patented. Among these solder alloys world consortia chose some alloys as strongly recommended alloys for industrial purpose. Ternary solder SnAgCu alloys based on tin, silver and copper are recommended for reflow soldering of surface mount devices (SMDs) by SOLDERTEC, JEITA (Japan Electronic Industries Development Association) and American NEMI (National Electronics Manufacturing Initiative).

The first consequence of lead-free soldering was higher temperature of soldering – reflow temperature of SAC 305 (SnAg3.0Cu0.5) commonly used by electronic industry is about 240°C higher than reflow temperature of eutectic lead-based alloy. Recommended peak temperature of reflow process is 235-245°C (depending on the producer of a solder paste). In fact solder lines had to be modified and adapted to carry out the soldering processes of electronic components on printed circuit boards (PCBs) at higher temperatures. Also components had to be adopted to withstand higher temperatures. Integrated circuit inside polymer package can withstand temperature about 260°C. Reflowing at high peak temperature above 260°C can damage the heat sensitive components (silicon chip) and cause warping the PCB. IPC/JEDEC J-STD-020 is the latest standard for the allowable peak temperature on the component body.

As well as solder lines or electronic components also soldering materials such as fluxes, pad or lead coatings had to be adapted to the new lead-free alloys. Nowadays VOC-free (Volatile Organic Compounds free), no-clean fluxes and ENIG (Electroless Nickel Immersion Gold), ImSn (Immersion Tin), HASL (Hot Air Solder Level) pad coatings are used. New material compositions (solder alloy/pad coating) caused changes within a solder joint structure eg different Intermetallic Compounds (IMCs) are formed during soldering process, more defects such as Kirkendall voids, micro cracks, gas pores are observed etc. Presence of defects or thick, uniform and continuous layers of IMCs can decrease fatigue strength of solder joints significantly [1-2]. In fact, although years of investigations and modifications, lead-free solder joints are still much less resistant to mechanical loading and fatigue failures than lead-based solder joints. Lead-free solder joints subjected to outer mechanical or environmental loading crack easily.

The current trends towards miniaturization of electronic devices as well as increasing their functional resources cause more and more leads (input and output pins) in increasingly smaller packages of integrated circuits. For this reason nowadays BGA (Ball Grid Array) packages designed for surface mounting are commonly used for a wide area of applications, including portable electronics such as notebooks, mobile phones, tablets as well as avionics, automotive industry etc. Depending on the construction, materials and ball size there is a lot of BGA package variants (eg CABGA – Chip Array Ball Grid Array, PBGA – Plastic Ball Grid Array, FBGA- Fine Pitch Ball Grid Array etc.).

The main advantage of BGA packages is much higher density of leads, (and smaller size as a consequence) compared to QFP (Quad Flat Package), PLCC (Plastic Leaded Chip Carrier) or packages designed for through hole mounting such as PGA (Pin Grigd Array), DIP (Dual-In-Line) etc. Thanks to very short distance between the package and the PCB thermal resistance between them is lower which affects more efficient heat dissipation. Other advantage is lower inductance of leads (electrical conductor between the package and the PCB) which affects less distortion of signals in high-speed electronic circuits.

The main disadvantage of BGA packages is fact that solder joints are much less resistant to thermal stress caused by differences of coefficient of thermal expansion between the package and the PCB or to mechanical stress caused by flexing or deformation of the PCB subjected to vibration or mechanical shocks. Under thermal or mechanical stress solder joints of BGA packages fracture easily. It is a common problem of graphic processor units mounted in notebooks.

Due to the problem of BGA inspection (solder joints are under the package) and the problem of manual soldering (positioning problem) any service works over the BGA packages are much more difficult and expensive than in case of non-BGA packages. Advanced equipment for inspection such as 2D X-ray imaging systems, 3D mico-CT X-ray imaging systems or special optical microscopes are needed. Also some dedicated equipment for soldering such as rework stations with infrared lamps or hot air, thermocouple and vacuum device for lifting are necessary.

Accelerated reliability tests

The essence of accelerated reliability tests is to accelerate degradation process of objects and the assessment of object reliability in field conditions. Based on the results of accelerated reliability tests, life cycle of objects can be estimated. Depending on loading or combinations of loadings, accelerated reliability tests vary in degradation effect eg the rate of creep, fatigue, electromigration etc. Acceleration factor of accelerated reliability tests was introduced in order to standardize accelerated reliability tests. Acceleration factor is a unitless parameter that relates a life time of objects at an accelerated stress level to the life time at the use stress level. Acceleration factor can be calculated on the basis of known model or determined experimentally. If acceleration factor of applied accelerated reliability test is known, life time of the tested objects in real field conditions can be calculated from the measured life time in testing conditions.

Test specimens

Lead-free SnAg3Cu0,5 (known as SAC 305) solder joints of BGA 256 daisy chains were the object of the study. SAC 305 was chosen because it is commonly used by electronic industry as a substitute of lead contained alloys. BGA 256 package is also used commonly by industry as a package for integrated circuits.

The BGA daisy chains were soldered to pads of test PCBs during reflow soldering process which was carried out in professional reflow oven in nitrogen atmosphere. Test PCBs were two sided, six layers PCBs made of epoxy resin reinforced with glass fabric. The thickness of PCBs was 1 mm. Copper solder pads were coated by electroless nickel immersion gold coating (ENIG) with a thickness of 4-5 µm (Ni) and 0,09-0,12 µm (Au). ENIG coating was chosen because it is commonly used int the case of fine pitch (as BGA 256) electronic assembly thanks to its great flatness.

In order to connect tested chains of solder joints to the event detector D-sub connectors (TH-through hole assembly) were applied. According to previous study D-sub connectors are much more resistant to noises caused by vibration in comparison to edge connectors, Centronics connectors etc.[1]. The holes for TH assembly, drilled in PCBs were metallized during manufacturing process. The connectors were soldered manually after reflow process and protected in order ensure high resistivity to vibration and temperature cycles (TH solder joints must be much more reliable than tested BGA solder joints). Both sides of a test PCB and X-ray imaging of both BGA soler joints and TH solder joints are presented in Figure 1.

According to Figure 1 each test PCB includes just one BGA daisy chain mounted in the middle. According to previpous investigations done usig 2D scanning laser vibrometer the middle of a PCB clamped along its opposite edges (as presented in Figure 7) and subjected to vibration is the most displacement area [1].

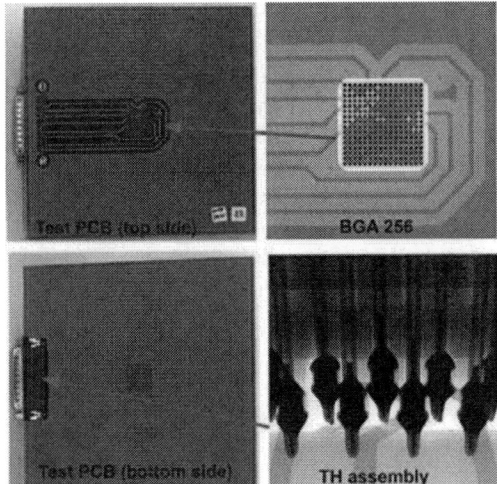

Figure 1: Both sides of a test PCB, 2D X-ray imaging of BGA 256 and TH connector

Every BGA daisy chain which is soldered to a test PCB includes four fully independent chains of solder joints. Each chain consists of 64 solder joints connected in series. The location of outer, two middle and inner chains is presented in Figure 2. The location of the chains is not accidental. The top side layout of the PCB was designed in order to determine influence of solder joint location on its fatigue when subjected to combined thermal and mechanical loadings.

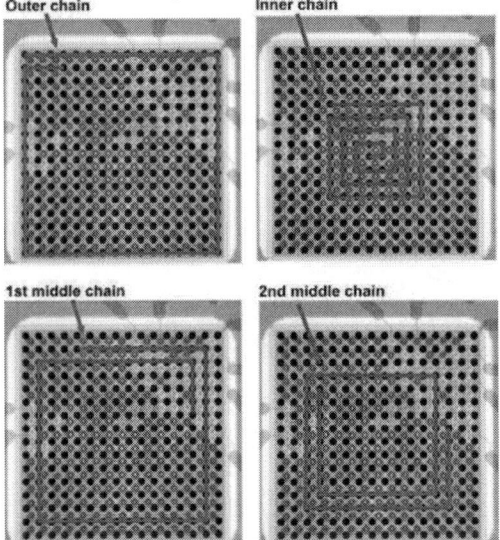

Figure 2: Chains of solder joints

After reflow process PCBs were subjected to 2D and 3D X-ray inspection. X-ray imaging of BGA solder joints (chip and PCB side view) is presented in Figure 3.

Figure 3: 2D X-ray imaging of solder joints (chip and PCB side)

3D X-ray imaging of BGA solder joints is presented in Figure 4

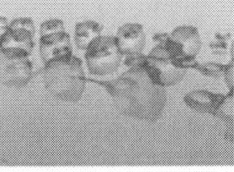

Figure 4: 3D X-ray CT imaging of BGA solder joints

All inspected PCBs met the standard requirements, no soldering defect has been observed. Twelve test PCBs were divided into three groups and subjected to three independent accelerated reliability tests after inspection.

Test conditions

Three accelerated reliability tests were carried out in research laboratory LIPEC of Wroclaw University of Technology. Climatic chamber combined with electrodynamic shaker was used in order to subject solder joints to temperature cycles combined with vibration. The equipment used during the reliability tests is shown schematically in Figure 5.

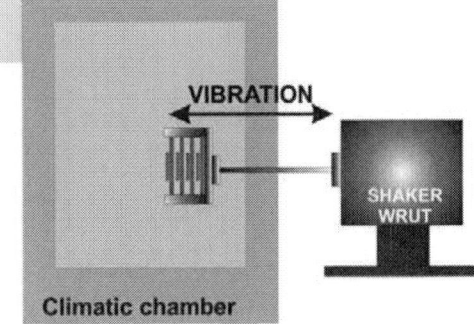

Figure 5: Construction of the climatic chamber and the electrodynamic shaker available in LIPEC laboratory

For each test a single temperature cycle was about 180 minutes long. Temperature changes inside the climatic chamber during a single temperature cycle are presented in Figure 6.

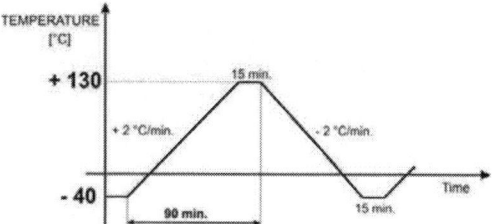

Figure 6: Temperature changes during a single temperature cycle

According to the graph presented in Figure 6. temperature varied from -40 to +130°C. Three reliability tests carried out differed just in mechanical loading (thermal loading was the same in each case). In all three cases mechanical loadings were fixed during the tests. During each test four PCBs were tested (four BGA 256 = 16 chains of solder joints = 1024 solder joints).

During the first and the second test solder joints were subjected to random vibration. In both cases the electrodynamic shaker was stimulated by white noise over a frequency bandwidth from 50 to 500 Hz, but they differed in root mean square of acceleration (G_{RMS}) and power spectral density (PSD) measured in the middle of the PCBs. Root mean square value of acceleration in the middle of tested PCBs was equal to 40 G_{rms} (PSD = 3,5 G^2/Hz) and 30 G_{RMS} (PSD = 2 G^2/Hz) in the case of the first and the second test respectively.

During the third test solder joints were subjected to sinusoidal vibration. In this case electrodynamic shaker was stimulated by sinusoidal signal with a frequency of 200 Hz. Root mean square value of acceleration in the middle of tested PCB was equal to 40 G_{rms}. The natural frequency of tested PCBs was 260 Hz.

Control and providing appropriate conditions of mechanical loading is crucial for comparative analysis. Therefore some measurements of acceleration were carried out before each reliability test. The acceleration was measured in the middle of test PCBs, precisely in the area where the BGA packages were mounted. The location of measurements is presented in Figure 7.

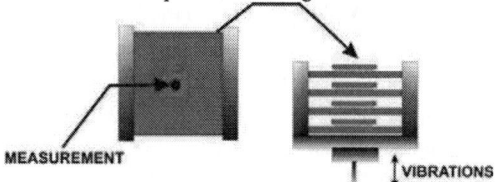

Figure 7: Location of acceleration measuremets

Measurements were performed using accelerometers in order to control the mechanical loading precisely. The information about acceleration of shaker head or holder is insufficient because printed circuit boards are made of FR4 laminate which is flexible material and its

deformation, acceleration and displacement during vibration strongly depend on temperature [1,4]. Previous investigations on the effects of substrate material on durability of solder joints confirm the concerns [6]. Moreover construction of a PCB (number of layers, layout design) can affects its flexibility, stress distribution and durability of solder joints in practice.

Event detection during reliability tests

Condition of all tested solder joints must be monitored continuously and simultaneously during accelerated reliability tests [1,5-6]. In order to detect temporary opens of electrical circuits caused by cracked solder joints subjected to mechanical loading, fast measurements of resistance changes must be carried out. In such study the sampling frequency must ensure detection of events lasting even several microseconds. Duration of the events is determined by the destruction process as well as frequency of mechanical loading (cracked solder joint becomes a vibrating mass).

In order to precisely detect time to failure of tested solder joints an event detector was applied. The multichannel event detector combined with data acqustion system was developed in LIPEC laboratory within the frame of PhD thesis [1] and successfully applied in many previous investigations of reliability of solder joints, electrically conductive adhesive joints or connectors [1-3].

The system and other equipment such as climatic chamber or electrodynamic shaker used for accelerated reliability testing are presented in Figure 8.

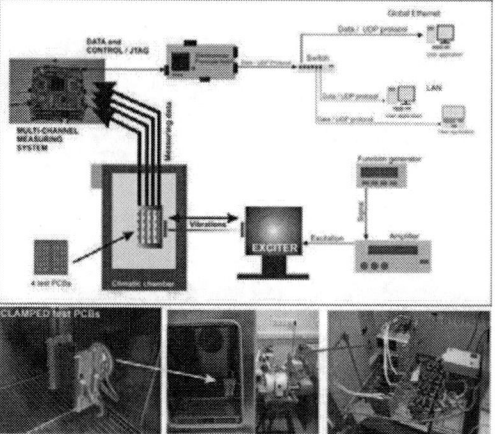

Figure 8 Equipment for accelerated reliability testing: view of clamped PCBs, climatic chamber, electrodynamic shaker and the event detector

The stationary version of the system is able to detect resistance changes of 128 channels simultaneously. The system measures resistance using four-probe method. The maximum sampling frequency is 64 Msps (Megasamples per second). The event detector is able to detect all temporary shorts and opens lasting several μs that occur during reliability test (which may take several weeks). Details such as time to failure, event duration, resistance value, number of events in specified time period and current status are stored by the data acquisition system. The event detector is connected to local area network (LAN) via the microcontroller that is responsible for data transmission (UDP protocol). [1, 4-5]

Developed event detection algorithm based on two resistance thresholds and counters in time and resistance domains was implemented in order to avoid noises. The algorithm is presented in Figure 9. Its functionality can be summarized as follows: the event is detected when resistance exceeds both thresholds and remains above both thresholds for a certain time (as presented in Figure 9).

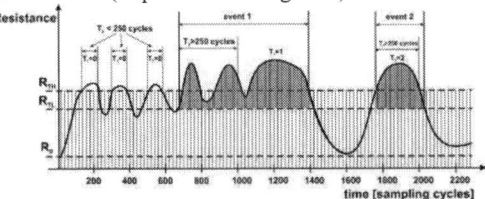

Figure 9: Conditions of event detection by the event detector

Collection of the data about condition of tested specimens without loosing any important data was possible thanks to application of advanced hardware level pre-processing and implementation of developed transmission protocols. These solutions enabled to reduce output stream from several GBps to the size available to store by hard drives of commercial computers.

During current study just 16 channels were monitored simultaneously (16 chains of solder joints). Sampling frequency was 1,25 Msps. It was reduced in order to detect temporary events lasting at leas 10 μs. The PCBs were connected to the system using 64 elastic shielded wires. The data was stored at hard drives of computers connected to LAN.

Results

The initial resistance of tested chains (initial resistance of 64 solder joints connected in series) was about 800 mΩ. According to applied algorithm the event was detected when the resistance exceeded 200% of the initial value (about 1,6 Ω). Detected time to first failure during 330-hour reliability test (110 temperature cycles) depending on mechanical loading is presented in Table 1.

According to the results time to failure strongly depends on both combination of combined loading (vibration level, type of vibration) as well as on localization of a solder joint within the frame of a ball grid array package.

Table 1: Time to failure detected by event detector during reliability tests

Failed chains	Number of temperature cycles to first failure (chain localization)		
	Random vibration 40 G_{rms}	Sinusoidal vibration 40 G_{rms}	Random vibration 30 G_{rms}
1st	1 (outer)	87 (outer)	1 (outer)
2nd	1 (outer)	97	16 (outer)
3rd	5 (outer)	99	18 (outer)
4th	6 (outer)	105 (outer)	40 (outer)
5th	26	105 (outer)	85
6th	40	105 (inner)	94
7th	47 (inner)	106 (inner)	-
8th	-	107 (outer)	-
9th	-	107	-

Temperature cycles combined with random vibration are much more destructive than temperature cycles combined with sinusoidal vibration, event if a root mean square value of acceleration was the same in both cases (acceleration was measured in the middle of PCBs and on the holder). Reliability tests shown also that outer chains of solder joints are the least reliable. Outer chains are the least resistant to fatigue because they crack considerably faster than other chains. The most reliable chains of solder joints were two middle chains. Outer and middle chains of solder joints are presented in Figure 2.

Data collected by the event detector is presented in Figure 10. An outer chain subjected to temperature cycles and random vibration was chosen as an example of failed chain. Plots of resistance changes and number of events (temporary shorts and temporary opens) detected and recorded are presented in Figure 10.

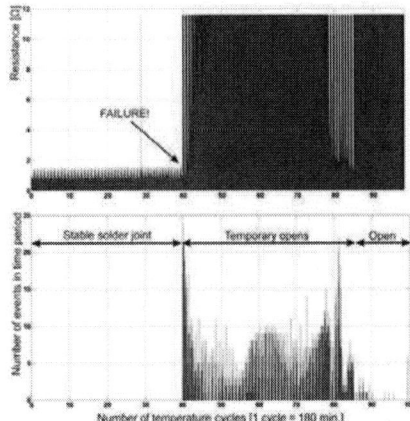

Figure 10: Changes of resistance and number of events detected by the event detector during reliability test of an outer chain (random vibration, 50-500 Hz, 30 G_{rms})

It can be seen that during first 40 temperature cycles all solder joints are in good condition. The number of events detected in constant time period is zero. Visible sinusoidal changes of resistance are caused by cyclic changes of solder joint temperature which is determined by temperature inside the climatic chamber. The resistance changes depend on temperature coefficient of resistance of applied solder alloy.

During the 40th temperature cycle first failure occurred. The failure is detected when any of tested 64 solder joint cracks and its resistance grows rapidly above specified thresholds. A cracked solder joint subjected to vibration is similar to a pin connector which is sensitive to vibration (its resistance changes because of vibration) so the event detector started detecting some temporary events. Cracked solder joints can still conduct electrical current but can also cause noises in electrical circuits when subjected to any mechanical loading. Total number of temporary opens detected within a specified time period was changing during the reliability test. It can be seen on the plot presented in Figure 10. The changes are periodical.

According to the plots just after the 95th temperature cycle the value of resistance still exceeded the measuring range but the number of events was reduced to 0. It means that permanent open (not sensitive to vibration) occurred. It is similar to previous investigations [1-3].

2D X-ray 3D and X-ray CT inspections were carried out in order to determine failure mode of failed chains. The analysis of solder joints, PCB, patches and BGA packages shown that cracks of solder joints along BGA metallization were the failure mode during reliability tests. Examples of cracks are presented in Figure 11. In most cases cracks were propagated across voids (voids were formed during reflow soldering process).

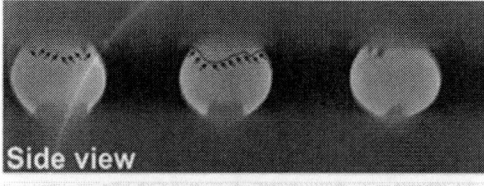
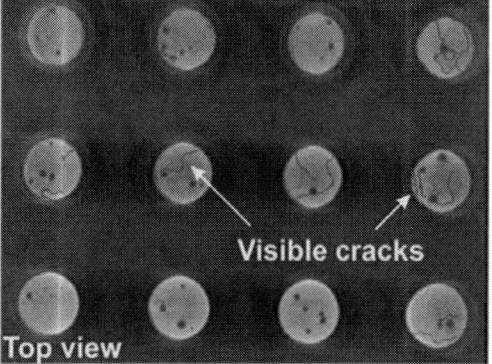

Figure 11: Cracks of solder joints along BGA metallization (some voids are visible)

Some cracks presented in Figure 11 were painted in order to make them more visible for readers of this paper.

Calculation of acceleration factors

According to the IPC standard acceleration factor (AF) from test to use condition can be based on the mean time to failure (MTTF) [10]:

$$AF = \frac{MTTF(Use)}{MTTF(Test)}$$

Use condition can be replaced by condition of other accelerated test in order to compare the results from different accelerated test conditions:

$$AF = \frac{MTTF(Test2)}{MTTF(Test1)}$$

Within the frame of current study three accelerated reliability tests that differed in mechanical loading were carried out. MTTF of each test can be calculated on the basis of detected time to failure. According to the time to first failure listed in Table 1, the only group of chains, which is damaged in 100% after reliability tests is a group of outer chains (outer solder joints presented in top left corner of Figure 2). Therefore only time to failure of outer chains will be used for preliminary estimation of acceleration factors. The mean time to failure of outer chains for each test is presented in Table 2.

Table 2: Mean time to failure of outer chains

Test condition (temperature cycles & specified vibration)	MTTF [hours]
Random vibration 50-500 Hz 40 G_{rms}	10
Sinusoidal vibration 50-500 Hz, 40 G_{rms}	303
Random vibration 200 Hz 30 G_{rms}	56

According to the results of previous investigations there is a strong interaction between temperature cycling and vibration which influences time to failure of solder joints subjected to combined loadings [3]. Nevertheless in the case of this study, temperature loading was the same in the case of each reliability test (the same temperature profile of temperature cycles) so the primary damage mechanism, creep-fatigue ought to be the same. Therefore interaction between temperature cycles and vibration will be skipped in order to simplify the preliminary calculations.

The first reliability test (combined temperature cycles with random vibration, 40 G_{rms}) was chosen as an accelerated reliability test (test1) for the calcultations. Results of acceleration factor calculation are presented in Table 3. It should be underlined that calculated acceleration factors are acceleration facors from test 1 to test 2 conditions, not from test to use conditions.

Table 3: Acceleration factor calculation results

Test1 (accelerated test)	Test2 (Reference test)	AF $\dfrac{MTTF(Test2)}{MTTF(Test1)}$
Random vibration 40 G_{rms}	Random vibration 30 G_{rms}	5,6
Random vibration 40 G_{rms}	Sinusoidal vibration 40 G_{rms}	30,3

According to the IPC standard a simplified fatigue relationship is often used to scale vibration levels, test durations, and derive acceleration transform if the loading is random vibration:

$$AF = \frac{PSD1}{PSD2}^{M} = \frac{time2}{time1}$$

A starting value of M for solder attachments is 3 to 4 [10]. Values of PSD level for both random vibration tests are presented in Table 4.

Table 4: PSD level of random vibration

Test condition	PSD [G^2/Hz]
Random vibration 50- 500 Hz, 40 G_{rms}	3,5
Random vibration 50- 500 Hz, 30 G_{rms}	2

The M parameter can be calculated as:

$$M = \log_{\frac{PSD1}{PSD2}} AF = \log_{\frac{3,5}{2}} 5,6 = 3,08$$

Calculated value is within range described in the IPC standard [10].

Conclusions

According to the results of the reliability tests of BGA daisy chains (BGA 256) outer chains of lead-free solder joints are much less resistant to fatigue under combined loading than other chains. Lead-free solder joint failures at the die edge are a common problem of GPUs (BGA packages) mounted in notebooks.

Cracked solder joints can still conduct electrical current in static conditions, but when are subjected to vibration they becomes the source of temporary shorts and opens.

Vibration loading parameters have a great influence on fatigue of lead-free solder joints subjected to combined loadings. Temperature cycles combined with random vibration are much more destructive for lead-free solder joints than temperature cycles combined with sinusoidal vibration, even if root mean square value of acceleration is the same in both cases. Random vibration is defined over a revelant frequency while sinusoidal over just one frequency, therefore random vibration loading is more destructive. When sinusoidal vibration was replaced by random vibration, the mean time to failure was decreased over 30 times. When root mean square value of acceleration (random vibration) was increased from 30 to 40, the mean time to failure was decreased over 5 times.

Acceleration factors from one test to the other were calculated. Calculated value of M parameter for a simplified fatigue relationship was within the range described in the IPC standard.

Further studies are in progress. Knowledge about acceleration factors (especially of combined loadings) will be important for the design of electronic equipment commonly exposed to heavy vibrations in various environmental loadings.

Acknowledgements

Thanks to Mr. Krzysztof Laskowski and Mr. Andrzej Mak – graduates who have participated actively in these research works.

Thanks to the company ELDOS and the company SONEL for their engagement in technological process.

These works were partially supported by The National Centre for Research and Development (NCBiR), Grant No. NR02-0001-10/2010

References

PhD dissertation
[1] P. Matkowski, "Thermo-mechanical analysis of microelectrocnic lead-free joints", PhD dissertation, Wroclaw University of Technology, Wroclaw, pp. 7-175, 2010.

Articles from conference proceedings
[2] P. Matkowski, J. Felba, "Influence of solder joint constitution and aging process duration on reliability of lead-free solder joints under vibrations combined with thermal cycling", Electronic Systemintegration Technology Conference (ESTC), Berlin, September 13-16, 2010.
[3] P. Matkowski "Reliability of SnAgCu solder joints during vibration in various temperature", 34th International Spring Seminar on Electronics Technology, International Spring Seminar on Electronics Technology (ISSE), High Tatras, May 11-15, 2011.
[4] P. Matkowski, R. Zawierta, J. Felba "Vibration response of printed circuit boards in wide range of temperature. Characterization of PCB materials", International Spring Seminar on Electronics Technology, International Spring Seminar on Electronics Technology (ISSE), Brno, May 13-17, 2009.

[5] P. Matkowski, K. Urbański, T. Fałat, J. Felba, Z. Żaluk, R. Zawierta, A. Dasgupta, M. Pecht, "Application of FPGA units in combined temperature cycle and vibration reliability tests of lead-free interconnections", Electronic Systemintegration Technology Conference (ESTC), London, September 1-4, pp. 1375-1379, 2008.

[6] R. Zawierta, P. Matkowski, K. Urbański, J. Felba, "Data visualisation in a fast data acquistion system for long-term reliability tests of microelectronic interconnections", Electronic Systemintegration Technology Conference (ESTC), London, September 1-4, pp. 275-278, 2008.

[7] Qi H., Ganesan S., Wu J. Pecht M., Matkowski P. Felba J., „Effects of Printed Circuit Board Materials on Lead-free Interconnect Durablility" International IEEE Conference on Polymers and Adhesives in Microelectronics and Photonics (Polytronic) Wrocław, October 23-26, pp. 140-144, 2005.

Standards
[8] IPC-A-600 "Acceptability of Printed Boards".
[9] IPC-A-610D "Acceptability of Electronic Assemblies".
[10] IPC-SM-785 "Guidelines for Accelerated Reliability Testing of Surface Mount Solder Attachments".

Thermo-Mechanical Fatigue Life Evaluation for Sn-Pb and Sn-Ag Solders

Noritake HIYOSHI*, Takamoto ITOH** and Masao SAKANE***

*Department of Mechanical Engineering, Ishikawa National College of Technology
Kita-chujyo, Tsubata, Ishikawa 929-0392, JAPAN
Phone +81-76-288-8091, Fax +81-76-288-8102, E-mail hiyoshi@ishikawa-nct.ac.jp
**Department of Mechanical Engineering, University of Fukui
Bunkyo, Fukui 910-8507, JAPAN
***Department of Mechanical Engineering, Ritsumeikan University
Noji-higashi, Kusatsu 525-8577, JAPAN

Abstract

This paper describes a new thermo-mechanical fatigue (TMF) machine developed for solders and discusses TMF characteristics of lead and lead-free solders. Isothermal low cycle fatigue (LCF) tests were performed using Sn-37Pb, Sn-3.5Ag and Sn-3.0Ag-0.5Cu solders at 253K and 353K and in-phase and out-of-phase TMF tests using the same solders at the temperature range of 253K-353K. There was no large difference in fatigue life between LCF and TMF tests for Sn-37Pb solder but there exited clear difference depending on temperature and phase shift for Sn-3.5Ag and Sn-3.0Ag-0.5Cu solders. This study proposed an energy based method for estimating the TMF lives of Sn-3.5Ag and Sn-3.0Ag-0.5Cu solders.

Key words: Solder, Thermo-mechanical fatigue, Temperature, Fatigue life estimation

Introduction

Solder joints in electronic devices undergo thermal fatigue damage caused by the mismatch of thermal expansion coefficient of different connecting parts [1]. There are many papers reporting low cycle fatigue, creep, creep-fatigue lives for solders at constant temperatures [2-7, 16-17] but there is no study reporting TMF lives, whereas TMF lives are essential for the quality assurance of solder joints under actual service conditions [8-15]. One of the reasons of a lack of report on TMF lives for solders results from no TMF machine available to TMF tests in the temperature range of 233K-398K [18-22]. This study presents a new TMF testing apparatus originally developed for solders and discusses TMF lives of Sn-37Pb, Sn-3.5Ag and Sn-3.0Ag-0.5Cu solders by comparing those with isothermal LCF lives.

Experimental Procedure

Figures 1 and 2 show the schematic and photograph of TMF testing apparatus developed by the authors. The apparatus is equipped with an electric hydraulic servo loading system and temperature control system. The temperature control system consists of an induction heater and a cold air generator. Figure 3 shows the induction heater coil and cold air outlet in a chamber to confine cold air. By using this system, TMF testing in the temperature range of 253K-423K is capable and that in the temperature range of 233K-423K is also capable in addition of liquid nitrogen.

Temperature of specimen was measured by K-type thermocouple attached to the center of the specimen. Heating and cooling rate of specimen was 1K/s.

Materials tested in TMF tests were Sn-37Pb, Sn-3.5Ag and Sn-3.0Ag-0.5Cu solders of which the chemical compositions and heat treatments are listed in Table 1. The specimen was a solid bar with 10mm in diameter and 10mm in gage length as shown in Fig. 4.

Specimens were annealed at $0.87T_m$ for one hour before testing to stabilize the microstructure of the solders, where T_m is a melting point of solder. Casting, machining and heat treatment of specimens were exactly followed to the testing standard of solders issued from the Society of Materials Science, Japan [19, 22]. An axial extensometer with linear variable differential transformer was used to measure the axial total strains. To avoid the cracking at the location of the extensometer rod tips, two small bumps of epoxy resin with grooves were formed on specimen surface and the tips of extensometer rods were placed on the grooves [19, 22].

The total strain (ε_{total}) measured by the extensometer is composed of the mechanical strain (ε_{mech}) and the thermal strain ($\varepsilon_{thermal}$). The thermal strain should be obtained before TMF testing because only the mechanical strain gives TMF damage to specimen. Free expansion heating and cooling cycling with no load for a few cycles were performed to obtain the thermal strain. TMF tests were conducted under mechanical strain control that

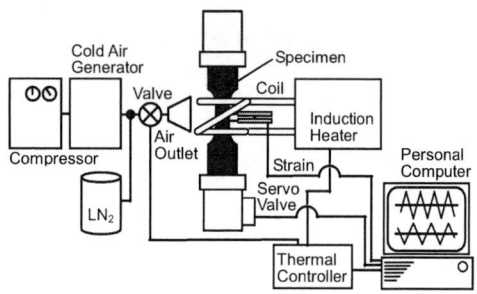

Figure 1: Schematic showing control system for thermo-mechanical fatigue testing.

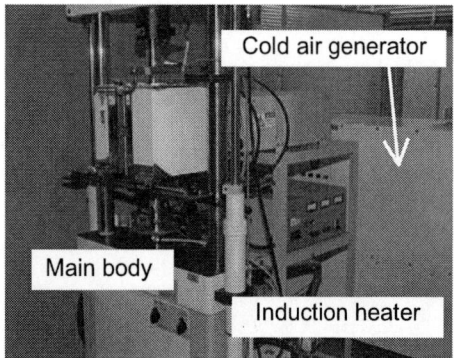

Figure 2: General view of thermo-mechanical fatigue testing apparatus.

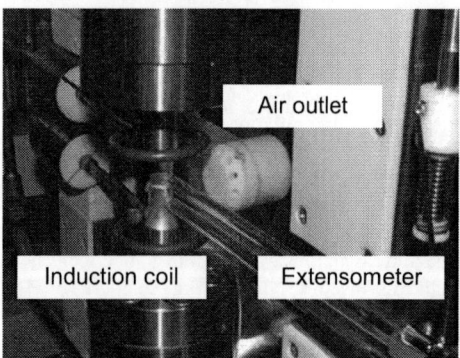

Figure 3: Induction heating coil and cold air outlet inside chamber.

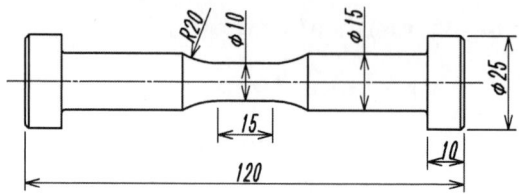

Figure 4: Shape and dimensions of the specimen (mm).

was calculated by subtracting the thermal strain from the total strain.

Two types of TMF tests were performed with triangle strain and temperature waveforms in the temperature range of 253K-353K. One was the in-phase test in which the mechanical strain waveform had no phase shift with the temperature waveform. The other was the out-of-phase test where the mechanical strain waveform had 180-degree phase shift with the temperature waveform. The tensile mechanical strain was applied at higher temperature in the in-phase test and the compressive mechanical strain at higher temperature in the out-of-phase test. The number of cycles to failure (N_f) was defined as the cycle of 25 percent tensile stress amplitude drop from that at a midlife.

Experimental Results and Discussion

Effect of Phase Shift on Stress-strain Hysteresis Loop

Figure 5 (a) shows hysteresis loops of three kinds of solders at a midlife in the in-phase TMF test. Asymmetrically shaped hysteresis loops are found in the test giving smaller tensile stress amplitude compared with the compressive stress amplitude. The peak tensile stress amplitude was +20.8MPa while the compressive peak stress amplitude was -30.9MPa for Sn-37Pb solder. The same trend, smaller tensile peak stress and larger compressive peak stress, was also found in Sn-3.5Ag and Sn-3.0Ag-0.5Cu solders. The smaller tensile stress results from the lower flow stress at higher temperature.

Table 1: Chemical composition and heat treatment of the materials tested (wt%).

	Sn	Pb	Sb	Cu	Bi	Fe	As	Cd	Ag
Sn-37Pb	63.08	Bal.	0.018	0.0006	0.0039	0.0004	0.0007	0.0001	—
Sn-3.5Ag	Bal.	0.0096	0.0077	0.0005	0.012	0.0061	0.001	0.0001	3.51
Sn-3.0Ag-0.5Cu	Bal.	0.0061	0.0054	0.50	0.0024	0.0006	0.0020	—	2.99

Sn-37Pb : 397K × 1h air cooling
Sn-3.5Ag : 430K × 1h air cooling
Sn-3.0Ag-0.5Cu : 430K × 1h air cooling

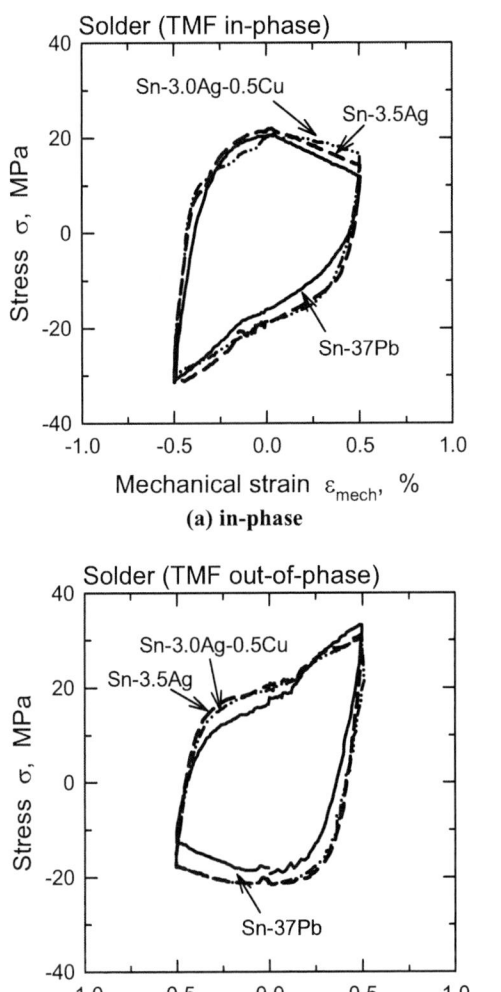

(a) in-phase

(b) out-of-phase

Figure 5: Hysteresis loops in TMF test at $N_f/2$.

Asymmetrical hysteresis loops are also found in the results of the out-of-phase TMF tests shown in Fig. 5 (b). The tensile peak stress was conversely larger than the compression peak stress in the out-of-phase TMF tests. The tensile peak stress and compressive peak stress are +33.2MPa and -19.2MPa for Sn-37Pb, +31.1MPa and -21.5MPa for Sn-3.5Ag, +30.4MPa and -21.3MPa for Sn-3.0Ag-0.5Cu, respectively. The difference in the peak stress between tension and compression results from the difference in the flow stress depending on temperature. The hysteresis loops shown in Fig. 5 (b) are the same in shape as those in Fig. 5 (a) if the formers rotate 180 degrees.

LCF and TMF Life Estimate

Table 2 and Fig. 6 correlate between the LCF lives and the TMF lives, where the total strain

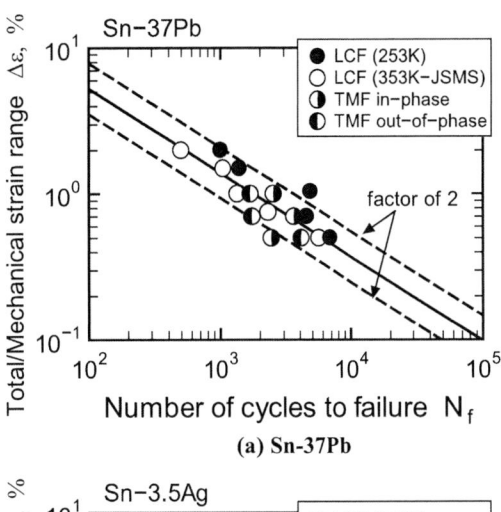

(a) Sn-37Pb

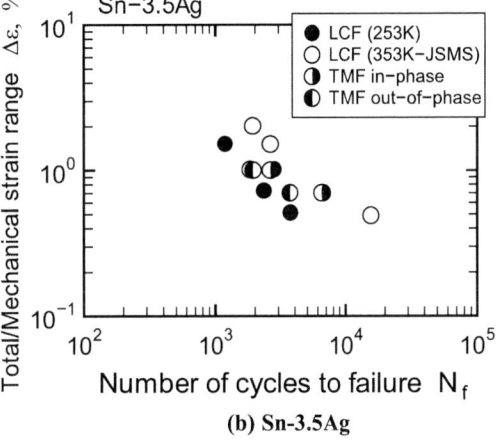

(b) Sn-3.5Ag

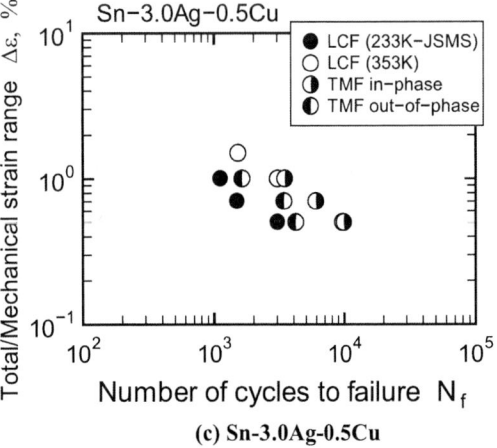

(c) Sn-3.0Ag-0.5Cu

Figure 6: Correlation of LCF and TMF lives with total / mechanical strain range.

range ($\Delta\varepsilon_t$) is used for LCF test and the mechanical strain range ($\Delta\varepsilon_{mech}$) for TMF test. Figure 6 (a) is the results in Sn-37Pb solder, where the solid line is drawn based on the LCF data at 253K and 353K and the dashed lines indicate a factor of two scatter band.

303

Table 2: Summary of LCF and TMF lives.

			Total / Mechanical strain range (%)				
			0.5	0.7	1.0	1.5	2.0
Sn-37Pb	LCF	253K [7]	6850	4490	4800	1400	1000
		353K [16]	5600	2300	1350	1050	500
	TMF	in-phase	2430	3600	2520	---	---
		out-of-phase	4100	1750	1700	---	---
Sn-3.5Ag	LCF	253K [7]	3730	2347	2794	1198	---
		353K [16]	15758	6650	1860	2622	1940
	TMF	in-phase	---	6500	2620	---	---
		out-of-phase	---	3700	1950	---	---
Sn-3.0Ag-0.5Cu	LCF	233K [23]	3028	1506	1121	---	---
		353K	9700	6000	3035	1540	---
	TMF	in-phase	10000	6000	3450	---	---
		out-of-phase	4200	3400	1650	---	---

Most of the LCF lives are correlated within a factor of two scatter band and the relationship between $\Delta\varepsilon_t$ and N_f is expressed as,

$$\Delta\varepsilon_t \times N_f^{0.57} = 80.0 \quad \text{for Sn-37Pb} \quad (1)$$

There is small difference in life between the in-phase and out-of-phase TMF tests. There is also small difference in life between the LCF and TMF tests. This trend indicates that the TMF lives of Sn-37Pb solder are estimatable by using the isothermal LCF lives expressed in Eq. (1).

Figure 6 (b) correlates the LCF and TMF lives for Sn-3.5Ag solder. The LCF lives at 253K are smaller than those at 353K. The smaller fatigue lives at 253K may result from the higher stress amplitude at 253K than that at 353K. The out-of-phase TMF lives were smaller than the in-phase TMF lives. The smaller out-of-phase TMF lives are attributed to the tensile mean stress in the tests as shown in Fig. 5. The out-of-phase TMF lives were almost the same as the LCF lives at 253K, the LCF lives at the lowest temperature in the TMF tests. On the other hand, the in-phase TMF lives were almost the same as the LCF lives at 353K, the LCF lives at the highest temperature in the TMF tests.

Figure 6 (c) correlates the LCF and TMF lives for Sn-3.0Ag-0.5Cu solder. The correlation of Fig. 6 (c) has the same trend as discussed for Sn-3.0Ag-0.5Cu as Sn-3.5Ag shown in Fig. 6 (b). The LCF lives at 233K were smaller than those at 353K. The out-of-phase TMF lives were smaller than the in-phase TMF lives. The out-of-phase TMF lives were almost the same as the LCF lives at 233K. The in-phase TMF lives were almost the same as the LCF lives at 353K.

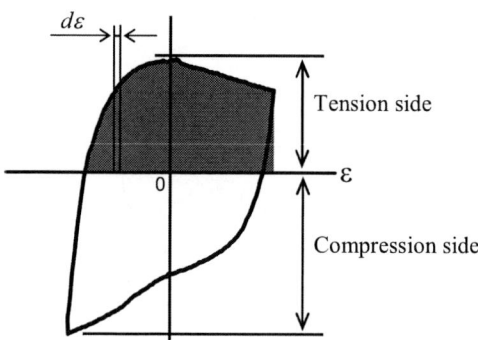

Figure 7: Calculation of strain energy in tension.

Strain Energy in Tension

Considering the results shown in Fig. 5, tension mean stress is more damaging to the compressive mean stress, so that the tensile mean stress should be take into account when correlating LCF and TMF lives. This study assumes that the tensile strain energy density relates to the LCF and TMF lives of the solders and proposes the density as a LCF and TMF correlating parameter shown in Eq. (2) and shaded area in Fig. 7.

$$E_{ten} = \int_{tension} \sigma(t)\,d\varepsilon \quad (2)$$

Figure 8 correlates the LCF and TMF lives of (a) Sn-3.5Ag and (b) Sn-3.0Ag-0.5Cu solders with the strain energy density, where the solid line is drawn based on all the LCF and TMF lives expressed by Eqs. (3) and (4) below and the dashed lines indicate a factor of two scatter band. All the LCF and TMF lives are collapsed within a factor of two scatter band.

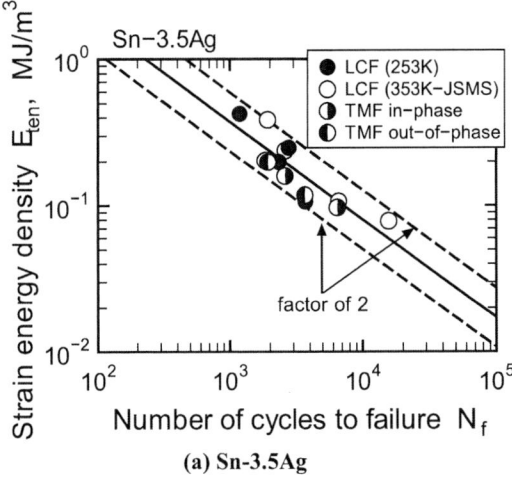

(a) Sn-3.5Ag

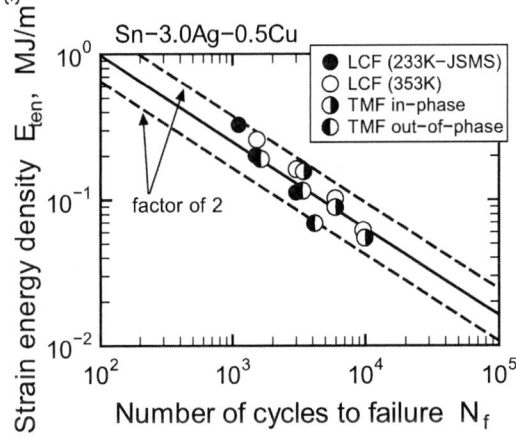

(b) Sn-3.0Ag-0.5Cu

Figure 8: Correlation of LCF and TMF lives with strain energy density.

$$E_{ten} \times N_f^{0.67} = 42.0 \qquad \text{for Sn-3.5Ag (3)}$$

$$E_{ten} \times N_f^{0.67} = 31.3 \quad \text{for Sn-3.0Ag-0.5Cu (4)}$$

Figure 9 compares the experimental LCF and TMF lives with those predicted by Eqs. (1), (3) and (4) for Sn-37Pb, Sn-3.5Ag and Sn-3.0Ag-0.5Cu solders. Most of the experimental lives fall into a factor of two scatter band for Sn-37Pb and all the data fall into narrower band for Sn-3.5Ag and Sn-3.0Ag-0.5Cu solders.

Conclusions

(1) Asymmetric hysteresis loops were found in the TMF tests for three kinds of solders. Tension peak stress amplitude was smaller than compression peak stress amplitude in the in-phase TMF test and reversed phenomenon was found in the out-of-phase TMF test.

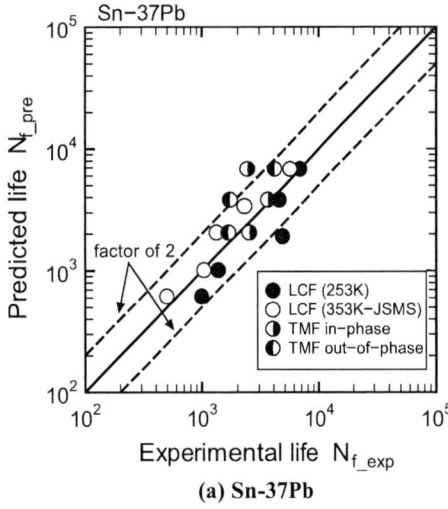

(a) Sn-37Pb

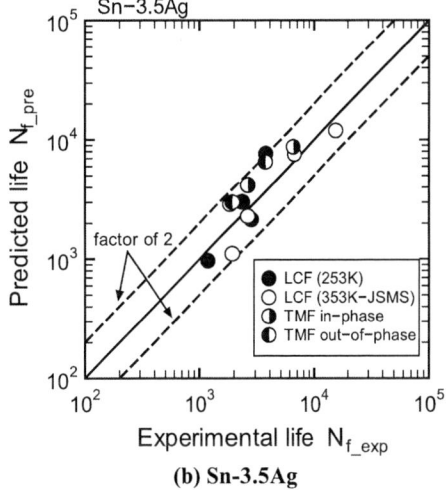

(b) Sn-3.5Ag

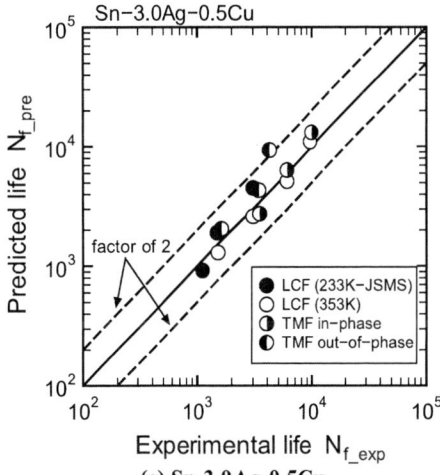

(c) Sn-3.0Ag-0.5Cu

Figure 9: Comparison of LCF and TMF lives between prediction and experiments.

(2) Since there was little difference in TMF and isothermal LCF lives in Sn-37Pb solder, the LCF lives are usable for estimating TMF lives.

(3) The out-of-phase TMF lives were smaller than the in-phase TMF lives for Sn-3.5Ag and Sn-3.0Ag-0.5Cu solders. The out-of-phase TMF lives were mostly the same as the LCF lives at the lowest temperature in the TMF test. The in-phase TMF lives were almost the same as the LCF lives at the highest temperature in the TMF test.

(4) Strain energy density in tension was a suitable parameter to estimate the LCF and TMF lives for Sn-3.5Ag and Sn-3.0Ag-0.5Cu solders.

References

Journal article

[1] T. Tsukada, H. Nishimura, M. Sakane and M. Ohnami, "Fatigue Life Analysis of Solder Joints in Flip Chip Bonding", Trans. ASME, Journal of Electronic Packaging, Vol. 122, No. 3, pp. 207-213, 2000.

[2] S. Yoshioka, S. Tani, M. Kumasawa and A. Inoue, "Low Cycle Fatigue Properties of Solder Material (36Pb62Sn2Ag) at Low Temperatures", Journal of the Society of Materials Science, Japan, Vol. 39, No. 442, pp. 908-913, 1990.

[3] T. Yamamoto, M. Sakane, M. Ohnami and T. Yamada, "Multiaxial Low Cycle Fatigue of 63Sn-37Pb Solder", Journal of the Society of Materials Science, Japan, Vol. 44, No. 503, pp. 1080-1085, 1995.

[4] T. Yamamoto, M. Sakane, M. Ohnami, Y. Tsukada and H. Nishimura, "Multiaxial Creep-Fatigue of 63Sn-37Pb Solder at Elevated Temperature", Journal of the Society of Materials Science, Japan, Vol. 46, No. 8, pp. 969-975, 1997.

[5] T. Yamamoto, M. Sakane, M. Ohnami, M. Yamashita and K. Shiokawa, "Multiaxial Low Cycle Fatigue of Five Kinds of Solders", Journal of the Society of Materials Science, Japan, Vol. 46, No. 12, pp. 1381-1388, 1997.

[6] H. Yamada, K. Saruki and K. Ogawa, "Influence of Test Temperature, Strain Frequency and Strain Hold on Fatigue Life of Eutectic Solder 63Sn-37Pb", Transactions of the Japan Society of Mechanical Engineers. A, Vol. 66, No. 647, pp. 1350-1354, 2000.

[7] N. Hiyoshi, A. Katoh, M. Sakane and Y. Tsukada, "Low Cycle Fatigue Lives of Sn-37Pb and Sn-3.5Ag Solders at Low Temperatures", Journal of the Society of Materials Science, Japan, Vol. 58, No. 2, pp. 155-161,2009.

[8] M. Miyazaki, S. Yoshioka and A. Hijikata, "Evaluation of Thermal Fatigue Strength in Lap Type Solder Joints", Journal of the Society of Materials Science, Japan, Vol. 30, No. 331, pp. 330-335, 1981.

[9] S. Ogashiwa, Y. Murakami and A. Inoue, "Thermal Fatigue of Pb-Sn-Sb-Cu Solders, Journal of the Japan Institute of Metals", Vol. 58, No. 8, pp. 952-958, 1994.

[10] Y. Uegai, S. Tani, A. Inoue, S, Yoshioka, S. Badono and A. Hijikata, "Evaluation Method of Thermal Fatigue Life for Surface-Mount Solder Joints by Mechanical Fatigue Test", Transactions of the Japan Society of Mechanical Engineers. A, Vol. 61, No. 584, pp. 729-735, 1995.

[11] M. Mukai, T. Kawakami, K. Takahashi, K. Kishimoto and T. Shibuya, "Effects of Hold-Time on Thermal Fatigue Life of Solder Joints", Transactions of the Japan Society of Mechanical Engineers. A, Vol. 3, No. 611, pp. 1594-1600, 1997.

[12] The First Division of the Committee on High Temperature Strength JSMS, "The Current Situation of Thermal Fatigue Testing and Low-Cycle Fatigue Testing at Elevated Temperature in Japan", Journal of the Society of Materials Science, Japan, Vol. 20, p. 439, 1971.

[13] The Committee on High Temperature Strength JSMS, "Standardization of Thermal Mechanical Fatigue Tests", Journal of the Society of Materials Science, Japan, Vol. 23, p. 219, 1974.

[14] M. Okazaki, "New Challenges for Thermo-Mechanical Fatigue2: Fundamentals of Thermo-Mechanical Fatigue", Journal of the Society of Materials Science, Japan, Vol. 56, No. 2, pp. 190-196, 2007.

[15] S. Taira, M. Fujino and R. Ohtani, "Collaborative Study on Thermal Fatigue Properties of High Temperature Alloys in Japan", Fatigue and Fracture of Engineering Materials and Structures, Vol. 1, pp. 495-508, 1979.

Book

[16] The Society of Materials Science, Japan, "Factual Database on Tensile and Low Cycle Fatigue Properties of Sn-37Pb and Sn-3.5Ag Solders", The Society of Materials Science, Japan, 2001.

[17] The Society of Materials Science, Japan, "Factual Database on Creep and Creep-Fatigue Properties of Sn-37Pb and Sn-3.5Ag Solders", The Society of Materials Science, Japan, 2004.

[18] The Society of Materials Science, Japan, "Standard Tensile Testing for Solders", The Society of Materials Science, Japan, 2000.

[19] The Society of Materials Science, Japan, "Standard Low Cycle Fatigue Testing for Solders", The Society of Materials Science, Japan, 2000.

[20] Japan Electronics and Information Technology Industries Association, "Standard of Japan Electronics and Information Technology Industries Association EIAJ ED-4701/100 Environmental and Endurance Test Methods for Semiconductor Devices (Life test I) ", Japan

Electronics and Information Technology Industries Association, 2001.

[21] Japan Electronics and Information Technology Industries Association, "Standard of Japan Electronics and Information Technology Industries Association EIAJ ED-4702A Mechanical Stress Test Methods for Semiconductor Surface Mounting Devices",

Japan Electronics and Information Technology Industries Association, 2003.

[22] The Society of Materials Science, Japan, "Standard Creep-Fatigue Testing for Solders", The Society of Materials Science, Japan, 2004.

[23] The Society of Materials Science, Japan, Solder data base-WG data, 2009.

Effects of Microstructure on Creep Deformation of Sn-3.5Ag Alloys

Sung Bum Kim[1], Jin Yu[2], and Won Jong Lee[2]

[1]Samsung Electronics Co.

San #16, Banwol-Dong, Hwasung-Si, Kyeonggi, 445-701

[2]Department of Material Science and Engineering, KAIST,

291 Daehak-ro, Yuseong-gu, Daejeon 35-701, Republic of Korea

Phone : +82-42-8841, Fax : +82-42-8840, and E-mail : jinyu@kaist.ac.kr

Abstract

Creep data of Pb-free solder alloys vary widely even for a given composition. Main culprits are differences in the solder microstructure. The role of β-Sn granular and particle size is particularly interesting as it is determined by the solidification rate of the solder from the melt. In the present work, creep tests were conducted under uniaxial tension by using the Sn-3.5Ag alloys with varying microstructures. The minimum strain rate ($\dot{\varepsilon}$) showed $d^{1.8\sim2.5}$ dependence where d is the granular size of the primary β-Sn phase. This is pretty much opposite to the grain size effect in creep where $\dot{\varepsilon}$ decreases inversely (D^{-2} and D^{-3} respectively) with the grain size (D). The inverse relationship originates from the diffusive matter transport through the bulk or along the grain boundary, but the granular effect observed in the eutectic solders has no resemblance to the former. Here, primary β-Sn granules are surrounded by walls of creep resistant eutectic regions which are mixtures of the β-Sn phases and Ag_3Sn precipitates, and the stress which drives the creep deformation of the eutectic phase increases with the granule size in a manner similar to that of the Hall-Petch relation.

Key words: Lead-free solder, Creep, Microstructure effect

Introduction

Wide scatterings of data [1-14] in the solder creep are thought to originate from two major sources; variations in the solder microstructure and the stress state. The former typically comes from the rapid solidifications of solders with small volume which yields substantially different microstructures from those of slowly cooled bulk alloys. Under equilibrium conditions, the eutectic Sn-3.5Ag solder typically shows a microstructure where needle shaped Ag_3Sn precipitates are embedded in the Sn matrix. However, upon rapid cooling from the melt, a complex nonequilibrium microstructure, which consists of the β-Sn dendrites and the eutectic phase, forms. The eutectic phase is mixture of intermetallic particles and the the β-Sn phase. According to Kobayashi et al.[8], the solder microstructure became finer with the cooling rate, i.e. β-Sn granule size, the volume fraction of the eutectic phase, and the size of intermetallic (IMC) particles in the eutectic region all became smaller. These microstructure factors affect the creep and other mechanical properties of solders profoundly. The solidification rate from the melt after the reflow process depends on the solder volume and the heat

dissipation through the packages, however solder joints generally have much faster cooling rate form the melt than bulk alloys and it would be misleading to predict the solder joint reliability based on creep data of bulk specimens.

In our previous work [2,28], variations of the cooling rate from the melt gave two orders of differences in the minimum creep rate ($\dot{\varepsilon}$) of Sn-3.5Ag solder, but little is known about the effects of microstructural constituents on the creep deformation rate.

In the present work, bulk specimens with various microstructures were prepared by controlling the cooling rate from the melt, and effects of microstructural parameters such as the β-Sn granule size and the IMC (Ag3Sn) particle size in the eutectic phase on $\dot{\varepsilon}$ were investigated. One advantage of using bulk specimen is the elimination of the IMC layers that form at the solder joint interfaces. Here, creep data can be solely ascribed to the microstructural characteristics of solders.

Experimental Procedure

Tensile creep specimens, which are schematically described in Fig. 1, were prepared by melting the Sn-3.5Ag solder alloy in an aluminum mold and quenching in water and two different oil baths to produce three solidification rates and thereby three different microstructures. The cooling rate of water quenched (WQ) and oil quenched (OQ-1 and OQ-2) specimens were approximately 37 K/sec, and oil quenched specimens had 3.7 and 1.6 K/sec, respectively. Creep tests were conducted at 373 ± 1 K under a constant load, and specimens were equilibrated for 2 hours at the test temperature before loading.

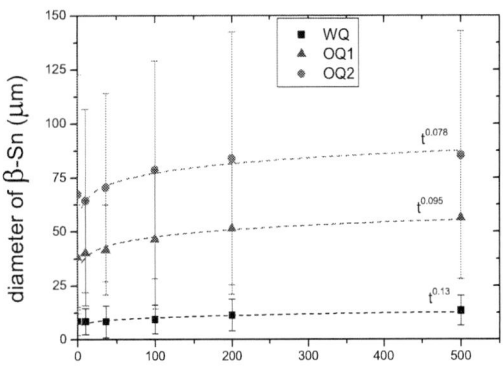

Fig.3 Kinetics of β-Sn granule growth as functions of the aging time at 373K.

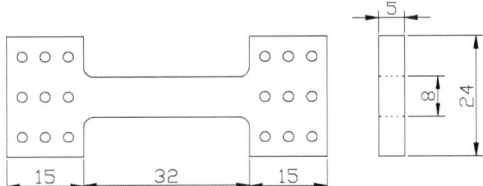

Fig. 1. Dog-bone shaped creep specimens. (in mm)

The average diameter of primary β-Sn granule of the three specimens is presented as a function of the ageing time at 373K in Fig. 3. It can be seen that β-Sn granule diameter (d) increases with the aging time (t) quite mildly at the given ageing temperature and is mostly determined by the quenching rate. Coarsening of Ag_3Sn particles in the eutectic phases during the ageing process is shown in Fig. 4 where the average radius of the IMC particles (r) is given as a function of the aging time at 373K. Note that the spherical Ag_3Sn particles coarsen more rigorously than the β-Sn phase.

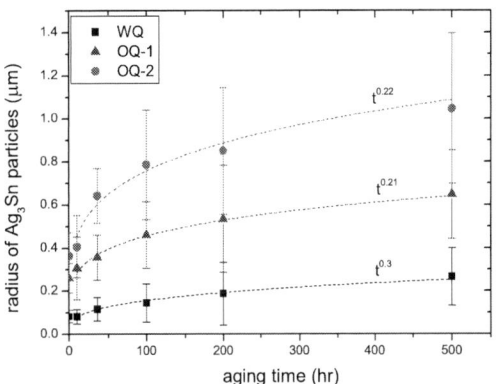

Fig.4 Growth kinetics of Ag_3Sn particles radius in the eutectic region as functions of the aging time at 373K

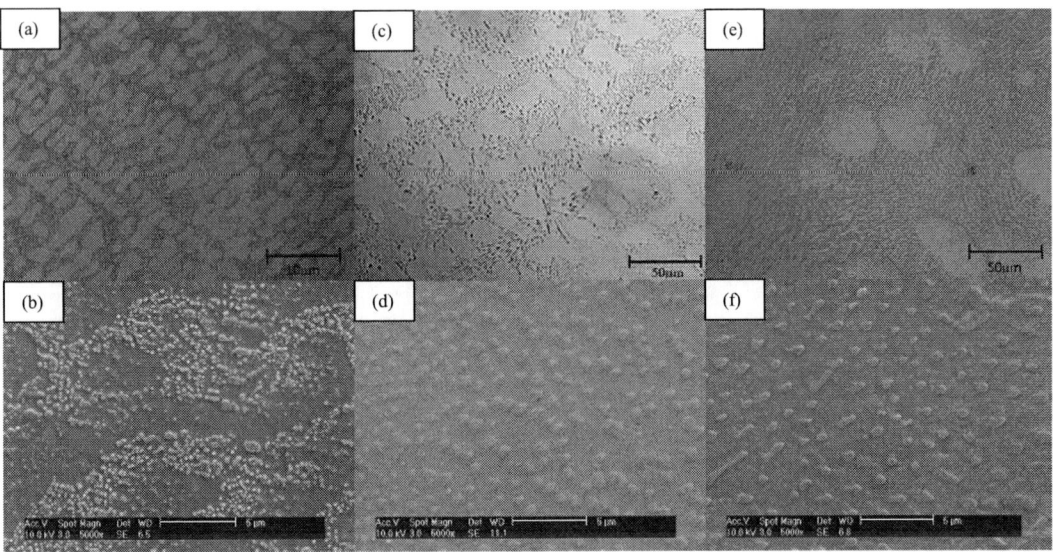

Fig.2. Optical and SEM micrographs of WQ (a, b), OQ1 (c, d) and OQ2 (e, f) specimens before creep tests. Micrographs on the right column were magnified 10 times more compared to those of on the left. Micrographs 2(d) and (f) were taken from the eutectic phase. Note opposite contrasts in Figs (a) and (b).

309

2. Creep Deformations

In Fig. 5, uniaxial tension creep curves of the three specimens are presented. It can be seen that the primary creep is quite limited in all case in contrast to creep curves of Pb-Sn eutectic solder. The lack of primary creep in the Sn based alloys seems to be related to the fast recovery and recrystallization of Sn grains at the creep temperature ($0.75T_m$).

The minimum creep strain rates ($\dot{\varepsilon}$) of the three specimens are plotted as functions of the applied tensile stress (σ) in Fig. 7 and expressed according to the power-law creep equation

$$\dot{\varepsilon} = B\sigma^n \qquad (1)$$

where B is the temperature dependent creep constant and n is the stress exponent.

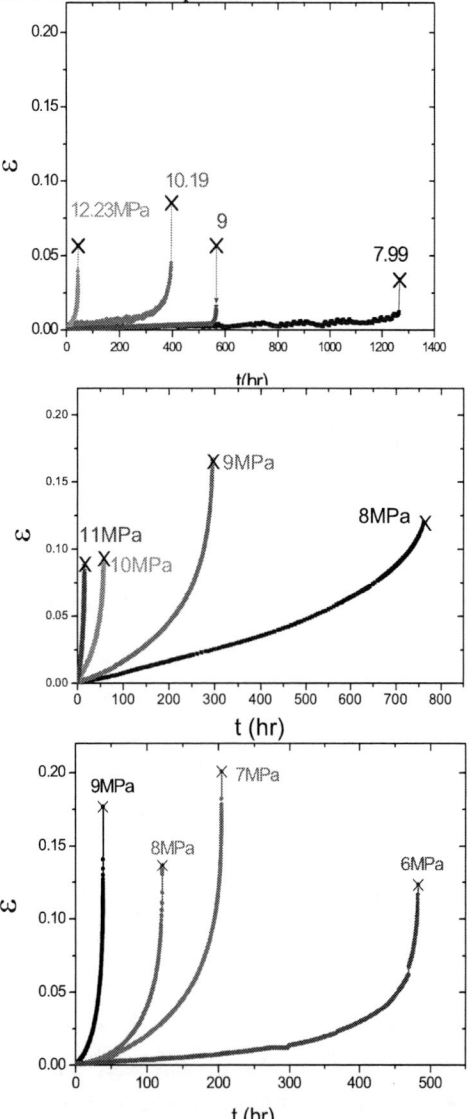

Fig.5 Typical creep curves of (a) WQ, (b) OQ-1 and (c) OQ-2 specimens of Sn-3.5Ag alloys under varying shear stress levels at 373K

If the stress exponent of all data are fitted to $n=11$, the creep constant B is found to decrease with the quenching rate and vary about two orders of manitude depending on the cooling medium. (The same data may also be fitted to dotted lines shown in Fig. 6 where the stress exponent n tends to decrease with the quenching rate; n is 9 for WQ and 12.5 for OQ2 specimen, respectively. Here, the stress exponent was assumed to be the same for all specimens for more rigorous discussion on the effects of microstructural constituents).

By comparing the creep data and the SEM micrographs shown in Fig. 6, one can ascribe the superior creep resistance of the WQ specimens to finely dispersed Ag_3Sn particles in the eutectic region. In eutectic alloys, it is well known that faster cooling rate ($\dot{T}$) results in larger amount of undercooling (ΔT) and finer microstructure .

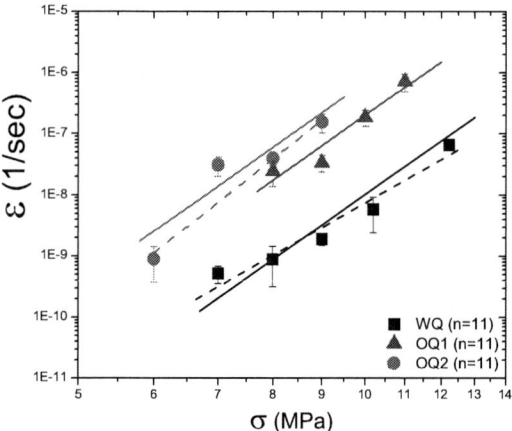

Fig.6 The minimum strain rates of Sn-3.5Ag solder at 373K as functions of the applied stress.

Rupture times of the three specimens are presented as functions of the applied stress in Fig. 7. When the rupture time (t_f) is described as $t_f = C\tau^{-m}$, all the rupture time data can fiitted reasonably using $m=8$. Note that m is smaller than n ($m \sim 2/3\ n$) and that variations of the rupture time with the quenching rate was smaller than that of $\dot{\varepsilon}$, which is consistent with the work of Joo and Yu [2]. Characteristics of rupture times are qualitatively consistent with those of the creep strain rates ; t_f increases in the order of OQ2, OQ1, and WQ, which is the order of the quenching rate.

Rupture strains (ε_f) of the three specimens are presented in Fig. 8. It can be seen that the creep ductility decreased as the solder microstructure became finer ; ε_f of the WQ specimens amounts to 1/3 of those of the OQ2 specimens. Since most of the rupture strain accumulated in the tertiary stage of creep which typically occurred after necking, direct correlations between ε_f and t_f were difficult to establish.

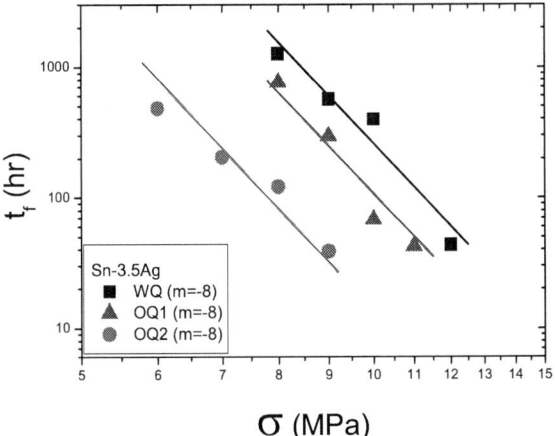

Fig. 7 Rupture times of Sn-3.5Ag solder at 373K as functions of the applied stress

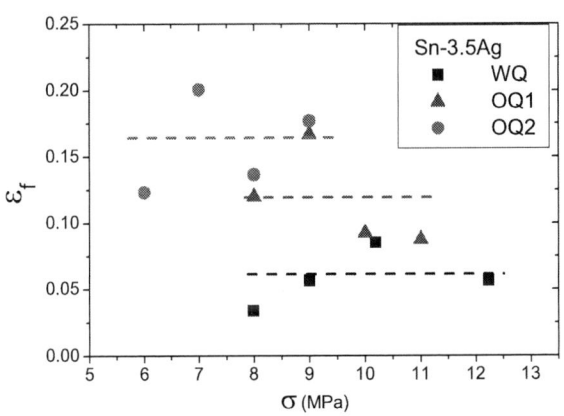

Fig.8 Variations of rupture with the applied stress.

In Fig. 9, the minimum creep strain rates are plotted as functions of the as-solidified β-Sn granule diameter (d) of the three specimens under the same stress levels, and it can be seen that $\dot{\varepsilon}$ increased with the β-Sn granule size as

$$\dot{\varepsilon} = \dot{\varepsilon}_0 + \text{const} \cdot d^{1.8 \sim 2.0} \quad (3)$$

At a first glance, this looks like a contradiction to the well-known grain size effect in creep where $\dot{\varepsilon}$ decreases inversely with the grain size (D). (in the Nabarro-Herring (NH) and Coble creep, $\dot{\varepsilon} \propto D^{-2}$ and D^{-3}, respectively) [22].

The boundary between the β-Sn and the eutectic phase is not the grain boundary joining two single crystals with different crystallographic orientations. It is rather a demarcation of regions with and without densely populated Ag$_3$Sn precipitate.

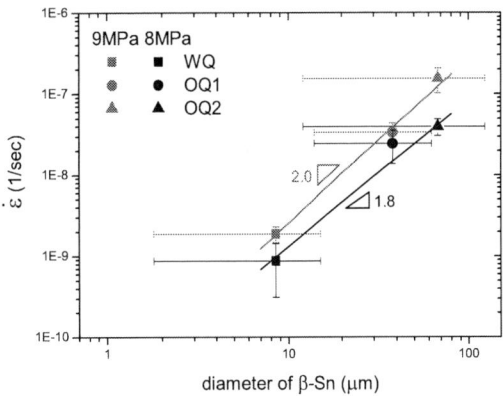

Fig.9 The minimum strain rates of Sn-3.5Ag as a function of β-Sn diameter.

Conclusions

Principal factors which affected the creep deformation rate of Sn-3.5Ag solder most significantly were solder microstructure, which changed the minimum strain rate by 2~4 orders of magnitude. The effect principally came from varying solidification rate. The minimum strain rate ($\dot{\varepsilon}$) showed $d^{1.8 \sim 2.5}$ dependence where d is the granular size of primary β-Sn. This is pretty much opposite to the grain size effect in creep where $\dot{\varepsilon}$ showed inverse dependence (D^{-1}, D^{-2}, or etc.) on the grain size (D).

References

1. A. Schubert, R. Dudek, H. Walter, E. Jung, A. Gollhardt, B. Michel, and H. Reichl, *Proc. of 52nd Electronic Components and Technology Conference* (Piscataway, NJ: IEEE, 2002), 1246.
2. D.K. Joo, and Jin Yu, *Proc. of 52nd Electronic Components and Technology Conference* (Piscataway, NJ: IEEE, 2002), 1221.
3. S.W. Shin and Jin Yu, *J. Electron. Mater.* 34, 188 (2005).
4. M.L. Huang, L. Wang, and C.M.L. Wu, *J. Mater. Res.* 17, 2897 (2002).
5. C.M.L. Wu, and M.L. Huang, *J. Electron. Mater.* 31, 442 (2002).
6. M.L. Huang, C.M.L. Wu, and L. Wang , *J. Electron. Mater.* 34, 11 (2005).
7. W.J. Plumbridge, C.R. Gagg, and S. Peters, *J. Electron. Mater.* 30, 1178 (2001).
8. Jin Yu, D.K. Joo, and S.W. Shin , *Acta Mater.* 50, 4315 (2002).
9. S.W. Shin, and Jin Yu, Jpn. J. Appl. Phys. 42, 1368 (2003).
10. J.W. Morris, Jr., H.G. Song, and Fay Hua, *Proc. of IEEE 53rd Electronic Components and*

Technology Conference (Piscataway, NJ: IEEE, 2003), 54.

11. F. Ochoa, J.J. Williams, and N. Chawla, *JOM* 55(6), 56 (2003).

12. J.P. Lucas, F. Guo, J. Mcdougall, T.R. Bieler, K.N. Subramanian, and J.K. Park, *J. Electron. Mater.* 28, 1270 (1999).

13. I. Dutta, *J. Electron. Mater.* 32, 201 (2003).

14. H. Mavoori, J.Chin, S. Vaynman, B. Moran, L. Keer, and M. Fine, *J. Electron. Mater.* 26, 783 (1997).

15. H. Riedel, *Fracture at High Temperatures* (Heidelberg : Springer-Verlag, 1987).

16. Steffen Wiese, Ekkehard Meuel, and Klaus-Juergen, *Proc. of IEEE 53rd Electronic Components and Technology Conference* (Piscataway, NJ: IEEE, 2003), 197

17. W.H. Chen, K.N. Chiang, and S.R. Lin, *J. Electron. Packaging* 33, 37 (2002).

19. I. de Sousa, D.W. Henderson, L. Patry, S.K. Kang, and D.-Y. Shih, Proceedings of the 56th Electronic Components and Technology Conference (Piscataway, NJ: IEEE, 2006),pp. 1454–1462.

20. S.K. Kang, D.-Y. Shih, D. Leonard, D.W. Henderson, T. Gosselin, S.-I. Cho, J. Yu, and W.K. Choi, JOM 56, 34(2004).

21. M. D. Mathew, Sashidhar Movva, Hong Yang, and K. Linga Murty, Proc. of Symp. of Minerals, Metals & Materials Society,(1999).

22. Josef Cadek, "Creep in Metallic Materials", Elsevier, 1988

Development and Characterization of 300mm Large Panel eWLB (embedded Wafer Level BGA)

Seung Wook Yoon, Yaojian Lin and Pandi C. Marimuthu

STATS ChipPAC Ltd. 5 Yishun Street 23, Singapore 768442

E-mail : Seungwook.yoon@statschippac.com

Abstract

This paper will highlight some of the recent advancements in 300mm eWLB large panel development. Compared to 200mm case, 300mm large panel has more warpage and process issues due to its area increase. Thermo-mechanical simulation shows 100~150% more warpage with 300mm large panel compared to 200mm. So various design parameters were studied to optimized warpage, such as dielectric materials and thickness, molding compound thickness etc. This paper also presents study of process optimization for 300mm eWLB and on overall warpage behavior in different process steps. Finally 300mm eWLB test vehicles are fabricated and tested in JEDEC standard test conditions. It also describes mechanical characterization, reliability data including component/board level, challenges encountered and overcome, and future steps.

Key words: embedded wafer level packaging (eWLB), Fanout WLP, 300mm large scale, process development, process optimization, reliability

I. INTRODUCTION

WLP applications are expanding into new areas and are segmenting based on I/O count and device. The foundation of passive, discrete, RF and memory device is expanding to logic ICs and MEMS. The WLP segment has matured over the past decade, with numerous sources delivering high-volume applications across multiple wafer diameters and expanding into various end-market products. With infrastructure and high volumes in place, a major focus area is cost reduction.

One of the most well known examples of a FO-WLP structure is eWLB technology[1]. This technology uses a combination of front- and back-end manufacturing techniques with parallel processing of all the chips on a wafer, which can greatly reduce manufacturing costs. Its benefits include a smaller package footprint compared to conventional leadframe or laminate packages, medium to high I/O count, maximum connection density, as well as desirable electrical and thermal performance. It also offers a high-performance, power-efficient solution for the wireless market[2]. Furthermore, next generation 3D eWLB technology enables 3D IC and 3D SiP (System-in-Package) with vertical interconnection. 3D eWLB can be implemented with through silicon via (TSV) applications as well as discrete component embedding.

II. eWLB TECHNOLOGY

eWLB technology is addressing a wide range of factors. At one end of the spectrum is the packaging cost along with testing costs. Alongside, there are physical constraints such as its footprint and height. Other parameters that were considered during the development phase included I/O density, a particular challenge for small chips with a high pin count; the need to accommodate SiP approaches, thermal issues related to power consumption and the device's electrical performance (including electrical parasitic and operating frequency).

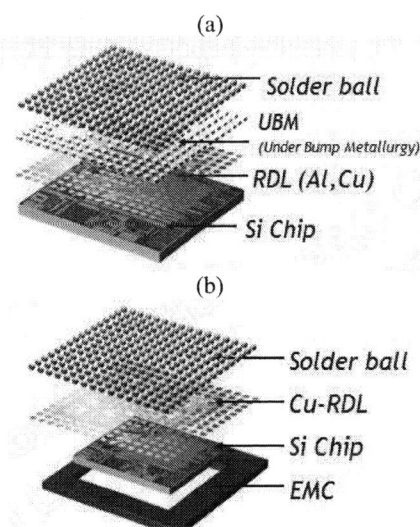

Figure 1. Structural comparison of (a) FI-WLP and (b) eWLB (FO-WLP)

The obvious solution to the challenges was some form of WLP. But two choices presented themselves: Fan-out or Fan-in. FO-WLP is an interconnection system processed directly on the wafer and compatible with motherboard technology pitch requirements. It combines conventional front- and back-end manufacturing techniques, with parallel processing of all chips. There are three stages in the process. Additional fab steps create an interconnection system on each die, with a footprint smaller than the die. Solder balls are then applied and parallel testing is performed on the wafer. Finally, wafers are sawn into individual units, which are used directly on the motherboard without the need for interposers or underfill.

Figure 1 shows the structural comparison between FI-WLP and eWLB. eWLB does have thick Cu-RDL so UBM is not required. (UBM is optional for eWLB). Schematics of redistribution layer (RDL) are compared for single-RDL and double-RDL as shown in figure 2.

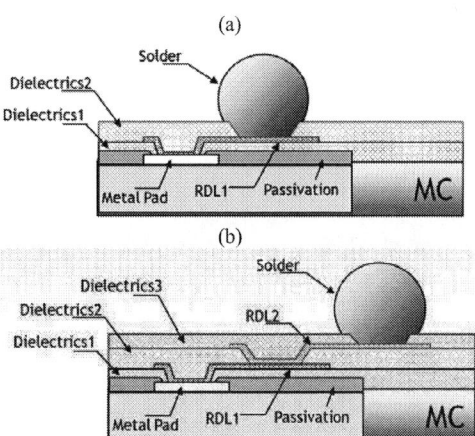

Figure 2. Schematics of eWLB RDL structure; (a) single-layer RDL and (b) double-layer RDL

Advantage of eWLB Technology

The current BGA package technology is limited by the organic substrate capability. Moving to eWLB helps overcome such limitations and also simplifies the supply chain. Building the substrate on the package itself, allows for higher integration and routing density in less metal layers. eWLB is a next generation platform that will support future integration, particularly for wireless devices and this packaging technology has a number of important features. Transition to eWLB packaging technology enables a significant reduction in recurring costs by eliminating the need for expensive substrates. The advantage of eWLB packaging can be summarized in Table 1. BGA packaging also faces a challenge with technology nodes beyond 65nm as the device performance density drives the need for flip chip. But advanced flip chip nodes drive fine pitch combined with weaker low-k dielectric structures resulting in flip chip becoming narrower in terms of packaging process margin,. In addition, there is a big trend in being environmentally friendly, driving lead

free and halogen free, or green, material sets. With ultra low-k and interconnects pitch becoming smaller and smaller and with the shift to lead free materials, the technical limitations faced by the packaging industry are becoming more challenging. eWLB technology provides a window for packaging next generation devices in a generic, lead-free/halogen free, green packaging scheme.

Table 1. Advantage of eWLB packaging.

1. The smallest and thinnest package other than fan-in WLCSP
2. Excellent electrical and thermal performance
• *Great for high frequency application*
• *Excellent for RF and mixed signal due to low parasitics compared to any laminate-based packages*
• *The lowest thermal resistance*
3. High density routing is easily implemented in RDL
4. No ELK damage issues for advanced Si nodes devices
5. Proven low cost path using a batch process & simple supply chain
6. Path to the flexible 3D packages – any array patterns on the top
7. Scalable technology to a larger panel production – Lower cost

III. 300mm large panel eWLB
A. Challenge of 300mm large scale eWLB

For 300mm eWLB, there is area increase of more than 230% times compared to 200mm as shown in Figure 3. So warpage is most critical for larger scale wafer handling as like in wafer fabrication process. Warpage affects wafer handling, processability, throughput as well as process stability thus yield so it is critical to optimize and well control its warpage behavior. Figure 4 shows clear warpage difference between 200mm and 300mm eWLB wafers. With computational simulation work for 300mm eWLB, it showed more than 2 time warpage than 200mm case. Figure 5 shows warpage change with different carrier thickness in 300mm eWLB. It clearly showed trend of less warpage with optimized carrier thickness.

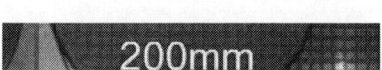

Figure 3. Wafer size difference between 200mm and 300mm eWLB wafers.

314

(a)

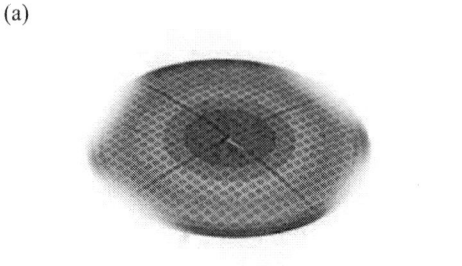

(b)

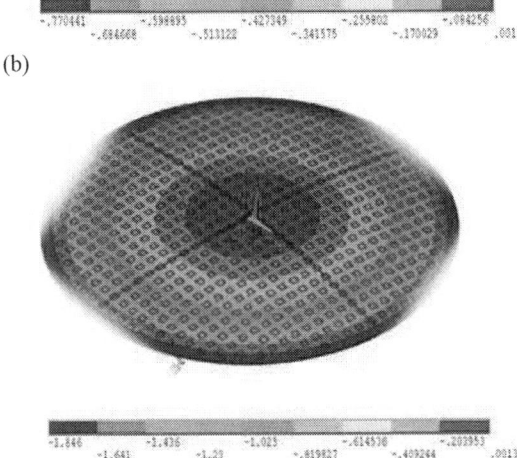

Figure 4. Warpage of (a)200mm and (b)300mm eWLB with thermo-mechanical simulation

(a)　　　　　(b)

(c)

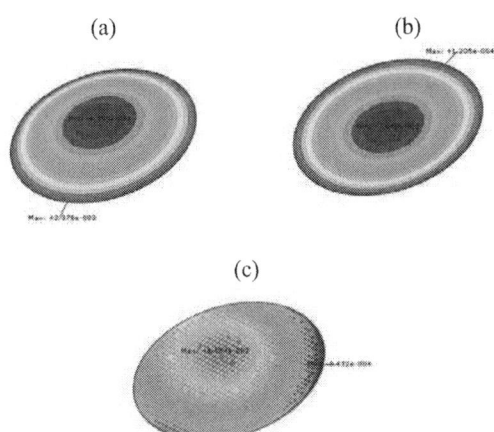

Figure 5. The warpage change with carrier thickness of wafer level molding in 300mm

To optimize these warpage behaviors, various material/process parameters were studied. Each material has different physical properties as like CTE (coefficient of thermal expansion), Young's modulus, and Poison ratio. So combination of each physical properties critically affect overall eWLB

wafer warpage. After basic thermal-mechanical simulation study of these parameters with several DOE (Design of experiment), key parameters were identified such as dielectric materials and thickness, molding compound thickness etc. Based on those parameters, in-depth simulation was carried out for several combinations of each parameter.

Figure 6 and 7 show the warpage behavior with different materials sets. With different set of materials, it showed significant warpage behaviors with maximum difference of 1000um (1mm).

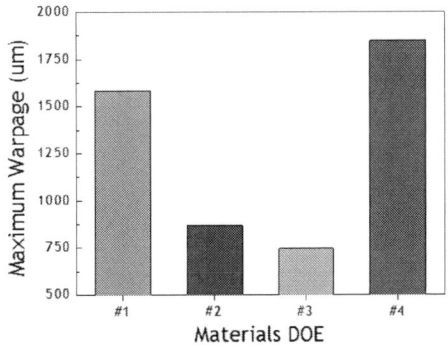

Figure 6. Computational mechanical warpage simulation data with different material DOE of 300mm eWLB.

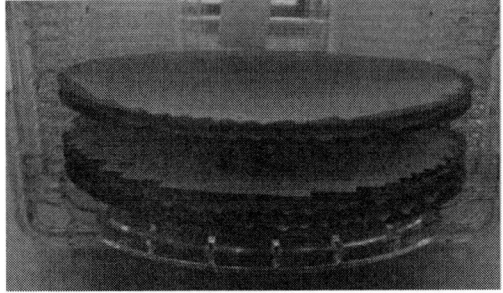

Figure 7. 300mm eWLB wafer warpage with different materials set DOEs.

This warpage behavior is very critical for overall process flow, manufacturability and overall yield. But reliability is another key challenge for products. So reliability evaluation was also carried out for different material DOE sets to investigate materials in 300mm eWLB packages. With comprehensive study of reliability study, final material set was selected for final test vehicle fabrication.

B. Comparison of process variation between 200mm and 300mm.

With test vehicles of 5x5mm die in 8x8mm eWLB, each major process variables are monitored and compared. Figure 8 shows the die displacement or movement after wafer level molding and compared with 200mm and 300mm carriers. As shown in figure, 300mm shows quite small die shift

315

as similar as 200mm case after process optimization and stabilization. Cu plating thickness, dielectrics thickness, ball shear strength are key comparison variables to investigate 300mm eWLB process stability as compared to 200mm eWLB.

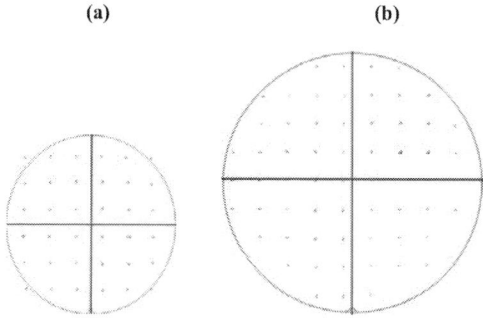

Figure 8. Comparison of die shift and movement after wafer level molding in (a) 200mm and (b) 300mm eWLB carriers.

300mm eWLB has larger standard deviation than 200mm mainly due to 230% times area, but measured average value and its standard deviation values in 300mm eWLB is close to 200mm in most cases. These shows the stability of 300mm eWLB process developed and established from 200mm baseline.

Table 6. Summary of eWLB Component level and Board level reliability test result

Moisture Sensitivity Level	MSL1 @ lead free condition (260°C)
Temperature Cycling	-40°C/125°C, 850 cycles
	-25°C/100°C, 1000 cycles
High Temperature Storage	150°C, 1000 hrs
Unbiased HAST	130°C/85% RH, 96 hrs
Temperature Humidity Bias Test	85°C/85%/5V, passed 1000 hrs
TC on Board	-40°C/125°C, 2 cycles/hr, passed 500 cycles
Multiple Solder Reflow	5x, 10x and 20x reflows with minimal reduction in bump shear strength
Drop Test	Passed JEDEC drop test for 8 x 8mm, 183 balls (0.5mm pitch)

C. Component level and Board level reliability test result

For 300mm eWLB package's reliability tests, test vehicle were prepared with 8x8mm and 183balls with 5x5mm die size. JEDEC standard reliability tests were carried and eWLB passed all reliability conditions. For board level reliability, JEDEC TCoB (temperature cycle on board) and drop test were carried out and it also successfully passed all test requirements as shown in Table 6. Figure 9 shows

board level drop test performance and it showed first failure was over 100 drops in industry standard test conditions.

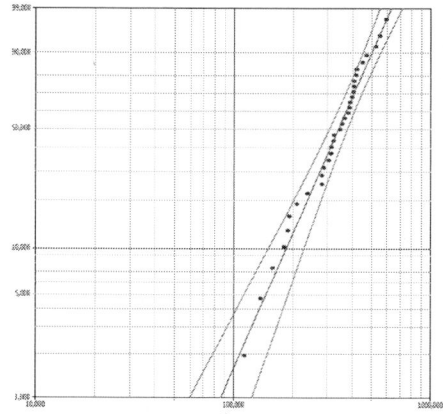

Figure 9. Weibull plot of drop test results of eWLB from 300mm carriers.

D. Next movement for large scale eWLB; panel approach

Significant cost and productivity advantages can be achieved with the larger scale reconstituted wafer eWLB format as compared to the existing WLB wafer format due to higher efficiency and economies of scale. For 12"x12" square panel, it has more than 30% more area compared 12" wafer because square panel can save corner space.

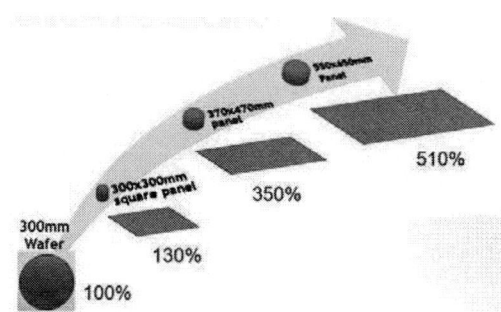

Figure 10. Potential trend of area increase with panel scale approach; more throughput with lower cost and economies of scale

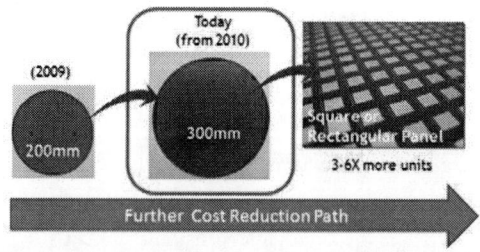

Figure 11. Scale of economics of eWLB; moving to large scale; panel approach for further cost reduction

Economies of scale arise when the cost per unit falls as output increases. Economies of scale are the main advantage of increasing the scale of production and becoming 'big'. Firstly, because a large business can pass on lower costs to customers through lower prices and increase its share of a market. Secondly, a business could choose to maintain its current price for its product and accept higher profit margins.

Based on Fan-out WLP costs comparison study from SavanSys Solutions and TechSearch [3], for all die sizes in the 6x6mm and 8x8mm eWLB packages, there is a significant cost advantage. However, the cost of fan-out WLP is much higher for larger packages. This is largely due to the fact that fan-out WLP technology is a semiconductor process, as opposed to flip chip and wire bond packaging which is primarily a printed circuit board (PCB) process. PCB processes use a large fabrication panel compared to the wafer used for a semiconductor process. For small packages, the wafer versus panel size difference is not as significant as with large packages. So moving to large scale eWLB with panel approach, eWLB would be on further cost-effective solution with inline batch process of fab technology.

V. CONCLUSION

Advanced packaging plays a crucial role in driving products with increased performance, low power, lower cost and smaller form factor. The industry requires innovation in packaging technology and manufacturing to meet current demands and the ability to operate equipment in high volume with large throughput. eWLB technology is an enhancement to standard WLPs, allowing the next generation of a WLP platform due to its fan-out capability.

eWLB is a low-cost solution with batch process and larger area utilization such as 12" and panel approaches. The ability to integrate passives like inductors, resistors and capacitors into the various thin film layers, active/passive devices into the mold compound and 3D vertical interconnection opens additional design possibilities for new Systems-in-Package (SiP) and 3D stacked packaging. Moreover, 3D eWLB technology provides more value-add in performance and promises to be a new packaging platform that can expand its application range to various types of devices as well as 3D TSV integration for true 3D SiP systems. For further cost reduction approach after 300mm eWLB, scaling-up such as panel approach would be next steps to move. It provides breakthrough productivity, compatible process for advanced Si node devices as well as functional Integration /combination of different node devices (32nm, 28nm or 22nm) with RF, discrete or memory devices.

Electronic product differentiation today is driven by ever-expanding functionality, feature sets, multi-functionality and faster communications. At the same time, consumers have made clear their desires for feature-rich products in compact form factors to enable maximum portability. eWLB technology is successfully enabling semiconductor manufacturers to provide the smallest possible, highest-performing semiconductors as cost-effective packaging solution.

ACKNOWLEDGEMENT

Authors appreciate Mr. Chow Seng Guan, for thermo-mechanical simulation of warpage behavior, design and characterization group in Statschippac LTD.

REFERNCE

[1] M. Brunnbauer and Thorsten Meyer, "Embedded Wafer Level Ball Grid Array (eWLB)," IMAPS Device Packaging Conference 2008, 17-20 March 2008, Arizona, US (2008)

[2] Graham pitcher, "Good things in small packages," Newelectronics, 23 June 2009, p18-19 (2009)

[3] Chet A. Palesko, Amy J. Palesko and E. Jan Vardaman, "COST COMPARISON FOR FLIP CHIP, WIRE BOND, AND WAFER LEVEL PACKAGING," Proceedings of IWLP2010, Santa Clara, US (2010)

Advanced Thermal Management Materials for Heat Sinks used in Microelectronics

Mathias Ekpu, Raj Bhatti, Ndy Ekere, and Sabuj Mallik

Manufacturing Engineering Research Group, School of Engineering,
University of Greenwich at Medway, Chatham, Kent, U.K.

Phone: +442083313873, E-mail: em52@gre.ac.uk

Abstract

Thermal management materials used for heat sink in laptop computers is reviewed in this paper. In laptop computers, heat sink plays a vital role of dissipating heat from the system. Heat sink is a vital component in a laptop computer because it dissipates the heat generated by the system. The overall efficiency, cost, and size of the system could be influenced by the heat sink device. Four selection criteria; thermal conductivity, coefficient of thermal expansion, density, and cost were abducted for selecting a laptop computer heat sink material in this paper. An ideal heat sink material exhibits high thermal conductivity, low coefficient of thermal expansion, low density, and low cost. Aluminium and copper are mainly used for laptop computer heat sinks. Aluminium is used where weight reduction is needed while copper is used when weight is not considered as a major factor. However, the high demand for materials with low coefficient of thermal expansion and high thermal conductivity is due to the use of ceramic materials for substrates. Aluminium and copper fall short of the required coefficient of thermal expansion hence the need for advance composite materials. Based on the four selection criteria for a laptop computer heat sink material, Al/SiC has superior property potentials and is recommended as a near optimum material for this purpose.

Key words: Laptop computer, heat sink, thermal management materials, thermal conductivity, coefficient of thermal expansion

Introduction

Increased heat from electronics component is a major issue facing surface mount technology (SMT) due to miniaturisation of such components [1-3]. Heat sink is used to remove heat generated by such electronics components. A heat sink can simply be described as an object that disperses or remove heat from another object. Heat sinks are commonly used in computers and many other microelectronics applications [4, 5]. In computers, a heat sink is mainly attached to a chip to help prevent the chip from overheating. Heat sink cannot be neglected in the production of recent computers because it is an essential component for the removal of heat from the system. Robert [5] stated that a computer heat sink is a thermal conductor that carries heat away from the computer's central processing unit (CPU) into fins that provide a large surface area for the heat to dissipate throughout the rest of the computer. This will result in the cooling of the heat sink and processor respectively. Fans are attached to most heat sinks to assist in the airflow of the system.

Heat sinks have been used over the years for only large computers because the heat generated by the processors was a problem. Presently, with the production of fast processors and miniaturisation of computer CPU, heat sink has become relevant because the system tend to overheat without a cooling mechanism.

Heat generally could be transferred in three different ways conduction, convection, and radiation [6, 7]. In conduction, heat is transferred in solid, and therefore the way heat is transferred in a heat sink [5]. Conduction is the transfer of thermal energy from one object to another through the movement or interaction between atoms or molecules of the matter. When two objects of different temperature come in contact (for example: a heat sink and a computer's chip during operation), the warm fast atoms from the chip crashes into the slow cold atoms of the heat sink. The atoms from both bodies continue to move until there is a state of equilibrium in the system. The heat transfer in a computer's CPU obeys Fourier's law which states that the rate of heat conduction (Q_c) is proportional to the transfer area (A) and the temperature gradient (dT/dx) [8, 9]. The equation is represented as:

$$Q_c = KA \, (dT/dx) \qquad (1)$$

Where K is the thermal conductivity (this measures the ability of the material to conduct heat).

Metal materials are usually used to produce heat sink that functions as a thermal conductor that removes heat from the CPU. However, there are advantages and disadvantages of the different metal materials used for heat sink production. The thermal conductivity of the metal material is an issue to consider. It is stated that the higher the thermal conductivity of the metal material the better it is at dissipating heat.

Aluminium is the most commonly used material for heat sinks because of its lightweight, lower cost, manufacturability, and the existence of infrastructure [10]. The thermal conductivity of aluminium is about 220 W/mK [11-14]. The thermal conductivity number for aluminium is therefore 220, which refers to aluminium ability to conduct heat. Therefore, the higher this number the more heat the material is capable of conducting. In addition, copper is increasingly been used for heat sinks production because of its high thermal conductivity value of about 400 W/mK [11]. Its major disadvantage over aluminium is that it is three times heavier and more expensive. It is important to note that when a heat sink is attached in a computer, it exerts additional level of stress on the motherboard. The motherboard must be designed to accommodate such level of stress. Hence, it is important to use a light material with high thermal conductivity for the production of heat sinks. Figure 1 shows a common design of a computer heat sink with fan attached [15].

Heat sinks are similar to other electronics components in the computer industries that require change due to miniaturisation. Rajiv *et al* [16] stated that the development, processing, and characterization of advanced composite materials for heat sinks are presently aided by novel approaches to materials design and synthesis based upon the fundamental understanding of the processing-structure-properties-performance relationships for a range of materials. The low coefficient of thermal expansion of ceramic substrate materials which ranges between 3 and 7 x 10^{-6} K^{-1} (Al_2O_3 or AlN) used by the electronics industries is a trigger for the need of advanced composite heat sink materials [4]. These advanced composite materials have different issues to be addressed due to the type of manufacturability or production processes [17]. The powder metallurgical approach is used for the recent development of advanced composite materials. The major method involved in this process of densification is sintering.

The purpose of this paper is to review thermal materials that may be applicable or important to the production of a laptop computer's heat sink. This study outlined the different materials used for a computer heat sink and the relevant thermal properties and manufacturing processes. The cost consideration of heat sink materials is presented in this paper. This paper addresses four criteria of material selection procedures for laptop computer heat sink materials. Finally, the conclusions drawn from this review is presented.

Figure 1: A common computer heat sink design [15]

Thermal Management Materials

Recent development in the production of computer components in electronics industries requires the need for materials that can dissipate heat effectively from the system. In other to maintain reliability of such products, thermal management should be a priority for the manufacturers of computer electronics components [18-22]. Materials with high thermal conductivity are the choices for laptop computer heat sink. The use of materials with low coefficient of thermal expansion for chips production in laptop computers meant that heat sinks should have low CTE as well. Miniaturisation of laptop computer components means that the weight (density) of heat sink materials has to be considered. Table 1 presents some materials with thermal properties and corresponding densities.

Aluminium Alloys

The choice of aluminium as heat sink material for laptop computer is attributed to its thermal conductivity (220 W/mK), low cost, and the ease of forming and fair machinability [10]. The drawback for aluminium heat sinks used in laptop computer is that secondary processing is inevitable. Aluminium in is pure state has a poor die casting capabilities [17], hence the need for the addition of a little percentage of silicon to allow it slip out of a mould [10]. This causes non-homogeneity in the internal structure of the aluminium which further reduces the thermal conductivity by a third. The thermal resistance in casting is increased due to poor material contact resulting from porosities within the casting forms. The porosity in the cast aluminium can be reduced by a method called Hot Isostatic Pressing, where pouring and pressing is done by hand at low pressure rather than high die cast pressures. The finish product has half the conductivity of raw aluminium and is 2 to 4 times more expensive than the standard die casting. However, in applications such as avionics, laptop computers, and portable electronics equipment where weight is a limitation it could be economically advantageous.

319

Table 1: Thermal properties of materials and the corresponding densities

Material	Thermal conductivity- W/mK	Coefficient of thermal expansion- $10^{-6}K^{-1}$	Density Kg/m^3	Ref.
Aluminium	220	22-24	2700	[11, 12-14]
SiC/Aluminium	170-220	6.2-7.3	3000	[11, 23]
Boron/Aluminium	145	13-15	2700	[11]
Copper	400(390)	16-17	8960	[11, 12]
Cu-coated Graphite/Cu	>400	2.8-3.5	5300	[11]
Copper/Molybdenum	170-210	5.7-6.0	10080	[11]
Copper/Tungsten	180-200	6.5-8.3	8400	[11]
Gold	315	14	19.32	[21]
Molybdenum	142	4.9	10.22	[21]
Tungsten	155	4.5	19.3	[21]
Diamond	2000	0.9	3.51	[21]
Beryllium oxide	260	6	3	[21]
Aluminium nitride	320	4.5	3.3	[21]
Silicon carbide	270	3.7	3.3	[21]

Copper Alloys

Copper is starting to gain relevance as heat sink material because of its high thermal conductivity (400 W/mK). For the possibility of casting and machinability of copper, impurities are needed [17]. These impurities reduce the thermal conductivity of copper drastically. Brass and bronze are not commonly used materials for thermal dissipation, but could form part of copper impurity [10]. Pure copper slug is the most common copper heat sink material and is used as a heat spreader on chip carriers [10]. Additional cooling is achieved by the mounting of heat sink on the top of the slug to increase the surface area. To avoid oxidation the slug is often plated. In corrosive environments where strength is required or abrasion can occur, bronzes are sometimes used for large heat sinks [10, 24, 25]. It is important to note that a material that is less expensive than Copper but more conductive than Aluminium would be ideal for a laptop computer heat sink.

Zinc Alloys

Zinc is often overlooked as a laptop computer heat sink material despite its thermal conductivity of about (112 W/mK) compared to that of pure aluminium that is (220 W/mK) [10]. This is mainly overlooked in weight sensitive equipment, but in other electronic cooling applications, for example medium to high volume, zinc alloys may have additional advantage over aluminium. Free porosity is maintained in the casting process of zinc despite the addition of alloys to improve its strength, durability, and cast-ability [10, 17]. This causes a reduction in the degree of loss thermal conductivity

than aluminium and copper. In addition, the cast thermal conductivity outperforms that of most cast aluminium and copper alloys.

Composite Materials

Erich *et al* [26], studied heat sink materials with tailored properties for thermal management. This research was based on processing advanced composite materials (refer to as metal matrix composites) for heat sinks production. Erich *et al* [26] developed different composite materials using powder metallurgical approach. The thermal properties and fabrication of composites are influenced by the properties and proportions of the matrix and the reinforcement [27]. The weight fraction and volume fraction are two factors that could be used to express the fabrication process and property calculation of composite materials respectively. In addition, the law of mixture is used as a guide during the development of advanced composite materials [27, 28]. The law of mixture formula is given as:

$$X_c = X_m v_m + X_f v_f \qquad (2)$$

Where X_c is an appropriate property of the composite, v is the volume fraction, and the subscripts m and f are the matrix and fibre reinforcement respectively.

The following subsections discuss some composites and ceramics materials that could be used for a laptop computer heat sink production.

Copper / Carbon Fibre

A copper-Carbon composite material exhibits high thermal conductivity abilities. It is therefore a good candidate for a laptop computer heat sink material. Erich *et al* [26] stated that the problems common with Cu-C composite when consolidated using hot pressing are the homogenous dispersion of the carbon fibres in the copper matrix and the weak interface between copper and the carbon fibre. The anisotropic material properties exhibited by Cu reinforced by carbon fibres using hot pressing [12, 13, 29] is due to the anisotropy of the carbon fibres. This results in the orientation of the fibres in x-y direction corresponding to the transversal hot pressing direction. It was reviewed that there was high thermal conductivity and low coefficient of thermal expansion along the fibre axis, and high coefficient of thermal expansion and low thermal conductivity parallel to the fibre axis. A further characterization of the thermal diffusivity of the Cu-C fibre using laser flash method in the x-y and z-direction was conducted by Erich et al [26]. The pitch type carbon fibres are recommended for Cu-composite materials because of the high thermal conductivity of about 100-1000 W/mK [26].

Cu-C composite material developed by copper coated carbon fibre using electrochemical coating process exhibited better distribution of the fibres in the matrix. This further improves the mechanical and thermal characteristics of the Cu-C material matrix composite. Erich et al [26] stated that 20 % more thermal diffusivity was achieved using the coated fibre approach. This resulted to a higher thermal conductivity in the z-direction of the fibre axis.

Copper / Silicon Carbide

Silicon Carbide has a thermal conductivity of about 200-300 W/mK with a coefficient of thermal expansion of 4.5 x 10^{-6} K^{-1} [30]. Silicon carbide particles are used for reinforcements of copper based composites. This is to improve the performance of copper based materials used for heat sink production. Copper silicon carbide is prepared using hot pressing [14, 31] and its major advantage over copper carbon composite is the isotropic behaviour exhibited. However, its major setback is its bad machinability. Cu-SiC has weak mechanical interface and weak thermal interface when compared with copper carbon composite materials. However, a more sophisticated fabrication process is required to help solve this machinability problem.

Copper / Diamond

Copper-diamond composites are showed to have good characteristics of a heat sink material [26, 32]. Diamond has a thermal conductivity of about 2000 W/mK and a coefficient of thermal expansion of about 0.9 x 10^{-6} K^{-1} [21]. Diamond composite is prepared by mixing and hot consolidation, putting into account the issue of the interface. Cu-diamond has a higher thermal conductivity than pure copper

and exhibited a coefficient of thermal expansion of between 10-11 x 10^{-6} K^{-1}.

Ke et al [32] studied the thermal conductivity of spark plasma sintering (SPS) consolidated Cu/diamond composites with Cr-coated diamond particles. Ke et al [32] stated that the major problem in the development of Cu/diamond composites is the achievement of a well bonded interface between the copper and the diamond. Two ways of solving the interface problem has been given as matrix alloying and particle surface metallization (metal coating) [32]. It has been reviewed that there has been a significant improvement in the interfacial bonding of Cu/diamond composites by the use of different alloys containing active elements capable of establishing chemical interactions. A novel sintering technique using spark plasma sintering (SPS) is gaining attention for the development of high performance metal matrix composites (MMCs) [32]. The introduction of chromium to the Cu/diamond composite showed a higher rate of densification than only Cu/diamond. A scanner electron microscope (SEM) image of the microstructure of the fracture surfaces of uncoated and coated Cu/diamond composite containing 50% volume particles sintered at 920 °C using chromium is shown in Figure 2 [32].

The presence of large cracks in Figure 2a confirms the weak interfacial bonding between copper and diamond. In Figure 2b it is showed that diamonds are closely embedded into the copper matrix and no clear pores and cracks can be found. This shows a good interfacial adhesion which is as a result of the addition of chromium-coated composite. This interfacial adhesion will improve the chances of Cu-diamond being used as a laptop computer heat sink material. However, Cu-diamond is expensive, and still suffers from poor machining and shaping problems.

Aluminium Nitride / Copper

Kwang-Min et al [22] studied thermo-mechanical properties of AlN-Cu composite materials prepared by solid state processing. The pulse electric current sintered mechanical alloying is used for the production of AlN-Cu composite. Kwang-Min et al [22] performed different experiment on the production of AlN-Cu composite using temperatures of 25 °C, 200 °C, 450°C, and a copper composition of 40 vol. %, 60 vol. %, and 80 vol. % respectively. The coefficient of thermal expansion achieved for AlN-40 vol. % Cu, AlN-60 vol. % Cu, and AlN-80 vol. % Cu were 7.9, 9.6, and 14.5 ppm / °C respectively. The thermal conductivity achieved is shown in Table 2.

Aluminium Nitride-Copper did not prove to be a material of choice for laptop computer heat sink

Figure 2: SEM image of the fracture surfaces of (a) uncoated and (b) coated Cu/diamond composites containing 50 vol. % particles sintered at 920 °C [32]

Table 2: Thermal conductivities of AlN-40 vol. % Cu, AlN-60 vol. % Cu, and AlN-80 vol. % Cu at temperatures of 25 °C, 200 °C, and 450 °C respectively [22]

	Pure Copper	Aluminium Bronze	Bronze	Red Brass
Conductivity	386 W/mK	83 W/mK	26 W/mK	6 1 W/mK
Composition	100% Cu	95% **Cu, 5% AL**	75% Cu, 25% Sn	**85% Cu, 9% Sn,** 6% Zn

due to the low thermal conductivity exhibited. However, different materials have been researched in other to improve the thermal conductivities of materials for thermal management. This includes Aluminium nitride reinforced by aluminium (AlN-Al), Aluminium nitride – molybdenum ceramic matrix composites (AlN-Mo), Silicon carbide – aluminium (SiC-Al), and Tungsten – copper composites (W-Cu) [33 – 39].

Aluminium-Silicon carbide (Al-SiC)

Aluminium-Silicon carbide exhibits a thermal conductivity of about 170-220 W/mK and a low coefficient of thermal expansion of about 6.2-7.3 x 10^{-6} K^{-1} [11, 23]. Al-SiC is produced using pressure infiltration casting (PIC) [23]. Development in pressure infiltration casting has allowed the production of products with tight tolerances, high thermal conductivity, and tailored coefficient of thermal expansion, reduced weight, and high stiffness [40]. Al-SiC is a good material for the production of a laptop computer heat sink because it has a reasonable thermal conductivity, low coefficient of thermal expansion, and light weight.

Cost Consideration

Determination of the cost of materials could be complex sometimes, depending on the complexity of parts, the amounts and costs of constituents and processes, production run size, total

annual production, and development of the technology [23]. Aluminium cost about £1351 per tons [41] and copper cost about £4565 per tons [41]. It is known that when composite materials are newly introduced into the market it is quite expensive. But as time goes on it becomes cheaper with increase in volume. In other to place a cost on newly introduced composite materials the following formula is suggested by Kevorkijan [42]:

$$C = X_1C_1 + X_2C_2 + C_3 \qquad (3)$$

C is the composite material and C_1, C_2, and C_3 are the cost of matrix alloy, the cost of reinforcement, and the production cost respectively. X_1 is the weight fraction of matrix alloy and X_2 is the weight fraction of the reinforcement in the composite.

The used of composite materials in some laptop computer applications shows that composite materials could be cost effective in the near future. However, the type of manufacturing process composite materials undergoes will largely affect its overall cost.

Selecting a Material for a Laptop Computer Heat Sink

Materials have different attributes such as density, strength, Young's Modulus, cost, resistance to corrosion, thermal conductivity, coefficient of thermal expansion, and so on [4, 31, 43-45]. Finding the material of choice for a laptop computer heat sink with tailored properties is vital. The smaller a component becomes the more the requirement for laptop computer heat sink material changes. The two most important properties for a heat sink are the thermal conductivity and coefficient of thermal expansion. Table 3 shows some selected materials and four selection criteria considered for a laptop computer heat sink.

A good laptop computer heat sink is expected to have the following property requirement due to miniaturisation and ceramic materials (with CTE of between 3 and 7 x 10^{-6} K^{-1}) used for the production of substrates:

a. Thermal conductivity: A material with high thermal conductivity is required
b. Coefficient of thermal expansion: The lower the CTE the more likely it is to match the substrate's CTE
c. Density: The weight of materials is fast becoming a major concern in the development of lighter laptop's computer
d. Cost: The cost of materials for a laptop computer heat sink cannot be overlooked in most cases, since making profit is the main target of laptop computers manufacturers. However, cost could be overlooked if high performance and durability is the main focus for laptop computer production. In addition, low cost material is always preferable for a laptop heat sink except stated otherwise.

Selecting a material from the list of available materials in Table 3 that exactly meets the requirement stated for a good laptop computer heat sink could be very difficult. However, a near optimum material could be selected based on the selection criteria. Diamond is a good choice of material for a laptop computer heat sink because it meets the entire thermal requirement and density. However, the cost to acquire a diamond is very high and therefore not a material of choice for manufacturers that require to make profit from its products. Materials such as copper and its composites are expensive and heavy. These are not chosen in laptop computer applications where weight reduction is necessary. Aluminium is mainly chosen by laptop computer heat sink manufacturers because it is cheap and has some thermal capabilities. However, aluminium suffers from high coefficient of thermal expansion which makes it difficult to function efficiently because substrates are produced with materials of low coefficient of thermal expansion. Aluminium/Silicon Carbide is a material with promising property for a laptop computer heat sink. Al/SiC has a reasonable thermal conductivity and a low coefficient of thermal expansion (30% less than Aluminium) to match that of ceramic substrates. It has a weight similar to aluminium and a little more expensive than aluminium. Therefore, Al/SiC is a near optimum material recommended for laptop computer heat sink production.

Table 3: Materials and selection criteria for laptop computer heat sink

Material	T.C W/mK	C.T.E 10^{-6}K^{-1}	Density Kg/m^3	Cost
Aluminium	Medium	High (not acceptable)	Low	Low
Copper	High (a)	High (not acceptable)	High (not acceptable)	High (not acceptable)
Zinc	Low (b)	Low	Low	Low
Copper/carbon Fibre	High (a)	Medium	High (not acceptable)	High (not acceptable)
Aluminium-Silicon carbide	Medium	Low	Low	Medium
Copper / Silicon Carbide	Medium	Low	High (not acceptable)	High (not acceptable)
Copper / Diamond	High (a)	Medium	High (not acceptable)	High (not acceptable)
Gold	High (a)	High (not acceptable)	Low	High (not acceptable)
Diamond	High (a)	Low	Low	High (not acceptable)
Copper/Molybdenum	Medium	Low	High (not acceptable)	High (not acceptable)

Key:

▭ : Medium/Average and is acceptable
▲a: High and is acceptable
▲ : High and is not acceptable
⬭ : Low and is acceptable
⬭b: Low and is not acceptable

Conclusion

Advanced thermal materials have been reviewed in this paper to be important to the efficiency of a laptop computer heat sink. Most laptop computers fail because of increased temperature within the system. An ideal heat sink exhibits high thermal conductivity and low coefficient of thermal expansion. Aluminium and copper are the two most commonly used material for laptop computer heat sink. It is important to note that the drift to advanced composite material for heat sink production is as a result of increased heat in microelectronic devices. Based on the selection criteria, aluminium showed outstanding qualities but Al/SiC has better qualities and is the optimum material of choice for a laptop computer heat sink. Machinability of newly developed composite materials is an issue affecting the production of new material for heat sink purposes. Other issues like homogeneity and isotropy of materials are currently being investigated to improve the quality and performance of advance composite materials.

Acknowledgements

The Author is grateful to the School of Engineering, University of Greenwich for providing the bursary for this research work and to MERG researchers for the support given.

References

[1] M.F. Ashby, Y.J.M. Brechet, D. Cebon, L. Salvo. Selection strategies for materials and processes. Materials and Design. Vol. 35, No. 1, pp. 51-67, 2004.

[2] Prasad P. Ray. Surface Mount Technology: Principles and Practice. 2nd ed. Massachusetts USA: Kluwer Academic. pp. 3-723, 2002.

[3] Nicholas Braithwaite, Graham Weaver. Materials in action series: Electronic Materials. London: Butterworth Scientific Ltd. pp. 11-82, 1990.

[4] P. G. Reddy, N. Gupta. Material Selection for Microelectronic Heat Sink: An Application of the Ashby Approach. Materials and Design. Vol. 31, No. 1, pp. 113-117, 2010.

[5] Robert Hartle. (2010). How heat sink works. Available: http://computer.howstuffworks.com/heat-sink.htm/printable. Last accessed 27th January 2011.

[6] Peter J. Ogrodnik. Fundamental Engineering Mechanics. 2nd ed. Harlow Essex: Pearson Education Limited. pp. 1-213, 2006.

[7] G. F. C. Rogers and Y. R. Mayhew. Engineering Thermodynamics: Work and Heat Transfer. 3rd ed. Essex: Longman Group Limited. pp. 5-601, 1983.

[8] Michael J. Moran and Howard N. Shapiro. Fundamentals of Engineering Thermodynamics. 6th ed. Asia: John Wiley and Sons ltd. Chapter 2, pp. 44-61. 2010.

[9] Yunus A. Cengel and Michael A. Boles. Thermodynamics: An Engineering Approach. 6th ed. New York: McGraw-Hill. pp. 92-97, 2008.

[10] K. P. Keller. Cast heat sink design advantages. IEEE intersociety conference on thermal phenomena. Vol. 1, No. 1, pp. 112-117, 1998.

[11] C. Gallagher, B. Shearer, G. Matijasevic. Materials Selection Issues for High Operating Temperature (HOT) Electronic Packaging. IEE. Vol. 1, No. 1, pp. 180-189, 1998.

[12] M. B. Dogruoz, M. Arik. On the conduction and convection heat transfer from lightweight advanced heat sinks. IEEE transactions on components and packaging technologies. Vol. 1, No. 1, pp. 1521-1528, 2010.

[13] M. B. Dogruoz, M. Arik. An investigation on the conduction and convection heat transfer from advanced heat sinks. IEEE. Vol. 1, No. 1, pp. 367-372, 2008.

[14] R. Kandasamy, W. Xiang-Qi, A. S. Mujumdar. Transient cooling of electronics using phase change material (PCM)-based heat sinks. Applied Thermal Engineering. Vol. 28, No. 1, pp. 1024-1057, 2008.

[15] Build-Your-Own-Computer.Net. (2010). What's the Best CPU Cooler?. Available: http://www.google.co.uk/imgres?imgurl=http://www.build-your-own-computer.net/image-files/cpu-heatsink-basic.jpg&imgrefurl=http://www.build-your-own-computer.net/computer-cooling.html&usg=__oIPSTS_RhOm. Last accessed 1st January 2011.

[16] Rajiv Asthana, Ashok Kumar, Narendra Dahotre. Materials Science in Manufacturing. London: Elsevier Inc.. pp. 1-607, 2006.

[17] Lyndon Edwards, Mark Endean. Materials in action series: Manufacturing with materials. London: Butterworth Scientific Ltd. pp. 11-415, 1990.

[18] X. L. Xie, Y. L. He, W. Q. Tao, H. W. Yang. An experimental investigation on a novel high-performance integrated heat pipe-heat sink for high-flux chip cooling. Applied Thermal Engineering. Vol. 28, No. 1, pp. 433-439, 2008.

[19] B. Zalba, J. M. Marın, L. F. Cabeza, H. Mehling. Review on thermal energy storage with phase change: materials, heat transfer analysis and applications. Applied Thermal Engineering. Vol. 23, No. 1, pp. 251-283, 2003.

[20] R. Kandasamy, W. Xiang-Qi, A. S. Mujumdar. Application of phase change materials in thermal management of electronics. Applied Thermal Engineering. Vol. 27, No. 1, pp. 2822-2832, 2007.

[21] D. D. L. Chung. Materials for thermal conduction. Apllied Thermal Engineering. Vol. 21, No. 1, pp. 1593-1605, 2001.

[22] K. Lee, D. Oh, W. Choi, T. Weissg¨arber, B. Kieback. Thermomechanical properties of AlN–Cu composite materials prepared by solid state processing. Journal of Alloys and Compounds. Vol. 434-435, No. 1, pp. 375-377, 2007.

[23] Carl Zweben. (1999). High Performance Thermal Management Materials. Available: http://www.electronics-

cooling.com/1999/09/high-performance-thermal-management-materials/. Last accessed 4th November 2010.

[24] A. Sommers, Q. Wang, X. Han, C. T'Joen, Y. Park, A. Jacobi. Ceramics and ceramic matrix composites for heat exchangers in advanced thermal systems - A review. Applied Thermal Engineering. Vol. 30, No. 1, pp. 1277-1291, 2010.

[25] Seri Lee. How to select a heat sink. Advance thermal engineering. Vol. 1, No. 1, pp. 1-5, 1996.

[26] E. Neubauer, P. Angerer, G. Korb. Heat Sink Materials with Tailored Properties for Thermal Management. IEE. Vol. 1, No. 1, pp. 258-263, 2005.

[27] F. L. Matthews and R. D. Rawlings. Composite materials: Engineering and science. 2nd ed. Cambridge: Woodhead Publishing Ltd and CRC Press LLC. pp. 10-23, 1999.

[28] G. Piatti. Advances in composite materials. Essex: Applied science publishers Ltd. pp. 75-89, 1978.

[29] H. Yin, X. Gao, J. Ding, Z. Zhang. Experimental research on heat transfer mechanism of heat sink with composite phase change materials. Energy Coversion and management. Vol. 49, No. 1, pp. 1740-1746, 2008.

[30] Th. Schubert, B. Trindade, T. Weißgarber, B. Kieback. Interfacial design of Cu-based composites prepared by powder metallurgy for heat sink applications. Materials Science and Engineering. Vol. 475, No. 1-2, pp. 39-44, February 2008.

[31] Michael F. Ashby. Materials selection in mechanical design. 2nd ed. Oxford (UK): Butterworth-Heinemann. pp. 5-400, 1999.

[32] K. Chu, Z. Liu, C. Jia, H. Chen, X. Liang, W. Gao, W. Tian, H. Guo. Thermal conductivity of SPS consolidated Cu/diamond composites with. Journal of Alloys and Compounds. Vol. 490, No. 1, pp. 453-458, 2010.

[33] C. Troadec, P. Goeuriot, P. Verdier, Y. Laurent, J. Vicens, G. Boitier, J.L.Chermant, B.L. Mordike. AlN dispersed reinforced aluminum composite. Journal of the European Ceramic Society. Vol. 17, No. 15-16, pp. 1867-1875, 1997.

[34] A. A. Khan, J. C. Labbe. Aluminium nitride—molybdenum ceramic matrix composites: Characterization of ceramic—Metal interface. Journal of the European Ceramic Society. Vol. 16, No. 7, pp. 739-744, 1996.

[35] A. Ureña, M.V. Utrilla, J. Rams, P. Rodrigo, M. Ferrer. Electroless multilayer coatings on aluminium–silicon carbide composites for electronics packaging. Journal of the European Ceramic Society. Vol. 25, No. 13-15, pp. 3983-3986, 2007.

[36] Dongdong Gu, Yifu Shen. Effects of processing parameters on consolidation and microstructure of W–Cu components by DMLS. Journal of Alloys and Compounds. Vol. 473, No. 1-2, pp. 107-115, 2009.

[37] S. Eghtesadi, N. Parvin, M. Rezaee, M. Salari. Mechanically induced driving forces in preparing W–Cu nanocomposite. Journal of Alloys and Compounds. Vol. 473, No. 1-2, pp. 557-559, 2009.

[38] Y. Guo, J. Yi, S. Luo, C. Zhou, L. Chen, Y. Peng. Fabrication of W–Cu composites by microwave infiltration. Journal of Alloys and Compounds. Vol. 492, No. 1-2, pp. 75-78, 2010.

[39] L. Shu-dong, Y. Jian-hong, G. Ying-Li, P. Yuan-dong, L. Li-ya, R. Jun-ming. Microwave sintering W–Cu composites: Analyses of densification and microstructural homogenization. Journal of Alloys and Compounds. Vol. 473, No. 1-2, pp. 5-9, 2009.

[40] Matt Napier. (2010). Metal Matrix Composites by Pressure Infiltrated Casting. Available: http://www.matweb.com/reference/MetalMatrix Composite.aspx. Last accessed 9th November 2010.

[41] London Metal Exchange. (2010). Non-ferrous metals. Available: http://www.lme.com/non-ferrous/index.asp. Last accessed 9th November 2010.

[42] Kevorkijan, V. "Development of Al MMC composites for automotive industry", Presented at the Syposiam on Deformation and Structure of Metals and Alloys, 26 – 27 June 2002, Belgrade, Yugoslavia.

[43] Michael F. Ashby, David R.H. Jones. Engineering materials 1: An introduction to properties, applications, and design. 3rd ed. Oxford (UK): Elsevier Ltd. pp. 1-410, 2006.

[44] Michael F. Ashby, David R.H. Jones. Engineering materials 2: An introduction to microstructures, processing, and design. 3rd ed. Oxford (UK): Elsevier Ltd. pp. 3-379, 2007.

[45] Michael F. Ashby. Material selection in mechanical design. 3rd ed. Oxford (UK): Elsevier Ltd. pp. 1-470, 2008.

Emerging Nanotechnology-based Thermal Interface Materials for Automotive Electronic Control Unit Application

K. C. Otiaba, N.N. Ekere, R. S. Bhatti, S. Mallik and E. H. Amalu

Electronics Manufacturing Engineering Research Group
School of Engineering at Medway, University of Greenwich, Chatham Maritime, Kent, ME4 4TB, UK

Email: k.c.otiaba@gre.ac.uk

Abstract

The under-hood automotive ambient is harsh and its impact on electronics used in electronic control unit (ECU) assembly is a concern. The introduction of Euro 6 standard (Latest European Union Legislation) leading to increase in power density of power electronics in ECU has even amplified the device thermal challenge. Heat generated within the unit coupled with ambient temperature makes the system reliability susceptible to thermal degradation which may result in catastrophic failure if not efficiently dissipated. Previous investigations show that the technology of thermal interface materials (TIMs) is a key to achieving good heat conductions within a package and from a package to heat sinking device. With studies suggesting that conventional TIMs contribute about 60% interfacial thermal resistance, innovative technology is required to improve the thermal performance of TIMs. A review of emerging nanotechnology in TIMs shows that carbon nanotubes (CNTs) and carbon nanofibres (CNFs) when used as the structure of TIM or TIM filler could improve the overall thermal and mechanical properties of TIMs. Hence, CNTs/CNFs have the potentials to advance thermal management issues in ECU. This search identifies chemical vapour deposition (CVD) as a low cost process for the commercial production of CNTs. In addition, this review further highlights the capability of CVD to grow nanotubes directly on a desired substrate. Other low temperature techniques of growing CNT on sensitive substrates are also presented in this paper.

Key words: Thermal Management, Electronic Control Unit, Thermal Interface Materials, Thermal Resistance, Carbon nanotubes.

Introduction

The operation of integrated circuits (ICs) at elevated temperature is a major cause of failures in electronic devices and a critical problem in developing more advanced electronic packages [1]. One such example of an electronic device is the ECU whose function in automotives has increased and is expected to further rise in the foreseeable future. The introduction of the recent European Union legislations (Euro 5 and 6 standards) [2] to put more stringent limits on pollutant emissions from road vehicles are amongst the major contributing factors of further ECU performance improvement and significant increase in ECU operating temperature.

Researchers seem to have responded to this heat rise in ECU with the suggestion of improved heat sinking material/device [3]. The extreme cost constraints being undergone by the automotive industry might have limited the feasibility of manufacturing real highly finished surface area resulting in interstitial air gaps between the heat sink base and heat source. The thermal resistance associated with this interfacial air gap have a detrimental impact in the overall heat dissipation from electronic devices. This interfacial thermal resistance is in series with the resistance of any heat sink and cannot be removed or reduced even by employing advanced cooling techniques on the side of the heat sink [1]. Hence, thermal interface materials (TIMs) need to be applied between contact surfaces to enhance heat conduction to the outside heat sink.

In ECU, TIMs are typically found between the printed circuit board (PCB) and the housing base plate as shown in Fig. 1. This is because the path of heat removal from an ECU involves conduction across the interface of the PCB surface, through a TIM, into the case or housing (heat sink) and then convection to the environment as delineated in Fig. 1c. Teertstra et al [4] reported that by using TIM, thermal resistance can be reduced by approximately a factor of five. Nevertheless, with researchers suggesting that conventional TIMs including polymer-based TIMs contribute about 60% interfacial thermal resistance in many electronic assemblies that employ a TIM to mount a heat sink [5,6], innovative technology is required to improve the thermal performance of TIMs.

The unique properties of one dimensional structure and materials have gained much attention in recent years for thermal management applications. Among such materials, carbon-nanotubes (CNTs) and carbon nanofibres (CNFs) seem promising TIMs owing to their special structural, mechanical and more importantly thermal properties [7-10]. The inherent thermal conductivity properties [7,11] of CNTs are excellent, and the ability to fabricate them in a controlled manner has been instrumental in realizing their potentials. CNTs appear promising to address the key concerns emanating from the use of polymer-based TIMs (such as dry-out over time, pump-out during cyclic thermal loading and non-uniform applications) and solder-based TIMs (like solder voids and fatigue failure) as discussed in reference [12].

This paper reviews the implications of the emerging nanotechnology in TIMs which could be applicable to automotive ECU.

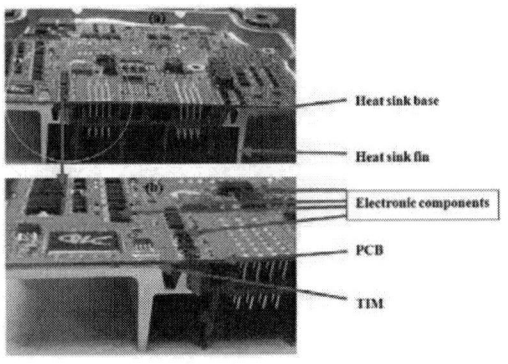

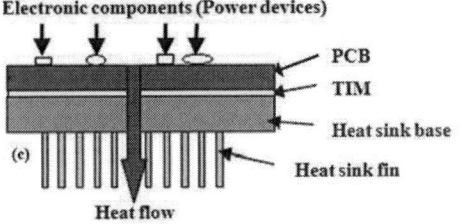

Figure 1: cross-section image of an ECU (a) showing ECU heat sink base and heat sink fin (b) showing TIM, PCB and electronic components (c) schematic diagram of ECU showing heat flow path.

Nanotechnology in TIMs

Since the discovery of Multi-Walled CNT (MWNT) in 1991 by Iijima [13] and Single-Wall NanoTube (SWNT) in 1993 by Iijima and Ichihashi [14] and Bethunes et al. [15], significant effort has been devoted to understanding and characterizing their properties [7]. The most interesting properties of CNTs are the ballistic transport of electrons and the extremely high thermal conductivity along the tube axis [7,16]. Also, phonons propagate easily along nanotubes [17]. Reported values of thermal conductivity are shown to be as high as 3000 W/mK [8,18,19] and 3500 W/mK [20] for a MWNT and SWNT, respectively, at room

temperature. Carbon nanofibres grown from CVD were measured to have a thermal conductivity in the range of 35 W/mK which increased to 2000 W/mK following annealing at 3000 °C [21,22]. Many of these values are comparable or even higher than that of diamond [23], giving them the greatest thermal conductivity of any known material. Many research works have already been published and patents filed on CNTs and CNFs potentials as TIM, some of which have been referenced in this paper [5,9,10]. Unsurprisingly, recent findings have shown that CNT-based interfaces can significantly conduct more heat than comparable state-of-the-art commercial TIMs at the same temperatures as shown in Fig. 2. CNTs/CNFs can be employed as TIM fillers or TIM structure.

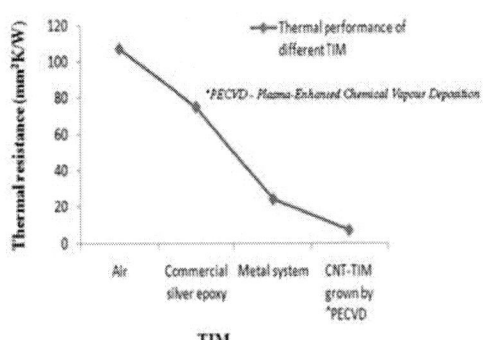

Figure 2: Measured total thermal resistance of different TIMs [24]

CNTs as Fillers

The discovery of high thermal conductivity in CNTs and CNFs has emanated suggestions that they could be employed as highly heat conductive fillers to improve thermal conductivity of TIMs. Many works [25-27] have been carried out in this regard. Remarkable improvements in the thermal conductivity of the TIM were achieved in each application. For instance, Hu et al. [28] employed the combination of CNTs and traditional thermal conductive fillers for TIMs. They accomplished a thermal conductivity value seven times that of the base material, approximately double the thermal conductivity of the equivalent TIM composite with only conventional fillers. Though the thermal conductivity of TIM could be improved by CNTs inclusion, references [1,29] suggest that the potential heat conduction of CNTs is not fully optimised when employed as fillers. The low efficiency could be as a result of firstly, the random dispersion of CNTs, which means that heat conduction is only through few portions of CNTs by effect. Secondly, heat is not directly conducted from one side to the other via CNTs. As a result of CNTs' small diameter, CNTs are discontinued by

327

other lower thermal conductive fillers or the base fluid which could deteriorate the thermal performance of the CNT composites [1]. The evolution of aligned CNT array as a better choice for TIM's basic structure is therefore not surprising.

CNT Arrays

The use of CNT arrays in an aligned manner has attracted much interest in recent times because CNT possess the highest value of thermal conductivity along their axis [30]. In this approach, vertically aligned CNTs is placed in the interface between the metallic heat sink such as aluminium and microelectronic devices. The overall heat conduction is determined by the thermal conductivity of the CNTs themselves and the thermal conductance at the two surfaces at the two ends (electronic components and heat sink devices) of the CNTs. Many works [1,11,29,31] propose that CNT arrays when employed as interface materials offer advanced thermal management due to their ability to significantly aid heat conduction with relatively high effective thermal conductivities (~ 80 W/mK). However, to fully realise the unique thermal properties of aligned CNTs array, more research is ongoing. The most challenging issue is attaining thermal contact between surrounding surfaces and vertically oriented CNTs as shown in Fig. 3 [5,32]. The contact resistance could be as a result of the variations in the length of the tubes resulting in poor mechanical contacts. Though improved mechanical contacts can be achieved by increasing attachment pressure, other key concerns like constriction effects and acoustic mismatch at the contact points may not be addressed with increased pressure [1]. Nonetheless, a few practical solutions exist to reduce the effect of the total contact thermal resistance without an overall increase in attachment pressure [1]. Firstly, is the combination of other conventional TIMs with high wetting properties, including phase-change-materials (PCMs) [33,34] and thermal greases. Xu and Fisher [33] have reported the least resistance values of 19.8 mm^{2}K/W and 5.2 mm^{2}K/W for copper-silicon interfaces with dry CNT arrays and PCM-CNT arrays, respectively, under moderate pressure. Secondly, is the growth of vertically oriented CNTs on both of the contact surfaces to form an interwoven mesh (cross-talk interface) [11] CNT array interfaces have been reported to produce thermal resistances as low as 8 mm^2 °C/W and 4 mm^2 °C/W (similar to that of a soldered joint), for arrays grown on one side [35] and both sides [31] of the interface, respectively, under moderate pressure. Thirdly, the CNT transfer process demonstrated by Zhu et al. [36] is promising as regards reducing thermal contact resistance. They employed a conventional solder

reflow process (peak temperature at 250 °C) for the assembly of open-ended CNT structures aligned to a eutectic tin-lead solder paste deposit printed on a copper substrate. The superb mechanical bonding strength on the CNT and solder interfaces should effectively assist in the reduction of the thermal contact resistance. They reported a thermal conductivity and thermal resistance of 81 W/mK and 0.43 cm^{2}K/W, respectively, for the assembly with CNT height of ~180 μm. An insulating layer like mica could be used to provide electrical isolation if required [37].

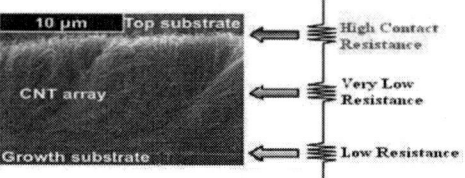

Figure 3: Contact resistance for a one sided CNT array [38]

CNT Mechanical Properties

CNTs, besides their outstanding thermal properties, are known to offer extraordinary mechanical properties [39,40]. Works [40,41] have shown that nanotubes can sustain large strain deformations such as twisting and buckling without showing signs of fracture, they have the intriguing capability of returning to their near original, straight, structure following deformation as depicted in Fig. 4. Such behaviour is highly unusual and could play a significant role in increasing the energy absorbed during deformation of CNT-filled composites during high temperature loading [39]. Ajayan and Zhou [39] reported that despites CNT's high elasticity and high conductivity, it is one of the strongest materials and often robust in most harsh environments. CNT array [42,43] and CNT-based polymer composites [44,45] have been largely studied for their mechanical strength and viscoelastic properties.

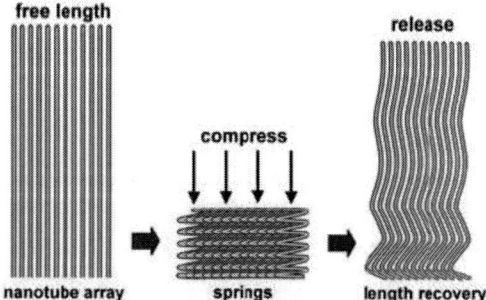

Figure 4: Compression testing of aligned CNT films. A schematic illustration shows a CNT array compressed to folded springs and then regaining the free length upon the release of compressive load [46].

Vertically aligned CNT arrays create parallel paths across mating surfaces with each path containing one CNT and two junctions at top and below surfaces. Hence, maximises thermal conductivity of CNT array and enhances temperature stability (no pump-out). Suhr et al. [43] showed that long, vertically aligned MWNTs manifested viscoelastic behaviour analogous to that observed in soft-tissue membranes under repeated high compressive strains. The mechanical response of the CNT arrays showed preconditioning, characteristic viscoelasticity-induced hysteresis, nonlinear elasticity and stress relaxation, and large deformations, under compressive cyclic loading. Additionally, they [43] did not observe any fatigue failure at high strain amplitudes up to half a million cycles. Cao et al. [46] reported that free standing films of vertically aligned CNTs exhibited super-compressible foam-like performance. Under compression, the CNTs collectively formed zigzag buckles that could fully unfold to their near original length upon release of load resulting in a strong cushioning effect (see Fig. 4). Compression stress-strain curves also manifested large hysteresis, indicating substantial energy loss (damping) which might be as a result of the friction between the CNT surfaces when in motion [46]. References [42,47] have also reported good mechanical responses and significant level of energy dissipation of CNT arrays under high-strain rate deformation. Nonetheless, in the as-grown state, aligned CNT films are only held together by weak van der Waals forces [48] and the substrate resulting in fragile adhesion between the tubes. This leaves pure CNT arrays with low resistance to shear and prone to splitting [49] when a stress is applied. No wonder the recent exploitation of CNT-based composites (with interest in matrix materials such as polymers, ceramics and metals).

CNT-polymer composites enhance stress distribution of the composite film through load transfer from the polymer to the CNT network [49]. Some experiments [50,51] have provided valuable insights into the mechanical characteristics of CNT-reinforced polymers, though with conflicting reports at times on the interface strength in CNT-polymer composites. Teo et al. [49] demonstrated a hybrid film consisting of CNTs and amorphous diamond. This unique structure referred to as carbon-based nanomattress encompassed extraordinary mechanical and viscoelastic properties. It exhibited good thermal performance (high temperature stability) and could protect components from mechanical vibration and wear [49]. Nanotube reinforcements have also resulted in an increase in the toughness of the nanotubes-based ceramic matrix composites by absorbing energy during their highly flexible elastic behaviour [52,53] under compressive loads.

The aforementioned studies indicate that CNT arrays and composite materials could potentially be employed as intrinsic damping materials coupled to its thermal management capabilities. These unique properties of CNTs are crucial for CNT-based TIMs especially for the high temperature and harsh under-hood automotive ambient applications. Future research should be focused on understanding the long term reliability and performance degradation of CNT-based TIMs when exposed to cyclic temperature loading. Such data seems scarce in literature.

CNT Fabrication

Among the various means of CNT synthesis, Chemical Vapour Deposition (CVD) appears the most promising method for industrial-scale production [7,36,54]. This can be attributed to CVD's price per unit ratio and its capability to grow nanotubes directly (well aligned) on a desired substrate [7,29,36,55]. However, direct synthesis of CNT array interfaces is a concern. This is because the temperatures at which these CNT array thermal interfaces are grown (>800 °C) may be too high for temperature-sensitive packages employed in the assembly processes of standard electronic device. Apparently, the electrical and mechanical performance of most metal contacts and interconnects degrades when exposed to high temperatures up to 450 °C for more than a limited time [54]. Cola et al. [55] proposed an insertable CNT array/foil as a viable method to apply CNT arrays to an interface without exposing mating materials to adverse CNT growth temperatures. This technique seems promising considering the low thermal resistances (10 mm^2 °C/W) achieved under moderate pressure. Zhu et al. [36] developed a CNT transfer process which is similar to flip-chip technology as delineated in Fig. 5. They employed a conventional solder reflow process (peak temperature at 250°C) for the assembly of open-ended CNT structures aligned to a eutectic tin-lead solder paste deposit printed on a copper substrate. This CNT transfer technology enabled the separation of the high temperature CNT growth and the low temperature CNT device assembly. References [54,56] reported other progress in low temperature growth of CNT arrays. These low temperature synthesis approaches appear advantageous for their ability to be incorporated into existing manufacturing processes and their good thermal interface conductance. For instance, CNT arrays can seamlessly be grown at low temperatures on sensitive substrates such as aluminium [3] that may be of interest to the automotive industry.

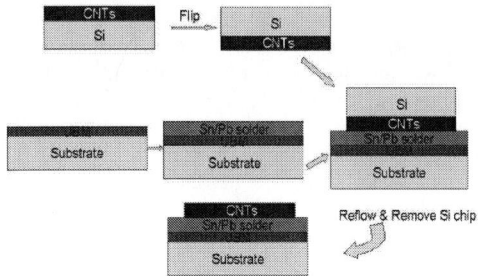

*UBM – under bump metallisation
*Si - silicon

Figure 5: Schematic diagram of the "CNT transfer technology" for assembling aligned open-ended CNT films (Zhu et al.) [36]

Summary

Concerted research and development is required to meet the thermal management challenges of current and next generation ECUs. This paper presents a review of the emerging nanotechnology in TIMs that could be applicable to automotive ECU. The discovery of CNTs potentials as TIMs owing to their high thermal conductivity values in axial direction stands out as the most recent development to improve TIMs' efficiency. However, in order to realize the promise of high thermal conductivity of CNTs with expected reliability, further practical approaches and extensive modeling must be found to characterize the performance degradation of CNT based TIMs. Undoubtedly, the development of affordable synthesis techniques is also vital to the future of carbon nanotechnology in automotive industry.

As with every other technology in its infancy, CNT application as TIMs may recently seem more complex and costly than polymer-based TIMs; however, it should be noted that CNT concept as TIM has good potentials and could lead to various creative/novel ideas that would be favourable in terms of cost and manufacturability. Indeed, there is much development space for the CNT assembly as TIMs.

References

[1] X. Jack Hu, Antonio A. Padilla, Jun Xu, Timothy S. Fisher, Kenneth E. Goodson, "3-Omega Measurements of Vertically Oriented Carbon Nanotubes on Silicon", ASME, Journal of Heat Transfer, Vol. 128, 2006.

[2] Summaries of EU legislation. Euro 5 and Euro 6 standards: reduction of pollutant emissions from light vehicles. [Online] January 7, 2010. [Cited: November 9, 2010.]
http://europa.eu/legislation_summaries/environment/air_pollution/l28186_en.htm.

[3] Sabuj Mallik, Ndy Ekere, Chris Best, Raj Bhatti, "Investigation of Thermal Management Materials for Automotive Electronic Control Units", Applied Thermal Engineering, Vol. 31, pp. 355-362, 2010.

[4] P. Teertstra, M. M. Yovanovich and J. R. Culham, "Calculating interface resistance", Electronics Cooling, Vol. 3, No. 2.pp. 24 - 29, May, 1997.

[5] Farhad Sarvar, David C. Whalley, Paul P. Conway, "Thermal Interface Materials - A Review of the State of the Art", Electronics Systemintegration Technology Conference, Dresden, Germany: IEEE, pp. 1292 – 1302, 2006.

[6] S.S. Tonapi, R.A. Fillion, F.J. Schattenmann, H.S. Cole, J.D. Evans, G.G. Sammakia, "An Overview of Thermal Management for Next Generation Microelectronic Devices", SEMI Advanced Manufacturing Conference: IEEE, pp. 250-254, 2003.

[7] David Gerke, Carbon Nanotube Technology, NASA 2009 Body of Knowledge (BoK), California Institute of Technology, Jet Propulsion Laboratory, Pasadena, California, 2009.

[8] Quoc Ngo, Brett A. Cruden, Alan M. Cassell, Megan D. Walker, Qi Ye, Jessica E.Koehne, M. Meyyappan, Jun Li and Cary Y. Yang, "Thermal Conductivity of Carbon Nanotube Composite Films", Materials Research Society, Vol. 812, 2004.

[9] Helen F. Chuang; Sarah M. Cooper; M. Meyyappan; Brett A. Cruden, "Improvement of Thermal Contact Resistance by Carbon Nanotubes and Nanofibers", Nanoscience and Nanotechnology, Vol. 4 (8), pp. 964-967, November 2004.

[10] Quoc Ngo, Brett A. Cruden, Alan M. Cassell,Gerard Sims, M. Meyyappan, Jun Li and Cary Y. Yang, "Thermal Interface Properties of Cu-filled Vertically Aligned Carbon Nanofiber Arrays", Nano Letters, Vol. 4 , No 12, pp. 2403-2407, 2004.

[11] ScienceDaily. Science News. [Online] May 2, 2006. [Cited: 11 4, 2010.]
http://www.sciencedaily.com/releases/2006/05/060502171803.htm.

[12] K.C. Otiaba, N.N. Ekere, R.S. Bhatti, S. Mallik, M.O. Alam, E.H. Amalu, "Thermal Interface Materials for Automotive Electronic Control Unit: Trends,Technology and R&D Challenges", In press, Microelectronics Reliability. 10.1016/j.microrel.2011.05.001.

[13] Sumio Iijima, "Helical Microtubules of Graphitic Carbon", Nature, Vol. 354, pp. 56-58, 1991.

[14] Sumio Iijima and Toshinari Ichihashi, "Single-shell Carbon Nanotubes of 1-nm Diameter", Nature, Vol. 363, pp. 603 – 605, 1993.

[15] D. S. Bethune, C. H. Klang, M. S. de Vries, G. Gorman, R. Savoy, J. Vazquez & R. Beyers, "Cobalt-catalysed Growth of Carbon Nanotubes with Single-atomic-layer Walls", Nature, Vol. 363, pp. 605 – 607, 1993.

[16] Hoenlein W., Krepul F. "Carbon Nanotubes Applications in Microelectronics", IEEE. Vol. 27, pp. 629 – 634, 2004.

[17] Baughman, R. H., Zakhidov, A. A. "Carbon Nanotubes - the Route Toward Applications", Science, Vol. 297, pp. 787 – 792, 2002.

[18] Matthew Weisenberger, Project Facts, Carbon Materials, [Online] 2009. [Cited: November 6, 2010.] http://www.caer.uky.edu.

[19] P. Kim, L. Shi, A. Majumdar and P. L. McEuen, "Thermal Transport Measurements of Individual Multiwalled Nanotubes", Physical Review Letters, Vol. 87, 2001.

[20] Eric Pop, David Mann, Qian Wang, Kenneth Goodson and Hongjie Dai, "Thermal Conductance of an Individual Single-Wall Carbon Nanotube above Room Temperature", Nano Letters, Vol. 6, pp. 96-100, 2006.

[21] J. Heremans and C. P. Beetz, Jr., "Thermal Conductivity and Thermopower of Vapor-grown Graphite Fibers", Physics Review B, Vol. 32, pp. 1981–1986, 1985.

[22] J. Heremans, I. Rahim and M. S. Dresselhaus, "Thermal Conductivity and Raman Spectra of Carbon Fibers", Physics Review B, Vol. 32, pp. 6742–6747, 1985.

[23] Taejin Kim, Mohamed A. Osman, Cecilia D. Richards, David F. Bahr, and Robert F. Richard, "Molecular dynamic simulation of heat pulse propagation in multiwall carbon nanotubes", Physics Review B, Vol. 76, 2007.

[24] K. Zhang, Y Chai, M. M. F. Yuen, D. G. W. Xiao and P. C. H. Chan, "Carbon Nanotube Thermal Interface Material for High-brightness Light-emitting-diode Cooling", Nanotechnology, Vol. 19, 2008.

[25] Jong-Jin Park, Minoru Taya, "Design of Thermal Interface Material With High Thermal Conductivity and Measurement Apparatus", ASME, Vol. 128, 2006.

[26] Kai Zhang Guo-Wei Xiao Wong, C.K.Y. Hong-Wei Gu Yuen, M.M.F. Chan, P.C.H. Bing Xu, "Study on Thermal Interface Material with Carbon Nanotubes and Carbon Black in High-Brightness LED Packaging with Flip-Chip", Lake Buena Vista, FL. Electronic Components and Technology Conference (ECTC): IEEE, pp. 60 – 65, 2005.

[27] E. S. Choi, J. S. Brooks, D. L. Eaton, M. S. Al-Haik, M. Y. Hussaini, H. Garmestani, D. Li, K. Dahmen, "Enhancement of Thermal and Electrical Properties of Carbon Nanotube Polymer Composites by Magnetic Field Processing", Journal of Applied Physics, Vol. 94, pp. 6034-6039, 2003.

[28] Xuejiao Hu, Linan Jiang and Kenneth E. Goodson, "Thermal Conductance Enhancement of Particle-filled Thermal Interface Materials Using Carbon Nanotube Inclusions", Thermal and Thermomechanical Phenomena in Electronic Systems, ITHERM '04: IEEE, pp. 63 – 69, 2004.

[29] K. Zhang, Matthew M.F. Yuen, N. Wang, J.Y. Miao, David G.W. Xiao, H.B. Fan. "Thermal Interface Material with Aligned CNT and Its Application in HB-LED Packaging", Electronic Components and Technology Conference (ECTC): IEEE; pp.177-82, 2006.

[30] S. Shaikh, L. Li, K. Lafdi, J. Huie, "Thermal Conductivity of an Aligned Carbon Nanotube Array" Carbon, Vol. 45, pp. 2608–2613, 2007.

[31] Baratunde A. Cola, Jun Xu, Changrui Cheng, Xianfan Xu, Timothy S. Fisher, and Hanping Hu, "Photoacoustic Characterization of Carbon

331

Nanotube Array Thermal Interfaces", Applied Physics, Vol. 101, 2007.

[32] Patrick K. Schelling, Li Shi, and Kenneth E. Goodson, "Managing Heat for Electronics", Materialstoday, Vol. 8, pp. 30-35, June, 2005.

[33] Jun Xu, Timothy S. Fisher, "Enhancement of Thermal Interface Materials with Carbon Nanotube Arrays", International Journal of Heat and Mass Transfer, Vol. 49, pp. 1658–1666, 2006.

[34] Shadab Shaikh, Khalid Lafdi, Edward Silverman, "The Effect of a CNT Interface on the Thermal Resistance of Contacting Surfaces", Carbon, Vol. 45, pp. 695–703, 2007.

[35] Placidus B. Amama, Baratunde A. Cola, Timothy D. Sands, Xianfan Xu and Timothy S. Fisher, "Dendrimer-assisted Controlled Growth of Carbon Nanotubes for Enhanced Thermal Interface Conductance", Nanotechnology, Vol. 18, 2007

[36] Lingbo Zhu, Dennis W. Hess, C.P. Wong, "Assembling Carbon Nanotube Films as Thermal Interface Materials", Electronic Components Technology Conference (ECTC): IEEE, pp. 2006-2010, 2007.

[37] Greg Bischak, Christine Vogdes, "Thermal Management Design Criteria and Solutions" IEEE. pp. 188-193, 1998.

[38] Baratunde A. Cola, "Carbon Nanotubes as High Performance Thermal Interface Materials", Electronics Cooling, [Online] April 2010. [Cited: November 9, 2010.] http://www.electronics-cooling.com/2010/04/carbon-nanotubes-as-high-performance-thermal-interface-materials/.

[39] Pulickel M. Ajayan and Otto Z. Zhou, "Applications of Carbon Nanotubes", Topics in Applied Physics, Vol. 80, pp. 391-425, 2001.

[40] B. I. Yakobson, C. J. Brabec and J. Bernholc, "Nanomechanics of Carbon Tubes: Instabilities beyond Linear Response", Physical Review Letters, Vol. 76, pp. 2511 – 2514, 1996.

[41] S. Reich and C. Thomsen, "Elastic Properties of Carbon Nanotubes under Hydrostatic Pressure", Physical Review B, Vol. 65, 153407, 2002.

[42] H.J. Qi, K.B.K. Teo, K.K.S. Lau, M.C. Boyce,W.I. Milne, J. Robertson, K.K. Gleason,

"Determination of Mechanical Properties of Carbon Nanotubes and Vertically Aligned Carbon Nanotube Forests using Nanoindentation", Journal of the Mechanics and Physics of Solids, Vol. 51, pp. 2213 – 2237, 2003.

[43] J. Suhr, P. Victor, L. Ci, S. Sreekala, X. Zhang, O. Nalamasu & P. M. Ajayan, "Fatigue Resistance of Aligned Carbon Nanotube Arrays under Cyclic Compression", Nature Nanotechnology, Vol. 2, pp. 417 – 421, 2007.

[44] Chenyu Wei, Deepak Srivastava, Kyeongjae Cho, "Thermal Expansion and Diffusion Coefficients of Carbon Nanotube-Polymer Composites", Nano letters, Vol. 2, pp. 647–650, 2002.

[45] D. Qian, E. C. Dickey, R. Andrews and T. Rantell, "Load Transfer and Deformation Mechanisms in Carbon Nanotube-polystyrene Composites", Applied Physics Letters, Vol. 76, pp. 2868 – 2870, 2000.

[46] Anyuan Cao, Pamela L. Dickrell, W. Gregory Sawyer, Mehrdad N. Ghasemi-Nejhad, Pulickel M. Ajayan, "Super-Compressible Foamlike Carbon Nanotube Films", Science, Vol. 310, pp. 1307 - 1310. 2005.

[47] S.Dj. Mesarovic, C.M. McCarter, D.F. Bahr, H. Radhakrishnan, R.F. Richards, C.D. Richards, D. McClainb and J. Jiao, "Mechanical Behavior of a Carbon Nanotube turf", Scripta Materialia, Vol. 56, pp. 157–160, 2007.

[48] A. J. Du and Sean C. Smith, "Van der Waals-corrected Density Functional Theory: Benchmarking for Hydrogen–nanotube and Nanotube–nanotube Interactions", Nanotechnology, Vol. 16, p. 2118, 2005.

[49] Edwin H. T. Teo, Wendy K. P. Yung, Daniel H. C. Chua, and B. K. Tay, "A Carbon Nanomattress: A New Nanosystem with Intrinsic, Tunable, Damping Properties", Advanced Materials, Vol. 19, pp. 2941–2945, 2007.

[50] P. J. F. Harris, "Carbon Nanotube Composites", International Materials Reviews, Vol. 49, pp. 31-43, 2004.

[51] P. C. P. Watts and W. K. Hsu, "Behaviours of Embedded Carbon Nanotubes during Film

Cracking", Nanotechnology, Vol. 14, pp. L7 – L10, 2003.

[52] Jing Sun, Lian Gao and Xihai Jin, "Reinforcement of Alumina Matrix with Multi-walled Carbon Nanotubes", Ceramics International, Vol. 31, pp. 893-896, 2005.

[53] Chang Sekyung, Doremus Robert H., Ajayan Pulickel M., Siegel Richard W., "Processing and mechanical properties of C-nanotube reinforced alumina composites", Ceramic Engineering and Science Proceedings , Vol. 21, pp. 653-658 , 2000.

[54] A. V. Melechko, V. I. Merkulov, T. E. McKnight, M. A. Guillorn, K. L. Klein, D. H.

Lowndes, M. L. Simpson, "Vertically Aligned Carbon Nanofibers and Related Structures: Controlled Synthesis and Directed Assembly", Applied Physics, Vol. 97, 2005.

[55] Baratunde A. Cola, Xianfan Xu, and Timothy S. Fisher, "Increased Real Contact in Thermal Interfaces: A Carbon Nanotube/foil Material", Applied Physics Letters, Vol. 90, 2007.

[56] Bojan O. Boskovic, Vlad Stolojan, Rizwan U.A. Khan, Sajad Haq and S. Ravi P. Silva, "Large-area synthesis of Carbon Nanofibres at Room Temperature", Nature Materials, Vol. 1, pp. 165 – 168, 2002.

3D LTCC and Flexible Structure Interconnections
Based on Galvanized Layers

Michal Nicák, Josef Šandera

Department of Microelectronics, Faculty of Electrical Engineering and Communication,
Brno University of Technology, Technická 10, 61600 Brno, Czech Republic

+420 604148614, xnicak00@stud.feec.vutbr.cz, sandera@feec.vutbr.cz

Abstract

Interconnections have an important role in packaging as they not only provide electrical connection but they also act as mechanical attachment between connected substrates and are supposed to be as reliable as possible. This paper describes design, construction and some recent laboratory tests of 3D stacked structures, where different materials with different coefficient of thermal expansion are combined together in one package. Goal of this work is to develop reliable 3D stacked structure intercronnection method, which will allow connection between LTCC, Al_2O_3 and also flexible substrates. Vacuum evaporated and galvanically enhanced layers in connection with solder balls are used to create interconnections between substrates.

Key words: LTCC, 3D, Interconnection, structure, packaging

Introduction

Why to perform research into 3D structures? Immediately after first printed circuit boards were introduced, it was obvious, that planar boards are not going to be able to provide enough interconnection density to satisfy rising demands. This led to new term "3D packaging". Focus on third dimensional design presents obvious way to create more and more complicated systems in smaller and more compact design. Both miniaturization and densification of electronic interconnections together are the inevitable future of not only packaging, but whole microelectronic world.

This project succeeds previous research of soldered 3D structures containing "dimpled" LTCC substrates, which was presented during ISSE2010 [7] and MSMF6 [8] conferences. The main goal of current phase is to develop method for design and implementation of lead-free soldered stacked three-dimensional structures (packages) based on substrates interconnected together by solder balls and solder paste in connection with vacuum evaporated and galvanically enhanced copper conductive layers. This method uses mentioned vacuum evaporated and galvanically reinforced copper layers as a base for pad construction. These pads can be realized on many different substrates from different materials including several types of ceramic and also plastic substrates. When compared to standard process, where conductive thick-film paste is screen printed on ceramic substrates and burned, this method has some advantages in thermal shock reduction and also pads created this way do not incline to leaching during reflow soldering process.

Multi substrate ceramic 3D structures created this way are base for further development of universal package, which might be useful as a compact carrier for complete sensor system containing conductive circuitry, soldered SMD, bonded chip devices and other technologies together. Similar package also can accommodate for example high power output LEDs utilizing good thermal conductivity and high thermal stability of ceramic materials.

Experiment setup

At the very beginning, the layout of test substrates was designed. Layout is inspired by standardized BGA test dummies and uses doubled daisy chain circuit with 64 soldering pads. Each pad diameter is 0.6mm and pitch (center-to-center distance) is 2.54mm (two times the standard 1.27mm). Wires connecting pads are 0.3mm wide. Test patterns of both connected substrates are displayed on figure 1.

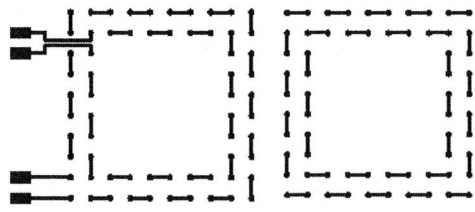

Figure 1: Layout of test boards

Main part of this experiment is based on ceramic materials (Al_2O_3 and Heraeus zero-shrink

LTCC). These are used because of their well-known structural stability over wide temperature range, good thermal management for dissipation and electrical insulation. Low temperature co-fired ceramic also adds possibility of creating highly integrated multilayer substrates to be used for the future package development. Initial tests were performed on quite affordable 0.65mm thick 2x2" Al_2O_3 substrates because many durable dies were needed during process development and tuning.

As process was improved enough on alumina, then LTCC substrates started to be used. It was decided to use special Heraeus HeraLock HL 2000 tapes. This material has unique parameters and is known for its almost zero shrinkage in x and y axes during firing process. Material shrinks only in z axis by about 30%. This means much easier structure design and also manufacturing process when compared to standard LTCC materials which reduce their dimensions in all axes and mostly with different ratio in different axes. The main advantage speaking for LTCC is of course the possibility of machining raw ceramic material into desired shape before firing. On the other hand one HL2000 layer is only 100μm thick and extremely fragile after firing. Because of this usually three layers were combined together to improve mechanical parameters and ensure satisfactory sturdiness.

Conductive copper patterns are realized in several steps. At first after cleaning, the whole substrate has to be covered with conductive surface finish. Physical vapor deposition method is used to do that. In vacuum chamber Cu particles are melted in crucible. As high current (50A) passes through, crucible is heated and molten copper evaporates. Hot copper vapors then condense on cooler target substrates, so their surface is covered by very thin Cu layer then. Created surface Cu layer is approximately 5μm thick, which is just enough to provide good electrical conductivity needed for next steps. Corner to corner measured resistivity has to be under 1Ω.

When substrate surface is covered by copper, application of solid photosensitive resistive layer follows. Whole substrate is covered by photoresist and only windows corresponding to desired pattern are opened by photo-development process (Fig. 2).

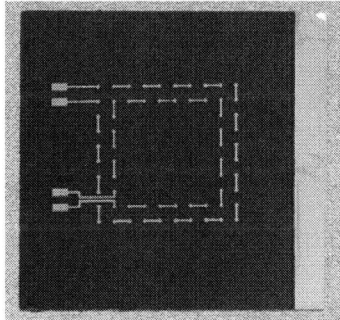

Figure 2: Substrate with photoresist

In next step the galvanic process using copper sulfate in solution with sulfuric acid is performed. Five minutes and 0.3A current are needed to enhance copper layer up to 30μm.

Resistive layer is then stripped down and desired pattern is revealed (Fig. 3).

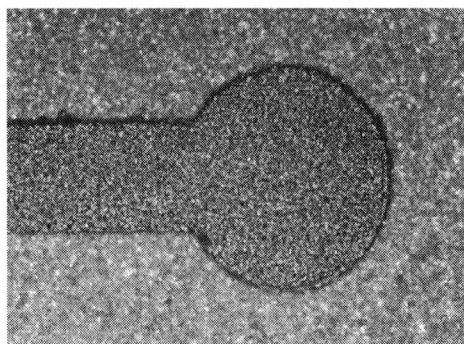

Figure 3: Detail of galvanically enhanced pad

Surrounding thin Cu layer is then removed by short etching. After that only main enhanced pads remain on substrate (Fig. 4).

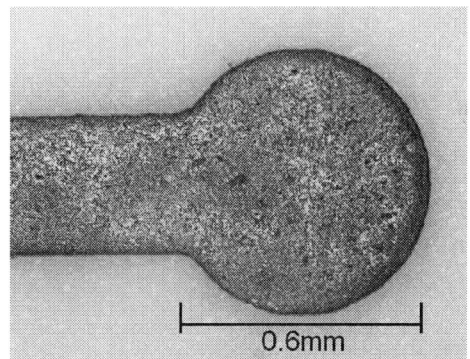

Figure 4: Detail of resultant pad after etching

This method of conductive layer creation has some advantages over standard thick film paste burning process. The main advantage is, that temperature remains at low lever during all process steps, which reduce thermal stressing. Also this means, that the same process can be applied on a wide range of materials including flexible plastic substrates which mostly are only able to withstand soldering temperatures. Soldering then also is the highest thermal point achieved during whole structure creation process. Copper layer also has an advantage as the surface is well wettable and solderable, especially when compared to LTCC conductive Ag pate characteristic. Leaching, which occurs during soldering, when Ag pads are used is also highly reduced, when soldering pads are being created out of copper.

Experiments

Initially a layer of non-solder mask was applied on substrates to achieve better behavior of solder and also better shaped joints during vapor phase reflow.

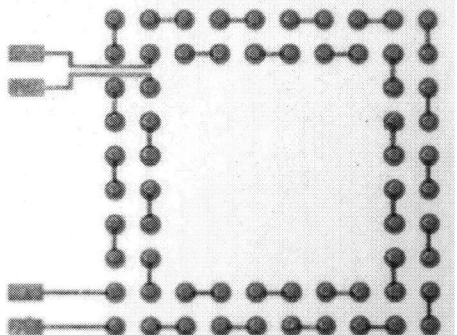

Figure 5: Detail of test substrate with non-solder mask and printed SAC305 solder paste

Solder balls used during soldering process were Senju lead-free SAC305 (96.5% Sn, 3% Ag and 0.5% Cu) type with 0.76mm in diameter. Balls were placed between two stencil-printed 100µm thick layers of SAC305 Cobar solder paste (Fig. 5). Structure is then assembled together with the use of FRITSCH microplacer rework station, where both upper and lower patterns can be easily matched to have the same orientation. Position of each pad on upper surface covered with solder paste then exactly matches position of lower pad, where solder ball is sitting on printed layer of solder paste. Detail of assembled structure just before reflow is depicted on figure 6, where both printed paste layers with solder ball placed between them are depicted.

Figure 6: Detail of assembled 3D structure before reflow

Whole structure assembled together was then reflowed in vapor-phase laboratory oven Asscon Quicky 300, where in Galden vapors uniform temperature across soldered structure and almost inert atmosphere with reduced oxygen content are present. This helps to minimize solder oxidation during reflow process.

Results

Several groups of soldered structures were completed during experimental part and research still continues with slightly modified design.

Soldered structures were inspected by various methods. The daisy chain pattern provides the easiest way to check if structure is conductive. Then a close optical inspection using BGA test station Ersascope was performed utilizing it's unique ability to look under package and to check internal solder joints. On figure 7 there is a detail of solder joints after reflow process. Ball, non solder mask circles and also wire connecting pad are clearly visible.

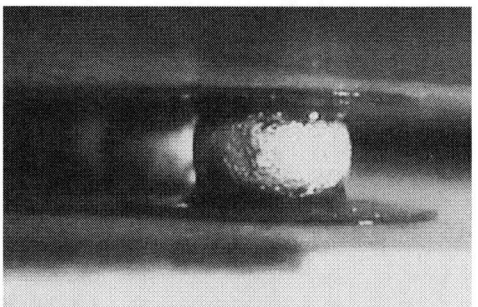

Figure 7: Detail of soldered 3D structure after reflow

To check structure of solder joints closely microsections of several soldered structures were made. One of these samples is described on figure 8, where both upper and lower ceramic substrates, galvanically enhanced copper layer and solder joint are visible.

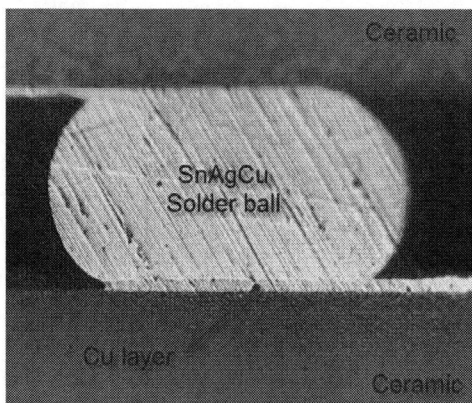

Figure 8: Detail of microsection of soldered 3D structure

Performed experiments show, that described process is usable to create desired structures. In next phase the process is going to be tuned and more structures are going to be created.

Recently performed and future experiments

There is also possibility to connect technology described in previous paragraphs with prior experiment with 3D dimpled LTCC structures described in [7] and [8]. 3D model visualization of this structure is described by figure 9.

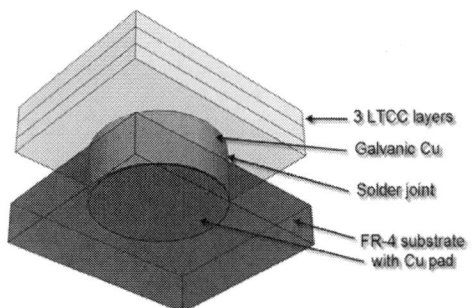

Figure 9: 3D model of recently tested structure

On this figure there are three laminated LTCC layers with round dimple created in third of them, where as a conductive connection a paste layer is buried under copper layer, which acts like part of thermo-mechanical stress buffer. This allows to achieve greater reliability and longer life, when substrates with different coefficient of thermal expansion are interconnected. The Cu layer also helps to effectively reduce leaching.

Conclusion

This research in it's early phase already shows that it is possible to build 3D stacked and interconnected package multi-substrate structures. These structures can combine the ceramic's advantages of thermal, insulating and mechanical properties with other substrates in one compact interconnected package. Possibility of one step reflow process also features way to thermal cycles count reduction.

Research still continues and is now focused mainly on Heraeus HL2000 LTCC substrates. There is also a plan for ANSYS simulations examining described structures using creep fatigue models. Long term reliability of used interconnection is going to be tested by accelerated thermal cycling using thermal cycling chamber.

Acknowledgement

Funding for this research was obtained through grant projects from Czech Ministry of Education (MSM 0021630503 MIKROSYN "New Trends in Microelectronic Systems and Nanotechnologies"), FRVŠ 2738/2011 project "Workplace for solder joint strength testing" and FEKT-S-11-5 project "Research of excellent technologies for 3D packaging and interconnection".

References

[1] Tummala, R., "Introduction to System-On-Package", The McGraw-Hill, 2007, ISBN 0071459065

[2] Mattox, D. M., "Handbook of Physical Vapor Deposition (PVD) Processing" S.E., Elsevier Inc., 2010, ISBN 978-0-8155-2037-5

[3] Tummala, R., "Fundaments of Microsystem packaging", The McGraw-Hill, 2001, ISBN 0071371699

[4] Lautzenhiser, F., Amaya, E., "HeraLockTM 2000 Self-constrained LTCC Tape", p. 43-49, Proceedings of the 2002 International Conference on Advanced Packaging and Systems (ICAPS), Reno, Nevada, March 12-13, 2002

[5] Lautzenhiser, F., Amaya, E., Barnwell, P., Wood, J., "Microwave module design with HeraLock[TM] HL2000 LTCC", Proceedings of IMAPS 2002, Denver, USA, ISBN 0-930815-66-1

[6] Kangasvieri, T., "Surface-mountable LTCC-SIP module approach for reliable RF and millimetre-wave packaging", University of Oulu, Tampere, Finland, 2008, ISBN 978-951-42-8921-7

[7] Nicak, M., Šandera, J, Szendiuch, I., "Contribution to realization of 3D structures", In 33rd International Spring Seminar on Electronics Technology (ISSE), 2010. Warsaw, Poland: IEEE Xplore digital library, 2010. s. 156-159. ISBN 978-1-4244-7849- 1.

[8] Nicák, M., Švecová, O., Šandera, J., Pulec J., Szendiuch, I., "Reliability and Simulation of Lead-Free Solder Joint Behavior in 3D Packaging Structure", Key Engineering Materials Vol. 465, 2011, Key Engineering Materials, Switzerland: Trans Tech Publications, 2011. s. 491-494, ISBN 978-3-03785-006-0, ISSN 1013-9826

Solder-TIMs (Thermal Interface Materials) for Superior Thermal Management in Power Electronics

Karthik Vijay

Indium Corporation of Europe, Milton Keynes, UK
kvijay@indium.com   ph.+44 (0) 7584 643 677

Abstract

The use of High Power devices are increasing exponentially given the need for increased current switching and conversion rates. High power devices are used in diverse end applications including traction (trains, powertrain management in an Electric Vehicle), LEDs (high brightness, high power), Diode Pumped Solid State Lasers, computing and graphics. Higher power translates to more heat being generated, and this heat needs to be removed from the power die and dissipated out of the package. Keeping the power die cooler is the key to increased functionality of the device and extending the life of the device. The choice of the right Thermal Interface Materials (TIMs) used to transport heat away from the power source is therefore crucial in preventing the power die from overheating.

This paper discusses the work done on Solder-TIMs that have been developed for high-power applications and how they compare to thermal grease. Solder-TIMs as the name indicates are solid solders (typically 100% Indium / Indium-containing alloys) that are very soft versus typical thermal greases that are metal-based (Ag particles) held in a silicone base. With higher power and higher heat dissipation needs, regular grease is not able to make the cut and superior performing Solder-TIMs are better suited to take the heat away from the die due to the following reasons: (a) Thermal grease has a low bulk thermal conductivity of 3-12 W/m.K compared to 100In that has a very high bulk thermal conductivity of 87 W/m.K; (b) During device usage over time, thermal grease tends to pump-out and migrate away from the center of the die (due to the diaphragm effect) and this means that the center of the power die gets hotter and this could lead to premature failures. On the other hand, there is no pump-out with a Solder-TIM. 100In is extremely soft (4X softer than lead) and this softness helps fill the interface gaps thus reducing thermal interface resistance. In addition, over time, the malleability of the solder helps fill the interface gaps even better. So *thermal interface resistance with a Solder-TIM decreases over time* as opposed to *thermal grease where the thermal interface resistance increases over time*; (c) Over time, grease tends to bake-out and dry (becomes powdery), thereby increasing thermal resistance and reducing heat-dissipation effectiveness. With Solder-TIMs, there is no bake-out; (d) Grease is messy when applying versus a solid solder that can be packaged in tape & reel and picked & placed.

The Solder-TIMs tested included (a) Heat-Spring® which is a foil made of In/In-containing alloys, with a proprietary *altered surface for reduced thermal interface resistance*. The Heat-Spring® needs only compression force, does not need to be melted/reflowed, does not need a flux and therefore eliminates voiding associated with flux and reflow, does not need any special substrate metallization; (b) Development Solder-TIM material which has solder particles (that are In-containing) held in a polymer base. This material also needs only compression force and does not need to be melted/reflowed. The Solder-TIMs were compared to industry-used thermal greases.

Testing regimes for Solder-TIMs included (i) Bake test: at 90 deg C for 3000 hours; (ii) Power Cycling: 1000-5000 cycles, 0-50W; (iii) Thermal Cycling: -55C/+125 C, 1000 cycles and for 2000 hours; (iv) HAST: 85C/85% RH for 1000-2000 hours. In all the tests, Solder-TIMs consistently outperformed thermal grease by achieving low thermal interface resistances especially over time and preventing the power die from overheating.

Keywords: Solder-TIMs, Heat-Spring®, Thermal Resistance, Thermal Grease, Pumpout

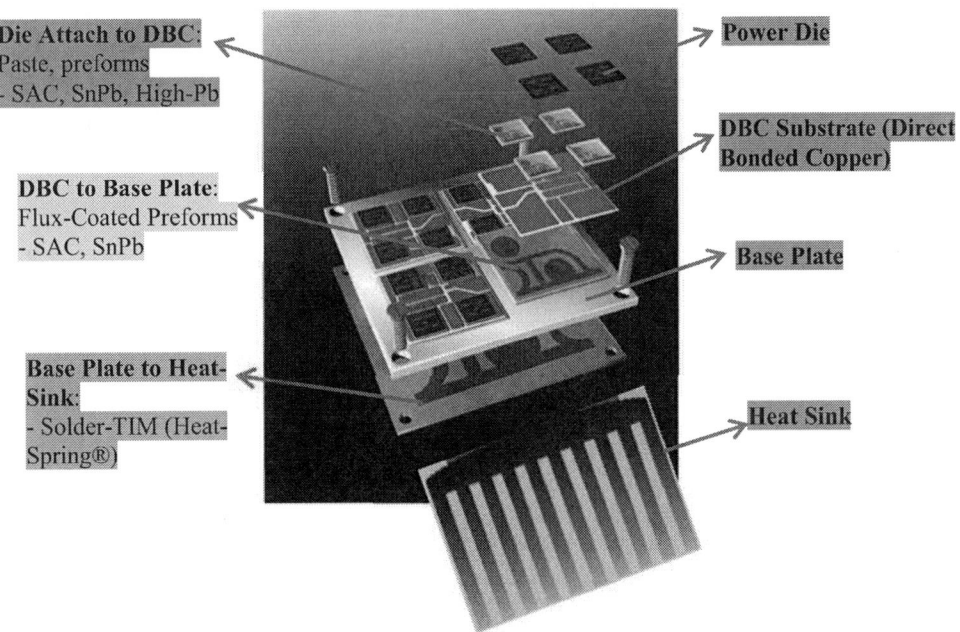

Die Attach to DBC:
Paste, preforms
- SAC, SnPb, High-Pb

DBC to Base Plate:
Flux-Coated Preforms
- SAC, SnPb

Base Plate to Heat-Sink:
- Solder-TIM (Heat-Spring®)

Power Die

DBC Substrate (Direct Bonded Copper)

Base Plate

Heat Sink

Figure 1. Schematic of IGBT Power Module and Materials Used

Power Devices and Materials Used

A power device controls or switches large electrical currents. IGBTs (Insulated Gate Bipolar Transistors) and MOSFETs (Metal Oxide Semiconductor Field Effect Transistors) would be typical power devices that operate in the 1KW-1MW range. Typically, a power device consists of a power die (Si/Si-Carbide); DBC (Direct Bonded to Copper – made of Cu/AlN) substrate; baseplate that acts as a heat-spreader (made of Cu/AlSiC) (depending on the device design, thermal and reliability requirements, a baseplate may or may not be used); and a heat-sink.

Typical interconnect material at the interfaces are: (a) Solder paste between the die and DBC. Paste is reflowed in a vacuum furnace and the flux residue is cleaned afterwards to enable wirebonding. Ultra-low residue pastes have been developed to leave very low post-reflow residue (1-3% as opposed to standard 35-50%) that could eliminate cleaning and allow for direct wirebonding. Solder alloys used include Sn/Ag/Cu (melts at 221 C), Sn/Pb (melts at 183 C), high-Pb (melts at 300+ C). With higher operating temperatures (> 175 C) and the move to Pb-free, higher-temperature Pb-free solders (Transient Liquid Phase Bonding techniques) as well as non-solder options (silver sinter) are being looked at; (b) Flux-coated solder preforms between the DBC and baseplate which are reflowed in a vacuum furnace. A

uniform flux-coating eliminates the variability associated with adding flux manually and a solder preform eliminates the variability associated with cutting ribbon manually to form the desired shape. Solder alloys used include Sn/Ag/Cu (melts at 221 C), Sn/Pb (melts at 183 C); (c) TIMs between the baseplate and the heat-sink. The TIMs considered are Solder-TIMs and thermal greases.

So heat needs to be transported from the power die to the DBC though the solder; from the DBC to the baseplate through the solder; and from the baseplate to the heatsink through the TIM; and from the heatsink to the ambient air. **Figure 1** shows the schematic of an IGBT power module and the materials used.

Solder-TIM Heat-Spring® Vs Grease

Solder-TIMs as the name indicates are solid solders (typically 100% Indium / Indium-containing alloys that are very soft) versus typical thermal greases that are metal-based (Ag or Al particles) held in a silicone base. A Heat-Spring® is a Solder-TIM made of In/In-containing alloys, with a proprietary *altered surface for reduced thermal interface resistance*. The Heat-Spring® needs only compression force, does not need to be melted/reflowed, does not need a flux and therefore eliminates voiding associated with fluxes and reflow. Solder-TIMs are better suited to take the heat away from the die due to the following reasons: (a) Thermal grease has a low bulk thermal conductivity of 3-12 W/m.K compared to 100In that has a very high bulk thermal conductivity of 87 W/m.K; (b) During device usage over time, thermal grease tends to pump-out and migrate away from the center of the die (due to the diaphragm effect) and this means that the center of the power die gets hotter and this could lead to premature failures. On the other hand, there is no pump-out with a Solder-TIM. 100In is extremely soft (4X softer than lead) and this softness helps fill the interface gaps thus reducing thermal resistance. In addition, over time, the solder creeps a little thus filling the interface gaps even better. So *thermal interface resistance with a Solder-TIM decreases over time* as opposed to *thermal grease where the thermal resistance increases over time*; (c) Over time, grease tends to bake-out and dry (becomes powdery), thereby increasing thermal resistance and reducing heat-dissipation effectiveness. With Solder-TIMs, there is no bake-out. The performance testing of the Heat-Spring® versus thermal grease is listed below.

1. Thermal Resistance at T=0 - Heat-Spring® Vs Grease

Figure 2 shows the comparison in thermal resistances (as per ASTM D5470-06) at t=0 for different TIMs for a 1 sq.cm die- Heat-Spring® (75µ thick), 2 types of industry-proven thermal greases (silicone-base; 50µ thick film) and a regular Indium foil (75µ thick). The Heat-Spring® had the lowest thermal resistance as the compression force increases beyond 3 bar, due to its softness (fills gaps) and higher bulk conductivity.

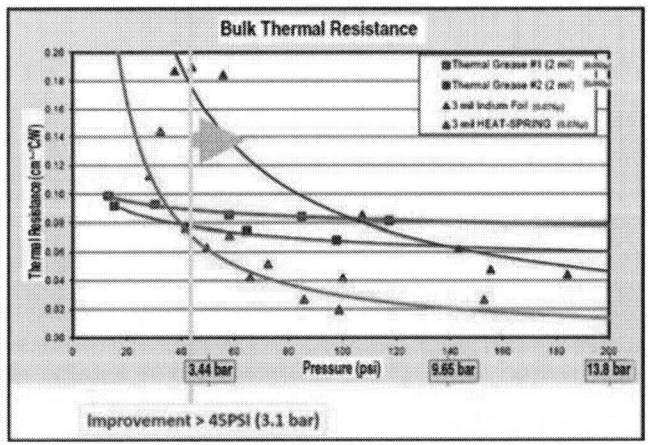

Figure 2. Comparison in Thermal Resistance between Heat-Spring®, 2 sets of Thermal Greases and a regular flat In foil

2. Power Cycling, Temperature Distribution & Thermal Resistance - Heat-Spring® Vs Grease

Power Cycling consisted of 1000 cycles for 0-50W, 3 minutes power-on and 2 minutes power off, with the TIMs under clamped conditions (4.8 bar). **Figure 3** shows the temperature distribution for thermal grease at T=0 and after 1000 power cycles. Due to the pumpout action of grease, the die becomes hotter over cycling, as thermal grease migrates away from the center of the die. **Figure 4** shows temperature distribution for the Heat-Spring® at T=0 and after 1000 power cycles. After power cycling, *the die actually becomes cooler* because the Heat-Spring® surface alteration and solder creep fills the interface gaps better over time and helps achieve uniform temperatures. **Figure 5** shows the comparison in thermal resistances under clamped conditions (4.8 bar) with power cycling over time between the Heat-Spring®, thermal grease and a phase-change TIM (PCM). The Heat-Spring® had the lowest thermal resistance due to its lower interface resistance and higher bulk conductivity.

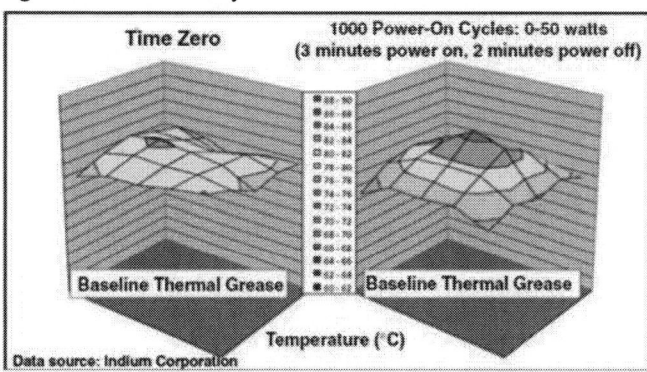

Figure 3. Temperature Distribution with Thermal Grease before and after Power Cycling

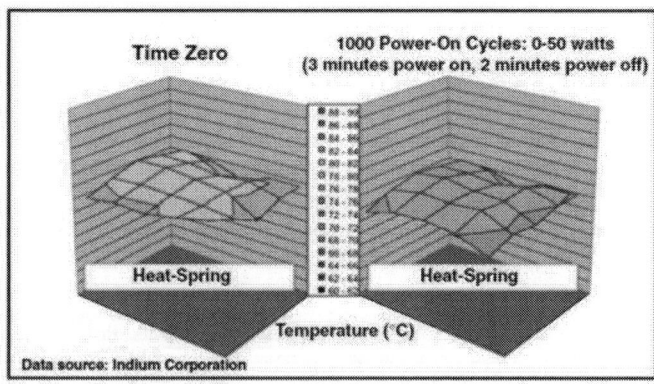

Figure 4. Temperature Distribution with Heat-Spring® before and after Power Cycling

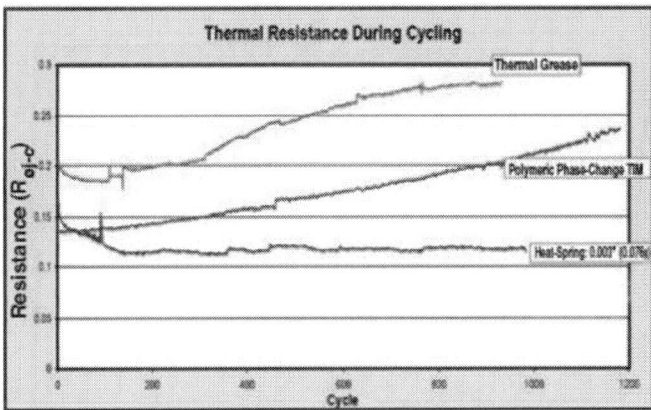

Figure 5. Comparison in thermal resistances under clamped conditions with power cycling over time for Heat-Spring®, thermal grease and a polymer phase-change (PCM) TIM

3. Bake Testing @ 90 deg C for 1500, 3000 hours - Heat-Spring® Vs Grease

Figure 6 shows the temperature distribution for a 1 sq.cm die under clamped conditions (4.8 bar), where the power module is baked at 90 deg C for 1500 and 3000 hours. With thermal grease, the temperature increased over time due to migration away from the center of the die and pump-out. With the Heat-Spring®, the temperature distribution was very uniform over 3000 hours due to its lower interface resistance and higher bulk conductivity.

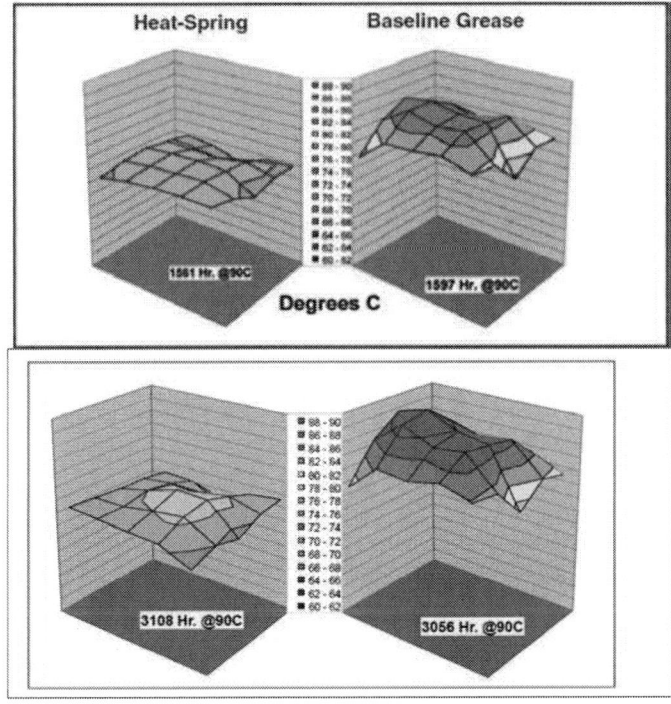

Figure 6. Bake Test@ 90 deg C- Temperature Distribution with Heat-Spring® Vs Thermal Grease at 1500 & 3000 hrs

4. Heat-Spring®- Change in Thermal Resistance and Solder Thickness Over Time for:

a. **T= 0–1000 hours:** under clamped conditions (4.8 bar) and at a temperature of 100 deg C, **Figure 7** shows the change in thermal resistance and thickness over 1000 hours for a 1 sq.cm die.

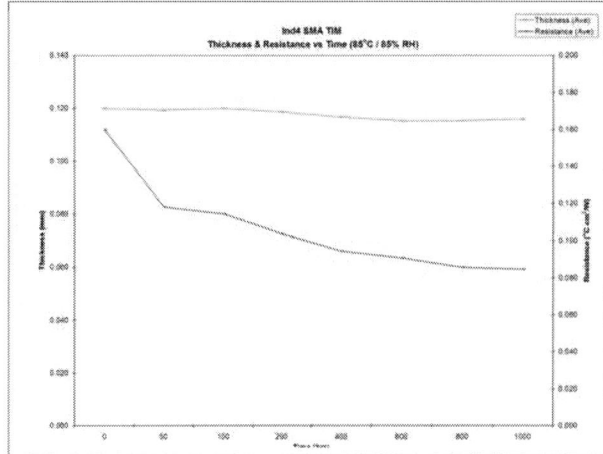

Figure 7. Heat-Spring® - Change in Thermal Resistance and Thickness over 1000 hours

b. **Thermal Cycling:** Under clamped conditions (4.8 bar), the power module with the Heat-Spring® and 1 sq.cm die, was thermally cycled between -10C to +95C for 1000 cycles. **Figure 8** shows the change in thermal resistance and thickness over the 1000 thermal cycles.

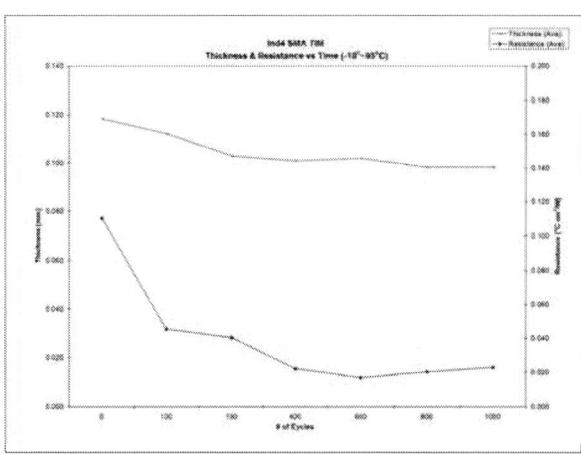

Figure 8. Heat-Spring® - Change in Thermal Resistance and Thickness during Thermal Cycling

c. **HAST Testing:** Under clamped conditions (4.8 bar), the power module with the Heat-Spring® and 1 sq.cm die, were subject to 85 deg C and 85% RH over 1000 hours. **Figure 9** shows the change in thermal resistance and thickness for the HAST test over a 1000 hours.

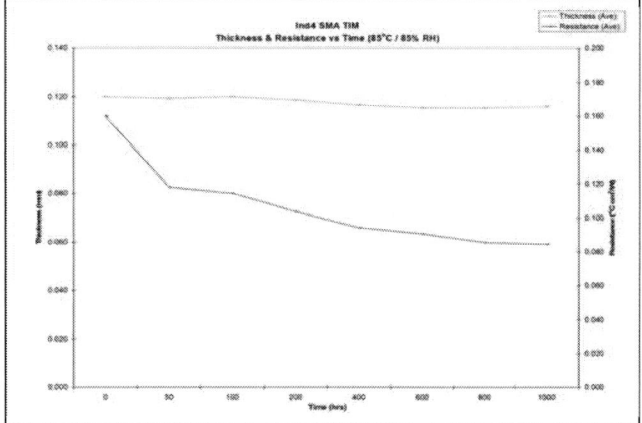

Figure 9. Heat-Spring® - Change in Thermal Resistance and Thickness during HAST test (85 deg C/ 85% RH)

In all the above tests, it was observed that the average thermal resistance reduced significantly and the solder thickness was fairly uniform.

Development Solder-TIM Material : HS-SC
A Solder-TIM product in development called HS-SC with solder particles (In-containing) in a polymer base was also studied. HS-SC like the Heat-Spring® is designed to work under compression force. There is no melting or reflow. This is printable/dispensable. **Figure 10** shows the comparison in thermal resistances (as per ASTM D5470-06) for a 1 sq.cm die at t=0 for the HS-SC (50µ thick bondline) vs Heat-Spring®   (100µ thick). A lower thermal resistance was achieved with the HS-SC material.

Further testing was done to compare the thermal resistance of the HS-SC with (a) couple of industry used thermal greases; (b) industry standard Phase Change Material (PCM) with metal fillers. **Table 1** lists the test regimes for the TIMs. The thermal test vehicle consisted of a Si die and a Copper heat spreader with the TIMs sandwiched in-between. The

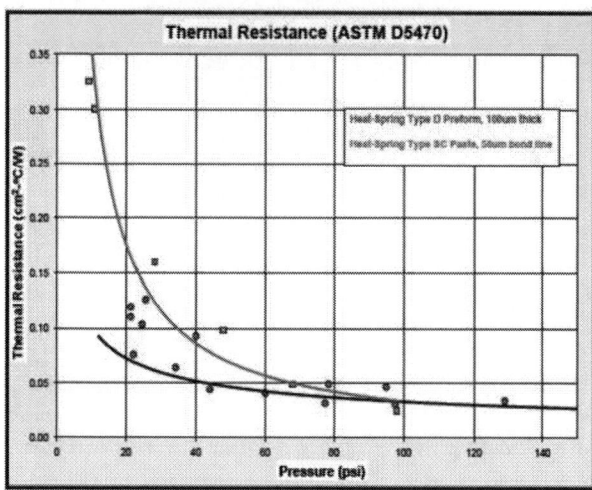

Figure 10. Compression Curves and Thermal Resistance Comparison between HS-SC (Development Material) and Heat-Spring®

thickness of the HS-SC was 100 µ and the thickness of the thermal greases and PCM was 50µ. Clamping force was 4.8 bar.

	Test methods	Cycle time (h)	Cycle #	Time (h)	Days
1	TTV, baked at 90 °C			2000	84
2	Temperature Cycling, -55 to 125 °C	1	1000	1000	42
3	85 °C, 85% humidity Chamber			2000	84
4	TTV power cycling, 50W, 3 min up, 2 min cool-down	0.08	5000	80	15

Table 1. Test Regimes for testing HS-SC, Thermal Grease 1 , Thermal Grease 2, Polymer Phase Change Material (PCM1)

1. HS-SC – Bake test @ 90 deg C for 2000 hours
Figure 11 shows that the HS-SC had the least thermal resistance and was the most stable when compared to Grease 1, Grease 2 and PCM1; over 2000 hours. **Figure 12** shows the pumpout with Grease after 2000 hours @ 90 deg C. **Figure 13** shows the HS-SC maintaining continuous coverage on the Si and Heat-Spreader with no pump-out. **Figure 14** shows the PCM degradation over time because of metal surface oxidation, and coalescence into discontinuous phases.

2. HS-SC – Temperature Cycling (-55C / +125C) for 1000 cycles
Figure 15 shows the thermal resistances after temperature cycling. HS-SC had the least thermal

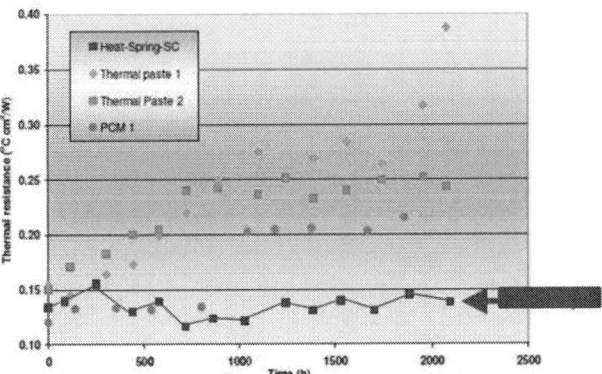

Figure 11. Bake-Test @ 90deg C for 2000 hours- Thermal Resistance Comparison

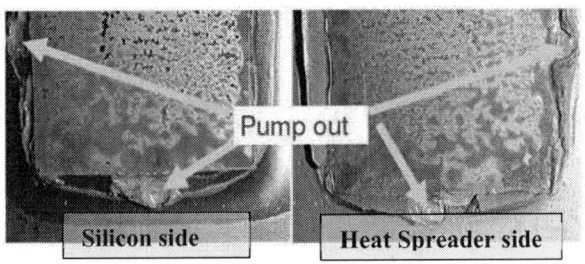

Figure 12. Bake Test @ 90 deg C for 2000 hours - Pumpout with Thermal Grease

Figure 13. Bake Test @ 90 deg C for 2000 hours - HS-SC exhibiting no pumpout and maintaining uniformity

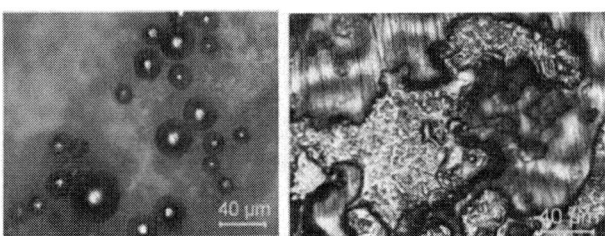

Figure 14. Bake test- Initial structure of Phase Change Material (on left). PCM exhibits metal oxidation and discontinuous aggregate phases (on right) after bake at 90 deg C for 2000 hours

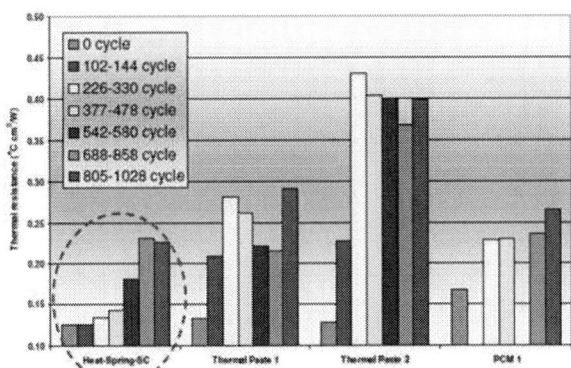

Figure 15. Temperature Cycling (-55C/+125C), 1000 cycles - Thermal Resistance Comparison

resistance and was the most stable when compared to Grease 1, Grease 2 and PCM1. Pumpout was seen with thermal grease as shown in **Figure 16**. The HS-SC was very uniform over the surfaces. The PCM experienced metal filler oxidation and formed discontinuous phases.

Figure 16. Temperature Cycling Test- Pumpout with Thermal Grease

3. HS-SC – Humidity Testing (85%RH, 85 deg C) for 2000 hours

Figure 17 shows that the HS-SC had the least thermal resistance and was the most stable when compared to Grease 1, Grease 2 and PCM1; over 2000 hours. **Figure 18** shows the pumpout with Grease after humidity exposure. **Figure 19** shows the HS-SC maintaining continuous coverage on the Si and Heat-Spreader with no pump-out. The PCM degraded with humidity due to metal surface oxidation.

4. HS-SC – Power Cycling Testing (0-50W), 5000 cycles

Figure 20 shows that the HS-SC performed best and had the least thermal resistance over 5000 cycles, and was the most stable when compared to Grease 1,

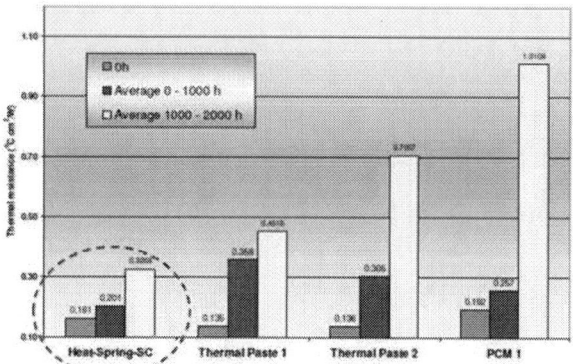

Figure 17. Humidity Testing (85%RH/855C), 2000 hours - Thermal Resistance Comparison

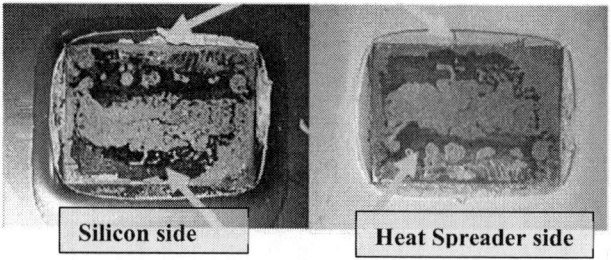

Silicon side | Heat Spreader side

Figure 18. Humidity Test (85%RH, 85C)- Pumpout with Thermal Grease

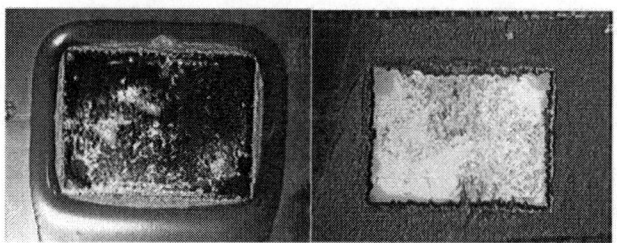

Figure 19. Humidity Test (85%RH, 85C) - HS-SC exhibiting no pumpout and maintaining uniformity

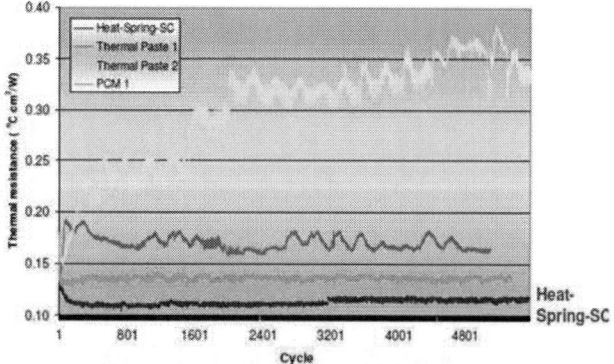

Figure 20. Power Cycling (0-50W), 5000 cycles- Thermal Resistance Comparison

Grease 2 and PCM1. **Figure 21** shows the pumpout with Grease due to power cycling. The HS-SC was very uniform over the surfaces as shown in **Figure 22**. The PCM experienced metal filler oxidation and degraded over the cycles.

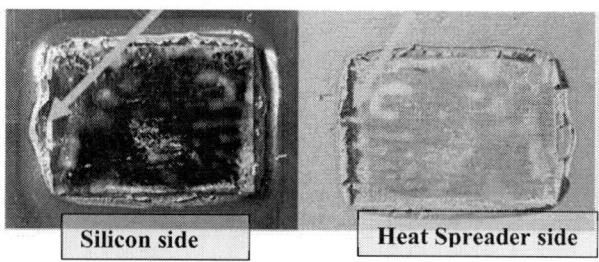

Silicon side | Heat Spreader side

Figure 21. Power Cycling Test (0-50W, 5000 cycles) - Pumpout with Thermal Grease

Silicon side | Heat Spreader side

Figure 22. Power Cycling Test (0-50W, 5000 cycles) - HS-SC intact, exhibits no pumpout and maintains uniformity

Conclusions

The Heat-Spring® Solder-TIMs consistently outperformed the thermal greases and achieved low thermal interface resistance especially over time. The three attributes of the Heat-Spring® that helps reduce thermal interface resistance are: (a) softness of the Indium solder that helps fill the gaps and reduce interface thermal resistance; (b) the proprietary surface alteration that helps increase contact area and reduce interface thermal resistance; (c) the high thermal bulk conductivity at 87 W/m.K (as opposed to that of grease at 3-12 W/m.K) that's helps to transport heat away faster. In addition, the Heat-Spring® is not prone to pumpout or bakeout as grease does. Pumpout makes the die hotter and causes performance degradation. These results were evident in the bake, power cycling and thermal cycling tests. The Heat-Spring® being a solid solder lends itself to high-volume automation in tape & reel packaging. The Heat-Spring® needing only compression force lends itself well to high-power modules where there is no constraint on the clamping force between the baseplate and the heatsink.

The HS-SC Solder-TIM is still in the development phase and all the initial results show that it can outperform thermal greases and Phase Change Materials. The HS-SC did not exhibit pumpout like grease. The polymer base of the HS-SC was able to protect the solder particles from oxidation as opposed to the PCM that experienced performance degradation due to metal filler surface oxidation. More testing is underway with the HS-SC to further characterize the material.

References

Jarrett, R.N., Merritt, C.K., Ross, J.P., Hisert, J., "Comparison of Test Methods for High Performance Thermal Interface Materials", 23rd IEEE Semi-Therm Symposium, San Jose, CA; March 20-22, 2007.

Saums, D.L., "ASTM D 5470-06 Thermal Interface Material Test Stand", DS&A LLC, 2006.

ASTM: Standard Test Method for Thermal Transmission Properties of Thermally Conductive Electrical Insulation Materials, Designation D 5470-06, ASTM International 2006

Samson, E., Machiroutu, S., Chang, J.Y., Santos, J., Hermeding, A., Prasher, R., Song, D.W., "Interface Matrial Selection and a Thermal Management Technique in Second Generation Platforms Built on Intel©Centrino Mobile Technology", Intel Technology Journal, Vol. 09, Issue 01, February 2005.

345

Low-Pressure Sintering of Silver Micro- and Nanoparticles for a High Temperature Stable Pick & Place Die Attach

J. Kähler, N. Heuck, G. Palm, A. Stranz, A. Waag, and E. Peiner

Institute of Semiconductor Technology, TU Braunschweig University of Technology,
Hans-Sommer-Straße 66, 38106 Braunschweig, Germany

Phone: +49-531-391-3783, Fax: +49-531-391-5844, and E-mail Address: j.kaehler@tu-bs.de

Abstract

Die attach using a pick & place process based on pressure-assisted Ag sintering was described for the assembly of hybrid circuits designed for operation up to 250 °C and above. Silver pastes with different ratios of nano- and micro-particles were analyzed regarding porosity, Young`s modulus, adhesion, electrical, and thermal conductivity in dependence of the sintered layer thickness. Temperature load capability was demonstrated by shear tests of bonded devices after temperature cycling between 30 °C and 250 °C. Commercial flip-chip devices were successfully bonded on substrates using pressure-assisted silver sintering for the first time.

Key words: Electronics packaging, nanoparticles, semiconductor device packaging, silver sintering, flip-chip

1 Introduction

High temperature applications put several demands on the joining layer. First, the mechanical stability must be given at room-temperature as well as operating temperature (e.g. at least 250 °C for 500 h in deep drilling applications). For the evaluation of the adhesion properties of the sintered layers, the American military standard for chip-substrate contacts (MIL-STD-883E, method 2019.5) was applied. Furthermore, a low Young's modulus for reducing stress during heating and cooling caused by thermal mismatch between chip and substrate, a high electrical, and a thermal conductivity are desired [1].

For die attach of power electronics the sintering of silver particles was introduced by Schwarzbauer and Kuhnert in 1991 [2]. In this process a porous silver layer serves as "glue" between chip and substrate. A hydraulic 50 tons press is generally applied to achieve the required pressure of typically 40 N/mm² [3-5]. However, the standard method faces some drawbacks: first, the high process pressure bears the risk of cracking or surface damaging of the joining partners. Second, the relative high process temperature of 250 °C does not allow the combination with standard processes for die attach (e.g. bonding with conductive adhesive). Third, the high-load press is associated with elaborate handling of the die. Therefore, several scientific groups have been working on the optimization of the sintering process: one approach to change the properties of silver powders is the combination with sinter additives. In doing so, a nearly pressure-free process was introduced [6]. However, for achieving shear strengths of more than 10 N/mm² long process time of 20 min and high

temperatures of 300 °C are required. Another proposal is to use nanoparticles [7-10]. It was demonstrated that the necessary process pressure or temperature could be reduced as well. But nanoparticles tend to the formation of agglomerates and cracks appear in layer thicker than 25 µm [6]. In addition, the fabrication of nano silver pastes is quite expensive. Our approach is to use a mixture of micro- and nanoparticles (micro-nano paste) [11]. For example, we achieved high shear strengths of (37.82 ± 14.9) N/mm² for a process at 200 °C and 40 N/mm² for 2 min. In order to meet the positioning requirements of hybrid assembly, we combined pressure-assisted sintering with high positioning accuracy of pick & place die attach, which is achieved using a modified flip-chip bonder.

In this contribution we describe the performance of this novel joining technique and the adhesive silver layer, i.e. the micro-nano paste in comparison to pastes comprising either pure nano- or micro-particles regarding porosity, Young`s modulus, electrical and thermal conductivity for different sintered layer thicknesses, low process pressure (12 N/mm²), and temperature (230 °C), respectively. As a result, we were able to demonstrate for the first time a pressure-assisted silver sintering flip-chip die attach of bumped devices (chip-area: 1 mm²) using micro-nano silver paste.

2 Analysis of Silver Pastes

First, we determined the porosity of different silver pastes in dependence of the layer thickness. The porosity P of the sintered layers was calculated according to eq. 1 by relating the density of the sample to the density of bulk silver ($\rho_{bulk} = 10.45$ g/cm³@20 °C). Silver pastes comprising mixtures of various ratios of nano- and micro-particles were screen-printed on microscope slides and afterwards compressed at 12 N/mm² and 230 °C for 2 min (see fig. 1, upper and lower row, respectively).

$$P = 1 - \frac{\rho_{sample}}{\rho_{bulk}} \qquad (1)$$

The micro and the micro-nano paste show smooth surfaces of the printed area (a, b) while a formation of dry channels was observed for the nano-paste (c). After sintering, the surface morphology did not change.

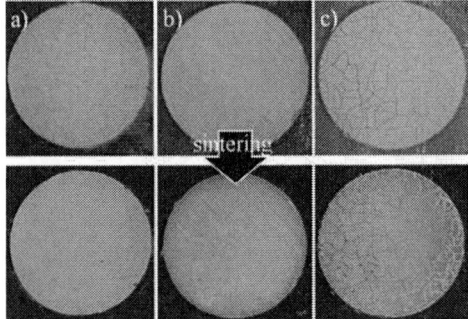

Figure 1: Silver layers before (upper) and after sintering (lower). a) Micro paste, b) micro-nano paste, 1:1 mixture, and c) nano paste.

The thicknesses of the sintered silver layers were measured by a digital length gauge (Heidenhain MT 25, reproducibility: ± 0.5 µm). For the porosity analyses only layers of uniform thickness (deviation < ± 10 %) were taken into account. This is important because the deviation of the thickness determined the uncertainty of the porosity. Finally, the mass of the samples was determined by an analytical balance (Kern ABT 120-5DM, reproducibility: ± 0.01 mg).

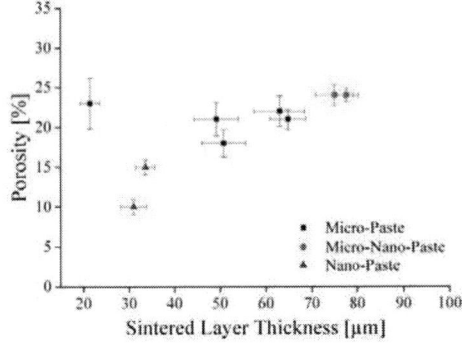

Figure 2: Porosity of sintered silver layers (t = 2 min, T = 230 °C, p = 12 N/mm²)

In general, the determined porosities are in good agreement with literature (e.g. [4]: Porosity of 30 % for micro powder sintered at 20 N/mm² for 3 min and 250 °C). We found different porosities

depending on paste mixture and thickness of sintered silver layers (see fig. 2). Interestingly, thin layers of micro-paste could not be compressed in the same manner as nano-paste. The reason for it is the larger particle-size: 10 % of the particles of the micro-paste are bigger than 8.5 µm (in comparison: nano-particles < 100 nm). There are only a few particles one above the other for thin layers resulting in a higher porosity. An exemplary cross section of a sintered layer is shown in fig. 3.

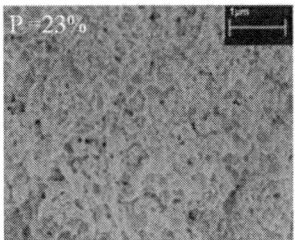

Figure 3: SEM pictures of cross sections from a sintered silver layer (Micro-Nano-paste, porosity: (23 ± 1.5) %)

The Young's modulus of the sintered layers can be calculated directly from the porosity P using eq. 2 [12] ($E_{Ag} = 80$ GPa, $\nu_{Ag} = 0.38$), which was reported for silver layers with low porosities (< 40 %) before [1][13].

$$E_{porous\ layer} = E_{Ag} \frac{3(3-5P)(1-P)}{9-P(9.5-5.5\nu_{Ag})} \qquad (2)$$

The determined Young's moduli (e.g. E = 50 GPa for P = 21 %) are in good agreement with other studies on the porosity of sintered silver layers ([3]: E = 54.8 GPa for P = 21 % and [4]: E = 40-55 GPa for P = 15 %).

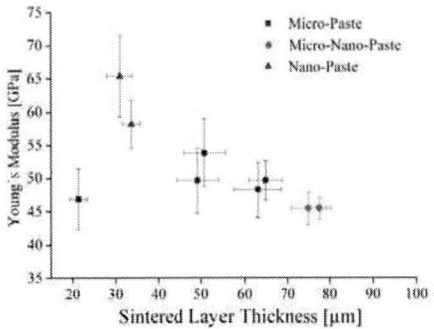

Figure 4: Young's modulus of sintered silver layers (t = 2 min, T = 230 °C, p = 12 N/mm²).

In the following, the influence of the layer thickness on the adhesion will be investigated. It is well known that a potential drawback of joining layers with a high Young's modulus is that they are less effective in relief of thermal stress induced by temperature cycling between chip and substrate having different coefficients of thermal expansion (CTE) [3]. Another important factor is the layer thickness between the joining partners, which has an influence on both stress and adhesion. It was suggested that for thicker layers the distribution of forces at the edges is better and thus stress is reduced [5]. For an evaluation of the micro, nano, and micro-nano silver pastes, we bonded small-area (Fairchild BC817, chip-area: 0.62 mm^2, metallization: Au) and large-area (chip-area: 4 mm^2, metallization: Ti/Pd/Au) silicon dice on Al$_2$O$_3$ ceramic substrates (metallization: Ti/Pd/Au, [14]) using the same sintering parameters as before (t = 2 min, T = 230 °C, p = 12 N/mm^2). In doing so, we chose different layer thicknesses and achieved an accuracy of ± 7 µm. After bonding, we performed initial shear tests at room temperature, i.e. without additional temperature treatment of the bonds, assuming an inaccuracy of the shear tester of ± 10 % (see fig. 5).

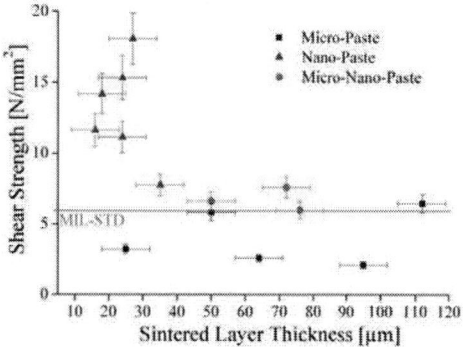

Figure 5: Shear test at RT without temperature treatment of the small-area die

The adhesion of bonded devices varies depending on paste mixture and thickness of sintered silver layers. This is in agreement with the lower porosity of the sintered nano-paste compared to the other pastes resulting in higher cohesion strength of the layer [15]. Another reason is that due to the smaller particle size and, with it, the larger surface-to-volume ratio of the nano particles, the contact area to the substrate surface is increased, which is beneficial for a good adhesion to the chip and substrate metallization [16] (surface area: micro-paste: 1.15-1.75 m^2/gm, nano-paste: 8-11.5 m^2/gm). The generally low shear strengths can be explained by the extremely smooth backside metallization of the die.

Subsequently, the adhesion of bonded dice was analyzed after 50 temperature shock cycles (i.e. one cycle: 1 min at 250 °C, 1 min at 30 °C) showing shear strengths which are higher than without annealing. After temperature treatment, all die attaches fulfill the requirements of the American military standard (MIL-STD-883E) for chip-substrate contacts (see fig. 6). While the adhesion of the nano-paste remains constant, the shear strength of bonded devices using micro and micro-nano paste was significantly increased. One reason could be

that the sintering process was not finished. The micro-particles are coated (e.g. fatty acids) in order to prevent agglomeration. The coating locally dissolves at a product-specific temperature initializing the sintering, whereas the nano-particles are not coated. Besides, this effect might be caused by strain hardening effects [17].

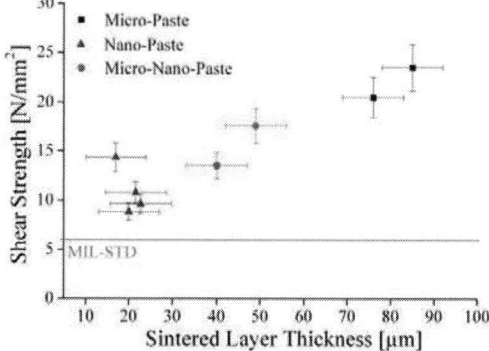

Figure 6: Shear test at RT after temperature treatment of the small-area die (50 cycles from 30 °C to 250 °C).

The bonded large-area chips could not be sheared off the substrate independent of the paste and its thickness (see fig. 7). In all cases the adhesion of the sintered layer was too strong and sometimes the silicon die broke at an applied shear-force of 50 N. Even after the shock-test, the adhesion was not impaired.

Figure 7: Shear test at RT of the large-area chips (micro: a)/b), micro-nano: c), and nano: d)).

For the measurement of the electrical conductivity σ of the sintered layers, the same samples as for the porosity measurements were used.

$$\sigma = \frac{1}{\rho} = \frac{I \times F_2 \times F_{31}}{4.532 \times t \times V} \qquad (3)$$

The four-point probe, which is commonly used to measure the semiconductor resistivity, was applied according to eq. 3 [18]. The equation is valid for thin layers, where the sample thickness (t) is smaller than a half of the distance of the measuring tips (probe spacing $s = 1$ mm).The value F_2 is the wafer diameter correction factor (here: $F_2 = 0.775$

for s = 1 mm and a sample width D = 6 mm) and F_{31} the boundary proximity correction factor versus normalized distance (d = 1.5) for non-conducting boundaries (here: F_{31} = 0.95). The measurement were conducted at room temperature (T = 300 K).

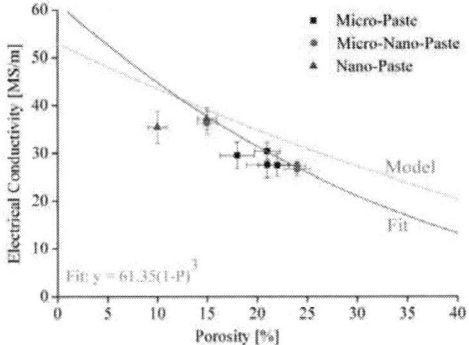

Figure 8: Electrical conductivity of sintered silver layers depending on porosity

In order to confirm the determined electrical conductivities, the results were fitted according to eq. 4 based on a study about comparable sintered micro powder [4]. Here, σ_m is the electrical conductivity of pure silver, which is 61.35 MS/m. For the sintered pastes we found a good fit with the model, although the values are in general a little bit lower (see fig. 8).

$$\sigma_{Model} = \sigma_m \times 0.86 \times (1 - P)^{1.86} \quad (4)$$

The model above has the drawback that the intersection with the y-axis (i.e. 0 % porosity) does not correspond to the electrical conductivity of pure silver. Therefore, we modified the model and used the following eq. 5 as fitting curve:

$$\sigma_{Fit} = \sigma_m \times 1.0 \times (1 - P)^{3.0} \quad (5)$$

The thermal conductivity λ can be calculated according to the Wiedemann-Franz law (see eq. 7), which validity was proofed for porous silver layers before [4]. The proportionality constant L, known as the Lorenz number, is equal to 2.44×10^{-8} WΩK^{-2}.

$$\lambda = \sigma \times L \times T \quad (7)$$

The values are in good agreement with [4] (λ = 250 W/mK at 100 °C for 15 % porosity for comparable sintered silver micro powder). For example, we found a thermal conductivity of (222.5 ± 13.4) W/mK for micro-paste (thickness: (64.8 ± 3.9) µm, porosity: (21 ± 1.3) %) and (272.3 ± 25.6) W/mK for nano-paste (thickness: (33.6 ± 2.1) µm, porosity: (15 ± 0.9) %). For comparison, the thermal conductivity for bulk silver is 429 W/mK.

3 Pressure-assisted Silver Sintering Die Attach

The novel bonding method is based on sintering of silver particles. For safe and easy handling they are dissolved to a paste using solvents such as terpineol and butanol. In a second step, the paste is deposited either on the chip or substrate (e.g. stencil printing).

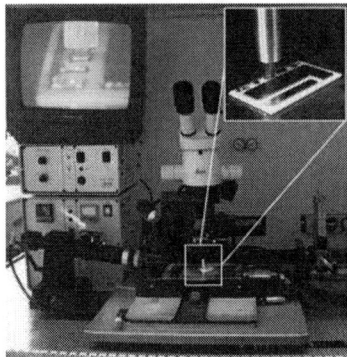

Figure 9: Flip chip bonder (Finetech electronic, Fineplacer-145 "Pico") used for pick & place die attach

The silver sintering process requires oxide-free noble metal finishes of chip and substrate (e.g. gold). The setup for the bonding process is shown in fig. 9. First, the device is picked up by a special handling device with an intake (diameter: 200 µm) for vacuum suction, which was designed in different materials (e.g. high performance plastic PEEK). After that the chip and substrate are positioned via a vision alignment system. The actual die attach is performed by a manual bonding force module, which allows a variation of the applied contact forces between the joining partners ($\geq$ 3 N/mm^2). The substrate is heated by a heating plate to the process temperature ($\geq$ 200 °C).

Figure 10: Bonded transistor (Fairchild, BC817) on Al$_2$O$_3$ gold thick-film ceramic

A field of application for the novel die attaches method is the bonding of small-area devices ($<$ 1 mm^2). For example, we bonded transistors (BC817, chip-area: 0.62 mm^2) at 20 N/mm^2 and 250 °C for 2 min on gold thick-film ceramics and tested the adhesion at room-temperature, at 225 °C, and after temperature-cycling (100 hours at 250 °C, 50 cycles from -40 to 200 °C) at 225 °C (see fig. 11). The adhesion was increased after temperature treatment demonstrating the high temperature capability of the pick and place sintering method.

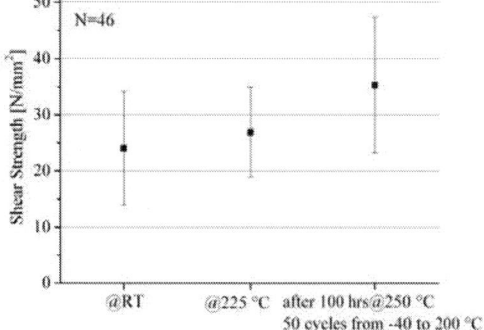

Figure 11: Shear strengths of bonded transistors (Fairchild, BC817) on Al₂O₃ gold thick-film ceramic. The data points (@RT, @225 °C) each correspond to 12 and the thermal cycling test to 22 bonded devices.

4 Pressure-assisted Silver Sintering Flip Chip

The flip chip die attach was performed using quad comparator test dice (National Semiconductor, LM139A) with a chip-area of approximate 1×1 mm². For the 14 Al contact pads gold stud bumps were realized (see fig. 12). The Al₂O₃ ceramic substrate was structured in a photolithographic lift-off process using a high temperature stable metallization (Ti/Pd/Au, [14]).

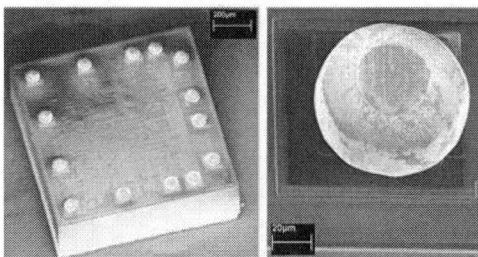

Figure 12: Test chip (LM139A) with gold stud bumps

A critical process step because of the small structures is the coating of the die or substrate with silver particles. First, we tried to depose the paste on the gold stud bumps (see fig. 13) applying the so called "transfer-method" [19]. In doing so, we stencil-printed the paste onto a special carrier foil and dried it. Subsequently, silver powder was transferred from the foil to the device in a print process below the sintering temperature ($t = 2$ min, $P = 12$ N/mm², $T = 150$ °C). After that the superfluous material should be removed from the die with the help of an adhesive foil in order to prevent short circuits between the bumps. We tested different silver layer thicknesses of 5, 8, and 35 μm.

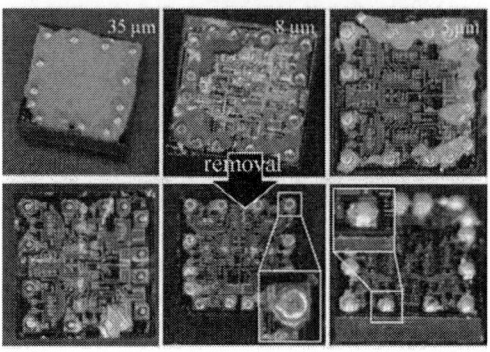

Figure 13: Test chip (LM139A) coated with silver powder

At a thickness of 35 μm the transferred layer was too compact (see fig. 13). As a consequence, it was removed in the second step in one piece without coating the gold bumps. However, for silver layers of 5 and 8 μm, which was below the height of the bumps (30 ± 3 μm), a coating was successful. The adhesion of the silver to the gold bumps was strong enough to withstand the pull force of the adhesive foil, whereas the cohesion strength of the superfluous material was too low. As a result, there was no electrical connection between the single bumps.

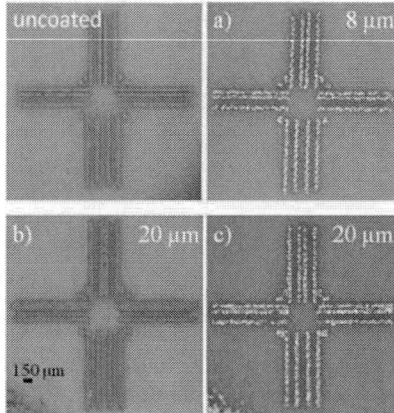

Figure 14: Metalized substrates coated with silver powder

Another possibility is the coating of the metalized substrate. The "transfer-method" could be successfully applied for a silver layer thickness of 8 μm. In this case, the silver powder could be removed from the ceramic surface without residue. Metalized structures as small as 110 μm with a distance of only 40 μm could be coated with silver powder (see fig. 14, a)). However, with increasing layer thickness (20 μm, b)) and with it cohesion strength, an after-treatment of the samples was necessary to remove the silver powder from small structures in order to prevent electrical short circuits between interconnects (see fig. 14, c)). In this case, the superfluous material was removed using fine needles under the microscope.

After the coating process the devices were successfully bonded on substrates using pick and place silver sintering at 250 °C and a pressure of 12 N/mm^2 for 2 min using the micro-nano paste (see fig. 15). In doing so, the high positioning accuracy of the novel die attach method could be demonstrated for both approaches (i.e. coating of the die and substrate).

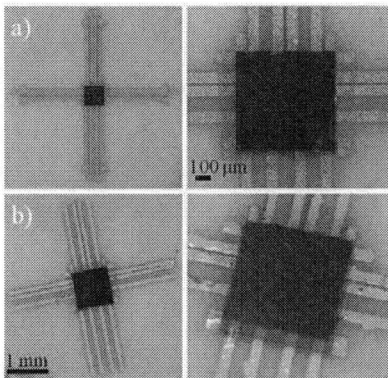

Figure 15: Pictures taken after pick and place silver sintering of the bumped devices: a) bumps, b) substrate coated with silver powder

Both devices passed the function-test successfully. According to the datasheet, different voltages (0, 1, and 2 V) were supplied between the pads 3 and 4. After that, the voltage gain was measured between the outputs (1, 2) and (13, 14), respectively. For the coated bumps (see fig. 15 a)) we measured voltages of 24/30.4/29.4 mV between (1, 2) and 7.1 mV/177.5 V/197.4 V between (13, 14). For the second approach (see fig. 15 b)) we determined a voltage gain of 14.8/17.5/17.08 mV between (1, 2) and 4.6 mV/183.07 V/188.05 V between (13, 14).

Figure 16: Pictures taken after shear test of bumped devices: a) bumps, b) substrate coated with silver powder

Finally, shear tests were performed with the flip chip bonds at room temperature (see fig. 16). In both cases fracture occurred at the interface between the gold stud bumps and the aluminum bond pads. The adhesion to the sinter layer was stronger than the connection to the aluminum pads which failed at 2.8 N (see fig. 16 a)) and 4.5 N (see fig. 16 b)), respectively. Taken into account the small contact area of each bump (between 70 and 90 µm) this

leads to shear strength in the range of (52 and 31 N/mm^2 for a)) and (83 and 50 N/mm^2 for b)).

5 Conclusion

Silver pastes containing nano- or microparticles as well as mixtures of both were investigated as adhesives for pressure-assisted sinter joining of silicon dies to ceramic substrates. We found different porosities depending on paste mixture and thickness of sintered silver layers. In addition, we determined the electrical and thermal conductivity of the bonds depending on porosity. With the help of the micro-nano paste even thick layers of 80 µm were realized without cracks. Temperature load capability was demonstrated by shear tests of bonded devices after temperature cycling between 30 °C and 250 °C. A novel sintering method was introduced allowing a precise pick & place bonding of small-area chips ($\leq$ 1 mm^2) flip-chip devices.

6 Acknowledgements

This contribution was performed within the collaborative project "gebo" (Geothermal Energy and High Performance Drilling) funded by the Ministry of Science and Culture, Lower Saxony (MWK) and Baker Hughes, Celle, Germany. The authors are grateful to the valuable technical assistance by J. Arens and F. Grosser. The authors also want to thank C. Schnöing, (Baker Hughes, Celle) for sample preparation and shear tests.

References

[1] N. Heuck et al., "Swelling Phenomena in Sintered Silver Die Attach Structures at High Temperatures: Reliability Problems and Solutions for an Operation above 350 °C", Proceeding of International Conference on High Temperature Electronics (HiTEC), Albuquerque, New Mexico, USA, May 11-13, 2010.

[2] H. Schwarzbauer et al., "Novel Large Area Joining Technique for Improved Power Device Performance", Transactions on Industry Applications, Vol. 21, No. 1, pp. 93-95, 1991.

[3] M. Thoben, "Zuverlässigkeit von großflächigen Verbindungen in der Leistungselektronik", PHD-Thesis, Bremen, 2002.

[4] C. Mertens, "Die Niedertemperatur-Verbindungstechnik der Leistungselektronik", PHD-Thesis, Braunschweig, 2004.

[5] J. Rudzki, "Aufbaukonzepte für die Leistungselektronik mit der Niedertemperatur-Verbindungstechnik", PHD-Thesis, Braunschweig, 2006.

[6] W. Schmidt, "Novel Silver Contact Paste Lead Free Solution for Die Attach", Proceedings of the 2010 International Conference on Integrated Power Electronics Systems (CIPS), Nuremberg, Germany, March, 16-18, 2010.

[7] M. Knoerr et al., "Reliability Assessment of Sintered Nano-Silver Die Attachment for Power

Semiconductors", Proceedings of the 2010 Electronics Packaging Technology Conference (EPTC), Singapore, December 8-10, pp. 56-61, 2010.

[8] J. G. Bai, J. Yin, Z. Zhang, G.-Q. Lu, "High Temperature Operation of SiC Power Devices by Low-Temperature Sintered Silver Die-Attachment", Transactions on Advanced Packaging, Vol. 30, pp. 506-510, 2007.

[9] T. G. Lei et al., "Low-Temperature Sintering of Nanoscale Silver Paste for Attaching Large-Area (> 100 mm^2) Chips", Transactions on Components and Packaging Technologies, Vol. 33, No. 1, pp. 98-104, 2010.

[10] D. Wakuda et al., "Room-Temperature Sintering Process of Ag Nanoparticle Paste", Transactions on Components and Packaging Technologies, Vol. 32, No. 3, pp. 627-632, 2009.

[11] J. Kähler et al, "Die-Attach for High-Temperature applications using Fineplacer-Pressure-Sintering (FPS)", Proceedings of Electronics System Integration Technology Conferences (ESTC), Berlin, Germany, September 13-16, 2010.

[12] G. Ondracek, "On the Relationship Between the Properties and the Microstructure of Multiphase Materials. Part III: Microstructure and Young's Modulus of Elasticity", Z. Werkstofft., Vol. 9, pp. 31-36, 1979.

[13] G. Ondracek, "On the Relationship Between the Properties and the Microstructure of Multiphase Materials. Part III: Microstructure and Young's Modulus of Elasticity", Z. Werkstofft., Vol. 9, pp. 96-100, 1979.

[14] J. Kähler et al, "Chip-Substrat-Kontaktierung optoelektronischer Bauelemente mittels Fineplacer-Verbindungstechnik", Proceedings of Micro-Nano-Integration (MNI), Erfurt, Germany, March 3-4, 2010.

[15] D. N. Lee, et al., "Plastic Yield Behaviour of Porous Metals", Powder Metallurgy, Vol. 35, No. 4, pp. 275-279, 1992.

[16] S. Joo et al., "Interfacial Adhesion of Nano-Particle Silver Interconnects for Electronics Packaging Application", IEEE Electronic Components and Technology Conference, pp. 1417-1423, 2008.

[17] F. W. Bach et al., "Moderne Beschichtungsverfahren", A Wiley-Interscience Publication, Weinheim, Chapter 2, pp. 15-21, 2005.

[18] D. K. Schroeder, "Semiconductor material and device characterization", A Wiley-Interscience Publication, third edition, New Jersey, Chapter 1, pp. 1-60, 2006.

[19] G. Palm, "Verfahren zur Befestigung von elektronischen Bauelementen auf einem Substrat", DE Patent 102004056702B3, March 2, 2006.

Low-temperature, photonic approach to sintering the ink-jet printed conductive microstructures containing nano sized silver particles

Tomasz Falat[1], Jan Felba[1], Bartosz Platek[1], Tomasz Piasecki[1], Andrzej Moscicki[2], Anita Smolarek[2]

[1]Wroclaw University of Technology, Faculty of Microsystem Electronics and Photonics, Wroclaw, Poland
[2]Amepox Microelectronics, Lodz, Poland

Tomasz.Falat@pwr.wroc.pl

Abstract

Since the late eighties of the last century, digital ink-jet technology has been widely explored by the electronic industry in developing new manufacturing processes. Nowadays, ink-jet printing technology is increasingly being used to create the electrically conductive structures on flexible substrates. Ink for printing conductive microstructures is typically based on noble metals of nano-sized dimensions – mainly Ag.

The structures are noncondcutive just after the printing. To obtain good electrical conductivity, they need an additional energy, mainly supplied during a heating process. Unfortunately relatively high temperature (~250 °C) used for thermal sintering process can not be applied to most polymer substrates. Current paper focuses on low-temperature, photonic sintering process. The influence of different parameters like pulses number or distance between sample and lamp on the sintering process was analyzed.

Key words: flexible electronics, ink-jet printing, nanosliver, photonic sintering

Introduction

Packaging of today's miniaturized electronics is based on conductive microstructures and contacts with dimensions in the range of tens of micrometers. Such lines, patterns or arrays of dots are possible under the condition that both special technologies and materials are applied. Since the late eighties of the last century, digital ink-jet technology has been widely explored by the electronic industry in developing new manufacturing processes [1]. Ink-jet technology needs a special liquid, usually called ink, which should satisfy at least the following three requirements: has very low viscosity, can be treated as a "true solvent" without component separation during high acceleration, and is able to make electrically conductive structures. If a suspension is used as the fluid for printing, the "conductivity" condition requires applying electrically conductive particles, and the "true solvent" demands particles with dimensions as small as possible, not higher than tens of nanometers. Ink for printing conductive microstructures is typically based on noble metals of nano-sized dimensions because of the chemical inertness in ambient atmosphere and good electrical conductivity – mainly Ag.

We are able to produce the silver particles with very narrow size distribution. The smallest controllable size is 4 nm, and size distribution window of few nanometers (e.g. particle size: 4-10 nm) [2]. The silver nanoparticles can be produced with controlled size up to 60 nm. On the basis of this nano sized filler, the ink for ink-jet printing technology is formulated [3] and microstructures can be printed with the use of a professional drop-on-demand printer.

The structures are noncondcutive just after the printing. To obtain good electrical conductivity, they need an additional energy, mainly supplied during a heating process. In our previous work [4] the properties of the structures made by ink-jet printing with the use of the ink containing the nano-sized silver particles were investigated. It was shown that resistivity of printed structures after thermal sintering process was only about three times higher than the value of the bulk material. Initial tests with the 4-10 nm silver particles showed that the resistivity of structures is about 4.5 $\mu\Omega\cdot$cm, while the bulk resistivity of silver is about 1.55 $\mu\Omega\cdot$cm. Also different electrical tests proved similarity between a bulk silver and the structures made by ink-jet printing with the use of nano-sized silver particles.

Unfortunately relatively high temperature (~250 °C) used for thermal sintering process can not be applied to most polymer substrates. Therefore for the flexible electronic substrates, the polyimides (like DuPont Kapton®) are commonly used since they are resistant to such temperatures. However, the polyimide foils are much more expensive than those made from other polymers like polyesters. Therefore many different ongoing research are devoted to lowering the temperature of sintering process.

One of the most promising method of sintering electrically conductive structures made of ink containing nano sized silver is applying photonic energy by pulsed light. The main advantages of this technique is low temperature (the whole process would be conducted in room temperature) and short process time which is also a requirement for large scale fabrication. Moreover, the linear flash lamp would be used in reel-to-reel technology.

Within the paper the low-temperature, photonic sintering of ink-jet printed electrically conductive structures made of ink containing 40% b.w. of nano sized (4-10 nm) silver particles was investigated.

Samples preparation

Samples in the form presented in Figure 1. were printed on polymer substrates by using the professional ink-jet, drop-on-demand material deposition system Fujifilm Dimatix DMP-2800 series.

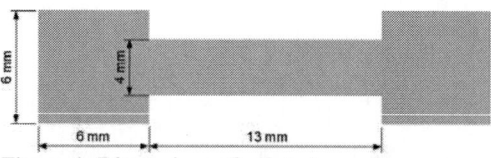

Figure 1: Dimensions of printed samples

Printer was equipped with cartridge having a 16 nozzles of 22.5 μm diameter. The drops of volume about 10 pl were jetted with the frequency of 5 kHz. The drop spacing were set on 20 μm (1270 dpi). The results of 3D profilometer measurements (see Fig. 2) done in our previous work [4] indicates, that for the same parameters of printing, the mean thickness of printed layer is about 400 nm.

Figure 2: 3D representation of the printed structure

After printing, the samples were only dried for 10 minutes at a temperature of 110 °C.

As the substrates the two different polymers were used: polyester foil (DuPont Melinex®) - 50 μm thickness and polyvinyl chloride (PVC) foil – 200 μm thickness. Both substrates were used as delivered without any chemical or mechanical treatment.

Photonic sintering

Sintering is a method for making the solid object from the powder usually by heating them but below the melting point. The nanoparticles has a lower melting temperature than bulk materials so lower temperature is also required to sinter them. Nevertheless the temperature above 200 ºC used for typical heat process of sintering silver nanopraticles is too high for most polymer substrates. Therefore some different sintering methods, besides sintering in oven, are developed e.g. arc discharge, laser or pulsed light.

Within the current investigations, the professional benchtop XENON SINTERON 500 system was used to photonic sintering. The energy was delivered in the form of light pulses - 800 J per pulse. The pulse duration was about 500 μs. The 4.2" xenon flash lamp made of Clear Fuse Quartz was used. This type of lamp provided a broad spectrum of light from deep UV to IR.

Samples were placed on the aluminum table below the flash lamp. The distance between the table and the lamp surface were controlled. The influence of distance between samples and lamp as well as number of pulses on the sintering process were investigated. For the multi-pulse sintering, about three seconds period between each pulse were applied to allow time for substrate and lamp cooling.

Results

In the first step the influence of distance between the sample and flash lamp surface on the impedance behavior of sintered samples were investigated. The samples printed on Melinex® foil were sintered with one pulse only, but the distance between sample and lamp was varied in the range of 1" to 1.5".

The changes of impedance after sintering process were measured by using Agilent 4294A impedance analyzer. The obtained results were presented below in Bode diagrams (in Fig. 3 and 4 - magnitude |Z| and phase angle Θ, respectively).

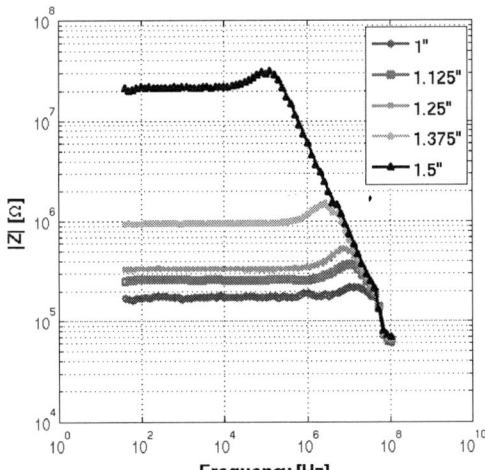

Figure 3: Magnitude of impedance |Z| of sintered structures – the influence o distance between samples and lamp

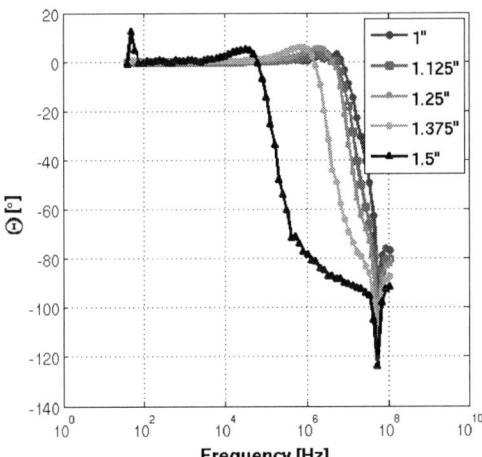

Figure 4: Phase angle Θ of sintered structures – the influence o distance between samples and lamp

As it can be seen o the above results the distance between sample and flash lamp influence strongly on the impedance. It is caused by increasing the amount of absorbed energy by samples when they are brought closer to the light source.

Moreover it is worth to notice, that the layers after sintering has a resistance nature (flat shape of impedance magnitude and phase angle equal 0 for lower frequencies; above ~10 kHz the influence of parasitic capacitance is visible).

The best results were obtained for the lowest distance between lamp and sample equal 1 inch. In this case, the average resistance of sintered strips was about 170 kΩ, while for the 1.5" distance, the measured resistance was about 21 MΩ.

Based on the above results, one could draw a conclusion, that to obtain the satisfied value of resistance after photonic sintering, the samples should be as close the flash lamp as possible. On the other hand, decreasing the distance between sample and lamp would damage the substrates, when the absorbed energy will be too high. It is due to increasing the surface power density (i.e. amount of energy in time per unit area) and therefore increasing the local temperature on sample surface. Nevertheless, it seems that increasing the peak power of single pulse would improve the sintering process and thus the electrical resistance of sintered structures, but to prevent the damage of structures and substrates the absorbed energy should be reduced. It requires a reduction of pulse time. Unfortunately, the used device was not equipped with the system of pulse time adjustment.

The other idea to improve the process without possible damaging of substrates is using multi-pulse sintering. It was tested by using PVC foil as substrate with ink-jet printed structures. The samples were placed on table underneath the flash lamp at the distance of 1.5 inches. The three sets of samples were exposed to 1, 2 and 3 pulses, respectively. Each set consisted of three stripes. The obtained results – the mean value of measured resistance are presented in Fig. 5.

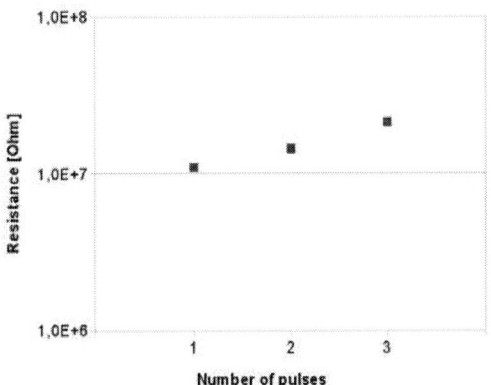

Figure 5: Influence of number of pulses on resistance of photonic sintered structures.

The obtained results indicates, that increasing the number of pulses does not improve the electrical conductance of sintered samples. The resistance of tested strips after one, two and three flash pulses remains on the high value - above tens of mega ohms. Moreover these values obtained for structures printed on PVC substrate is in the same range as those obtained for structure printed on polyester foil (Melinex®) sintered from a distance of 1.5".

It is worth to indicate that the lowest value of resistance obtained within photonic sintering is about five order of magnitude higher than the resistance of samples sintered with using standard convection heat method [4]. It is mainly caused by non-continuous, highly porous structure of layers after photonic sintering. The example – microscopic picture of surface of sample sintered by using flash lamp is presented in Fig. 6.

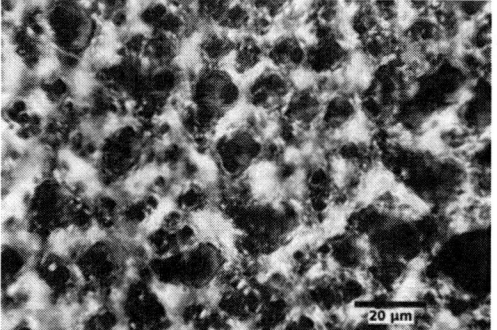

Figure 6: Surface of sample after photonic sintering

This picture illustrates well that the layer sintered by using flash light is highly porous and that the voids among the sintered silver has a diameters in the range of few µm. Such highly defected structure does not occur while sintering process is conducted by using convection heat.

Conclusions

Within the current paper the influence of sintering process parameters like flash pulse number or distance between sample and lamp on the sintering efficiency for two different substrate types was investigated.

The best result (the lower measured resistance) was obtained, when the distance between the test samples and flash lamp was very low. It was caused by increasing the surface power density while the distance to light source decreases. Obtained results also showed that increasing of pulse number does not influence on the resistance of tested samples. Moreover the final resistance of stripes exposed on flash lights was much higher than those obtained in heating process. It is probably caused by the high porosity of structures sintered by using flash light.

Nevertheless it seems that the optimization of equipment and process parameters e.g. by increasing the power with reduce the pulse time would improve the sintering and thus the electrical conductance of created layers.

Acknowledgements

These works were supported by The National Centre for Research and Development, Grant No. NR02-0001-10/2010.
The authors would like to thank Mr. Peter Schullerer and Mr. Thomas Feusse from Polytec GmbH, Waldbronn, Germany, for the opportunity of using SINTERON system and for help in sintering process.

References

[1] Felba J. and H. Schaefer H., "Materials and technology for conductive microstructures" in: "Nanopackaging: nanotechnologies and electronics packaging". Ed. James E. Morris, Springer, New York, 2008

[2] Mościcki A., Felba J., Sobierajski T., Kudzia J., Arp A. and Meyer W., "Electrically Conductive Formulations Filled Nano Size Silver Filler for Ink-Jet Technology", Proc. of 5th International IEEE Conference on Polymers and Adhesive in Microelectronic and Photonics, Polytronic2005, Wroclaw 2005, p.40-44

[3] Mościcki A., Felba J. and Dudziński W., "Conductive Microstructures and Connections for Microelectronics Made by Ink-Jet Technology", Proc. of 1st Electronics Systemintegration Technology Conference, ESTC2006, Dresden 2006, p.511-517

[4] Felba, J. et al, "Properties of Conductive Microstructures Containing Nano Sized Silver Particles", Proc. of 11th Electronics Packaging Technology Conference, EPTC2009, Singapore, 2009, p.879-883

Influence of different type protective layer on silver metallic nanoparticles for Ink-Jet printing technique.

Andrzej Mościcki[1*], Anita Smolarek[1], Jan Felba[2], Tomasz Fałat[2]

[1] *Amepox Microelectronics, Ltd., Lodz, Poland*
[2] *Wrocław University of Technology, Faculty of Mikrosystem Electronics and Photonics, Wrocław, Poland*
[*] *amepox@amepox.com.pl*

Abstract

Very quick development of the new techniques for making printed circuits (R2R, Ink Jet) makes possible for using as a base substrate - flexible plastic foils. This solution are giving very strong requirements for printing process what is connected with substrate foils parameters. One of the most important problem is as low as possible processing temperature for obtaining electrical conductivity for printed structures. At the present moment takes that the temperature of the processing should not be higher as 150C. The most popular conductive fillers for printing applications are silver metallic nanoparticles [1]. As well know, nanosize silver doesn't exist without protection layer around each particles. Nanosilver must be encapsulated by a protective layer of organic type dispersants to avoid aggregation. When metallic nanoparticle inks are heated, the dispersants should be effectively removed, allowing the remaining active metallic nanoparticles to be successfully sintered [2]. In the article we'll present several types of nanosilver's with different coating shells, for example polymer type, carboxylate and others. The results of quantitative investigations will be introduced for various type of coating and their dynamics of removing in the function of the temperature and time.

Introduction

The miniaturization in the electronics sector is a real progress in producing active and passive components, as well as in the new technologies of assembling microelectronic circuits. Additionally, the important matters are the purity of all used materials.

In the case of electrically conductive materials the most widely used metal is silver. In the most cases the silver filler is added to the various types of conductive adhesives, pastes or paints.

The most interesting with production point of view we have two methods – R2R and Ink Jet. For both methods pastes for printing (we'll call inks) will have similar properties. But in this case we'll be working with the ink-jet printing technology which is used year by year wider for microelectronic packaging [3], especially for printing on flex type of substrates. There are special demands for this technology - the most important is low viscosity and very homogenous structure like full molecular fluid parameters.

For its implementation, the liquid has to contain quite high of percentage conductive filler with silver size as close as possible to "atomic" size range. As well known, is very important that each particle of nanosized filler must to be protected by special kind of protection layer. It is necessary to avoid re-coagulation. Protection layer is built during nano Ag production and kind of it depends on technology used to silver powder production process.

The protection material rounded each silver particle is necessary for keeping uniform and stable ink parameters. Unfortunately, the layer is the main barrier for easy nanoparticles sintering process, and obtaining very low resistivity printed structure value. Also, removing protection layer should be at as low as possible temperature what is a key for using low-cost flexible substrates such as PE, PET, etc. (Fig.1.).

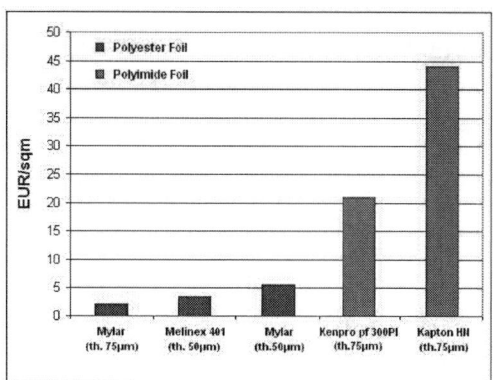

Fig.1. Cost comparison for plastic substrates.

Metallic Nanosilver Production

Is a several methods for making nAg [4,5,6,7,8,9], but in our work we choose the chemical reaction process [10], because this will give possibility of protection layer changing, without change of the method silver powder production and the structure of particles themselves.

For testing types of protective layer for sintering temperature, silver salt was reduced by three different reducing agent in precursor of solution, which was added to protect the synthesized Ag from agglomeration (Fig. 2). In this way the nanosilver with different type of protective layers were obtained, namely:

- nanosilver with carboxylate coating (Ag1),
- nanosilver with amine type coating (Ag2),
- nanosilver with polymer coating (Ag3).

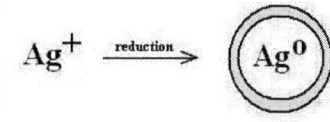

Fig. 2. Scheme of preparation silver nanoparticles with different protective coating.

For all types of the basic nanosilver Ag1, Ag2 and Ag3 SEM pictures and Malvern Zetasizer examinations were carried out for their dimensions and size distribution (Fig.3a,b,c, and 4a,b,c).

a)

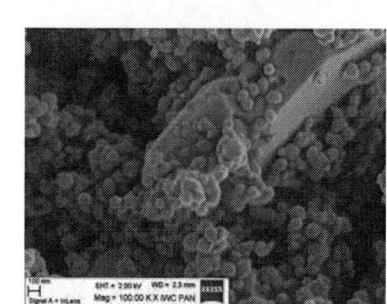

b)

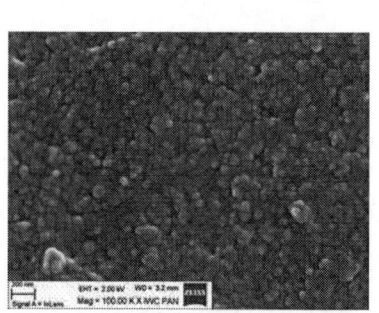

c)

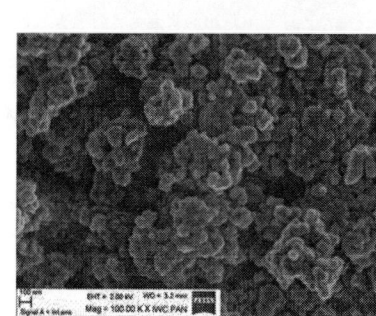

Fig. 3. SEM pictures of nanosilver a) Ag1,

b) Ag2, c) Ag3.

a)

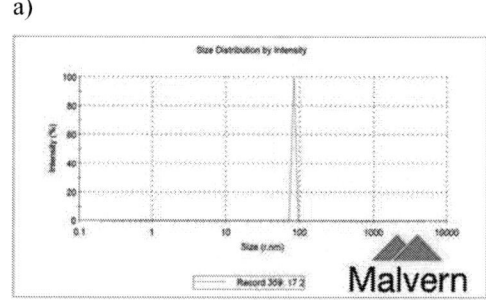

b)

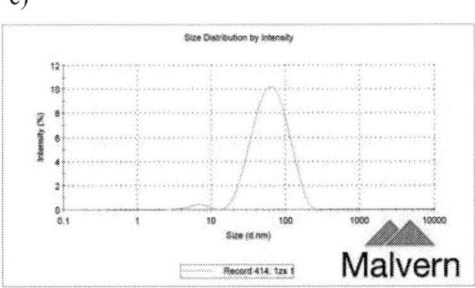

c)

Fig. 4. Size distribution of nanosliver a) Ag1, b) Ag2, c) Ag3.

Nature of Protective Layers

As we've described earlier, silver nanoparticles were prepared by using polymer, carboxylate and amine type coating. The method by which a polymer is added to the medium be calls steric stabilization (Fig. 5a).

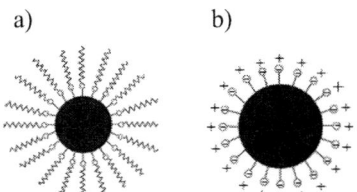

Fig. 5. Polymer (steric) (a) and charged (electrostatic) (b).

One end of such a polymer can attach to the silver particles in suspension, while the other end points away from the particle, resembling a tail. In this way each particle is surrounded by protective layer. Since these polymers will not bind to each other, they will prevent the silver particles from agglomerating.

Other types of protective layers stabilize a nanoparticles in a suspension by electrostatic

stabilization (Fig. 5b). This is method by which ions are attached to the silver particles to create a unified charge. This charge around particles is known as a zeta (ζ) potential, measured in mili-volts.

Each kind of protection layer was built during nano Ag production process and has exist on nanosilver all time up to end of fully curing/sintering process.

Removing Protection Layer Tests

For dependency of silver nanoparticles protective layers on sintering temperature, series of thermal testes were done. In this case, the most important results of studies were presented:

- the purity analysis for Ag1, Ag2 and Ag3 samples,
- the dynamics of removing the protective coating at 150 °C and 230 °C as a function of time for Ag1, Ag2 and Ag3 samples,

a)

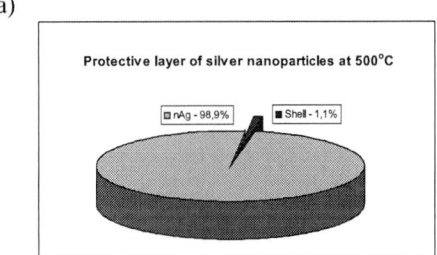

b)

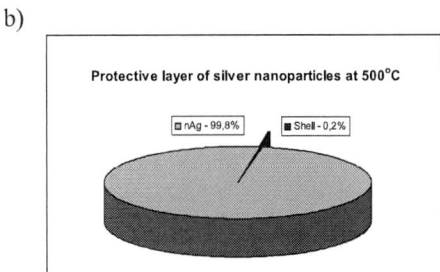

c)

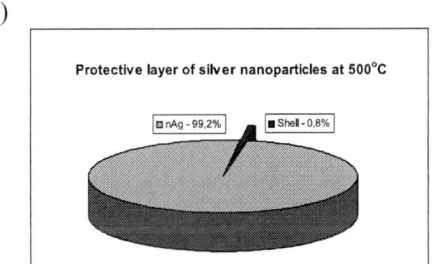

Fig. 6. Results of purity analysis for a) Ag1, b) Ag2, c) Ag3.

For finding total amounts of protective layers, the purity test were made by weight method. This tests were done by thermal removing all organic chemicals (protective coating) at 500 °C for 1h. It is very low content of protective layers as is illustrated on Fig. 6, and it is depend about method of making nAg particles. As a results, maximum protection material is less than 1.5 % of total nanosilver mass for Ag1 with carboxylate coating. In the best case the amount of protective material like tertiary amine was only about 0.2 % of total Ag2. While the Ag3 was included 0.8 % of polymer type of protective layer.

When is well known total amount of protective shell, it is interesting speed of removing it as a function of time in a constant temperature (Fig. 7). All types of nanosilvers has been tested at temperature (150 °C and 230 °C) and time up to 60 min.

a)

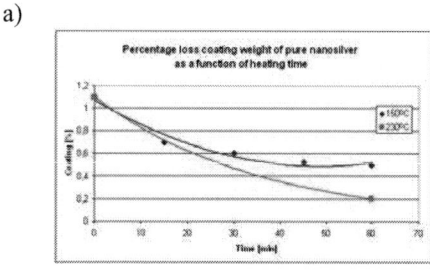

b)

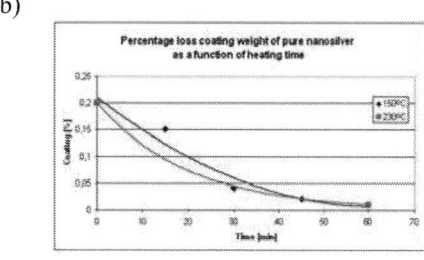

c)

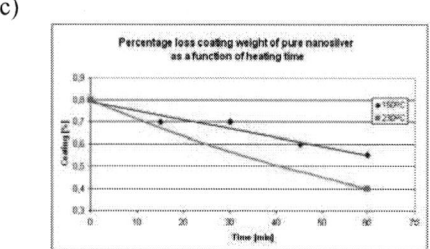

Fig. 7. The speed of removing the protective coating at 150 °C and 230 °C as a function of time for a) Ag1, b) Ag2, c) Ag3.

Obtained results show that weight loss of each coatings at the required temperature depends on the heating time of nanosilver. Thermal studies of nanosilver specimens sintered by heating at temperatures 150 °C every 15 minutes during 1 hour and 230 °C after 1 hour give us very important informations. As it is shown on above graphs, whole percentage content of protection materials on each tested nanosilver were marked in time equal 0 min and were 1.1 %, 0.2 % and 0.8 % for Ag1, Ag2 and Ag3, respectively. It was observed that the loss weight coating curves vs. time has the similar tendency for both at 150 °C as well as at 230 °C. The graphs show that the bigger loss of protective layer is by heating at higher temperature and in longer time. It is connected with the materials of which the shells are built of. In the case of Ag1 it was 0.5 % and 0.3 % carboxylate layer at 150 °C and 230 °C, respectively (Fig. 7a). Similarly the polymer layer has existed on Ag3 for the whole heating time up to 230 °C (Fig. 7c). The situation has changed when we used amine type of shell (Ag2). In this case the removing almost the whole of protection layer was already observed at 150°C (Fig. 7b).

It is very clear, that some parts of organic ingredients inside tested protection shells will be removed in high temperature as quick, as temperature will be higher.

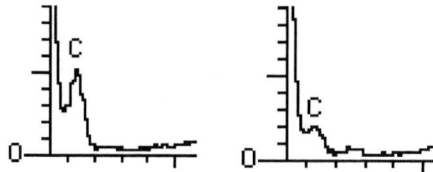

Fig.8. Fragments of EDX spectra indicated the carbon presence in nAg structure; on the beginning (left) and after the heating process (right).

When protection shell will be represented by carbon, than EDX spectrums during thermal process (Fig.8) will give us information about loosing volume of organic protection layer.

We should note, that the minimal amount of remained protection layer does not bother for obtaining good stable conductive properties. It is probably responded for adhesion properties to substrate.

For determination resistivity of printed layer. thickness of 4 times printed structure was measured by TEM. For number of measurement as a example is showed on Fig.9.

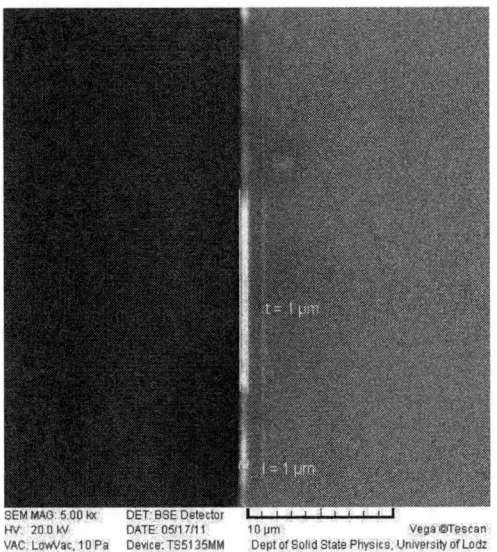

Fig.9. Picture of cross-section printed structure.

Average thickness is 1 μm, so after accounting resistivity ρ from the simple formula:

$$\rho = \frac{R \cdot d \cdot h}{l},$$

where R – measured resistance, d, h and l – respectively width, thickness and length of measured lines, obtained result is:

$$\rho = 2.9 \times 10^{-5} \ \Omega cm$$

Conclusions

The synthesis of nanosilver's with different protective coatings by chemical reduction was described. The maximum protection material was about 1.1 % for nanosilver with carboxylate coating and in the best case 0.2 % for nanosilver with amine type coating. The research was carried out in comparison to commercially available inks, which contain about 0.5-3 % protective layers.

In this paper we have demonstrated dependency of silver nanoparticles protective layers on sintering temperature. After "removing" process at 150 °C at 1 hour – sintering nanoparticels start to appear. This is necessary process for obtaining good and stable electrical conductivity of inks printed on low cost flexible substrates as an example PET, PPS, PC and others.

The SEM and Zetasizer have revealed that the particles dimension were in range 50-100 nm and each kind of studied nanosilver with carboxylate, tertiary amine and polymer protective materials were uniformly distribution.

This described our work is a part of company R&D for development new family of low temperature sintered inks for Ink Jet and R2R technology.

Acknowledgments

Authors would like to thanks Polish Academy of Sciences in Warsaw for SEM analysis of samples and University of Łódź for EDX analysis.

References

[1] Mościcki A., Felba J., Sobierajski T., Kudzia J., Arp A., Meyer W. „Electrically Conductive Formulations Filled Nano Size Silver Filler for Ink-Jet Technology", IEEE Polytronic Conference, Wrocław, 2005.
[2] Felba J., Nitsch K., Piasecki T., Tesarski S., Mościcki A., Kinart A., Bonfert D. „Properties of Conductive Microstructures Containing Nano Sized Silver Particles", IEEE Nano Conference, Genoa, 2009.
[3] Mościcki A., Felba J., Gwiaździński P., Puchalski M. „Conductivity improvement of microstructures made by nano-sized silver filled formulations", Polytronic 2007-6th International Conference on Polymers and Adhesives in Microelectronics and Photonics.
[4] Hsien-Hsueh Lee, Kan-Sen Chou, Kuo-Cheng Huang, „Inkjet printing of nanosized silver colloids" Nanotechnology 16 (2005) 2436–2441.
[5] Dongjo Kim, Jooho Moon, „Highly Conductive Ink Jet Printed Films of Nanosilver Particles for Printable Electronics", Electrochemical and Solid-State Letters, 8 (11) J30-J33 (2005).
[6] Ming Chen, Wen Hua Ding, Yang Kong, Guo Wang Diao, „Conversion of the Surface Property of Oleic Acid Stabilized Silver Nanoparticles from Hydrophobic to Hydrophilic Based on Host-Guest Binding Interaction", Langmuir 2008, 24, 3471-3478.
[7] Steve Lien-Chung Hsu, Rong-Tarng Wu, „Synthesis of contamination-free silver nanoparticles suspensions for micro-interconnects", Materials Letters 61 (2007) 3719–3722.

[8] Joseph Lik Hang Chau, Ming-Kai Hsu, Ching-Chang Hsieh, Chih-Chun Kao, „Microwave plasma synthesis of silver nanopowders", Materials Letters 59 (2005), p. 905.

[9] C.H. Bae, S.H. Nam, S.M. Park, Applied Surface Science 197–198 (2002) 628.

[10] Starowicz M., Stypuła B., Banaś J., „Electrochemical synthesis of silver nanoparticles", Electrochemistry Communications 8 (2006) 227–230.

Additive Photolithography Based Process for Metal Patterning Using Chemical Reduction on Surface Modified Polyimide

David E. Watson, Jack H.-G. Ng, and Marc P.Y. Desmulliez

MIcroSystems Engineering Centre (MISEC), School of Engineering & Physical Sciences,
Heriot Watt University, Edinburgh, EH14 4AS, Scotland UK

Tel:+44(0)131 451 3783, Email:degw1@hw.ac.uk

Abstract

This article presents progress towards the direct metallisation of polyimide via silver ion impregnation and nanocluster formation using a novel solvent free photoresist prior to electroless silver plating. Such a process is envisaged to play a major part in the fabrication of a wide range of electronic devices, such as RFID, flexible electronics, biomedical applications, OLED, e-paper and flip-chip bonding applications by providing a quicker, greener and more cost effective methods of polymer electronics production. A new method involving the chemical reduction of silver ions is presented here providing a comparison to previous UV photoreduction experiments, and a further test for electrolessly plated microstructures.

Key words: Electroless plating, Metallisation, Polyimide, Polymer Electronics, Manufacturing Processes

Background

Research and commercial activities in plastic electronics are gowing exponentially and there are many different approaches being taken to provide the next generation of more efficient and higher resolution production technologies [1-7]. A recent review of the different techniques currently researched or commercially available highlights the need for a lower cost solution to mass production of application specific components [1]. Other important factors include the speed of production and the minimum achievable feature size. Our group is currently investigating an alternative manufacturing method, based on the "bottom up" approach; an additive process that not only reduces processing time by eliminating the metal etching step, but also circumvents the problems of subtractive methods where large quantities of metal waste are produced.

Previous work carried out by our group utilised UV light to photoreduce silver ions embedded in a polyimide substrate, resulting in a seed layer consisting of clusters of silver nanoparticles for subsequent electroless silver plating [1-3]. Here we apply a chemical reduction method [8,9] for the creation of the seed layer micropatterns using a novel photoresist with UV exposure, which provides a useful comparison to the equally well-researched direct photoreduction process for micropatterning [1, 10], with respect to the formation and density of the silver nanoclusters. This article also examines the quality of plating on this alternative means of silver ion reduction / seed layer formation, both in terms of resistance and adhesion.

Fabrication Process

The fabrication steps are outlined in Figure 1 and described below.

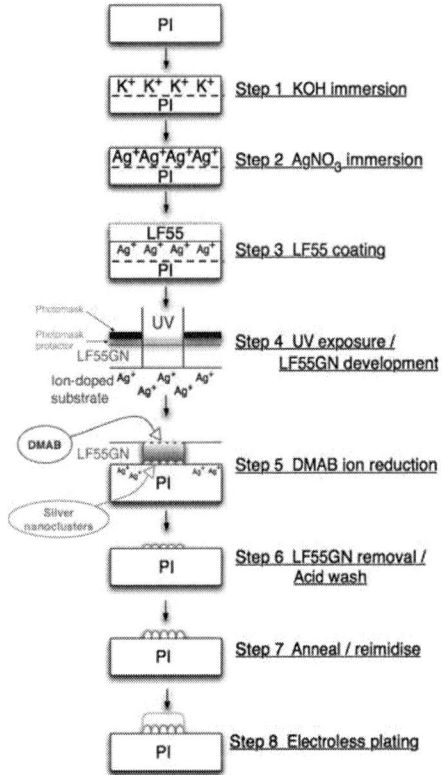

Figure 1: Direct metallisation and electroless plating process.

All chemicals were procured from Fisher Scientific, UK. For the substrate, Kapton 200 HN (DuPont, USA) polyimide of thickness 50μm was selected. Following rinse cleaning with acetone, isopropyl alcohol and submersion in an ultrasonic bath of deionised water, the substrate was treated with 1M potassium hydroxide solution at 50°C for 5 minutes to hydrolise the imide groups within the polymer. The substrate was them immersed in 100mM silver nitrate solution for 15 minutes to allow the silver cations to exchange with their potassium counterparts in the substrate, resulting in silver ions electrostatically bonded to the carboxyl groups within the polymer chains.

The samples were then spin coated with LF55GN liquid photoresist (MacDermid Printing Solutions, USA) on a Spin150 spin coater (SPS Europe) at 2000 rpm for 55 seconds. Negative photoresist LF55GN was chosen because it contains no solvents and requires only a jet spray of water to develop. This photoresist was originally designed for flexographic printing applications; here we have adopted it for microengineering fabrication. It requires no further chemical treatment for stripping the crosslinked resist, only mechanical peeling is sufficient to remove the photoresist. This is of particular importance for the surface modified polyimide substrate as the first stage of the process cleaves the imide ring within the polymer chain, leaving it vulnerable to attack from more aggressive developers. With no drying required, the photoresist remains in a viscous liquid state so a protective acetate film was placed between the resist and the photomask to keep the latter clean. The covering of the top surface prevents the presence of oxygen at the surface of the resist, which would which would prohibit crosslinking of the photopolymer. Exposure was carried out using a roughly collimated UV arc lamp delivering 200 mJ of energy and patterning was achieved with an acetate photomask (JD Phototools, UK),. The film was then developed by thorough rinsing in deionised water using a spray gun.

Reduction of the silver ions was carried out by submersion in 50mM dimethylamine borane (DMAB) solution for 5 minutes before rinsing in deionised water. Prior to resist removal, the structure was then subjected to a further 200mJ exposure. This causes the resist to assume a less sticky, more rubbery form, thus making it easier to handle and remove. The substrate was then submerged in dilute sulphuric acid for 15minutes to remove any unreacted silver ions before annealing for 30 minutes at 250°C. This step has the dual effect of reimidising the polymer and also further developing the silver nanoclusters.

Finally the samples were electrolessly plated for 8 minutes to achieve high conductivity and given a post-plating bake at 150°C for 40 minutes. For the bath, ESM500 (Polymer Kompositer AB, Sweden)

was selected for its slightly acidic chemistry as alkali solutions are prone to attack the polyimide.

Results and Discussion

Previous results indicated that, while large UV energy doses (approximately 3000 times greater than the energy dose supplied here) and long annealing times (3-4 hours at 300°C) can have a positive effect on silver nanoparticle formation and agglomeration, they may also have a negative, degradative effect on the polyimide substrate and therefore a trade off must be made [3].

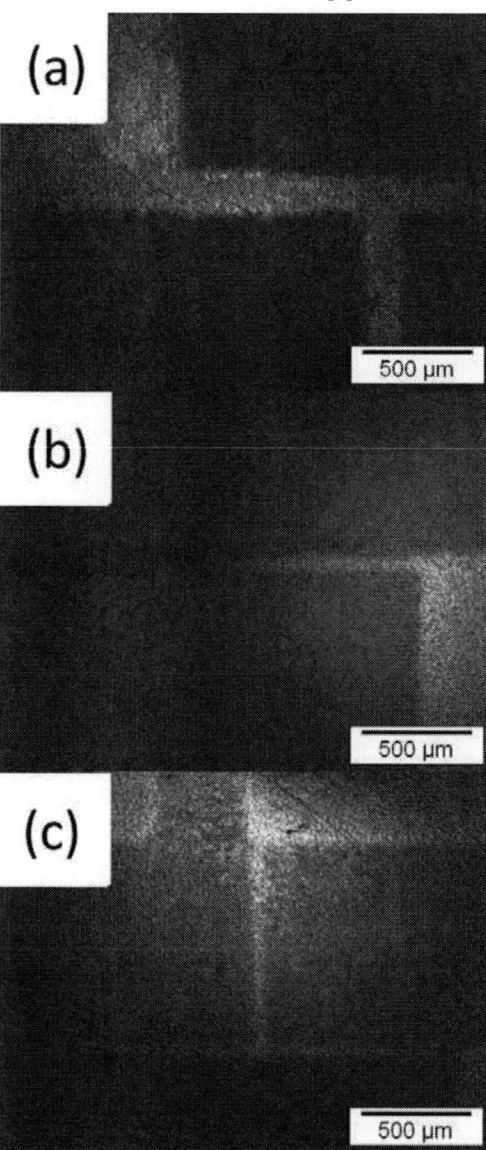

Figure 2: Microscope images of different linewidths (μm): (a) 250, (b) 100, (c) 50.

To investigate this effect on the outcome of the plating, an alternative, chemical ion reduction method was employed to eliminate the effects of UV irradiation. This was also the first time such a photoresist has been used as a mask for reduction so naturally the linewidths achieved and any deviation from the mask width is of interest. The linewidth deviation, height profile and any substrate degradation were characterized by optical microscopy using an Olympus SZX10 and white light interferometry with the Zygo Viewmeter 5620 while the adhesion was tested by the standard Scotch tape test.

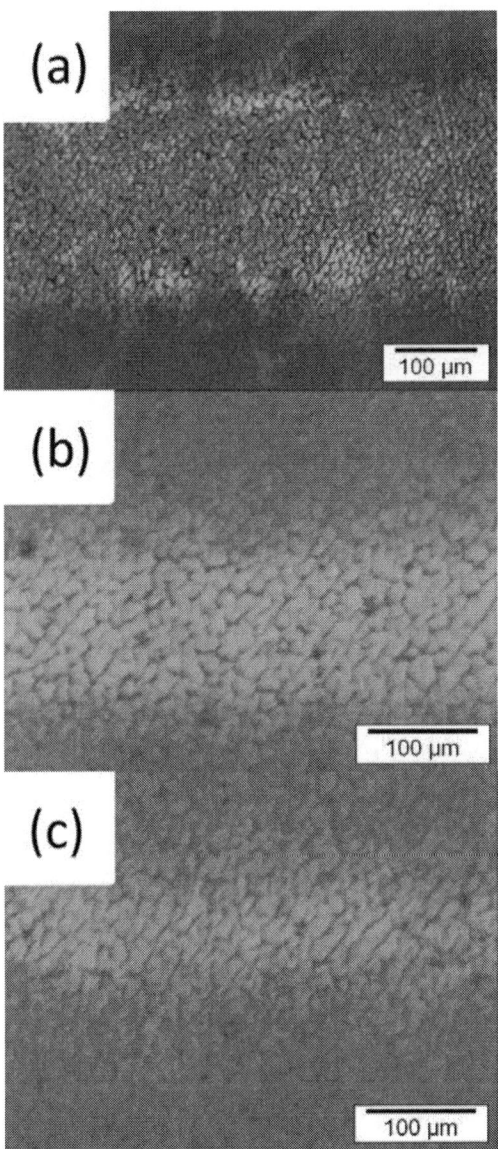

Figure 3: Linewidths at higher magnification (μm): (a) 250, (b) 100, (c) 50.

The photomask used consists of test structures with linewidths of 500, 250, 100, 50 and 15μm. As expected, the 500μm seed layers were very well formed and are not shown here. Of more interest are the smaller linewidths and Figure 2 shows microscope images of the silver seed layers of different linewidths after reduction and resist stripping. It can be clearly seen that the 250 and 100μm structures are very well patterned and there is evidence of reduction activity on the 50μm structures. No 15μm lines were visible and so these images have been omitted. This could be due to the swelling effect exhibited by the LF55GN liquid resist during development. The silver circles in apparently random places in these images are most likely caused by the acetate film applied on top of the resist before first exposure to protect the photomask from becoming dirty and contaminated with photoresist. The completely smooth application of this layer is an intricate process and any air holes left allow oxygen in, which, as mentioned previously, hinders the crosslinking process. Although the DMAB-developed lines are clearly visible, the seed layers are slightly bigger than the linewidths of the mask at this stage as seen in Figure 3.

Rather than a clear divide between plastic and metal being visible, there is more of a gradual transition, with the density of silver nanoparticles decreasing from the centre of the line. The average deviation from the mask dimensions is shown in Table 1

	Linewidth (μm)				
Mask	500	250	100	50	15
Pre-plating	560	313	133	72	-
Post-plating	582	330	148	87	-

Table 1: Linewidth deviation in the different samples before and after plating.

This could be down to the DMAB partially diffusing into the polyamic acid where the silver ions are contained. This conclusion is strengthened when a sample was overexposed during the post-resist-development step. Figure 4 shows how this second UV exposure has caused silver to form in an approximately circular shape around the areas of substrate that were in contact with the DMAB. Another possibility is the partial detachment of resist from substrate during resist development. This is hypothesised because the water jet spray is very powerful and has been seen to dislodge partially cured resist in previous tests.

Figure 4: Over exposure during post-development resist hardening step.

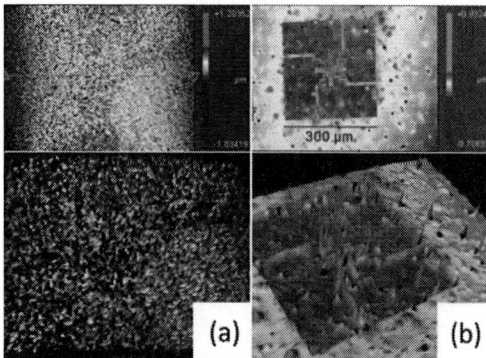

Figure 5: Zygo surface profiles of test structures via different reduction methods: (a) chemical (DMAB), (b) photo (UV).

Figure 5 compares the surface profiles of the DMAB reduced test structures and their UV photoreduced counterparts. It is clear from the absence of recessed pattern on the chemical sample that there is no degradation of the polymer under the silvered areas as previously reported with UV exposure [3]. This confirms the previous hypothesis that UV irradation, extended high temperature heat treatment or both degrade the substrate.

Figure 6 shows microscope images of the same test structures from Figures 2 and 3 after plating. As can be seen from Figure 6, the plating was not always uniform. This is most likely due to a combination of a very thin film of LF55GN resist remaining and sub-optimal plating conditions. Even thought it is water soluble, it is a very jelly-like and tacky substance and therefore it is not a trivial exercise to remove it. There are many plating parameters to monitor and the process is still being tailored to a polyimide substrate. An example of unremoved resist can be seen in Figure 7 below.

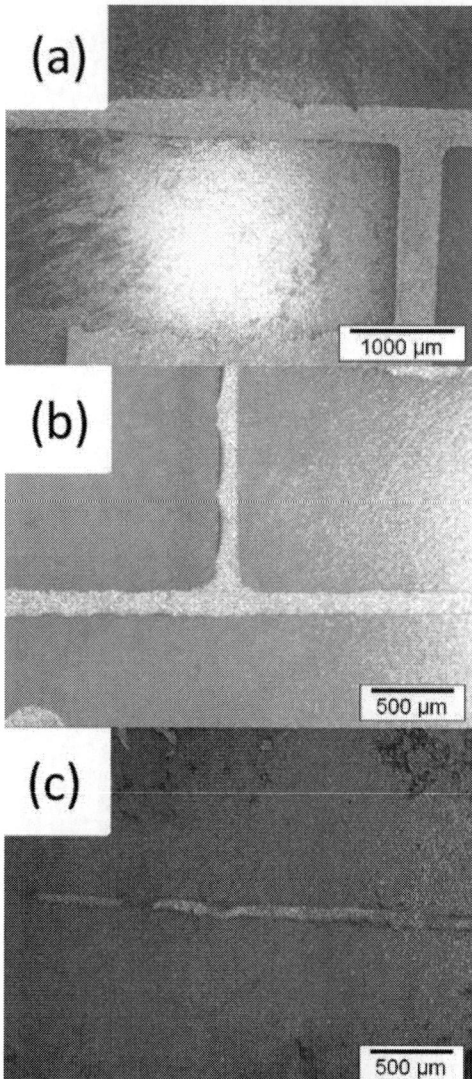

Figure 6: Plated test structures, linewidths (µm): (a) 250, (b) 100, (c) 50.

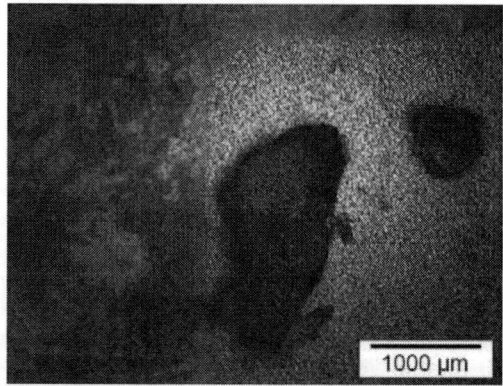

Figure 7: Residual resist possibly causing plating defects.

Despite some occasional defects, smooth larger area plating was achievable, as can be seen in Figure 8. Unsurprisingly the linewidths after plating increased. Figures 9 and 10 show microscope images and Zygo profiles of the plated lines while Table 1 summarises deviation from the mask specified widths.

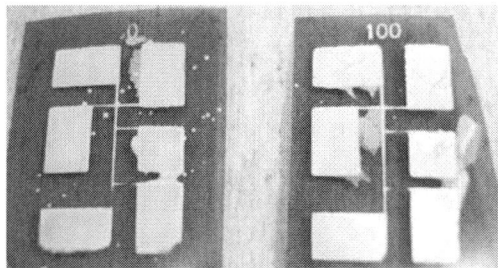

Figure 8: 100µm plated test structure.

As can be seen from Table 1 and Figure 9, the linewidths have increased approximately in proportion to the plating height. The nonlinear relation can be explained by the partial seed layer outwith the trackwidth, visible in Figure 3. Zygo measurements for the 50µm linewidths were highly anomalous and so do not appear here.

Of great importance to flexible electronic component manufacturing is obviously the conductivity of the metal tracks created. A basic 2 point probe testing along the longest path of the test structure (10mm) gave slightly different resistances depending on the linewidth as summarized in the Table 2 below.

Also of high importance is the adhesion of the metal layers onto the substrate. Peel tests were carried out on the structures using the Scotch tape method. Figure 11 shows the before and after results of a typical test on the bond pad of a structure. While the adhesion is good, a small amount of silver is removed.

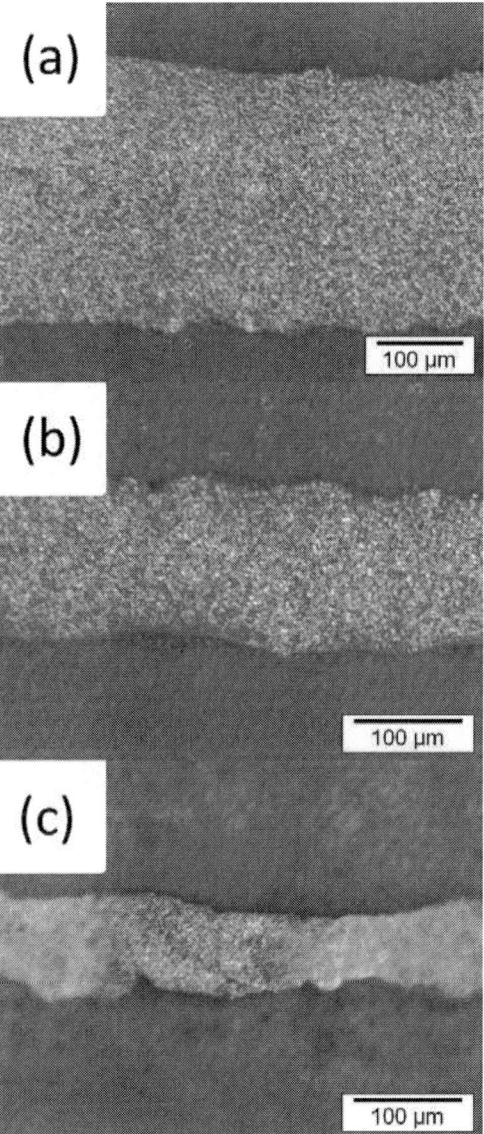

Figure 9: Plated linewidths (µm): (a) 250, (b) 100, (c) 50.

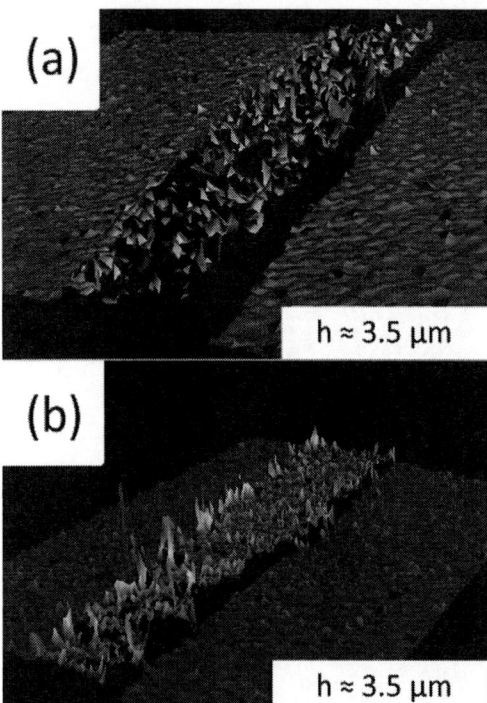

Figure 10: Zygo images of plated lines (µm): (a) 250, (b) 100.

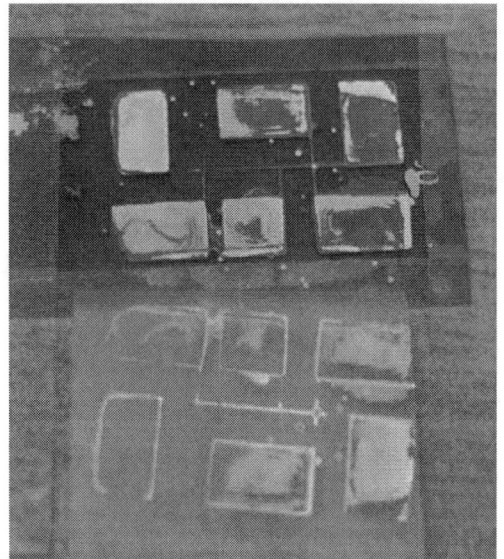

Figure 11: Peel test of plated 100µm test structure.

Linewidth (µm)	500	250	100	50	15
Resistance (Ω)	1.4	2.4	9.9	-	-

Table 2: Resistances of test structures

Conclusions

We have reported an alternative additive manufacturing process for the creation of flexible electronic components that is fast, cheap and generates very few hazardous waste products. Chemical reduction in the form of DMAB is highly effective at producing a very good seed layer for subsequent electroless plating; work is underway in achieving smaller resolution features. We have demonstrated robust conductive tracks of 100µm linewidth and partially formed tracks of 50µm. The LF55GN photoresist used is designed for high aspect ratio structures and although its swelling during development will provide a natural limit, it can still be very useful because of its purely water-based development characteristics. Furthermore, we have demonstrated that our process is ideal for producing seed layers that have good adhesion to subsequent electroless plating. The phenomenon illustrated in Figure 4 is also of note as it shows that a combination of both photo- and chemical- reduction processes may provide the solution to the next generation of higher resoltion flexible electronic manufacturing processes.

Acknowledgements

The author would like to acknowledge Anders Remgård of Polymer Kompositer AB and the financial support from the British Engineering and Physical Sciences Research Council (EPSRC).

References

[1] J.H.-G. Ng, M.P.Y. Desmulliez, M. Lamponi, B.G. Moffat, "A direct-writing approach to the micropatterning of copper onto polyimide", Circuits World, Emerald Group Publishing Limited, Vol. 35, N.2, pp.3-17, 2009.

[2] J. H.-G. Ng, M.P.Y. Desmulliez, K.A. Prior, D. Hand, "UV direct Patterning of metal on polyimide", IET Micro & Nano Letters, Vol.3, N.3, pp.82-89, 2008.

[3] D.E. Watson, J. H.-G. Ng, J. Sigwarth, J. Bates, M.P.Y. Desmulliez, "Silver nanocluster formation using UV radiation for direct metal patterning on polyimide", Proceedings of 3rd Electronic System-Integration Technology Conference (ESTC), 10.1109/ESTC.2010.5642829 , 2010

[4] S.M. Bidoki, J. Nouri, A.A. Heidari, "Inkjet deposited circuit components", J. Micromech. Microeng. Vol. 20, N.5, 2010.

[5] S.S. Yoon, D.O. Kim, S.-C. Park, Y.K. Lee, H.Y. Chae, S.B. Jung, J.-D. Nam "Direct metallization of gold patterns on polyimide substrate by microcontact printing and selective surface modification", Microelectronic Engineering. 85, 1, 136-142, 2008.

[6] S.-C. Park,Y.-B. Park, "Effect of Temperature / Humidity Treatment Conditions on Interfacial Adhesion Energy between Inkjet-Printed Ag and Polyimide", Japanese Journal of Applied Physics. 48, 8, pp. 08HL02-08HL02-6, 2009.

[7] D.S. Thompson, L.M. Davis, D.W. Thompson, R.E. Southward, "Single-Stage Synthesis and Characterization of Reflective and Conductive Silver‚àíPolyimide Films Prepared from Silver(I) Complexes with ODPA/4,4'-ODA", ACS Applied Materials & Interfaces. 1, 7, 1457-1466. 2009.

[8] S. Ikeda, H. Yanagimoto, K. Akamatsu, H. Nawafune, "Copper/Polyimide Heterojunctions: Controlling Interfacial Structures Through an Additive-Based, All-Wet Chemical Process Using Ion-Doped Precursors", Advanced Functional Materials, 17: 889–897. doi: 10.1002/adfm.200600527, 2007.

[9] W. Su, P. Li, L. Yao, F. Yang, L. Liang, J. Chen "Direct Patterning of Copper on Polyimide by Site-Selective Surface Modification via a Screen-Printing Process", ChemPhysChem, 12: 1143–1147, 10.1002/cphc.201100040, 2011.

[10] K. Akamatsu, S. Ikeda, H. Nawafune, "Site-Selective Direct Silver Metallization on Surface-Modified Polyimide Layers", Langmuir, 19 (24), 10366-10371, 2003.

Copper Makes a lot of Cents

Chris Flowers and Martyn Owen

CSR, Cambridge Business Park, Cambridge, UK

chris.flowers@csr.com

Abstract

Transitioning from gold to copper wirebonding can provide significant cost saving for the semiconductor industry. However the move to copper wirebonding is not without risk and this paper will address the measures of end product quality that are essential for reliable products in the commercial sector.

Key words: Copper, wirebond, reliability, bHAST, IMC

Introduction

Historically, gold has been the dominant material used for wirebond interconnect, but over the last 5 years the gold price has increased from $500/oz to in excess of $1500/oz for a variety of reasons associated with the equity and commodity markets [1]. During this time, wirebond material and Out Sourced Assembly and Test (OSAT) suppliers have worked to indentify an alternative material which could be used as a replacement for gold.

In 2006, Levine and Deley reported a comparison of gold and copper wirebonding stating that potentially copper wire has a superior lifetime providing the manufacturing challenges are overcome [2].

Following this, in 2009, ASM Pacific Technology (Charles J. Vath III, M. Gunasekaran, Ramkumar Malliah) reported wire pull data from a variety of different bondwire types after extended high temperature baking [3]. They, too, concluded that both copper wire pull and ball shear results are comparable to, or better than, gold and indicated that the presence of a thin Palladium (Pd) coating on the copper wire may be beneficial.

Also in 2009, at ECTC, Nippon (Uno, Terashima, and Yamada) presented data to show that Pd coated copper wire has greatly improved storage life and floor life, compared to uncoated copper wire. In the same paper Nippon explained that the use of Pd coating allows the formation of a uniform, symmetrical Free Air Ball (FAB) in a nitrogen only environment [4]. This avoids the need for expensive forming gas facilities inside the wirebond manufacturing site.

The nitrogen environment is required because copper is more reactive than gold and it is important that the copper wire is not allowed to oxidise. If the copper wire oxidises during the formation of the FAB then it will result in non-symmetrical ball formation, and the ability to bond the wire is diminished, leading to Non Stick On Pad (NSOP).

If designed and manufactured correctly, reliability concerns such as wirebond failure can be controlled to provide a reliable end product. This has been shown to be achievable in both Quad Flat-pack No-lead (QFN) and Ball Grid Array (BGA) wirebond packages, covering multiple Complementary Metal Oxide Semiconductor (CMOS) wafer process nodes with varying bond pad structures and materials.

In order to optimize the copper wirebond quality it is necessary to carefully select the mold compound used to encapsulate the die and wirebonds. Parameters such as its glass transition temperature (Tg), percentage water absorption (PCT) and halogen ppm count, all need to be considered.

Key quality metrics for copper wirebonded devices include:

- Inter-Metallic Compound (IMC) percentage coverage under the wirebond ball,

- Thickness of the remaining bondpad metal (Aluminium) under the wirebond ball on the IC bond pad

- Wirebond ball shear strength

The relationship between these key quality metrics and the reliability of the devices, will be covered in this paper.

In order to assess the reliability of the devices, the industry standard tests defined in JEDEC JESD47 (Stress-Test-Driven Qualification of Integrated Circuits) are used. The most stringent JEDEC reliability test has been found to be Biased Highly Accelerated Stress Test (bHAST). Results for the reliability tests will be presented, and the failure modes will be discussed.

Experiment

In order to transition Cambridge Silicon Radio's (CSR's) QFN and BGA products from gold wirebonding to copper wirebonding, 8 QFN assembly lots and 4 BGA assembly lots have been subjected to qualification testing. The qualification procedure included assessment of:

- Assembly quality and yields
- Reliability (Jedec JESD47)
- Test yields and fail mode pareto analysis
- Constructional analysis

The 12 assembly lots were split over 5 assembly runs (QFN: A, B, C and BGA: A, C) using different assembly process conditions in order to investigate the relationship of the key quality metrics to the bond strength and reliability performance of the end product.

The same copper wire type was used for all of the assembly lots: Palladium coated copper wire. Two different types of mold compound were used: one for BGA, plus one for QFN devices (see details below).

The measurement of the quality metrics was performed by two independent laboratories in order to reduce measurement error. This was crucial as copper wire is eroded by the standard etchants used for removing the mold compound (de-capsulation), so copper wirebonded device de-capsulation requires special techniques compared to the established gold wire de-capsulation methods.

Careful selection of a mold compound, with optimal material properties, is crucial to ensuring the quality and reliability requirements are met. Some of the key attributes of the mold compound that help to ensure the reliability are shown below:

Mold compounds have their water absorption measured by exposing the mold compound to raised temperature, pressure and humidity, and measuring the increase in mass. A lower value of water absorption is desirable, as it indicates that the material is more resistant to the diffusion of water molecules.

As the ambient temperature approaches the Tg of the mold compound, the bonds inside mold compound undergo a change that allows more water vapour to be absorbed. Although different mold compounds can have different rates of water absorption, having a high Tg mold compound helps to reduce the amount of water that is absorbed.

If significant concentrations of chlorine, or other halogens, exist within a mold compound and moisture allows these chlorine ions to become mobile, then the bond pads, under and around the wirebond balls, can become corroded. This corrosion process is accelerated by applying an electrical bias. An example of bond pad corrosion is shown below.

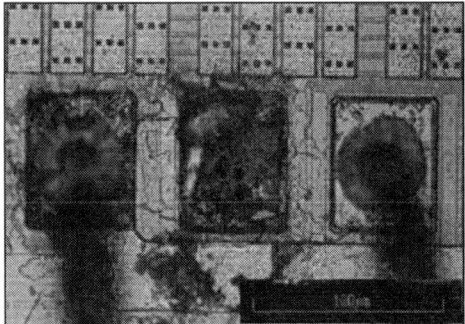

Figure 1: Typical bond pad corrosion.

Chlorine ions can react with the Cu-Al IMC to create Aluminium oxide, with the final result that the bond becomes open-circuit see Figure 2.

$$2Cu_4Al + 6Cl^- \rightarrow 8Cu + 2Al^{3+} + 3Cl_2$$
$$2Al^{3+} + 3O^{2-} \rightarrow Al_2O_3$$

Figure 2: Reaction for corrosion of the IMC

Although Chlorine is the most common halogen impurity found in mold compounds, the reaction can also be caused by other halogen impurities such as Bromine or Fluorine.

CSR worked closely with the OSAT and mold compound supplier to select the two mold compounds that met the requirements for the BGA and QFN devices. After evaluating a number of improvements to the initial materials, the two mold compounds were fixed, and not changed during the assembly.

bHAST simulates the ageing of a product in the field by using accelerated humidity and temperature conditions :

- 130°C/85%RH/96 hours or
- 110°C/85%RH/264hours

The combination of high temperature and high humidity drives the moisture into the packaging materials, such as the mold compound. When the moisture penetrates as far as the Integrated Circuit (IC) bond pads it promotes bond pad corrosion, especially in the presence of halogen ions. The electrical bias accelerates the bond pad corrosion by increasing the rate at which the halogen ions are attracted to specific bond pads.

Measurement of the key quality metrics is essential, so that the reliability performance can be correlated to these metrics. The metrics can then be used in process quality monitoring and control. The key quality metrics that were identified were:

1. IMC coverage
2. Aluminium pad thickness under the ball bond
3. Wirebond ball shear test

These quality metrics were measured in four bond pad locations on each die.

In the X-Y plane (parallel to the die surface) a key parameter is IMC Coverage (see Figure 3). This is a measure of the percentage area of the underside of the wirebond ball that has formed an intermetallic join between the copper wire and the aluminium pad.

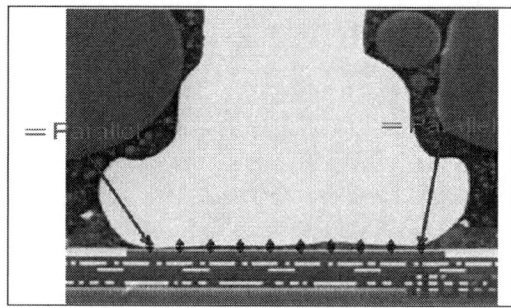

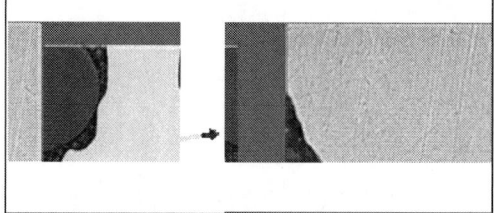

Figure 4: Aluminium pad thickness remaining measurement illustration.

Standard wirebond ball shear testing was performed, to measure the strength of the join between the copper wirebond ball and aluminium bond pad.

Figure 3: IMC coverage photograph

In the Z-plane (perpendicular to the die surface) the thickness of the Aluminium bondpad underneath the ball bond is a key parameter. As part of the quality assessments we measured the thickness of the Aluminium underneath the ball bond in 10 equally spaced locations underneath the ball bond as shown in Figure 4. Measurements were made across the area where ball bond surface is approximately parallel with the die surface.

Results

The results for the reliability legs are contained in Figure 5.

130nm QFN Wirebond Related Fails			
Assembly Run	A	A	A	A
Package	6x6x0.9/40L	8x8x0.9/68L	7x7x0.9/48L	8x8x0.9/68L
HTOL	80/80 pass	80/80 pass	80/80 pass	
MSL3	311/311 pass	308/308 pass	231/231 pass	77/77 pass
TCT	77/77 pass	77/77 pass	77/77 pass	
TST	77/77 pass	77/77 pass	77/77 pass	
PCT	77/77 pass	77/77 pass	77/77 pass	
BHAST 96h	80/80 pass	77/77 pass	>>>>	77/77 pass
HTS 150'C 1000h	77/77 pass	77/77 pass	77/77 pass	
BHAST 192h	40/40 pass	2/77 fails	>>>>	77/77 pass
BHAST 288h	25/25 pass			
HTS 150'C 1500h	61/61 pass			
HTS 175'C 500h	77/77 pass			
HTS 175'C 1000h	62/62 pass			
HTS 150'C 1000h	231/231 pass			

(a) Wirebond related fails of QFN assembly lots from assembly run A.

130nm QFN Wirebond Related Fails			
Assembly Run	B	C	C	C
Package	6x6x0.9/40L	6x6x0.9/40L	6x6x0.9/40L	8x8x0.9/68L
HTOL	80/80 pass	80/80 pass		
MSL3	311/311 pass	236/236 pass	231/231 pass	231/231 pass
TCT	77/77 pass	77/77 pass	77/77 pass	77/77 pass
TST	77/77 pass	77/77 pass	77/77 pass	77/77 pass
PCT	77/77 pass			
BHAST 96h	4/80 fails	85/85 pass	80/80 pass	80/80 pass
HTS 150'C 1000h	77/77 pass	77/77 pass	77/77 pass	77/77 pass
BHAST 192h				
BHAST 288h				
HTS 150'C 1500h				
HTS 175'C 500h				
HTS 175'C 1000h				
HTS 150'C 1000h				

(b) Wirebond related fails of QFN assembly lots from assembly runs B and C.

130nm BGA Wirebond Related Fails			
Assembly Run	A	A	C	C
Package	8x8x1.2 169B	6.5x6.5x1 72B	8x8x1.2 169B	6x6x1 84B
HTOL	77/77 pass			77/77 pass
MSL3	231/231 pass	154/154 pass	231/231 pass	154/154 pass
TCT	77/77 pass	77/77 pass	77/77 pass	77/77 pass
TST	77/77 pass		77/77 pass	
PCT		77/77 pass		
BHAST 96h	77/77 pass		77/77 pass	77/77 pass
HTS 150'C 1000h	77/77 pass		77/77 pass	
BHAST 192h	77/77 pass			

(c) Wirebond related fails of BGA assembly lots from assembly runs A and C.

Figure 5: Reliability results indicating wirebond related fails.

As shown in Figure 5 (a) after passing the JEDEC requirements of 96 hours bHAST, parts from QFN assembly run A were subjected to a further 96 hours. Automatic Test Equipment testing of the parts after the 192 hours of bHAST revealed two failing devices.

When QFN assembly run B, Figure 5 (b), was tested after 96 hours bHAST four failing devices were found.

An example of the relationship between reliability and the IMC distribution is shown in Figure 6. The lowest IMC % distribution (indicated in red) had the poorest reliability performance. The blue distribution showed good reliability performance and the green showed exemplary performance.

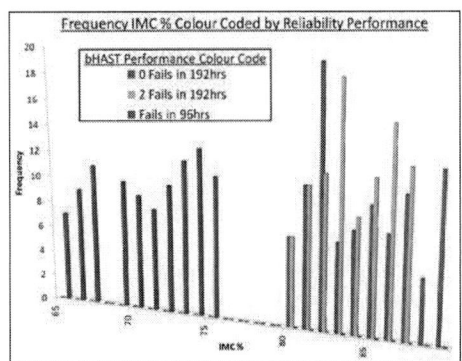

Figure 6: Frequency of IMC coverage percentage, colour coded by reliability performance.

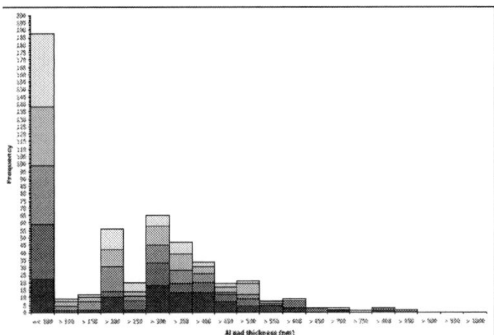

Figure 7: Distribution of the aluminium pad thickness remaining, for devices from an assembly run with good reliability (QFN assembly run C).

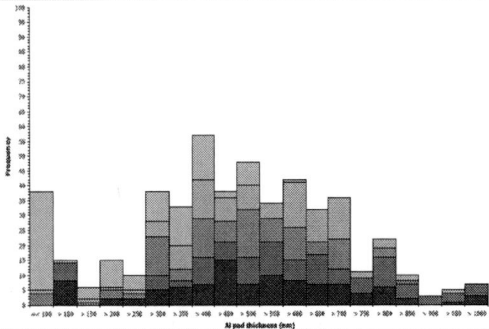

Figure 8: Distribution of the aluminium pad thickness remaining, for devices from an assembly run with bad reliability (QFN assembly run B).

Figure 7 shows the distribution for an assembly run that passed bHAST and Figure 8 shows a distribution for an assembly run that failed bHAST. By comparing Figure 7 and Figure 8 it can be seen that the run that failed bHAST has a much wider distribution and includes results with much thicker remaining bond pad under the wirebond ball.

The higher strength of the IMC interface for copper wire compared to gold wire means that during wire pull test the majority of failures are at the neck of the wire ball bond. This means that it is not possible to use wirebond pull test to analyse the quality of the copper wirebond IMC. In order to analyse the strength of the joint between the copper ball bond and the aluminium pad, ball shear must be used. Ball shear strength for QFN assembly runs A, B and C is shown below in Figure 9. QFN assembly run C is colour-coded red as it has the poorest reliability performance. This distribution also has the largest standard deviation and lowest ball shear strengths in comparison to the other QFN assembly runs with higher reliability performance, that are colour coded in green and blue.

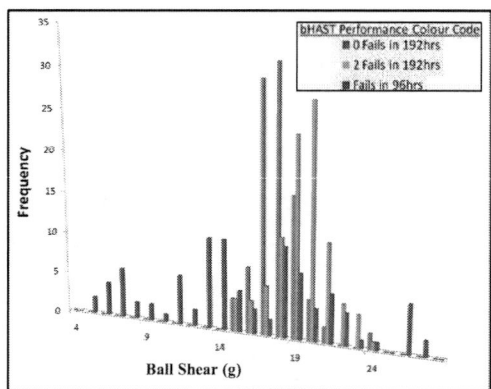

Figure 9: Distribution of ball shear strength for QFN assembly runs A (blue), B (red) and C (green).

Discussion

High Temperature Storage (HTS) is a key test for gold wirebonding, due to Kirkendall voiding, but the IMC growth is much slower for copper wire than for gold wire. So HTS does not simulate a reliability risk for copper wirebonding, as quickly as for gold wirebonding.

From the reliability work it is clear that bHAST is the most stringent test. As mentioned previously, this is due to the susceptibility of the IMC to moisture driven corrosion. Four bHAST failures were found from one of the Assembly Runs. Three of the failing units were open circuit, between the QFN leadframe and die pad, and the other unit was short circuit.

The open circuit devices were de-capsulated and it was found that the open circuit was

due to lifted wires at the die bond pad i.e. the joint between the wirebond and the die pad had failed.

An EDX was performed on one of the pads which suffered bond lift, and on one of the good bond pads (as a control) – see Figure 10 and Figure 11. The comparison between the EDX plots showed a significantly higher concentration of oxygen and slightly higher concentration of fluorine on the open circuit bond, in comparison to the good bond pad. This data supports the corrosion mechanism failure mode theory that was explained earlier in the paper.

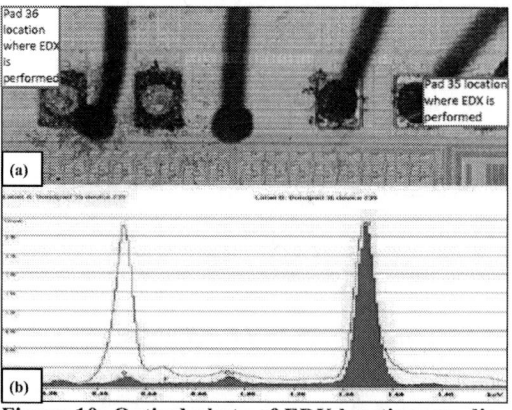

Figure 10: Optical photo of EDX locations on die (a) and EDX chart (b) of a failing bHAST device.

Failure analysis of the other open circuit units revealed similar EDX results, with high concentrations of oxygen. But rather than fluorine being detected, chlorine was detected instead.

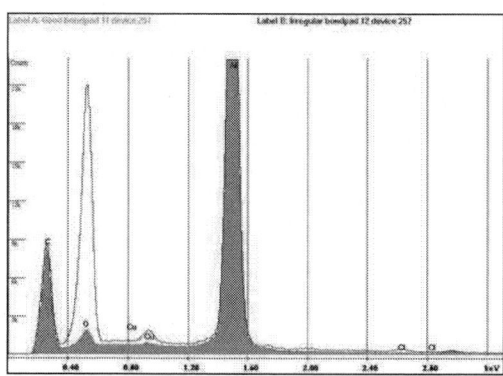

Figure 11: EDX chart of a failing bHAST device.

The short circuit device was found to have die surface damage that was not related to the wirebonding.

To ensure that the de-capsulation process was not influencing the EDX results, we also analysed a failing pad using cross-section and SEM. The SEM revealed the bond lift and corrosion on the pad that was open circuit.

374

As indicated in the results section, the assembly run that failed bHAST also showed:

1. The lowest IMC coverage distribution
2. The widest distribution of aluminium pad thickness remaining under that ball bond, as well as the largest aluminium pad thickness remaining.
3. The widest wirebond ball shear strength distribution, including the lowest ball shear strengths.

The IMC coverage and the remaining aluminium pad thickness are key attributes of the wirebond process set up. The wirebond ball shear strength is dependent upon the IMC coverage and remaining aluminium pad thickness, amongst other variables. This means that the ball shear strength can be easily used as an immediate indicator of the copper wirebond quality.

Conclusion

The most stringent test for copper wirebonding has been shown to be bHAST, as this can result in accelerated IMC corrosion. Reducing the level of corrosion by careful selection of the mold compound is critical to achieving good reliability.

The copper wirebond IMC coverage and the depth that the ball bond is driven into the aluminium pad on the die are crucial to the quality of the ball bond to die pad joint. Both of these key quality metrics have been shown to be correlated to the reliability results.

We have seen a correlation between the assembly run that failed bHAST and the assembly run with the lowest ball shear strength. This shows that the ball shear strength is a good indicator for the copper wirebond key quality metrics.

Acknowledgements

CSR would like to acknowledge MASER Engineering B.V. and Advanced Semiconductor Engineering Inc.

References

[1] GOLD PRICE PTY LTD 2002-2011.
Spot Gold Price Chart. [Internet]
Available from:
<http://www.goldprice.org/spot-gold.html>
[Accessed: June 2011]

[2] M.Deley and L.Levine, "Copper Ball Bonding Advances for Leading Edge Packaging", Proceedings of the 2005 Semicon, Singapore.

[3] Charles J. Vath III, M. Gunasekaran, Ramkumar Malliah, "Factors Affecting the Long Term Reliability of Copper Wire Bonding", Proceedings of the 2009 EPTC, Singapore.

[4] T.Uno, S.Terashima, T.Yamada, "Surface-Enhanced Copper Bonding Wire for LSI", Proceedings of the 2009 Electronic Components and Technology Conference (ECTC), San Diego, United States.

Current Industry Adoption of Fine-Pitch Cu Wire Bonding
and Investigation Focus at iNEMI

Grace O'Malley[1], Peng Su[2], Haley Fu[1], Martin Bayes[3] and Masahiro Tsuriya[1]

[1]gomalley@inemi.org, [1]haley.fu@inemi.org, [1]m.tsuriya@inemi.org, International Electronics Manufacturing Initiative
[2]pensu@cisco.com, Cisco Systems, Inc.
[3]MBayes@dow.com, Dow Electronic Materials

ABSTRACT

Gold (Au) wire bonding is one of today's most common and well understood first level interconnects. However copper (Cu) wire bonding is increasingly being used for a wide variety of components, predominantly driven by the cost benefit of Cu. As the application moves from mostly consumer products to high-reliability electronic systems, the long term reliability of fine-pitch Cu wire bonding must be better understood. A collaborative project has been launched by iNEMI (International Electronics Manufacturing Initiative) member companies to assess the current adoption landscape of Cu wire bonding in the industry, and to understand the critical factors that impact component reliability. An industrial survey has already been conducted by the project. The results of the survey not only provide a clear overview of the general application status of Cu wire bonding, but also illustrate key technical concerns such as reliability performance, materials and process capability (or optimization), and qualification standards requirements.

INTRODUCTION

Cu bond wires are increasingly being used for a wide variety of components, from consumer applications to high-reliability electronic products. Recent published papers have shown that Cu wire bonding requires more rigorous bonding process control and stricter packaging material selection. Despite the positive impact of these improvements, reliability still needs to be collectively assessed by the industry in a quantitative manner. Furthermore, for component qualification purposes, standard reliability test methods and durations established for Au wire device may not be appropriate for Cu. Extended reliability evaluation may be needed to validate today's commonly used BOMs (bill of materials) for Cu wire bonded devices.

iNEMI launched a collaborative project on Cu Wire Bonding Reliability in 2010. At this stage, fifteen companies are participating in this project.

The project includes two phases. In the first phase, member companies outlined and conducted a survey on the industry-wide conversion status, as well as key reliability concerns for Cu wire bonding. Based on the findings from this survey and published literature, it was determined that reliability performance of Cu wire bonding needs to be further evaluated with experimental work, so that the effects of key factors such as packaging material selection on reliability performance can be better understood. The project team has designed an experimental matrix that includes realistic packaging material variations for both leadframe and substrate-based packages. During the 2nd phase a series of accelerated tests will be performed on components made with these material variations and lifetime data will be collected. Based on the accelerated tests and failure analysis data, the effects of packaging material properties will be better understood. The effectiveness of current accelerated test methods and durations will also be determined.

SURVEY METHODOLOGY

The project team conducted an initial survey in September 2010, and a follow-up survey in December 2010 to clarify some issues from the initial results. Representatives from more than 40 major companies responded to the survey. These companies covered a full spectrum of the industry, ranging from material and interposer suppliers, semiconductor manufacturers, assembly houses, to OEMs with total annual revenue of over 300B USD.

32 suppliers, 20 OEM/EMS, and also several industry organizations and research institutes responded, providing good coverage of end-product segments including consumer/portable, Netcom, computers and medical, automotive. The survey included 20 questions covering Cu wire adoption status, reliability, process, materials, failure analysis and technical concerns.

SURVEY RESULTS

This section summarizes the major observations from the survey.

1. Preferred Wire Type

The survey asked what the preferred type of wire material is.

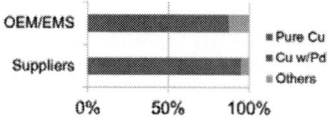

Fig.1 Preferred Wire Type

Response from suppliers and OEM/EMS were different. OEM/EMS preferred Palladium (Pd) coated Cu wire, while Suppliers preferred pure Cu wire. The choice of wire type was influenced by both reliability performance and cost considerations. Some commented that Cu wire with Pd shows better reliability test result than bare Cu wires, while others reported opposite results.

2. Cu Wire Adoption Status

The survey asked the current application status of Cu wire bonding for different device types.

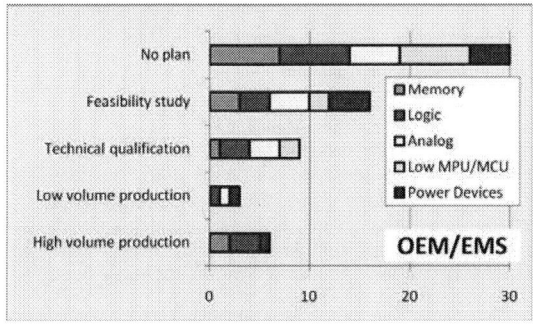

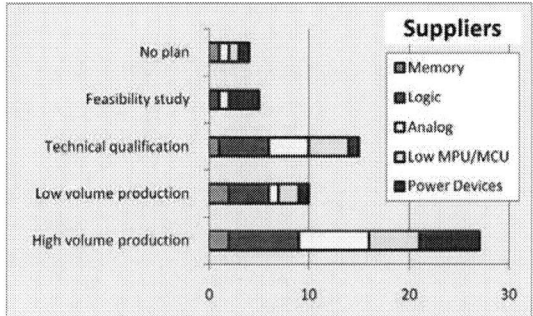

Fig. 2 Adoption Status by Device Technology

Many OEM/EMS have no adoption plan at this time, but some are conducting feasibility studies. A small percentage reported being in production for memory, logic and power devices.

The responses from semiconductor suppliers and assembly houses were counted in the graph, but material or equipment suppliers were not included. Most suppliers are in technical qualification or production stage. Cu wires are used for all device types.

Adoption status by market segment was also assessed for OEM/EMS.

Memory, Logic and Power devices are in volume production for some Consumer and Portable computers and also some High-end server products.

No production was reported for Netcom, Medical, and Automotive products and there were no plans reported for Military or Aerospace.

3. Cu Wire Implementation Status

The survey asked for what package types were there plans to adopt Cu wire bonding technologies.

Copper wire usage in all package types is in "planned" or "already adopted" phases, but varys from company to company. xBGA, xQFN and xQFP packages have higher adoption rates.

Results showed more cautious views of implementation by OEMs than suppliers, which seems most likely related to reliability concerns.

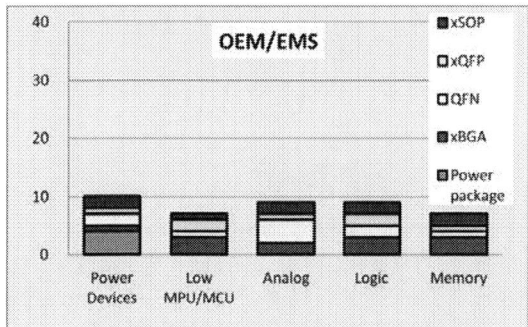

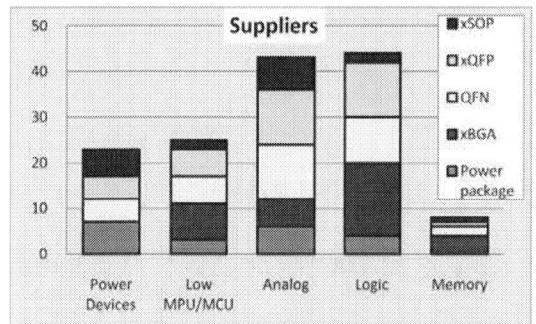

Fig. 3 Implementation Status by Package Type

4. Bond Pad Pitch & Wire Diameter

The survey asked what the smallest bond pad pitch and wire diameter were for current and future production.

Currently copper wire diameters from 1μm down to 0.7μm diameter of Cu wire are in use for pad pitchs of 40μm to 55μm. There is a trend towards use of more fine wire diameters and smaller pitches. 0.6μm wire diameter is being considered for pitches ranging from 20μm to 45μm.

5. Main Concerns:

The survey asked for information on the main concerns with Cu wire bonding.

In-service reliability and unproven historical performance were the major concerns expressed by the OEM/EMS community. Suppliers had concerns in these areas as well, but to a lesser degree. Suppliers had additional concerns about the throughput and yield.

When "in-service reliability" was selected, the major reason was a lack of use history (80%), followed by "unknown or poor correlation of JEDEC testing with in-field performance" (20%).

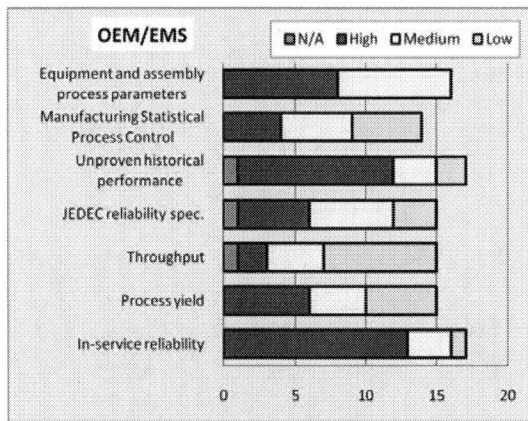

Fig. 4 Concerns with Cu Wire Bonding

6. Failure Analysis (FA) Techniques

The survey asked whether there are any issues for FA techniques.

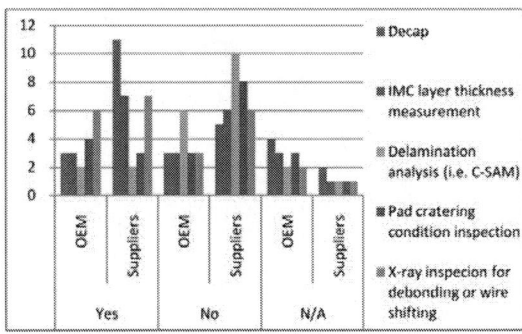

Fig.5 Concerns with Failure Analysis Techniques

Suppliers indicated that Decap, IMC thickness measurement and X-ray inspection of debonding / wire shifting are issues. X-ray inspection for debonding / wire shifting is an issue for OEM/EMS. C-SAM analysis for delamination was not reported to be an issue for either suppliers or OEM/EMS. This data indicates that suppliers and OEM/EMS have seen challenges for FA.

7. Failure Modes

Reported failure modes were similar from both OEM/EMS and Suppliers – with Ball lift, IC damage and 2nd bond defects being the major failure modes.

OEM/EMS also cited wire break as a common failure mode.

The responses for the 'ball lift' failure mode most likely referred to ball lift defects during the bonding process, rather than interfacial cracking after accelerated testing.

8. Surface Finish

The survey asked about the preferences for both chip pad finish and interposer surface finish.

Chip Pad Finish: There was no specific preference indicated from suppliers, while OEM/EMS may not have the information required to make comments at this time.

Interposer Surface Finish: A wide range of surface finishes were suggested. There was no clear consensus amongst suppliers or OEM/EMS. For Flex / Rigid Substrate: Au finish is preferred. For Ceramic Substrate: No clear preferred finish. For Leadframe: Ag was preferred by suppliers, but OEMs had no clear preference. The variability of OEM/EMS inputs may have been limited by small sample sizes. Suppliers' inputs showed similar surface finishes to those presently used for Au wire bonding.

9. Mold Compound

The survey asked whether new specifications for mold compounds were required for Cu wire bonding.

45% of component suppliers are adopting materials with lower halogen content and more neutral pH. Others reported not special requirements for materials such as mold compound. 32% of suppliers stated no performance advantage was observed for materials with nominally better properties.

Some OEMs may not have sufficient information on package BOMs to allow them to clearly specify material property requirements.

10. Reliability Test

The survey asked what reliability tests are most difficult to pass for the current production process.

The survey results showed that suppliers see biased HAST as the most difficult reliability test to pass, and that OEM/EMS perceive biased HAST and thermal cycling as equally difficult. HAST and thermal cycling are commonly used for package qualification. OEMs often have difficulty passing these items in qualification and would be aware of such failures.

65% of respondents responded to this survey question based on actual test results, while the responses from the remaining 35% were based on what they expected would be difficult to pass.

The question also asked whether current JEDEC standard accelerated test methods were sufficient for system application requirements.

Responses showed that both suppliers and OEM/EMS have great uncertainty about the suitability of current tests. There are concerns on test duration, temperature range, and whether current JEDEC standard tests can properly detect corrosion. A majority of suppliers believe different or new tests are needed for Cu wire device qualification.

The word 'sufficient' in the question may have been interpreted differently by different companies. Some may have interpreted it as 'sufficient to generate failures', while others may have interpreted it as 'sufficient to ensure product reliability'.

SURVEY SUMMARY

The purpose of this survey was to understand key issues and concerns regarding the adoption of Cu wire bonding.

- 58 individuals from over 40 global leading companies responded to this survey, covering a wide spectrum of the industry, ranging from material suppliers to OEMs with total annual revenue of over 300B USD.
- The results showed that 20% of the OEM/EMS surveyed use Cu wire for either high volume production (HVP) or low volume production (LVP), and 55% of suppliers are in either HVP or LVP phase.
- Main concerns for OEMs about factors that would slower the implementation progress were reliability and unproven historical performance Suppliers had similar concerns, but to a lesser degree.
- Establishing appropriate failure analysis techniques is challenging Decap, IMC thickness measurement and X-ray inspection are most difficult ones based on suppliers' input. OEM/EMS mentioned that X-ray inspection for debonding/ wire shifting is an issue.
- Biased HAST was rated as the most difficult reliability test to pass. Test duration, temperature range, and whether current JEDEC standard can properly detect corrosion were the major concerns.
- Overall, there was strong consensus that the reliability performance of Cu wire bonding needed to be further evaluated with experimental work.

NEXT STEP

iNEMI is now initiating the 2nd phase of the Cu wire bonding reliability project. Experiments have been designed to investigate the technical issues highlighted in the survey results. Phase 2 will:

- Assess reliability performance of components with Cu bond wires and compare to components with Au bond wires. 3 bonding wire types will be studied: pure Cu and Pd coated Cu, vs. an Au wire control.
- Identify key packaging material properties that impact reliability performance, and provide guidelines on packaging material selection. 2 mold compounds and 3 surface finishes will be evaluated on 2 package types (BGA and leadframe package).
- Evaluate results from several temperature-humidity level combinations and attempt to derive a lifetime model based on the results from these tests. 5

conditions for Biased-HAST, 1 condition for HTS, 1 condition for AATC, and MSL3 are planned.
- Perform failure analysis and identify test methods and material properties that impact lifetime during such tests.
- By using a test-to-failure methodology, assess the effectiveness of standard reliability test methods and durations so as to better address reliability risks for Cu wire bonded devices.

The project phase 2 is expected to be completed in approximately 1 year from the time of project launch.

AKNOWLEGEMENT

The authors acknowledge the great contributions of Copper Wire Bonding Reliability Phase 1 project team members: Agilent, Amkor, ASE, Atotech, Celestica, Cisco, Dow Chemical, HP, Heraeus, IBM, IST, Lenovo, NMC, Plexus, and STATSChipPAC.

Use of Harsh Wire Bonding to Evaluate Various Bond Pad Structures

Stevan Hunter[1], Bryce Rasmussen[1], Troy Ruud[1], Guy Brizar[2], Daniel Vanderstraeten[2], Jose Martinez[3], Cesar Salas[3], Marco Salas[3], Steven Sheffield[3], Jason Schofield[3], Kyle Wilkins[3]

[1]ON Semiconductor, Pocatello, ID, USA; [2]ON Semiconductor, Oudenaarde, Belgium;

[3]Brigham Young University Idaho student intern.

Ph. 208-233-4690, FAX 208-234-6796, stevan.hunter@onsemi.com

Abstract

IC bond pad structures having Al metallization and SiO$_2$ dielectric have been traditionally designed with full plates in underlying metallization layers, connected by vias. In addition, pads having bond over active circuitry (BOAC) which are much more sensitive to pad cracks, are likely present in the same IC. Cracks in the pad dielectric weaken the bond reliability and may cause electrical leakage or shorts to circuitry under the pad. Cracks are more likely to occur during Cu wire bond due to higher bonding stress as compared to Au alloy wire bonding. Experimental data from bonding with 1mil Au or Cu wires reveals dramatic differences in pad robustness against cracking, depending upon the underlying metal structures and patterns. A "harsh" Au wire bond recipe is also developed to produce the stress effects of Cu wire bond in experiments without having to upgrade older bonding equipment for Cu wire. Cratering test after wire bond is used to evaluate pad cracking. Ball shear testing followed by a cratering test further reveals pad cracking tendencies. Design principles for increased pad robustness to cracking are developed based on the data. Reliability data verifies the effectiveness of the design principles. Proper design of interconnects beneath the pad can greatly increase pad robustness to cracking, allowing much more margin in bonding stress, enabling the option of Au or Cu wire bond on the same IC without pad cracking.

Key words: wirebond, circuit under pad, BOAC, Au wire, Cu wire, bond pad

Introduction

Gold (Au) wire bonding often has very little process margin because the pad structure is fragile. Wire bonder recipes are often intended to avoid issues with the pad structure, such as too much pad aluminum (Al) displacement, bond lifting, dielectric cracking, divots in the dielectric, and cratering of the silicon (Si). Additionally for circuit-under-pad (CUP) (or bond-over-active-circuitry (BOAC)) designs, deformation of Al in the interconnects below the "pad window" or subtle shifts in the semiconductor device electrical behavior should be avoided. Deformation and cracking are found together on traditional pad structures that have experienced too much stress. But deformation in the Al of CUP pads is detrimental even if there are no cracks, due to local increase in electrical resistance and degraded electromigration reliability.

Traditional pad structures are not sufficiently robust to tolerate a switch to the higher mechanical stress needed for successful copper (Cu) wirebond

without significantly increasing pad Al thickness or making other processing changes that add cost or require unacceptable design or manufacturing tradeoffs. Successful Cu bonding on highly sensitive CUP pads is a big challenge.

Bond pads we will consider have a Si substrate and 2 or more layers of Al-based metallization. Metallization includes thin titanium-nitride (TiN) barrier layers above and below the Al (actually Cu –doped Al with 0.5% Cu) conductor film. The bond pad metal is the exposed Al of the thin "metal top" (MT) layer within the pad window, where the wirebond makes contact. The pad window is surrounded by the passivation films covering the die surface. In these experiments the pad structure of interest includes all of the features enclosed by the drawn pad window, meaning everything physically within or below the pad window opening down to the Si substrate. These are the structures expected to receive highest stress during bonding on the pad Al.

The traditional bond pad structure has sheets of metal in all metal layers across the entire pad window. Tungsten (W) plug vias in the dielectric layers between the metal levels electrically connect all of the levels to the pad.

Mechanical stress from both wafer probe and wirebond can cause pad damage, with wirebond expected to exacerbate a weakness or crack already present from probing. This is a serious reliability concern, but will not be considered here; most of these experiments use unprobed pads to avoid any pre-damage. Fig.1 shows examples of cracks from harsh palladium-doped Au (AuPd with 1% Pd) wire bonding and even harsher Cu wire bonding on a traditional pad structure that does not have top vias in the pad window. "Parenthesis" shape cracks near the ball contact edge are typical, with cracks initiating perpendicular to the direction of ultrasonic vibration. A crack across the middle of the pad may also occur, often associated with concealed damage from wafer probe (not shown).

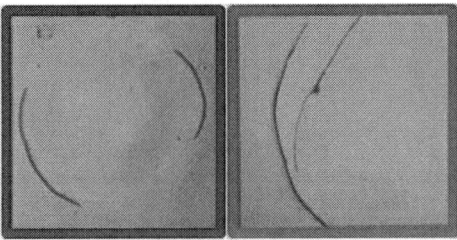

Figure 1. Pad cracks after: (left) 1 mil AuPd (1% Pd) wire ball bond, (right) 1 mil Cu wire ball bond

Experiments

This project explores the bond pad structure's robustness to cracking through the use of harsh ball bonding using Au wire, intended to simulate the higher mechanical stress which may actually occur during a Cu wirebond process. "Harsh" Au ball wirebond recipes were developed to purposely cause nearly 100% of traditional pads to crack, as observed in the usual "cratering test" (etching to remove the bond ball and pad Al, then microscope inspect for damage). Au wire diameter is 1mil, and pad dimensions are generally 75um x 75um, with pad Al thickness of 1um or less for most tests, and up to 3um for one test. A few designs were also bonded with AuPd and Cu wires to confirm that the Au wire "harsh" bonding is sufficiently similar.

Wafers were fabricated in 2-level metal up to 5-level metal processes in 4 different CMOS technologies having Al / SiO_2 interconnect. Various experimental pad structures were fabricated along with the "traditional" pads, then all were mechanically stressed by standard or harsh Au wirebond. Experimental pad structures consist of various metal layout densities within the pad window in the underlying layers, and a patterned polysilicon layer. These are to simulate circuitry under pad in all layers. Some structures include ESD protection devices beneath as well, and all can be electrically tested to detect electrical shorts from pad to underlying circuitry. For convenience in referring to the interconnect metal layers in the bond pad stack, we designate the top metal as layer "MT" or Metal-Top, layers below as Metal-Top-minus-one "MT(-1)", Metal-Top-minus-two "MT(-2)", and so forth. Some example MT(-1) patterns are shown in Fig. 2.

Figure 2. MT(-1) experimental designs, with slots and holes

Slots or holes were used to lower the pattern density of MT(-1), with more variety than those shown here. Other MT(-1) designs placed dummy metal fill in the pad window, or left out the MT(-1) completely in the pad window. Similar design variations in MT(-2) and MT(-3) were included as well.

A sample of at least 3 die were analyzed for each pad structure and bonding condition, each die typically having 30 to 40 pads bonded. Bonded pads were sample tested for bond pull strength and wire ball shear, and all pads were inspected after cratering etch. Additional data was obtained on smaller sample sizes by etching away the barrier to get a clear view of damage to underlying features, or scaning electron microscope (SEM), polished cross section SEM (XSEM), or cross section by focused ion beam (FIB). TiN barrier and top dielectric films *bending* on top of a deformed underlying metal layer is detectable by a "ripple effect" seen optically in microscope inspection, having an appearance similar to ripples on a pond. Ripple effect is easily observed (but in poor detail) at low magnifications. We were able to adjust lighting and sample tilt sufficiently to observe ripple at higher magnifications on standard microscopes, though the photos don't reveal as much detail. Ripple is best observed using a differential interference contrast (DIC) microscope. Wire pull strength test (PST) and ball shear test (BST) data is gathered on smaller sample sizes.

Harsh bonding recipes for Au ball bond produced the desired results by increasing the ultrasonic power and reducing the stage temperature. The pancake shaped bond often becomes more flattened, usually with larger diameter than targeted for production. We assume that the lower pad temperature causes less of the ultrasonic energy to be absorbed at the bonding interface or dissipated across the pad Al, causing more of the energy to transfer into the top SiO_2 and into the underlying Al of MT(-1). Later in the experimentation, a second capillary style was tried, forming a more bell shaped bond that appeared to cause even more cracking in harsh recipes due to the increased downforce and ultrasonic energy coupling at the ball's outer edge. Fig. 3 shows some example Au ball shapes resulting from harsh bonding recipes.

Figure 3. Example bond ball shapes in 1 mil Au harsh bonding recipes, (left) tall ball, (middle) flattened ball, (right) bell-shaped ball

Results

Traditional pads are highly damaged in the harsh bonding conditions, showing strong ripple and serious cracking. Bonding stress includes the dynamic downward force of the ball combined with ultrasonic vibration, with the highest stress concentrated under the ball contact edge in our harsh recipes. Extra care had to be taken in the

sample preparation and cratering etch sequence so as to not pull apart the pads by laterally stressing the wires, and not etching away all the underlying pad metal Al through the cracks causing the top SiO_2 to break off before cratering inspect.

Fig. 4 shows an FIB cut across a crack, from harsh bonding. Some MT(-1) Al was etched through the crack during crater etch. An important thing to note is the difference in MT(-1) deformed Al thickness on each side of the crack.

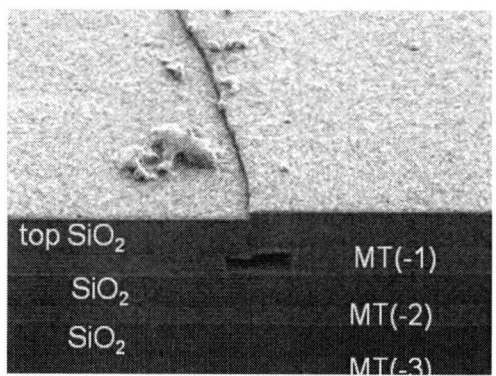

Figure 4. FIB cross section through a pad cracked in harsh bonding, after crater test.

Standard Au wire bonding typically caused 10% to 50% cracking to the weakest traditional pads (pad structures with top vias and full sheets of metal in all interconnect layers below the pad window). Without top vias under the pad window, 0% to 20% cracked pads is typical in these experiments. There was no cracking response for standard bonding in any experimental pad design. The ripple effect is always observed on bonded traditional pads, cracked or not, top vias or not, optimized or harsh bonding -- indicating that bonding consistently deforms MT(-1) Al into local "valleys" and "hills" (see Fig. 5).

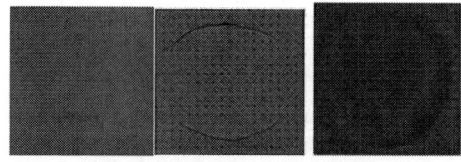

Figure 5. Standard bonding cracks in traditional pad structures: (left) no top vias, (middle) with top vias, (right) ripple example

Harsh Au wire bonding: 60% to 80% of traditional pads crack, and 90% to 100% when there is an array of top vias in the pad window. Traditional

382

pads showed strong ripple effect, whether cracked or not, whether top vias or not. Barrier lifting also occurred on some of the traditional pads with top vias. Fig. 6 shows examples of cracked pads from harsh bonding (in cratering test) and the ripple view.

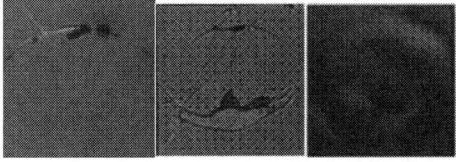

Figure 6. Harsh bonding cracks in traditional pad structures: (left) no top vias, (center) with top vias, (right) ripple example

Every experimental pad structure showed less cracking, reduced or nonexistent ripple effect, and no barrier lifting or SiO_2 divots as compared to traditional pads. No crack was found that was not accompanied by strong ripple effect in that location. Pads having greatly reduced MT(-1) pattern density improved significantly, with 0% to 13% cracking overall. Fig. 7 shows a XSEM of a standard Au ball bond on a cracked experimental pad having full sheets of metal, illustrative of the method used to analyze in detail (we can see Au-Al intermetallic (IMC) formation, pad metal displacement (PMD), bending and cracking of SiO_2 layers, and deformation of Al in metallization layers).

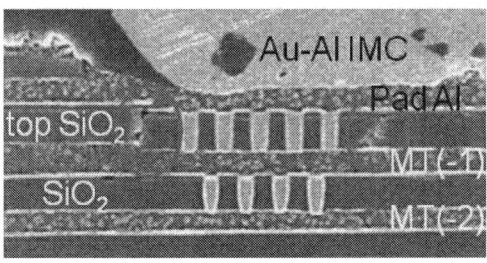

Figure 7. XSEM of cracks near the ball contact edge in an experimental pad design having full sheets in metal layers, but top vias only near the pad window. Al deformation in MT(-1) and top SiO_2 cracking occur near the ball contact edge.

Full metal sheets: Pad structures containing a full metal sheet across the pad window in any underlying layer showed ripple effect, decreasing in magnitude as the effective top dielectric thickness increased: MT(-1) full sheet pads cracked as much as the traditional pads regardless of the pattern in metals below MT(-1) with strong ripple effect observed, MT(-2) full sheet (with

patterned or absent MT(-1)) in the pad window) had less than 10% of the pads cracked and reduced ripple effect, MT(-3) full sheet (with patterned or absent MT(-1) and MT(-2) layers in the pad window) had less than 1% of the pads cracked and weak ripple effect observed. Fig. 8 shows a cross section SEM of standard bonding on an experimental pad design having full sheets of metal in all layers. Note how the crack near the ball contact edge initiates on the bottom side of the top SiO_2, where the MT(-1) Al transitions from a slight "valley" and slopes up into a "hill".

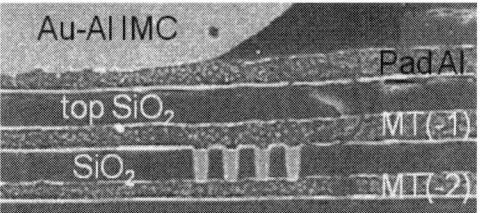

Figure 8. XSEM of a crack in a pad structure having full sheets in metal layers.

Fig. 9 shows a cross section SEM of a harshly bonded pad, showing huge deformation in the Al of MT(-1), and including cracks below MT(-1) in the SiO_2 as well. Note the drastic difference in MT(-1) Al thickness due to the mechanical stress. Sufficient stress reached the MT(-2) to deform it as well, initiating cracks in the SiO_2 above it.

Figure 9. XSEM of cracks near the ball contact edge (just out of view to the right) from harsh bonding on a traditional-style pad structure.

Fig. 10 is another example of cracks from harsh bonding occurring over MT(-1) near the ball contact edge. This experimental pad design has reduced pattern density in MT(-1) and underlying metal layers, but does have MT(-1) and top vias in the region of the ball contact edge. The cracks initiate from the bottom side of the top SiO_2. There may be a tendency for cracks to form above metal feature edges as seen here, but this has not been confirmed for harsh bonding. These cracks will not necessarily be visible in cratering test if they don't break the surface of the SiO_2 and break the TiN on top.

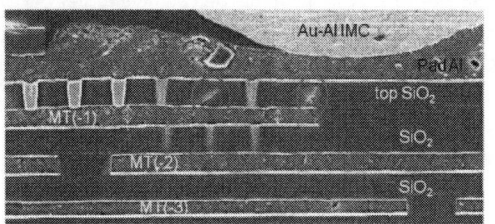

Figure 10. XSEM of cracks in an experimental pad design

The role of MT(-1): MT(-1) patterning plays the most important role in determining robustness to cracking for harsh bonding on thin pad Al, if full metal sheets are removed from underlying levels in the pad structure. Some experimental pad structures simulate CUP applications with MT(-1) patterns of 4um wide metal "busses" and 57% pattern density in the pad window. These pads showed 5% to 13% of pads cracking from harsh bonding (some of which may be attributable to a full metal sheet in an underlying layer), with slight ripple effect detectable above the MT(-1) metal busses. MT(-1) patterns with <50% uniform density across the pad had even fewer cracks and no ripple, *unless there was a full sheet of metal beneath* in MT(-2) or MT(-3), which caused both cracks and ripple. MT(-1) patterns with <50% density *and* limited maximum metal width between spaces, slots or holes consistently showed 0% cracks and no ripple. No obvious interaction of cracks and MT(-1) layout was observed.

Effects of MT(-2) and MT(-3): For pad structures having reduced density MT(-1) across the pad window, cracking is already greatly reduced, and there appears to be little dependence on MT(-2) or MT(-3) *as long as they are not full metal sheets*. The presence of MT(-2) features and vias below are beneficial in reducing cracks from harsh bonding. For pad structures having MT(-1) absent in the pad window, up to 93% pattern density in MT(-2) has been shown to be robust to cracking from harsh bonding.

For CUP applications, MT(-1) absent, and MT(-2) bus patterns of up to 4um metal width and overall pattern density of 57% is also robust.

Effects of MT(-4), poly, Si devices: Patterned MT(-4) and polysilicon layers were included in the tests, but no effect was observed. No detectable shift in Si device parameters was observed in any test.

Effect of Pad Al thickness: Harsh bonding recipes caused a high percentage of traditional pads to crack for pad Al thickness of 0.55um and 0.8um. Table 1 shows additional data from a technology having thicker pad Al options, on traditional pads. The percentage of cracked pads reduces as pad Al thickness increases.

Bond Type	1um pad Al	3um pad Al
Standard Au bond	8%	0%
Harsh Au bond	18%	2%

Table 1. % of traditional pads cracked for two pad Al thicknesses

For the experimental pads without a full sheet of metal beneath the pad window, no cracks or ripple were observed for 1um or thicker pad Al.

Other relevant tests: PST and BST data for both standard and harsh Au wire bonding indicate that pads robust to cracking perform as well or better than traditional pads in comparative tests. PST *followed by cratering test* is not usually done, because it is known that traditional pads are too fragile. For the experimental pads, PST *followed by cratering test* showed no cracking or craters. BST *followed by cratering* is not usually done because traditional pads are fragile. In these experiments, 99% of 120 traditional pads were cracked after standard bonding followed by ball shear *then cratering test*; while no experimental pads out of 697 showed cracks, divots, or craters.

Discussion

Cracking and ripple effect together help show how the application of high stress to a bond pad structure containing brittle SiO_2 dielectric film over a ductile Al film leads to cracking. This is caused by the dielectric bending in conformance to the Al valleys and hills (We ignore the thin TiN films for simplicity). Fig. 11 demonstrates a case for a harshly bonded traditional-style pad, where large tensile stress on the top side of the top SiO_2 film bending over the Al "hill" caused a crack to initiate, resulting in a divot that pulled out in the cratering etch.

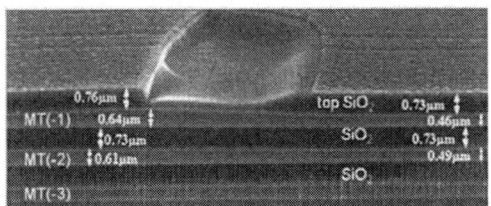

Figure 11. FIB cross section of a pad structure having full metal sheets, after harsh bonding and cratering etch. The MT(-1) "hill" is 1.4x the thickness of the neighboring "valley".

The presence of top vias further weakens the SiO_2, creating the weakest pad structure of any that were tested. It becomes clear that such cracks may be prevented if the SiO_2 doesn't bend significantly. This may be accomplished by lowering the pattern density, especially in MT(-1) and limiting the width of SiO_2 above MT(-1) features (in other words, limit the MT(-1) width between spaces, slots or holes) in the region of high bonding stress, and eliminating the use of full sheets of metal in underlying levels of the pad structure. Increasing pad Al thickness may be used to greatly reduce cracking, but this is not a complete solution for harsh bonding.

Bond pads can be redesigned for improved robustness to cracking, replacing traditional pads, and facilitating a switch to Cu wirebond without the requirement of thick MT. Cu wire bond replacing Au not only reduces cost, but lower electrical resistance and stiffer wire behavior during packaging are significant advantages. Also, Cu's much lower rate of intermetallic formation at the bond interface can lead to improved reliability as compared to Au wirebond at higher operating temperature.

CUP pads can be designed by following the simple principles of reduced metal pattern density and limited metal width between spaces, slots, or holes. This will facilitate successful design of free-form CUP circuitry in all interconnect layers beneath the pad window to produce a robust pad structure, and enables the concept of Cu wire bond on CUP pads without the need for thick MT.

Additional positive results were later obtained using actual AuPd wirebond (more bonding stress in this recipe than for Au) and Cu wirebond (even more bonding stress). Figure 12 shows the cratering test cracking data for sets of 50 die of

each selected pad design, bonded with AuPd wire or with Cu wire. The pads most robust in harsh Au bonding are the same designs that showed no cracking in AuPd and Cu bonding tests.

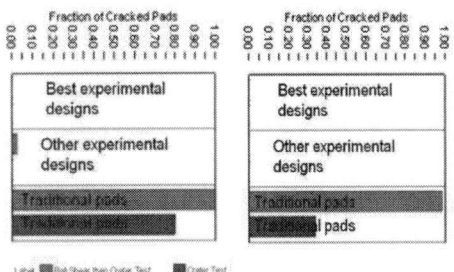

Figure 12. (left) 1mil AuPd wirebond results, (right) 1mil Cu wirebond results; including crater test after ball shear.

PST and BST testing were done to compare 240 "robust" pads to 60 traditional pads ball bonded with 1mil AuPd wire. All samples passed, with no statistical difference in values observed. PST and BST testing was also done to compare 360 "robust" pads to 60 traditional pads ball bonded with 1mil Cu wire. All samples passed, with no statistical difference in values observed.

Reliability Test Results

Reliability testing was first done for one lot of Au wire bonded parts in a technology having 0.55um pad Al thickness, comparing traditional pads (with and without top vias) to a pad design having 4um wide parallel metal busses and 57% pattern density in MT(-1). Harsh bonding this time was with 1.2mil Au wire, which caused cracking on many traditional pads but not on the experimental pads. Assembly was in 20-pin SOIC plastic packages. Though parts with cracks were stressed in parallel with parts having no cracks, no electrical changes in leakage or shorting were detected over 1khrs high temperature operating life (HTOL), 2000hrs high temperature storage life (HTSL), 2000cyc temperature cycling (TC), or 200cyc thermal shock (TS) -55C to 150C. As well, special tests were done such as TC first, then HTSL, and vice versa, all with passing results and no degradation found. A few plastic packages began to come apart, causing invalid "opens" fails after 1500cyc TC, verifying that the stress was indeed harsh. One possible conclusion is that cracks already present from bonding do not appear to propagate further or at least don't cause new fails. It is clear that standard reliability stress tests don't seem effective in propagating or detecting cracks in pads, and thus it may be assumed that products with cracked pads might be commonplace…

Based on the success of "robust pads" in every result, two new CUP pad designs were fabricated in another technology for reliability stressing: 4-level metal, with 0.8um pad Al thickness. One design with MT(-1) absent under the pad window, has 4um wide metal busses with 53% overall metal density in MT(-2), including circuitry in MT(-3), various electrical nodes, and full ESD protection circuitry beneath. The other design added MT(-1) circuitry to the previous design, with an overall MT(-1) pattern density of 43%, no top vias, but adding some via connections to the appropriate MT(-2) nodes from MT(-1).

These CUP pads were subjected to a barrage of harsh stresses, including harsh probing, harsh Au wire bonding, harsh AuPd bonding, and Cu bonding, again confirming previous results with no issues found. Reliability testing was done after assembly in 20-pin SOIC plastic packages, with parallel testing of 300 parts with 1mil Au wire and 300 parts with 1mil Cu wire bonding. 80 parts with each wire type were stressed in Biased HAST, HTSL, and TC. Standard and extended reliability testing was completed, followed by PST, BST, and cratering tests. No reliability fails or issues were discovered (see Table 2). Of special interest in this case was the PST, BST, and cratering test results comparing after stress and at various read points with the initial test values. No degradation in performance and no issues of any kind were discovered in these additional tests.

Rel Test	Stress	Duration	Result
MSL2	Moisture preconditioning, 3 x 260C		PASS
Biased HAST	130degC 85% RH,	215hrs	PASS
HTSL	175degC	1000hrs	PASS
TC	-65C to 150C, air to air	2000x	PASS
PST	Bond pull strength, 5 die	Initial, 2 read points, and after stress	PASS
BST	Wire Ball Shear, 2 die	Initial, 2 read points, and after stress	PASS
Cratering	Etch away ball and Al, all pads	Initial, after stress, and after each PST and BST	PASS

Table 2. Rel test data for two CUP designs, including Au and Cu wirebond

Our confidence in standard reliability test stressing to detect the presence of cracks is low. "Before stress" and "after stress" PST, BST, and cratering tests and follow up analyses are helpful to learn about cracks and weak bond pads.

Conclusion

Harsh bonding experiments on various pad structures, analyzed by cratering test, aid in our understanding of the pad cracking mechanism and how to prevent cracks. Traditional pad designs having full sheets of metal under the pad window, and especially those with top vias, are the weakest in terms of pad cracking from bonding tests. The importance of underlying Al-based metal structures, their pattern density, and limiting the metal width between spaces, slots or holes is shown. These principles can easily be used in both successful pad design and successful CUP pad design. "Robust" pad designs are demonstrated to be resistant to cracking and other damage from harsh bonding. Ripple effect exists for all cracks found, and is present whenever there is a sufficiently wide metal feature in the pad structure. Ripple effect is not observed in robust pads. Sample CUP pad designs, having circuitry in all interconnect levels below the pad window and Si devices beneath, were reliability tested, demonstrating no pad cracking, and positive results overall for both Au and Cu wirebond on thin pad Al.

Acknowledgements

We recognize the significant contributions of the engineers and technicians of the Pocatello Prototype Operation, and Materials Analysis Lab, especially Lynda Pierson; also reliability test engineers and technicians in ON Calamba, Philippines, and assembly engineers and technicians in ON Carmona, Philippines. Special thanks to Jim Workman and Craig Christensen.

Flipchip bonding of thin Si dies onto PET foils: possibilities and applications

Jeroen van den Brand [1], Roel Kusters [1], Maarten Cauwe [2], Daan van den Ende [3], Muge Erinc [3]

[1] Holst Centre/TNO, HTC31, Postbus 8550, 5605 KN Eindhoven, the Netherlands
[2] Imec Gent, Technologiepark Zwijnaarde, Grote Steenweg Noord, B-9052 Zwijnaarde, Belgium
[3] TNO Science and Industry, De Rondom 1, Eindhoven, the Netherlands

Abstract

Low cost large area flexible electronic products are expected to be used in a wide range of applications and in large quantities in our society. Examples of this include sensor packages added to food or conformal intelligent patches that monitor a patient's well-being. Because of their large area, the preferred substrate material for these applications will be low cost materials like polyesters (PEN/PET). Intelligence or communicative capabilities are preferably added to these devices by integrating the chips directly on the low cost foil itself. To maintain the flexibility of the package and not to add too much to the thickness, the Si chip needs to be integrated into the product as a bare, thinned die. Flip chip bonding is currently the most mature, widely available technology to integrate these thin chips. The low temperature stability of the PET foils however puts serious constraints on the materials and the process. The current paper specifically addresses the challenges associated with this. Initial results from a finite element model will be discussed. The model is being developed to understand the influence of the bonding process and material parameters on the final stresses and warpage of the chip. Additionally, lifetime and flexural test results will be discussed of ultrathin chips bonded on Cu and Ag-based screen printed circuitry. Finally, some applications of the technology will be shown: a microcontroller integrated on a Cu-PET foil and a supply chain monitoring tag.

Key words: thin chip, flip chip, PET, large area electronics

Introduction

A new class of low cost, large area electronics is starting to emerge. Examples include cheap sensor packages attached to food packaging to measure the ripeness of food, smart patches that monitor the patient well-being and smart active or passive RFID tags.

The substrate material will not be polyimide as commonly used in flexible electronics. By using polyesters like PEN (polyethylene naphtalate) or PET (polyethylene teraphtalate), the substrate costs can be reduced by a factor of 5-10 [1]. A disadvantage of polyesters is however that they are considerably less thermally stable. PEN has a glass transition temperature (T_g) of ~130 °C and PET of ~85 °C while poly(imide) has a Tg of ~350 °C) [1]. This limitation in thermal stability excludes many well established processes for making electronic products and/or renders the use of existing processes much more challenging.

Printing technologies will play a key and pivotal role in manufacturing of these large area low cost electronics devices. Despite the fact that also quite advanced functionality can be made with printing, it will still for some time be needed to integrate the intelligence and communication capabilities of the device as a Si-based chip. To keep the flexibility of the device and not to add too much to its thickness, this chip will need to be integrated

as a bare, thinned Si chip. Because of the low temperature stability of polyesters it is necessary to interconnect the chip using an adhesive.

A wide variety of different types of conductive circuitry types are being considered, developed and evaluated for these devices [2-8]. The circuitry can be printed using technologies like screen printing or inkjet printing by using (mostly Ag-based) conductive inks. Of these technologies, screen printing is likely the most mature. The technology is however limited in resolution to a pitch of around 150 µm. The circuitry can also be made using more traditional technologies like etching or plating of Cu through photolithographic processing. This has the advantage over printing that much finer line widths and spacings are possible and additionally that a huge installed base of equipment is already available.

The current paper specifically addresses the interconnection of ultrathin Si chips on both screen printed and etched Cu circuitry on PET foils using anisotropic conductive adhesives. Flip chip technology is chosen as the bonding method because it is existing (for example used in passive RFID tags) and has a large installed base of manufacturing equipment.

A finite element model has been developed which allows studying the influence of the various bonding parameters and material properties on the

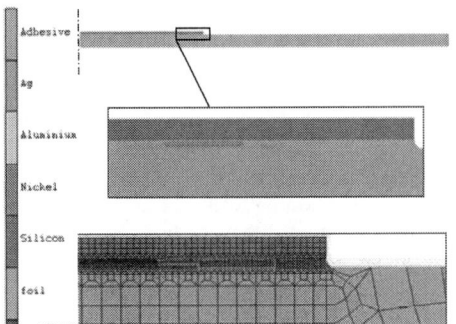

Figure 1. Cross section of modeled geometry.

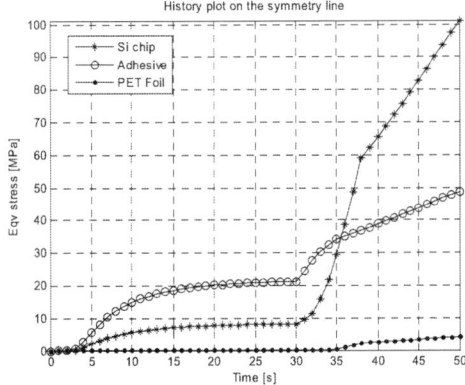

Figure 2. Stress evaluation during bonding. Stresses given as equivalent von Mises stresses. At t=0s, a temperature of 120 C is applied to cure the adhesive. At t=30s, the heating is switched off and the system is cooled down to room temperature until t=50s.

stresses in the system and the warpage (i.e. out-of-plane bending) of the ultrathin chip due to the bonding process. First results of this model will be discussed. Furthermore, it will be shown that flip chip bonding of ultrathin chips on low cost PET foils can be done reliably: a good lifetime and flexural stability has been achieved. Finally, it will be shown that the technology is ready to be used for actual applications in the field of large area low cost flexible electronics. Two examples will be shown of actual applications of the technology. The first example is a fine pitch bare chip Texas Instruments MSP430 microcontroller integrated onto a Cu-PET foil. The second example is a smart RFID food monitor tag which is being developed in the framework of the Catrene project Pasteur [9]: a sensor chip and an RFID chip on a PET foil with screen printed circuitry and battery.

Experimental

The PET foil that was used for the screen printed circuitry was a 120 μm thick thermally stabilized foil from Agfa. The conductive structures were printed using Dupont 5025 Ag-filled screen printing paste on a DEK Horizon 03i screen printer. Curing of the paste was performed in a conventional oven

by heating the substrates for a period of 20 minutes at 120 °C.

The PET-Cu foils that were used for the fine pitch circuitry were obtained from Hanita Coatings. They consist of a 12 μm thick Cu layer on a 50 μm thick thermally stabilized PET foil. Etching of the Cu was performed using traditional photolithographic processing. First a mild acid cleaning and micro etching was performed. Then, a dry film resist was laminated, the substrates were exposed (7s at 10 mW/cm2), developed and spray etched in CuCl₂. Finally, stripping using NaOH was performed and anti-tarnish was applied.

Bonding on the screen printed circuitry PET foils was investigated using a 2 x 2 mm IZM28 daisy chain test chip having 20 IO's, a pitch of 300 μm and a thickness of 20 μm. Bonding on the Cu circuitry PET foils was investigated using a 5 x 5 mm IZM42 daisy chain test chip having 168 IO's, a pitch of 50 μm and a thickness of 20 μm.

For both circuitry types, a dedicated fan-out circuitry was made which allowed accurate 4 point (4p) measurements to be performed so as to obtain only information on the bump-circuitry contact resistance.

The flip chip bonding was performed using a Dr. Tresky T3200 semi-automatic bonder. First, the anisotropic conductive (Delo AC163) adhesive was dispensed on the foil at the bonding position of the chip. Then, the preheated teflon-coated bonding tool with the attached chip was aligned and pressed down with the appropriate force onto the foil. Bonding of the daisy chain test chips was performed at a bonding temperature of 120 °C (as measured in the adhesive) and a bonding time of 30 seconds. For the bonding on the screen printed circuitry a bonding force of 1 N was used. Bonding of the larger daisy chain test chip on the Cu-PET foil was done at a slightly higher bonding force of 2 N.

Results and discussion

Finite element modeling of bonding process

Bonding of the chip is done at an elevated temperature and under pressure. As a result of the bonding process, stresses will be introduced into the system. Si chips having a thickness below 25 μm are flexible. As a result, these stresses can cause warpage (i.e. out of plane bending) of the thin chip. A too high warpage should be prevented as it can lead to damage of the internal structures in the chip.

A generic finite element model has been developed which can simulate the full chip bonding process so as to better understand the influence of the various bonding parameters and material properties on the internal stresses and the warpage of the chip.

The geometry, boundary conditions and material parameters of bonding the IZM28 test chip on a PET foil with screen printed circuitry have been implemented into the model, see the Experimental section. Figure 1 shows a cross section of the model.

Curing of the adhesive was modeled using the cure kinetics model from Kamal and Sourour [10] where the different parameters have been fitted on experimental data obtained from the adhesive.

Figure 2 shows the stresses in the system (shown as equivalent von Mises stresses) in each layer (silicon chip, adhesive layer and the foil) as a function of time, starting from the moment the temperature is applied during chip bonding (t=0). Nodes have been selected for each layer from the middle of the cross section.

Initially, the system is assumed to be in a stress-free state. When the temperature is applied at t=0, the adhesive is in a liquid state. As a result, the foil and the chip can expand freely with respect to each other due to heating. They will do so differently: the Si chip has a low CTE of 3 ppm/°C while the PET foil has a much higher CTE (CTE < T_g: 25 ppm/°C, CTE > Tg: 60 ppm/°C, Tg=85 °C)

During curing, the adhesive will transform from a liquid to a solid state, thereby fixing the chip and the foil to each other. Figure 2 shows that the stress levels in the chip and adhesive rise due to the cure shrinkage. The foil will absorb the stresses by deformation due to its lowered modulus at elevated temperature.

At t=30 seconds, the heating is stopped and the system is cooled down to room temperature until t=50 seconds. As a result of the cooling, the Si chip, the adhesive and the foil will shrink. The foil and the chip can however no longer do so freely with respect to each other as they are fixed through the cured adhesive. The stresses rapidly increase, see Figure 2. This causes out-of-plane bending of both the foil and the chip. For the considered bonding conditions, the predicted chip warpage was around 10 μm (measured as the vertical upward movement of the chip center with respect to the edge of the chip). This value is acceptable [11, 12] and corresponds well with measurement values obtained earlier for this chip on this foil type [13].

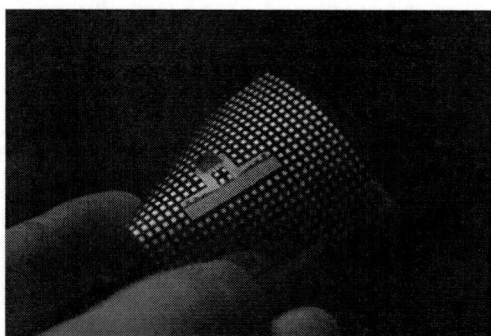

Figure 3. Photograph of 20 μm thick, 50 μm pitch daisy chain test chip bonded on Cu-PET foil

Further investigations have been performed to study how the different material and bonding parameters have an influence on the stresses in the system and warpage of the chip. This will be discussed in detail in a future paper. The results indicate that the used bonding *temperature* is the main factor that determines the stresses in the system and thus also the final chip warpage. Based on this, it can be concluded that for these low cost foils the bonding temperature should be kept as low as possible.

Initial bonding test results

Flip chip bonding experiments of the 20 μm thick daisy chain test chips were performed on both the Ag-based screen printed and on the Cu circuitry PET foils, see the Experimental section. A bonding temperature of 120 °C was used, which is the lowest possible temperature for this adhesive. For both types of circuitry a good and reproducible low contact resistance could be achieved. For the screen printed circuitry, an average bump-circuitry resistance of 63.5 ± 2.9 mOhm was obtained while for the Cu circuitry a comparable contact resistance of 64.0 ± 3.7 mOhm was obtained. Each value is the average of 8 bonded chips, having 8 measurable 4p contacts per chip. As an illustration, Figure 3 shows a photograph of the 20 μm thick, 50 μm pitch daisy chain test chip, bonded on a PET-Cu foil.

Lifetime testing

Accelerated Humidity Testing (AHT, 60 °C/90% RH) and temperature shock testing (-40 °C to +120 °C, 30 minutes cycle time) were performed to evaluate the lifetime stability of the bonds.

A total of 8 chips, each having 8 measurable contacts were prepared for each circuitry type and for each test. A protective cover layer or coating was not applied onto the chips, thus mimicking a worst case scenario for the humidity testing. The obtained humidity test results are shown in Figure 4 for the screen printed circuitry. Very similar results were obtained for the Cu-based circuitry. These results are therefore not shown.

Plotted are both the average circuitry-bump resistance (4p, Ohm) with the corresponding standard deviation (left axis) and the number of open contacts (right axis).

The accelerated humidity test results (Figure 4 top) show stable resistances up until around 100 hours of exposure. After this, the average resistance and the spread can be seen to increase. This increase is caused by a limited number of the contacts. At 500 hours of exposure, specifically these contacts showed failure. A microscope investigation showed evidence of corrosion of the bonding pads at the failed contacts.

The other contacts showed a stable resistance and a low spread for the remainder of the exposure period.

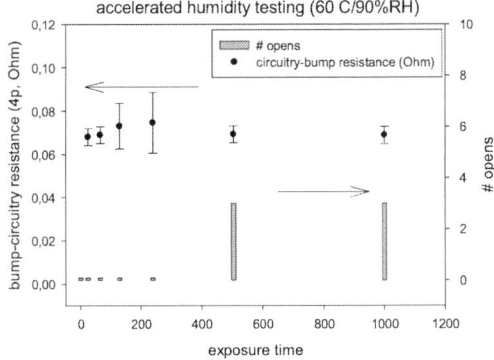

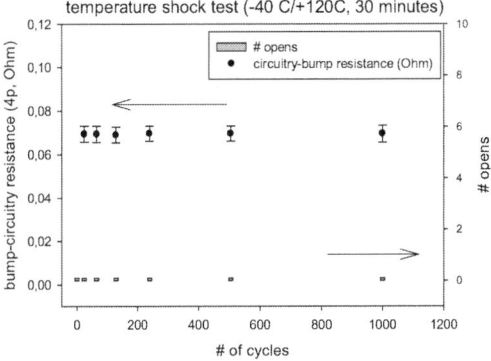

Figure 4. Top: accelerated humidity test results (AHT), Bottom: temperature shock test results (TST)

Overall, it can be concluded that the AHT are reasonably good. Further improvement of the results can be achieved by covering the chip by a protective coating or layer so as to protect it from moist. This is current being evaluated.

The temperature shock tests (-40 °C to 120 °C, 30 minutes cycle time), see Figure 4 bottom, were performed to evaluate thermal stability. The obtained results are very good: up to 1000 cycles no significant increase in resistance is observed and none of the contacts failed. This despite the fact that the T_g of the PET is traversed in each cycle.

Flexural testing

When a thin layer system is bended, it will experience tension (elongation) at its outer surface and compression (shortening) at its inner surface. Along the cross section, the strain gradually changes from compression to elongation. For pure bending, there is a position where the system neither experiences tension nor elongation. This position is denoted the 'neutral line' or 'neutral plane'. The exact position along the cross section strongly depends on the moduli and thicknesses of the different layers in the system. Because thin chips are brittle, they are preferably placed at this neutral line.

A surface mounted thin chip, like considered in the current work, will not be located at the neutral line of the system. For maximum flexibility of the

system, it will therefore be preferred to cover the chip with an additional layer so as to move the neutral line to the position of the chip.

Finite element modeling and analytical calculations were performed to study the stresses that occur upon bending. The modeled system consisted of a base foil of a given thickness with an ultrathin Si chip mounted on top using a thin layer of adhesive. This system was virtually bended to a given radius by applying three point bending conditions. The same system but with a laminated top foil, having the same thickness as the base foil, was additionally modeled. The chip is then placed in the center of the laminate and should thus experience less strain, thereby improving the flexural reliability.

The results of the calculations are shown in Figure 5. Plotted is the maximum equivalent von Mises stress in the chip as a function of the reciprocal of the bending radius. Results are shown for both a 50 and a 120 μm thick base PET foil with and without a laminated cover PET foil of the same thickness.

The results show that for the 120 μm thick base foil, the stresses are roughly decreased by a factor of 2 as a result of covering it with an additional layer. For the 50 μm thick base foil, the differences are considerable less: a roughly 20% decrease in stress is observed as a result of laminating a cover foil. This difference can be explained by the fact that the thinner the base foil and the cover foil, the larger the contribution of the chip thickness to the total stack will be. In any case, the calculations clearly show that it is advisable to apply a covering layer on a surface mounted chip as it gives a significant decrease in the maximum stresses that the chip will experience.

Experiments were performed to also practically study the influence of the cover layer. Samples were prepared where 8 ultrathin chips were mounted on a 120 μm thick PET foil. Mounting was performed with the grinding and polishing lines on the chip both parallel and anti-parallel to the planned bending

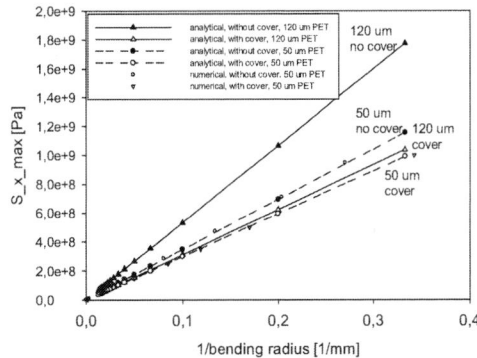

Figure 5. Analytical and numerical results of 3 point bending of an ultrathin chip bonded onto different PET foils, with and without a cover foil.

391

direction. Also another set of the same samples was prepared where a 120 μm cover foil was laminated. Subsequently, the samples were bended down to a diameter of 25 and 10 mm by using a custom-built flex tester.

Both types (with and without cover layer) could survive bending over a diameter of 25 mm. Also continued bending for up to 500 cycles resulted in no broken chips. The 10 mm bending diameter was more critical. The sample without a cover layer did not survive this diameter: 100% of the chips were broken already after the first cycle. The sample with a cover layer could survive this bending diameter with no broken chips for a limited number of cycles. Continued exposure up to 500 cycles however also here resulted in some broken chips where especially those mounted with the polishing marks perpendicular to the bending direction showed failure. This indicates that fatigue plays a role with the polishing marks acting as stress initiators.

Application of the technology

It has been shown above that bonding of ultrathin Si chips on low cost PET foils is feasible and results in

Figure 6. Top: MSP430F1611 bare die microcontroller bonded on Cu-PET foil. Bottom: bottom view microscope photograph of chip bonded to foil (looking through foil)

a good lifetime and flexural reliability. Currently, the process is being applied in several applications that are under development at Holst Centre. As an illustration, two of these applications will be discussed.

MSP430 microcontroller integrated onto a PET foil

Many applications require a microcontroller to be integrated to allow data processing / analysis or driving of the system. A commonly used, general purpose, low power and low cost microcontroller is the Texas Instruments MSP430. The subtype MSP430F1611 was selected for integration on a PET foil. This 16 bits microcontroller runs at 8 MHz and has for example on-board memory, clock and AD convertors. The size of the chip is 5 x 5 mm, it has 64 IO's, a pad size of 75 μm and a minimum pitch of 100 μm. Thinning and electroless Ni-Au bumping of the chips was performed in-house.

Integration of the microcontroller on a PET-Cu foil was performed successfully using the adhesive and settings discussed above. The microcontroller showed full functionality, also for a prolonged period. As an illustration, Figure 6, top side shows a photograph of the microcontroller driving a bare die 50 μm thick LED, which was mounted onto one of the IO's of the chip using an isotropic conductive adhesive. Figure 6, bottom shows a bottom view microscope photograph, looking through the foil onto the internal circuitry of the microcontroller. The microscope photograph shows a good alignment of the chip with respect to the circuitry on the foil. Currently work is being performed to also integrate a radio chip onto a PET foil.

Food monitoring tag

A good application example of a future low cost smart RFID tag is being developed in the Catrene project Pasteur which is led by NXP [9]. In this project, a wireless sensor platform is being developed to monitor the environmental conditions of perishable goods in the supply chain between producer and consumer.

The platform is based on a CMOS-based sensor chip which can measure different gases (O_2, CO_2, ethylene) but also temperature and humidity. By adding an active tag (e.g. including a battery) with this sensor chip to packaging solutions (crates, containers, boxes etc.) the quality of the product can be guaranteed more effectively throughout the supply chain.

The active RFID tag in its current form is shown in Figure 7. It consists of a PET foil with screen printed circuitry, antenna and battery (left side, not present in photograph).

Two Si chips need to be mounted on the PET foil: an RFID chip and the sensor chip. Again, flip chip attachment was used for both chips. For the RFID chip, the exact process, adhesive and conditions as discussed above have been used. The

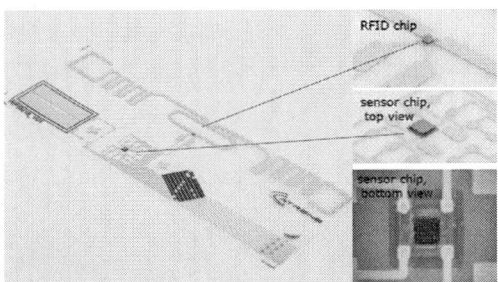

Figure 7. Photograph of smart RFID tag being developed in Catrene project Pasteur [9]

sensor chip has its functionality on the bottom side. Thus, when mounted face down, a cavity needs to be present in the foil to allow passage of gases. The assembly procedure that was used is as follows. First, a cavity was laser-machined in the foil using a 355 nm Nd:YAG laser source. Then, the anisotropic conductive adhesive was carefully dispensed around the perimeter of the cavity. Finally, the chip was flip chip mounted. The sensor chip attached to the PET foil is shown in Figure 6, bottom right hand side. Both chips were mounted successfully and showed appropriate functionality.

Conclusion

Flip chip bonding of ultrathin Si chips on low cost PET foils has been investigated on two different types of low cost electronic circuitries. Based on the experimental results it can be concluded that for both types of circuitries, a reproducible, low contact resistance and a good lifetime can be achieved. Flexural test experiments and mechanical calculations show that for small bending radii it is preferred to cover the chip with an additional layer as this can lead to significant decrease of the stress level upon bending. A finite element model has been developed to understand how the different material properties and bonding parameters influence the stresses in the system and the warpage of the ultrathin chip. First results indicate that the main stresses in the system are introduced upon cooling down of the system. Additionally, mainly the used curing temperature determines the stresses in the system and thus warpage of the chip. Finally, two examples have been shown where the developed technology has been used successfully: a bare die microcontroller and a food monitoring tag.

References

[1] Fjelstad, J.: 'Flexible Circuit Technology' (BR Publishing Inc, 2007, 3rd edn. 2007)

[2] van Osch, T.H.J., Perelaer, J., de Laat, A.W.M., and Schubert, U.S.: 'Inkjet printing of narrow conductive tracks on untreated polymeric substrates', Adv. Mater., 2008, 20, (2), pp. 343-+

[3] Ko, S.H., Pan, H., Grigoropoulos, C.P., Luscombe, C.K., Frechet, J.M.J., and Poulikakos, D.: 'All-inkjet-printed flexible electronics fabrication on a polymer substrate by low-temperature high-resolution selective laser sintering of metal nanoparticles', Nanotechnology, 2007, 18, (34), pp. 8

[4] Wang, J.Z., Zheng, Z.H., Li, H.W., Huck, W.T.S., and Sirringhaus, H.: 'Dewetting of conducting polymer inkjet droplets on patterned surfaces', Nat. Mater., 2004, 3, (3), pp. 171-176

[5] Chung, J.W., Ko, S.W., Bieri, N.R., Grigoropoulos, C.P., and Poulikakos, D.: 'Conductor microstructures by laser curing of printed gold nanoparticle ink', Applied Physics Letters, 2004, 84, (5), pp. 801-803

[6] Sankir, N.D., and Claus, R.O.: 'An alternative method for selective metal deposition onto flexible materials', J. Mater. Process. Technol., 2008, 196, (1-3), pp. 155-159

[7] Pique, A., Mathews, S.A., Pratap, B., Auyeung, R.C.Y., Karns, B.J., and Lakeou, S.: 'Embedding electronic circuits by laser direct-write', Microelectron. Eng., 2006, 83, (11-12), pp. 2527-2533

[8] van den Brand, J., Kusters, R., Barink, M., and Dietzel, A.: 'Flexible embedded circuitry: A novel process for high density, cost effective electronics', Microelectron. Eng., 87, (10), pp. 1861-1867

[9] 'Catrene project Pasteur'

[10]Sourour, S., and Kamal, M.R.: 'Differential Scanning Calorimetry of epoxy cure - isothermal cure kinetics', Thermochim. Acta, 1976, 14, (1-2), pp. 41-59

[11] Banda, C., Johnson, R.W., Zhang, T., Hou, Z.W., and Charles, H.K.: 'Flip chip assembly of thinned silicon die on flex substrates', IEEE Trans. Electron. Packag. Manuf., 2008, 31, (1), pp. 1-8

[12] Wu, A.T., Tsai, C.Y., Kao, C.L., Shih, M.K., Lai, Y.S., Lee, H.Y., and Ku, C.S.: 'In Situ Measurements of Thermal and Electrical Effects of Strain in Flip-Chip Silicon Dies Using Synchrotron Radiation X-rays', J. Electron. Mater., 2009, 38, (11), pp. 2308-2313

[13] Brand, J.v.d., Kusters, R., Heeren, M., Remoortere, B.v., and Dietzel, A.: 'Flipchip bonding of ultrahin Si dies onto PEN/PET substrates with low cost circuitry'. Proc. ESTC 2010, Berlin2010

Effects of different combinations of environmental tests on the reliability of UHF RFID tags

Kirsi Saarinen, Laura Frisk and Leena Ukkonen

Department of Electronics, Tampere University of Technology
P.O. Box 692, FIN-33101 Tampere, Finland

Phone +358 40 849 0613
Fax +358 3 3115 3394
Email address: kirsi.saarinen@tut.fi

Abstract

Accelerated environmental tests can be used to study the effects of environmental stresses on reliability. Typically environmental tests are used in parallel so that only one test is performed for test samples and new test samples are used in another test. However, different tests one after another for the same test samples may describe the operational environment better and give more reliable results in a short time. In addition, the different stresses may accelerate the effects of another stress, and such behaviour can be perceived if combinations of environmental tests are used. In this study, passive ultra high frequency (UHF) radio frequency identification (RFID) tags were tested with different combinations of environmental tests. Some of the tags were kept first in a bending test, in a constant humidity test or in a temperature cycling test, and then subjected to a humidity cycling test. In addition, as a reference some of the tags were tested only in the humidity cycling test. Changes in the performance parameters of the RFID tags were examined during testing. The tags which were held first in the constant humidity test failed significantly faster than the other test combinations. Consequently, on the basis of this study, it is important to study the effect of sequential environmental tests on the same test samples.

Key words: Environmental testing, radio frequency identification, reliability

1. Introduction

Radio frequency identification (RFID) is an emerging technology in the field of identification and security [1]. With RFID tags it is possible to identify objects individually and reliably [2]. An RFID system consists of a reader and a tag. The reader transmits radio wave signals to the tag, which replies by backscattering the modulated signal. [2,3] The tag consists of an antenna which receives and backscatters radio frequency signals and of a microchip which controls the operation. Passive tags take their operating energy from the radio waves emitted by a reader and have no internal energy source. [3]

RFID tags can be effectively used in various applications. For example, they can be used to identify and track products during manufacturing, distribution, and shipment of a device. [4] Ultra high frequency (UHF) passive RFID tags are especially suitable for these purposes because they are inexpensive, compact, mechanically robust, and their read range is several meters [3]. Tags are used successfully in various industries, such as aeronautics, automotive and medicine [5], [6]. Additionally, there is potential to use them even more widely and diversely [6]. For example, the tags

may be used as sensors which detect the environmental conditions around them [5].

Due to their numerous different applications, RFID tags may be exposed to various environmental conditions during their lifetime. Different stresses from the environment may affect the reliability of a tag. The environmental stresses may, for example, include high temperature or humidity levels, changes in the temperature or humidity, and mechanical vibration or bending [7]. Accelerated environmental tests can be used to study the effects of such environmental stresses on the reliability of tags. [8]

Typically the environmental tests are ran in parallel, so that only one test is performed for the test samples and new test samples are used if another test is conducted. However, RFID tags may be exposed to various stresses in their operational environment. Testing the stresses separately may not yield reliable data, and thus different tests one after another for the same test samples may describe the operational environment better. In addition, the different stresses may accelerate the effects of another stress, and this kind of behaviour can be perceived if combinations of the environmental tests are used. By studying the combinatory effects of different stresses, environmental test methods giving more reliable results in a short time may be developed.

In this study, a passive UHF RFID tag was tested with different combinations of environmental tests. Parts of the tags were first held in a bending test, in a constant humidity test or in a temperature cycling test, and after that they were subjected to a humidity cycling test. In addition, as a reference some of the tags were subjected only to a humidity cycling test. Changes in the performance parameters of the RFID tags were examined during testing to ascertain differences between the different combinations of the environmental tests.

2. Experimental

2.1 Test samples

The test tag was a passive UHF RFID tag with a dipole antenna. The substrate was a thin flexible polyethylene terephtalate (PET) with aluminium wiring on top of it. The wiring was attached on the substrate with a thin adhesive layer. The thickness of the substrate was 50μm and the thickness of the wiring was 9μm. The antenna was a dipole antenna length approximately 7 cm. An illustration of the tag is presented in Figure 1.

Figure 1: Test tag

The chip in the middle of the tags was a small radio frequency integrated circuit (RFIC). The size of the chip was about 590μm x 590μm and its thickness was 150μm. In each corner of the chip was a nickel bump. The sizes of the nickel bumps were 76μm x 76μm and their heights were 15-25μm. The RFIC was attached on the antenna with a commercially available epoxy-based anisotropic conductive paste (ACP). The conductive particles in the ACP were nickel particles average 1.2μm in diameter. Bonding parameters were chosen according to the recommendations of the ACP manufacturer for this kind of structure.

2.2 Test methods

The test tags were tested with different combinations of environmental tests. Three different combinations of tests were made and 15 tags with each combination were tested. The tags were held first in a bending test (series 1), in a constant humidity test (series 2), or in a temperature cycling test (series 3), and after that they were tested in a humidity cycling test. As a reference, 15 tags (series 4) were tested only in the humidity cycling test. Altogether 60 tags were tested.

In the bending test the tags were tested for 250 cycles with a special bending tester designed for the mechanical testing of RFID tags. The physical bonds and mechanical durability of the test structure

were tested. The tags formed a loop which travelled through the bending tester with small diameter rolls with constant web tension. In addition, during the test dynamic pressure was tested in a nip between the rolls. Web tension in the test was 250N/m and nip tension was 300N/m. Velocity of 20m/min was used.

In the constant humidity test the tags were kept 168 hours in an environment with temperature 85°C and relative humidity 85%. In the temperature cycling test the temperature varied from -40°C to 80°C. Soak time in the extreme temperatures was 15 minutes. Ramp rate for cooling was 3K/min and for heating 5K/min. Thus the total cycle time was 68 minutes. The tags were tested in the temperature cycling test for 200 cycles.

In the humidity cycling test humidity changed from 10%RH to 85%RH and temperature from 25°C to 85°C. The exposure time at both humidity levels at 85°C was 2 hours. The transition time from high humidity level to low humidity level was 40 minutes, while the transition from the low level to high level was 3 hours and 20 minutes. The times were different because the temperature was dropped to room temperature for a while, when the humidity level was raised from the lower level to higher level. This was done to increase the harshness of the test. The total cycling time was 8 hours. The cycle profile is shown in Figure 2. The tags were tested for 420 cycles in the humidity cycling test.

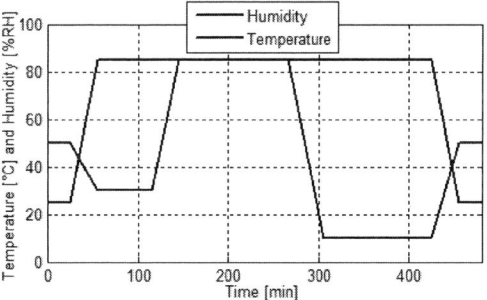

Figure 2: Profile of a cycle in the humidity cycling test.

To observe the changes in the tags during testing, the performance parameters of the tags were measured before testing, after the first tests, and after every 84 cycles in the humidity cycling test. They were tested within a frequency band from 800MHz to 1GHz using customized RFID measurement equipment. The core operations of this system are performed with a vector signal analyzer. Threshold power, which is the minimum power transmitted from the reader antenna to activate the tag and to receive a response to ID query at a certain frequency and a distance, was measured as a function of frequency using 1MHz steps. The measurements were carried out in a compact measurement cabinet where the distance between the reader antenna and the tag was 0.45m. The gain of the measurement antenna was 8.5dBi. The threshold power sweeps after different periods of testing were compared for any significant changes in the performance parameters of the tags. The tags were deemed to have failed when the threshold power was doubled in their optimal operating frequency.

3. Results

No significant changes in the performance parameters of the tags from series 1, 2 and 3 were seen after the first tests. Tags from test series 1 were held first in the bending test, tags from test series 2 in the constant humidity test and tags from test series 3 in the temperature cycling test. In the humidity cycling test most of the samples failed during the 420 cycles of testing. In order to study the reliability data statistically a two-parameter Weibull distribution was used. This is a versatile statistical method which has been used to study the reliability of ACA connections [9,10].

For the two-parametric Weibull distribution, the probability density function is:

$$f(t) = \frac{\beta}{\eta}\left(\frac{t}{\eta}\right)^{\beta-1} e^{-\left(\frac{t}{\eta}\right)^{\beta}} \qquad (1)$$

where β is a shape parameter and η is a scale parameter. The shape parameter β helps to define the shape of a distribution. When $\beta < 1$, the distribution has a failure rate that decreases over time. On the other hand, when $\beta > 1$, the failure rate of the distribution increases over time. With $\beta = 1$ the failure rate is constant. The scale parameter η is a life-point value and gives the age at which 63.2% of the test samples fail [11].

The Weibull plot for the results in the humidity cycling test is shown in Figure 3. The shape parameters (β) and the scale parameters (η) of Weibull distributions, and the numbers of failed samples (F) and censored samples are listed in Table 1.

There was no significant difference in the shape parameters of the different test series. All the shape parameters were over one and thus the failure rate of the distributions increased over time.

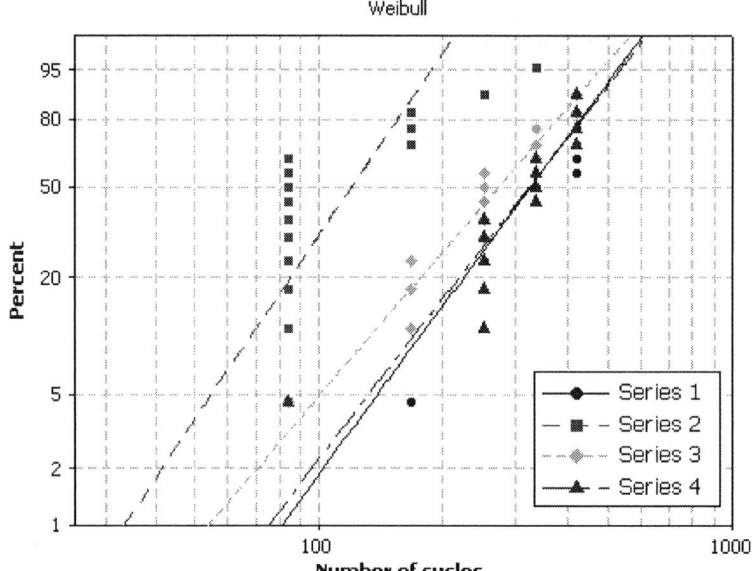

Figure 3: Results of the humidity cycling test.

Table 1: Shape and scale parameters of the Weibull distributions.

	Tests	β	η	F	C
Series 1	Bending test + humidity cycling test	3.041	367	13	2
Series 2	Constant humidity test + humidity cycling test	3.291	134	15	0
Series 3	Temperature cycling test + humidity cycling test	2.600	314	14	1
Series 4	Humidity cycling test	2.893	367	14	1

Test series 2 had a significantly smaller scale parameter than the other test series. The tags in test series 2 were held in a constant humidity test before the humidity cycling test. Consequently, the exposure to the high humidity and high temperature seem to significantly impair the reliability of the tags in a humidity cycling test.

In the constant humidity test the PET substrates of the tags were exposed simultaneously to high temperature and humidity. This started a hydrolysis of ester groups in amorphous parts of the material and thereby changed the properties of the PET [12]. Random chain scission during the hydrolysis slowly made the PET substrates brittle [12]. In this study the exposure time to the high humidity and temperature was relatively short and no visible changes were observed after the test. However, it is likely that the hydrolysis started and the mechanical properties of the PET changed slightly. In addition, the glass transition temperature (T_g) of the PET substrate was exceeded when the temperature was held at 85°C in the constant humidity test. Consequently, the PET material changed to a softer form [13] and it was more prone to dimensional changes. It is possible that these slight changes in the dimensions and in the material properties of the PET decreased the reliability of the tags from series 2 compared to the other test series.

The scale parameters of series 1 and 4 were equal. Consequently, the bending test seems to have no significant effect on the reliability of the tags in the humidity cycling test. The scale parameter of the test series 3 was only slightly smaller than in test series 4. Thus it is possible that subjection to the temperature cycling test may slightly impair the reliability of the tags in the humidity cycling test. However, the difference is so small that it may also be caused by the deviation, as the test series contained only 15 tags per series.

In the temperature cycling test the materials expanded and contracted as the temperature fluctuated. The dimensions of the polymers changed more than those of metals and silicon. Most of these dimensional changes were elastic and recovered after testing. However, when the T_g temperature of the PET substrate was exceeded at the upper limit temperature of the test, the PET softened [13], and it is possible that plastic changes in the dimensions also occurred. The plastic changes may have impaired the reliability of the tags from series 3.

4. Conclusion

A passive UHF RFID tag was tested with different combinations of environmental tests. Some of the tags were held first in a bending test, in a constant humidity test, or in a temperature cycling test, and then subjected to a humidity cycling test. In addition, as a reference some of the tags were tested only in a humidity cycling test. Changes in the performance parameters of the RFID tags were examined during testing to ascertain differences between the different combinations of the environmental tests.

The exposure to constant humidity test before the humidity cycling test impaired the reliability of the tags in the humidity cycling test significantly. On the other hand, bending testing before the humidity cycling test had no effect on the reliability of the tags in the humidity cycling test. The exposure to temperature cycling test before the humidity cycling test seemed to slightly impair the reliability of the tags in the humidity cycling test. However, the difference was so small that it may also be due to experimental error.

This study demonstrated the importance of studying combinations of different environmental tests one after another on the same test samples. Although exposure to certain environmental stresses may not immediately cause a device to fail, it may impair the reliability when the device is later exposed to another environmental stress. In future studies, the effects of different test conditions on the failure mechanisms will be studied and further different test combinations will be tested.

Acknowledgements

We would like to thank TEKES (the National Technology Agency of Finland) and the following companies which supported this work: UPM Raflatac Oy, ABB Oy / Drives, Vaisala Oyj and Elcoflex Oy.

References

[1] P.V. Nikitin and K. V. S. Rao, "Theory and measurement of backscattering from RFID tags", IEEE Transactions on Antennas and Propagation, Vol. 48, No. 6, pp. 212–218, 2006.

[2] R. Want, "An Introduction to RFID Technology", IEEE Pervasive Computing, Vol. 5, No. 1, pp. 25-33, 2006.

[3] D. Dobkin, "The RF in RFID - Passive UHF RFID in Practice", WJ Communications, San Jose, 504p., 2007.

[4] K.V.S. Rao, P.V. Nikitin and S.F. Lam, "Antenna Design for UHF RFID Tags: A Review and a Practical Application", IEEE Transactions on Antennas and Propagation, Vol. 53, No. 12, pp. 3870-3876, 2005.

[5] K. Domdouzis, B. Kumar and C. Anumba, "Radio-Frequency Identification (RFID) applications: A brief introduction", Advanced Engineering Informatics, Vol. 21, No. 4, pp. 350-355, 2007.

[6] B. Nath, F. Reynolds and R. Want, "RFID Technology and Applications", IEEE Pervasive Computing, Vol. 5, No. 1, pp. 22-24, 2006.

[7] P. Viswanadham and P. Singh, "Failure Modes and Mechanism in Electronic Packages", Chapman & Hall, England, 370 p., 1998.

[8] E. Suhir, "Accelerated Life Testing (ALT) in Microelectronics and Photonics: Its Role, Attributes, Challenges, Pitfalls, and Interaction

with Qualification Tests", ASME Journal of Electronic Packaging, Vol. 124, No. 3, pp. 281-291, 2002.

[9] W. Kwon, M. Yim, K. Paik, S. Ham and S. Lee, "Thermal Cycling Reliability and Delamination of Anisotropic Conductive Adhesives Flip Chip on Organic Substrates With Emphasis on the Thermal Deformation", Journal of Electronic Packaging, Vol. 127, No. 2, pp. 86-90, 2005.

[10] J. Liu and Z. Lau, "Reliability of Anisotropically Conductive Adhesive Joints on a Flip-Chip/FR-4 substrate", Journal of Electronic Packaging, Vol. 124, No. 3, pp. 240-245, 2002.

[11] W.D. Brown, "Advanced Electronic Packaging – With Emphasis on Multichip Modules", IEEE Press, New York, 791p., 1999.

[12] B. Fayolle, X. Colin, L. Audouin and J. Verdu, "Mechanism of degradation induced embrittlement in polyethylene", Polymer Degradation and Stability, Vol. 92, No. 2, pp. 231-238, 2007.

[13] M. Kutz, "Handbook of Materials Selection", John Wiley & Sons, New York, 1497p., 2002.

A Dynamic Bending Method in CSP Package Validation for Portable Electronics Application

Jeffrey ChangBing Lee*, Graver Chang, Cherie Chen, Jandel Lin

IST-Integrated Service Technology Inc.,Taiwan

19, Pu-Ding Rd., Hsin-chu 30072, Taiwan, R.O.C

E-mail:Jeffrey_lee@istgroup.com

Abstract

In the study, instead of electrical resistance monitoring in JEDEC JESD22-B111 for mechanical shock testing of handheld electronics, a proprietary strain-controllable dynamic bending method is applied to verify the effects of lead free SnAgCu solder ball with different dopant on the board level reliability of green TFBGA package.

The result shows the strain-controllable dynamic bending method can identify the weakness of IC package on board by way of the setting of the low strain to high strain gradually. Various failure modes from interfacial IMC crack in component and PCB side, pad crater due to core material adhesion issue of HF PCB and solder bulk itself are scrutinized by advanced FA application of SEM/EDX, polarized OM, respectively. The SnAgCu solder with 4th element dopant to compare with SAC105 will demonstrate certain improvement in solder joint reliability. On the contrary, the HF PCB will downgrade certain percent solder joint reliability due to its higher stiffness, as well the larger package size and thermal aging precondition present similar behavior.

Keywords: Halogen free, Lead free, Strain-controllable bending method, Drop, Fine pitch,

Introduction

Halogen free compliance beyond RoHS has got highly attention by branding customer to raise the environmental ranking in Greenpeace. The supply chain in ODM and OEM for mobile product, Notebook and Desktop computer will be drive to meet the low halogen requirement of Br and Cl in the flame retardant of plastic material based on ongoing IPC-JEDEC JSTD-709 guideline. The plastic material related to IC packaging reliability consists of molding compound, substrate, die attach and underfill and so on. Moreover, the board level reliability effect from PCB and solder ball will also play more and more important roles to validate IC packaging integrity due to SnAgCu solder involvement and the HF dielectric materials conversion of PCB.

General speaking, the normal SnAgCu solder is stiffer than SnPb solder, as well the HF PCB is stiffer than conventional FR4 PCB. Combined these two factors with the higher peak reflow temperatures for lead-free assembly, could induce more failures in the substrate and PCB site caused by weaker cohesive energy between the halogen free dielectric materials and copper, which is also called pad cratering (1).

The SnAgCu solder ball with 3 or 4 %Ag content has been adapted in thermal cycling concerned product and 1% Ag in higher strain rate drop concerned mobile product due to its unique creep behavior with Ag content during various strain condition (6~10). The low Ag SAC with more cost effective advantage is considered to be modified with minor dopant addition to compromise the both performance by way of finer grain formation and interfacial IMC growth mitigation. Therefore, there are number of proprietary solder candidates in the industry to approach the issue by different solder company. To date, there is still no best answer to resolve the problem due to too complicated reliability testing and various used condition(2).

The popular JESD22-B111is adapted to verify the board level drop performance of IC package for handheld electronic products. The higher strain rate applied to identify weakness of IC package integrity often generate over killing situation so as not to identify correctly failure mode for following improvement, especially in advanced IC package due to its more and more complicated 3D integrated structure.

The dynamic bending method developed by Motorola is adapted to verify the IC package integrity in board level performance due to its strain-controllable design (5). The drop strain can be adjusted from low to high step by step to result in failure mode evolution among package, solder joint and PCB, so that the weakness in whole package integrity can be identified for improvement.

Experiment

Two TFBGA with Green compliance (LF+HF) BOM and two types of PCB (FR4 and Halogen Free) with solder mask define (SMD) and OSP finish are adapted as test vehicle in the study, as described in table 1 and 2 respectively. The solder ball in 14X14 contains SAC105 and low Ag percentage SAC with 2 different dopants named Mn and X in order to investigate the dopants effect on the board level performance. The HF PCB with phenol curing agent is designed with close pad opening as that in packaging substrate in order to eliminate geometry effect of solder joint, which will be able to focus on metallurgical analysis only. The 15x15 TFBGA package with SAC105 can be applied to understand package dimension effect on the solder joint life.

Table 1. The IC package profile.

Package Size (mm)	I/O	Ball Size (mm)	Pitch (mm)	Surface Finish	Solder Ball Material
14*14	457	0.3	0.5	NiAu	SAC105 SACMn SACX
15*15	468	0.3	0.5	NiAu	SAC105

Table 2. The PCB profile (75.0 x 40.0 x 1.08mm ,Length x Width x Thickness, SMD)

Type	Curing Agent	Pad opening (±25µm)	SolderMask opening (±38µm)	Solder Paste opening (±6.5µm)
FR4	Dicy	0.38	0.3	0.305
HF	Phenol	0.38	0.3	0.305

The one selected IC packages are subjected to surface mounting on the PCB at 235 °C (+5/-3 °C) with dwell time over 217°C of 70 seconds (+/- 10 seconds) with SAC387 no clean solder paste.

The dynamic 4 bending method with strain-controllable design by the adjustment of drop height with fixed ball weight, as shown in the figure 1.The mounted package will face down to withstand the drop strain from low to high level defined by previous strain-drop height calibration curve, then the solder joint failure number and location can be recorded after dye and pry analysis, A proprietary statistical calculation is adapted to get the life of first solder joint to fail in strain level unit after weibull analysis.

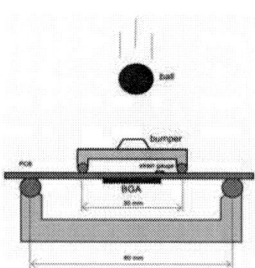

Figure 1. High strain rate 4 point dynamic bending setup.

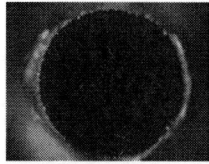 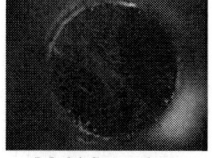

Single Sn grain   Multi Sn grain

Figure 2. The polarized light image comparison of Sn grain structure

Furthermore, the mounted HF PCB with 14X14 package

are subject to 125°C/250hrs thermal aging prior to test to drive the interfacial IMC growth and solder grain boundary evolution. After Z-axis cross-section at 50% height of solder joint for bottom view, a polarized OM is adapted to check the ratio of single Sn grain transforming to multi Sn grain at 64 solder joint near the package edge before and after thermal aging as defined in figure 2. The multi grain number will be ignored but account solder joint number with multi grain presence (3,4).

Result and discussion

1. Solder ball and PCB effect in 14X14 TFBGA on the life of solder joint

The table 3 showed the life result from different solder ball and PCB. The life in strain level of first joint to fail is that SACX>SACMn>SAC105 whatever PCB type, while HF PCB will reduce certain strain level whatever any solder ball type.

Table 3. The effect of solder ball and PCB on the joint life (unit : strain)

PCB / Solder ball	FR4	HF
SAC105	0.61%	0.50%
SACMn	0.63%	0.60%
SACX	Over 1.0%	0.80%

In FR4 PCB category, the X dopant in SAC solder presents highest strain level and the Mn dopant has a little improvement compared to SAC105. The failure distribution over different bending strain in figure3 shows most IMC failure in component side for SAC105 and SACMn, while there is no solder joint failure but board pad crater at highest 1.1% strain in SACX. In other words, X dopant in SACX does play a critical function in modifying the interfacial IMC and solder bulk micro-structure to prevent solder joint from failure during mechanical stress. Eventually, the failure will occur in PCB itself.

In HF PCB category, the strain level reduces certain level in each solder ball type, but the ranking is still identical as that in FR4 group. The failure distribution is transferred some to the IMC in board side in the 3 solder ball types, which illustrate the HF PCB will induce more stress from component to PCB due to its higher flexure modulus, as shown the result tested by 3 point bending method in figure4.The SACX still has higher failure ratio in PCB itself, so that the joint life can be still maintained at highest level.

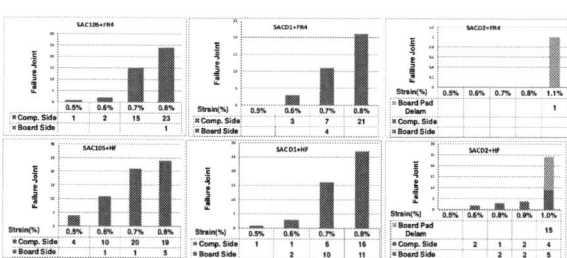

Figure 3. Failure mode mapping (up: FR4, bottom: HF

PCB)

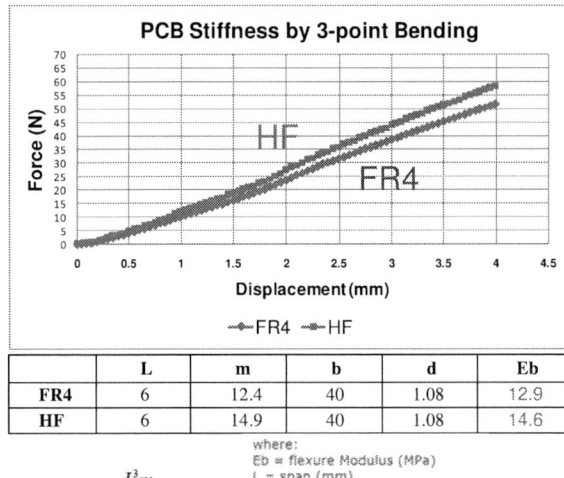

Figure 4. PCB flexure modulus measurement by 3 point bending test

	L	m	b	d	Eb
FR4	6	12.4	40	1.08	12.9
HF	6	14.9	40	1.08	14.6

$$E_b = \frac{L^3 m}{4bd^3}$$

where:
Eb = flexure Modulus (MPa)
L = span (mm)
b = width of beam (mm)
d = depth of beam (mm)
m = slope of stiffness curve (N/mm)

The typical failure modes during bending test in all legs are explained in figure 5. Lower strain level generates less and single failure mode and higher strain will generate more and mixing failure modes.

Mode I is the flat solder crack happening inside the IMC layer between NixSny and (CuNi)xSny in the component side, which is majority of the solder joint failure on the FR4 PCB. Due to solder mask define (SMD) in both component and PCB side and similar pad opening, the result could explain that the interfacial strength of NiSn+CuNiSn will be worse than that of Cu3Sn + Cu6Sn5 under the dynamic bending test. Therefore, the fracture will be easier to happen in the weakest point of solder joint due to its more brittle characteristics. The Mn dopant in SACMn does not function well to change the failure mode compared to SAC105, but X does, that is to illustrate the unique solder ball design will enable more robust solder joint life.

Mode II is the irregular solder crack happening inside the IMC layer between Cu3Sn and Cu6Sn5 in the PC board side. Its ratio is raised when converting to stiffer HF PCB even if the interfacial strength is stronger than in component side.

Mode III is the pad cratering happening in the core material of PCB during the bending to illustrate the weakest point will be in PCB itself. The SACX still maintain high ratio mode III failure to own highest strain level even if in HF PCB.

Generally in halogen free PCB, the adhesion between polymer core material and copper will be weaker than in conventional FR4 due to its higher filler content design in the resin in order to increase flame retardancy and decrease CTE

in alpha 2 above Tg, as well reduce moisture uptake. Hence, mode III is also popular in HF PCB.

Mode I IMC Crack on component side

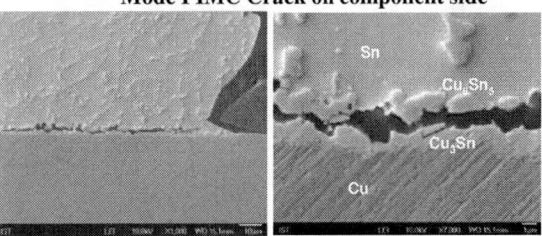

Mode II IMC Crack on board side

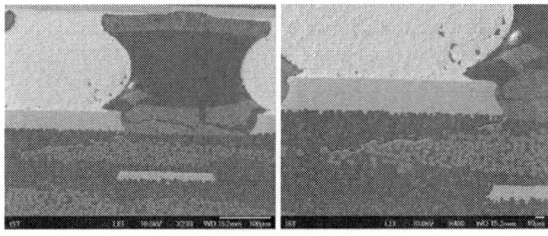

Mode III PCB pad crater
Figure 5. The typical failure modes during bending test

2. Thermal aging effect in 14X14 TFBGA and HF PCB on the life of solder joint

The precondition with thermal aging 125°C/250hrs prior to bending testing is to simulate actual service condition when using different SAC solder. The IMC growth after aging in component and PCB side will be different due to discrepancy characteristic of SnNi and SnCu IMC system. As shown in figure 9, there is no apparent IMC growth in component side, but certain IMC growth in PCB side, particularly from Cu3Sn growth in SACMn and SACX. The inter-diffusion rate between Sn and Ni to form IMC will be slower than that between Sn and Cu. The Mn and X dopants do not function well to mitigate interfacial SnCu IMC growth in PCB side, particularly in X dopant.

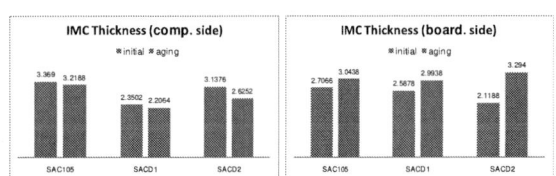

Figure 6. IMC thickness comparison over thermal aging

Figure 7 and 8 show the SEM picture of cross-section in component and PCB side. There are a number of Cu6Sn5 precipitates distributed in the solder joint, less Plate like Ag3Sn is not found due to low Ag composition in the three SAC balls. The SnNi interfacial IMC presents flat microstructure and SnCu interfacial IMC presents irregular microstructure due to spalling Cu6Sn5 on the Cu3Sn layer as shown in figure 9. After thermal aging, there is no apparent IMC precipitates increasing and coarsening inside the solder bulk observed.

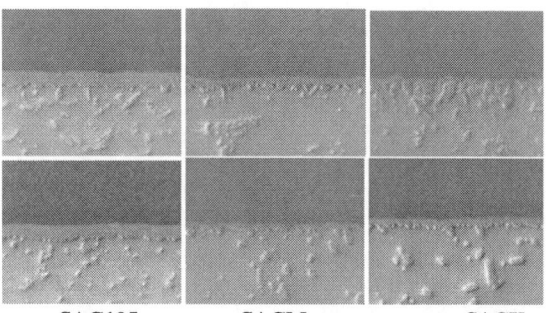

SAC105 **SACMn** **SACX**

Figure 7. IMC at component side (up: no aging, bottom: aging)

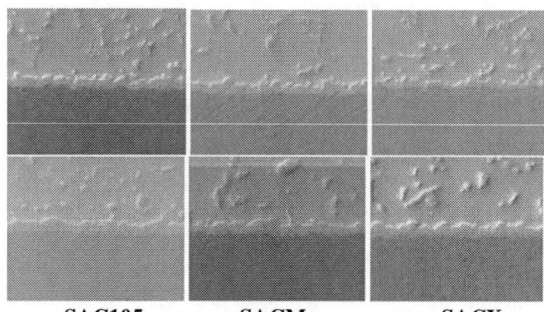

SAC105 **SACMn** **SACX**

Figure 8. IMC at board side (up: no aging, bottom: aging)

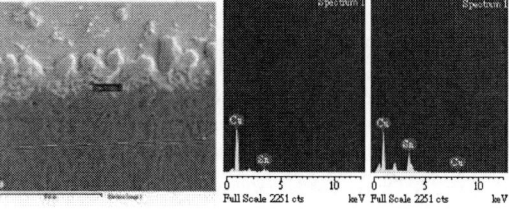

Figure 9. EDX for SACMn at board side after aging

The table 4 showed the life result from thermal aging. The life in strain level of first joint to fail reduce certain percentage after aging, especially in SACX although the life ranking after aging is still SACX>SACMn>SAC105. The life in SACX reduce down 32.5% after aging to illustrate its potential risk if apply it in real service after period of time. SACX's result after aging is not like in initial stage so perfectly. Figure 10 show the failure mode mapping before and after aging in each solder ball type. The mode II will be dominant failure after aging to accommodate with the evidence of the interfacial SnCu IMC growth in HF PCB side. The mode III disappears

in SACX to explain well the significant reduction in strain level, also explain the weakest point in interfacial SnCu IMC once again.

Table 4. The effect of thermal aging on the life of first joint to fail (on the HF PCB)

	No Aging	Aging	Reduction
SAC105	0.50%	0.46%	8%
SACMn	0.60%	0.52%	13.3%
SACX	0.80%	0.54%	32.5%

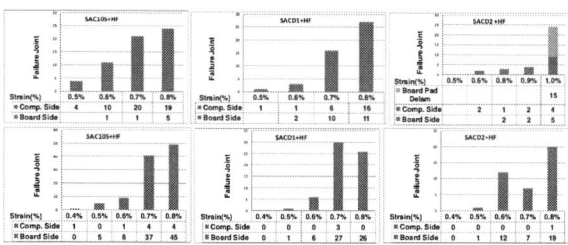

Figure 10. Failure mode mapping (up: no aging, bottom: aging)

Figure 11 show the polarized light image before and after thermal aging at 50% cross-section from the bottom view. There are numbers of single grain transforming to multi grains after thermal aging in each solder ball type. The grain number and size inside multi grain will be ignored but only calculate the multi grain ratio of total joints. The multi grain is generated as a result of single Sn grain phase transformation under thermal excursion such as reflow at SMT and thermal aging or thermal cycling.

The multi grain ratio before thermal aging is SAC105>SACX>SACMn, there is certain increase after thermal aging in each of solder ball type. Most significant increase is from SACX, next is SAC105. There is less change in SACMn. as shown the comparison in figure 12. More multi grain presence means more grain boundary in the solder bulk so as to induce easily crack propagation once it is initiated. Although it could not be adapted to explain exactly the strain level difference among the three solder ball types, it could be one of the reasons for SACX about most significant reduction and major failure mode conversion from PCB crater to interfacial CuSn IMC crack after aging.

On the other hand, the solder joint recrystallization degree of outer I/O matrix are more evident than inner solder joint, and the solder joint recrystallization degree on the solder bulk near the IMC layer is higher than that in the middle of solder bulk. According to the cross-section analysis result, it can be found the fatigue fracture on the solder balls happen mostly on the solder bulk near the interfacial IMC, and recrystallization area also exhibit apparent intergranular fatigue crack. These phenomena can explain why the SACX joint life decreases obviously than other low Ag solder.

The non-linear or linear interaction about both creep

402

fatigue behavior between solder bulk with multi grain and interfacial IMC in SnNi and SnCu system will be investigated further by way of stress simulation after acquiring their related mechanical property. The micro-hardness measurement can be tested by nano-identer.

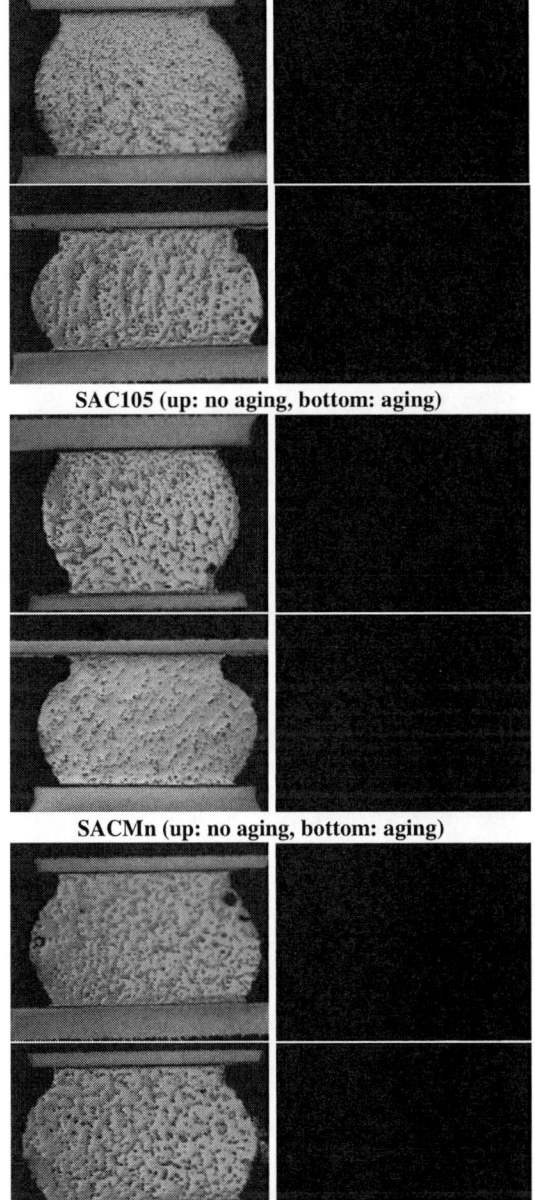

SAC105 (up: no aging, bottom: aging)

SACMn (up: no aging, bottom: aging)

SACX (up: no aging, bottom: aging)

Figure 11. The polarized light image before and after thermal aging. (At 50% X-S from the bottom view)

3. Package dimension effect on the life of solder joint

As shown in figure 13, there is no apparent strain level

difference in 15X15 and 14X14 package, which can illustrate similar solder joint constrain around the package corner in the 2 package size. Due to different package layout in the 2 packages, only the joint failure from package corner can be comparative. The test in smallest package size with 10X10 is underway.

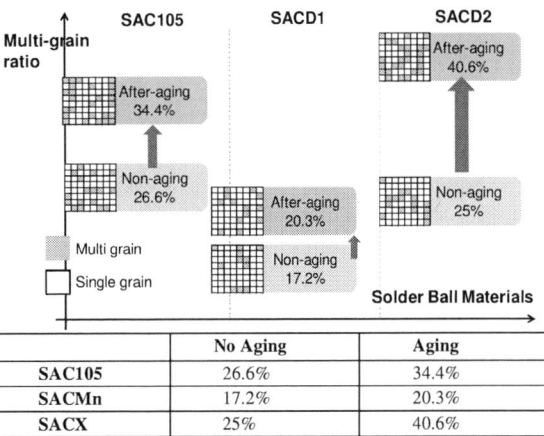

	No Aging	Aging
SAC105	26.6%	34.4%
SACMn	17.2%	20.3%
SACX	25%	40.6%

Figure12. The Multi grain transformation ratio versus thermal aging

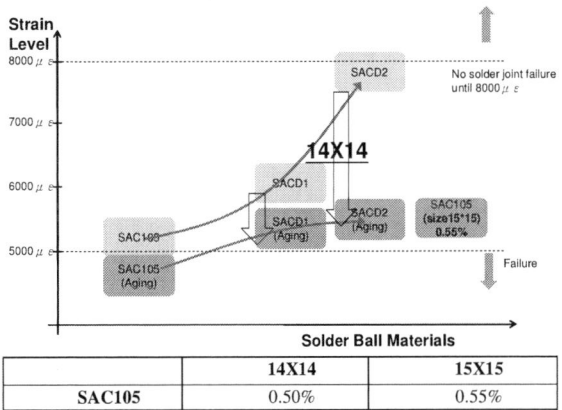

	14X14	15X15
SAC105	0.50%	0.55%

Figure13. The effect of package size on the life of first joint to fail (on the HF PCB)

Conclusions

Compared with the popular high strain rate 1500 Gs, 0.5 millisecond duration, half-sine pulse drop in JEDEC JESD22-B111, the strain-controllable dynamic bending method will be better to identify the failure mechanism in more detail from material and package design viewpoint. Control from lower strain to higher strain will gradually develop different failure mode for material improvement. It will be helpful when developing new IC packaging integrity due to new material introduction, especially for advanced fine pitch device such as 3D IC, WLCSP, and so on. Moreover, the halogen free conversion for consumer electronics will be inevitable requirement due to environmental concern. It will be paid more attention in negative reliability contribution when designing into product. .

Acknowledgement

Authors wish to thank the following individuals for their assistance in package sample assembly from Xiang-Kai.Meng in Freescale China and DOE discussion from Anthony Gallagher and Matt Brown in Motorola.

References

1. Mudasir Ahmad, et al, "Comprehensive methodology to characterize and mitigate BGA pad cratering in printed circuit boards", Volume 22, issue 1, SMTA Journal, 2009,pp.21-28.

2. Weiping Liu, Ning-Cheng Lee, et al, "Achieving High Reliability Low Cost Lead-Free SAC Solder Joints Via Mn or Ce Doping", ECTC, 2009,pp.994 -1007.

3. Bize Zhou, et al, "Characterization of SAC solder microstructure evolution during thermal cycling with statistical and in-situ methods", IMAPS, , 2009, pp.158-165.

4. Tae-Kyu Lee, et al, "Grain orientation and microstructure evolution in SnAgCu solder joints as a function of position in ball grid array package", IMAPS, , 2009, pp.142-149.

5. Min Ding and Adriana Porras, "Aging effects on dynamic bend test performance of Pb-free solder joints on NiAu finish".Volume 19, Issue 4, SMTA Journal, 2006, pp.11-17.

6. Ranjit S Pandher, et al, Drop shock reliability of lead free alloy-Effect of micro additive, ECTC, 2007, pp. 669-676

7. Jeffrey C.B Lee et al, The solder joint characterization in green WLCSP. *Proc. 54th Electr. Comp. Technol. Conf.*, 2004, pp.1914-1920.

8. Jeffrey C.B.Lee, et al,"One solution for IMC failure in CSP package. 6th IPC/JEDEC international lead free conference, Singapore, Aug.18-20, 2004.

9. Jeffrey C.B.Lee, et al," The SnAgCu solder joint integrity in WLCSP for green conversion APM, Irvine, Mar, 2005.

10. Sung K.Kang, et al, " Formatopn of Ag3Sn plate in SnAgCu alloy and optimization of their alloy composition", ECTC, New-Olean, May.2003, pp.64~70

Data analysis techniques for real-time prognostics and health management of semiconductor devices

Thamo Sutharssan, Stoyan Stoyanov, Chris Bailey, Yasmine Rosunally

Computational Mechanics and Reliability Group, School Computing and Mathematical Sciences, University of Greenwich, Old Royal Naval College, Park Row, London SE10 9LS

Tel: +44 (0)20 8331 8669, Fax: +44 (0)20 8331 8665, Email: t.sutharssan@gre.ac.uk

Abstract

Prognostics and health management (PHM) has emerged in the last few years as one of the most efficient approaches in failure prevention, predicting reliability and remaining useful life of various engineering systems and components. The diagnostics of unusual performance trends typically is associated with the requirement for a continuous monitoring of system's behaviour using data from sensors. Often it is necessary to monitor the system in real-time, especially in the case of safety critical applications, to predict when a fault will occur and/or to assess the remaining useful life. In this paper we present an investigation on the suitability of a number of data analysis algorithms to realise real-time diagnostics, prognostics and health monitoring of engineering systems. The focus is on the following two techniques for data-driven PHM: (1) Euclidean Distance (ED) and (2) Mahalanobis Distance (MD). These techniques are implemented into a real-time operating platform for health management and tested using data from high power light emitting diodes (LEDs). These LEDs are tested for the PHM of semiconductor devices/solid state lighting systems and include monitoring of current, temperature and light intensity. We also demonstrate the real-time PHM capability of these algorithms by programming the associated sensor data manipulation and numerical computations. Although for the application reported in this work the real-time aspect is not essential, the actual prognostics techniques can be applied in other applications where this capability might be necessary.

Key words: Prognostics, Real-Time Health Monitoring, Mahalanobis Distance, Light Emitting Diodes

Introduction

Prognostics and health management (PHM) is an engineering process of failure prevention, and predicting reliability and remaining useful life-time. It has emerged in the last few years as one of the most efficient approaches in failure prevention and predicting reliability and useful life time of various engineering systems and components. PHM of engineering systems has become very important as a malfunction or failure may cause severe damage to the system, environment and users, and may result in significant repair on un-scheduled maintenance costs. Anomaly detection and failure prevention can be achieved effectively by monitoring the key performance parameters continuously. It will be more accurate if the parameters are monitored in real-time, especially in the case of critical applications. Many safety critical systems and mission critical systems consist of electronic hardware and software that control the electronic hardware and also interact with the user. Most of these electronic hardware devices use thousands of individual semiconductor components to perform their operation. Malfunction or failure of any individual semiconductor component, electronic hardware or software module independently affects the system as a whole.

Health of a system is defined as the extent of deviation or degradation from its expected typical operating performance [1]. This extent of deviation or degradation of the expected typical operating performance has to be determined accurately to prevent the failures. It is also necessary to determine the parameters which deviate or degrade with time, or which are good indicators of failures.

There are various approaches available to assess the reliability of systems and predict the remaining useful life using PHM. Data-driven PHM is one of the approaches which does not need any system model and depends only on operational and failure data sets. The principal advantage of the data-driven approaches is that they can be applied very quickly given there are enough performance data from sensors. Availability of run-to-failure data sets for a particular system or component is the main disadvantage of data-driven PHM [2], as running a system or a component to failure might be time consuming and expensive [2].

Real-time computing is required to be not only fast and high performance computing but also very deterministic and reliable. It must meet typically requirements for 24x7 operations and for an extended period of time with high reliability.

Real-time computing can be classified into two categories: 1) periodic real-time computing 2) non-periodic real-time computing. Periodic real-time computing will have continuous deterministic tasks to response whereas the non-periodic computing will have to respond to certain events deterministically. Complexity of a real-time computing or system varies from low cost microcontrollers to high end microprocessors and Field Programmable Gate Arrays (FPGA). A real-time PHM system will need periodic real-time computing to acquire the sensor data periodically and will also require non-period computing to respond to some events based on the periodic computing, for example to provide an early warning.

In this paper we focus the investigation purely on data-driven PHM approaches. Health monitoring and failure prognostics for high-power light emitting diodes (LEDs) are used as the application to test and demonstrate these techniques. The PHM is based on monitoring sensor data for the input current and board temperature of the LEDs under accelerated voltage load tests. The main focus is on applying and testing of two particular distance measure based techniques, Euclidean Distance and Mahalanobis Distance, used to realise a data driven PHM. A demonstration of the real-time prognostics capability of these algorithms by programming the associated numerical data manipulation and computations is provided. The tests are set using the NI PXI real-time data acquisition and automation hardware platform. Real-time data acquisition and the prognostics algorithm computations are successfully realised and also demonstrated.

Distance Measure Algorithms for Data Driven PHM

Distance measure techniques are capable of providing the best estimate of the state of a system in real-time using a number of monitored performance parameters. These parameters are device or system specific and must be identified in each instance. For example, in the case of the semiconductor devices extensive tests are necessary to determine the key monitoring parameters as well as their deviation from normal values associated with a "healthy" state of the system. It is also important to determine the safe and failure threshold values of these deviations or degradation of the typical operating performance.

A distance measure is the physical distance between data points and can be used to determine the level of similarity between them. The underlying principle is to create a cluster which is cantered with respect to the typical operating performance values. Deviation in the operating performance is then determined by the distance from the centre point. In relation to PHM, such value defines the health of the system. Threshold values or outliers for the safety operation and failures modes are determined by

extensive testing of the system that generates "healthy" performance data. These strategies have been applied successfully in different data-driven PHM approaches. The evaluations of the contours centred on the typical operating performance data are useful in determining the different states of the system. In the case of three parameters, the contours will be replaced by the different surfaces centred on the typical operating values. Introduction of soft and hard margins for the safety and failure threshold values enable the condition based maintenance (CBM) to be performed in an effective way. Distances greater than these threshold values should be considered as failures or will indicate a system in failure mode. Thus, the distance measures are important means by which the state of the system in real-time can be assessed very effectively.

Euclidean Distance

Euclidean distance (ED) is the physical distance between two data points and it is the most commonly used distance measure in many different fields. This is simply a distance examines the root of squares differences between any data sets i.e. it can be in any dimensions. The straight line Euclidean distance measure between the typical operating performance data and the current operating performance data gives the straight deviation or degradation of the system. Using this evaluation, the health of the system can be assessed. Once the outliers or threshold values of ED for the safety and failure operating performance is defined in the Euclidean space or in the Cartesian coordinate system, the state of the system can be observed every time a new set of data becomes available.

Let us consider typical operating performance data set $X = \begin{bmatrix} C_1 T_1 \end{bmatrix}$ where C_1 and T_1 are the mean values of key monitoring parameters C and T, and current operating performance data set $Y = \begin{bmatrix} C_2 T_2 \end{bmatrix}$ where C_2 and T_2 are the mean values of key monitoring parameters C and T The Euclidean distance of the current state of the system is given by the following equation [6]:

$$ED_{current} = \sqrt{\left(C_2 - C_1\right)^2 + \left(C_2 - T_1\right)^2}$$

The $ED_{current}$ is proportional to the current deviation or the degradation of the systems and acts as a good indicator of the current health of the system.

Mahalanobis Distance

Mahanobis distance (MD) is another physical distance measure [4]. Although similar to the Euclidean distance in some respect, the Mahalanobis distance takes into account the actual correlations of the data. Since it is based on the correlation of the data sets, Mahalanobis gives a better estimation.

Since the health of the system is defined as the deviation from expected typical operating performance, Mahalanobis distance is useful in determining the similarity/distance between the typical operating performance and unknown current operating performance and it can also be used to predict or forecast the remaining useful life of a systems. Since MD brings the multidimensional characteristics into one scale , the MD estimate of the system gives better knowledge of the system and therefore monitoring individual sensor data is not necessary [5].

If typical operating performance data set is denoted as $X = \begin{bmatrix} C_1 T_1 \end{bmatrix}$ where C_1 and T_1 are the mean values of key monitoring parameters C and T, and current operating performance data set $Y = \begin{bmatrix} C_2 T_2 \end{bmatrix}$ where C_2 and T_2 are the mean values of key monitoring parameters C and T, then Mahalanobis distance of the current state of the system can be expressed as [6];

$$MD_{current} = \sqrt{(C_2 - C_1) \times Cov_{CT}^{-1} \times (T_2 - T_1)}$$

where Cov_{CT}^{-1} is called the inverse covariance matrix of key monitoring parameters under the expected typical operating conditions. MD calculation can be formulated using the following formula [6]:

$$Cov_{CT} = \begin{bmatrix} \sigma_C^2 & \rho_{CT}\sigma_C\sigma_T \\ \rho_{CT}\sigma_C\sigma_T & \sigma_T^2 \end{bmatrix}$$

where σ_C^2 and σ_T^2 are the variance of monitoring parameters C and T and $\rho_{CT}\sigma_C\sigma_T$ is the covariance of C and T. Using these variables MD can be formulated as follows [6];

$$MD_{current} = \sqrt{\left(\frac{C_2 - C_1}{\sigma_C}\right)^2 + \left[\left\{\left(\frac{T_1 - T_2}{\sigma_T}\right) - \rho_{CT}\left(\frac{C_2 - C_1}{\sigma_T}\right)\right\}\frac{1}{\sqrt{1 - \rho_{CT}^2}}\right]^2}$$

Data Driven PHM – an application to LEDs

Light Emitting Diodes

Light emitting diode or LED is an optoelectronic device which consists of p-type region, n-type region and the p-n junction. When the current passes through this p and n region radiactive recombination process takes place in which recombination of electrons and holes, releases energy in the form of photos. Power LED has evolved from simple indictor light into future lighting source. Applications of LED Lighting are continuously increasing as it has several advantages

including energy efficient (typically 85%) and very long life time. Different materials are used to make different LEDs. *Aluminium Gallium Arsenide* (AlGaAs) is used to make red LEDs and blue LEDs are made from *Indium Gallium Nitride* (InGaN) and *Aluminium gallium phosphide* (AlGaP) is used to make green LEDs. Right combinations of these three are used to generate different colours including white LEDs. A schematic cross-section of a LED assembly with typical construction is shown in Figure 1.

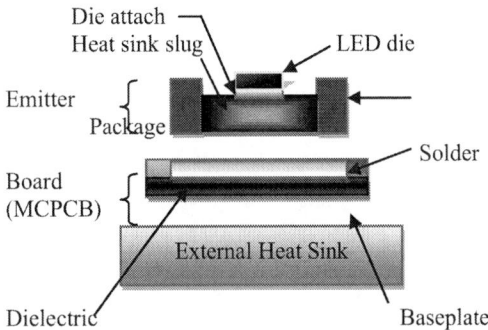

Figure 1: Cross section of LED assembly

Previous research in performance of LEDs has shown that depreciation of lumen up to 30% will have little effect if the depreciation is appropriately gradual. Therefore, the useful life of a LED for general lighting is given by the time in which the luminance of the led is 70% or higher of the initial light level. According to Alliance for Solid-State Illumination Systems and Technologies (ASSIST) recommendations time taken to depreciate 30% of the initial lumen is the useful life of general purpose lighting and time taken to depreciate 50% of the initial lumen is the useful life of decorative purpose lighting. But some specific application may require less than 30% of depreciation such as medical purpose lighting and safety purpose lighting. Typical life time of a general purpose LED under use conditions is around 50,000 hours (more than 5 years) and maximum of 100,000 hours (more than 11 years) which is very much higher than any other lights.

There is no specific life time model for LEDs. Main cause of the failures in the LEDs is heat generated at the p-n junction. Under the forward bias condition the p-n junction carries a current which is almost exponential function of the applied voltage which means if there is an increase in applied voltage, current through the p-n junction increases exponentially (Shockley Equation). Current increase will cause the temperature to increase dramatically which means the heat generated in the p-n junction increases with the current. Typically the life of the LED is affected by the current in the p-n junction and the board temperature. Therefore, any accelerated test can use the current or the temperature as the stress parameter [7].

Experiment Setup and Data Collection

The primary objective of this experiment is to test above mentioned data-driven PHM approaches for real-time PHM systems. National Instruments' PXI system is used to implement these algorithms and National Instruments' data acquisition system was used to collect the data from the Luxeon Star LEDs which is product of Philips Lumileds Lighting Company (see Figure 2). Secondary objectives are to develop a methodology to test the LED under the accelerated voltage conditions and prognosis of LEDs.

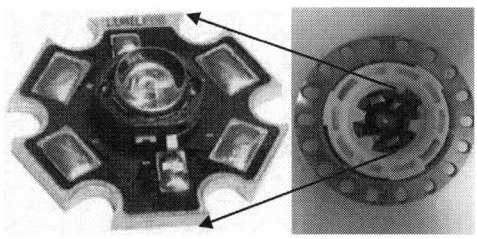

Figure 2: Luxeon Star LED from Philips

Since the input current play a crucial role in determining the brightness as well as the life of the LEDs, it is the most important parameter to be monitored and it is also used in the calculation of MD values. Another important parameter is the temperature of the P-N junction of the LED as it has shown that the heat at the junction determines the performance of the LEDs in terms of light output, spectral shift and light output degradation over the time [3]. Measuring the junction temperature is very difficult and therefore in this paper we consider corresponding 'near board temperature' which is the temperature at a point in the heat sink near to the LED board, to calculate the MD values.

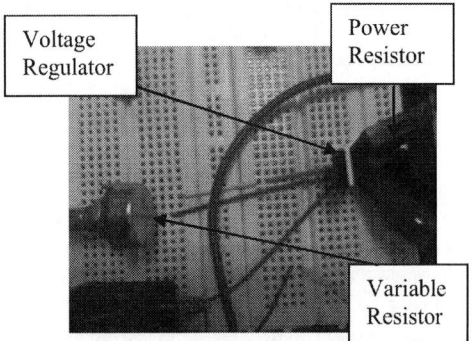

Figure 3: Test circuit

Test circuit is designed to handle the high current using the voltage regulator (3-terminal adjustable regulator LM317AT) and power resistor (MHP 100), and variable resistor was used to control the input voltage. This set up is shown in Figure 3. Power resistor is used to measure the current in the circuit by measuring the voltage across the power resistor. A thermocouple is attached with the LED's board to measure the board temperature (NI readymade J type) and photodiode (TSL250R-LF) is used to measure the output light intensity inside a spherical light integrator. These sensors are connected to the data acquisition NI PXI system (see Figure 4).

Figure 4: NI PXI System

Initially, Luxeon Star LEDs were tested under two conditions to observe the effect of the heat and temperature in the LED's life. First, the test was carried out without a heat sink. In this case, when the current increases to 500mA the LED starts to fail very quickly as the LED board temperature increases to maximum of 144°C. Second test was carried out with a heat sink which is a Star LED holder. In this case, current needs to be increased to 1A to accelerate the failure very quickly as the 'near board temperature' increases to maximum of 82°C. Both tests were carried out under different voltage conditions and under room temperature condition. In both tests, the input current, output light intensity and board temperature were recorded.

Test to Failure Specification

A run-to-fail accelerated voltage test for high power LEDs is designed to provide data for the PHM techniques. The accelerated load condition is realised through applying higher than the required operational voltage to an individual LED. Changing the applied voltage cause the increase in the current through the p-n junction.

The elevated voltage level has a maximum value of 3.99V which is also the typical maximum forward voltage of the LED. Nominal operational voltage that is required for the LEDs is 3.42V. Different accelerated voltage tests are defined and applied to the LEDs as detailed in the next sections.

Data Analysis and Results

For "healthy" operating, both the MD and ED values have been found in the range 0-10. This range has been observed in all undertaken tests as the range that corresponds to the LED luminance of 100%. Figure 5 shows the MD values for a LED which was operating under the typical expected operating conditions (i.e. voltage 3.42 V). MD values are calculated from the input current and the near board temperature. Light output data was used to find out the state of the LED, i.e. whether the LED is failed or not. Failure of a LED in this work is defined as the 30% or more reduction in the light intensity. In most cases such a failure of a LED is not physically observable as it still produces the light but below this failure level (i.e. 70%).

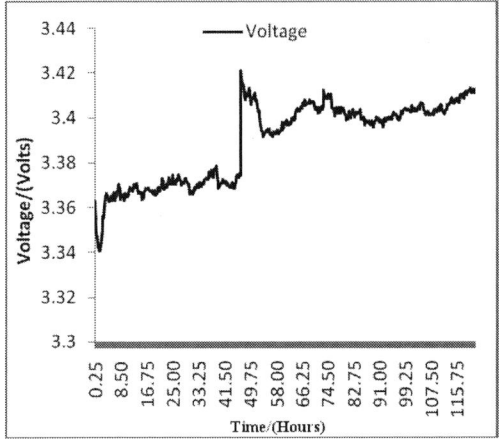

Figure 6(a): Applied voltage profile for typical expected operating condition

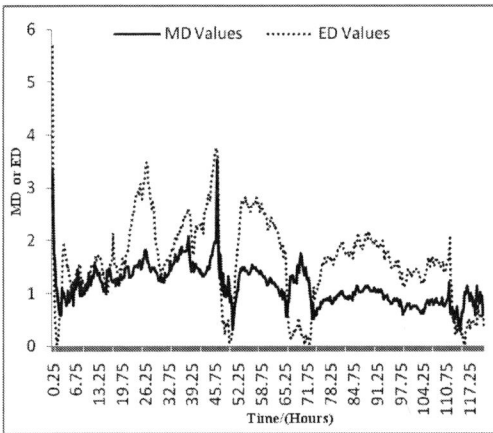

Figure 5: MD and ED values for typical expected operating performance

Based on this experiment and the data supplied by the manufacturer the typical operating conditions can be defined as follows. At the nominal applied voltage of 3.42V and input nominal current or forward bias current of 0.350A, the photodiode output voltage corresponds to light output of the LED of 1.18V and near board temperature of 42.7°C.

The results obtained from the distance measure techniques show that they can be used to estimate the health of the LED in real-time provided

all important parameters are considered for the distance measure. Figures 6(b) shows the graphs of MD and ED values for a given applied voltage profile shown in the Figure 6(a). A failure parameter that can take only two values depending on the LED's light output, 0 and 1, is used to indicate the health of the LED. A value 1 defines a healthy LED with luminance above 70% and is associated with a functional device (photodiode voltage is higher than 0.83V). If the value of the failure parameter is zero, then this is an indication of a failed LED where the light level has dropped below the critical limit. The value of the failure parameter over time is plotted in Figure 6(b) as dotted line, and the value is indicated on the right vertical axis of the graph.

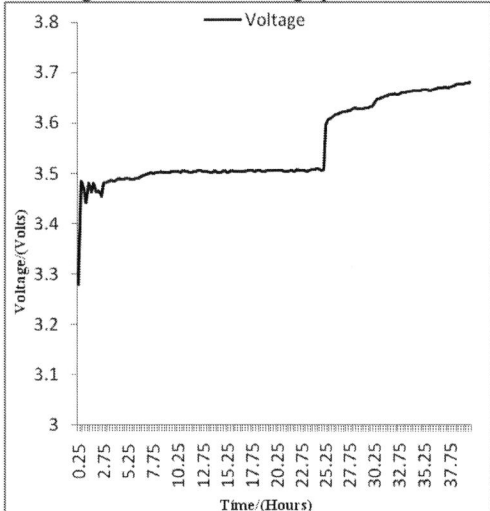

Figure 6(a): Applied voltage profile

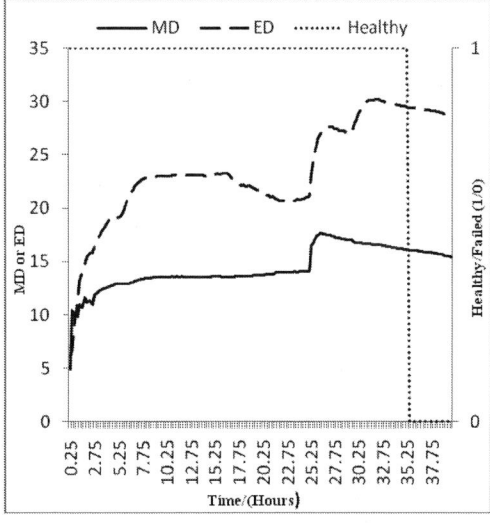

Figure 6(b): MD and ED graphs with the healthy indicator

Applied voltage load profile shown in Figure 6(a) is set so that the voltage amplitude is almost 3.5V. This is higher than the typical forward voltage of 3.42V required for the device As a result, the light output initially increases to 1.46V which is almost 126% of the typical operating light performance. Relative light output is steadily increased to 133% in 25 hours. After 25 hours, the applied voltage is further increased to 3.6V and the LED has started to fail as the light output started to decrease with time. At time 35 hours the light level drops below 70%. This result also shows that from the start of the operation the ED and MD values are above the typical maximum value of 10 which means LED was operating outside its typical expected performance. When MD value reached 15 and ED reached 25, the LED has started to degrade and after approximately 10 hours the device has failed.

From the above experiments and graphs, LED's health levels can be defined in terms of MD as follows:

1. Typical expected operating region which correspond MD values 0 to10.
2. Accelerated damage region which is MD values 10 to 15.
3. Failure threshold limit is MD equal 15 or above.

In terms of ED, health level of the LED can be defined as follows:

1. Typical expected operating region which correspond ED values 0 to10.
2. Accelerated damage region when ED values are in the range from 10 to 25.
3. Failure threshold limit based on ED is higher than 25.

Validation

To validate the above result, two more LED accelerated voltage tests are performed and data collected from the attached sensors. Figure 7(a) shows the applied voltage profile of the first LED. Figure 7(b) shows the corresponding MD and ED graphs for the LED under test and the applied voltage profile is shown in the Figure 7(a). In this case the applied voltage level was set to the LED's typical forward voltage of 3.42 initially but then increased to 3.5V at time 4 hours. The corresponding MD and ED values increased to 5 and above 5 respectively. The voltage applied to the LED is then increased again by 0.1V in three steps at times 24, 28 and 30 hours. Both MD and ED values increase above 10 indicating that the LED has started to degrade. Indeed, the measured light intensity increased to 121% following the first voltage increase.

In this case MD reached its threshold values of 15 in 29 hours. ED reached its threshold value of 25 also at time 29 hours. At this point the light

intensity peaks at 136% and then the light output started to decrease down 70% of the typical operating value. This decrease was indicated by the drop to 0.83V of the photodiode output. The LED actually failed in 33 hours. Both distance measure techniques that were tested were capable of providing early warning for the unusual performance and also have predicted accurately the time of failure. Note that in this application the failure of a LED is not directly observed as it is defined as a percentage drop in the light level, hence detectable through measurement only. Therefore, the availability of these data driven techniques is an efficient way not just for an early warning to failure but also indicating the time of failure itself.

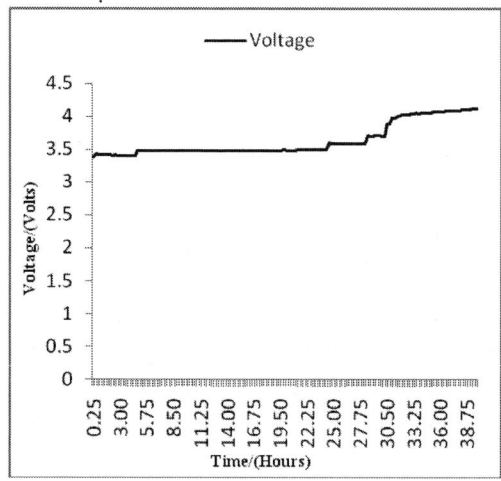

Figure 7(a): Applied voltage profile for the first LED

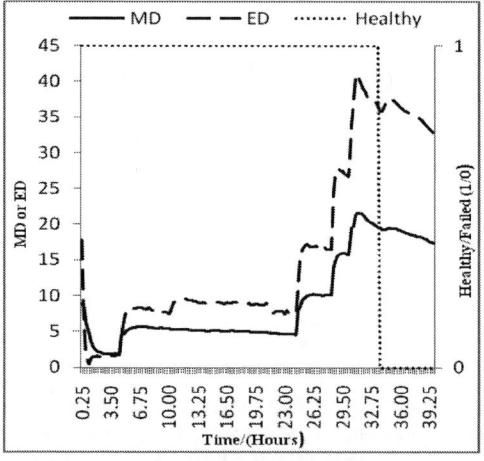

Figure 7(b): MD and ED graphs for the first LED with healthy indicator

Remaining Useful Life Estimation

In order to estimate the remaining useful life time of the LED or any system in real-time, appropriate extrapolation techniques can be used. If the LED operates under the typical expected operating conditions, then the expected useful life time can be calculated based the expected life time of a typical LED. But if MD is out of the safety range which is MD values 0 to 10 in our case and MD values are moving towards the failure threshold value of 15 with time, and then remaining useful life time of the LED can be estimated using suitable state extrapolation techniques.

Our experiment shows the life time of a LED depends on the applied voltage which increases the current exponentially, and surroundings temperature of the LED. Heat generated in p-n junction depends on the input current. So the input current is the key deciding parameter of the life of the LED.

Conclusion and Future Work

This paper has presented the application of distance measure techniques for real-time PHM of semiconductor devices and systems based on monitoring of key operational parameters. For the purpose of PHM, ED or MD measures can be defined as health distance measures of the particular system. PHM based on these techniques has been developed and demonstrated for LEDs. The data driven PHM using distance measure approach has shown good capabilities in predicting the deterioration of the high power LEDs.

In the aspect of the real-time computing ED measure can be implemented in many low cost microcontrollers where as the MD measure may need relatively bigger computing power. Both algorithms will be benefitted from the floating point unit (FPU) which is part of a real-time computing specially designed to do the floating point arithmetic easily. Both distance measure technique have been implemented in National Instruments' PXI real-time system. In the mean time these algorithm can be implemented very easily using available C compliers for many microcontrollers. Nowadays real-time operating systems (RTOS) and C compliers are available to most of the microcontrollers and this makes the implementations easier.

Future work will focus on combining the investigated in this work data driven techniques for PHM with Physics-of-Failure (PoF) predictive models. This is very important as incorporating a PoF approach into real-time PHM will allow also to capture the underlying failure mechanisms.

References

[1] Vasilis A. Sotiris, Peter Tse and Michael Pecht, "Anomaly Detection Through a Bayesian Support Vector Machine" Journal of Transaction on Reliability, Vol. 1, No. 1, June 2010

[2] Saxena, A., Goebel, K., Simon, D. and Eklund, N., "Damage propagation modeling for aircraft engine run-to-failure simulation", International Conference on Prognostics and Health Management, Denver, CO, October 6-9, pp. 1 – 9, 2008.

[3] Lalith Jayasinghe, Yimin Gu and Nadarajah Narendran, "Characterisation of thermal resistance coefficient of high power LEDs", Proceedings of SPIE 6337, 63370V Sixth International conference on solid state lighting, 2006.

[4] Elizabeth A. Cudney, Kioumars Paryani and Kenneth M.Ragsdell "Applying the Mahalanobis-Taguchi System to Vehicle Ride" Vol. 1, No. 3, pp 251-259, Fall 2007.

[5] Elizabeth A. Cudney, David Drain, Kioumars Paryani and Naresh Sharma "A Comparison of the Mahalanobis-Taguchi Systems to a Standard Statistical Method for Defect Detection" Vol. 2, No. 4, pp 250-258, winter 2009.

[6] R. De Maesschalck, D. Jouan-Rimbaud and D. L. Massart, "The Mahalanobis Distance", Chemometrics and Intelligent Laboratory Systems, Vol. 50, Issue 1, pp 1-18, 4 January 2000.

[7] Nadarajah Narendran, Yimin Gu, Lalith Jayasinghe, Jean Paul Freyssinier and Yiting Zhu, "Long-term performance of white LEDs and systems", Proceeding of First International Conference on white LEDs and solid state lighting, Tokyo, Japan, pp 174-179, Novembr 26-30, 2007

Planar Thick Film Inductor Characterization

Jiri Pulec, Ivan Szendiuch

Faculty of Electrical Engineering and Communication; Brno University of Technology
Technická 10, 61600 Brno, CZ
ji.pulec@post.cz, szend@feec.vutbr.cz

Abstract

This paper describes experiment where model of thick film inductor was created, simulated and verified. For simulation, the FEM (Finite Element Method) was used. In this paper are shown graphical outputs of simulation. Next part of this work was experimental verifying of obtained results. For this, testing sample was designed, manufactured and measured. Measured values were statistically evaluated.

Keywords: Planar inductor, Inductance, Magnetic field, Numerical modeling

Introduction

Although in contemporary technology inductors are receding into the background, they are still used in some field, e.g. in communication technologies, military and generally for lumped elements. Hereat, on-chip and off-chip technologies are used in implementation. Geometry and materials are crucial for functionality of this device. When operating, planar inductor generates magnetic field which can influence adjacent devices and shape of this field also influences this inductor itself (skin effect, reflections). This paper deals with simulation of magnetic field generated by planar inductor. Knowledge of shape of this field is helpful when designing these devices, it is possible to simulate device and design it to operate effectively. Results of these simulations are also way to detect potential inductive coupling to adjacent parts of system (which is undesirable) and design can be matched to be compliant with respect to these aspects. In addition, numerical methods are useful to determine the planar inductor inductance because there is possible to integrate magnetic flux density over surface.

Scope of interest

For purpose of thick film inductor chcaracterization, it is necessary to solve Maxwell's equation, namely Ampere's law, or equation known as Biot –Savart law. But because of complexity of considered structures without any obvious symmetry, it would be very ambitious. So electromagnetic field is simulated and viewed by method of FEM (Finite Element Method), which is method approximating solution of differential equations. This method is implemented in computational system ANSYS, which is used for this purpose. Geometry was created, materials and initial conditions were adjusted in order to simulation of magnetic field of inductor and determine of its shape. This shape is viewed by means of vector field, thus it is possible to determine the direction of this field. Next, surface integration of this field and surface integration of current density was made to determine overall inductance of inductor. The usual method to inductance determination is as follows:

$$L = \frac{\Phi}{I}$$

where Φ denotes magnetic flux generated by an inductor. L is the value of inductance and I is current passing through inductor.

Next, it is possible to express Φ as an integral of magnetic flux density B over surface of the inductor, that is:

$$\Phi = \int_{\Sigma} B \cdot dS$$

where Σ is the overal area of inductor and B.dS is overal magnetic flux in element dS of an area. Obviously, overal magnetic flux is obtained by integrating of values elementar magnetic flux over entire area of inductor.

In each elemental area dS, magnetic flux is generated by the current passing thorugh conductor from which the inductir is created. Here is possible to determmine the magnetic flux density.

Usual form of Biot – Savart law is:

$$dB = \frac{\mu}{4\pi} \frac{I \cdot ds \cdot \sin\varphi}{r^2}$$

where I denotes passing current, μ is a permeability, ds denotes the conductor element, r is the distance and φ is the angle between axis of conductor and flowline of this conductor and point where magnetic flux density is calculated.

Inductance of inductor is described as realation between generated magnetic flux and current passing through inductor which is generating this magnetic flux. Obvoiusly, using FEM, we can obtain value of magnetic flux by surface integrating of magnetic flux density.

Experiment

Inductor described and simulated in this article has rectangular shape, with outer dimensions 20,5 by 20,5 mm. Trace width is 0,5 mm, and its thickness of simulated was originally 70 μm but later reduced to 70 μm. Inductor is realized on alumina substrate by thick film technology and placed in air ambient. On Fig. 1 is shown geometry of simulated inductor. It is realized by thick film technology using silver conductive paste. Thickness of this pattern after cofiring is approximately 10 microns. Therefore, thickness of simulated model had to be reduced.

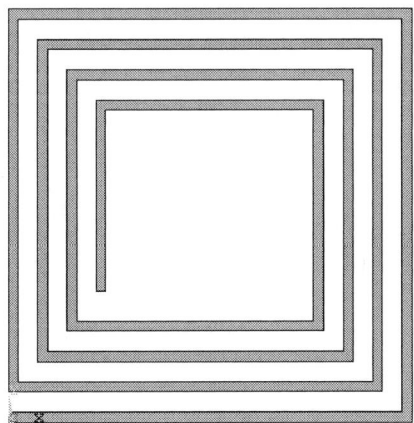

Figure 1: Geometry of planar inductor

For solution, model must be excited by current passing through conductor. This is made by setting of initial conditions where current 5 amps was set. Model was then simulated and in the plane of bottom of inductor, results were obtained. On Fig. 2 is displayed distribution of magnetic flux density in this plane in direction tormal to this plane.

On Fig. 2, line shows the area where integration is performed. The surface integral was applied and result was related to an applied current (5A), so inductance was calculated on 0,133 μH.

But on FEM, there is also possible to display vectors of magnetic field – unlike on Fig. 2 where only magnitude in direction normal is displayed. On vector field is possible to see vectors of magnetic flux density or of a magnetic intensity. This is displayed on Fig. 3 and Fig. 4.

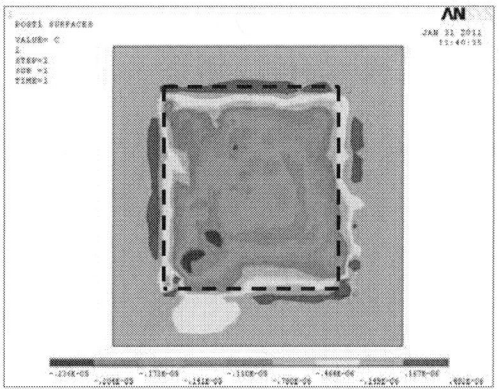

Figure 2: Magnetic flux density in direction normal to plane

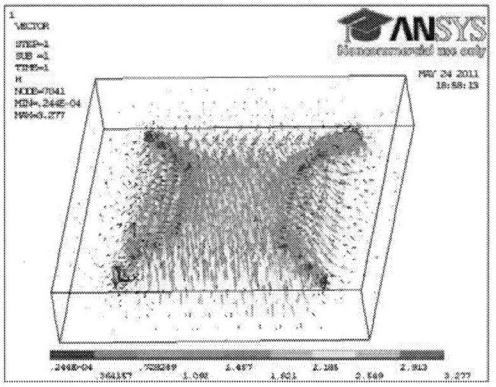

Figure 3: Vector plot showing intensity of magnetic field

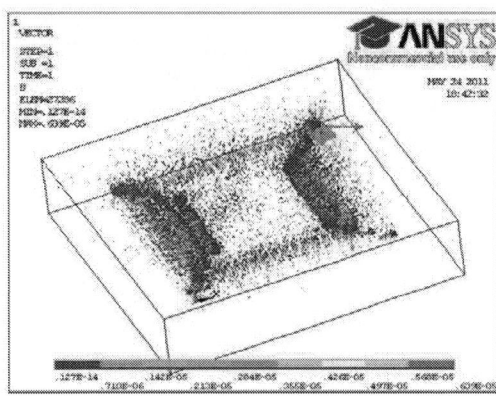

Figure 4: Vector field representing magnetic flux density

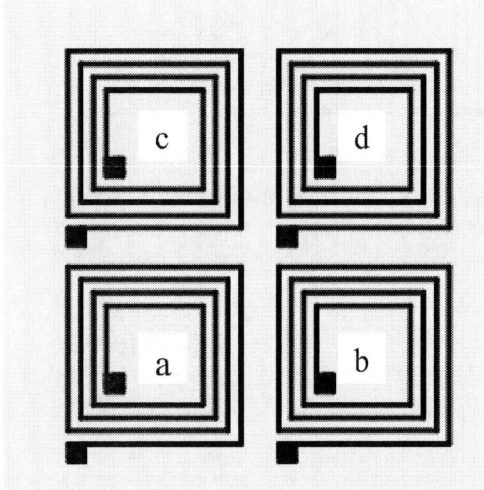

Figure 5: Layout of set of inductor for thick film technology

Next, for verifying this method of planar inductor characterization, sample with real inductors was made on alumina substrate using silver conductive paste. The thickness of film is 10 microns with 4 micron roughness. On Fig. 5 is layout of substrate with inductors.

As the substrate is here alumina ceramic with dimmensions 50 x 50 mm and 1 mm thickness. Inductors in this set are labelled as "a", "b", "c" and "d". two such sets were made and results were statistically evaluated in table.

Results

Set of capacitors made on substrate was measured, where values of capacitance were listed into Tab. 1.

Sample 1	a	0,124
	b	0,103
	c	0,122
	d	0,141
Sample 2	a	0,129
	b	0,132
	c	0,122
	d	0,102
Average value		0,122
Standard deviation		0,013

Tab. 1. Measured values of inductance of inductors

Conclusion

As we can see, average value of inductance is 0,122 µH, which is 0,011 µH lower value than value computed. But this difference is lower than is value of standard deviation.

Simulationg of inductance is thus method for reliable determinig of value of inductance. In addition, here is also possibility of displaying of detailed distribution of magnetic intensity and magnetis flux density vectors. This is very useful when designing devices with coupled fields; e.g. transformators or other. Displaying of vectors of magnetic intensity and magnetic flux density is also way to reliable design of devices, where the electromagnetic compatibility is crucial. Here is possible to show, where potential (undesirable) interferences will occur. So, FEM modeling in an useful tool, for component design.

References

[1] Hoffmann, K. Planární mikrovlnné obvody. Praha: ČVUT,2007. 2. ed. 145 p.ISBN 978-80-01-13705-8

[2] Svačina, J. Mikrovlnná technika. Praha :SNTL,1987. 2. edition. 218 p.

[3] Soutor, Z., Šavel, J., Žůrek, J., Hybridní integrované obvody. Praha: SNTL. 1982. 369 p. ISBN 04-532-82

[4] Conn, D. R., Naguib, H. M., Anderson, C. M. Mid/Film for Microwave Integrated Circuit. IEEE Transaction on Components, Hybrids, and Manufacturing Technology, Vol. CHMT-5, No. 1, March 1982

[5] Ilgenfritz, R. E. , Mogey, L. E. , Walter, D. W. A High Density Thick Film Multilayer Process for LSI Circuits, IEEE Transactrions on Parts, Hybrids, and Packaging, Vol. PHP-10, No. 3, September 1974

Acknowledgement

This research was supported by the grant from the Czech Ministry of Education (MSM 0021630503 MIKROSYN „New Trends in Microelectronic Systems and Nanotechnologies")

Laser Patterning of Thin Films for Luminescence Applications

Thomas Höche[1], Michael Lorenz[2], Alexander Müller[2],
Marius Grundmann[2], and Kai Mittwoch[3]

[1] Fraunhofer Institute for Mechanics of Materials IWM, Walter-Hülse-Straße 1, 06120 Halle, Germany

Phone: +49 345 5589 197, Fax: +49 345 5589 101, E-mail: thomas.hoeche@iwmh.fraunhofer.de

[2]Universität Leipzig, Faculty of Physics and Earth Sciences, Linnéstraße 5, 04103 Leipzig, Germany

[3]3D-Micromac AG, Technologie-Campus 8, 09126 Chemnitz, Germany

Abstract

Fresnoite, $Ba_2TiSi_2O_8$, possesses a very intense and broad photoluminescence band similar to that of the most intense oxide scintillator materials. As such, fresnoite is to be considered a well-suited material for photonics applications, including phosphors. Complete miscibility with the isotypic fresnoite compounds $Ba_2TiGe_2O_8$ and $Sr_2TiSi_2O_8$ as well as doping with rare-earth ions facilitates colour-space design. In order to qualify fresnoite as a material for photonic applications, fresnoite thin films were synthesized on a-plane sapphire, upon which, in comparison to growth on silicon (100), quartz, and MgO (100), the texture is most pronounced. Infrared-laser direct writing in amorphous fresnoitic films on sapphire allows a spatially resolved crystallization and an enhancement of the luminescence intensity by about three orders of magnitude. The latter fact can be utilized for UV-sensitive marking and wavelength converters.

Key words: Thin-Film Patterning, Photoluminescence, Structure-Property Relationships, Optically Active Materials

Introduction

After the first report on the mineral fresnoite, $Ba_2TiSi_2O_8$ (BTS), back in 1965 [1], the research interest in synthesized crystalline fresnoite and the corresponding glass is still increasing due to their unique application potential. Mainly due to its acentric structure, the latter is based on a very special combination of pronounced luminescence, pyro-, piezo-, and ferroelectric properties, as well as non-linear optical properties, such as second-harmonic generation, of fresnoite. The blue-green fluorescence and the long decay time of the luminescence that are typical for fresnoite were reported early by Blasse and attributed to deep Ti^{4+} centers [2], cf. Fig. 1. In contrast, a blue-white luminescence was found at melt-quenched fresnoite glass ceramics [3].

The preparation of glass-ceramics by ceramming of fresnoitic glasses has been used for more than 20 years. Laser-induced crystallization of optically non-linear fresnoite crystallites inside a bulk glass, however, was only demonstrated recently [4]. The essentially isostructural compounds $Ba_2TiGe_2O_8$ (BTG) and $Sr_2TiSi_2O_8$ are completely miscible with $Ba_2TiSi_2O_8$ (STS) [5]. Various routes to the synthesis of fresnoite bulk materials have been pursued, but only very few reports on the growth of fresnoitic thin films are known. Polycrystalline BTS thin films on Si (100) were magnetron sputtered and annealed and the IR absorption modes were correlated to

particular bonds of fresnoite [6]. Optically nonlinear BTG thin films were grown by RF magnetron sputtering, and their second harmonic generation was investigated [7].

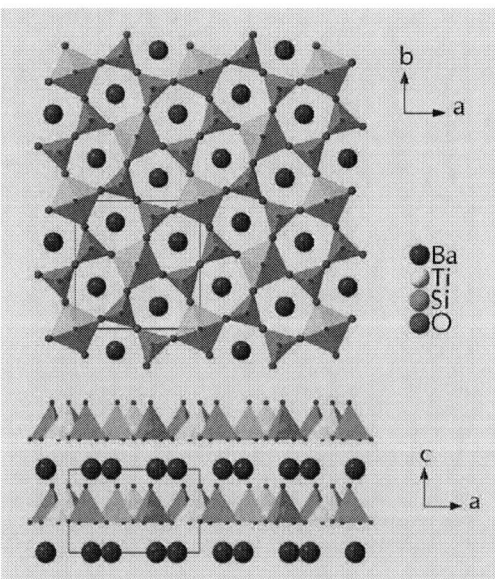

Figure 1: Crystal structure of $Ba_2TiSi_2O_8$.

In this contribution, by means of pulsed laser deposition (PLD), a different growth route to obtain thin fresnoitic films is followed, and the potential of fresnoite for photonic applications is demonstrated. Not only is its photoluminescence intensity in the visible spectral range comparable to that of other established scintillator materials, including phosphors, but also the photoluminescence properties can be controlled by local conversion of glassy fresnoitic thin films grown onto a-plane sapphire into crystallized fresnoite patterns. Laser direct writing of laterally defined crystallized structures with considerably enhanced luminescence intensity is demonstrated, and applications of this approach are discussed.

Materials synthesis

A BTS single crystal was grown by automated Czochralski pulling with radio-frequency induction heating. Experimental conditions are detailed in [8].

Prior to BTS thin film growth, the BTS PLD target was prepared by spark-plasma sintering of oxide powder pellets. The corresponding powders used for compacting polycrystalline BTS pellets (1" diameter, 4 mm thickness) were obtained by ceramming of stoichiometric glasses and grinding them to a grain size of several microns. The bulk density of the targets was greater than 90%.

For photoluminescence (PL) measurements, pellets were pressed from commercial $MgWO_4$ powder (99.9% purity) and from crystallized and grinded BTS powder and sintered for 6 h at 1.100°C and for 3 h at 1.150°C, respectively.

Thin film growth and crystallization

BTS thin films were grown on 10 x 10 mm^2 a-plane sapphire substrates by PLD using a KrF excimer laser operated at a wavelength of 248 nm and 2 J/cm^2 energy density on the target.

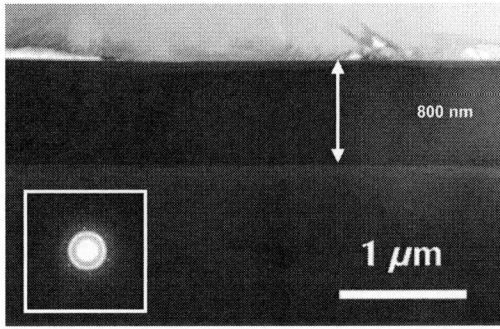

Figure 2: Cross-sectional TEM micrograph of the as-grown BTS thin film deposited onto a-plane sapphire. The electron diffraction pattern in the lower left corner proves the thin film to be fully amorphous (diffraction spots are attributed to the substrate).

The substrate temperature and the oxygen partial pressure $p(O_2)$ for PLD film growth were precisely controlled between room temperature and 700°C, and 1.6 x 10^{-2} mbar to 3.0 x 10^{-4} mbar, respectively. According to transmission electron microscopy (TEM) cross-section imaging (Fig. 2), film thicknesses ranged between 400 nm and 950 nm. Post-synthesis crystallization of the X-ray amorphous films obtained by PLD was achieved by (a) thermal annealing with quartz lamps in vacuum at about 1,000°C and 700 mbar O_2 or N_2 atmosphere for 30 minutes (Fig. 3), and (b) laser direct writing using short pulse CO_2 lasers with a maximum output power of 25 W, pulse length of ~ 1 µs, and repetition rate of 10 kHz.

Figure 3: Cross-sectional TEM micrograph of the BTS thin film deposited onto a-plane sapphire after annealing. The inset in the lower left corner illustrates the (001) BTS texture.

Using the CO_2 laser, patterns were written to large area BTS thin films on 3-inch diameter sapphire wafers. To control the fluorescence intensity and lateral resolution of the laser written structures, the laser output power as well as the scanning speed and the number of subsequent scans were varied.

TEM

The microstructure of BTS thin films was studied at cross-sections in a transmission electron microscope (HITACHI H-8100, acceleration voltage: 200 kV, point resolution: 0.23 nm). Cross-sectional samples were obtained by laser cutting of ca. 1.8 mm wide stripes, forming of "sandwiches" with the thin films aligned face-to-face, embedding the sandwiches into alumina tubes, cutting disks, grinding the disks to 100 µm thickness, one-sided dimpling to a residual thickness of ca. 20 µm, both-sided, low angle ion-beam etching (2.5 kV Ar$^+$ ions, beam current ca. 5 µA), and selective carbon coating [9].

Photoluminescence, transmission spectroscopy

Photoluminescence (PL) measurements were performed using the frequency-tripled light ($\lambda = 265$ nm) of a mode-locked Ti:Sapphire laser (pulse length 200 fs, repetition rate 76 MHz), details are given in [10].

Laser-written structures were recorded without spectral discrimination using a digital camera equipped with a macro objective. The samples were illuminated using a mercury spectral lamp, filtered by a Schott UG5 filter to select deep UV radiation (250 to 400 nm) for excitation. The integration time was 30 s, the camera sensitivity ISO 1,600.

Fresnoite Luminescence

In Fig. 4, room-temperature PL spectra of three fresnoite bulk samples (single-crystalline [(100)-oriented], polycrystalline and glassy) are juxtaposed to two high-intensity oxide scintillator materials, viz. polycrystalline $MgWO_4$ and a phosphor coating used in commercial fluorescent lamps. The absolute intensities of all luminescence spectra in Figure 4 are directly comparable within a factor of 2 as they have been measured under equal or very similar excitation and detection conditions. Polycrystalline BTS shows luminescence intensities well comparable to those of established commercial scintillator materials.

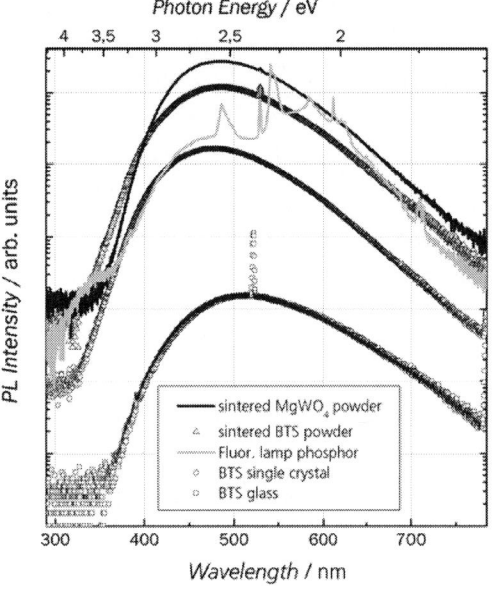

Figure 4: Room-temperature photoluminescence spectra of single-crystalline BTS, sintered, polycrystalline BTS and glassy BTS. Three typical scintillator materials are plotted for reference: polycrystalline $MgWO_4$ and phosphor coating of a commercial fluorescence lamp. The sharp peak at 530 nm visible for the BTS glass is attributed to the second order diffraction of the excitation light. Note the logarithmic PL intensity scale covering more than 5 orders of magnitude.

While pure $MgWO_4$ is known to exhibit a very high luminescence quantum efficiency up to $\eta = 1$, the room-temperature PL signal of the polycrystalline BTS pellet is found to be only slightly weaker. The intensity of the polycrystalline BTS is

fully comparable with that of the fluorescent-lamp phosphor. This indicates a comparably efficient luminescence mechanism in fresnoite and the established phosphors, clearly making fresnoite a very promising material for photonics applications.

Concerning the significantly weaker PL intensity of the BTS single crystal, it has to be taken into account that the measured luminescence intensity has to be weaker due to total internal reflection at the sample surface. Therefore, the luminescence from the BTS pressed powder sample is 7.5 times stronger than the signal from the BTS single crystal in Fig. 4, indicating that light coupling out is much more efficient for the powder sample. The photoluminescence of single- and polycrystalline BTS is peaking at a photon energy of ~ 2.59 eV.

As shown elsewhere [10], annealed BTS thin films show spectral maxima within 20 meV of this value as well. Spectral shapes of BTS spectra in Figure 4 are generally very similar to those of other materials, such as the structurally very different polycrystalline $MgWO_4$, exhibiting its spectral maximum at 2.54 eV. Single-crystalline ZnO, possessing its near-band-edge maximum at 3.26 eV in the UV range, is clearly different, reaching a comparable peak height as the BTS single crystal. The typical green luminescence band of ZnO at 2.5 eV, often attributed to oxygen vacancies, is observed in the visible spectral range.

Laser Direct Scribing into Fresnoite Films

Using laser annealing with a short-pulsed CO_2 laser (P < 25 W, pulse length: 1.0 ms), it is possible to locally crystallize thin amorphous BTS films and to form patterns of a locally enhanced luminescence intensity (compare PL intensity of BTS glass and BTS polycrystal in Fig. 4).

Sapphire proofed a particularly well-suited substrate since the ablation threshold is significantly lower than that of fresnoite. Already well below the fresnoite ablation threshold, clear changes in the film structure became visible (Fig. 5).

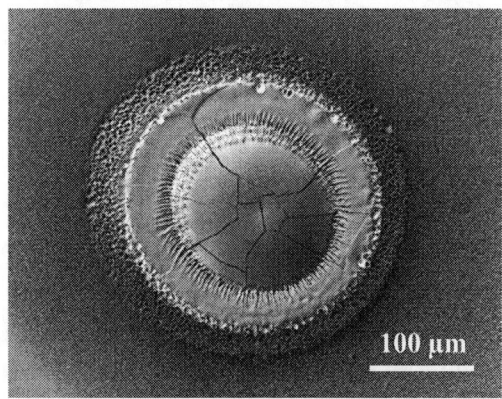

Figure 5: Scanning electron micrograph of an area crystallized by a single CO_2 laser shot.

In order to improve the spatial resolution of local crystallization, the laser power was further reduced to generate the pattern shown in Fig. 6. By careful adjustment of the laser irradiation parameters, somewhat dimmer patterns, possessing much more uniform fluorescence, can be generated. A spatial resolution of a few tens of microns can be readily achieved.

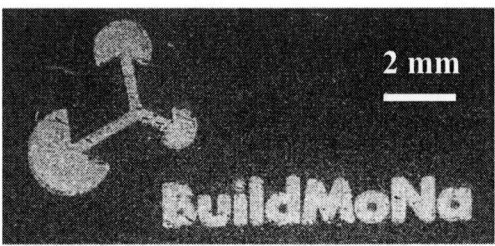

Figure 6: Luminescence of laser-scribed pattern in a fresnoite thin film on sapphire under illumination with a filtered mercury lamp. For local crystallization, an attenuated CO_2 laser (3 W output power, 10 kHz repetition rate) was applied at a mark speed of 200 mm·s^{-1}. For enhancement of luminescence homogeneity, 24 passes of a 80 μm – 100 μm diameter laser spot were applied. Adapted from [10].

Conclusions

The recipe given above, represents only one among various possibilities to locally vary fluorescence intensities in fresnoitic thin films.

Other approaches include:
(i) local ablation of the luminescing film using laser fluences above the ablation threshold of the thin film but below that of the substrate, followed by bulk crystallization of the remaining thin-film pattern,
(ii) localized ablation of an already crystallized thin film using "cold" ablation (ultra-short laser pulses) [11].

Both these approaches render the transition between luminescing and non-luminescing areas smoother. In general, emission characteristics of underlying emitters (e.g. LEDs) can be taken into account by measuring the latter and adapting the fluorescence pattern accordingly, or geometric shapes of the converter can be designed.

Color-space design can be accomplished by synthesis of solid-solution thin films, e.g. containing a blend of $Ba_2TiGe_2O_8$, $Ba_2TiSi_2O_8$, $Sr_2TiSi_2O_8$, or doping the aforementioned mixed compositions with rare-earth elements like Yb, Tm, Nd, Er, Sm etc.

Coupling out can be enhanced by using PLD targets with surplus SiO_2, resulting in a less uniform microstructure characterised by crystalline dendrites in an amorphous SiO_2 matrix [12].

Last but not least, a use of CO_2 lasers is not mandatory, but other IR lasers, such as diode-pumped solid-state lasers, could be used as well. Besides higher spatial resolution, the latter might possess the advantage that the substrate might become transparent, facilitating selective ablation and enabling laser micromachining even through the substrate. This might proof beneficial for industrial applications such as the fabrication of security features against product piracy, individualized branding of products, or frequency converters for UV LEDs.

Acknowledgements

We are indebted to Z.J. Shen (Stockholm University) and G. Ramm for the spark-plasma sintering and powder synthesis of PLD targets, respectively. We thank H. Hochmuth for synthesis of the BTS thin films and G. Benndorf for help with the PL measurement setup as well as R. Uecker (Institute for Crystal Growth Berlin) for growth of BTS single crystals. B. Rauschenbach is acknowledged for granting access to the SEM, TEM, and preparation facilities at Leibniz Institute for Surface Modification (IOM) in Leipzig, where microdiagnostics was performed when TH was still working with IOM. Financial support by Deutsche Forschungsgemeinschaft within SFB 762 "Functionality of oxide interfaces", of the Graduate School "Leipzig School of Natural Sciences - Building with Molecules and Nano-objects" BuildMoNa Grant No. GS185/1, and of the European Social Fund in the framework of "ESF in Sachsen" is kindly acknowledged.

References

[1] J. T. Alfors, M. C. Stinson, R. A. Matthews, and A. Pabst: "Seven new Barium Minerals from Eastern Fresno county, California," Am. Mineral., Vol. 50, pp. 314-340, 1965.

[2] G. Blasse, "Fluorescence of compounds with fresnoite ($Ba_2TiSi_2O_8$) structure," J. Inorg. Nucl. Chem., Vol. 30, pp. 2283-2284, 1968.

[3] A. Halliyal, A. S. Bhalla, S. A. Markgraf, L. E. Cross, and R. E. Newnham, "Unusual Pyroelectric and Piezoelectric Properties of Fresnoit ($Ba_2TiSi_2O_8$) Single Crystals and Polar Glass-Ceramics," Ferroelectrics, Vol. 62, pp. 27-38, 1985.

[4] Y. Dai, B. Zhu, J. Qiu, H. Ma, B. Lu, S. Cao, and B. Yu: "Direct writing three-dimensional $Ba_2TiSi_2O_8$ crystalline pattern in glass with ultrashort pulse laser," Appl. Phys Lett. Vol. 90, Art. 181109, 2007.

[5] Th. Höche, C. Rüssel, and W. Neumann: "Incommensurate Modulations in $Ba_2TiSi_2O_8$, $Sr_2TiSi_2O_8$, and $Ba_2TiGe_2O_8$," Solid State Commun., Vol. 110, No. 12, pp. 651-656, 1999.

[6] M. K. Zhu, Y. H. Yang, X. H. Li, B. Wang, H. Wang, and H. Yan: "Preparation of $Ba_2Si_2TiO_8$ thin films by magnetron sputtering," Microelectron. Eng., Vol. 66, pp. 745-749, 2003.

[7] R. Ogawa, H. Masai, Y. Takahashi, H. Mori, T. Fujiwara, and T. Komatsu: "Fabrication of $Ba_2TiGe_2O_8$ thin film with optical nonlinearity," Jpn. J. Appl. Phys, Vol. 46, pp. 7145-7147, 2007.

[8] P. van Aken, Th. Höche, F. Heyroth, R. Keding, and R. Uecker: "Insights into Oxygen-Cation Bonding in Fresnoite-Type Structures from O-K-Electron Energy-Loss Spectra and *Ab Initio* Calculations of the Electronic Structure" Phys. Chem. Minerals, Vol. 31, pp. 543-552, 2004.

[9] Th. Höche, J.W. Gerlach, and T. Petsch: "Static-Charging Mitigation and Contamination Avoidance by Selective Carbon Coating of TEM Samples" Ultramicroscopy, Vol. 106, pp. 981-985, 2006.

[10] A. Müller, M. Lorenz, K. Brachwitz, J. Lenzner, W. Skorupa, K. Mittwoch, M. Grundmann, and Th. Höche: "Spatially Resolved Laser Crystallization of Fresnoitic Thin Films: Evolution of Texture and Photoluminescence" CrystEng-Comm, doi: 10.1039/C1CE05265A, 2011.

[11] D. Ruthe, K. Zimmer, and Th. Höche: "Etching of $CuInSe_2$ Thin Films - Comparison of Femtosecond and Picosecond Laser Ablation" Appl. Surf. Sci., Vol. 247, pp. 447-452, 2005.

[12] Th. Höche, R. Keding, R. Hergt, and C. Rüssel: "Microstructural Characterisation of Grain-Oriented Glass-Ceramics in the System $Ba_2TiSi_2O_8$ - SiO_2," J. Mater. Sci. Vol. 34, No. 1 pp. 195-208, 1999.

Effect of Substrates Surface Condition on the Morphology of Silver Patterns Formed by Inkjet Printing

Z. Xiong, C. Liu

Loughbourough University, Leicestershire UK, LE11 3TU

+44-07507147130, Z.Xiong2@lboro.ac.uk

Abstract

Preheating is considered to be an effective method to improve the resolution of inkjet printing conductive tracks. However, although preheating can accelerate the evaporation of the droplets, it also changes the surface condition of substrate, which affects the spreading behaviour of the droplets. In this study, four types of commonly used substrates are employed to investigate the effect of preheating on inkjet printed silver patterns. The result shows that coffee stain effect becomes more obvious as the substrate temperature increases. Optimal preheat temperature exists for each substrate, which is >80°C for glass, 30-40°C for polyimide, 60-70°C for FR4 and 50-60°C for hydrophobic paper. Mathematic model has been established to discuss the factors in this process. During the track printing stage, overlap rate was also studied to achieve continuous and homogenous tracks.

Key words: Silver track, inkjet, surface condition, high resolution

Introduction

Inkjet technology has been demonstrated for applications in Printed Electronics (PE) in the past decade. The direct writing of metal patterns in its data-driven nature is particularly attractive. To obtain optimal conductive patterns (e.g. using Ag inks) from ink-jet printing, preheating of the substrates is usually applied [1][2]. High substrate temperature can change the course of deposition and ink/substrate interactions that can be associated with the evaporation of the solvent in the inks. Once the droplets land on the substrate surface, the spreading of the deposited inks is inevitable, and it is expected that the preheating can suppress such spreading. However, the increase of temperature also leads to an increase in surface energy of the substrate surface. Subject to the materials and surface properties, the interactions and wetting behaviour of the ink droplets with different substrate can be drastically changed. Therefore, optimal preheating temperatures should exist to suit different types of substrates. In this study, the effect of substrate type and preheat temperature on the formation of the inkjet printed patterns is discussed to provide an insight into the major factors that can determine the quality and robustness of the printed features.

Experimental

Nano silver ink with a content of 20% silver (from Printed Electronics Ltd) is used in the ink-jet printing with a Microfab Jetlab® 4 inkjet system. Waveforms of the dispensing device (40μm orifice and 60μm orifice) in the jetting process are firstly studied, thus the processing parameters can be

achieved for the optimum droplet formation to eliminate potential deviation and satellite droplets. Four types of substrates commonly used in the industrial applications including glass, FR4 Epoxy Glass Cloth, polyimide (PI) and hydrophobic paper are utilised to investigate the droplet/substrate interactions and the spreading behaviour of ink droplets on substrates with different surface conditions. Dataphysics OCA 20 is used to measure the surface energy of the four substrates. Prior to the printing, the substrates are preheated to a temperature from 20°C to 80°C. Both single droplet and formation of continuous tracks by multiple deposition of droplets are investigated, then the printed deposits are treated by a post-sintering process at 200°C for 20 min. Optical microscope is employed to observe the printed patterns, and their dimensions such as droplet diameter and track width are measured by Zygo NewView 5000. The cross-sectional profiles of such sintered features are also analyzed to understand the behaviour of evaporation and surface wetting characteristics as preheat temperature increases.

Results and discussion

The optical top views and cross section profiles of the droplets on glass substrate with different preheated temperatures are shown in Fig 1. It is obvious that once the temperature increases, not only the diameter of the dots is getting smaller, but also the shape becomes more regular. Since the surface of the glass cannot be absolutely flat and homogeneous, the irregular outline of the dots on substrate with low temperature is supposed to be caused by the spreading of the droplets before a steady state obtained. However, the growth of substrate temperature leads to a higher evaporation speed, resulting in suppression for spreading. The evaporation rate of the droplet can be calculated as following [3]

$$M = \frac{A(m+nv_a)(P_w - P_a)}{H_v} \quad (1)$$

Where A is the surface area, m and n are constants, v_a is the air velocity over water surface which is zero in this study, P_w is the saturation vapour pressure at present temperature, P_a is the saturation vapour pressure of air dew point, and H_v is the latent heat of vaporization of water. Considering the time between the end of impact stage and steady state is often longer than the time of evaporation t in practical situation, droplet can be supposed to keep spreading during the whole evaporation process. Thus the spreading time is

$$t_0 = t = \frac{Q_0}{M} = \frac{Q_0 H_v}{mA(P_w - P_a)} \quad (2)$$

Where Q_0 is the total mass of the solvent and according to Antoine formula

$$P_w = e^{A_1 - \frac{B_1}{T + C_1}} \quad (3)$$

It can be concluded that the spreading time decreases as the raise of temperature, which is independent on the property of substrate. Besides, the edge of the dots is found to getting dark from optical observation. This is because of, according to the cross section profile, the coffee stain effect which causes a material enrichment on the edge of the dots. The severity of the effect is proportional to the evaporation speed.

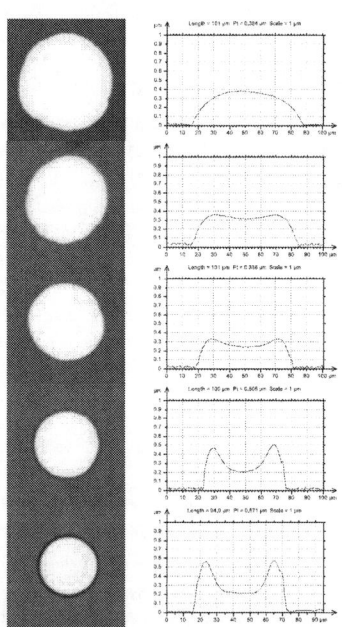

Figure 1: Optical top views and cross-section profiles of droplets on preheated glass substrate (top to bottom is 20°C, 35°C, 50°C, 65°C and 80°C)

Although preheat of the substrate can accelerate evaporation of the droplet and accordingly suppress the spreading time, it cannot always decrease the dot size. The temperature growth of the substrate also increases the surface energy which offers better wettability between the droplet and substrate. In this case, the contact angle in steady state will decrease leading to a larger contact area. Therefore, subject to substrates with different surface properties, the temperature growth can either diminish the dot size (when the change of spreading time dominates) or increase the dot size (when the trend of spreading dominates). The situation on four commonly used substrates is shown in Fig 2.

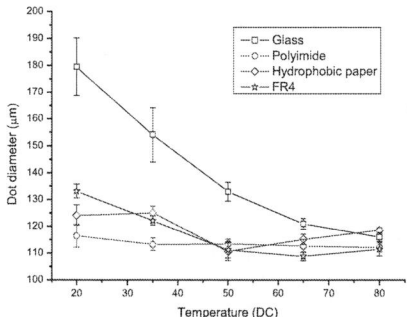

Figure 2: Dot sizes on four substrates under different preheat temperatures

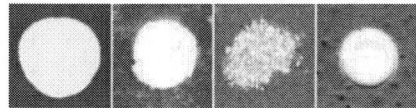

Figure 3 Optical top view of dots on (from left to right) glass, hydrophobic paper, FR4 and polyimide at 20°C

As the experimental result suggested, to pursue the highest printing resolution, the optimal preheat temperature in this experiment is >80°C for glass, 30-40°C for polyimide, 60-70°C for FR4 and 50-60°C for hydrophobic paper. For all the substrates, the variation of dot diameters became smaller as the temperature increased, which indicates more uniform dot morphology is obtained.

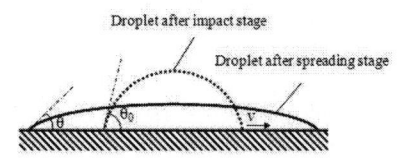

Figure 4: Schematic model of droplet spreading process

The final dot diameter can be theoretically estimated using a mathematics model, shown in Fig 4. Assuming the diameter of the contact area between droplet and substrate after impact phase is D_0, the spreading speed of the contact line is v and the spreading time is t_0, the dot diameter after spreading will be

$$D = D_0 + 2\int_0^{t_0} v\,dt \qquad (4)$$

The instantaneous spreading velocity of the contact line is [4]

$$v = 2K^0\lambda\sinh\left[\frac{\gamma_{LV}(\cos\theta_0 - \cos\theta)}{2nk_BT}\right] \qquad (5)$$

Where K^0 is the frequency of molecular displacements, λ is their average length, γ_{LV} is the surface tension of the liquid, θ_0 is the present contact angle, θ is the contact angle of steady state, n is the number of adsorption sites per unit area, k_B is Boltzmann's constant and T is the temperature. According to M. J. de Ruijter's research [5], the formula can simplify to

$$v = \frac{1}{\zeta_0}\gamma_{LV}(\cos\theta_0 - \cos\theta) \qquad (6)$$

ζ_0 is a friction coefficient per length of the contact line. γ_{LV} can be estimated with Eotvos rule

$$\gamma V^{2/3} = k(T_c - T) \qquad (7)$$

V is the molar volume of substance. T_c is the critical temperature and k is a constant valid for almost all substances. Using the Gibbs adsorption isotherm, thermodynamics and a molecular interaction model, the contact angle of steady state can be described as [6]

$$\cos\theta = 1 + C(T_c - T)^{a/(b-a)} \qquad (8)$$

Where C is an integration constant, and a and b are temperature-independent constants from a balance of intermolecular forces. The model predicts the general experimental trends of θ with respect to T. Based on formula (4) and (5), both the surface tension of the droplet and the contact angle of steady state decrease when the temperature rises. However, the decrease of the contact angle of steady state can lead to an increasing spreading speed, while the decrease of droplet surface tension has a negative effect on it. Since a, b and C are related to the substrate properties (e.g. surface energy and surface roughness), temperature influences the spreading speed of contact line in different ways according to the substrate type.

Table 1: Contact angle and surface energy of the four substrates

Substrate	Contact angle (°)		Dispersive	Polar	Total
	Water	Diiodomethane			
Glass	35.7	61.0	14.42	46.37	60.79
Hydrophobic paper	77.4	40.3	39.45	4.40	43.85
FR-4	122.7	80.3	19.77	0.75	20.52
Polyimide	80.4	31.9	40.59	2.14	42.73

In this study, without preheating the substrate, the dot diameter on glass, hydrophobic paper and polyimide is propotional to the surface energy of the substrate (shown in Table 1). However, the size of the dots on FR4, which should be the smallest in theory, is second to the one on glass. This is because, instead of a static situation, the droplets impact the substrates with a speed around 2 m/s during the inkjet process. In the impact phase, the liquid recoils towards the impact line of the first collision and climbs along the impact line, driven by the surface tension [7]. Nevertheless, the contraction of the dots on FR4 has been prevented owning to the high surface roughness, resulting in an irregular outline shown in Fig 3.

The acceleration of the solvent's evaporation significantly contributes to the decrease of dot size when heated. However, small range of spreading occurred while the droplets were landing on polyimide. The dot size was slightly decreased at the early stage of preheating. When the substrate was heated up to around 35°C, a balance generated between the evaporation of the solvent and the increasing spreading speed kept the dot size staying at the same level. For FR4 and hydrophobic paper, minimal dot size can be achieved through preheating in a certain range of temperature. Beyond the dividing point, the increase of spreading speed became dominant and the dot size turned to increase.

Dot spacing is an important parameter in printing tracks. Continuous track can only be obtained with suitable dot spacing value. With variant conditions of ink and substrate, the optimal dot spacing will be different. Results of dot spacing study on glass are shown in Fig 5. An overlap rate of 27% is comparatively suitable in this process. For polyimide, FR4 and hydrophobic paper, the optimal overlap rate is 32%, 33% and 28% respectively. When the dot spacing is too large, dots may not connect to each other. When the dot spacing is set too small, on the other hand, bulge will be generated.

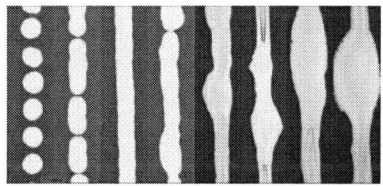

Figure 5 Printed silver tracks with different dot spacing values (left to right is 170 μm, 125 μm, 110 μm, 100 μm, 80 μm, 70 μm, 60 μm and 50 μm. The average dot diameter is about 150 μm)

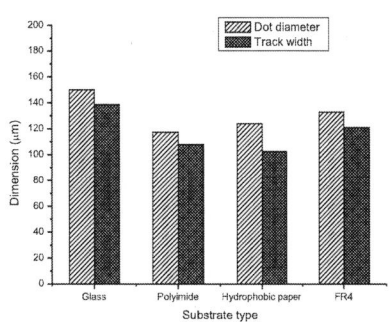

Figure 6 average dot diameters, track width of 1 layer on different substrates

According to Fig 6, the widths of the tracks are normally smaller than the related dot diameters. This can be explained using the model in Fig 7. Instead of spreading in all directions on the substrate as the dot behaves, a horizontal uniform distribution was first achieved when the track was printed. As a result, the contact angle between droplet and substrate in the interface of two droplets was larger than other area, leading to a preferential spreading trend. Thus relatively linear track outline can be obtained even though the original droplet outline is undee.

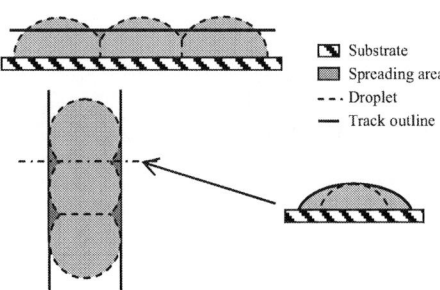

Figure 7 schematic illustration of droplet spreadingSucceeding Pages

Summary

In summary, evaporation of the droplets can be accelerated by preheating, resulting in a more obvious coffee stain effect and more uniform dot diameters. Although preheat can diminish the time of spreading, the upgraded wetting property can increase the spreading speed of the contact line as well. According to the experimental results of this study, to pursue the highest printing resolution, the optimal preheat temperature in this experiment is >80°C for glass, 30-40°C for polyimide, 60-70°C for FR4 and 50-60°C for hydrophobic paper. The results are supposed to provide a general guidance to the practice of inkjet printing industry. A mathematics model has been developed to explain the roles of surface tension, roughness and temperature in this process. In the track printing stage, the optimal overlap rate for the four substrates are 27%, 32%, 33% and 28% respectively. The width of the track is normally smaller than dot diameter due to the nonuniform spreading trend in different area of the printed droplets.

References

[1] S-H Lee, K-Y Shin, J Y Hwang et al., "Silver inkjet printing with control of surface energy and substrate temperature", Journal of micromechanics and microengineering, vol. 18, 075014, 2008.

[2] Dong Jun Lee, Je Hoon Oh, "Shapes and morphologies of inkjet-printed nanosilver dots on glass substrates", Surface and Interface Analysis, vol. 42, pp. 1261-1265, 2010.

[3] Charles C. Smith, George Lof and Randy Jones, "Measurement and analysis of evaporation from an inactive outdoor swimming pool", Solar Energy, Vol. 53, No. 1, pp. 3-7, 1994.

[4] T. D. Blake and A. Clarke, "Contact angle relaxation during droplet spreading: comparison between molecular kinetic theory and molecular dynamics", Langmuir, vol. 13, pp. 2164-2166, 1997.

[5] M. J. de Ruijter, T. D. Blake and J. De Coninck, "Dynamic wetting studied by molecular modelling simulations of droplet spreading", Langmuir, vol. 15, pp. 7836-7847, 1999.

[6] Adamson, A. W., "Potential distortion model for contact angle and spreading II. Temperature dependent effects", Journal Colloid Interface Science, vol. 44, pp. 273-281, 1973.

[7] Hsuan-Chung Wu, Weng-Sing Hwang, Huey-Jiuan Lin, "Development of a three-dimensional simulation system for micro-inkjet and its experimental verification", Materials Science and Engineering A, vol. 373, pp. 268-278, 2004.

Embossed Ceramic Reflectors with Nano Dispersive Coatings for Compact Optoelectronic Systems

W. Buß *, H. Richter**, Th. Bartnitzek***, W. Brode****, A. Heymel**** and
Th. Walther*****

(*) Fraunhofer Institute for Applied Optics and Precision Engineering IOF, Jena, Germany;
(**) Fraunhofer Institute for Ceramic Technologies and Systems IKTS, Hermsdorf, Germany;
(***) VIA electronic GmbH, Hermsdorf, Germany;
(****) SIEGERT TFT GmbH, Hermsdorf, Germany;
(*****) LUST Hybrid-Technik GmbH, Hermsdorf, Germany

Phone: +49 3641 807 441, Fax: +49 3641 807 603, E-mail: wolfgang.buss@iof.fraunhofer.de

Abstract

Modern optoelectronic elements used in harsh environments like high ambient temperatures and UV-radiation are often equipped with ceramic housing and substrates which include integrated optical elements. This reflectors or beam shaping elements are realized by dead-mold casting or embossing technologies. The present paper describes a combined embossing, infiltration and coating process for realizing smooth reflector surfaces into Low Temperature Cofired Ceramics (LTCC) substrates. The measurement results of the reflection coefficients are comparable with full-metal aluminum-surfaces. The main fields of application so far have primarily been grooves for edge-emitter UV-LEDs with forward scattering reflector areas.

Key words: LTCC reflector, embossing, infiltration, coating, LED beam shaping and alignment

Introduction

Optoelectronic modules used in sensor systems, image processing, or communication devices, are often equipped with more and more complex micro-optical components, enable the effective shaping and alignment of emitted beams as well as the selection of incoming light with respect to wavelength, polarization, and acceptance angle.

Figure 1 LTCC chip reflectors and embossing tools

To meet the demand for low cost optical elements, high precision replication techniques like injection molding, hot embossing, reaction molding, etc. have been developed. Among some special glasses which can be hot embossed during a comparatively expensive method, the commonly applied materials for these processes are polymers. These materials

are limited in their thermal load and their long-term stability is relatively low. In the strongly expanded area of LED-illumination - due to its temperature stability - LTCC as a rugged substrate material plays an important role, enabling the transport of thermal power dissipation of the LEDs on small space. Currently, more and more UV-LEDs are used in illumination modules. Because LTCC is more insensitive towards radiation than organic material it can be used as a reflector material instead of metals which have an unwanted high electrical conductivity

Objectives

Previously [1] the techniques of cold embossing for realizing optical structures in green LTCC-foil have been presented. But the roughness of the sintered reflector areas was much too high.

The goal of the present work was to find a proceeding for setting up a microstructure gradient in glass-ceramics. This gradient should generate a structure refinement in the near surface region of the LTCC. The procedure should be realized by means of infiltration of nano-dispersive solutions and a subsequent cofiring step. Furthermore, it was required to embed the complete process into the common LTCC-substrate preparation run.

Infiltration of LTCC green-foil

A lot of different type-similar and identically nano-dispersive solutions have been tested in regard of their suitability. The application technologies as well as the drying processes of the solutions have been varied and analyzed as well as the clamping procedure and the sintering conditions of the infiltered probes.

The infiltration of green LTCC-laminates using a polymeric sol by means of dip-coating and a following sinter process at 850°C led to best surface qualities.

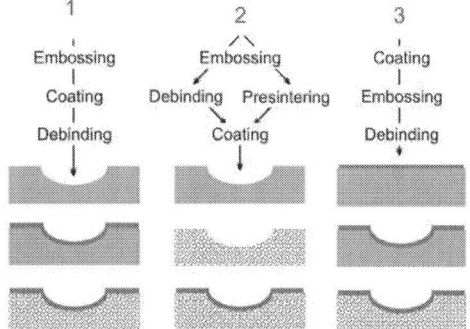

Figure 2: Variation of infiltration process

At first, a simple sequence of the process steps infiltration – embossing – sintering was chosen. But this does not lead to the requested smooth surfaces. That's why the molding process of the LTCC reflector areas has been replaced by an interlaced sequence of technologies. Figure 3 shows exemplary the manufacturing steps for realizing LED grooves.

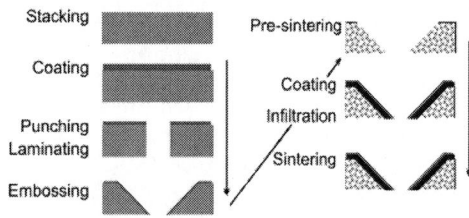

Figure 3: Manufacturing of LED grooves

Multistage Infiltration, Cofiring and Debinding

Already the one-step infiltration of green LTCC with the following embossing and sintering process steps has shown a considerable smoothing of the surface – figure 4b. But there are also low ripples visible which arise by reason of unevenly shrinkage of the reflector areas during the embossing process. Figure 4c shows a LTCC surface after the two-step infiltration procedure with a pre-sintering step in between. As a result there were significant smoother partial areas but also deep craters, emerging evident from the debinding process during the final sintering. But after aluminum-coating these sintered probes show the best reflection coefficients because the macroscopic defects do not influence the integral reflection ability. Figure 4d shows a LTCC-probe after a supersaturated second infiltration step. The final surface is characterized by macroscopic rejections of a closed surface of injection material.

These surfaces are not usable for reflectors because the reflectance is irregular and not reproducible.

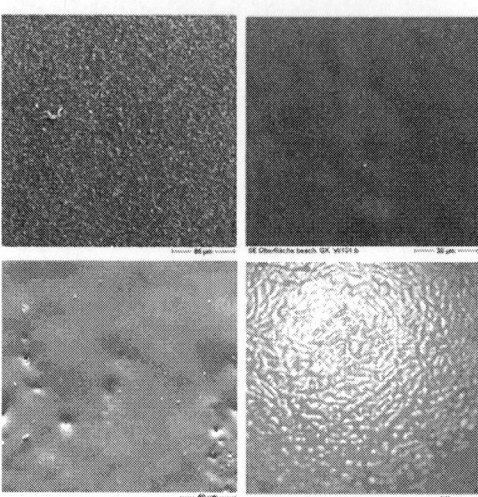

Figure 4: Surface images (after sintering) of different infiltration procedures, 4a: substrate without infiltration, 4b: one-step Infiltration, 4c: two-step infiltration with pre-sintering, 4d: supersaturated two-step infiltration

The accurate coordination of consistency of the infiltration material as well as the speed of the dip-coating procedure in connection with the sinter regime are finally crucial for realization of smooth reflector areas. But the technologies for surface leveling mentioned here are ineffective without precision molding tools with optical smooth embossing areas.

Embossing Tools and Techniques

The transfer of the optical structures into the LTCC green-foil stacks - as already shown in [1] – has been realized in parallel with the lamination process. During this procedure about 15 to 20 green-foils with about 0.1mm thickness, which may be structured e.g. by contacting paths or vias, are pressed into a flat raw substrate.

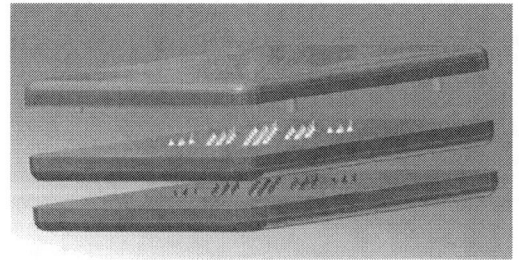

Figure 5: Combined punching and embossing tool 8", counter plate in the top, punching plate in the middle and embossing plate in the bottom; every plate was fabricated from stainless steel, 20mm thick.

426

In contrast to previously published technologies [1], a punching and an additional embossing step were used to mold the 2 mm deep reflector grooves for side-emitting LEDs. The volume of the LED-groove is too big for supplanting the LTCC green-material only by an embossing step.

The counter plate – figure 5 - is characterized by punching slots in the positions of the punch pins integrated in the punching plate. Moreover, it is equipped with stainless pins for threading green LTCC foils and also for mechanical guiding the plates during the pressure process of lamination.

After stamping out 70% - 80% of the final reflector volume in the first step, subsequently the different forms of the reflector devices will be molded under pressure using the embossing plate instead of the punching plate.

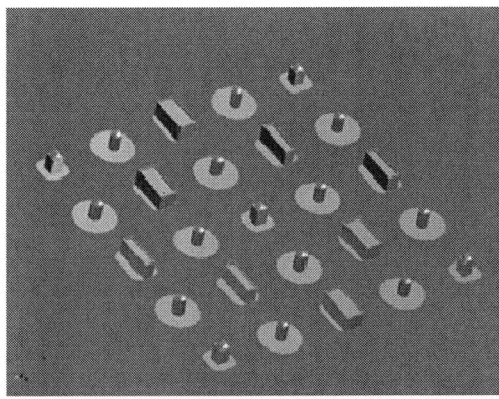

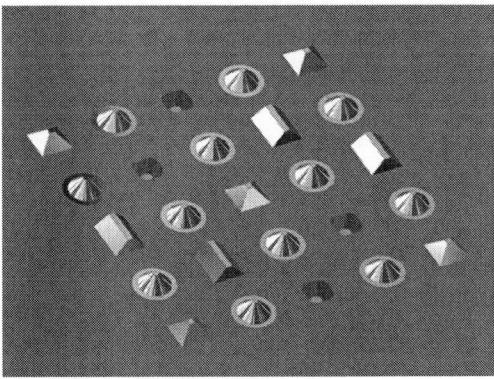

Figure 6: Top: punching plate with inserted tools for punching round, square and rectangular holes into the LTCC stack; Bottom: embossing plate with inserted embossing tools for molding pyramidal, conical and prismatic reflector surfaces

The punching pins are made from hardened steel with a Ni-coating for better demolding of the laminates from the tools. The embossing-tool-inserts comprise a hardened steel body and a Ni-coating, too. The molding surfaces of the punching-inserts have been grinded and polished up to <10 nm rms if

there are molded straight areas. Alternatively, they have been formed individually by ultra-precision single-point diamond turning.

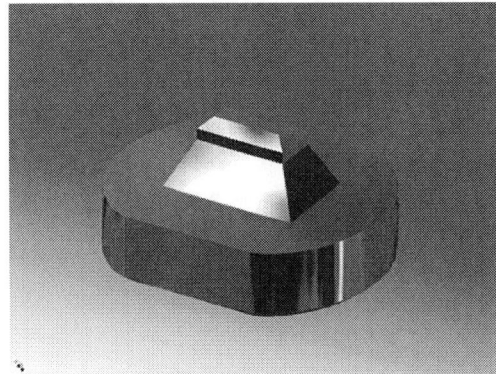

Figure 7: Embossing-tool-insert with polished working areas for molding prismatic reflector chips

A representative measured deviation profile of an oblique area relative of a planar surface is +/- 5µm; which meets a radius of curvature of 13,3m. On the same profile a roughness rms-value of 20,8nm was measured.

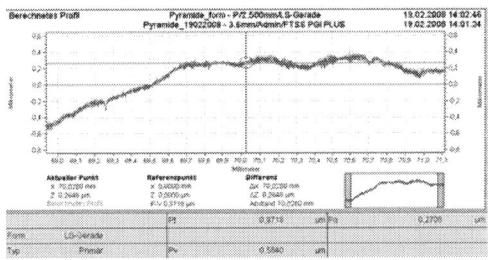

Figure 8: Deviation profile of an oblique embossing area related to a straight area

Final Coating

Because of the remaining volume scattering of the molded LTCC reflector areas, a final metal coating is essential. A 200 nm aluminum layer acts as reflector layer, and for passivation 160nm SiO_2 have been sputtered. This layer system has no leveling performance with regard to the reflector area. All the existing bumps of the infiltered and sintered areas will be transferred 1:1 outward.

Complete processed probe surfaces with two infiltration steps, pre-sintering, and final-sintering achieve approximately reflection coefficients like solid aluminum bodies with polished external faces (Figure 10).

The LTCC chip-reflectors have been molded as composite into a 2.5"x2.5" laminate. Before aluminum coating, the surrounding substrate areas have to be grinded after sintering in order to remove raised edges or corrugations. The molded reflector areas are not affected by this mechanical processing.

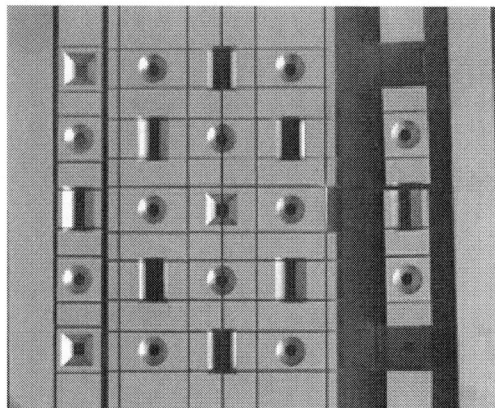

Figure 9: Al-coated composite of 15 different chip reflectors after dicing

Roughness and Reflectance Measurement

Optics uses the parameter "integral roughness" rms or rq synonymous with the average surface roughness along a defined test track as criterion of surface quality. So areas with surface roughness rms <50 nm are "mirrors" and <30 nm rms "good mirrors" with low distortion and acceptable scattering. For the striven purpose – the realization of LTCC reflector structures for LEDs or other optical emitters – partial smooth areas with low micro roughness are of interest. Long-wave bumps or partial damages like cracks - which in mirrors would induce image distortions – have no or low influence to wished simple reflector function. But the cracks influence the rms-value and so this parameter is not meaningful alone for the evaluation of the surface quality of embossed, infiltrated and coated LTCC surfaces. So in addition to the surface profile measurement, the spectral resolved reflection measurement and white-light interferometric images have been taken in order to characterize the probe surfaces.

For measuring the reflection coefficient a spectral photometer LAMBDA 900 was used.

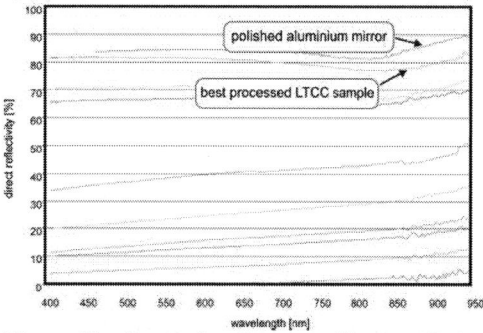

Figure 10: Spectral reflectance (direct reflex) of selected samples and different processing levels

Figure 10 shows, that the best measured sample has only a slightly worse reflectance than an aluminum mirror.

The angle-resolved measurement of the reflection ability of different probes has been carried out on a goniometer to determine the scattering distribution depending of the optical wave length

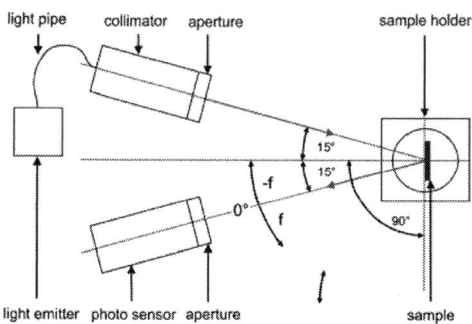

Figure 11: Goniometer measuring station for angle-resolved detection of scattered reflected light.

The divergent beam of a point light source will be collimated into a quasi-parallel beam. An aperture defines the size of the measurement spot. The incidence and reflection angles are 15°. The detection unit with an entrance pupil is moveable in a plane concentrically to the scattering center. To measure the small scattering light signals a Lock-In µV-meter was used.

A polished aluminum mirror was applied as reference object for the quality assessment of the reflector samples. The intensity distributions are characterized by a high ration of direct reflected light. Within an angle tolerance of +/- 7° the measured values are below 10%.

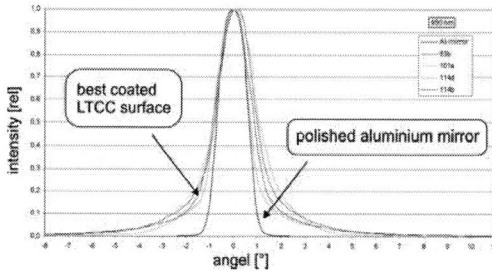

Figure 12: Scattering distribution of reflected light (950nm) from processed LTCC surfaces compared with an aluminum mirror

Measurements were performed at 528 nm and 950 nm. The relative small part of scattering light follows indirectly to the reflectivity of the different LTCC samples. In the infrared (IR) region, reflectors with a short-wave micro roughness show a lower scattering then in the visible (VIS) region, while samples with long-wave damages and partial smooth areas are characterized by nearly equal values of scattering reflection.

Generalized the molded LTCC-reflectors are favored in IR-applications, especially if a direct reflection with low stray light portion is required. In the VIS or ultra violet (UV) region a considerably proportion of stray light should be noted.

LTCC Chip Reflectors

As demonstrators LTCC chip reflectors with contact bumps for fixing have been designed and realized. It can be inserted in the same way on optoelectronic PCBs like SMD resistors or capacities

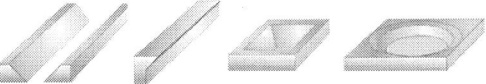

Figure 13: Basic types of chip reflectors

Currently in simple optoelectronic PCBs, mechanical elements for light shaping, deflection or shadowing are applied which are made of metal or plastics. [3] Unfavorable is however, that these components have to be assembled by special techniques. The here presented LTCC chip reflectors have contact bumps on the bottom for reflow-soldering or for conductive gluing.

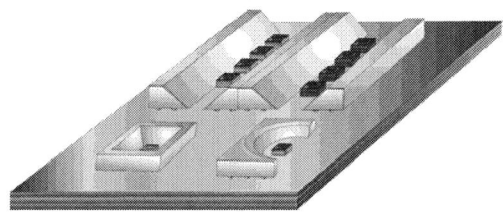

Figure 14: Possible configurations of LED chips and LTCC reflector chips

Rectangular, circular and line-shaped reflectors with 45° tilted functional areas have been constructed as demonstrators which are moldable with simple embossing tools. These elements can be separated easily by dicing and assembled by pick-and-place processes.

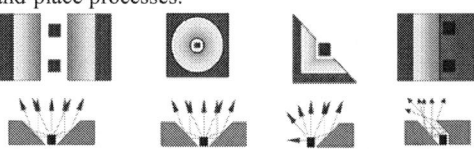

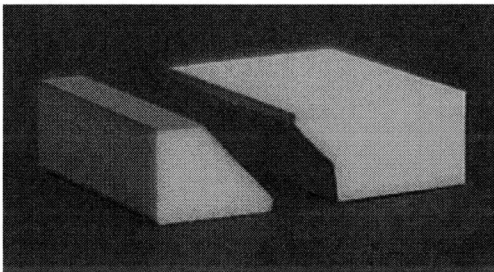

Figure 15: LTCC reflectors and LED chips and their optical action; realized deflection elements [2]

Summary

The realization of smooth glass ceramic surfaces by a combined embossing and infiltration process was shown. The molding technology can be used for forming non-expensive integrated or single-element reflectors with high thermal and irradiation stability that are usable for light emitting, light deflection or light shaping modules where forward scattering is desired or harmless.

Acknowledgements

The work was financially supported by German Bundesministerium für Bildung und Forschung (BMBF), Förderinitiative Innovative regionale Wachstumskerne, Verbundvorhaben Nanostruktur-technologie für funktionelle keramische Materialien und Schichten- fanimat nano grade, Fkz.: 03WKF23G

The authors gratefully acknowledge the collaboration with Felix Kraze, Petra Puhlfürß, Beate Pawlowski, Franz Bechtold, Reinhard Pelke.

References

[1] W. Buss, P. Schreiber, W. Brode, A. Heymel, F. Bechtold, E. Müller, K. Kaschlik, B. Pawlowski, T. Bartnitzek, "LTCC-Based Optical Elements for Opto Electronic Applications", Proceedings of EMPC 2005, Brugge, Belgium, June 12-15.

[2] F. Kraze, "Untersuchungen an geprägten und infiltrierten Oberflächen hybrider LTCC-Chipbauemente bezüglich ihrer Eignung für Reflektoranwendungen in optoelektronischen Schaltungen", Bachelor Thesis, University of Applied Science Jena, August 04 2008.

[3] D. J. Mooney, „Light reflector and barrier for light emitting diodes", U.S.Patent 7,086,754 B2, August 08, 2006.

Dispensing process with additional ultrasonic energy

M. Bursik, I. Szendiuch, and J. Jankovsky

Department of Microelectronics, FEEC, Brno University of Technology, Czech Republic,
phone: +420541146095, email: bursik@feec.vutbr.cz

Abstract

This paper deals with the dispensing of viscous materials, focused especially on thick film pastes and thixotropic materials. The main aim of our solution is to reduce the viscosity of paste and thus ensure better conditions for the deposition during deposition process. The basic classification of dispense systems into groups depends from the discharge of material. There are two main principles; the first is based on pressure actuation and the other on the screw principle. For our new developed head we have used the pressure system which allows better application of additional ultrasonic energy. New principle of ultrasonic energy application at the capillary outlet is based on piezoceramics resonator. During generation of the appropriate harmonic signal the resonator will vibrate and transferees the kinetic energy to capillary walls and thus causes the additional movement of paste. The field of printed media in the capillary section is more homogenized. If we keep constant pressure we can control the speed of material in capillary very sensitively. For dispensing systems this new principle gives good possibility for the achievement of increasing resolution for the standard materials. Our experimental work has verified decreasing of minimal required pressure by the application of additional ultrasonic energy. This paper describes the new principle and the construction of the ultrasonic dispensing head for precise dispensing of thixotropic materials including thick film pastes. Printing method can be optimize by analyze of printed profile, which is necessary part of precise option of the ultrasonic power unit.

Key words: ultrasonic, dispensing, viscosity, thick film

Introduction

Thick film technology and the deposition of viscous materials are long-proven technologies. Dispensing process of thick film pastes is also designed for special applications, prototype production parts and samples development. Process for the deposition of viscous materials is very sensitive for input parameters of dispense unit. In thick film technology we are using the pressure principle, which allow us reduce the minimum volume amount.

Parameters of the dispenser unit

The shape of the deposited layers should remain within specific dimensional intervals. This parameter is related to the viscosity of thick film pastes and the dispensing process optimalization. Process parameters, such as pressure and its dynamic changes during the dispensing head movement over the printing motive trajectory, are linked with the pressure unit using to handle the process. Pressure is controlled in direct relation with the time or displacement speed. Viscous materials are divided into materials changing the viscosity depending on temperature level and materials which has constant viscosity. Especially thick film pastes changes its viscosity during the deposition through the screen/capillary. This characteristic feature is called thixotropy and it is necessity from the principle for successful paste deposition by the dispensing device. By changing the viscosity, in conjunction with other parameters of dispensing mechanisms, we are able to targeted affect the amount of the paste, both horizontally and in the thickness of the layer. The flatness and contours of the shape deposited are related to consequential operations to dispensing process. The operations such as leveling, drying and sintering of thick film pastes are dedicated to the final adjustment process of components content inside the structure of the layer and its curing.

The sensitivity of dispenser follows from a simple distribution of the chamber linked with the dispensing part containing the nozzle. The ratio of dispenser chamber and the dispensing nozzle diameter determines the pressure sensitivity of the dispensing unit.

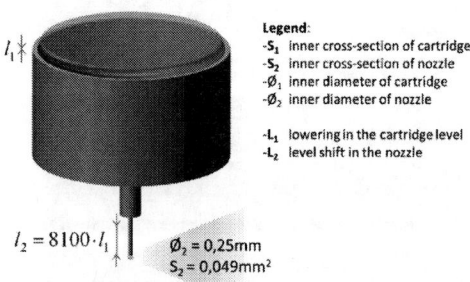

Legend:
- S_1 inner cross-section of cartridge
- S_2 inner cross-section of nozzle
- $Ø_1$ inner diameter of cartridge
- $Ø_2$ inner diameter of nozzle

- L_1 lowering in the cartridge level
- L_2 level shift in the nozzle

$l_2 = 8100 \cdot l_1$

$Ø_2 = 0,25mm$
$S_2 = 0,049mm^2$

Fig. 1 Dimensions of dispensing systems

Problems related to the shape of printed motive, is reflected especially in printing of the curved lines. Together with the increasing resolution should be considered the growing influence of radius in the corners. The possibility of the filtration of defects is through the software correction of the dispensing head movement trajectory. Together with this correction method is necessary to optimize the dynamic control of individual parameters such as especially pressure and the material characteristics. The shape of the applied layer shows the **Chyba! Nenalezen zdroj odkazů.**.

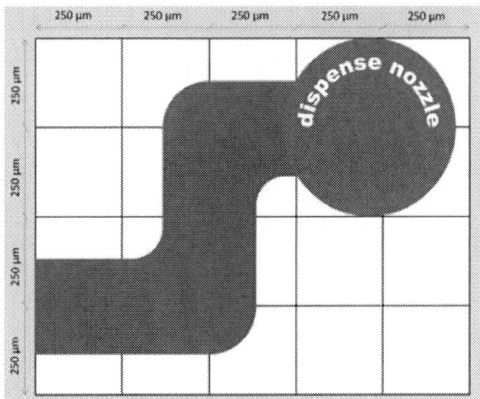

Fig. 2 Printed layer of viscous material

Principle of dispensing head with the additional energy

When we talk about the viscosity and its changes during deposition, we find its mainly affect to the homogeneity of deposited layer. The friction forces of the inner mechanical parts during the dispensing process results a change of the viscosity value resulting from the heat affected zones. By the controlled oscillating motion can be friction and its consequences artificially produce and thus optimize the viscosity in defined areas. To this purpose it is possible to use an ultrasonic head as a source of the mechanical oscillations.

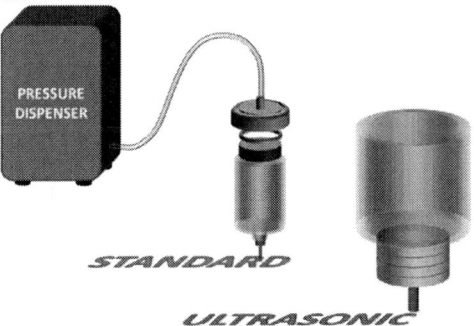

Fig. 3 Standard and ultrasonic system

Construction of the ultrasonic head

The arrangement of dispensing unit with the ultrasonic head must be optimized according to the requirements for the efficiency and the transfer method of the oscillations from piezoceramic rings to the mechanical parts of dispensing unit. Selected design is composed from the static part, a dispenser cartridge. To this part is fixed piezoceramic resonator, when with increasing distance from this nodal point grows up the displacement amplitude of ultrasonic system. Ultrasound system parameters depends on the number, power and performance of piezoceramic ring themselves. **Chyba! Nenalezen zdroj odkazů.** shows the functional prototype of head with the ultrasonic unit to change the viscosity by the mechanical oscillations. In **Chyba! Nenalezen zdroj odkazů.**b is shown the dispensing head with the preloaded piezoceramic system. Interconnection of positive and negative poles of resonator is realized by using the copper rings rotated to each other about 90 °.

a)   b)

Fig. 4 Prototype head,
a) Piezoceramic resonators, b) Complete preload system

Technological parameters of ultrasonic head

Exciter electronic signal should be optimize according to the requirements to dynamics of the oscillating motion. The frequency of excitation signal is selected by evaluation of the frequency characteristics of resonant system. Mechanical characteristics of piezoceramic rings and their mechanical connection related with the preload of the system. Preload the system is necessary for the optimal working characteristic setup of piezoceramic tension in relation to their transient dimensions. Because this crystal changes its thickness, comes up for shrink and stretch.

The default fixed position is membrane consists to the bottom of dispensing unit cartridge, which is the base for the first piezoceramic ring. Each pole is connected to the circuit through the copper electrodes. A copper electrode interconnects the piezoceramic system too. Electrical interconnection is parallel, while the system is physically interconnected in series. Sum of the oscillations amplitude of each of resonator is then the maximum displacement with the orientation corresponding with the polarity of the alternate signal. The change in thickness of piezoceramic ring is dependent to amplitude of the excitation signal.

Depending on the displacement range should be chosen the initial tension of system and design optimal material with the corresponding mechanical properties. During oscillation the ring system should not be relieved to such an extent that has swung the ring from the axis. The positive polarity signal may crack the rings result from excessive tightening of the system.

Functional tests

Prototype dispensing unit with ultrasonic head was designed for the input exciter signal with amplitude $160V_{p-p}$. This test was performed using the generator with the amplitude $16V_{p-p}$. There were possible only the monitoring of changes in the volume of the paste printed at a certain time interval (**Chyba! Nenalezen zdroj odkazů.**). First tests confirm the transmission of ultrasonic oscillations on the capillary by contact test. Consequently was tested the size of dispense batch, but these results using a low excitation signals are not usable to evaluation of the process.

Fig. 5 Dispense nozzle with incidence of ultrasonic

The task of the first prototype compilation was to verify the transmission of mechanical oscillations to the target area in the described configuration.

Principle of the ultrasonic head

Chyba! Nenalezen zdroj odkazů. shows a comparison of dispensing head with additional

ultrasonic energy and the standard nozzle. Paste inside the nozzle and its viscosity interact with the inner dispenser surface. This indicates the necessity of optimalization of these dispensing process parameters and their ability to controlled adjusting.

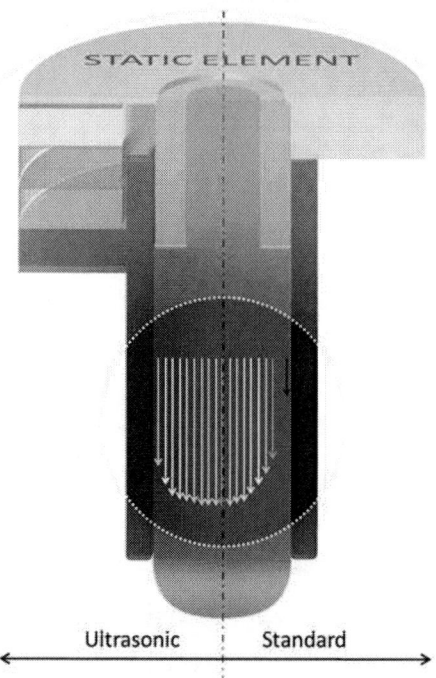

Fig. 6 Principle of ultrasonic head

Further work

Priority now is to optimize the excitation signal and its automatic tuning. This is due to changes in weight and absorbing ratios of the whole system. By increasing the amplitude of the signal gets higher mechanical amplitude. Prototyping with defined inner surface of dispensing system is necessary to perform simulations of material flow curves (**Chyba! Nenalezen zdroj odkazů.**).

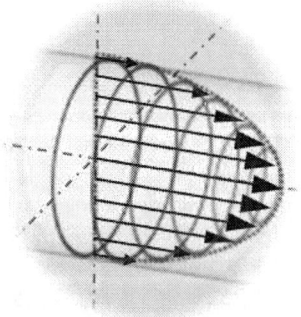

Fig. 7 Flow curves in capillary

Statistical evaluation of batch volume with different factors optimized dispensation classified the factors according to relevancy. Based on statistical evaluation will develop a systematic evaluation the impact of various factors on the characteristic parameters of printed layers.

Conclusion

Thick film materials including polymers have wide spectrum of use in modern electronics, both for non-conventional applications and for sampling of hybrid integrated circuits. That is why perfection of deposition process is important and gives new opportunities for non-traditional solutions. The new dispensing principle should enable us to achieve better resolution and the possibility to create various shapes.

Acknowledgements

Funding for this research was obtained through grant project from the Czech Ministry of Education MSM 0021630503 MIKROSYN „New Trends in Microelectronic Systems and Nanotechnologies") and the Grant projects MP180S14001 (2A-1TP1/143) "Research of new mechatronic structures MEMS useable for pressure measurement" and FEKT-S-11-5 "Research of excellent technologies for 3D packaging and interconnection".

References

[1] I. Szendiuch, M. Buršík, J. Hladík "Deposition of Thick Film Fine Line Patterns by Direct Writing", 42nd International Microelectronics and Packaging Society (IMAPS) 2009, USA, San Jose November 1, p. 10-15, 2009.

[2] J. Kadlec, "Optimalization of ultrasonic transmitter and receiver", Master thesis, Vol. 2, p. 40, Brno 1994

[3] P. Gregor, "Optimalization of ultrasonic wirebonding process", Master thesis, Vol. 3, p. 98, Brno 2011

[4] S. Yang, J.R.G. Evans "A dry powder jet printer for dispensing and combinatorial research", Volume 142, p. 219-222, April 30, 2004, ISSN: 0032-5910

[5] XS. Lu, SF. Lu, J.R.G.Evans, "Microfeeding with different ultrasonic nozzle design", Volume 49, p.514-521, June 2009, ISSN: 0041-624X

[6] H. Wang, T. Ritter, W. Cao, K. K. Shung, "High frequency properties of passive materials for ultrasonic transducers", IEEE, Transactions on ultrasonics, ferroelectrics, and frequency control Vol. 48, January 2001, p. 3

[7] J. W. Phair, M. Lundberg, A. Kaiser, "Leveling and thixotropic characteristics of concentrated zirconia inks for screen-printing", Rheol Acta, Vol. 48, p. 121-133, 2009

3D TSV Mid-End Processes and Assembly/Packaging Technology

Seung Wook YOON, HSIAO Yung Kuan
DZAFIR Shariff, Andy Yong Chang Bum, Won Kyung Choi, *Y.C. Kim, *G. T. Kang,
and Pandi C. Marimuthu

STATS ChipPAC Ltd. 5 Yishun Street 23, Singapore 768442
* STATS ChipPAC Korea Ltd. Ichon, Kyunggi-Do, Korea 467-701
** STATS ChipPAC Inc.47400 Kato Road Fremont, CA 94538 USA
Seungwook.yoon@statschppac.com

Abstract

Increasing demand for new and more advanced electronic products with a smaller form factor, superior functionality and performance with a lower overall cost has driven semiconductor industry to develop more innovative and emerging advanced packaging technologies.

One of the hottest topics in the semiconductor industry today is a 3D packaging using Through Silicon Via (TSV) technology. Driven by the need for improved electrical performance or the reduction of timing delays, methods to use short vertical interconnects have been developed to replace the long interconnects found in 2D packaging. The industry is gearing up to move from technology path finding phase for TSV into commercialization phase, where economic realities will determine the technologies that can be adopted. Choosing the right process equipment and materials, combined with innovative design solutions addressing thermal and electrical issues will be the key success factors.

This paper addresses TSV middle-end process as well as TSV assembly/packaging process. Latest developments in the key elements of 3D Si integration such as backside via reveal and thin wafer handling. The status of "bridge" technologies such as interposers and TSV substrates as an interim play prior to full productization of the active Si TSV approach is reviewed with specific examples of configurations approaching volume production in real products. For TSV assembly/packaging, microbump bonding and reliability characterization will be discussed. TSV Packaging challenges and experimental results will be presented for CTC (Chip-to-Chip), CTW (Chip-to-Wafer) bonding with ultra fine pitch microbump interconnections in this paper.

Integrating TSV and IPD technology in an eWLB design delivers clear advantages such as advanced heterogeneous system integration, higher electrical performance and reduced form factor packaging. The ability to integrate TSV and IPD with eWLB technology opens up a wide range of possible design configurations for SiP and 3D packaging at the silicon level.

Key words: Through Silicon Via (TS), packaging and assembly, microbump, Cu column, Chip-to-Chip, Chip-to-Wafer, 3D integration, TSV, IPD, eWLB (embedded wafer level BGA)

I. INTRODUCTION

The packaging and assembly industry imitated the laminated multilayer circuit board by stacking packages. Package-on-package (PoP) and flex-based origami packages result substantially higher densities without significant infrastructure disruption.

One of the hottest topics in the semiconductor industry today is 3D Packaging using Through Silicon Via (TSV) technology. Driven by the need for improved performance and the reduction of timing delays, methods to use short vertical interconnects have been developed to replace the long interconnects found in 2D packaging. The industry is moving past the feasibility (R&D) phase for TSV technology into the commercialization phase, where economic realities will determine the technologies that can be adopted. Low-cost, high aspect ratio, reliable via formation and via filling technologies are the need of the hour. Choosing the right process equipment and materials with innovative design solutions addressing thermal and electrical issues will be the key winner. As functional integration requirements increase, assembly and wafer fabrication companies are looking to 3D TSV technology, which allows stacking of LSIs thereby enabling products to be made smaller with more functionality. 3D technology realizes miniaturization by 300-400% compared to conventional packaging.

3D integration is progressing on three fronts starting with package-level (die, package stacking), wafer level (die-to-wafer bonding, fan out WLP) and more recently at the Si level (TSV) as shown in Fig.1.

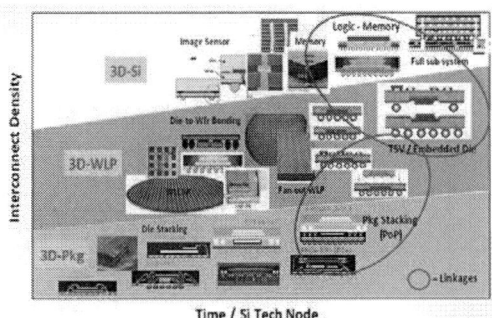

Figure 1. 3D Integration- The solution space – 3D at package level, wafer level & Si level

The demand for high density and multifunctional microelectronics leads to the development of 3D and wafer level packaging, which provides an optimal solution for the shortened interconnects, increased performance and functionality, miniaturization in size and weight, integration of hetero-generous technologies and complex multi-chip systems as well as reduced power consumption. Such packaging technology normally requires the use of ultrathin devices (less than 100μm in thickness). The key benefits from thinned wafers include improved heat dissipation and reduced electrical resistance which offers better flexibility for 3D stacking. However, it brings up a challenge for assembly and packaging; thinning and handling ultrathin semiconductor devices in both front-end and back-end processes due to its fragility and tendency to warp.

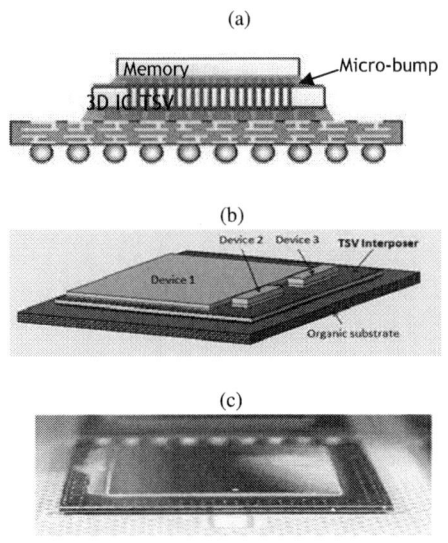

Figure 2. 3D integration packaging with microbump and TSV, (a) 3D-TSV, (b) 2.5D TSV interposer and (c) Super thin TSV PoP (Package-on-Package)[2]

II. TSV Mid-End TECHNOLOGY

"Mid-end" process flow that occurs between the wafer fabrication and back-end assembly process. Mid-end processes support the advanced manufacturing requirements of 2.5D and 3D TSV as well as wafer level packaging, flip chip and embedded die technology.

Flip chip and wafer level packaging are important drivers of mid-end processing in addition to the anticipated growth in 3D solutions utilising TSV technology, particularly with the integration of memory and logic devices at advanced technology nodes. The initial markets that are expected to embrace 2.5D and 3D TSV technology are mobile applications and high performance processors for the computing segment.

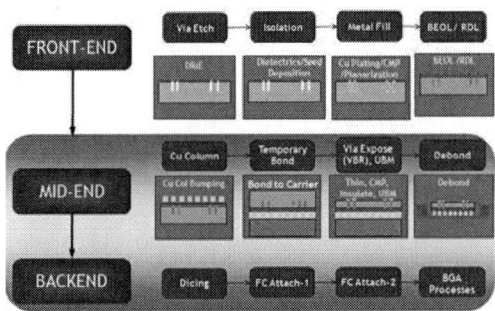

Figure 3. Mid-End process in overall TSV process flow.

One of the first implementations of TSV technology is in the form of silicon interposers used to bridge 2D silicon designs into more advanced and efficient 3D configurations. Often referred to as the 2.5D technology, TSV interposers are an immediate and practical approach to die-level integration using the capabilities of TSV technology. TSV interposers provide flexibility for the integration of die from different semiconductor technology nodes and deliver advantages in miniaturisation, thermal performance and fine line/width spacing in a semiconductor package.[2]

2.5D interposers are one of candidates that would be useful for ELK devices and high performance applications. TSV interposers also have the potential to replace advanced substrates due to advantages in thermal performance, precise dimension control, fine line width/spacing, embedded passives as well as overall thickness. There should be cost/performance crossover between Si interposer and advanced substrates in the near future.

2.5D TSV interposer approach is an efficient and practical approach to solving die integration challenges. Many microsystem device applications that will have to move to wafer-level packaged silicon form factor devices, thereby also facilitating further integration using silicon TSV interposers.

Thin wafer handling

There are many ways to overcome current issues to get better process compatibility and repeatability of handling TSV thin wafers for packaging process. Because the TSV wafer is thinned down to 40~50um thickness, handling of this wafer for packaging process is another challenge in order to meet the different and varying needs, such as temperature and chemical stability, temporary adhesion strength et al. For example, polymer bonding/debonding process is based on high temperature, so it may not suitable for solder bumped wafer. And 12" wafer applications have more challenges than 8" applications and need to optimize the process and materials for larger area bonding/debonding.

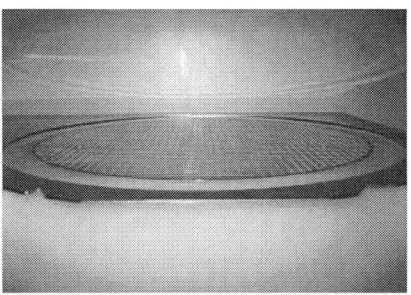

Figure 4. 12" TSV wafer after successful debonidng after mid-end process with temporary carrier.

Backside Via Reveal Process

As shown in Fig.3, TSV is to be reveal to backside for 3D vertical interconnection after front-end TSV formation. With temporary bonding/debonding system, TSV wafer from fab is to be backgrounded and Si etched to expose Cu via with fab process. Fig.5 shows the successfully exposed Cu via on backside with 12" TSV wafer. There was TOF SIMS (Time-of-Flight Secondary Ion Mass Spectroscopy) analysis for Cu contamination on Si wafer during CMP process and verified non-detectable Cu content after chamical composition analysis along whole 12" TSV wafer.

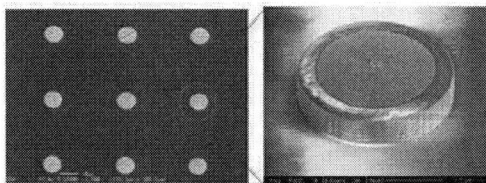

Figure 5. SEM micrographs of backside TSV via reveal with CPM and Si etching process in 12" 3D TSV wafer.

III. 3D TSV ASSEMBLY AND PACKAGING

Compared to conventional flipchip process, TSV assembly process is more complex due to TSV wafer as well as microbump. As shown in Fig. 6, there are additional materials, like as additional encapsulation in between bump and flipchip die or bump and TSV die. There are quite critical

challegenges for assembly view point both in materials and assembly process.

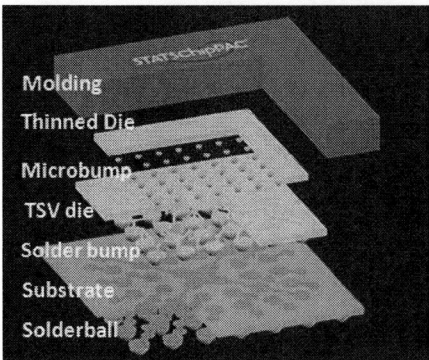

Figure 6. Schematics of 3D TSV assembly and packaging.

In advanced 3D stacking technologies, one of the important steps is to develop and assembly fine pitch and high density solder microbumps. Solder microbumps for flip-chip interconnections allow high wiring density in the Si-carrier, as compared to organic or ceramic substrates, and enable high-performance signal and power connections. [2]

(a)

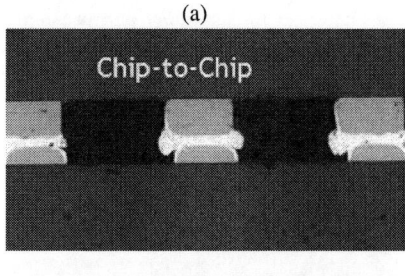

(b)

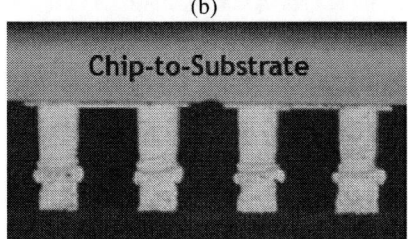

Figure 7. SEM micrograph of 40/ 50μm pitch of (a) chip-to-chip and (b) chip-to-substrate with Cu column bump.

Flip chip package assembly was carried out to investigate the bonding quality and interconnections with microbump flip chip as shown in Fig. 7. After flip chip die fabrication with bump, the die attachment was carried out with thermo-compression flip chip bonders. Several DOE (design of experimental)s were carried out to find optimized process conditions as functions of time, stage temperature, pickup tool temperature as well as pressure. Fig. 9 shows x-ray image after thermocompression successful flipchip bonding without void.

40μm and 50μm both microbump test vehicles were sent to JEDEC standard reliability tests. Reliability samples passed MSL-3 with 3x reflow process at Pb-free 260°C peak temperature. All samples passed unbiased HAST and HTS reliability tests. There was no failure found after 1000 T/C.

(a)

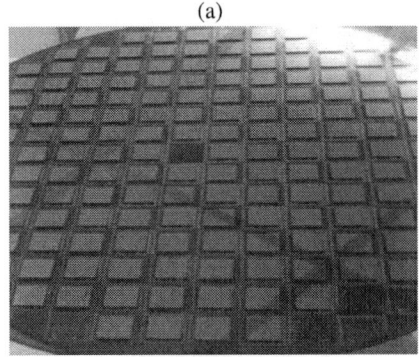

(b)

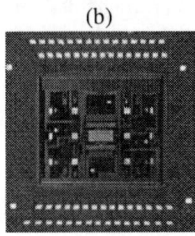

Figure 8. Micrographs of (a) Chip-to-Wafer bonding and (b)Chip-to-Substrate assembly with encapsulation in small gap of microbumps.

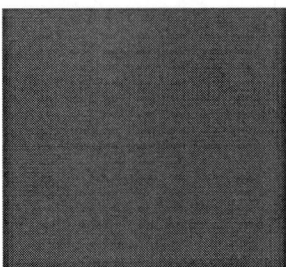

Figure 9. X-ray micrographs of thermocompression bonded package of Chip-to-Substrate assembly

Thin Die Warpage

Package warpage is also a challenge in the assembly process and reliability. It showed after reflow process, the thin flip chip die warpage difference was 60 um at room temperature between before-reflow and after-reflow. This serious warpage causes higher stress in TSV structures thus catastrophic failure will happen at the package level. Therefore, it should be studied more and characterized with further experiments.

IV. 3D WAFER LEVEL INTEGRATION WITH eWLB, TSV AND IPD TECHNOLOGY

Demand for through silicon via (TSV) is being driven by the need for 3D stacking to shorten interconnection length, increase signal speed, reduce power consumption and reduce power dissipation. Increasing demand for new and more advanced electronic products with a smaller form factor, superior functionality and performance with a lower overall cost has driven the semiconductor industry to develop more innovative and emerging advanced packaging technologies.[3]

TSV is typically not a packaging solution by itself. TSV uses only back-end manufacturing techniques such as bonding, fine pitch bumping, back grinding and thin wafer handling. Final packaging is required to connect the device to the PWB. Due to assembly constraints, the choice of the final package solution could affect the entire TSV process flow. This final packaging could be a BGA package, a fan-out WLP type, an embedded die in substrate (EDS), or other. Here, it is interesting to notice how complementary 3D IC configurations with TSV and 3D packaging can be. In effect, 3D eWLB(embedded Wafer Level BGA) can enable designs to fully benefit from the 3D IC integration and can reduce the package footprint with more aggressive design rules than BGA packages.

A truly seamless wafer level integrated 3D packaging module that will incorporate aspects of 3D stacking, as well as Si package with IPD, actives in 3D eWLB packaging with TSV, flip chip, and microbump as well as 3D WLPs.[4]

By combining eWLB technology with our Through Silicon Via (TSV) and Integrated Passive Device (IPD) technology we are achieving new levels of heterogeneous integration in a wide range of design configurations including small die, large die, multi-die, multi-layer and stacked packages. eWLB is a powerful wafer level technology that has the design flexibility to accommodate an unlimited number of interconnects and is unconstrained by die size. This has allowed eWLB to meet the relentless form factor requirements and performance demands of the mobile and handheld market. Next generation eWLB technology includes single die, multi-die, ultra thin, System-in-Package (SiP) and three dimensional (3D) packaging, all with superior electrical and thermal operating characteristics. With its many technology attributes including a solid integration platform and a revolutionary structure which makes it inherently one of the thinnest package profiles in the industry, eWLB appeals to an increasingly broad range of market applications.

Although active and stacked ICs are a highly functional and important component of the overall system, they are only one set of components; many other components including other actives, passives, power systems, wiring, and connectors must be considered in a complete system. As a result, there is a need to think at module and system levels and this need is largely met by the current technology domain in the areas of through silicon vias (TSVs), 3D stacking, and wafer level packaging. There should be further study on integration, focusing on TSVs, 3D stacking and 3D eWLB with better electrical and thermal performance, greater system reliability, and reduced form factor and overall cost. It will go far beyond this to realize a truly seamless wafer level integrated 3D packaging module as shown in Fig. 10, that will incorporate aspects of 3D stacking, as well as Si package with embedded passive, actives in 3D eWLB packaging with TSV, flip chip, and microbump as well as 3-D WLPs.

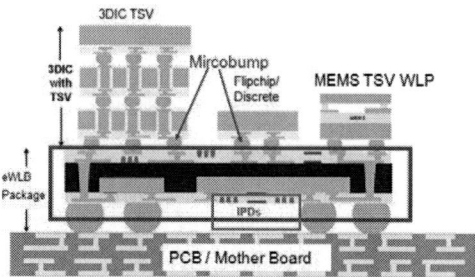

Figure 10. Total solutions for 3-D packaging with eWLB(FO-WLP) and TSV technology. [3]

V. CHALLENGE AND FUTURE WORK OF TSV TECHNOLOGY

There is no question that 3D TSV will be adopted, but the timing for mass production depends on how the TSV technology compares in terms of cost with existing technologies. CIS image sensors for camera modules are already in mass production. For other applications, the adoption time is longer than originally expected, as is quite common with the introduction of new technologies. TSV technology itself is not a packaging technology apart from a few exceptions. 3D TSV and 3D packaging do not have to be considered as competitors but more as complementary areas. In recent years, the semiconductor industry has expressed some growing interest in these ideas and put some significant efforts in allowing the emergence of these new breakthrough technologies. Still, some challenges remain ahead for a wide adoption including cost, a shift in the design method paradigm, system co-design, new CAD tools, new architectures, and more new challenges. While progress is being made, design, thermal, reliability and testing issues remain a barrier to TSV adoption in some applications.

There should be further study and understanding of TSV cost-ownership analysis, value-chain and

systematic analysis of competitive advantages of TSV in business perspectives in order to move into high volume manufacturing. The industry is still lacking real test strategies to evaluate the reliability of these TSV interconnects. TSV engineers and researchers have been very focused on developing 3D TSV processes, but testing of these interconnects is still at its early stage. This is a very similar situation to SoC, SiP and WLP. A test strategy and methodology have to be defined in order to decide which type of measurements need to be made at which level (wafer, microbumping, Chip-to-Wafer/Wafer-to-Wafer bonding, packaging/assembly, final test) and which test infrastructure needs to be set up.

Technical feasibility alone does not enable high volume production. Proven materials, process, and equipment capability is essential for manufacturing and high volume production with quality and reliability. Compatibility of materials/processes in the total assembly flow is also quite critical concerns in process integration process.

2.5D/3D TSV is not a single technology element and productivity benefits and reliability can be realized at several levels. As shown in Fig. 11, Industry-wide alliance necessary across design, fab, assembly, reliability, characterization and test etc.

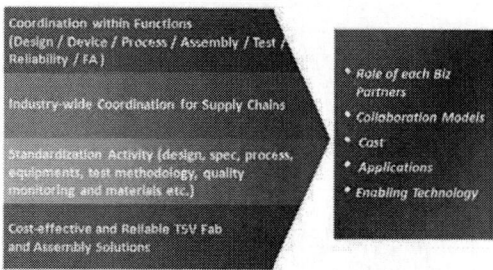

Figure 11. Diagram of challenges of 2.5D/3D TSV technology.

Table 1. Requirement of 3D TSV industrial standardization.

Scope	Areas
Design Enablement	EDA 3D tool compatibility. Chip-to-chip interface. Chip-to-Package Interaction
Manufacturing	DFM, process & material specs
SiP Integration, Quality & Test	Supply chain hand-off, shipping methods, incoming/outgoing specs. KGD strategy, Quality monitoring, Test hardware
Reliability Assurance	Lifetime/reliability criteria (TSV & micro bump) & thermal/thermo-mechanical attributes

Major challenge of TSV 3D is probably not technical, but more that TSV is disruptive and emerging technology. Traditional CMOS scaling and predictive or planned roadmaps by shrinking transistor dimensions may not be the only one way to go. Whole semiconductor industry needs to be convinced of the additional values of 2.5D/3D TSV integration approach. Things are moving, the momentum is still there, but the first products must show up in market to make 3D TSV a reality thus move further mainstream market and applications. 3D TSV must help to reduce overall semiconductor cost, improve performance or reduce form factor and reduce the time-to market.

VI. CONCLUSION

In this paper, 3D integration, 2.5D TSV interposer technology and TSV packaging/assembly issues were discussed. For successful implementation of TSV technology to microsystem products, TSV process and TSV packaging/assembly both should be well prepared and established with close collaboration and clear understanding.

TSV technology enables the integration of semiconductor die fabricated in different technology nodes with diverse testing requirements. The short vertical TSV interconnections through the silicon wafer achieve greater space efficiencies for a smaller form factor and higher electrical performance. Passive devices such as resistors, capacitors, inductors, filters and baluns can consume 60% to 70% of available space in a system, subsystem or SiP package. Integrating TSV and IPD technology in an eWLB design delivers clear advantages such as advanced heterogeneous system integration, higher electrical performance and reduced form factor packaging. The ability to integrate TSV and IPD with eWLB technology opens up a wide range of possible design configurations for SiP and 3D packaging at the silicon level. This is an effective approach to system partitioning which offers an overall outstanding system performance.

REFERENCE

[1]. Flynn Carson , Kazuo Ishibashi , Seung Wook Yoon, Pandi Chelvam Marimuthu, "Development of Super Thin TSV PoP", IEEE CPMT Symposium, Japan, 24-28 Aug 2010. (2010)

[2]. Knickerbocker, J. U. et al, "Development of next generation system-on-package (SOP) technology based on silicon carriers with fine-pitch interconnection," IBM J. Res. Dev. Vol. 49, No. 4/5 (2005), pp. 725-754 (2005)

[3]. Newsletter on 3D Packaging, 3D IC, TSV, WLP & Embedded Technologies, MARCH 2009, No.10, Yole Development (2009)

[4]. Seung Wook YOON, Meenakshi PADMANATHAN, Andreas BAHR, Xavier BARATON and Flynn CARSON "3D eWLB (embedded wafer level BGA) Technology: Next Generation 3D Packaging solutions," San Francisco, IWLPC 2009 (2009)

Ultra Fine Line Print Process Development
for Silicon Solar Cell Metalisation

T. Falcon, S. Clasper.

DEK Solar, 11 Albany Road, Weymouth, DT4 9TH, UK.

Tel: +44 1305 760760, +44 1305 760123, Email: tfalcon@dek.com

Abstract

One of the principal drivers for the continuing miniaturisation of frontside conductors on silicon solar cells is the trend toward higher emitter sheet resistances. This trend is apparent both in mainstream cell production, and with the introduction of selective emitter technologies. High R_{sheet} emitters make more efficient use of the blue end of the light spectrum, thereby enhancing cell efficiency. However, this benefit comes with the downside that the higher R_{sheet} emitter impedes the flow of electrons through the semi conducting silicon, and therefore reduces current collection. Using a greater number of finer conductors with reduced conductor spacing can mitigate these resistance losses with no additional shadowing loss, thereby acting as an enabler for high R_{sheet} emitter technologies. With today's mainstream cell manufacturers printing lines below 100µm wide, the sub 50µm printed line becomes the next logical milestone in the miniaturisation of frontside conductor geometries. This paper will describe the path to that milestone and beyond.

Key words: Screen Printing, Metalisation, Ultra Fine Line.

Background

More than 80% of silicon solar cells are metalised by 3 separate screen printing steps. On the rear of the cell either 2 busbars (or several smaller areas) are printed with silver paste to form solderable contacts. This silver paste is dried, and then the remainder of the rear side is printed with an aluminium paste. The aluminium paste is then dried before the wafer is flipped over in preparation for the next printing stage.

The sun facing (or front side) of the solar cell is often viewed as having the most critical metalisation step. This is typically an "H pattern" grid, with 50 to 80 fine lines and 2 or 3 much wider busbars. Both the fine lines and the busbars are printed from silver paste in a single screen printing operation. The fine fingers collect electrons from the sun facing emitter layer and transport them to the busbars. The purpose of the busbars is simply to collect current from the fingers and to provide a solderable surface so that the cells can be strung together in series with a solder coated copper ribbon.

Once all three printing stages have been completed the wafers are passed through a co-firing furnace to form the positive and negative connections with the silicon wafer.

The front side grid conductors are one of the main areas of focus for increasing solar cell efficiencies. The reasons for this are listed below:

i) The printed silver lines and busbars cover about 7% of the sun facing surface of the wafer, effectively blocking it from the sun. Printing narrower and taller lines (high aspect ratio) allows more light to hit the wafer surface, thereby enhancing cell efficiency. The lines must be taller if they are to be made narrower in order to maintain (insofar as is possible) the same cross sectional area and therefore the same electrical conductivity.

ii) At the time of writing the cost of silver has doubled in the last 12 months to over US$1200/kg. The 2nd greatest cost in cell manufacturing after the silicon is the printed metalisation pastes, consequently they are a primary area of focus for cost reduction. Efficient use of the frontside silver paste is a high priority for many major cell manufacturers.

iii) The majority of cells are made with a boron (P-doped) base and a phosphorus (N-doped) emitter. The standard level of emitter doping has recently been 50 to 70 Ohm/square. This is a compromise between doping at a high level to achieve a low contact resistance with the metalisation which also gives a low sheet resistance (R_{sheet}), and doping at a lower level to give a good conversion efficiency at the blue end of the spectrum. Printing more and narrower grid lines at a finer pitch enables the use of higher sheet resistances, and therefore a better blue response from the cell. Industry forecasts indicate that sheet resistance will increase year on year, almost doubling by 2015 [1] (fig. 1).

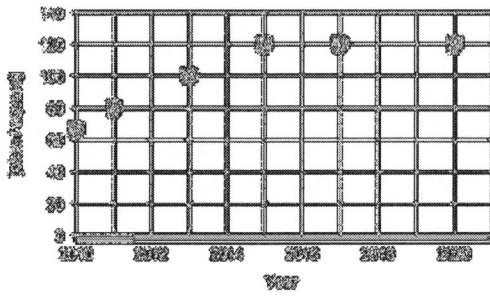

Figure 1: Emitter Sheet Resistance Forecast.

Screen Printing the Front Side Grid.

Historically (2000 to 2008) front side grid metalisation pastes have been formulated for their electrical performance with rather less regard for the aspect ratio of the fired conductor. Typical line widths were 100 to 150µm with fired heights of 10 to 15µm. This gives an aspect ratio (H/W) of 0.1 to 0.15. However, in the last 3 years paste suppliers have made rapid and significant advances in printed conductor aspect ratio, whilst simultaneously improving their product's electrical characteristics (lower contact resistance and better conductivity). In high volume manufacturing today printed line widths of 70µm with fired heights of 22µm are amongst the best possible with a single print process. This gives a much more favourable aspect ratio of >0.3.

Industry forecasts predict frontside conductor width will decrease year on year, almost halving between 2011 and 2015 [1] (fig. 2).

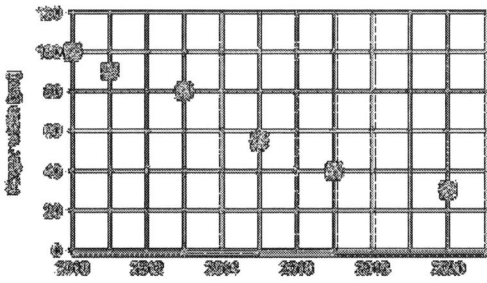

Figure 2: Finger Width Reduction Forecast.

Screen technology has also advanced rapidly in order to meet the challenge of high aspect ratio fine line printing. Historically screen meshes with 30µm wire diameter were the norm. More recently, with the introduction of high tensile alloys, screens with 18µm wire diameter are common and these are preferred for sub 80µm lines.

Objective.

The purpose of this work is to explore several potential methods for printing sub 50µm lines and to assess their printing performance. The potentially suitable methods are listed below. These are followed, in the next 3 sections, by some background information on each:

i)   Ultra-fine wire mesh screens.
ii) Single layer stencil.
iii) Double layer "hybrid" screen.

Ultra Fine Wire Mesh Screens

There is a general rule for screen printing fine lines that states the finest printing aperture's width should be a minimum of 3 times the mesh's wire diameter. Also, due to some post print spreading and slumping of the paste, the printed line will typically be anywhere from 5 to 20µm wider than the screen aperture through which it was printed. Consequently, in order to print sub 50µm lines, mesh screen apertures with dimensions from 30 to 45µm should be tested. To facilitate this, the finest commercially available screen meshes were sourced and these have wire diameters of 13 to 14µm (fig. 3). A mesh with 11µm wire was listed by a Japanese manufacturer but was not available when requested.

Figure 3: 30µ aperture in 380 mesh screen.

Single Layer Stencils

Screen printing equipment is generally capable of stencil printing too. Scientists at the Energy Research Centre of the Netherlands (ECN) have been exploring this possibility for several years and have recently developed the process to a high standard [2] (fig.4). However, it should be noted that it is not possible to print the full H grid using a single layer metal stencil in much the same way as it is not possible to stencil print a continuous letter "O". The simple problem being that the middle falls out. ECN circumvented this problem by printing the busbars first with a mesh screen, and then printing the fine grid lines in a second print operation. Clearly the disadvantage of this is that an additional dryer and printer are required in each metalisation line. However the advantages of reduced silver consumption (by de-coupling the high aspect ratio grid lines from the now lower aspect ratio busbars) and the higher cell efficiencies possible with high aspect ratio gridlines, generally outweigh the cost of the additional equipment.

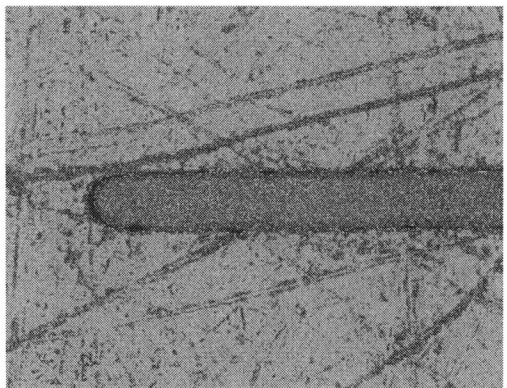

Figure 4: 30µ aperture in metal stencil.

Hybrid Screen

Whilst screen printing has been the mainstay of silicon solar cell metalisation for many years, there are some inherent inefficiencies with the process, particularly with regard to transferring the maximum amount of paste to the narrowest possible printed line. (The percentage of paste transfer is later referred to as transfer efficiency). The nature of a mesh screen is such that the printing apertures are partly occupied by the wire mesh. This reduces the potential transfer efficiency on 2 counts;

i) The wires in the apertures have considerable surface area and some paste will adhere to these surfaces in preference to transferring to the wafer.

ii) The wire is occupying aperture volume which could otherwise be occupied by transferrable paste.

To overcome the issues with wire mesh in the apertures, DEK Solar invented and are further developing a 2 layer hybrid screen [3]. This screen uses a high tensile steel foil with relatively large openings for the top layer (squeegee side), and a photo-imageable polymer to define the printing aperture and add thickness to the base layer (fig. 5). The presence of a base layer also facilitates the use of fine "bridges" in the steel to hold the stencil together and stop the 'middle falling out'. This enables the whole frontside H grid to be printed in a single operation.

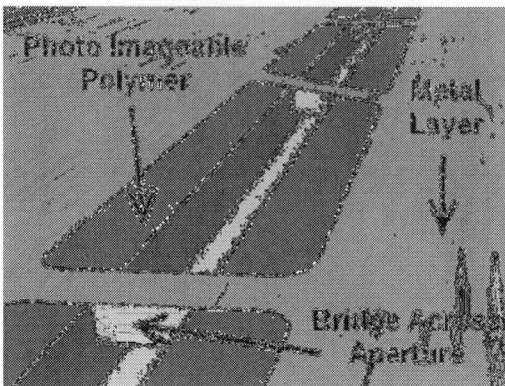

Figure 5: 3D sketch of the hybrid screen.

Screen and Stencil Preparation.

i) Ultra-fine wire mesh screens were fabricated in-house from the finest commercially available meshes. Several experiments to optimize screen tension on these relatively fragile meshes were required before a completed and tensioned framed mesh was ready for emulsion coating. The final tension was approximately 14N/cm.

The test pattern chosen for these mesh screens included apertures from 20 to 40µm wide in 5µm increments. Each aperture size had 5 blocks of apertures in a column and each block contained 20 apertures each 20mm long (fig. 6).

Further experiments to optimise the emulsion coating thickness were also required. These were performed in combination with print tests and exposure tests to ensure good definition on the 30 to 40µm apertures.

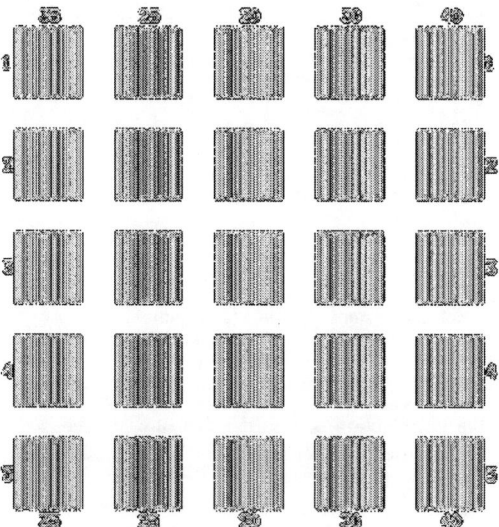

Figure 6: Mesh screen print test pattern.

442

ii) Single layer stencil
This was electro-plated from nickel at DEK Singapore to a thickness of 30µm. A 3 busbar design was chosen to minimise the length of the apertures. Only the fine lines were included in the stencil with no busbar apertures. The fine line apertures were 30µm wide.

iii) Double layer "hybrid" screen.
Earlier iterations of this screen were based around a 60µm wide aperture. For this experiment the relative geometries of screen remained the same but were reduced by 50%. This gave an aperture width of 40µm and an overall screen thickness of 40µm.

Table 1 below gives some dimensional details on the various screens & stencils tested.

	Wire Dia.	EOM	Metal Thickness	% Open Area
590 Mesh	13µm	12µm	N/A	49
380 Mesh	14µm	12µm	N/A	62
Hybrid	N/A	20µm	20µm	85
Stencil	N/A	N/A	30µm	100

Table 1: Screen / Stencil Measurements

Printing Experiments

Essentially these required printing all the screen / stencil variants on the same machine with the same paste and the same process parameters in order to compare the results.

To ensure the printing paste was used in a "steady state" rheologically, 10 wafers were printed in rapid succession for each of the tests, with the 10th wafer being inspected and measured.

Printing was performed on a DEK Solar Eclipse printer with 80 shore 'A' hardness polyurethane squeegees mounted at 60 degrees. Print speed was an industry typical 200mm/s and the print gap was 1.2mm in all cases. Print pressure varied a little between the various screens and stencils, with the metal foil based screens / stencils typically requiring 2 to 3 kgs pressure whereas the mesh screens performed better in the 4 to 5kg range.

The results included here are all from the same industry typical screen printing paste. This paste being one of the top selling products from one of the top three vendors. Some later experiments not included here used other pastes potentially more suited to the particular type of screen or stencil being tested. These results will be presented in a later paper.

Measurement of the printed lines was performed using a Nikon VMR 3020 video microscope equipped with 3D topographical capability by multi-point video auto-focus.

Data was analysed and 3D images generated using Digital Surf's MountainsMap Topography XT.

Results

	Mean Width	Mean Height	Min Height	Aspect Ratio
590 Mesh	51.7	10.9	9.1	0.21
380 Mesh	54.0	14.3	12.8	0.24
Hybrid	59.2	17.7	15.8	0.30
Stencil	N/A	N/A	N/A	N/A

Table 1: Results from 40um wide apertures.

	Mean Width	Mean Height	Min Height	Aspect Ratio
590 Mesh	42.0	8.95	7.48	.21
380 Mesh	45.3	10.9	9.21	.24
Hybrid	N/A	N/A	N/A	N/A
Stencil	41.5	18.9	17.4	0.46

Table 2: Results from 30um wide apertures.

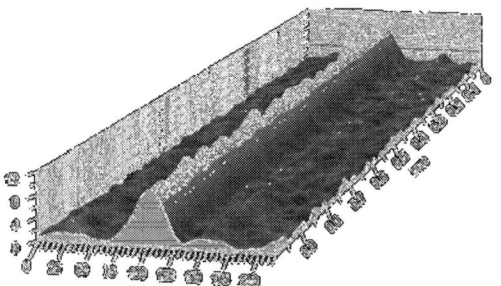

Figure 7: 590 Mesh Screen, 30µm Openings.

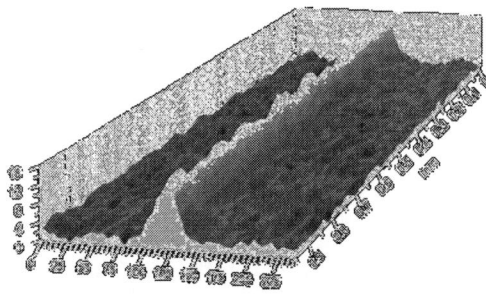

Figure 8: 380 Mesh Screen, 30µm Openings.

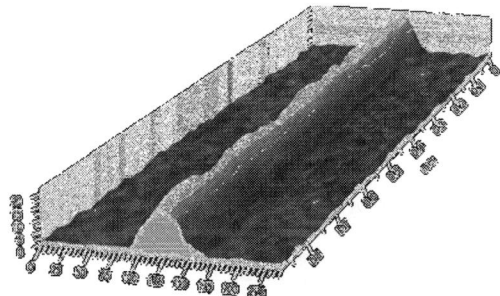

Figure 9: Hybrid Screen, 40µm Openings.

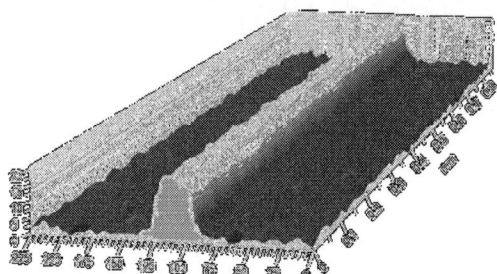

Figure 10: Stencil with 30µm Openings.

Observations.

The shape of the printed line is one indicator of its potential for low resistance conductivity, the area of the cross section giving a good idea of the quantity of bulk silver available to conduct.

The 3D images above illustrate that the mesh screens give a triangular profile with numerous peaks and troughs. Any silver in the peaks is effectively wasted as it contributes little to the line's conductivity. The lowest, narrowest part of the line will of course give the highest electrical resistance.

The hybrid screen gives a more rounded, fuller profile which could be described as being part way between a triangle and a hemisphere. The peaks and troughs are gentler and much closer together.

The stencil print gets much closer to the ideal cross section, being a slightly rounded rectangle. Peaks and troughs are almost non-existant indicating that the electrical conductivity will be highly uniform along the entire length of the line.

Conclusions

All screen / stencil options printed continuous lines, however, one remark made in an earlier paper originally referring to mesh screens only, also holds true for this work, (now the concept is extended to include stencils too.....) "Percent open area of the particular screen meshes had the largest influence on aspect ratio" [4].

The open areas of the hybrid screen and of the stencil tested here are significantly higher than any mesh screens, and open area seems to be a good indicator for printed line aspect ratio and uniformity in all cases, not just for the mesh screens.

As printed conductor line width becomes increasingly narrow, printing with mesh screens is still quite possible down to 40µm, with higher open area screens giving the highest printed line aspect ratio possible with a single print process.

In order to gain the maximum paste transfer efficiency and final printed line aspect ratio, stencil printing is preferred. However, this is only possible with an additional printer and dryer in each metalisation line, so that the busbars can be printed separately. The hybrid screen would seem to be a good compromise as it requires no additional equipment.

In some ways it is unfair to compare the print performance of a single step process with that of a 2 stage process, especially considering the recent high volume implementation of a "Print on Print" process using two mesh screen printing stages to print high aspect ratio frontside conductors. Further work is therefore planned to develop the Print on Print process to deliver sub 50µm lines, allowing it to compete on a level footing with the stencil print process.

Further work is also planned to develop the hybrid screen such that it will deliver high aspect ratio fingers and with lower paste heights for the busbars.

It should also be recognised that it may be quite possible to achieve more favourable results for each screen / stencil type by using a paste optimised specifically for it. Further work is also planned to explore this area.

Acknowledgements

The authors would like to acknowledge the staff at DEK Screens and DEK Singapore's electro-formed stencil experts for their invaluable help.

References

[1] International Technology Roadmap for Photovoltaics (ITRPV.net) Results 2010, 2nd Edition.

[2] B. Heurtault, J. Hoornstra, "Towards Industrial Application of Stencil Printing for Crystalline Silicon Solar Cells", Proceedings of the 25th European Photovoltaic Solar Energy Conference 2010. pp. 1912-1916.

[3] T. Falcon, "Aspect Ratio Improvements for Printed Frontside Conductors on Silicon Solar Cells", Proceedings of the 24th European Photovoltaic Solar Energy Conference 2009. pp. 1338-1346.

[4] D. Buzby, A. Dobie, "Fine Line Screen Printing of Thick Film Pastes on Silicon Solar Cells", Proceedings of the 41st International Symposium on Microelectronics, November 2008.

Mechanical Problems of Manufacturing Processes for Photovoltaic Modules

S. Wiese [1], F. Kraemer [1], R. Meier [2], and S. Schindler [2]

[1] Saarland University, Chair of Microintegration and Reliability, Saarbruecken, Germany
[2] Fraunhofer Center for Silicon Photovoltaics, Halle (Saale), Germany

Phone: +49-681-302-71309, Fax: +49-681-302-64872 and E-mail: s.wiese@mx.uni-saarland.de

Abstract

The paper describes thermo-mechanical problems of the soldering process, which is used to interconnect single solar cells with each other, in order to create a photovoltaic module. The interconnection wires need to conduct high currents, because of the demands for high power modules that operate at low voltages. Therefore the resistance of the wiring is a very critical factor, because it influences strongly the total module efficiency. However significant mechanical problems are created by increasing the width and height of the rectangular wires, because the brittle silicon solar cells are prone to breakage. Therefore innovations in interconnection design are necessary to release mechanical stresses during the assembly process.

Key words: Reliability, Photovoltaic Modules, Mechanical Stresses, Copper Ribbon, Cell String

Introduction

The lifetime of photovoltaic modules is a crucial factor for the profitability of a solar energy system. Longer lifetimes can either increase the profitability, e.g. when the price for generated electricity is fixed by a feed in law, or they will the decrease the average price for the generated electricity under non subsidized conditions. In contrast to the lifetime of conventional electronic products, which is defined by the occurrence of a functional failure, the reliability of photovoltaic modules is characterized by a power derating. Standard modules are designed to have a power derating of -20% in a operational period of 20 years. The power derating characteristics will be guaranteed by the module manufacturer. If the module exceeds the specified power derating figures, the manufacturer is obliged to exchange any module which does not longer fulfill the specified power figures. Therefore knowledge about the expected power derating is important for the profitable production of photovoltaic modules.

Assembly Process for PV Modules

Figure 1 shows a cross section trough a typical PV module. It consists on a sandwich structure. The first layer is the front glass, which provides the mechanical stability of the module. Usually it is a tempered high strength glass, because

it needs to protect the module against hail storms or large snow loads. It is optimized for low iron content to minimize light absorption. The second layer is a soft encapsulate sheet that adheres the solar cells to the front glass.

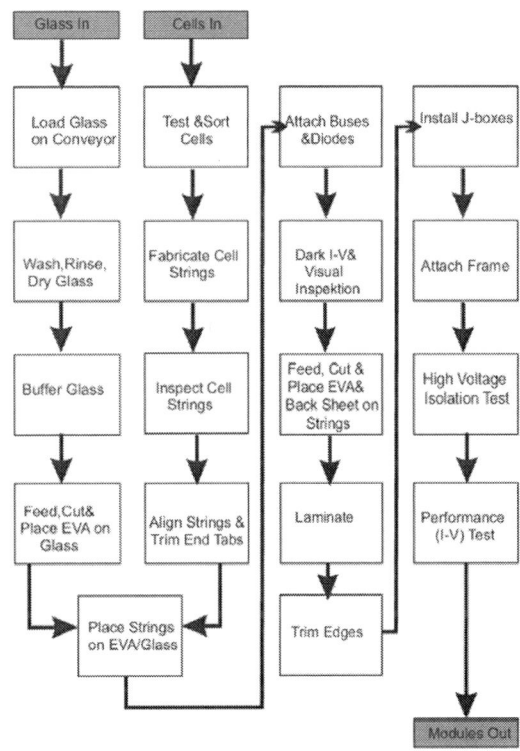

Figure 2: Process Flow Module Assembly for Crystalline Silicon Solar Cells [1, 2]

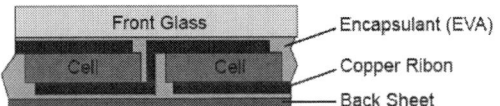

Figure 1: Schematic of a Photovoltaic Module (Cross Section View)

The transparent encapsulate is made most commonly from ethylene-vinyl acetate copolymer (EVA), because it combines the necessary optical, mechanical and adhesion properties. The same encapsulate sheet is used on the back side of the solar cells to adhere the back sheet to the sandwich assembly as the final layer. Several materials can be used for the back sheet, such as safety glass, or special tailored polymer like Tedlar or Mylar [3].

Figure 2 shows the standard processes flow for the assembly of cells into a module. It starts with two different branches for front glass and solar cells. In the solar cells branch individual cells were connected to each other by copper ribbons. This manufacturing step is called tabber/stringer process. It puts out cell stings, which were layed up onto a front glass/EVA-sheet-stack. Then several cell strings were connected to each other by bus ribbons. After that the whole assembly is covered by a second EVA-sheet and the back sheet. Subsequently the whole stack is assembled through a vacuum lamination process. Final process steps like framing, inpection, mounting of J-boxes complete the module assembly.

Critical process steps

Soldering of individual cells into a string and the vacuum lamination of the assembly are critical steps, concerning the mechanical integrity of the module. During the soldering process the cell and the wires heat up and expand. After soldering, when the assembly cools down the cell and the wires start to contract. Since the coefficients of thermal expansion differ from each other, there is a different amount of contraction between the Cu and the Si, which builds up stresses in the assembly. Since the solder thickness (5 … 20 µm) is relatively small compared to the thickness of the copper wire (130 … 200 µm) and the thickness of the silicon solar cell (180 µm), there is nearly no release of mechanical stresses in this assembly.

One of the approaches to reduce the thermally induced stresses during the soldering of the copper ribbons targets the systematical analysis of specific factors. The most significant factors are:

- the copper ribbon material properties

- the copper ribbon geometry

- the solder alloy creep properties

FE-Analysis were carried out, to find out the contribution of each of these factors to thermally induced stresses during the soldering of copper ribbons on solar cells [4].

FE-Analysis of the soldering process

The investigation of thermal mechanical loading during the soldering process was carried out by 3-D FEM simulation on the ANSYS simulation software package. The specific goal of these FEM-simulations was to analyze the mechanical stress, that builds up in a single cell interconnection during cooling down phase of the soldering process.

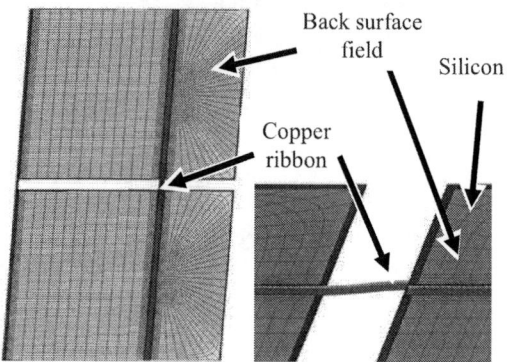

Figure 3: left) 3-D FE-model of the investigation of the string-ribbon soldering process, backside view; right) Detailed view of an interconnection [4]

The finite element model represents the currently used solar cell type (quadratic shape with 156 mm edge length and an overall thickness of 170 µm). Those cells are connected via 3 busbars. For symmetry reasons the model consists of two quarter cells which are connected by one and a half copper ribbons (shown in figure 3). The solar cell consists of several material layers (wafer = silicon, thick film pastes for the bus bars = silver, and thick film paste for the the back surface field = aluminum). The bus bars were simulated with an average thickness of 10 µm and the back surface field was assumed to have a thickness of 20 µm.

Table 1: Material models for FE-Analysis and specific properties at room temperature [4]

Material	E [GPa]	CTE [ppm/K]	Yield Stress [MPa]	Material model
Al Paste	6.0	11.9	28.3	Bi-linear temperature dep. elastic-plastic
Copper (initial)	85.7	17.0	95.1	Bi-linear temperature dep. elastic-plastic with isotropic hardening
Silver paste	7.0	9.8	64.0	Bi-linear temperature dep. elastic-plastic
Silicon	130.0	3.5	-	temperature dep. elastic

The material models applied in this investigation are listed in table 1. The FEM-model assumed the following boundary conditions: Both cells were clamped at their corners, so that the cells are able to warp [4].

The temperature profile assumed a constant cooling rate. The start temperature was set to the melting point SnPb solder (183°C). The final temperature was set to room temperature (20°C). In this first attempt, the solder layers between copper and bus bars were not represented in the model, since they were assumed to have little effect on the stress relief.

Mechanical properties of copper ribbons

The material properties of the copper ribbons are crucial for the FEM-simulation results. Therefore extensive experimental investigations were carried out, in order to provide a reliable material model for the interconnector material. The goal of the experiments was to perform all tests on true copper ribbons as used for the assembly of PV module were chosen as specimens (Figure 4). The most used copper materials are OFHC-Cu, OF-Cu 99,99% and Cu-ETP 99,9%. The solder coatings are either leaded (Sn63Pb37, Sn62Pb36Ag2) or unleaded (Sn100, Sn96.5Ag3Cu0.5, Sn96.5Ag3.5, Sn42Bi58). An uncoated copper ribbon with a rectangular cross section of 2.0 mm x 0.1 mm was used for the experiments, in order to represent different types of copper ribbons [5].

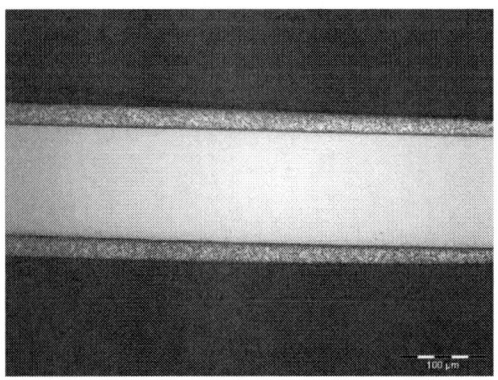

Figure 4: Cross-section through a copper ribbon (as delivered by the manufacturer; light microscopy): rectangular copper core with solder coating [5]

Tensile tests at room temperature were carried out on a universal material testing machine (Zwick). The clamping of the thin ribbon into standard wedge grips turned out to be difficult. In order to improve gripping forces sand paper was put between the wedges and the sample. To avoid any errors due to slipping of the sample during the test, strain measurement was realised via a video extensometer. The tensile tests were carried out as

monotone tests and pulsating load tests (see figure 5) [5].

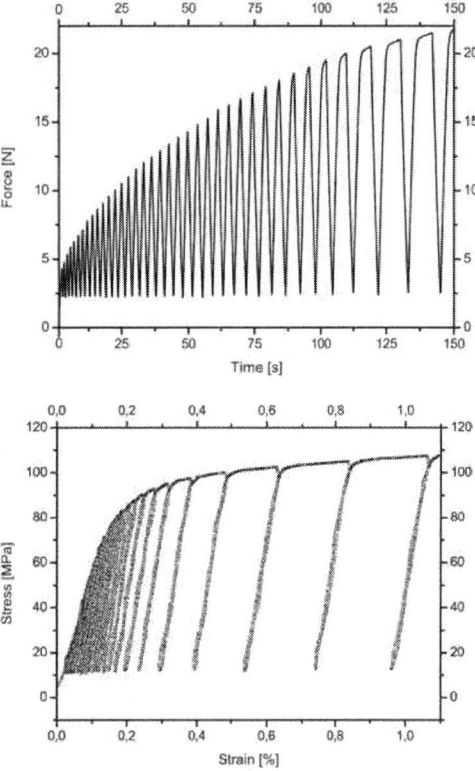

Figure 5: Loading profile and resulting stress-strain-curve of a pulsating load tensile test [5]

Basing on the results of the experimental investigations on copper ribbons, a parameter set was defined, which contains experimentally gained and theoretical values for the Young's modulus and the yield stress of the copper ribbons (see Table 2). Stresses calculated by FEM simulation were analyzed at room temperature. First principal stresses and third principal stresses in silicon were applied as result criteria (see figure 6) [4].

Table 2: Matrix of investigated combinations of Copper Young's Modulus and Yield Stress [4]

Yield Stress [MPa]	30	60	95	120
Young's Modulus [GPa] 67	x		x	x
86	x	x	x	
121		x	x	x
192	x		x	x

447

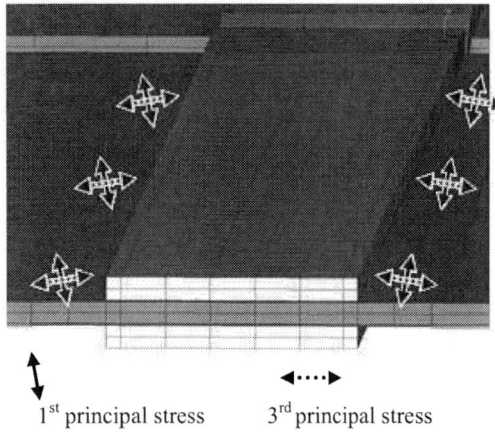

1st principal stress   3rd principal stress

Figure 6: Orientation of first and third principal stress in silicon along the busbars [4]

It was found, that first and third principle stress have their maximum values beneath the edge of the front side busbar (FS-BB). At this position, the first principal stress is oriented through the silicon wafer thickness, while the third principal stress is oriented perpendicular to the length direction of the FS-BB (figure 6). Thus, both stress orientations may cause the shell cracks of silicon, as seen in production.

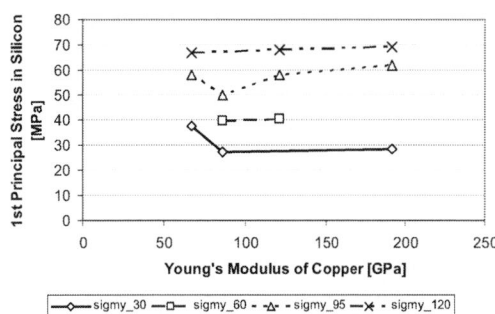

Figure 7: First principal stress in silicon underneath the frontside busbar in dependence of the Young's modulus of the copper ribbon [4]

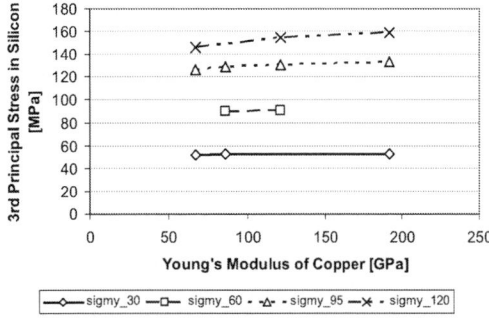

Figure 8: Third principal stress in silicon underneath the frontside busbar in dependence of the Young's modulus of the copper ribbon [4]

Figure 7 and figure 8 depict the relations between the Young's modulus of copper and the first and third principal stress in silicon after the assembly was cooled down from the SnPb eutectic point (183°C) to room temperature (20°C). It becomes evident that the copper Young's modulus has only little effect on the stress in silicon.

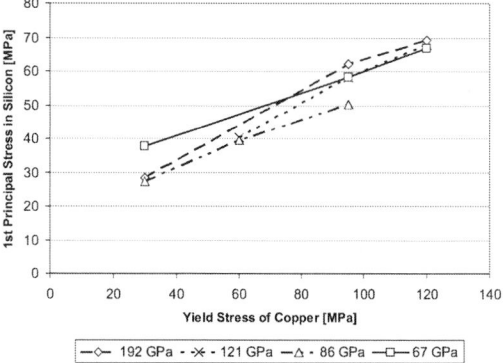

Figure 9: Influence of the copper ribbon yield stress on the 1st principal stress in silicon solar cells after soldering in dependence of the yield stress of the copper ribbon [4]

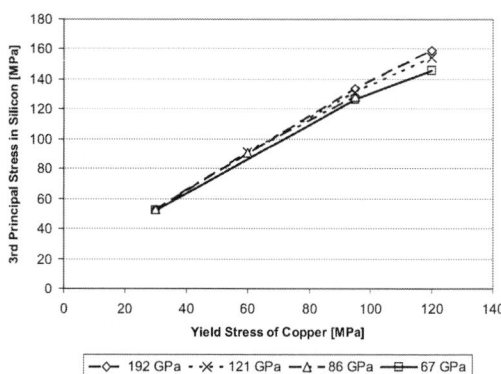

Figure 10: Influence of the copper ribbon yield stress on the 3rd principal stress in silicon solar cells after soldering in dependence of the yield stress of the copper ribbon [4]

Figure 9 and figure 10 show the relations between the yield stress of the copper and the first and third principal stress in silicon after the assembly was cooled down from the SnPb eutectic point (183°C) to room temperature (20°C). It can be seen that the yield stress has a significant influence on the thermo-mechanical stress in the silicon solar cell. The third principal stress in silicon increases by about factor three, when the yield stress is varied between 30 MPa to 120 MPa. Therefore it can be concluded, that the copper material of the interconnectors always deforms plastically, when the whole assembly is cooled down from soldering temperature to room temperature. This is also the reason, why the Young's modulus of the copper ribbons has nearly no influence on the thermo-mechanical stresses that were build up in the whole assembly [4].

Geometry of copper ribbons

Basing on commercially available copper ribbons, a parameter set for interconnector geometry was defined, which contains several cross section geometries for copper ribbons combined with corresponding and front side busbar (FS-BB) widths (see Table 3).

Table 3: Geometry: Copper ribbons cross sections (width x height) and frontside busbar (FS-BB) width in millimeters [4]

Ribbon type	Type 1	Type 2	Type 3	Type 4
Ribbon x-section (H x W) [mm]	1.7 x 0.20	1.7 x 0.15	2.0 x 0.15	1.2 x 0.28
FS-BB-Width [mm]	1.5 or 2.0	1.5 or 2.0	1.5 or 2.3	1.5

Stresses calculated by FEM simulation were analyzed at room temperature. First principal stresses and third principal stresses in silicon were applied as result criteria. Figure 11 and figure 12 show the results und the FEM calculations.

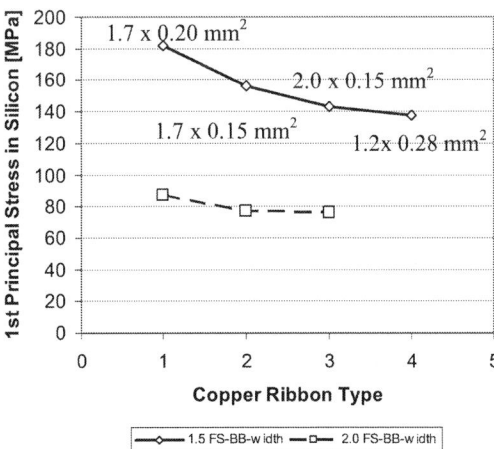

Figure 11: Influence of copper ribbon cross section and the front side busbar width on the first principal stress in silicon solar cell after soldering [4]

It can be seen, that the silicon stress reduces when the front side busbar width is larger than the copper ribbons width. Wider busbars reduce the first principal stress by nearly 50%, while the third principal stress is only reduced by 10 – 15%.

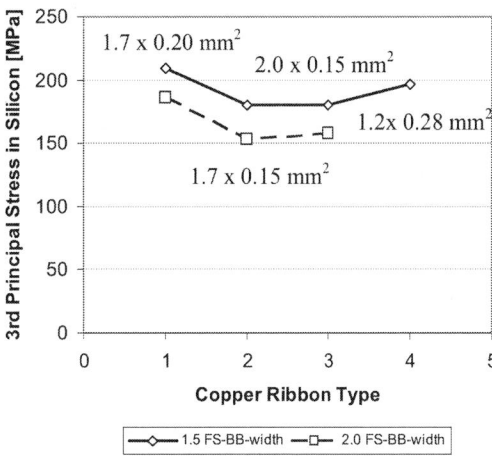

Figure 12: Influence of copper ribbon cross section and the front side busbar width on the third principal stress in silicon solar cell after soldering [4]

Solder Material

The solar cell assembly remained basically unchanged in comparison to the previous investigations. The model applied the $1.7x0.2mm^2$ copper ribbons on the 1.5mm wide FS-BBs. This configuration created the highest mechanical stress in silicon (see figure 12 and figure 13). The cooling down phase was assumed to be 60s for all solder alloys, whereas the starting temperatures were set according to the individual melting temperatures. The following starting temperatures were applied for several solder alloys: Sn63Pb37 = 183°C; SnAg1.0Cu0.5 = 210°C; SnAg3.5Cu0.75 = 220°C.

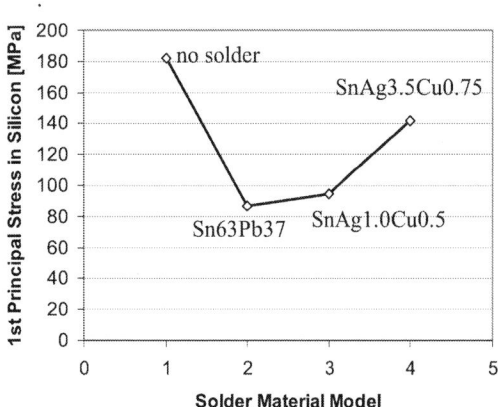

Figure 13: First principal stress in silicon underneath the front side busbar in dependence of the solder alloy and soldering temperature [4]

Figure 13 shows the first principal stress in the silicon. Solder creep clearly reduces the stress in silicon. The first principal stress created in the reference without the assumption of solder creep is about two times higher than in case of SnPb solder.

The analog effect can be observed for the third principal stress (see figure 14). However the amount of stress reduction is not as large as for the first principle stress.

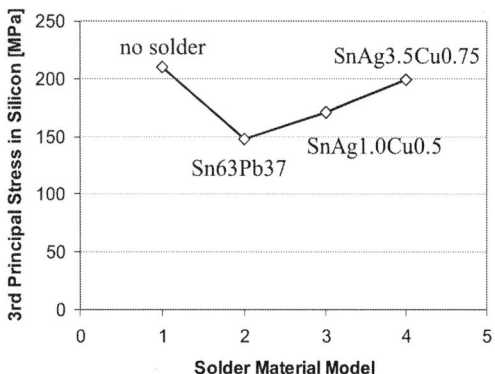

Figure 14: Third principal stress in silicon underneath the front side busbar in dependence of the solder alloy and soldering temperature [4]

Conclusions

The soldering of solar cells into strings causes significant thermo-mechanical stresses. FE-analysis were conducted to determine the specific contribution of several factors to the thermally induced stresses during the soldering of copper ribbons on solar cells.

The yield stress of the copper ribbons contributes significantly to the amount thermally induced stresses into the silicon solar cells, while the Young's modulus has nearly no influence on the stress level in the silicon.

The specific geometry (width and thickness) of the copper ribbon has only a minor influence on stresses during the soldering process. However the relation between copper ribbon width and front side busbar width has a significant influence on the thermally induced stresses during the soldering process.

Solder creep can help to relief stresses during the soldering process. On the other hand there will be higher stresses by the use of SnAgCu-solders because of their higher melting point compared with SnPb solders.

Acknowledgments

This work was supported by the FhG internal Programs under Grant No Attract 692131 and the German Federal Ministry of Education and Research under Grant No 03SF0337E.

References

[1] Wiese, S.; Meier, R.; Kraemer, F.; Bagdahn, J.: "Constitutive Behaviour of Copper Ribbons used in Solar Cell Assembly Processes", Proceedings 10th EuroSimE Conference 2009, 28-30 April 2009, Delft, pp. 44-51.

[2] Nowlan, M.: "Development of Automated Production Line Processes for Solar Brightfield Modules", Subcontract Report, NREL/SR-520-43190, April 2008

[3] Mrig, L.: Photovoltaic module reliability workshop, Sheraton Inn Lakewood/Lakewood, CO, October 25-26, 1990

[4] Wiese, S.; Kraemer, F.; Betzl, N.; Wald, D.: "Interconnection Technologies for Photovoltaic Modules – Analysis of technological and mechanical Problems", Proceedings 11th EuroSimE Conference 2009, 26-28 April 2010, Bordeaux

[5] Wiese, S.; Meier, R.; Kraemer, F.: "Mechanical Behaviour and Fatigue of Copper Ribbons used as Solar Cell Interconnectors", Proceedings 11th EuroSimE Conference 2010, 26-28 April 2010, Bordeaux

Development of an intelligent integrated LED system-in-package

A.W.J. Gielen [1], P. Hesen [2], F. Swartjes [3], H. van Zeijl [4], F. Boschman [5], J-E. Bullema [1], R.J. Werkhoven [1], S. Koh [4]

1: TNO, De Rondom 1; 5612 AP Eindhoven; The Netherlands
Tel +31 88 866 55 36, Fax +31 88 866 88 47, Email sander.gielen@tno.nl
2: Philips Lighting, Eindhoven, The Netherlands
3: NXP, Nijmegen, The Netherlands
4: TU Delft, Delft, The Netherlands
5: Boschman Technologies, Duiven, The Netherlands

Abstract

Intelligent LED lighting systems can save up to 80% of energy compared to traditional lighting systems. In order to provide these intelligent lighting products at reaseonable costs, integration and miniaturisation are important steps. To this end we have designed a LED system-in-package that combines LEDs, driver ICs, and passives into a single package. In future we will expand this package with other functions, for example presence detection technology and wireless communication. Due to the heat dissipation in the LEDs and driver ICs, the thermal load in the package is considerable, challenging the reliability of the system-in-package. Furthermore, to be economically viable, the lifetime of the LED SiP is much longer than traditional electronics. The design lifetime of LED luminaires is typically 25 to 50 thousand hours. An initial assessment of the thermal performance of our LED SiP shows that the thermal management is challenging the reliability limits, and we expect that with optimization the thermal loads can be accommodated at reaseonable levels. Also our first failure calculations indicate that our required lifetime is feasible. Finally, the fabrication of our SiP is based on readily available manufacturing techniques. Future work will focus on further optimizing our design.

Key words: LED, system-in-package, integrated system, energy saving.

Introduction

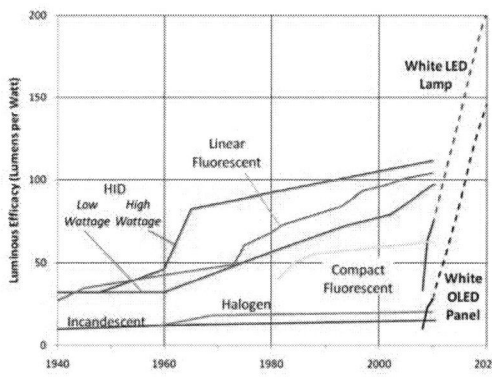

Fig. 1: Luminous efficacy of LEDs and OLEDs compared to other light sources [4].

Europe has committed itself to reduce primary energy consumption by 20% compared to projections for 2020. Energy efficiency is the most cost-effective way of reducing energy consumption while maintaining an equivalent level of economic activity. Improving energy efficiency also addresses the key energy challenges of climate change, energy security and competitiveness [1]. The energy used for lighting ranges from 15% to 25% depending on the application [2] and typically 19% of the global

energy produced is consumed in lighting [3]. LED lighting holds huge potentials for saving energy due to its high efficiency and future efficacy potential (see Fig. 1).

Furthermore, intelligent lighting systems can significantly contribute to the energy saving by sensing the lighting needs, for example depending on environmental light and presence of people, and consequently delivering the optimum lighting [18]. Integrating intelligence or smartness in systems is a successful strategy in the ICT industry, where highly functional compact systems are built using integration and packaging technologies like system on chip (SOC) and system in a package (SiP).

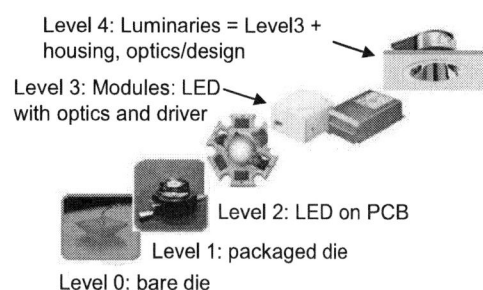

Fig 2: The SSL system value chain.

Contrary to light sources that are developed today, the novel LED lighting technology is based on semiconductor technologies. Therefore, it allows the use of integration and packaging technologies from the IC industry to build smart lighting devices. The integration approach can be used at different levels (see Fig 2). In this work we propose a SiP technology to integrate LED lighting technology at level 3.

These smart systems must be delivered against competitive prices in order to quickly penetrate the market. Therefore, cost-reduction for these smart luminairies is important and many R&D efforts are targeted at lowering the price. In Fig. 3 the relative manufacturing costs of a LED luminaire are shown including the expectations for the future.

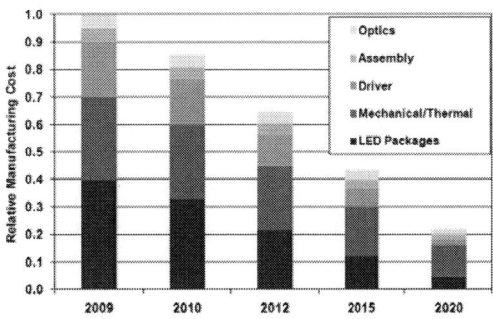

Fig. 3: Relative manufacturing costs of LED luminaires [4].

The US Department of Energy solid-state-lighting manufacturing has established that a 10 time cost reduction by 2020 was not aggressive enough - the industry now needs to cut costs per lumen by a factor 8 by 2015 to enable a viable volume LED lighting market that competes with fluorescents. This means that the current average costs level of 18 USD/klm has to be brought down to an average costs level of 2.20 USD/klm [17].

A costs and material efficient approach, that supports the development of costs efficient smart LED lighting, consists of a system-in-package (SiP) design that uses multiple dies of different nature. In order for such a SiP to be competitive it needs to outclass a conventional smart LED system either in size, costs, performance, or lifetime, or preferably on all aforementioned aspects. The trade-off between costs and performance including the future targets are shown in Fig. 4.

Another additional advantage of a LED SiP is embodied in the supply chain: If the package contains known good LEDs, drivers, and other components for intelligence, then the production will have higher yields. Furthermore, it opens up the possibility to completely remove the PCB from the luminaire product and put the LED SiP directly on the heatsink, thereby removing the thermalthermal barrier the PCB may form.

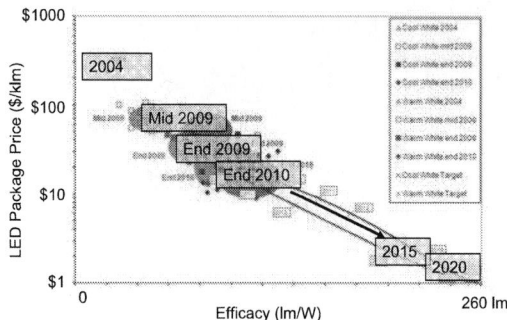

Fig 4: LED package prices Price-Efficacy tradeoff for LED Packages at 35 A/cm² [4]. The red fields represent warm white lights and blue fields the cool white lights.

The design approach

We target to replace a 60W/806 lumen incandescent light bulb and comparable LED retrofit bulbs with an intelligent LED SiP. The challenges that need to be solved are (1) the thermal management, (2) the thermo-mechanical reliability, (3) the internal interconnect (1st level) reliability, (4) the external interconnect (2nd level) reliability, (5) the incorporation of the passive components (resistors, capacitors, and inductors), and (6) the packaging both in terms of manufacturability and costs. Depending on the application the LED SiP may be connected directly to the mains (230 VAC) or a low voltage feed from an existing application infrastructure, for example kitchen hood lighting. Further requirements include a lifetime of at least 25 to 50 thousand hours for indoor conditions. The indoor conditions are specified in more detail in Table 1.

Table 1: LED luminaire user conditions for indoor.

Item	Indoor
Ambient temperature (degC)	10 – 35
Switching (cycles / day)	1 (general use) 10 (corridor / toilet)
Operating (hours / day)	4 (general use) 12 (office) 24 (commercial)
Moisture (degC / %RH)	20 / 60 (Europe) 40 / 85 (Asia)

In order to design and ultimately produce a LED SiP, Philips Lighting, NXP, Boschman Technologies,, TU Delft, and TNO, represented by the authors of this paper, joined forces in the ENIAC ESiP project.

With our wish lists in mind, a first concept design was drafted as shown in Fig. 5. This concept showed promising thermal behaviour at a package level, when using an optimal heatsink (Fig. 6): The junction temperature of the LED was only a few degrees higher than the heatsink temperature. Using this concept the different groups each worked on their field of expertise to improve the design and to understand the performance under more realistic conditions.

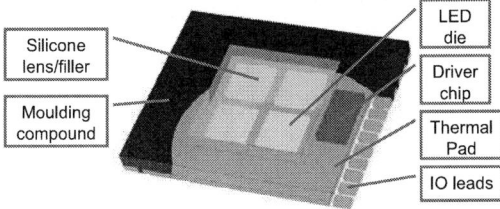

Fig 5: Initial concept of a LED SiP with integrated driver.

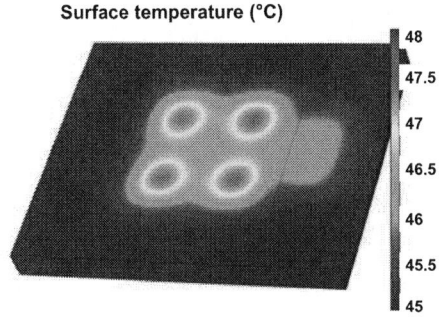

Fig 6: Temperature distribution of the concept when mounted on an ideal heatsink.

LED SiP design

Three producable designs were drafted of which one design is shown in Fig. 7. It consists of a QFN-like package with a large thermal copper pad. As there is no need for a large number of I/Os the number of I/Os is reduced and we can even consider putting the I/Os only on two sides or even on one side of the package. Four LEDs can then be placed on the thermal pad, leaving sufficient room for the driver dies and other passive components.

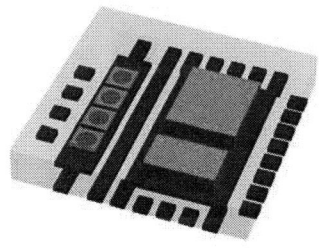

Fig 7: LED SiP package design. The gray area is black molding compound, except for a wedge on top of the LEDs with a clear (silicone base) molding compound.

As the light colour and the light output depend on the temperature the package must demonstrate a stable thermal behaviour. Current LED technology is not at the end of the learning curve and we may expect that LEDs' efficacy will improve beyond the current 100 - 150 lm/W range [5]. Future LED SiPs will profit from reduced heat production in the LED chip and make the SiP more attractive to reduce the packaging costs.

Thermal performance of the design

Also the electronic driver is a source of heat. In our concept we use the SSL2101 driver of NXP [12]. This driver is designed for controlling mains dimmable LEDs and is a fully integrated LED driver for lamps and luminaires up to 10W. By combining the driver function and the lighting function in a package, the challenge for the thermal management is increased by the heat dissipation in the driver. In particular, one must consider the noticeable costs of a heatsink in current LED luminaires (see Fig 3: mechanical / thermal solution is 25-30% of the costs of the luminaire). The overall thermal resistance, expressed in K/W, is commonly used as a performance indicator for a LED package. This performance indicator is closely related to the overal system design of the LED package. For die sizes in the 500 to 2000 micron range, the thermal resistance is typically in the range of 2 K/W to 12 K/W [6].

Although LED components are supposed to have a lifetime of over 50,000 hrs, ASSIST envisages LED components as the Methuselah of electronic components [7]. In practice, the useful lifetime of a LED is closely related to the LED junction temperature. Narenran [8] estimates that the life of a LED is reduced with 50% for every 10°C increase in temperature. Commercially available LEDs typically have a maximum junction temperature of 120°C for a lifespan of 25 to 50 thousand hours.

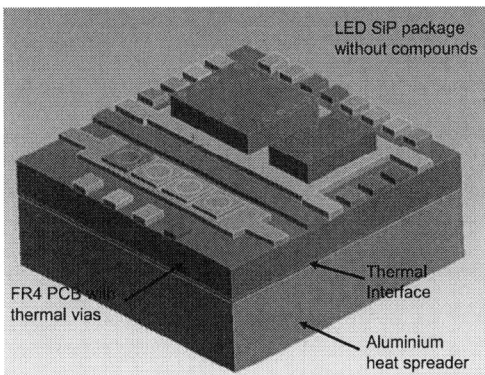

Fig 8: Concept model on top of a FR4 PCB, connected to an aluminium heat spreader using thermal interface material.

The thermal performance of the design was examined using a finite element model. To this end the package concept (Fig. 7) was mounted on an FR4 PCB with an aluminium heatsink underneath it. The connection of the PCB to the heatsink is made using a thermal interface material (Fig. 8). This material provides electrical insulation while providing a reaseonable thermal conductivity. This PCB-heatsink configuration is used in commercially available LED retrofit lamps. The finite element division is shown in Fig 9, while Table 2 lists the material properties, boundary conditions, and loading conditions. It should be marked that we used quite pessimistic values for the heat dissipation in the LEDs and driver ICs in order to push the limits of our design. For example, the typical heat dissipation of the driver SSL2101 IC is 1W, while we used a value of 2.1W: We doubled the IC dissipation and added 0.1 W for dissipation in the passives that were not yet modelled.

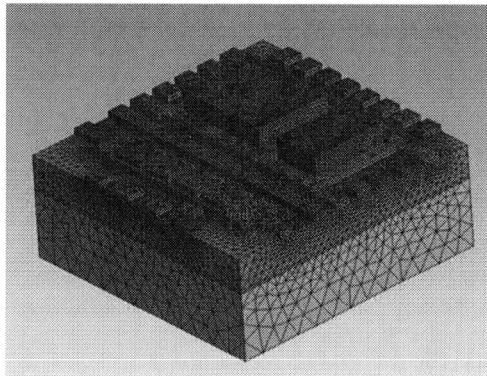

Fig 9: Finite element division of the model. Note that the elements of the black and clear molding compound are removed.

Table 2: Material properties, loading and boundary conditions of the thermal model.

Material	Thermal conductivity [W/mK]
Clear compound	0.2
Black compound	1.0
LED GaN	200.0
Copper	400.0
FR4 with vias	40.0
Thermal interface material	2.0
Aluminium spreader	100.0
Driver chip	4.0
Load	
LED heat dissipation	1.16 W per LED
Driver heat dissipation	2.1 W
Boundary conditions	
Convection @ top of package	5 W/(m² K)
Convection @ bottom of package [9, 10]	1000 W/(m² K)
No radiation	

The thermal results of the concept design are shown in Fig. 10 and Fig. 11. The temperature gradient between the package and the aluminium heat spreader shows clearly the presence of the thermal interface material, that is clearly limiting the heat flux. Currently the junction temperature of the LEDs will be around 142°C. This is clearly too high,

but the design was not optimized in any way yet, leaving room for improvement. For example, the temperature difference between the hot spots in the package and the FR4 PCB is only 20°C, which also supports our earlier conclusion on the concept design that the heat fluxes involved can be accommodated by the proposed package design and that the main bottle neck is in the board and heat sink.

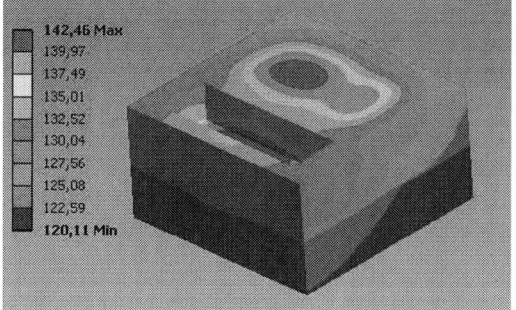

Fig 10: Temperature distribution of the LED SiP package mounted on FR4 and connected to an aluminium heatsink via thermal interface material.

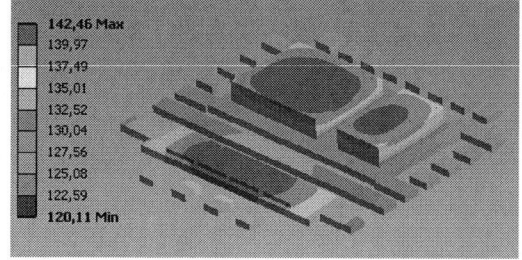

Fig 11: Temperature distribution of the LED SiP for the LED chips and the driver chips with the copper lead frame.

Initial lifetime estimates of the design

Although the previous section showed that the LED junction temperatures are not yet in the required range, the lifetime of the package as system is not solely determined by the LEDs. It is strongly influenced by the other components and interconnects that are present in the system. The LED driver electronics may be composed of about a hundred components. In case more functionality and intelligence is added to the system, then reliability calculations following MIL HDBK 217 show that the System in Package is likely to fail, long before the 50 thousand hrs lifetime due to the failure of specific electronic components other than the LED chip.

Still, integration of the driver electronics and other components into a single package, may remove vulnerable interconnects, such as the solder connection from the package to the PCB, or even remove the complete PCB. This also removes some of the limitations current retrofit lamp design face that are not well manufacturable with standard PCB technology [11].

A typical LED driver circuit is shown in Fig 12. Using the components of this circuit and some main characteristics of our SiP design, a straightforward life estimate was calculated using handbook FIT values from MIL HDBK 217. The calculations are summarized in Table 3.

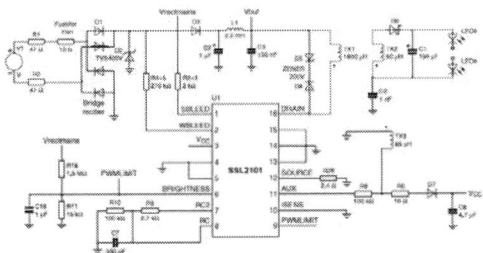

Fig 12: Fly-back LED driver circuit scheme for 230 AC mains connection with dimming function of NXP [12].

Although more advanced connections are possible, for the calculations we considered the interconnections of all the passive components to be solder joints within the package. Wirebonds are needed to connect the driver dies to each other and the LEDs to the driver and passives. From the scheme we counted 12 resistors, 7 capacitors, 3 inductors, and 6 supporting diodes. The LEDs, although being diodes as well, are considered separately.

For the lifetime estimates all the FIT values are summed. Subsequently, the mean-time-before-failure (MTBF) is calculated as:

$$MTBF = \frac{10^9}{\sum FIT}.$$

then the F10 and F50 limits are calculated from:

$$t_{F(x)} = -\ln(1-x) * MTBF$$

where x represents the failed fraction. The lifetime calculated with this method is valid if we consider electrical failures only.

In the LED industry the B50 (or F50) limit is used for prediction of the lifetime. In our case we would very well reach the targeted 50 khours. However, it is more likely, and also common in the electronics industry, to work with the F10 (or even tighter) limits. In that case we can get to a lifetime of over 25 khours, based on sudden electrical failures only. However, there are several potential failure modes for the LED System in package.

Table 3: Lifetime estimations for electrical failures only, using straightforward FIT calculations. The FIT values used here are estimates and are not based on experimental values.

Type	Number	FIT	FIT design
Solder joints	70	0.1	7
Wirebonds	20	0.01	0.2
LED	4	100	400
Driver	2	500	1000
Capacitors	7	2	14
Resistors	12	2	24
Inductors	3	2	6
Diodes	6	20	120
Substrate	1	130	130
		TOTAL	1701
		MTBF (hours)	587820
		t_F10 (hours)	26897
		t_F50 (hours)	176952

For the LED two types of failure are dominant: failures in the bulk of the LED material and failures at the interface. As stated before, the junction temperature is seen as most critical for these failures [13]. Yellowing of plastics is often caused by either temperature of the system or short-wavelength emissions. The location of the phosphors influences this type of degradation.

Han and Nahrendran [14] studied failure modes of stand alone LED drivers and noticed that electrolytic output capacitors in the driver were the 'weakest' link in the driver: failure of the driver was caused by elevated pin temperatures. A lifetime of 7,000 hrs at a pin temperature of 100°C was measured in this study.

Kan [15] proposes the so-called roller coaster model for LED degradation. Failure mechanisms occurring were, beside systematic wear-out failure and early failures, also so-called early wear-out failures in overal system degradation. According to Kan, a burn-in is required to come to a first stable light output for the LED system, where thermal management of the system remains one of the largest challenges.

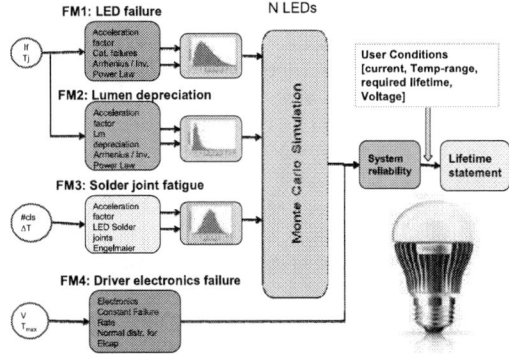

Fig 13: simulation scheme to predict the life time of the LED SiP using Monte-Carlo methods.

A more accurate calculation of the expected lifetime is obtained by Monte-Carlo simulation techniques (see Fig 13) or even more advanced Bayesan Belief Nets and Markov Chains, see [16] for more details on this approach for LED systems.

In short, Monte-Carlo and the advanced Bayesan Belief Nets with Markov Chains take into account not only the catastrophic failure behaviour of the LEDs, but also the lumen depreciation. Furthermore, this method not only considers the different failure modes for the computation of LED lifetime, it is also able to take into the account the interaction between these different modes. This enhances the realism of the prediction. This may be necessary since the failure modes in the LED system will influence the degradation rates. For example, the crack growth on the solder layer may increase the thermal resistance and this will in turn increase the junction temperature and hence, the degradation of the LED and LED lens.

Discussion, conclusions and future work.

We have designed a LED System-in-Package that consists of four LED dies, driver dies, and room for incorporation of passive components on a QFN-like package. For the package the design incorporates a black and a transparent moulding part that can be moulded using available technologies. As expected, the thermal management of the LED SiP is very challenging. However, our SiP is not yet optimized and we used pessimistic heat dissipation values for our assessment. We expect that with improving our design we will be able to control the LED and driver IC junction temperatures to acceptable values.

First lifetime estimates, based on electrical failures only, show that the design lifetime of 25 to 50 thousand hours can be achieved. In order to generate a more complete insight in performance and the lifetime of our SiP, we will employ a combination of methods: with finite element methods we will extract the thermo-mechanical behaviour to investigate the thermal and mechanical hot-spots. These will help us to reveal the (potential) failure modes and failure probability. This information, in combination with field data of known failure modes in comparable products, can then be fed into system-reliability approaches, such as the Bayesan Belief Nets and Markov Chains [16].

For our design we still have to solve several challenges, such as optimal placement and electrical routing of the LED dies, driver dies, and passive components in the package.

Beside the performance of the SiP itself, we need to investigate its behaviour in combination with a realistic heatsink. Furthermore, we want to explore if our SiP is able to make the PCB redundant, thereby eliminating costs and also vulnerable second level interconnects.

Finally, we will build a series of LED SiPs to demonstrate the manufacturability of our SiP and to validate our performance and lifetime preditions.

Acknowledgements

This work is supported by the ENIAC JU and the national authorities of the participating countries in the project ESiP. The authors thank Mark Müller and Richard van der Stam from ALSI for their valuable discussions on this project.

References

[1] Communication from the EU commission: "Energy efficiency: delivering the 20% target", Brussels, 13 November 2008, COM(2008) 772.

[2] A. de Almeida, et al., Characterization of the household electricity consumption in the EU, potential energy savings and specific policy recommendations, Energy Buildings (2011), doi:10.1016/j.enbuild.2011.03.027

[3] EIA 2011 data www.eia.gov.

[4] Solid-state-lighting reseach and development: Multi-year project plan, and manufacturing plan. US Department of Energy www.eere.energy.gov 2011 update.

[5] R. Lineback, "Solid-state lighting set to boost LED growth," LED Magazine, May 2006.

[6] E. David, Patent WO2010/126592.

[7] Lighting Research Centre, ASSIST program. http://www.lrc.rpi.edu/programs/solidstate/assist/index.asp

[8] N. Narendran, J. Bullough, N. Maliyagoda, and A. Bierman. What is useful life for white light LEDs?, Journal of the Illuminating Engineering Society 30 (1) : 57-67, 2001

[9] H. Ye et al. "Numerical modeling of thermal performance: natural convection and radiation of solid state lighting." Proceedings of 12th Int. Conf. on Thermal, Mechanical and Multiphysics Simulation and Experiments in Microelectronics and Microsystems, EuroSimE 2011.

[10] J. Jakovenko et al., "Thermal simulation and validation of 8W LED Lamp." Proceedings of 12th Int. Conf. on Thermal, Mechanical and Multiphysics Simulation and Experiments in Microelectronics and Microsystems, EuroSimE 2011.

[11] Z. Shook, "Lighting the way: LEDs in SMT production", Global SMT and Packaging, October 2010, pp. 38-40

[12] Application note AN10829 "SSL2101 dimmable high efficiency flyback design" rev02, 2009. www.nxp.com.

[13] N. Narendran, Y. Gu, J.P. Freyssinier, H. Yu, L. Deng, Solid-state lighting:failure analysis of white LEDs, Journal of Crystal Growth 268 (2004) 449–456

[14] L. Han and N. Narendran. 2009. "Developing an accelerated life test method for LED drivers." Ninth International Conference on

Solid State Lighting, August 3-5, 2009, San Diego, Proceedings of SPIE 7422: 742209.

[15] Y. Kan, "Degradation modelling white light tube lights", Master Thesis Eindhoven University of Technology, The Netherlands, August 2007.

[16] W. van Driel, et al. "LED system reliability", Proceedings 12th Int. Conf. on Thermal, Mechanical and Multiphysics Simulation and Experiments in Microelectronics and Microsystems, EuroSimE 2011.

[17] P. Doe, "LED fabrication roadmap targets packaged LED prices of $2.20/klm by 2015", LED Magazine, May 2011

[18] "New Philips LED-streetlights in Tilburg reacts on movement and saves up to 80% on energy consumption" (in Dutch). http://www.newscenter.philips.com/nl_nl/standa rd/about/news/press/20110117_licht_op_aanvra ag_in_tilburg.wpd

MultiChip Mems Sensor Packaging

Alastair Attard*, Mark Azzopardi*, Conrad Cachia*, Angelo Crobu**,Fulvio Fontana, Luca Maggi**, Mark Shaw, Federico Ziglioli

ST Microelectronics, Corporrate Packaging and Automation (CPA), Via Olivetti 2, Agrate, Italy. *CPA, Kirkop, Malta. **MEMS Buisiness Unit, Castelletto, Italy.

+39 0396035422,+3903964426930,mark.shaw@st.com

Abstract

In consumer applications the size and cost of the package are becoming increasingly important especially in the applications for the user interface for Gaming and Mobile telephones. The initial adoption of accelerometer devices has enabled human movement to become part of the man machine interface, simply tilting the device to change menu or navigate a web page. One of the key factors in the adoption of the accelerometer devices was the development of LGA style packages which opened up these new market segments to MEMS devices by the use of a low cost, low stress flexibile platform, adaptable to different form factors and chip dimensions. Recently following on from the success of accelerometer devices Gyroscopes devices are increasingly being adopted to add dynamic angular movement sensing putting two different sensors in different packages. In addition to these devices, digital compasses or magnetometeters and altimeters are now being introduced to include directional and height sensing functions. Increasing therefore more than one sensor is used within the same hand held device and instead of using separate device there is now a requirement for Multichip LGA based MEMS Packages. These packages must take into account the particular devices constraints of each individual device within the same package. The individual assembly building blocks for the various device requirements are described in detail, in particular the development of novel assembly techniques for 90° chip mounting and interconnection is outlined. The design of such packages is outlined and the problems encountered in packaging multiple MEMS devices in this style of package structure discussed.

Key words: MEMS, LGA, Accelerometer, Gyroscope

Introduction

The first wave of MEMS devices have now penetrated the main stream of consumer applications, and after initial acceptance by a few innovative players they are becoming ubiquitous in high end cell phones, gaming and tablet computers. These sensor devices have had such success due to a combination of innovative consumer electronics companies and component suppliers design of new MEMS devices suitable for high volume production [1,2]. In addition to the MEMS Silicon technology the adoption of the LGA package which can provide High volume and low cost, with the benefits of low stress and a flexible pin out and dimensions facilitated this expansion in the use of MEMS motion sensor devices[2].

In certain applications more than one sensor is being used within the same device so there is now requirement to integrate different MEMS sensors into the same package device to provide additional functionality. This is the case for Gyroscope and Accelerometers to have X,Y,Z and angular acceleration and also for Navigation systems with a Magnetometer and accelerometer. In some case a pressure sensor is also used for altitude

measurements. In the first step of integration this is done on a PCB board (iNEMO) but now devices are appearing where the integration has been done at the component level with the first Mems system in Package.

LGA package and substrate design

The LGA (Land grid Array) package [3,4] consists of a laminate substrate with a BT (Bismaleide. Triazine) resin Core, so far MEMS sensor devices have required only a top and bottom metal tracks with corresponding solder mask but as more complicated integrated designs develop these can be easily accommodated with extra layers. The exposed soldering pads (lands) and wirebond pads are suitably plated. One of the advantages of the LGA substrate design is it's flexibility with the ability to route the Cu tracks to permit different designs of sensor or ASIC to have the same pin out, or the same designs to have different pin outs with no change to the assembly process including even the wirebonding layout, only a different substrate design. This gives a large amount of flexibility which is not available in more traditional (e.g QFN) designs.

In the case of an accelerometer the sensor device is usually co-packaged with a custom ASIC controller this enables a more user friendly output for the customer either in analog or in digital format.

Depending on the size of die and package different configurations of sensor and ASIC are possible, they can either be bonded side by side or in a stacked mode (fig 1). The MEMS accelerometer is first chip bonded to the Substrate, The ASIC bonded to the top of the MEMS, the devices wirebonded, overmoulded the singulated.

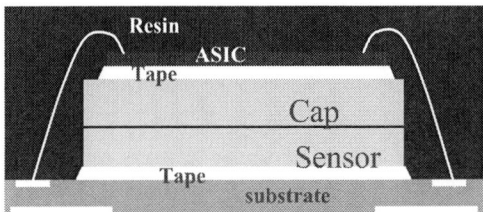

Fig 1 schematic of stacked MEMS accelerometer LGA

The side by side mode (fig 2) obviously has a larger footprint than the stacked devices but has advantages for lower profile devices, in that the separate die may not need to be backlapped to fit in the required package height. A stacked approach however with backlapped die will result in the smallest overall package size

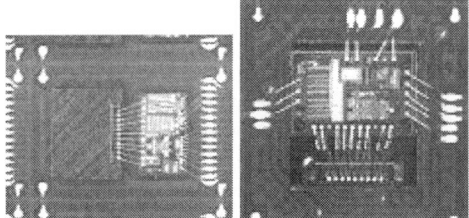

Fig 2 Photo of side by side bonded accelerometer and ASIC on LGA substrate and stacked version

The design of the substrate should be done in order to balance as much as possible the stress exerted on the MEMS device. The most important contribution from the substrate is due to the via placement, [5] with the preferred positioning of the vias showing a different stress profile and drift than layouts dictated only by routing requirements. The bond pad opening of the sensor device is normally placed opposite the transfer moulding gate or entry point for the moulding epoxy, this is to ensure that during moulding this area, which has the largest stress concentration within the glass frit used to seal the sensor and cap silicon, is protected. This poses little difficulty when assembling a single sensor device in the package but when multiple sensors are assembled it poses restrictions on the wirebond layout and substrate design. This often requires longer wires which need to be placed parallel to the

moulding compound flow to reduce wire sweeping effects.

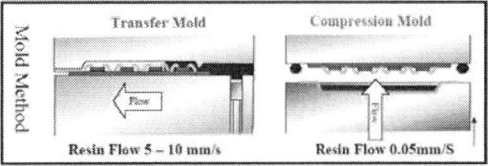

Fig 4 Transfer and compression moulding schematic

One way to avoid this is to move to compression moulding which used a dispensed granular resin directly into the mould cavity avoiding any large resin flow (fig4). This also has the advantage in that it minimizes any wirebond sweep from the transfer of the resin into the cavity. The compression moulding technique has further advantages in that instead of a substrate with separate moulded blocks to reduce the amount of warpage during moulding a single block cavity can be produced, this increases the number of devices produced on each substrate. It is also possible to increase the size optimising the number of strips produced at the supplier in each panel thus reducing cost the strip size. The via position effects not only the stress in the mems device but also the stress on the glass firt Simulations have confirmed that a non homogeneous distribution has an adverse effect on the stress distribution (fig3).

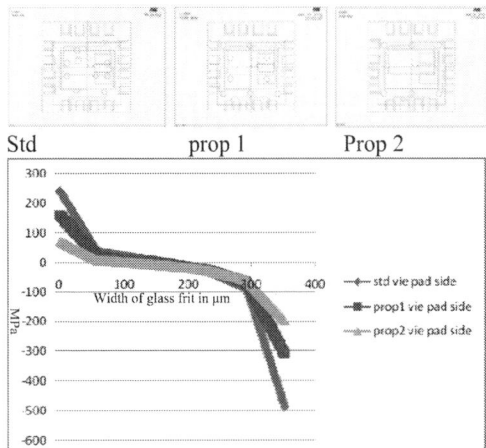

Fig 3 Stress due different via layout substrate designs for a LGA4x4 with Max stress in the sensor chip

The area of the glass frit is then minimized to keep the unproductive non sensor seal area as low as possible, The area of the glass frit is optimised for the stress conditions in moulding and reliability that the device is subjected to, compression moulding also uses much less pressure during the moulding process than conventional transfer moulding, this should mean that it should be possible to reduce the width of the glass frit used to hermetically seal the cap layer to the sensor. In order to test this, devices that failed the automated visual inspection for the glass frit width at wafer level were assembled and half moulded using the conventional transfer moulding and half moulded using compression moulding. Failures of the hermetic seal were observed on devices moulded using transfer moulding but no failures found on devices moulded using the compression moulding technique after moulding and subsequent stress testing. Work is now ongoing to optimize the glass frit dimensions with the compression moulding to determine the design rules and consequent size reduction and cost optimisation of the MEMS sensor chip.

Combinded Mems Sensor packaging

The combination of Gyroscopes and accelerometers in the same package can reduce the amount of space taken up on the PCB board by up to 50%.

Fig 5 MEMS combined Gyroscope and accelerometer

The package is relatively straight forward with two side by side stacks hermetically sealed sensor accelerometer or gyroscope and its relative ASIC stacked on top (fig 5). Smaller footprint 4 stack devices have been produced but this comes at the expense of an overall height increase. Some care is however needed as the increased percentage of silicon within the package gives rise to a different stress distribution within the package and on some early prototypes some die attach delamination was observed.

Magnetometer Packaging

The packaging of magnetometer devices consist of a X-Y magnetometer and a separate Z axis Magnetometer with an controlling ASIC. The peculiarity of the Packaging is that the Z axis device as it uses the same technology as the X-Y devices with the active part on the surface of the chip requires the surface of this Z axis to be at 90 degrees to the surface of the X-Y device.

One solution tried has been to mount the Z axis on a flexible part of a substrate and then bend this substrate 90 degrees to the base plane [6]. Another method is to bond the Z axis device at 90 degrees to the substrate on its side and then make the connections around the top corner. This requires the metallisation of the device to connect around the

corner edge [7]. However for a device fabricated on a silicon wafer this would require the devices to be metallised around the corner after dicing or using an intermediate ceramic part.

Side metallization on wafer

The first method evaluated was the possibility of metallising the side surface of the Z axis chip so that a conventional wirebond could be performed. This was done by dicing a groove in the wafer prior to the final dicing cut (fig 6). The devices were then masked and sputter coated, the coating then formed a pad around the 90 degree edge of the Z axis chip [8]. The final dicing process can then be made next to the groove cut.

Fig 6 Z axis chip showing bond pads sputtered in dicing groove

Whilst this technique enabled the first samples to be assembled (fig 7) it was very difficult to control the quality of the metallization within the groove and it was decided to look for an alternative method for mass production

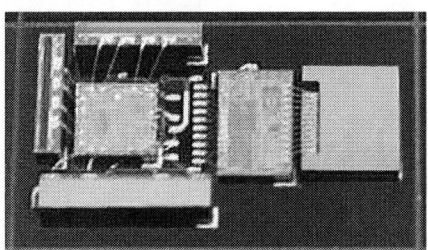

Fig 7 Layout on LGA7.5x4.4 package with wirebonded Z axis

In hard disk manufacture 90 degree interconnects are regularly used for the reading head, this is often done by stud ball wirebonds [9,10]. This technique is difficult to apply on the array strip LGA as either the strip would have to be tilted or the bond head would need to arrive at 45 degrees. In either case a new bonder would be required with either a completely new bond head or a new indexing system. A second problem is the need to have the bond pads at a height of less than the diameter of the stud bond in order to make the connection. If we assume the ball of roughly 2x the wire diameter using a standard 1mil wire this would mean that we would need to have the bond pad at no more than 25um from the pad on the substrate to make a good connection. This is difficult to do with

460

the dicing tolerances and the height of any die attach material (Fig 8).

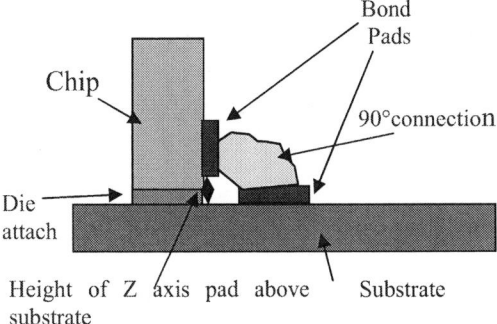

Fig 8 Schematic of Ball bond interconnection

Epoxy connections 400um pitch

Attempt were made to use jet dispensing to place a dot of conductive epoxy in the corner, the difficulty here is that the dot size had to be limited to eliminate any possibility of shorting between the connections of the Z axis, especially as the dot size has to be sufficiently large to cope with the height difference between the pads (fig 9).

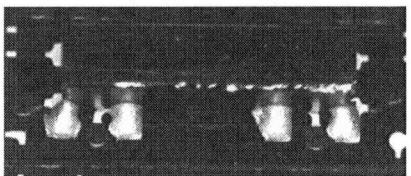

Fig 9 epoxy dispensed interconnections

The process can be assisted and the UPH improved by the fact that the movement of the dispense bond head during the stepping from one component in the strip to the next can be utilised to project the glue dot into the corner of the placed component.

There is however a problem when the pitch between the pads is lower than the required dot size. If improvements can be made it may however be possible to stack dots together to produce the connections.

Dispensed solder connections

Making the connections with solder has one big advantage that even if after dispensing the solder some of the solder extends too far connecting between two pads giving rise to possible shorting. The solder paste during reflow should automatically shrink back onto the wettable metallised pads (fig 10,11). This enables sufficient solder to be dispensed at the side of the die and during the reflow of the solder it wicks along the track and wets the pad on the 90° chip. Again the movement of the dispense bond head during the stepping from one component

in the strip to the next can be utlilised to project the glue dot into the corner of the placed component. The Pitch however for this process is still restricted by the size of the solder dot that can be dispensed. Even with the shrinkage onto the solder tracks the shrinkage can not compensate if the pitch is reduced too much as the dot will either not separate onto the two separate tracks leading to a short or unevenly separate leading to an open where not enough solder is on one pad.

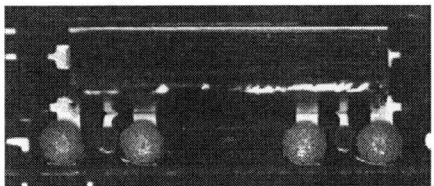

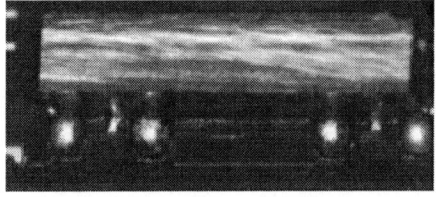

Fig 10 solder paste dispense and reflow interconnections

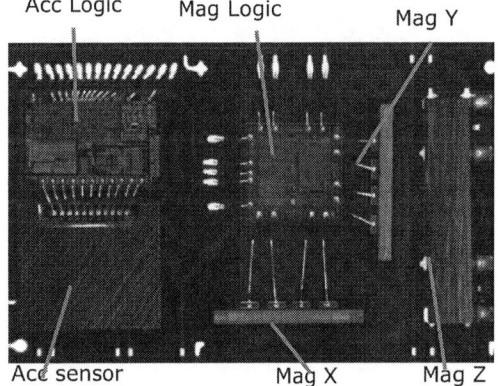

Fig 11 Layout in LGA7.5x4.4 package with dispensed solder connections

Reduced Pitch 90° Interconnections

A change in chip design for the 90° Magnetometer necessitated a reduction in the pitch size from the first prototypes produced to 200um. This increased the number of connections required to 8 and reduced the pitch size. This required an alternative approach for the 90° interconnect. Two different approaches have been evaluated, High definition screen printing and liquid solder jetting

Liquid solder jetting uses a technique of jetting instead of a solder paste a pre molten liquid drop of solder directly into the angle corner in order to make the connection [11]

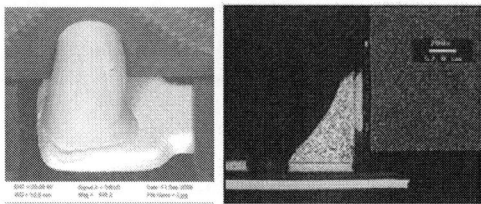

Fig 12 Solder jetting 90° interconnections

This produces a good connection at very fine pitch but like dispensing solutions requires every single connection to be made separately in sequence (fig 12) but with relatively low UPH

Screen printing

The final technique chosen for mass production involved screen printing solder directly onto the substrate prior to the 90° flip chip bonding, a small amount of solder is also dispensed on the opposite side to the 8 connections. The flux from the solder paste and the reflow of the solder on the opposite side maintains the chip in position avoiding any tombstone effect [12] (fig 13-16).

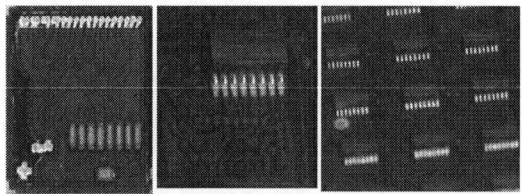

Fig 13 solder printing, die attach and reflow of interconnections

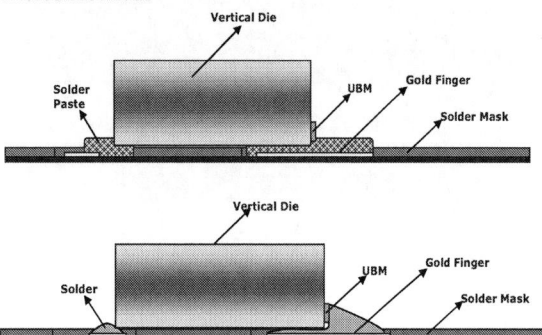

Fig 14 schematic of screen printed solder interconnections

Using the screen printing solution we have a technique that can make all the connections in parallel giving a high UPH. The technique is however limited by the pitch available on the stencil printing so it is difficult to shrink the chip any further

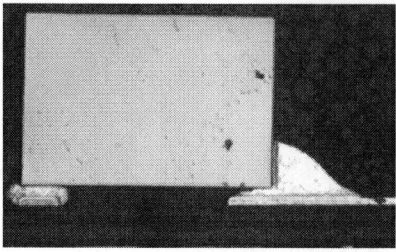

Fig 15 Section of screen printed interconnections

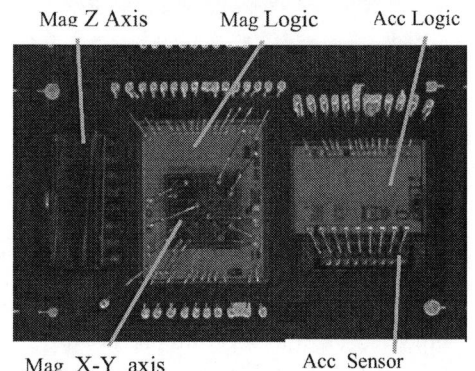

Fig 16 Layout in LGA3x5 package with screen printed Z connections

Electroless plating

The sensitivity of the Magnetometer device can be distorted by the presence of magnetic materials near the device. The only magnetic material used in the package is the nickel plating layer than covers the Cu tracks. Standard plating is electrolytic which is very magnetic, the alternative is to use electroless plating

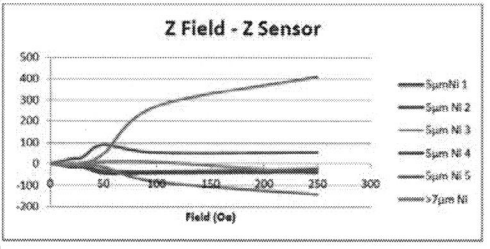

Fig 17 Comparison of Magnetic permeability of substrates with Enipig Nickel showing Z axis field shift in null field reading

The ideal would be to have a high Phosphorous content Nickel as with above 10 % of phosphorous electroless Nickel is essential non Magnetic but unfortunately plating solutions are not commercially available with such high P content. Standard Nickel plating with around 7% P still has some residual magnetic properties but if the thickness of the Nickel layer is reduced to a minimum this can meet the device requirements. The graph (Fig 17) shows the magnetic performance of the 3 axis sensor as a function of the nickel concentration in the substrate

Pressure Sensor Packaging

Three types of pressure sensor packages have been developed within ST. The Holed LGA (HLGA) (Fig 18) which used a holed silicon cap layer and film assisted moulding. This package uses the silicon cap as a barrier or dam to prevent the moulding compound covering the pressure sensor membrane [13].

Fig 18 HLGA pressure sensor

The second package developed the Rear hole LGA (RHLGA), (Fig 19) has the silicon pressure sensor flip chipped onto a substrate with a hole. The pressure is then sensed through this hole in the substrate [14].

Fig 19 RHLGA pressure sensor

The HLGA package has the disadvantage of using a large amount of silicon for the seal area with the cap and an extra silicon wafer for the cap. The RHLGA package requires the hole to be on the reverse side of the substrate. Both packages also suffer from the fact that as the moulding compound is in contact with the MEMS chip so some of the stress from the expansion of the moulding compound is transmitted to the MEMS devices, this can partially warp the MEMS membrane giving a false offset which requires compensation.

An alternative package recently developed uses a Cap over the stacked Mems devices, the holed cap LGA (HCLGA), (Fig 20).

This cap is fabricated directly from standard substrate core material and once assembled on the strip the devices can be singulated using standard processes as the holed cap is placed face down making a seal around the hole. This design using a substrate cap has the advantage over coined metal caps in that the top surface is flat over the whole

surface making standard pick and place operations easier

Fig 20 HCLGA pressure sensor

Even though there is no moulding compound involved the die attach material still has a large effect on the sensor performance (Fig21) and has to be carefully selected.

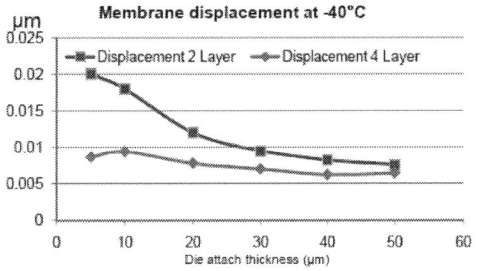

Fig 21 Simulation of membrane displacement with different die attach thicknesses and substrate thickness in HCLGA package

iNemo

The iNemo concept is to mix and match the various Mems sensor devices and assemble them together within the same package (Fig22).

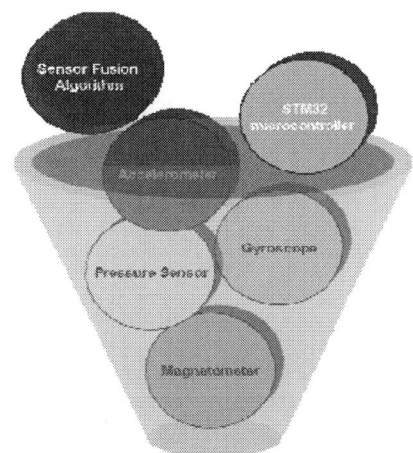

Fig 22 schematic of iNEMO concept

463

The devices have been produced with assembling the various components onto a PCB to produce a complete 10 axis which together with the sensor algorithms and application software produces a complete inertial navigation unit (Fig23).

Fig 23 iNEMO PCB Board

The first steps of combination sensors have been made with combi Accelerometers/Gyroscopes and accelerometer/magnetometers. The next step is the integration of all of the above mentioned sensors into the same package. The most integrated Mems package so far assembled in ST has been a prototype LGA combined Accelerometer Gyroscope, magnetometer for the European collaborative Minami project. This combined a 3 axis accelerometer, 3 single axis gyroscopes and 3 single axis magnetometers in an LGA 15x15 package (fig24). Problems were also encountered with wire sweeping with some long wires being parallel to the mould flow moved during the transfer moulding process, switching to a compression moulding process would improve this aspect.

Fig 24 X ray inspection of combined accelerometer, magnetometer and Gyro 9 axis LGA15x15 package

However in order to reduce the space occupied by combining the axis in single chips, single 3 axis Gyroscopes and 2 axis magnetometers are now available, and package a mixture of stacked and side by side dies will be required (Fig 25).

Fig 25 LGA6x6 concept of 9 axis package

Integration of a pressure sensor within this multichip package poses further problems. The easiest method would be to assemble an RHLGA type package with a hole on the multichip substrate, a HLGA package would also be possible but would suffer height restrictions due to the need for the wafer cap layer to be in contact with the moulding Film or The protrusion can either be made from a rubber compound to protect the sensor or a plastic film is inserted between a metal protrusion in the cavity and the sensor membrane [15]. The drawbacks of both the RHLGA and HLGA package in terms of package stress would however remain. Using a HCLGA package would require moulding around the substrate cap. Devices have been made using protective moulded caps [16] and trials have shown good adhesion between the a HCLGA cap and the moulding compound (Fig 26)

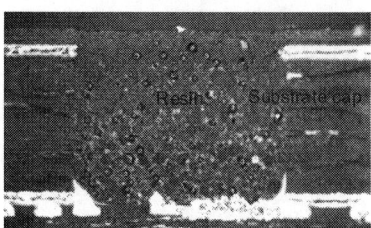

Fig 26 Over moulding of a package cap with film assisted moulding

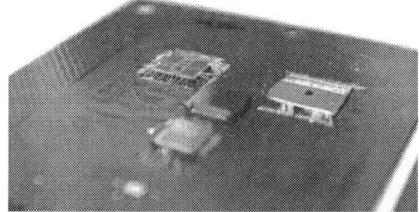

Fig 27 package with accelerometer, ASIC, memory and memory controller

Other components can also be integrated fig shows the integration of an accelerometer ASIC Memory and memory controller chip (Fig27) another example is the inclusion of passives.

Conclusion

It may be in the future that single chip solutions can be found to combine the sensors and or ASIC on the same chip. However as at present the combination of the Accelerometer devices with the ASIC for example does not yet make sense due to the different silicon usage of each part and the yield on the individual devices. The System in Package solution is seen as the best route forward for the immediate future of combined sensor systems as it allows a mix and match approach allowing the most appropriate sensors to be combined in the same package for a particular clients specification.

References

[1] Vigna B., "Physical Sensors Drive MEMS Consumerization Wave" The 13th International Micromachine / Nanotech Symposium July 26th, 2007

[2] Vigna,B MEMS EPIPHANY. MEMS 2009. IEEE 22nd International Conference on Micro Electro Mechanical Systems Page(s): 1 – 6

[3] Vigna Mems Move into Smart sensors Semicon Japan 9/8/2010

[4] Shaw et al High Volume Mems Packaging EMPC2007 p182-187

[5] US patent application 20100052144 Semiconductor Package substrate and method in particular for mems devices

[6] Report MPR-0907-801-1 www.chipworks.com

[7] Yoshinobu, H et al '6-Axis Motion Control Sensor Using Magneto-Impedance Sensor INTMAG.2006 Page(s):693

[8] http://www.linetec.com.tw/download/Microgate _6D_mode/microgate_6D_module_english.ppt

[9] King J Ball in corner www.palomartechnologies.com/.../Ball%20in% 20Corner%20-%20J%20King.pdf

[10] Ball Bumping application data sheet FK Delvotec

[11] Sb2 jet Laser solder jetting Systems Brochure Pactech http://pactech.com/files/PAC%20TECH_SB2-Jet%20SB2-SM%20SB2-M.pdf

[12] 3 D integrated compass package US Patent US2009/0072823

[13] Shaw et al 'Package Design of Pressure Sensors for High Volume Consumer Applications' ECTC 2008 p834 - 840

[14] Fontana MEMS Pressure sensors – new LGA Packagings EMPC2009 p 1-7

[15] Boschman F., "Film assisted molding technologies and applications for high volume MEMS and Sensor packages" EMPC2007 p208,209

[16] Saitoh Hiroshi European Patent EP1860694A1

Miniaturised Integrated Hybrid Microsystem Assemblies for Harsh Environment Applications

Steve Riches

GE Aviation Systems – Newmarket, 351 Exning Road, Newmarket, Suffolk, CB8 0AU, United Kingdom

T: 44 (0) 1638 663 381, F: 44 (0) 1638 660 628, E-mail: steve.riches@ge.com

Abstract

There is a growing requirement for distributed sensors to measure, monitor and control systems such as airframes, wind turbines, engines, power trains, human motion and seismic sensors. These type of devices need to be as unobtrusive as possible and miniaturisation of the sensor and its control circuitry is a key factor in achieving this. This paper reports on a method of combining the sensor element and the control electronics in a single hybrid packaged unit which can provide a step decrease in the size of the component.

The manufacturing processes developed within the project have addressed a rugged vacuum compatible package environment with co-located control electronics. For microsystem encapsulation, wafer level metal sealing techniques have been realised through bonding of the multi-sensor device to a silicon capping/interconnect wafer. The connections from the multi-sensor device have been designed using through silicon vias (TSV). An application specific integrated circuit (ASIC) has been designed and manufactured to provide the control electronics for the multi-sensor device and this has been connected through to the multi-sensor device using solder bump technology. The ASIC has also been designed to accommodate decoupling capacitors to minimise signal noise and parasitics and a flexible circuit to provide connections. The demonstration unit has a size of <10mm x10mm, which is a 15X reduction in size compared to equivalent packaged units.

The approach developed within the project has potential in applications for inertial measurement and also for microsystems involving fluidic and optical components.

Key words: Microsystem assembly, flip chip, miniaturisation.

Introduction

This paper has investigated novel packaging and assembly processes to service the growing need for 3D multi-functional nanosystems [1]. The manufacturing processes have demonstrated the concept of extreme miniaturisation of a 6 degree of freedom (DoF), low power, high performance nano-inertial measurement unit (nIMU) for precision position, orientation and load measurement in a volume of $1cm^3$, some 15x smaller than current inertial measurement systems.

The demonstrator used for the 3D integration manufacturing processes is a highly integrated, novel multi-sensor 2D inertial measurement platform based around a monolithic inertial silicon MEMS multi-sensor array on a $<1cm^2$ chip, as shown in Fig. 1. The generic manufacturing processes are applicable to a wide range of sensors, actuators and fluidic/optical nanosystems. This paper addresses the key aspect of integration of ASIC electronics to the MEMS multi-sensor array. As indicated by the dotted sectors in the schematic, it will be possible in the processes to allow for fluidic and optical ports allowing the assembly techniques to be used for microfluidic and hollow waveguide optical products.

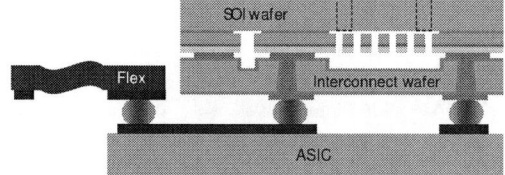

Figure 1. Schematic cross sectional diagram of multi-sensor demonstrator platform with fine pitch flexible PCB external connection.

The manufacturing processes investigated address a rugged, full hermetic and high vacuum compatible package environment. The areas of wafer level packaging and through silicon via (TSV) interconnect are described below. Wafer level metal seal bonding techniques have been developed for bonding the multi-sensor die to a stress matched silicon capping/interconnect wafer using silicon-gold eutectic bonding to polished highly doped polysilicon. The sealing process simultaneously form the required interconnection to the control/readout ASIC by means of through silicon vias (TSVs) in a specially designed interconnect/capping wafer.

This paper presents results of attaching demonstrator interconnect wafer devices directly to the ASIC electronics using underbump metallisation and advanced flip-chip surface mount device (SMT) bonding processes to minimise size compared to wire bonded devices. The position, routing and length of the interconnections to the ASIC has been optimised for resistance and capacitance as required for the particular feed/pick-off, with external passive components mounted onto the ASIC. This approach gives the flexibility to use ASICs and MEMS devices fabricated on different wafer sizes and sources. External connection is via a fine pitch (~250μm pitch) flexible circuit interconnect.

Assembly Overview

The assembly of the miniaturised microsystem consists of the following three main interconnect stages:
- ASIC to interconnect devices
- Passive component to ASIC
- Flexible circuit to ASIC

An overview of the assembly processes is presented in Table 1 and illustrated in Fig. 2.

Table 1. Overview of assembly processes

Process	Approach
ASIC UBM	Ni/Au
Joint to I/connect wafer	Au-Sn solder
Joint to passive components	Sn-Ag-Cu solder Pb-Sn-Ag solder
Joint to flexible circuit	ACF/ACP Sn-Ag-Cu solder Pb-Sn-Ag solder

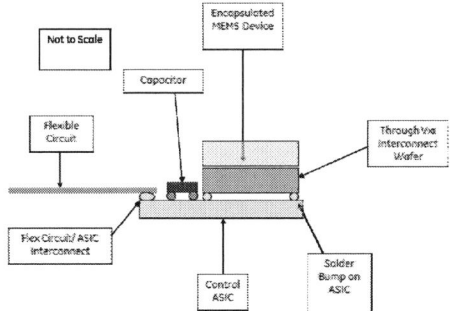

Figure 2: Overview of assembly processes

The first stage in the assembly process was to under bump metallise (UBM) the aluminium bond pad metallisation on the ASIC, using electroless Ni/Au plating.

ASIC to interconnect devices

The ASIC, as shown in Fig.3, was designed to drive and control the microsystem through the interconnect device.

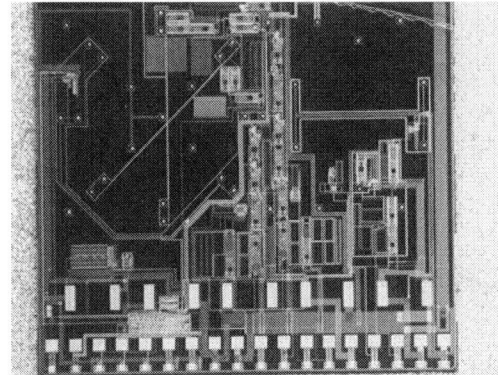

Figure 3: ASIC chip for microsystem drive and control

In order to replicate the ASIC in the assembly trials, a dummy ASIC, see Fig.4, with the same bonding pad co-ordinates was designed and manufactured alongside the real ASIC to establish bonding conditions and assess the environmental performance.

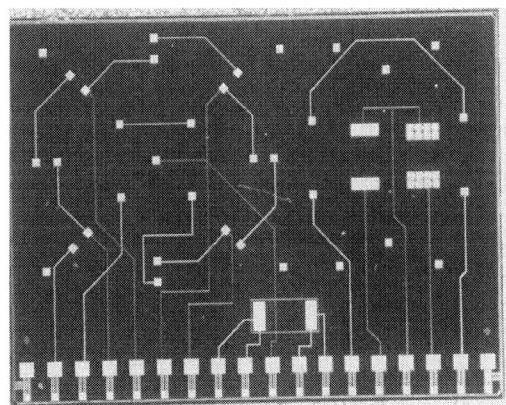

Figure 4. Dummy device to replicate ASIC for assembly trials

In order to connect to the ASIC, a dummy interconnect device with matching bond pad locations to the ASIC (dummy and real) was designed and manufactured with a Au thin film metallisation, as shown in Fig. 5.

The layout of the interconnect device and dummy ASIC was designed in a daisy chain pattern to allow electrical resistance measurements to be made along a string of bonded connections, see Fig. 6 for the tracking details on the dummy ASIC.

467

Figure 5: Interconnect device for bonding to ASIC

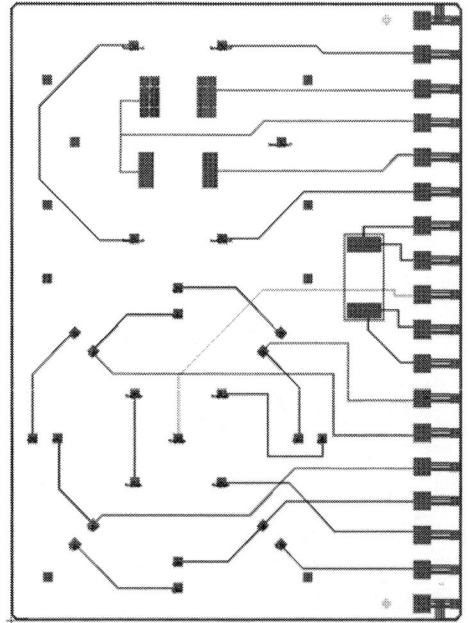

Figure 6: Layout of tracks on dummy ASIC

The ASIC and replicate ASICs were then Au-Sn solder bumped with 100 μm diameter solder balls on the ASIC to interconnect device bond pads.

The replicate ASIC devices were then aligned with the interconnect device and reflowed with two different weights applied. The resistance along two of the daisy chains was measured showing continuity through a number of joints, as summarised in Table 2.

Table 2. Resistance measurements on assembled daisy chain devices

Weight, gms	Resistance, Ω	
	Pin 2-6	Pin 12-16
0.5	2153	2426
1.0	2140	2383

Assembled devices were then submitted to high and low temperature storage and temperature cycling between the upper (250°C) and lower (-40°C) levels and resistance measurements along the daisy chains were taken after each test. The normalised values against the as-assembled results are presented in Table 3.

Table 3. Environmental tests on assembled daisy chain devices – normalised values

Thermal Cycling	2-6	12-16
As-assembled	1	1
3 hrs @250°C	0.96	0.95
15hrs @-40°C	0.97	0.95
3 hrs @ 250°C + 15 hrs @-40°C	0.95	0.96
3 hrs @ 250°C + 15 hrs @-40°C	0.95	0.97
3 hrs @ 250°C + 15 hrs @-40°C	0.95	0.97

Passive component to ASIC

The ASIC design allowed for decoupling capacitors to be mounted close to the ASIC to minimise issues with signal noise. Surface mount capacitors were reflow soldered using Sn-Ag-Cu or Pb-Sn-Ag solder paste at lower temperatures than the ASIC to interconnect device joints, as illustrated in Figure 7.

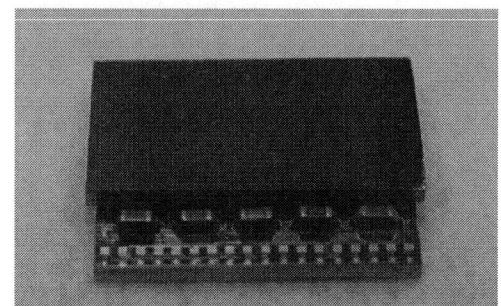

Figure 7: Surface mount capacitors soldered to ASIC

Flexible circuit to ASIC

The ASIC design also allowed for a flexible circuit carrying drive and control circuitry to be assembled along one edge of the ASIC. The flexible circuit had 18 connections on a 0.5mm pitch (0.25mm pad width, 0.25mm gap) and was joined to the bond pad metallisations on the ASIC using either solder or anisotropic conductive film/paste, as shown in Figure 8.

The flexible circuit joints to the ASIC were made with a thermode (hot bar), applying temperature and pressure to the circuit. The general heating from the thermode was not sufficient to affect the adjacent capacitors or interconnect device.

Figure 8: Flexible circuit attached to ASIC

Discussion

The reduction in size of a microsystem and control ASIC has been achieved through the elimination of a level of packaging and close integration of the control ASIC with the microsystem by flip chip assembly methods and through silicon vias. The use of temperature cascades during processing allows the subsequent assembly of capacitors and flexible circuits with lower temperature materials.

The robustness of the system within its intended environment needs to be proven, particularly in high vibration conditions, where underfill materials may be required to reinforce the connections. In principle the approach developed can be applied to other microsystems and miniature electronics. The robustness of the system would need to be demonstrated on a case by case basis.

The material and process costs associated with this process would also need to be assessed on a case by case basis. The advantage of elimination of a level of packaging may be outweighed by the engineering costs of processing at a wafer scale, particularly for small quantities. However, for larger quantities, the benefits of this approach will become more attractive due to the material saving per device.

Conclusions

This work has demonstrated a flip chip assembly approach for integrating microsystems with a control ASIC to achieve a size reduction of x15 compared to a conventionally packaged unit. The approach has also included close proximity mounting of decoupling capacitors and flexible circuit interconnect to provide power and signal inputs/outputs.

The approach developed within the project has potential in applications for inertial measurement and also for microsystems involving fluidic and optical components.

Acknowledgements

The author acknowledges the efforts of Nigel Collins of GE Aviation Systems – Newmarket for the assembly trials and the work of Alan Brown, David King, Chris Reeves and Andy Lewis formerly of QinetiQ, Malvern for the supply of materials, including the ASIC and interconnect device design.

This work was supported by UK Technology Strategy Board (www.innovateuk.org) under the project title UK-MINO.

References

[1] H Kuisma "3D WLP MEMS: Market Drivers and Technical Challenges", 3D Packaging, Issue 9, January 2009

Requirements for Microfluidic Sensors Based on Ceramic Technologies in High Temperature Chemical Processes

Samuel Hildebrandt, Klaus-Jürgen Wolter

TU Dresden, Electronics Packaging Lab; Helmholtzstr. 10, 01062 Dresden, Germany

Phone: +49 (0) 351 463-36426; Fax: +49 (0) 351 463-34531; E-mail: samuel.hildebrandt@tu-dresden.de

Abstract

After a short introduction into one actual development in the field of chemical reaction engineering, the requirements resulting from the application of sensors in a chemical reactor are shown. Physical quantities of interest when investigating and running chemical reactions are specified. As an example, the implementation of a multi-channel temperature sensor is described. Due to the chosen sensor technology, it was necessary to find a working combination of a chemically resistant steel and an electrically insulating glass paste. The compatibility of steel grade 1.4571 and Heraeus SD 1000 glass is shown.

Key words: Microfluidic Sensor, High Temperature, Chemical Reaction Engineering

Motivation

In the wide range of products made in the chemical industry, fine chemicals are a small but important subgroup. They are characterized by low volume production and precious products like fragrances or flavours. The usual way to produce them is a continuously or discontinuously operated stirred-tank reactor. However, this method shows several drawbacks. Reaction conditions are not homogeneous in the reactor: local temperature, velocity and the concentration of reactants or the catalyst may differ notably from the intended values. Different batches have to be blended to achieve constant quality. There is still a safety risk due to the amount of reactants in the range of hundreds or thousands of litres. The resulting safety restrictions (lower temperatures and / or lower pressures) may lead to a suboptimal efficiency of the reaction.

One approach to overcome these drawbacks is to use a large number of continuously operated microreactors in parallel. The continuous operation of chemical reactions is the key to a constant and increasing product quality. With closed loop control, unknown regions of conversion can be reached, helping new technologies to amortize faster. To obtain deeper understanding of reactions and to control the reactors precisely, microfluidic sensors are necessary.

Boundary Conditions

As described above, there are chemical reactions that are done at lower than optimal temperatures and pressures due to safety considerations. With smaller reactor volumes, this danger is reduced. In discussions with chemical reaction engineers, temperatures up to 350 °C and pressures up to 200 bar were regarded as reasonable.

Sensors have to withstand these parameters in operation, and even tougher conditions are applied during qualification. The temperature rules out almost all organic materials as well as many solder alloys. It is likely (and common) to separate the sensor itself and the necessary electronic circuits, so that the latter can be manufactured using standard technologies. For the sensor, ceramic materials seem to be a good choice.

The flow velocity in such a microreactor can reach 10 m/s and will likely cause abrasion on any exposed edges. In some cases, the conversion rate of the chemical reaction can be increased by a special flow regime. For example, several three phase reactions (gas + liquid + solid catalyst) show optimal results when a so called taylor flow is established in a small channel. This flow regime must not be disturbed by any sensor.

To avoid any influence on the chemical reaction, ohmic contacts to the fluids are not permitted. Finally, all sensors have to be resistant against as many chemicals as possible. At the moment, research is still ongoing to find out which reactions benefit from being transferred to continuously operated reactors. Therefore, especially in research, no one is able to specify a fixed list of chemicals.

Physical Quantities

One of the main parameters of a reaction is the temperature. At the first glance, that standard task seems simple. However, with the boundary conditions mentioned above, it gets more complicated. It is not allowed to insert anything into the fluidic channel since even a "small" sensor with a diameter of 0.5 mm would disturb flow and reaction considerably.

Close-knit is the demand for temperature distribution, allowing to get a hint at how fast the reaction runs. Knowing this helps to dimension the length of a reactor for a certain reaction.

It is also important to know and monitor the mentioned flow regime. In particular, it is necessary to differentiate between the liquid and the gaseous phase and to determine the length of each phase. This parameter shows directly the amount of conversed source materials. Furthermore, if the sensor is calibrated for example with a high speed camera, it might be possible to determine whether a certain (optimal) flow regime is established or not.

What counts in the end is the efficiency of the reaction. Of course, it is always possible to take a sample of the reaction product to the laboratory and perform a chemical analysis. However, it is desirable to build an online sensor for that task, so that the process efficiency can be used to control the reaction.

Example Implementation: Temperature Sensor

One of the major challenges when building a sensor for high pressure applications is a reliable fluidic connection between sensor and the rest of the system. One approach is to apply the sensor on a steel tube, preferably of a stainless steel type used in chemical reactors. The main advantage of such a solution is the applicability of standard tube fittings like the widely used Swagelok® system, which can easily handle the requirements regarding pressure and temperature. Another bonus is a certain mechanical robustness of the sensor given by the stiffness of the steel tube. This is necessary to prevent damage during the mounting process of the tube fittings.

Using thick film paste systems optimized for steel substrates and special screen printers for tubes, a variety of sensors can be implemented on steel tubes. Usually, the process starts by insulating the tube with at least two separately fired layers of a glass ceramic paste. Upon that glass film a sensor can be built.

[2] and [3] describe the realisation of a calorimetric flow sensor for liquids. The glass-on-steel-process was designed for heater applications first. Therefore, mainly conductive or resistor pastes with low sheet resistance were adapted to the insulation paste. As a result, sensor pastes can show a different behaviour on the insulated metal substrate compared to the alumina they were developed for. For example, [2] documents a "drop of the temperature coefficient of resistance from the nominal value of 2200 ppm/°C to 1490 ppm/°C as well as a slight increase in sheet resistance."

[1] describes a torque sensor based on printed resistance strain gauges. [1], [2] and [3] originate from the same research group and use an ESL paste system on the widely used steel grade 1.4301.

[4] uses the same steel grade for a oxygen sensor. The paste system is not mentioned and might include proprietary materials.

This work focuses on the "SD 1000 Dielectric Paste for Chromium-Steel" produced by Heraeus. In the datasheet, only one steel grade is listed: 1.4016, a stainless steel used for kitchen equipment and in applications with low corrosive liquids. The material is not suitable for (potentially high corrosive) chemical reactors. Furthermore, the cold forming ability needed for constructing tube systems is not given. SD 1000 is also compatible with the 1.4301 steel grade. This combination seems to be used in industrial applications; however, no confirmative publications were found. An own quick test showed that it is possible to coat a larger area of (24 x 12) cm² on a flat steel sheet without defects.

In our application, the steel grades preferred by the chemical reaction engineers are 1.4571 and 1.7458. Both are widely used in chemical reactors. The necessary compatibility tests are described in the following section.

The goal is to implement a multi-channel temperature sensor (see Fig. 1). A grid of 1.5 mm is chosen, resulting in 32 measurement channels on a length of just fewer than 50 mm. The outer diameter of the given tube is 6 mm with a wall thickness of 1 mm. Due to the wide temperature range of (0 ... 350) °C and the resulting change of mechanical stress, a thick film resistor might show high measurement errors. To avoid that, Platinum is chosen as sensor material. The resistance of one channel will be around 11 Ω at room temperature, increasing to about 23 Ω at 350 °C. By four-terminal sensing, this low resistance can be measured reliably.

One application for this sensor is testing and comparing different types of catalyst pellets or powders used in chemical reactions. The temperature gradient along the sensor gives a quick feedback about the intensity of the reaction. The sensor could also be useful for process control.

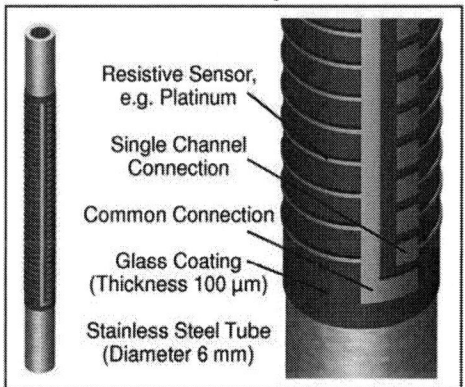

Fig. 1: Concept for a Multi-Channel Temperature Sensor on an insulated Steel Tube

Glass Isolation on 1.4571

Steel grade 1.4571 is a high performance stainless steel with high corrosive resistance. It is the material of choice for many chemical reaction systems. Fig. 2 shows an exemplary collection of tubes coated with one and two layers of SD 1000. The surface of the samples was analyzed intensely by optical microscopy - no defects or cracks were found.

Fig. 2: Steel Grade EN 1.4571: Upper Tube 1x coated, middle 2x, lower: initial state

To confirm the optical inspection, metallographic cross-sections were prepared (see Fig. 3 and 4). The surface of the steel tube is coated uniformly, without delamination or cracks. The images of the double coated tubes do not show any boundary, so the two layers bonded perfectly.

In summary, steel grade 1.4571 can be added to the compatibility list of the SD 1000 paste.

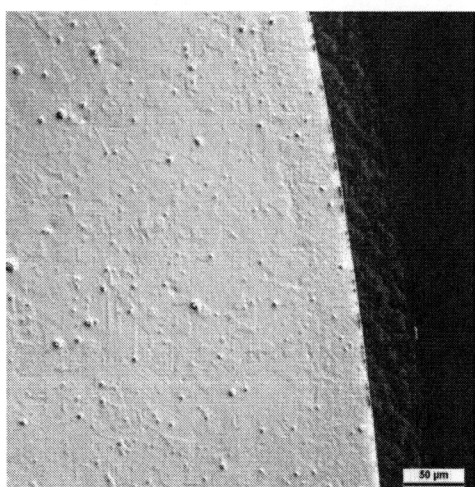

Fig. 3: Steel Type EN 1.4571: Coated 1x; Fired Thickness 40 μm

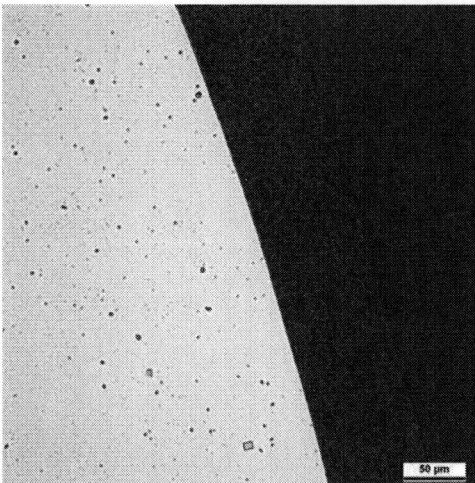

Fig. 4: Steel Type EN 1.4571: Coated 2x; Fired Thickness 83 μm

Glass Isolation on Steel Type EN 1.7458

Steel grade 1.7458 is also used for chemical reactors, and thus the SD 1000 compatibility test was also done on this material. All treatment was done together with the 1.4571 samples, so all tubes have seen the same coating, drying and firing cycle under the air atmosphere recommended in the datasheet. However, during the cooling phase after sintering, the glass layer started to chip off. A cross-section revealed that the glass paste had hardly sintered on the steel tube (see Fig. 6). In direct comparison to 1.4571, a more intense oxidation of the bare ends of the tubes is observed.

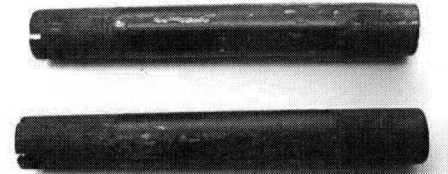

Fig. 5: Steel Type EN 1.7458: Upper Tube 1x coated, lower 2x

472

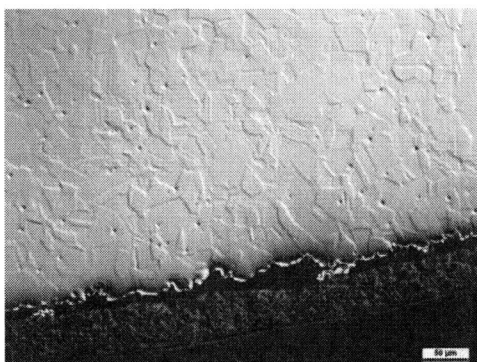

Fig. 6: Steel Type EN 1.7458: Glass Coating peels off (1x coated)

In summary, steel grade 1.7458 has to be regarded incompatible to the SD 1000 paste.

Further compatible materials

The composition of the two "new" compatible materials is similar. There are some more steel grades with comparable alloys like 1.4401 and 1.4541 (see Tab. 1). The author expects these materials to be compatible with SD 1000. However, no tests were conducted to prove this assumption.

Except 1.4016, all the other steel grades have similar properties: They are austenitic and their mean coefficient of thermal expansion is in the range of $(16.5 \ldots 17.5) \, 10^{-6}/K$ [5]. In contrast, 1.4016 is a ferritic steel with a CTE of $10.5*10^{-6}/K$.

Tab. 1: Steel Grades compatible with Heraeus SD 1000 *(italic: assumed to be compatible because of similar composition – not tested!)*

EN No.	EN Name Designation	AISI No.
1.4016	X6Cr17	430
1.4301	X5CrNi18-10	304
1.4571	X6CrNiMoTi17-12-2	316Ti
1.4401	*X5CrNiMo17-12-2*	*316*
1.4541	*X5CrNiTi18-10*	*321*

Further Challenges

For the presented temperature sensor, the fluidic connection is quite simple. There are however other measurement tasks that cannot be performed on the outside of a steel tube. It is necessary to develop either a permanent or a removable fluidic connection from steel tubing to ceramic substrates. Literature research and first tests have shown that the combination of high temperature and high pressure is quite challenging for a removable connection, since huge forces are required to make a gasket leak-proof. On the other hand, the difference in thermal coefficient of expansion and tensile strength between steel and ceramics do not make a permanent connection easier.

As mentioned in the beginning, the electrical connection at the sensor side is also challenging due to the high operation temperature of up to 350 °C. One approach is to use wire bonding and to encapsulate the junction with a special sealing glass. This process is used for single discrete Platinum temperature sensors. However, the necessary thick precious metal bonding wires are expensive. There seems to be a need for a cheap and practical, yet reliable high temperature multi-pole electrical connection technology.

Another point is the ageing of the sensor itself or its electrical connection. If it is not possible to eliminate these effects, they need to be at least predictable.

Acknowledgements

The author would like to thank the colleagues from TU Dresden (Chair for Chemical Reaction Engineering as well as Electronics Packaging Lab) for input on chemical processes and fruitful discussions.

Thanks also go to J. Kruse for the preparation of the metallographic cross sections.

References

[1] H. Hochstöger et al., "Piezoresistive Torque Sensor Realized in Thick Film on Tube Technology", Proceedings of the 28th International Spring Seminar on Electronics Technology (ISSE), Vienna, Austria, May 19-22, 2005.

[2] W. Smetana et al., "Thick Film Flow Sensors for Liquids Based on the Calorimetric Principle", Proceedings of the 2004 International Symposium on Microelectronics (IMAPS), Long Beach, USA, November 14-18, 2004.

[3] D. Güleryüz et al., "Aspects Concerning the Optimization of a Calorimetric Flow Sensor Built Up in Thick-Film Technology", Proceedings of the 28th International Spring Seminar on Electronics Technology (ISSE), Vienna, Austria, May 19-22, 2005.

[4] S. Ziesche et al., "Oxygen sensor for gas mixtures with combustible components", Proceedings of the XXII EUROSENSORS, Dresden, Germany, September 7-10, 2008.

[5] Euro Inox, "Stainless Steel: Tables of Technical Properties", Second Edition, 2007. ISBN: 978-2-87997-242-8. Online available at http://www.euro-inox.org/pdf/map/ Tables_TechnicalProperties_EN.pdf

Multilayer Technology as an Integration System for Ceramic Micro Fuel Cells

Adrian Goldberg[a], Uwe Partsch[a], Steffen Ziesche[a], Lars Fallant[a], Jakob Schöne[a], Holger Neubert[b], Johannes Ziske[b]

[a] Fraunhofer Institute for Ceramic Technologies and Systems IKTS, [b] TU Dresden, Institute of Electromechanical and Electronic Design

Winterbergstr. 28, 01277 Dresden, Germany

Phone +49 3512553-7783, Fax +49 351 2554-341

adrian.goldberg@ikts.fraunhofer.de

Abstract

Fuel cells are considered to be a possible alternative to battery technology for many years. But because of the too high production costs and the lack of long-term experience as well as the missing miniaturization of such systems there is no commercialisation. The aim of the Fraunhofer IKTS is the development of planar self-breathing µPEMFC system based on LTCC technology. By using this mass production technology togehter with multiple printed panels you can build highly compact and integrated micro fuel cell systems. The work presented here gives an insight into the development of µPEMFC system in various levels of integration and performance classes. They show the feasibility of miniaturization and cost reduction through integration of Balance of Plant (BoP) components and electronics in the bipolar plate of the fuel cell. Different sample systems give an outlook on possible approaches for further miniaturization.

Key words: LTCC, fuel cell, PEFC, USB charger, fuel cell charger

Motivation

In recent years the development of entertainment electronics goes to mobile computers such as tablet PC and smartphones with increasing processing power and screen resolutions. Special features for example digital camera, mp3-, video-player, mobile internet, GPS navigation leads to an increased energy consumption of these devices. In spite of increasing energy density of the Li-Ion technology the battery life of mobile devices is not extended. The desire for higher energy density and faster load times combined with a smaller volume is the first priority of the electronic industry. One possible approach for higher energy densities, especially for long time operation offer micro fuel cells (e.g. µDMFC, µPEMFC). The possibility of refilling a fuel cartridge offers an alternative to the long loading times of batteries. For many years the ceramic multilayer is used in the automotive and communication industry as a reliable and extremely robust material system. This technology offers the possibility to integrate all necessary components of a micro fuel cell system. The LTCC (Low Temperature Co-fired Ceramics) technology is a special multilayer ceramic technology which enables producing multiple printed panels similar to wafer manufacturing processes. The advantage is production of high volumes with parallel processes and lower costs.

For many years the IKTS has competence in the LTCC Technology and deals with the development of miniaturized systems in LTCC and HTCC. Because of the expertise in the development of ceramic base materials and conductive pastes the necessary operation experience for building a micro fuel cell completely available in-house. Since 2002 IKTS deals with the development of air breathing µPEMFC in LTCC and has built up various technology demonstrators. The work shown here presents an insight into the development of passive air breathing systems in LTCC and gives an outlook into possible development goals and technological challenges.

LTCC Technologie for Micro Fuel Cell Systems

The LTCC technology offers an integration platform for micro fuel cell systems. For example the electronic packaging to implement DC/DC converters, User Interfaces and actuator control is possible. Also complex channel geometries for providing reactions with fuel gases are possible to integrate. Due to the specific material properties of LTCC (hermetic, chemically stable and mechanically rigid) bipolare plates can be developed. If the micro fuel cells have to be positioned on the market they must be cost-effective and profitable. The general manufacturing process of a multilayer-based µPEMFC is shown in figure 1. At the beginning the still unsintered films are cutted into their processing dimensions (4x4, 6x6 or 8x8 inches). In the next step a punching machine punches electrical connections and stack labels in the tape (vias). In the following screen printing process metallisations are applied and the individual layers will be laminated (uniaxial or isostatic) as shown in the figure 1. By laser cutting the openings in the cathode side and the gas channels in the anode side were structured. After the sintering process and the following separation of the bipolar plates out of the multiple printed panels you get a monolithic ceramic fuel cell with the desired properties.

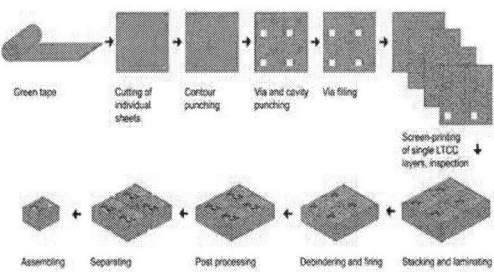

Figure 1: LTCC manufacturing process steps

High-speed 3D Structuring of Multilayer Ceramics

Because of necessary cost reduction a multilayer microstructuring equipment for large substrate sizes is needed. In cooperation with KMS Automation GmbH a new automated machine was developed which combines the ability for micro punching and laser micromachining for ablating and cutting (UV laser by Spectra-Physics, wavelength 355 nm, 20 W). The machine has an option for automated feeding of tapes or substrates. The prototype machine (figure 2) is built for Fraunhofer IKTS to test and manufacture multilayer components (LTCC, HTCC). With this prototyping machine structuring of 8x8 inch substrate in combination of ablating cavities in the range of 50 µm and punch vias in the range of 50 µm is possible [1]. Using this newly developed combination of laser and punching machine simultaneous punching and subsequent laser processing of substrates are possible.

Figure 2: Prototype machine, a combination of laser and punch

The automatic feeding system allows time-efficient working. Figure 3 shows an 8x8 inch substrate of a µPEMFC cathode array produced by this new technology.

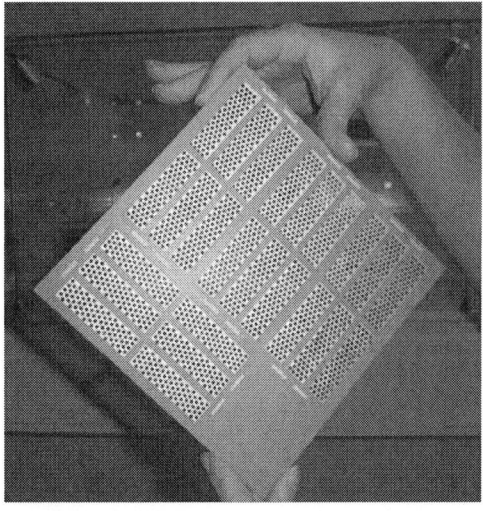

Figure 3: 8x8 inch substrate of a µPEMFC cathode array

Specifcation and Research Goals

According to the EU initiative of standardization of cell phone and smart phone charger interfaces [2] all the European cell phone manufacturers (e.g. Nokia, Apple) agreed to the micro USB standard 3.0 to be the standard charging port. This development could help the fuel cells to enter into the commercial market as universal charger systems. First prototypes of marketable system are already presented from Motorola, MYFC or Horizon.

Results of Research Aktivities for Ceramic µPEFC

Since 2002 the IKTS develops planar self breathing micro PEFCs. The goal was to develop passive, miniaturized and simply structured fuel cells in LTCC technology to build very compact and simple systems. The feasibility of PEFC production in multilayer ceramic has been shown since 2002 and was steadily improved (figure 4) [3].

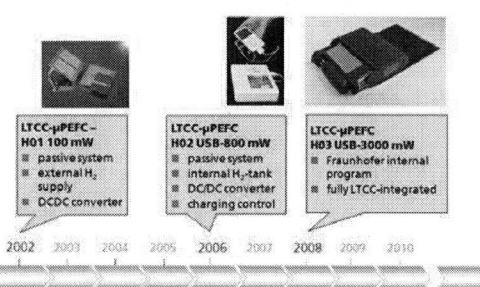

Figure 4: Timetable of micro fuel cell development at the IKTS

In early 2006, the IKTS presented a prototype of a fuel cell charger on international trade fairs and showed the feasibility of the system. The base of this unit was a 1 Watt LTCC multilayer fuel cell with a 5 volt USB 1.1 specified charge output (figure 4) [4].

In early 2008 a 3 watt USB 2.0 charging system was developed with all necessary components for the regulation and the control of the system (valves, charger interface, USB output voltage generator). In addition a user interface with a removable hydrogen cartridge system was implemented. The size of this system was 125x100x35 mm and the most space was used for the cartridge compartment and the valve implementation. The format of the planar fuel cell stack was 85x35x4 mm with a maximum power of about 3 watts a serial connection of 6 cells and a power density of 120 mW / cm^2. In this system metal hydride hydrogen storage was used. The system could be operated for 5 hours and had a capacity of 15 watts per hour [5].

The hydrogen cartridge was a commercially available product of the company Horizon. To extend the operating time the cartridge could be changed at any time by simple turning the cartridge out of housing. Figure 5 illustrates the buildup of the system.

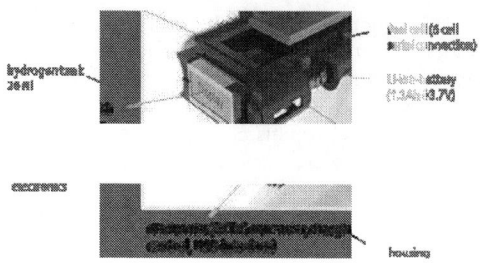

Figure 5: 3 watt USB 2.0 charging system

Since 2009, with the help of simulation tools such as Ansys and Comsol the geometry parameters were optimize. With the foundation project M3 of the Fraunhofer Gesellschaft the patent to minimize the internal resistance of an electrochemical cell was developed and with the new technological layout could be realized only by multi-layer technology [6]. A finite number of vertical through connection (vias) were introduced in the individual ceramic layers and connected to an inner current collector layer and afterwards lead to the external contacts. Advantages of this layout are the use of highly conductive internal silver layers and the parallelism of the power dissipation (uniform current distribution). This allows to minimize the use of expensive gold and additionally connects non contactable surface area to the outside. An example of the simulation results shows the figures 6a and 6b.

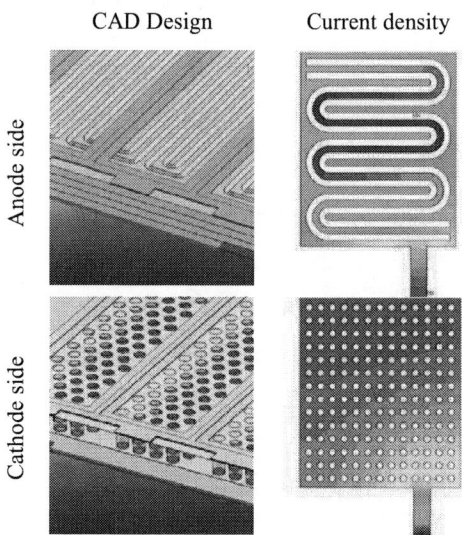

Figure 6a: Simulated top layer current collection

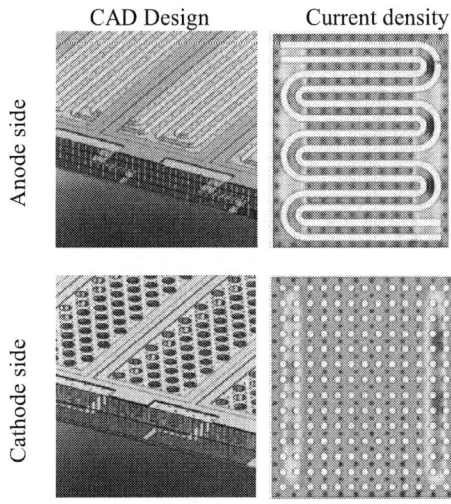

Figure 6b: **Multiple layer current collection with vias**

A further result of this project was a procedure to replace the gold layer on the top of cathode and anode sides [7]. By using polymer pastes after ceramic sintering a passivation with thin carbon layer was printed over the silver vias. In combination the two patents a very cost-reduced solution while minimizing the internal resistance was achieved. Therefore, the cost of a multilayer PEFC was reduced significantly whereas system integration and miniaturization was increased.

Next Step of Miniaturization and Integration LTCC μPEFC

At the moment the development of the LTCC μPEFC USB charger in the power range of 5 W with USB 3.0 specification starts. The target size of the system is 110x52x15 mm. Compared to the previous system the size is smaller by 1/3 while the energy density doubles. These properties are mainly caused by the integration of electronics directly on the anode side and the implementation of the control valves in the LTCC ceramics. An initial design for such a system be seen in figure 7. A key advantage of such packages is the close coupling between the metal hydride tank and the fuel cell. Because of the discharge of the hydrogen storage and the redox reaction inside the PEFC the stack is cooled simultaneously. Additionally overheating of the system is minimized. Using a quick-change coupling the metal hydride storage can be exchanged during operation time. In this situation the integrated battery supplies the system and the consumer load. With the special valve design (Normally Closed) and internal temperature and pressure monitoring the security of the system can be improved and the risk of failure minimized.

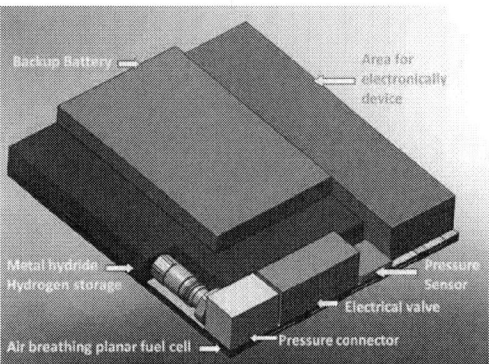

Figure 7: **New schematic design of a LTCC USB charger**

New μPEFC Stack Design

Current work is focused on the development of the LTCC μPEFC stack. The 6-cell stack is designed in a planar arrangement. The cathode is made completely passive and the supply of the anode with hydrogen is carried out by pressure from the hydrogen storage system. The current design (Figure 8) was designed by using the previous development with the internal power dissipation and the ability to use carbon electrodes.

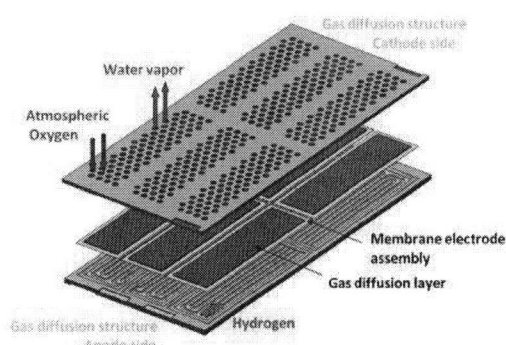

Figure 8: **New μPEFC stack design**

The dimensions of the stack comes to 106x51x2,5 mm. In experiment the good performance has already been proven and a maximum power density of 120 mW/cm^2 was achieved. The characteristic curve of the stack is shown in figure 9 with parameters in the following table 1.

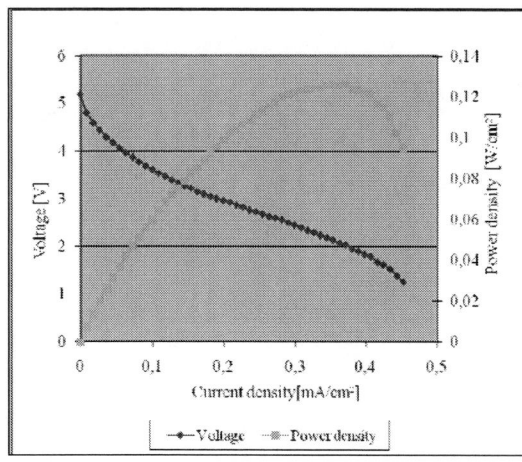

Figure 9: Characteristic of the new µPEFC stack

Table 1: Parameters of measurement

Parameter	Data / Discription
MEA	Gore 5710 (A510.1 ; K510.4) Anode: 0.1 Pt[g/cm²] ; Cathode: 0.4 Pt[g/cm²]; Thickness: 18µm
GDL	SGL Carbon Group (Anode: 35 BC / Kathode: 35 BA)
Number of fuel cell	6 cell, every 5.52 cm² (serial connection)
Gas composition	H_2 5.0 Linde
Gas pressure and flow	Anode flow mode: 100 ml/min Kathode air breathing: Atmospheric oxygen
Environment temperature	Room temperature (23°C)
Mounting force	0,8 N/mm²

The preparation was carried out in the newly evaluated 8x8 inch standard LTCC Technology (Figure 1). Every 8-inch substrate consists of four bipolar plates (Figure 3). After separation by laser the individual parts of the fuel cells were glued together with a vision controlled micro-dispenser (Asymtec Axiom X1000). The final assembly prressure was 0,8 N/mm^{2}. The serial interconnection of the cells was done with silver glue or soft solders at the edges. A ready manufactured fuel cell is shown in figure 10.

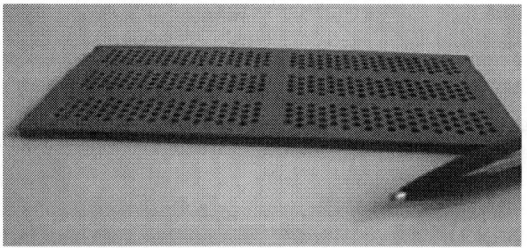

Figure 10: 6-cell µPEFC LTCC stack

Aktor integration

An important part of system miniaturization is the integration of valves on the LTCC interconnector. Gas supply, gasket seal and the positions system were integrated in the LTCC ceramic (figure 11). This special, extremely compact valve design allows the supply of hydrogen to fuel cells while providing an important safety feature. In case of an automatically power failure the valve closes and remains closed (normally closed function).

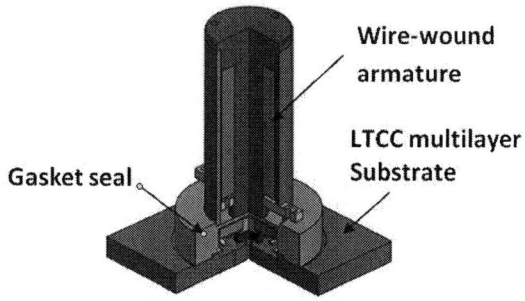

Figure 11: LTCC integrated valve

The design parameters of the fuel cell system were evaluated by FEM modelling. Already the first prototypes of the valve showed their functionality. In the next step the assembly is positioned on the LTCC substrate with subsequent characterization of properties such as leakage rate, maximum pressure, maximum flow rates of hydrogen and electric power consumption. Figure 12 shows the actuator of the valve.

Figure 12: The Actuator of the LTCC integrated valve

Electronic Integration on the Anode Side

The final step is the development of a control and evaluation unit as well as the power conversion and charging mangement for USB 3.0 devices. Efficiency is the main focus of the individual parts. Using a micro-controller all relevant systems (valves, sensors) can be controlled and displayed via a user interface. All necessary electronic components are implemented on the anode side of the fuel cell by SMD assembly. The development of electrical structure and circuit design of the controlling electronics will be the work for the next time.

Conclusion

The work on μPEFC systems and LTCC as an integration platform could make a contribution for miniaturization of fuel cells charging systems. In several prototypes the feasibility, ever-increasing functionality and power density were already demonstrated. The ongoing work shows again the increasing of the level of integration with the help of implementation of functions such as valves and electronics. Thus, the energy density per volume is increased and the overall system can be miniaturized (total volume is 86 cm^3). In later work the following sectors of development (stack, valves and electronics) are combined and assembled as one system.

[1] Annual Report Fraunhofer IKTS, 2009, Internet www.ikts.fraunhofer.de

[2] European Commission, Homepage 2011, http://ec.europa.eu/enterprise/sectors/rtte/chargers/index_en.htm, visit 05.2011

[3] T. Schirgott, "Entwurf, Aufbau und Charakterisierung einer PEM-Mikrobrennstoffzelle in LTCC-Technik", diploma thesis, 2004, Germany

[4] A. Goldberg, "Micro Proton-Exchange-Membrane Fuel Cell System in LTCC (Low Temperature Cofired Ceramics", ISSE 2006, Conference

[5] U. Partsch, "Ceramic Interface and Multilayer Technology for Micro Fuel Cells", CICMT 2010 Japan

[6] M. Stelter, U. Partsch, T. Rabbow, j. Schöne, A. Goldberg, M. Schneider, "Kontaktelement zur Stromaufnahme und Stromableitung in stromerzeugenden elektrochemischen Zellen", DE 102009037147A12011.02.17, 06.08.2009

[7] M. Stelter, U. Partsch, T. Rabbow, j. Schöne, A. Goldberg, M. Schneider, „Kontaktelement zum elektrischen Kontaktieren einer stromerzeugenden elektrochemischen Zelle sowie Herstellungsverfahren für selbige", DE 102009037144A12011.02.17, 06.08.2009

THE i-MODULE APPROACH:
TOWARDS IMPROVED PERFORMANCE AND RELIABILITY
OF PHOTOVOLTAIC MODULES

Jonathan Govaerts[1], Riet Labie[1], Jose Luis Hernandez[2], Mario Gonzalez[1], Guy Beaucarne[3],
Stefan Dewallef[4], Arvid van der Heide[5], Jan Vanfleteren[6,7], Ivan Gordon[1] and Kris Baert[1]

[1]imec, Leuven, Belgium
[2]imec, Leuven, Belgium, now at Kaneka, Leuven, Belgium
[3]Dow Corning, Seneffe, Belgium
[4]Soltech, Tienen, Belgium
[5]Photovoltech, Tienen, Belgium
[6]Cmst-imec/UGent (Centre for MicroSystem Technologies), Gent, Belgium
[7]also at Dept. ELIS, UGent, Gent, Belgium

Phone: +32-16-288675
Fax: +32-16-28
e-mail: jonathan.govaerts@imec.be

Abstract

Imec is developing a new concept, referred to as i-module (for interconnect-module), to integrate and interconnect back-contact solar cells into modules, which is based on embedding cells in silicone on top of a glass superstrate. This technology aims at an improved optical performance and reliability (through the use of silicones and low-temperature metallization), offering a module-level interconnection scheme (which implies interconnection only after the cells have first been fixed to the carrying superstrate).

First, the paper elaborates the process flow, comparing it with the current standard technology for manufacturing modules. Then, some first mini-modules are discussed, demonstrating proof-of-concept for this kind of packaging technology. Finally, the strategy for assessing the reliability of such modules is indicated and illustrated with first results. This reliability assessment will be key for the viability of the proposed technology, as PV modules typically need to survive over 20 years in harsh outdoor conditions, where it is subjected to a (cyclic) combination of moisture, temperature, mechanical loads, electrical stress and UV bombardment.

Key words: PV module, silicone embedding, back-contact cells

Background

Conventional module manufacturing for the current mainstream type of technology today, which is based on wafer-based crystalline silicon solar cells, typically consists of first interconnecting the separate cells into strings through soldering ribbons from one cell's contacts to the next (so-called tabbing/stringing) and then laminating those strings in between a glass front- and glass, metal or polymer backsheet. This requires soldering of standalone cells and handling of long strings of cells prior to lamination, two steps where (micro-)cracks may easily be introduced, especially in case of very thin cells, e.g. 100 μm and below. These cracks may possibly not directly affect the performance, but typically they will result in additional degradation of the module during ageing [1]. In addition, although the described mainstream technology is widespread and very mature for cells with contacts on both sides (requiring out-of-plane interconnection schemes),

there is not yet a lot of modules based on back-contact (BC) solar cells (i.e. cells that have both positive and negative contacts on the backside).

However, BC solar cells are expected to grow rapidly in market share, so module integration schemes for these are to be developed. The i-module (interconnect-module) approach we proposed [2] aims to minimize the risk on cracks by providing a module-level interconnection only after attaching the cells to the module substrate (glass). This is of course only possible if all of the cells' contacts are still available at this stage and thus is (only) suitable as a solution for back-contact cells.

While the concept and approach have been presented in different PV conferences already [2,3], the idea here was to present it to the packaging community and sollicit possible "microelectronics-packaging-inspired" contributions to this concept.

i-Module Concept and Process Flow

The basic process flow, shown in Figure 3, and described in more detail in [2] elaborates this conceptual approach. It uses coating technologies for the deposition of adhesive and encapsulant layers, via drilling technology for opening up the encapsulant above the cells' contacts and low-temperature metallization techniques. These choices could be compatible with the need for very high throughput processing demands (< 1 min/m2)[1] typical for PV manufacturing lines.

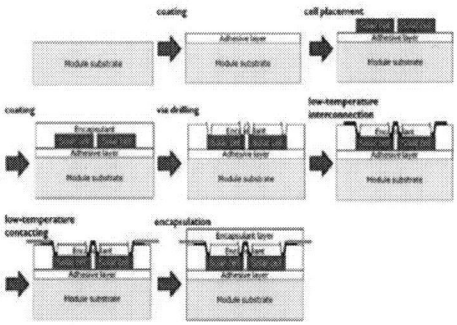

Figure 1: i-module schematic flow

Some potential advantages may be identified as compared to the more conventional approaches for manufacturing modules based on back-contact cells as e.g. described by SunPower [4] and ECN [5]. Wet coating as opposed to dry lamination could result in an increased speed and throughput, and allow for thinner cells as any uneven pressure distribution during lamination could result in breakage. The use of silicones as adhesive and encapsulant instead of EVA is furthermore considered to be beneficial in terms of optical performance [6,7], due to a reduced UV absorption, and reliability [7,8], because of a lower glass transition temperature and Young's modulus, a better UV stability, a reduced moisture take-up, and a higher heat and flame resistance. These last assets are interesting from both a processing (e.g. the silicone can withstand soldering temperatures) and safety point-of-view. And finally, of course the primary strength of the i-module concept remains its module-level interconnect metallization, as elaborated in the previous section. The i-module approach can also be extended to the so-called i2-module approach, which maintains the advantages of the i-module approach and in addition is compatible with the trend[2] to using ever thinner Si wafers. In this paper however, we focus on the i-module approach itself.

---

[1] From Table 2 in International Technology Roadmap for PV, ITRPV.net

[2] From Fig. 5 in International Technology Roadmap for PV, ITRPV.net

Processing Feasibility and Demonstrators

Obviously, one of the first steps in developing such a concept, is demonstrating its feasibility in terms of processing. To this end, different proofs-of-concept have been fabricated, most of which have been reported earlier [7]. Such demos have been made based on both industrial-type Metal Wrap-Through (MWT) cells and lab-type Interdigitated Back-Contact (IBC) cells.

Figure 2 shows an i-module demo based on 4 IBC cells. These cells measure 20 mm by 20 mm, have a thickness in the range of 100 micron, and were fabricated at imec [9]. They were first glued to a 3 mm thick BoroFloat glass using a silicone from Dow Corning, "PV6010 cell encapsulant"[3], and subsequently encapsulated with the same material. Holes were drilled with a CO_2 laser to provide vias to the busbars, and the cells are connected in series through the use of a low-temperature curable Ag paste from DuPont, "PV410". Outside contacts consist of standard tabbing ribbons that have been contact welded to the Ag paste interconnect metallization. Finally, the full stack is covered once more with the silicone "PV6010 cell encapsulant".

Figure 2: i-module demo based on lab-type IBC cells

Figure 3 shows another i-module demo, this time based on MWT cells and tabbing. For this, 156 mm by 156 mm, 180 micron thick MWT cells from Photovoltech [10] were diced into unit cells of 34 mm by 44 mm. 6 of these are glued to a 0.7 mm thick glass substrate (Corning EAGLE), using Dow Corning's silicone "PV6100 cell encapsulant"[3]. Then a silicone isolation layer "PV6150 cell encapsulant"[3] is manually dispensed around the MWT contacts at the back to prevent shunting during tabbing in the next step. Neighbouring cells are rotated 180°, so that the interconnecting strings can be soldered in a straight line. Two series of 3 cells are fabricated and connected in such a way that each cell can still be measured separately. After

---

[3] Datasheet available through Dow Corning at www.dowcorning.com

481

soldering the ribbons, the stack is encapsulated with the silicone "PV6150 cell encapsulant".

Figure 3: i-module demo based on industrial MWT cells

The electrical performance of one string and its individual cells is shown in Figure 4. The standalone reference is a cell that is soldered but not glued to glass. The shown measurements of cells 4, 5 and 6 are module-level measurements, as well as the series measurement (necessarily, as the interconnection is only carried out after the cells are attached to the glass superstrate).

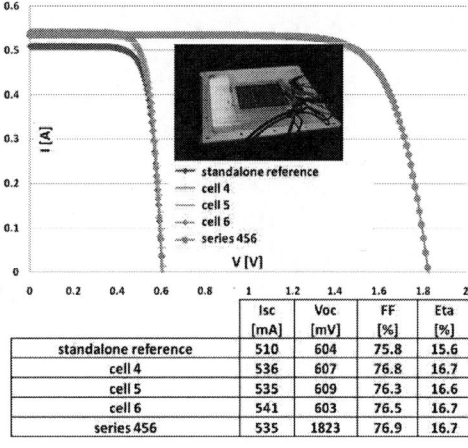

	Isc [mA]	Voc [mV]	FF [%]	Eta [%]
standalone reference	510	604	75.8	15.6
cell 4	536	607	76.8	16.7
cell 5	535	609	76.3	16.6
cell 6	541	603	76.5	16.7
series 456	535	1823	76.9	16.7

Figure 4: electrical performance of cell/string in i-module demo shown in Figure 3; the inset shows a picture of the IV simulator measurement setup

Further on, we are developing the technology for Cu plating as another low-temperature module-level metallization technique. This would solve two issues: the higher temperature of the soldering process soldering, and the use of the low-temperature Ag, which is considered to be less abundant and therefore less sustainable.

Reliability Approach and Preliminary Results

In the next phase, after demonstrating processing feasibility, reliability assessment is a very important step, especially in the case of photovoltaics, where module lifetimes above 20 years, in harsh outdoor conditions, are typically guaranteed for commercially available products. First steps have already been taken in this area, based on the following methodology. Starting from a failure mode and effect analysis (FMEA), simulations based on finite element modeling have been carried out to get a feel for the correlation between layer buildup and stress- and strain-levels inside the module. Being partially determined by material properties (next to dimensions), also material characterization is required for this modeling. To validate the FMEA, accelerated reliability testing is necessary, either failure-driven or standards-driven (for crystalline Si photovoltaics e.g. IEC61215). The results coming out of this testing can be analysed and compared to simulations and expectations from the FMEA. Next to this, it's also important to look at outdoor weathering for reliability testing as the outdoor situation is typically much more complex than the necessarily simplified accelerated reliability tests. Finally, all these results are fed back into the model in order to elaborate an ageing model, where we can correlate the lifetime of the module with the design and processing parameters.

Currently research and development on this has been started and progress has been made in several areas of the reliability assessment: simulations based on thermomechanical modeling, material characterization and reliability testing.

A first possible failure mode that may be considered is cracking of the cells. To evaluate this, simulations are carried out to get an estimation for the stress experienced by the silicon wafer, and compare it against the strength of such a wafer. It turns out that the stress in the silicon would be compressive and far below what silicon can withstand [3]. A second failure mode that may be simulated is deformation of the adhesive layer. According to simulations [3], the elastic strain may become significant, which might result in shearing or delamination of this layer, depending on its cohesive and adhesive strength. Also the significant change in displacement during thermal cycling, that this strain implies, may result in fatigue failures in the interconnects.

Other failure modes may be more closely linked with failure of the encapsulation. For this, accelerated testing schemes in a reliability chamber may be appropriate. So far, fully encapsulated, but not metallized, samples have been tested with conditions based on a commonly used standard in the PV area, IEC61215: 1000 hours damp heat 85%RH/85°C, 200 thermal cycles 85°C/-40°C and 10 humidity freeze cycles 85%RH/85°C > -40°C. The results here and so far show no significant difference between the silicone and EVA glued samples which may be considered as an indication for the viability of the encapsulation [2].

The interconnection scheme is another obvious source of failure [3]. Highly depending on the thermal and viscoelastic behaviour of the materials, their lateral dimensions and thicknesses, interconnect cracking and fatigue issues may be expected. First experiments have been carried out: in order to mimick a (plated) Cu metallization that would act as interconnect between the neigbouring cells in an i-module, we have glued a 9-micron-thin Cu foil as a bridge across two 156 mm by 156 mm dummy wafers, that in turn have been glued to a glass (module) substrate. During curing of the silicone adhesive, the thermal mismatch of the different components causes strains and stresses in the Cu foil. Depending on the stiffness of the encapsulant materials, this will result in either residual stresses in the interconnect or (out-of-plane) deformation. The idea was to check if this deformation (which can be calculated and simulated) could be observed, but it turns out to be rather difficult to discriminate this stress-induced deformation against the deformation that is already present in the foil due to handling and processing (as such thin standalone foils are very fragile). Therefore, more samples would be required to statistically confirm the significance before drawing any conclusive result here. To further assess these interconnects, the samples were then subjected to thermal cycles 150°C/-40°C to check for further deformation and possible fatigue failures. An out-of-plane spring-like topography is visible after 238 cycles, but no cracking is observed even after over 500 cycles. Figure 5 illustrates this experiment. All surface topography measurements were carried out at room temperature.

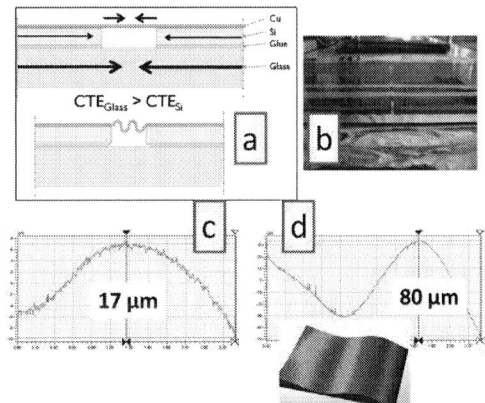

Figure 5: preliminary interconnect bridge experiment: schematic cross-section (a), prepared sample (b), and surface topography of the Cu foil before (c) and after (d) 238 thermal cycles

Apart from the above, more device-related degradation could gain in importance when more and more advanced cell and module technologies and concepts start to be developed and deployed. Examples of this are e.g. light-induced degradation of cells [11] and the (reversible) effects of externally-induced buildup of charges in the cell structure [12]. In the same field of functional device breakdown, you could also for instance consider diffusion of Cu as a threat to (minority carrier) lifetime and therefore possible performance killer of a solar cell device. Depending on the used barrier, expected temperatures and cell and module structure, the impact of Cu diffusion may be very minor or highly significant. For future modules, where Cu is likely to replace Ag as a more commonly available resource, this is a topic that has to be tackled. First experiments have been conducted in this area [13]. The properties of a Ti layer to act as a barrier for Cu diffusion in cells with a Cu metallization were investigated for different temperatures by checking the solar cell efficiency as a function of the applied thermal budget. A clear degradation of the cell efficiency was observed for an increased thermal budget, which may be attributed to the breakdown of the barrier and local Cu diffusion. The diffusion constant of Cu through the Ti barrier layer could then be derived from these experimental data. Furthermore, such cells were also incorporated into standard 1-cell modules (ribbon soldering and glass/EVA/transparent backsheet lamination), and subjected to damp heat (85%RH/85°C) conditions for 1000 hours. In total, 4 out of 5 1-cell modules survived the testing conditions while one sample showed a more than 5% reduction in fill factor after 740h of testing.

Finally, considering the complexity involved in simulating actual outdoor conditions, outdoor weathering remains an important test to validate any new module technology. For this reason, we have also exposed one of the very first demonstrator in the outdoor environment, already very early on to gain some early insight into possible failures. So far, after an exposure time of nearly 500 days, no clear and significant degradation has yet been identified, as indicated in Figure 6. The module has been very plainly put outside, roughly facing south and tilted again roughly around 45°. It is sometimes shaded by trees, and is left in open-circuit condition. The discoloration that is visible at the edges, is due to soiling at the open and unprotected sides (absence of a laminated backsheet and framing).

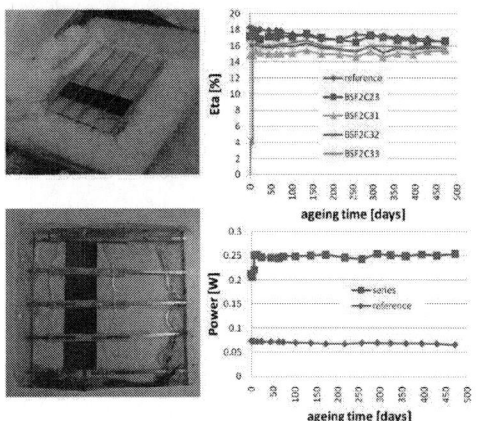

Figure 6: monitoring of the electrical performance of the i-module demo shown in Figure 2, subjected to outdoor weathering

Summary

In this paper the i-module concept for back-contact solar cells is presented. As a packaging concept, it aims at improving output performance and reliability of the photovoltaic devices inside.

The described process flow could potentially reduce the module assembly costs and improve the process in terms of mechanical yield and reliability. Regarding the latter, several tests have been carried out. Although some possible failure modes have been identified (e.g. delamination due to high stress, interconnect fatigue failure), so far no unacceptable failure appeared: most of the accelerated ageing tests have been succesfully passed, and an outdoor test is ongoing with no significant decay of performance after almost 500 days.

References

[1] M. Koentges et al., "Quantifying the risk of power loss in PV modules due to micro cracks", Proceedings of the 25th EUPVSEC, Valencia, 2010, pp. 3745-3752.

[2] J. Govaerts et al., "A novel concept for advanced modules with back-contact solar cells", Proceedings of the 25th EUPVSEC, Valencia, 2010, pp. 3850-3853.

[3] M. Gonzalez et al., "Thermo-Mechanical Challenges of Advanced Solar Cell Modules", Proceedings of the 12th EuroSimE, Linz, 2011.

[4] D. Rose et al., "Development and manufacture of reliable PV modules with >17% efficiency", Proceedings of the 20th EUPVSEC, Barcelona, 2005.

[5] P.C. De Jong et al., "Single-step laminated full-size PV modules made with back-contacted mc-Si cells and conductive adhesives", Proceedings of the 19th EUPVSEC, Paris, 2004.

[6] K.R. McIntosh et al., "An optical comparison of silicone and EVA encapsulants for conventional silicon PV modules: a ray-tracing study", Proceedings of the 34th IEEE PVSC, Philadelphia, 2009.

[7] B. Ketola et al., "Silicones for photovoltaic encapsulation", Proceedings of the 23rd EUPVSEC, Valencia, 2008.

[8] M. Kempe et al., "Acetic acid production and glass transition concerns with ethylene-vinyl acetate used in photovoltaic devices", Solar Energy Materials and Solar Cells, Vol. 91(4), pp. 315-329, 2007.

[9] J. Robbelein et al., "Industrial type passivation on interdigitated back junction solar cells", Proceedings of the 24th EUPVSEC, 2009.

[10] www.photovoltech.be

[11] J. Schmidt et al., "Light-induced degreadation in Cz silicon solar cells: fundamental understanding and strategies for its avoidance", 12th Workshop on Crystalline Silicon Solar Cell Materials and Processes, Breckenridge, Colorado, 2002.

[12] R. Swanson et al., "The Surface Polarization Effect in High-Efficiency Silicon Solar Cells", Proceedings of the 15th International PVSEC, Shanghai 2005, pp. 410-411.

[13] R. Labie et al., "Cu plated i-PERL cells: ageing and humidity reliability tests", Proceedings of the 26th EUPVSEC, Hamburg, 2011.

Improved Testing of Soldered Busbar Interconnects on Silicon Solar Cells

R. Klengel*, M. Petzold*, D. Schade**, B. Sykes**

*Fraunhofer Institute of Mechanics of Materials IWM (Halle, Germany)

**XYZTEC b.v. (Panningen, The Netherlands)

Robert.Klengel@iwmh.fraunhofer.de; +49 345 5589 159

Abstract

The expected life of a PV solar cell is 20-25 years. To ensure this the quality and reliability of the material and process parameters of solder joints that connect single cells into strings has to be assured. A standardized test method or clearly defined good-bad-criteria are not available for PV products. Thus a product, manufacturer or joining technology comparison is not possible. Due to the lack of standards and commercial available test equipment PV module manufacturers often implement their own test solutions and specifications. However, these mostly have limited scientific proof, are difficult to interpret and lack optimisation. Because there is no standardization these solutions cannot be used for comparative characterization of different manufacturers, cell types, base- and joining materials or joining technologies. The most commonly used test for solar cell ribbon interconnects is the Pull test, as used in micro-electronics. For thin, brittle, large area silicon solar cells substantial modifications of the test equipment and methodology are required to realize a suitable metrology. In addition the factors influencing the defect conditions in multi crystalline material, high strains around the solder connect, inhomogeneous contact interfaces has to be considered. Results show the development, optimization and evaluation of a test method for soldered bus bar interconnects on silicon a solar cell that is independent of product, manufacturer and cell assembly. A comparison between this new method, its results and the previously mentioned issues is made. The test method demonstrates how cell breakage during a test can be avoided by optimising how it is mounted and guided.

Key words: Photovoltaic, Solar Cell, Mechanical Testing, Quality Improvement, Solder Joint Characterization

Motivation

Reductions in the cost and performance of solar cell and module production are an ongoing requirement in the PV industry. There are two factors of essential importance. On the one hand the optimization of material properties with the goal to reach a high and long-term stabile efficiency. On the other hand reducing production costs, whilst ensuring the highest quality and reliability possible are required. These requirements place a high demand on test and inspection that enable investigation into the mechanical, electrical and micro-structural properties of a solar cell and how they are affected by the materials and processes used during its manufacture. Often basic methods are adopted from other industries like micro electronic or electronic packaging. But these were modified, enhanced and optimized for appliance in the PV sector.

Basics

Silicon solar cells are made from wafer disks with a thickness of 200 microns or less. They are manufactured by numerous etching, coating, screen printing, and firing steps. A metallization grid on the front side collects the light generated charge and is connected to bus bars. Flat copper ribbons, soldered to the bus bars, transfer the current from one cell to another in a series connection to so called strings. Several strings sealed in a lamination process create a module. [1]

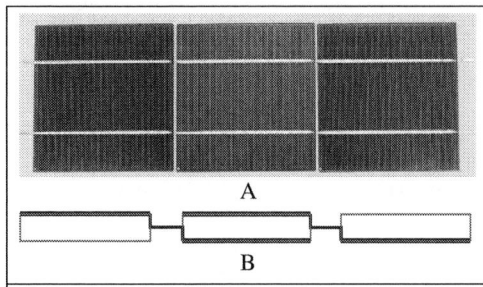

Fig. 1: A- Overview of a three cell string connected by two soldered copper ribbons each
B- scheme of the solar cell series connection by soldered copper ribbons

The soldering process is a well known interconnection technology in the electronic packaging industry. Low costs and high reliability are the main advantages of this type of bond. For quality control, there are many known tests and standards. For example, the pull test for electronic components is specified in DIN EN 61189 [2]. The challenge is whether these test conditions can simply be transferred to solar cell soldering. There are

485

differences between soldering of under bump metallization (UBM) and the under interconnector metallization (UIM) used in solar cell construction. Usually the size of an UBM is at most only a few square millimetres. The width of an UIM is between 1 and 3 mm with lengths up to 156mm. Therefore, the current density in an UBM is much higher than in an UIM. Voids for example are less problematic than thermo mechanical stress.

Fig. 2: Solar cell after pull test with common concept – due large area silicon disruptions no conclusion to the interconnection quality can be made

The different coefficient of thermal expansion between copper and silicon results in mechanical stress after the soldering process. In some cases, this leads to interruption of the thin grid finger at the front side of the cell [3]. Charge carriers, occurring around the interrupted grid finger, are not able to dissipate. The efficiency of the harmed cell and subsequently the efficiency of the whole module are reduced. Pulling ribbons from pre-damaged cells results in large area silicon tears. Micro cracks underneath the UIM are forced to spread along silicon grain boundaries and cause large area silicon disruptions. Instead of the UIM, the silicon wafer is tested [1]. Figure 2 shows a silicon solar cell after pull test using a common test concept.

One aim of the presented development is to restrain the spreading micro cracks during the pull test and avoiding the large area silicon tears what is indispensable for getting evaluable and reproducible results. Another important motivation is to provide test equipment which allows comparing the results indendent of cell geometry, condition of contact materials, soldering technology or manufacturer. Further goals were making the test easier to handle, reaching a higher throughput and getting clearer criteria to distinguish good or poor solder joints.

Principle of the Test Method

There are three main types of failure modes occurring through the pull test (Figure 3). The first and favoured is that the interconnection fails within the UIM. The pull forces between soldered and

unsoldered areas are clearly distinguishable from each other.

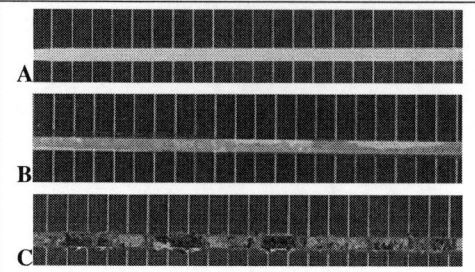

Fig. 3: Failure modes after pull test
A- no wetting, no solder joint formation
B- failing within the UIM, good quality
C- silicon disruptions

The second is that no wetting took place. The UIM remains unsoldered and there are nearly no measurably pull forces.

The third and most inappropriate one is that the silicon wafer underneath the UIM fails, causing large area silicon disruptions. When silicon has brittle fracture behaviour, nearly no pull forces are measured [1]. Adjustments in soldering conditions like low temperature and cooling rate may reduce this problem [4].

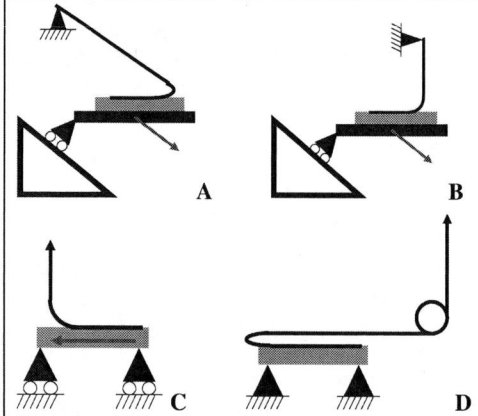

Fig. 4: Overview of some test concepts
A+B- cell mounting on an incline plane and stationary ribbon clamping lead to inconstant pull angles
C- flexible horizontal cell mounting and no static deviating point for the ribbon lead to overshooting and inconstant pull angles
D- inconstant bending stiffness of the ribbon lead to varying pull angles, for 180° deviating a very long ribbon is needed which is not available by taking sample from a string

The pull test of solder joints of electronic components is standardized in DIN EN 61189, including the pull angle of 90° and a velocity of 200 mm/min. Nevertheless, there are many different concepts in use. Figure 4 gives a view over some solutions.

Looking at the natural load case, the main direction of applied force is longitudinal to the solder ribbon. For that reason, a pull angle of 180° may seem to be more realistic. But it is not the task of the pull test to simulate a realistic load case. The pull test is used as process control for the soldering process as well as for characterisation of the solder joint stability as a quality criterion. Figure 5 images one advantage of a 90° pull angle.

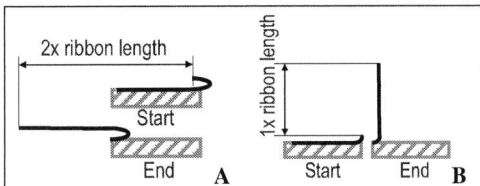

Fig. 5: Scheme shows one of the advantages of performing test with 90° pull angle
A- for 180 ° pull angle a long ribbons and place of the double cell lengths are needed
B- for 90° pull angle only the ribbon and place of the cell length are needed

Every micro crack can lead to failure of the silicon and afterwards to a brittle crack extension underneath the UIM. To restrain this crack extension, it was found that an opposing force applied at the point of brake prevented crack propagation. Single micro cracks will only lead to small silicon tears. Large area silicon tears will only occur if the cell has been badly pre-damaged.

Development of Test Equipment and Routine

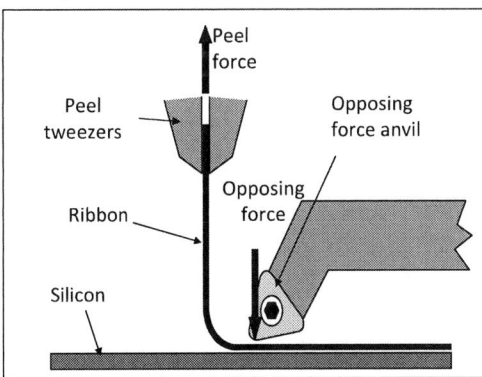

Fig. 6: Scheme of the function principle developed for the new solar cell pull test equipment

Since solar cells and module manufacturing is mass production, there are special requirements for measuring equipment. The main requirements for the ribbon pull test are stability, simplicity and high throughput. Additionally, it is necessary to generate reports that allow easy interpretation of the results. Calculated mean values, measured curves, and pictures of the fracture interface should be included.

A simple way to combine these requirements with an innovative measurement strategy is to upgrade a common measuring unit which has a flexible data base and onboard visual inspection. The XYZTEC Condor 250 multifunctional test machine was selected. The machine includes a vertical moving z-axis equipped with a force sensor and an x-y-stage for horizontal movement of the test piece.

According to DIN 61189, the pull angle should be 90°. The X-Y table moves the cell at the same velocity so that the point of breakage remains under the measuring head. So the measuring range equates to the ribbon length. The pull test with 180° causes a measuring range twice as long as the ribbon (Figure 5).

Due the point of breakage remains in the same position the opposing force can be easily applied by using a triangular anvil with rounded edges made from special low-wear material.

Pressing the anvil onto the cell may cause additional damage. Therefore the cell has to be pulled to the edge of the anvil and should not be fixed at the table in vertical direction. Figure 6 shows a scheme of the developed anvil and a ribbon being pulled. Figure 7 gives a picture of the complete machine and a detail of test situation.

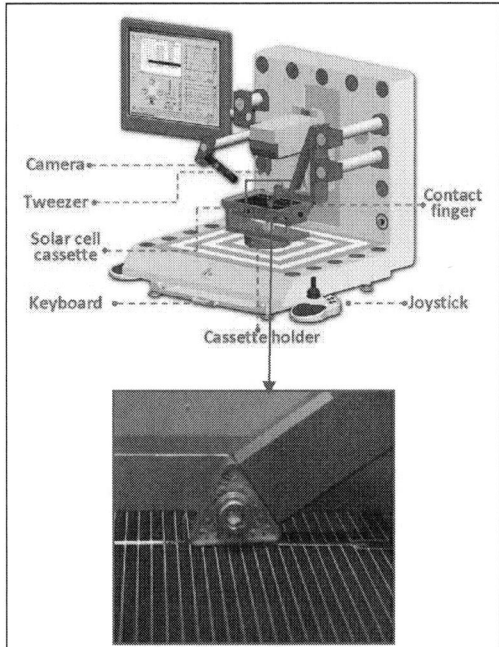

Fig. 7: Image of the universal test machine equipped with the new developed solar cell work holder for solder ribbon pull test

The cell is placed in a cassette that can be removed from the machine. Using two cassettes, one can be in test while the other is being reloaded. The following test procedure was developed:

1) Inserting the cell in the work holder
2) Fixing the work holder on the machine
3) Move to start position of first ribbon (automatically)
4) Clamp the ribbon in the pull tweezers
5) Starting the test of the first Busbar
6) After finish first test unclamp the ribbon (automatically)
7) Move to the second start position (automatically)
8) Repeat procedure from point 4)

After finishing the last ribbon, the work holder is moved automatically to the load position. For security demands, the whole work holder is pneumatically lowered while the table moves toward the next starting- or load position.

The test of one cell with three bus bars takes less than five minutes. The user has to clamp and unclamp the ribbon and to change the solar cells. While the test is ongoing, the next cell can be loaded to the second cassette. After finishing the test, the cassettes will be interchanged and the next test can be started immediately.

It is possible to test any kind of silicon solar cell. The work holder with the cell cassette as well as the test routine can be adapted to every cell geometry and condition like two or three bus bars, cell thickness, ribbon dimensions and soldering method.

While the test is ongoing, a camera looks at both the cell and the pulled ribbon. So the failure mode can be observed and easily documented. How the developed test machine looks like is shown in Figure 7.

Results of Test Development

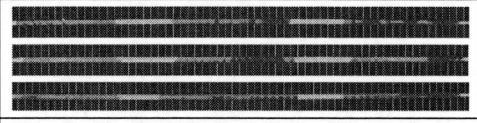

Fig. 8: Front side bus bars after pull test performed with the new developed test equipment – no silicon tears and clear distinguishable failing interface levels

Silicon solar cells were tested from several manufacturers, in varying conditions and different geometries. For all cells a pull test velocity of at least 5mm per second could be performed without any significant silicon disruptions. Figure 8 shows three front side bus bars after pull test without any damaged.

Due to the semi-automatic application flow the test is considerable accelerated and the handling is remarkable easier.

The dependence of the test results on secondary conditions like influences of varying friction depending on test velocity, influences of different ribbon material expansion, influences of solder material etc. were extensive investigated and will be published in near future.

Micro structural Correlation

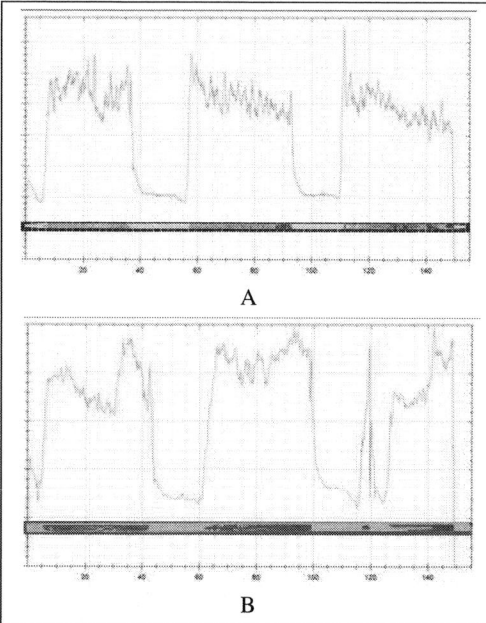

Fig. 9: Graph of pull test measurement (force over distance) with imaging of the corresponding failure interface
A- front side bus bar
B- back side bus bar

The graphical analysis of the measurement results and the images of the fracture interface allow correlating the pull force values with the corresponding failure mode. Figure 9 shows measurement graphs of a front side bus bar and a back side bus bar (both connected with soldered and unsoldered areas alternating) with the images of the correlating fracture interface.

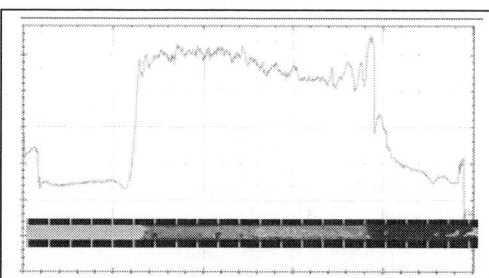

Fig. 10: Detailed graph (force over distance) of pull test measurement on a front side bus bar

In the detail of a measurement graph, pictured in Figure 10, it is visible that every single change in the fracture interface is reproducible in the progression of the curve. That indicates it seems to be possible to conclude to the solder joint condition by interpreting the measure curve. But to ensure this hypothesis and to give defined criteria for the curve interpretation, a lot more investigations are necessary.

Subject of actual research is the possibility to correlate the pull test measurement curves and fracture interfaces with concrete interface weakening like voids, non wetted areas or micro cracks etc. before performing the pull test. Therefore different non-distructive analyzes of soldered bus bar interconnects like scanning acoustic microscopy (SAM), X-ray inspection or electroluminescence were performed before the pull test. Afterwards the results of the preliminary investigation have to be correlated with the measurement curves and the images of the fracture interface. Figure 11 shows the solder joint interface of bus bars investigated with non distructive methods.

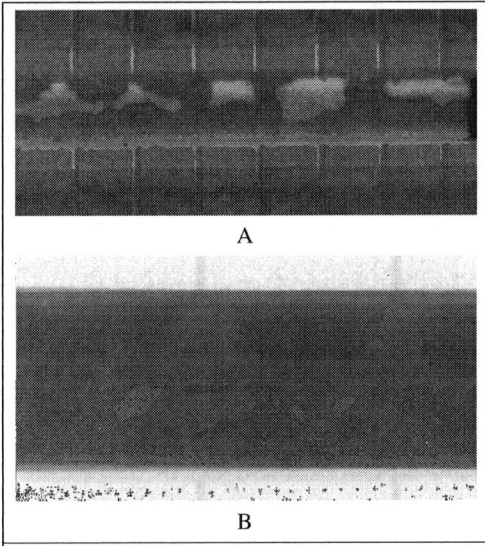

Fig. 11: Overview of used micro structural investigation methods for solder joint quality analyzes:
A- Scanning Acoustic Microscopy (SAM)
B- 2D-xray inspection

Conclusions

Improving measurement quality for testing solder joints on silicon solar cells includes many aspects. Firstly there is no standardization. Different test conditions like pull angle and test velocity implicates difficulties concerning comparability between different users. The second aspect is that it has to be assured that only the UIM is tested instead of silicon underneath. The disadvantage of common testers is that the cell is fixed while the ribbon is pulled. This can be improved by applying an opposing force on the ribbon at the point of test load application. Silicon disruptions will then not lead to large area tears, but only when extensive pre-cracks are present over the solder joint. [1]

With the developed test equipment and measurement routine it is succeeded to provide a reproducible, fast and easy handable method which is usable indendent from solar cell geometry, condition of contact materials, soldering technology or manufacturer.

It is succeeded to show that it is possible to correlate the measurement curve of the pull test with the fracture interface. This allows to interprete the results with statements in terms of the solder joint quality. To ensure these interpretation hypothesis further investigations were performed in present time using non-distructive test methods before the pull test. The sum of results from preliminary investigations, pull test measurement and analysis of fracture interface should allow defining criteria for rating the solder joint quality.

Acknowledgements

This work was partly supported by the German Ministry of Economics and Technology under contract "ATSOLOT" (FKZ01FS10010).

Many thanks to Tino Stephan, Marcel Mittag and René Härtel for the great assistance in this work.

References

[1]   J. Wendt, R. Klengel, D. Schade: "Improved quality test method for solder ribbon interconnects on silicon solar cells"; 12th IEEE Intersociety Conference on Thermal and Thermomechanical Phenomena in Electronic Systems (ITherm), 2.-5. June 2010 Las Vegas, Nevada USA

[2]   DIN EN 61189: Test methods for electrical materials, printed boards and other interconnection structures and assemblies. German DIN standard

[3]   J. Wendt, M. Träger, M. Mette: "The link between mechanical stress induced by soldering and micro damages in silicon solar cells"; 24th European Photovoltaic Solar Energy Conference, 21-25. September 2009, Hamburg, Germany

[4]   A.M. Gabor: „Soldering Induced Damage To Thin Si Solar Cells And Detection Of Cracked Cells In Modules"; 21st European Photovoltaic Solar Energy Conference, 4-8 September 2006, Dresden, Germany

Solidification Processes of SnCu, SnAg and SnAgCu solder alloys and interface reactions to charactize solar cell interconnections processes

Sebastian Schindler [1], Steffen Wiese [2]

[1] Fraunhofer-Center for Silicon Photovoltaics, Walter-Huelse-Str. 1, 06120 Halle
[2] Saarland University, Chair for Microintegration and Reliability

Phone: +49 345 5589 417, E-mail: sebastian.schindler@csp.fraunhofer.de

Abstract

Lead-contained and lead-free solder joints have been well studied in electronics packaging applications over the past years. Within the solar industry, establishing the Si crystalline solar cell interconnection process as an environment friendly and RoHS-conform process, more substantial investigations are necessary to prove the process reliability and characterize interconnection formation in the Ag-thickfilm and Cu-ribbon interfaces. This paper focuses on the evaluation of material parameters of the joining processes for solar interconnections and quantifies the solder formation of the typical PV interfaces and the solder compositions. Therefore, varying solder alloys (content gradiation of SnCu, SnAg compositions) are investigated. We used DSC (differential scanning calorimeter) measurements to characterize the melting and solidification behaviour with and without the interfacial reactions. Influecing factors like peak-temperature and cooling rate within the contextof PV manufacturing process will be discussed to convey a better understanding for the Ag-paste to solder to Cu-ribbon interconnecetion level for solar cell stringing. The results show that the conditions of solidification behaviour and microstructural properties are of significant importance for silicon-based solar cell interconnection processes.

Key words: photovoltaics, wettability, soldering, joint formation, process reliability

Introduction

Lead-contained solders were widely used because of their well know wetting behaviour and mechanical properties. Establishing the Si crystalline cell-based module manufacturing as an environment-friendly process more substantial income of the soldering process reliability and efficiency for photovoltaic applications is still necessary.

The common used interconnection method [1, 2] is the tabber-stringer-process, used world wide with a throughput >1,000 wafers per hour in full automated production lines. A screen-printing step provides the electrical contact interface on the solar cell wafer by printing and firing a thick film pastes (containing Ag, Al) on the surface and backside.

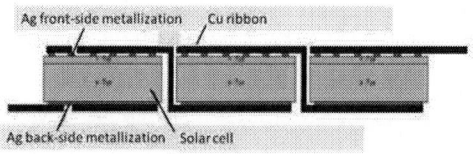

Figure 1: Interconnection scheme for serial solar cell interconnection for module assembly

The tabber-stringer-process is conducted with varying and relatively simple soldering and handling equipment (hot contact element, inductive, laser, hot air solder heads). Without technology modifications this method becomes impractical due to reduced wafer thicknesses [3] or new solar cell designs. Lead-free solders are almost not used, because of more distortion and different soldering process parameters and missing module reliability (up to 25 years).

The interconnection technologies in microelectronics [4, 5] have been studied and described very well, compared to PV interconnections and application sector.

Materials and methods

For this study different solder compositions (SnCu0.5, SnCu1.5, SnCu3.0, SnAg1.0, SnAg3.0, SnAg5.0, SAC305, Sn99.9) and standard commercial solar cells were used.

The differential scanning calorimetry (DSC) technique, ref. image 1, was used to observe the physical transformation from solid to liquid of the solder material. We measured the melting and solidification temperatures of solder spheres (70 µg, 5k/min rap rate).

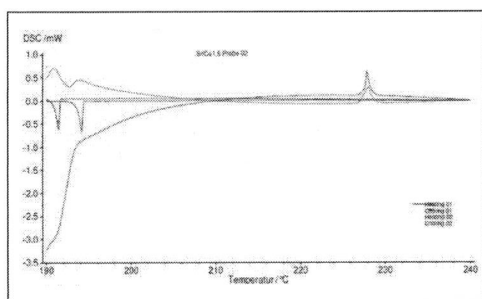

Figure 2: DSC curves of SnCu1.5, multiple heating and cooling runs to measure solder melting and solidification temperatures

The experiment were conducted with 2 heating and cooling cycles for each specimen, and 2 specimen for each solder composition (28 DSC runs in total) see figure 2.

The determination of the specific melting and solidification temperature, as a function of the Ag and Cu content allocates a better understanding for the required tabber stringer joining process parameters.

In further experiments SnAg solder depots on solar cell Ag-busbar specimen were melted, readjusting the stringing process on the cell side by laser soldering. Cross sections were made for IMC formation exposition and contact angle measurements as wettability characteristic.

Results

DSC measurements

The result of this DSC experiments are curves of heat flux versus the temperature and evaluated the exothermal difference from the baseline, fig. 1. The peak temperature points are the points of maximum heat flow from the baseline. The determination of the peak temperatures is usable as process parameter evaluation for contact formation and process control indicator.

The following diagrams (figures 3-6) show the average values of the DSC runs. Table 1 lists the temperature values and standard deviation (SD).

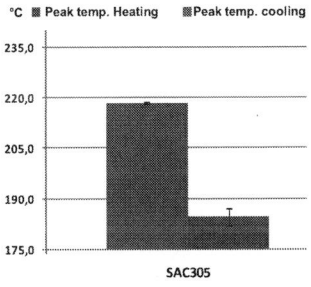

Figure 3: DSC measurement of melting and solidification temperatures of SAC305

The melting point of SAC305 of 218,5 °C, shown in figure 3, is within the expected band. The solder specimen solidified at a temperature of 184,5 °C. The temperature range between liquidus and solidus, ref. table 1, amounts 34 K.

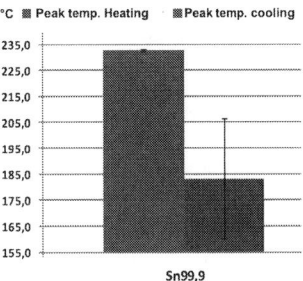

Figure 4: DSC measurement of melting and solidification temperatures of Sn99.9

The Sn99.9 specimen melted above 230 °C. The solidification temperatures decreased most for all measurements. The average values decreased by ca. 50 K to 183,3 °C. The Sn99.9 specimen showed a solidification between 206,5 – 160,1 °C, as the highest standard deviation too.

Figure 5 and 6 show the experimental results for the SnAg and SnCu solder alloys.

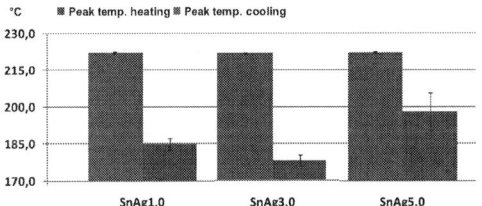

Figure 5: DSC measurements of melting and solidification temperatures of SnAg1.0, SnAg3.0, and SnAg5.0

Fig. 5, the SnAg solder alloys melted at the characteristic eutectic (SnAg eutectic 220,3 °C). The solidification temperature of SnAg1.0 is higher than SnAg3.0, plus 6,7 K. An increased Ag content increased the solidus point of SnAg5.0 (197,9 °C) by higher standard deviation, see table 1.

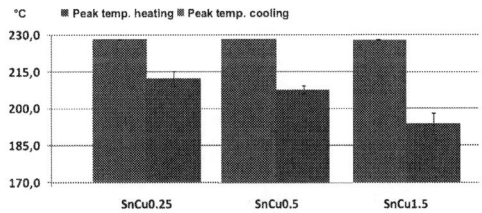

Figure 6: DSC measurements of melting and solidification temperatures of SnCu0.25, SnCu0.5, and SnCu1.5

For all SnCu compositions, shown in fig. 6, the melting point is in a small range at the SnAg eutectic of 226,8 °C, comparable like SnAg. The solidification temperature decreased with increasing Cu content. For SnCu1.5 a maximum of 34,2 K temperature range occurred between liquidus and solidus point, ref table 1.

Therefore the Cu ribbon for the serial cell connection marks as partner for Cu dissolution in the solder joint, affecting solidification process and IMC formation and contact realibility.

Table 1: Average values and standard deviation of DSC measurements of SAC305, Sn99.9 SnCu-x and SnAg-x solder alloys

Solder / [°C]	Peak temp. heating	SD	Peak temp. cooling	SD
SAC305	218,5	0,3	184,5	2,4
Sn99.9	232,7	0,2	183,3	23,2
SnAg1.0	222,1	0,3	184,9	2,4
SnAg3.0	221,9	0,3	178,2	2,3
SnAg5.0	222,1	0,3	197,9	7,8
SnCu0.25	228,2	0,1	212,1	3,0
SnCu0.5	228,3	0,1	207,7	1,7
SnCu1.5	227,9	0,2	193,7	4,3

Influence of Ag and Cu content on the solidification behaviour

The average values of the DSC measurements of the solder alloys are listed in table above. By comparison the temperature range between liquidus and solidus, Sn99.9 (Δ=49,4 K) is affected by the unspecified 0.1% element incorporation. The addition of 0.25% Cu decreased the temperature shift significant to Δ=16,1 K. Contrary, the experiments with higher Cu content (SnCu0.5, SnCu1.5) enlarged that range again, up to 20,6 K and 34,2 K respectively.

SnAg3.0 solidified at the lowest temperature of 178,2. SAC305 melted at the lowest temperature. Comparing ΔT_L-T_S of these solder alloys, a incorporation of 0.5% Cu decreased the values at 10 K.

Interface reactions of SnAg on cell interface

The second experiment readjusted the soldering process of SnAg alloys on standard Ag-thickfilm busbars. Compared to soldering process within the microelectronic industry, soldering processes for serial cell interconnection are reduced to a minimum. Time over liquidus for element dissolution and contact formation is within fractions of a second. Supporting process gases or controlled temperature ramps are not applied.

Figures 7 – 9 show exemplary results of the melting experiments on solar cell busbars.

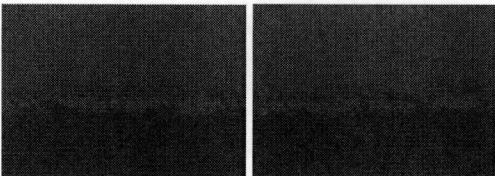

Figure 7: cell solder joint interface for SnAg1.0, left: 1 s, right: 2 s process time

Figure 7 shows the influence of the process time on formation on Ag_3Sn phase band between Ag thickflim paste and SnAg1.0 solder. After 2 s soldering temperature (right side) continues phase formation with starting needle growth is visible.

Figure 8: Ag_3Sn phase formation for SnAg3.0, 1 s

Comparing the interface reaction for Sn3.0 for short processing times of 1 s, a obvious needle growth at the busbar interface can be observed. Comparable to SnAg1.0 longer soldering times inducing a SnAg IMC band alongside the interface.

Figure 9: Significant Ag_3Sn needle formation at the busbar interface for SnAg5.0, 1 s

The last figure 9 shows the interface for SnAg5.0. At short soldering time, no continuous phase band is visible, a increased Ag_3Sn needle growth within the soldering joint can be observed.

Discussion and conclusion

The DSC measurements provided detailed results of the melting and solidification temperatures of SnAg, SnCu, SnAgCu solders. The influence of solder compositions on the required process temperatures and liquidus and solidus temperatures within the solder contact were evaluated.

The results show that Ag increases ΔT_L-T_S up to >40 K (SnAg3.0). An increasing Cu content increased ΔT_L-T_S for the investigated SnCu compositions. Regarding the Cu ribbon for the serial cell connection, the ribbon possibily marks as partner for Cu dissolution in the solder joint, affecting solidification process and IMC formation and contact realibility [6].

For SAC305 the Cu incorporation plays a signification role for melting and solidification temperatures. Lowest peak temperature heating (218,5 °C) compared to SnAg3.0 and SnCu0.5. And ref. to SnCu0.5, it promotes a higher solidification temperature.

By comparison SnAg-x and SnCu-x, a directed solder solidification from Cu-ribbon to Solder to Ag-metallized solar cell can be assumed, ref. cross section in fig. 1.

The cross sections of different soldering times show distinct Ag_3Sn needles within the solder joint for SnAg3.0 and SnAg5.0. For a continuous phase formation between busbar and solder longer process times are necessary, contrary to state-of-the-art machine setups.

More experiments of the soldering process have to be carried out to prove the contact formation and characterization of wettability and IMC formation at the interfaces.

These experiments will include investigations of varying fluxes and Ag-thickfilm cell metallizations.

These first results reveal parameters to ensure the quality and contact formation of solar cell joint interconnections and shows that the conditions of wetting behaviour and additional investigations on microstructural properties and IMC formation are important for PV module reliability up to 25 years.

References

[1] A. Goetzberger, V. Voß und J. Knobloch, Sonnenenergie: Photovoltaik, Stuttgart: Teubner, 1994

[2] Kreß, A.: Emitterverbund-Rückkontaktsolarzellen für die industrielle Fertigung, Dissertation, Universität Konstanz, 2001

[3] Gabor, M. et al: Soldering induced damage to thin Si Solar Cells and Detection of Cracked Cells in Modules. In: Proc. of 21st European Photovoltaic Solar Energy Conference, September 4-8, 2006, Dresden

[4] Vianco, P.: Handbook of Lead (Pb)-Free Technology for Microelectronic Assembly (New York: Marcel-Dekker, 2004), pp. 167–210

[5] Artaki, I.: Evaluation of lead-free solder joints in electronic assemblies, Journal of Electronic Materials, Springer Boston, Volume 23, Number 8, 1994, pp. 757-764

[6] S. Schindler, S. Wiese, " Investigation of wettability and interface reactions of Sn-Pb, Sn-Cu, Sn-Ag and Sn-Ag-Cu solders for solar cell interconnections", Electronic Systeme Integration and Technology Conference (ESTC), Berlin, Germany, 2010

3D Stacking Approaches for Mold Embedded Packages

T. Braun ([1]), K.-F. Becker ([1]), K. Piefke ([2]), S. Voges ([2]), T. Thomas ([2]), M. Töpper ([1]),
T. Fischer ([1]), R. Kahle ([2]), V. Bader ([1]), J. Bauer ([1]), R. Aschenbrenner ([1]), K.-D. Lang ([2])

(1) Fraunhofer Institute for Reliability and Microintegration
Gustav-Meyer-Allee 25, 13355 Berlin, Germany

phone: +49-30/464 03 244 fax.: +49-30/464 03 254 e-mail: tanja.braun@izm.fraunhofer.de

(2) Technical University Berlin, Microperipheric Center

Abstract

The constant drive to further miniaturization and heterogeneous system integration leads to a need for new packaging technologies which also allow large area processing and 3D integration with potential for low cost applications. Large area mold embedding technologies and embedding of active components into printed circuit boards (Chip-in-Polymer) are two major packaging trends in this area.

This paper describes the use of a novel S2iP (Stacked System in Package) interconnect technique using advanced molding process for multi chip embedding in combination with large area and low cost redistribution technology derived from printed circuit board manufacturing with a focus on integration of through mold vias and vertical interconnect elements for package stacking.

The use of compression molding equipment with liquid or granular epoxy molding compounds for the targeted integration process flow is a new technology that has been especially developed to allow large area embedding of single chips but also of multiple chips or heterogeneous systems on wafer scale, typically 8" to 12". The wiring of the embedded components in this novel type of SiP can be done using PCB manufacturing technologies, i.e. a resin coated copper (RCC) film is laminated over the embedded components. Also thin film redistribution technologies can be applied – using similar processes as used for WLP on silicon. For the 3D interconnection two technologies are evaluated: In a process flow similar to conventional PCB manufacturing vias are drilled using a UV laser after RCC lamination and are metalized in the same process as the vias for chip interconnection. A second via generation process uses vertical interconnect elements, e.g. silicon dies with well defined lead structures, assembled in the same process step as the embedded dies. Top and bottom surface of those vertical interconnect elements are exposed after mold embedding. Planar interconnection is then formed by thin film redistribution or after RCC lamination, the vertical interconnect elements are contacted using a μVia process.

Within this study the different approaches to vertical interconnection in a mold embedded wafer have been intensively evaluated on their processability. A strong focus was put on the process chain including chip placement on a temporary carrier - compression vacuum molding for embedding – RCC lamination or thin film redistribution – laser drilling processes for μVias & Thru Holes – metallization structuring – module singulation & 3D assembly. The feasibility of the entire process chain is demonstrated by fabrication of a Ball Grid Array (BGA) type of system package with two embedded dies and through mold vias (TMVs) allowing the stacking of these BGA packages. Reliability of the manufactured 3D stacks is evaluated by temperature cycling and is analyzed both non-destructively and destructively.

The paper depicts a final technology demonstrator where two BGAs are stacked on each other and mounted on a base substrate enabling the electrical connection of the stacked module, allowing the evaluation of the technology and the applied processes.

Key words: embedding; 3D packaging, stacking, wafer level molding

Introduction

Drivers for 3D packaging solutions are manifold and each requirement calls for different answers and technologies. Mainly miniaturization, but density and performance, simplification of design and assembly, flexibility and functionality and finally, cost and time-to-market have been found to be the core drivers for going 3D as well. Besides die and package stacking and folded packages, embedding dies is a key technology for heterogeneous system integration [1].

There are two main approaches for embedded die technologies: Wafer level integration, where dies are embedded into polymer encapsulants and 3D vertical integration, where dies are embedded into the substrate. For wafer level integration a lot of activities are running worldwide. Main drivers are here the

Embedded Wafer Level Ball Grid Array (eWLB) by Infineon [2] and the Redistributed Chip Package (RCP) by Freescale [3]. Singulated dies are assembled on an intermediate carrier and encapsulated by compression molding, forming a polymer wafer with embedded silicon dies. This "reconfigured" wafer is then released from the carrier. Using thin film technology, an electrical redistribution layer is routed on the wafer. Finally, the wafer is singulated by sawing into single packages. One trend in eWLB technology is at the moment a double sided eWLB packaging with integration of vias through the encapsulant by integration of preformed PCB based vias allowing the stacking of eWLB packages [4].

Another concept for 3D integration of active components is the Chip in Polymer (CiP) technology, introduced by Fraunhofer IZM, TU Berlin and Wuerth. It is based on embedding of ultra-thin dies into build-up layers of printed circuit boards (PCBs). The dies are bonded onto a core substrate using an adhesive; a resin coated copper (RCC) layer with thin Cu is used for the subsequent lamination. Interconnects are established by laser drilled micro vias followed by a PCB-compatible Cu plating [5, 6].

The combination of both concepts embedding into polymer by molding and redistribution by PCB technologies has the potential for highly integrated low cost packages and was successfully demonstrated for a 2-chip LGA package [7]. The direct integration of Through Mold Vias (TMVs) can be easily integrated in such packages as vias are a standard feature in the PCB manufacturing process and can be adapted for the proposed concepts of embedding into polymer by molding and redistribution by PCB technologies. A principle draft of a Package-on-Package assembly (PoP) based on a wafer level embedded package with PCB based redistribution technology is shown in Figure 1.

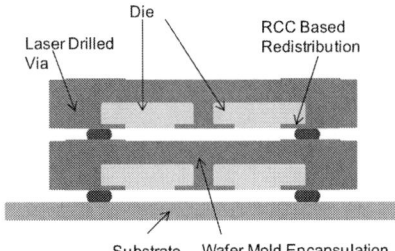

Figure 1: Schematic of a Package-on-Package assembly based on wafer level embedded package with PCB based redistribution technology and laser drilled TMVs

The goal of the thinfilm routing ontop of the embedded die in molded wafers is the manufacturability on existing bumping and redistribution lines which already exist world-wide [8; 9]. Only modifications of the wafer handling should be tolerable. 3D routing is realized by integrating

vertical interconnect elements (VIE), e.g. silicon dies with well defined lead structures, assembled in the same process step as the embedded dies.

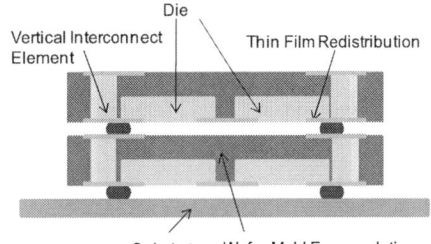

Figure 2: Schematic of a Package-on-Package assembly based on wafer level embedded package with thin film redistribution VIE

Top and bottom surface of those vertical interconnect elements are exposed after mold embedding. Planar interconnection is then formed by thinfilm redistribution. Advantage of this concept is that the 3D routing with these vertical interconnect elements is impedance controlled and therefore even high frequency applications can be targeted with this technological variant. The schematic of this approach is shown in Figure 2.

Process Flow

The process steps for both technology options described in detail below are summarized in Figure 3.

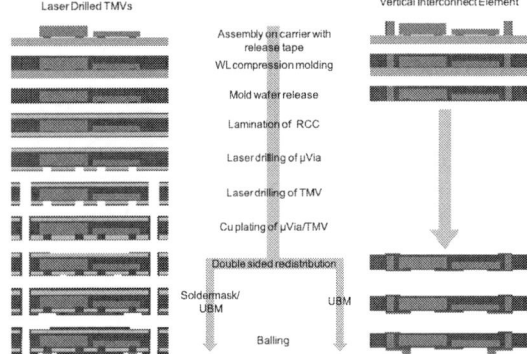

Figure 3: PoP manufacturing process flow based on WL embedded package with laser drilled TMVs (left) and VIE (right)

The general process flow starts with the lamination of an adhesive film to a carrier. This special adhesive film has one pressure adhesive side and one thermo-release side, i.e. by heating up the tape above a certain temperature, the thermo-release side of the tape loses its adhesion strength. On this carrier-adhesive film sandwich dies are precisely placed, the active side facing down towards the carrier. High accuracy is needed as die pads have to match with the redistribution layer. Molding is done by large area compression molding.

One option for chip redistribution is a low cost PCB based technology with RCC in combination with laser drilled vias for 3D routing. After lamination of the RCC film on both wafer sides in one step, μvias are drilled to the die pads and through mold vias in the same process step to connect to and bottom side. Next process steps are cleaning, palladium activation and copper plating. By plating both, via filling and die pad connection to the copper layer and the top copper layer to the bottom copper layer are achieved. Conductor line formation is done by laser direct imaging (LDI) in combination with a dry film resist and copper etching. Finally, a solder mask and solderable surface finish as NiAu and solder balls can be applied.

The thinfilm process flow uses low T curable polymers like Epoxy (JSR WPR 5100) and a metallization based on sputtering TiW/Cu and electroplated Cu for the RDL. Main concern of the processing is the reduction of the process temperature to stay below 200°C. The vertical interconnection elements are based on electroplated Cu on Si-Monitor-Wafers which are singulated and turned to 90° before molding. A schematic is shown in Figure 4:

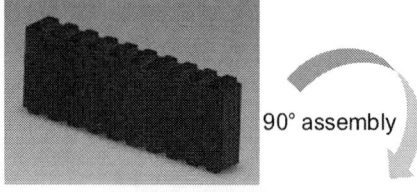

Figure 4:   Schematic of a VIE

Final process steps for both technologies described above are package singulation by sawing followed by 3D package assembly on main board.

Materials

There is a variety of compression molding compounds for embedded wafer level molding from different suppliers on the market available at the moment. For mold embedding technology materials

should have low chemical shrinkage, low cure temperature and match thermo-mechanical properties for low warpage of the molded wafer and low die shift after molding. Flow properties should allow homogeneously filling of 8" and large cavities. Additionally for through mold vias filler particle size and distribution has to be taken into account. Table 1 gives an overview on the selected compression molding compounds for this study is given. Materials were selected to give a broad cross section of materials concerning process parameters and final mechanical and thermo-mechanical properties. Additionally focus was put on filler particle sizes as this was assumed to be one of the most import criteria for via drilling into molding compounds. Hence, materials selected have maximum filler sizes from 25 μm up to 75 μm. 25 μm seems to be the smallest maximum filler size available at the moment as with smaller filler particles the filler content has to be reduced to keep the viscosity low for best processing. But this also means an increased CTE, leading to wafer warpage due to thermal mismatch.

For a better understanding of the flow and cure behavior of the compression molding and with that to optimize the compression molding process concerning e.g. voiding and minimizing flow marks and material inhomogeneities which may gain further importance with thinner and larger molded wafers/panels the materials were analyzed by Differential Scanning Calorimetry (DSC) and by viscosity measurements [10]. As a summery it can be noticed that there are clear differences in material behavior between liquid and granular compounds even for materials with close processing temperatures. In general the liquid compounds start to cure lower temperature but show also a lower viscosity at process temperature. The highest viscosity is detectable for material B, the granular compound with the smallest filler particles and the lowest filler content. Here the interaction of the surface of the small filler seems to be the major reason for the higher viscosity in combination with the lower filler content. In contrast the two liquid materials D and H with the largest filler show the lowest viscosity.

Table 1:Compression molding compounds selected, datasheet values

Material	A	B	C	D	E	F	G	H
Supplier	1	1	2	3	4	4	4	5
Type	Granule	Granule	Granule	Liquid	Granule	Granule	Liquid	Liquid
Filler Content	86%	85%	89,50%	89%	89%	86%	88%	90%
Filler Sieving Size	55 μm	24 μm	45 μm	75 μm	53 μm	30 μm	53 μm	75 μm
Gel Time	70 s @ 175 °C	60 s @ 175 °C	90 S @ 150 °C 33 s @ 175 °C	40 s @ 150 °C	30 s @ 175 °C	30 s @ 175 °C	250 s @ 120 °C 160 s @ 130 °C	
CTE $_1$	8 ppm/K	11 ppm/K	7 ppm/K	7,4 ppm/K	7 ppm/K	9 ppm/K	7 ppm/K	8 ppm/K
CTE $_2$	30 ppm/K	53 ppm/K	32 ppm/K	33 ppm/K	25 ppm/K	30 ppm/K	40 ppm/K	30 ppm/K
Tg	135 °C	155 °C	140 °C	160 °C	190 °C	190 °C	145 °C	138 °C
Flexural Modulus @ RT	26 GPa	19 GPa	24GPa	22 GPa	23 GPa	20,5 GPa	22 GPa	29,5 GPa
Mold Temperature	175 °C	175 °C	140 °C	120 °C	130 °C	130 °C	130 °C	120 °C
Inmold Cure Time	90 s	90 s	120 s	600 s	300 s	300 s	300 s	400 s
Post Cure Temperature	175 °C	175 °C	140 °C	150 °C	175 °C	175 °C	150 °C	150 °C
Post Cure Time	4 h	4 h	3 h	1 h	6 h	6 h	1 h	1 h

Resulting warpage after molding and carrier release has been measured contactless at room temperature with a CyberScan Vantage System. All materials were characterized as blank wafers without dies inside and selected materials were additionally analyzed as reconfigured molded wafers with 1400 dies (2x3 mm² and 250 μm thickness). Molded wafer thicknesses for all measurements were 400 μm. Warpage results of blank wafers are shown in Figure 5. Lowest warpage can be found for materials with low molding temperature, high glass transition temperature and low CTE. Hence, best results could be found for the granular compounds C, E and F.

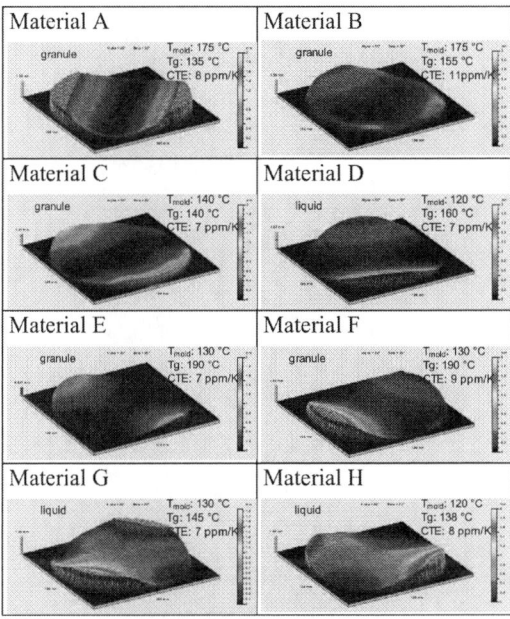

Figure 5: **Warpage of molded blank wafers with a thickness of 400 μm**

Warpage results from reconfigured mold wafers with selected molding compounds are depicted in Figure 6. For all materials a higher warpage compared to the blank wafers molded wafers is visible. But lower warpage is again found for materials with low molding temperature, high glass transition temperature and low CTE. As a summary from the warpage point of view the granular compounds with a low cure temperature showed a better performance as the liquids.

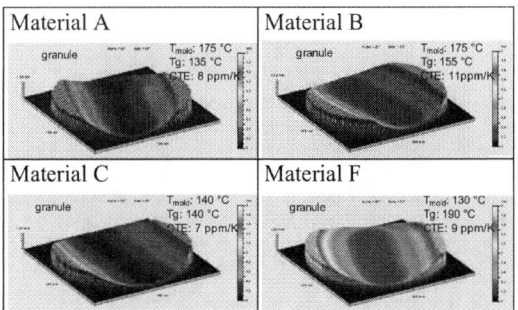

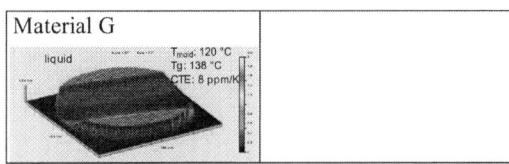

Figure 6: **Warpage of molded wafers with 1400 dies inside (2x3 mm², 250 μm thickness) and a thickness of 400 μm**

Experimental – Laser Drill Parameter Study

A test vehicle has been designed to optimize laser drill parameter for TMVs in the various compounds evaluated. In Figure 7 the layout of the test vehicle is shown. Within the test vehicle different TMV diameters from 50 μm up to 200 μm and different pitches from 150 μm up to 300 μm can be tested. Additionally mold layer and via metallization thickness can be varied. For every via combination 104 TMVs are connected by a daisy chain and can be electrically tested on their interconnect integrity initially and during reliability testing e.g. temperature cycling.

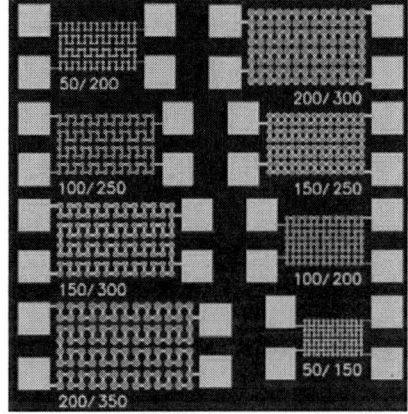

Figure 7: **Layout test vehicle for laser drill parameter study with via diameter and pitch variation**

For sample preparation 8" wafers are molded according the material data sheet recommendations for the different materials without embedded dies on a TOWA 120 t compression mold machine. A filled RCC material with 5 μm Cu and 35 μm resin thickness was selected as redistribution layer. Lamination was done on both sides under vacuum at 10 bar for 50 min at 185 °C. Through mold vias were drilled with an UV laser with an accuracy of ± 10 μm. TMV metallization was done by Cu plating after a cleaning and activation step of the via.

Mold wafer thickness for this investigation was 400 μm comparable to the final technology demonstrator presented below.

Drill parameter study was done with selected materials presented in Table 1. Materials with different filler particle sizes have been evaluated first as maximum filler size was assumed to be one of the import points for drilling through mold vias with small diameters and fine pitches for highest integration. During laser drilling the epoxy resin is ashed and the SiO_2 filler particles are pulled out with the vaporized resin but they do not melt.

Figure 8 shows two cross section of TMVs in material D with the largest maximum filler size of 75 µm demonstrating the limitation of materials with larger filler particles. That means on the one hand that for smaller via diameters in the range of the maximum filler size no reproducible and continuous via can be guaranteed (s. Figure 8, left). One the other hand also finer pitches are not achievable with materials with large filler particle as no mechanically reliable wall between two vias can be obtained where only one or two filler particle are left (s. Figure 8, right).

Figure 8: **Cross sections of TMVs of material D (left: 50 µm diameter and 200 µm pitch, right: 200 µm diameter and 300 µm pitch)**

However, materials with small filler particles as material B with a maximum filler particle size of 25 µm allow the fabrication of vias with a very precise and smooth via surface (s Figure 9). Here the materials have potential for small vias and fine pitches. Nevertheless, materials with finer fillers currently have the drawback of higher viscosities and lower filler content leading to a higher CTE.

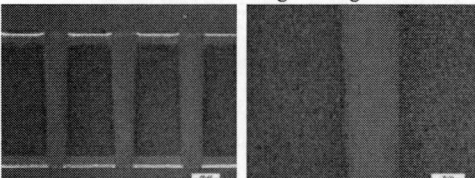

Figure 9: **Cross section of TMVs of material B (100 µm diameter and 250 µm pitch)**

Figure 10 summarizes the results from the test vehicle for material C which showed good moldability, low warpage in combination with medium sized filler particles with a maximum diameter of 45 µm. All combinations in diameter and pitch could be manufactured with a homogeneous Cu layer without any shorts. Even with this material TMVs with a diameter of 50 µm and a pitch 150 µm could be achieved.

Test vehicle of material C were submitted to temperature cycling tests -55 °C / 125 °C with a cycle time of 30 min. Test vehicles were intermediately electrical tested on their daisy chain integrity.

TMV Ø [µm]	Pitch [µm]	Material C max. filler size: 45 µm
50	200	
100	250	
150	300	
200	350	
50	150	
100	200	
150	250	
200	300	

Figure 10: Cross sections of TMVs of material C and D with different diameters and pitches

After 250 cycles only one failure occurred for a structure with 50 µm diameter and 200 µm pitch and after 500 cycles 1 failure for structure with 50 µm diameter and 150 µm pitch out of 15 test coupons. Cycling tests are ongoing at the moment.

Test vehicles will be analyzed on their failure mode destructively after test end. These first results are very promising and show the potential of this technology.

Demonstrator Manufacturing

The general process flow described above was used for the fabrication of a 2-chip BGA stackable package to demonstrate the feasibility of the proposed process flow.

Package layout

A 2-chip package 8x8 mm^2 with BGA pin-out has been designed for technology demonstration. The package consists of two daisy chain chips with a size of 2x3 mm^2. The daisy chain dies have peripheral pads with a size of 120 µm and a pitch of 200 µm. The chips are based on sputtered AlSiCu daisy chain pads on thermal Oxide passivation with 3,5 µm thick PI (s. Figure 11).

Figure 11: Test chip with daisy chain structure

Additionally, fiducial chips have been introduced for via drill and redistribution alignment on the molded wafer. First demonstrators presented below were fabricated with 250 µm thick dies.

The electrical wiring of the package was designed allowing the testing of the daisy chain die interconnection integrity as well as the TMVs/VIEs. By assembling two modules where the second package is rotated by 90° on top of the other on a main board the electrical testing of the entire stack is possible.

Reconfigured Wafer Molding

Before die placement, a thermo-release film was laminated on an 8" carrier. Release film was selected according to compression molding compound and the related molding temperature. A high speed chip assembly machine, Siplace CA3, was used for chip assembly with an accuracy of ± 15 µm. Compression molding was done on a compression mold machine 120 t press from TOWA. After wafer release from carrier die positions were measured with an optical measuring system Mahr OMS 600 to study the die shift of the different materials and to adapt the drill and wiring layout or the placement strategy according to the die placement accuracy and die shift during molding.

Vertical Interconnection Elements (VIE)

200 mm Si-Monitor Wafers with thermal Oxide are sputtered with a plating base (100 nm TiW /

300 nm Cu) and laminated with a 50 µm thick dry film Photo-Resist from DuPont (WBR 2100). The dry film is structured by UV-lithography using mask aligners (SüssMicroTec, MA 200) and developed using 1,2% K$_2$CO$_3$ in water. After electroplating of 50 µm Cu the resist is stripped and the wafers are diced. An example of a Cu VIE with 75 µm lines and spaces is shown in Figure 12.

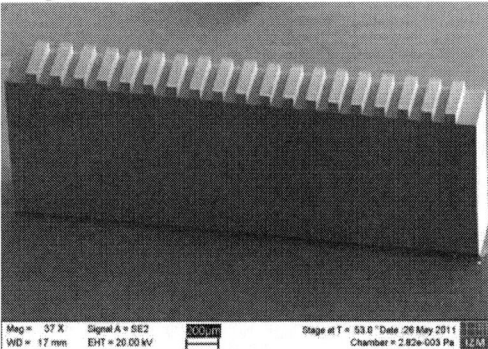

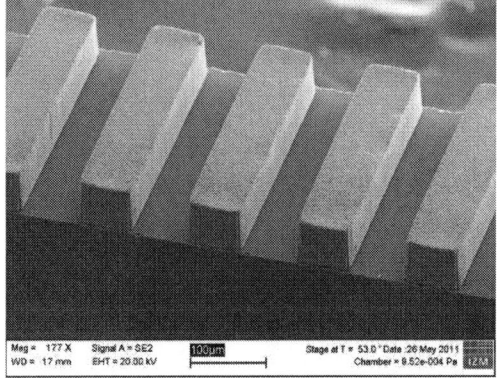

Figure 12: SEM pictures of vertical interconnect elements with 75 µm lines and spaces, top: overview, bottom: detail

First assembly and molding VIE trials have been successfully carried out. VIEs could be assembled by standard pick and place together with daisy chain dies with same placement accuracy and die shift after molding. Detailed picture of a reconfigured molded wafer with VIEs after carrier release is shown in Figure 13.

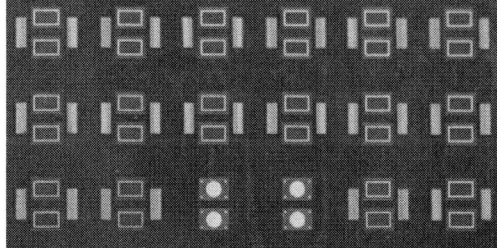

Figure 13: Detailed picture of a reconfigured molded wafer with VIEs after carrier release

First wafers were molded with a thickness of 400 µm. Backside opening of the VIE will be done in a first step by backgrinding to the Cu lines. Later work will also deal with backside opening of VIE during mold process. At the moment first redistribution tests are performed.

Laser Drilled TMVs

For the laser drilled TMV version the resulting package thickness was 400 µm. TMVs were realized with a diameter of 150 µm and a pitch of 300 µm. For first process evaluations material C and D were used as process flow should be demonstrated with a liquid and granular compound and both materials show a good overall processability.

For the PCB based redistribution technology the embedded dies need to be prepared with an under bump metallization (UBM) to avoid incompatibilities with the used PCB processing. Typically, NiPd pad reinforcement is applied to the wafer.

After the reconfigured mold wafer is released from carrier a filled RCC film with 5 µm copper and 35 µm resin was used as redistribution layer. Lamination was done in one step on both wafer sides under vacuum at 10 bar for 50 min at 185 °C. Through mold vias as well as µvias for chip connection were drilled in one step using a UV laser with an accuracy of ± 10 µm. Parameters from process setup of the test vehicle were used for TMV drilling.

Next step was via metallization by Cu plating. Here also the goal was to fill µvias together with TMVs is one process step. Cu structuring is conducted by dry film resist lamination, laser direct imaging (LDI) which allows layout adaptation. Finally an electroless NiAu surface finish, a soldermask and lead free solder balls were applied. Afterwards two singulated packages were soldered on a main board. Figure 14 shows two photographs of such an assembled stack.

Figure 14: Photograph of two stacked wafer level embedded packages assembled on a base substrate

Assembled modules were tested on their electrical integrity and for both molding compound used full electrical integrity could be achieved.

X-ray inspection was performed to analyze the interconnections as well as the alignment of the stack. Figure 15 depicts two x-ray images of a stack with wafer level embedded dies demonstrating the good alignment and interconnection of the entire stack.

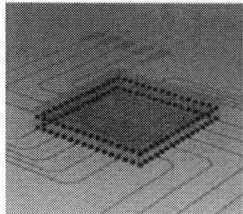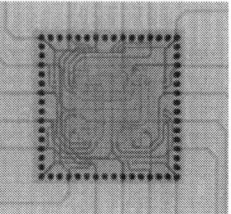

Figure 15: X-ray microscopy of two stacked wafer level embedded packages assembled on a base substrate

Cross sectioning was used to investigate µvia interconnects as well as through mold vias. Figure 16 shows an image of a cross section of two stacked modules demonstrating µvias and TMVs with a good integrity. All pads are well aligned connected with void free copper filled vias and homogeneously metalized TMVs.

Figure 16: Cross section of two stacked wafer level embedded packages

In summary the proposed process flow for stacking wafer level embedded dies by integration of through mold vias (TMVs) was successfully demonstrated.

Conclusion & Outlook

This paper describes the development of stackable wafer level embedding technology based on two technologies:

- large area compression molding and PCB based redistribution technology with a focus on integration of through mold vias (TMVs) for package stacking. TMVs are laser drilled through the embedding molding compound and metalized by Cu plating.
- large area compression molding and thinfilm redistribution in combination with vertical interconnect elements, e.g. silicon dies with well defined lead structures, assembled in the same process step as the embedded dies

An intensive study with eight different liquid and granular molding compounds on their molding processability has been performed.

A test vehicle for TMV evaluation taking into account molding compound type, mold thickness, via diameter, pitch and Cu thickness was designed and manufactured with different molding compounds. One key facture for TMVs are the SiO_2 filler particle in the molding as their size limits the minimum achievable diameter pitch. First reliability temperature cycling tests show promising results and the potential of laser drilled TMVs.

The overall process flow was successfully demonstrated for liquid and a granular molding compound for the fabrication of a stack consisting of two 2-chip package with BGA pin-out assembled on a main board.

First vertical interconnect elements were successfully manufactured and together with daisy chain dies embedded into a reconfigured mold wafer. Ongoing work deals with the redistribution of these wafers.

Further work will also deal with reliability testing of stacked modules assembled in the two 3D routing technologies described above.

Acknowledgements

The authors would like to acknowledge the support of the German ministry for education and research [BMBF] and of VDI/VDE Innovation + Technik GmbH in the MST LowCostTMV Project, (Contract No. W40043).

Additionally the authors want to thank M. Minkus and M. Stanjek for analytical support and L. Böttcher for electroless Ni bumping.

References

[1] T. Thomas, F. Yuwen Lin, K.-F. Becker, T. Braun, E. Jung, R. Aschenbrenner, H. Reichl; State-of-the-art of 3D SiP Technology; Proceedings of IMAPS Poland 2009.

[2] T. Meyer, G. Ofner, S. Bradl, M. Brunnbauer, R. Hagen; Embedded Wafer Level Ball Grid Array (eWLB); Proceedings of EPTC 2008, Singapore.

[3] B. Keser, C. Amrine, T. Duong, O. Fay, S. Hayes, G. Leal, W. Lytle, D. Mitchell, R. Wenzel; The Redistributed Chip Package: A Breakthrough for Advanced Packaging, Proceedings of ECTC 2007, Reno/Nevada, USA.

[4] Y. Jin, X.r Baraton, S. W. Yoon, Y. Lin, P. C. Marimuthu, V. P. Ganesh, T. Meyer, A. Bahr; Next Generation eWLB (embedded Wafer Level BGA) Packaging; Proceedings of EPTC 2010, Singapore.

[5] L. Boettcher, D. Manessis, S. Karaszkiewicz, A. Ostmann, R. Aschenbrenner, H. Reichl; Innovative embedded-chip QFN package realization; Proceedings of SMTA International Wafer Level Packaging Conference 2009, Santa Clara, California, USA.

[6] J. Kostelnik, K.-F. Becker, F. Ebling, J.-P.Sommer; E Noack; L Böttcher; KRAFAS – Einbetttechnologien für aktive Komponenten; PLUS; Band 10, 11.2008, pp. 2447-2456.

[7] T. Braun, K.-F. Becker, L. Böttcher, J. Bauer, T. Thomas, M. Koch, R. Kahle, A. Ostmann, R. Aschenbrenner, H. Reichl, M. Bründel, J.F. Haag, U. Scholz; Large Area Embedding for Heterogeneous System Integration; Proceedings of ECTC 2010, Las Vegas, USA.

[8] M. Töpper, D. Tönnies; Microelectronic Packaging (Chapter 25), Semiconductor Manufacturing Handbook (H. Geng, Ed.), McGraw-Hill, 2005, ISBN: 0-07-144559-5.

[9] M. Töpper, T. Fischer, V. Bader, K.-D. Lang, N. Matsuie, T. Motobe, T. Minegishi, M. Knaus; Ultra Low Temperature PBO-Polymer for Wafer Level Packaging Applications, Proceeding of ICEP 2011, Japan.

[10] T. Braun, K.-F. Becker, S. Voges, T. Thomas, R. Kahle, V. Bader, J. Bauer, K. Piefke, R. Krüger, R. Aschenbrenner, K.-D. Lang; Through Mold Vias for Stacking of Mold Embedded Packages; Proceedings of ECTC 2011, Lake Buena Vista, USA.

Chip embedding technology developments leading to the emergence of miniaturized system-in-packages

D. Manessis[1], L.Boettcher[2], S. Karaszkiewicz[2], A. Ostmann[2], R. Aschenbrenner[2], and
K-D. Lang[1]

[1] Microperipheric Research Center, Technical University Berlin (TUB)
[2] Fraunhofer Institute for Reliability and Microintegration (IZM), Berlin
Gustav-Meyer-Allee 25, Berlin 13355, Germany
E-Mail: manessis@izm.fhg.de

Abstract

At PCB manufacturing level, 50 μm thin chips have been embedded with pitches up to 200 μm in up to 18"x24" panels using face-down technology for chip assembly. This paper describes in detail the merits and shortcomings of both face-down and face-up embedding. In addition, it shows the further developments in chip embedding technologies to incorporate chips with even smaller pitches up to 50μm. Embedding of small pitch chips has been realised with concurrent developments in accurate chip positioning, plating methods and chemistries and ultra fine line patterning. The results in this paper show the emergence of a new prototype Embedded chip-QFN package with contact pads at 400μm pitch and a total number of 84I/Os with dimensions of 10mmx10mm. The embedded chip in the QFN package is 5mmx5mm in size and has a peripheral pad configuration at 100μm pitch. All Embedded chip-QFN packages have been manufactured in 10"x14" panels at prototype level. This paper also shows the latest developments in semi-additive processes for copper structuring of chip embedded packages with pitches lower than 100μm. Ultra fine line structuring technology has been developed up to 15μm L/S copper structures using innovative 2μm copper base foils. Qualitative analysis using acoustic microscopy and shear testing of the QFNs provides evidence of good resin adhesion and package mechanical robustness. Furthermore, this study shows promising results for embedding of chips with pitches dwon to 50μm, introducing a new "vialess" face-down embedding approach where direct contacts are established between copper pads and functional copper foil. Developmental work on "vialess" embedding technology is still ongoing.

Key words: chip embedding, semi-additve, ultra fine line structuring, vialess ultra fine pitch embedding.

Introduction

Electronics systems have witnessed an unprecedented miniaturization trend driven by the continuous pursuit of increased functionality, optimum exploitation of substrate area for even larger integration of components. Conventional technologies for assembly of component packages on susbtrates (SMD, wire-bonded chips, Flip chips) are still the locomotive for electronics systems intergration but progressively yield the packaging roadway to embedding technologies which have achieved to bring the component packages into the substrates and thereby letting free space for further mounting of components on top of the substrates. Embedding of active and passive components has become a very dynamic technology with great potential in many prospective industrial markets [1,2]. The technological front has rapidly advanced from a 2D system-in-package (SiP) integration to a 3D-SiP module integration to keep up with miniaturization trends. In the new packaging roadmaps towards a "More than Moore" direction, chip and component embedding turns out to be a very promising and feasible technology route for

even higher 3-dimensional homogeneous and heterogeneous integration of components. Embedding technologies offer the advantages of direct contact to the chips without use of long wires or solder bumps and thus improved electrical performance, capability of 3D-stacking, reduced package thickness if thin components are available, and enhanced thermal performance for components assembled on thermal interfaces and heat sinks, as in the case of embedding power components. The emergence of embedding packages steadily changes the packaging value chain and thus sets new roles for all players in the whole packaging value system. The initial value share of substrate suppliers for production of FCBGA devices was about 20% but now with embedded technologies the substrate suppliers can also perform the assembly of componenets before embedding and thus count for 55% of the embedded package value [2]. However, it should be underscored, that the shift to embedding technologies marks also necessary adaptations to the supplain chain which will potentially burden the value system. For instance, the necessity of RDL layer for chip pitch enlargement that makes chip componenents compatible with existing embedding capabilities or copper pad deposition should be definitely accounted for before the shift in embedded packages [4-7]. Initial applications for embedded packages will be low cost, low pin counts applications such as analog and power devices (DC/DC converters, Power MOSFETS etc.) [2]. There are forecasts for a half billion dollars extended market by 2015 [2].

A comprehensive overview of embedding technologies for passive and active components is given in the literature [2-8]. Among the most recent technologies, an Integrated Module Board (IMB) technology was developed by Helsinki University of Technology [5] and industrialised by Imbera with face-down assembly of the chips, and on the other hand the chip-in-polymer technology by Fraunhofer IZM using a face-up embedding concept [6-10]. This paper discusses in detail the technology developed in the EU-funded project "HERMES" whose scope is to further develop all embedding technologies at advanced R&D level and more importantly to bring embedding technology in real manufacturing PCB production with a concrete goal of embedding components in 18"x24" PCBs. This paper intends to address all pertinent technological challenges arisen with embedding of chips with pad pitches up to 100μm employing FhG-IZM "chip face-up" approach and in addition to highlight the first results obtained for embedding ultra fine pitch chips at 50μm pitch using a "vialess" face-down embedding technology.

CHIP EMBEDDING TECHNOLOGIES

Technology overview & comparisons

There are different ways of embedding an active Si chip on an organic or inorganic substrate. The die can be either placed face-up like a die that would be assembled for wire bonding of the pads or with the face-down like a conventional flip chip as illustrated in Figures 1a and 1b, respectively. Both techniques use standard printed board processing for the embedding of the chips but different approaches for the assembly of the chips.

The technology developed at Fraunhofer IZM basically focuses on the face-up approach in combination with the formation of laser micro via to the chip bond pads. Figure 1(a) illustrates the process flow for IZM technology on Copper substrate. The specific process flow in Figure 1(a) can yield a package with both top and bottom contacts and direct accessibility of the chip backside. Very thin chips in the range of 20μm-50μm can be assembled and laminated in thin resin layers resulting in very thin embedded packages.

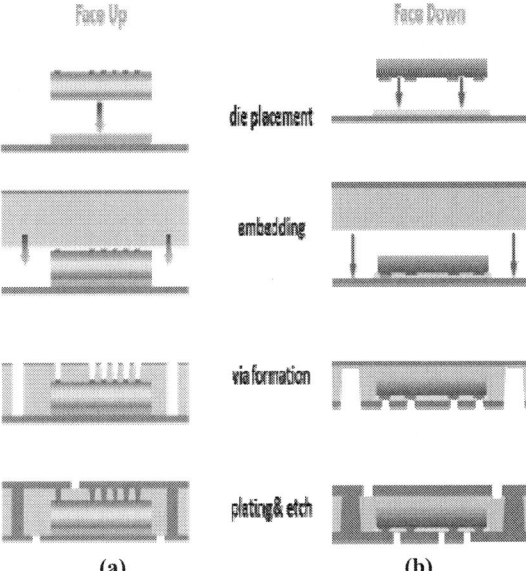

Figure 1. Process sequence for chip embedding with face-up (a) and face-down (b) assembly approaches.

The face-down approach as illustrated in Figure (b) has been initially developed by Imbera and has been the technology of choice for Asian manufacturers of embedded packages [5]. In this technology approach, a copper foil has been pre-structured with drilled holes which help for the face-down alignment of the die. A non-conductive adhesive is applied prior to die bonding. After die placement and curing of the die adhesive, either Resin-coated-copper (RCC) foils or epoxy resin prepregs can be vacuum laminated for the final embedding of the die. The microvias to the pads of the embedded chip are made by laser drilling and the electrical contact is realized by electrolytic deposited copper in the drilled via. Finally the copper is structured realising the needed routing structure for the substrate. These substrates can be processed further like an inner layer of a PCB if additional routing layers are required. Based on the face-down embedding concept, new alternative technologies have been developed by Wuerth Electronic like the "Laser cavity" where cavities in PCB substrates are opened and chips are assembled in a flip chip fashion [3].

The main difference between the two approaches is the die assembly. In the face-down technology, the flip chips are assembled with non-conductive adhesives on copper substrates. By the face-up technology, in addition to non-conductive adhesives (NCAs), conductive adhesives and solder materials can be used and thus offering the opportunity to establish a "direct" silicon backside electrical and thermal contact without the drilling of vias. In this respect, the face-up approach can be very beneficial for the embedding of vertical IC (e.g. Power MOSFET) where the backside drain contact

503

can be formed for advanced thermal dissipation in combination with the electroplated microvias.

However, it should be mentioned that the face-up technology with microvias necessitate very accurate placement of the chip and poses a lower limit of about 100µm for the chip pad pitch that can be contacted with laser microvias. The face-down assembly of the chips can reach finer pitch capabilities (<100µm) on the predrilled substrate with most of the new assembly machines and thereby decoupling the process from laser microvia pitch limitations [9]. As evidenced, both technologies can offer complimentarily chip interconnections without the use of costly and fragile RDL layers for pitch enlargement. However, the common pursuit in both technologies is even more precise die placement and bonding for better microvia alignment in combination with high speed die placement at high yield. The following sections in this paper elucidate both technology approaches.

FACE-UP EMBEDDING TECHNOLOGY

Face-up chip embedding technology has been employed by IZM for the development of a miniaturised system-in-package in a standard QFN format. The embedded chip has a pitch of 100µm and the resultant QFN package a standard pitch of 400µm. The main process steps and details are discussed below.

QFN package development

Wafer preparation

Laser drilling of microvias and the PCB metallization process is not compatible with Al of semiconductor chips. Therefore, a further layer of 6-8 µm Cu is applied electrolytically to the bond pads of the chips to be embedded. The copper pads have a size of 85µm. Other more cost-efficient pad metallisations such as electroless Ni/Pd can be optimised for microvia plating [8,9]. Passivation layers can be tested for their fragility as well as for their adhesion with the RCC laminate layers. It is worth noting that chips with very small pitches originally designed for wire bonding can also be compatible with face-up embedding after an application of a redistribution layer provided by subcontractor services. In addition, if very thin embedded modules are to be manufactured, the wafers should be thinned down to a desirable 50 µm or 20µm thickness.

Chip placement and bonding

Placement accuracy is extremely crucial for chip embedding. The process tolerances for sequential die bonding, via drilling and Cu structuring (Figure 1a) have to be very low in order to achieve an acceptable yield. One of the requirements is that the machines for these three process steps should use the same alignment fiducials on the core substrate. Prior to chip placement, die ejection from a dicing blue tape takes place and should be optimised according to the adhesiveness of the tape, size and ultimate thickness of the chip. Adhesiveness of the tape is regulated by UV exposure and is to be reduced for component pick-up and placement without endangering the chip integrity. By using a Datacon Evo placement and bonder machine a placement accuracy of ±10µm at 3σ is achieved on 18"x12" panels with a placement speed of 1000 components/hr [8].

Chip bonding has been developed by using printable pastes and die attach films (DAF). Stencil printing allows a precise control of volume and location of the adhesive paste, which is rather a problem for dispensing. Electrically conductive Ag-filled pastes or solder pastes can be used. Another method is the use of a non-conductive die dicing attach film (DDAF) of 20µm thicknes. It is a UV dicing tape which has two layers, a conventional UV dicing foil and the 20µm adhesive layer on top. Wafers are mounted on the adhesive layer. The dicing blade has to cut the silicon and the top layer of the tape. In the picking process this layer remains at the chip and serves as adhesive. Die attach films have shown superior adhesive coplanarity compared to printed pastes, which is extremely important for precise epoxy thickness over the chip after RCC lamination, and for avoiding chip cracking. For the QFN development a DDAF film was used.

Lamination process and embedding materials

Both for face-up and face-down embedding technologies depending on chip thickness, different combinations of epoxy resin materials can be used as shown in Figure 2. For thin chips, in the range of 20µm-50µm, a resin-coated-copper (RCC) layer can be laminated. Temperature and pressure profiles should be adjusted carefully to promote epoxy adhesion at all interfaces and avoid chip breakage. A thickness of 15-20 µm over the chip surface is desirable for the subsequent microvia opening and filling. The overlaying Cu layer serves as the base for package routing. Curing of the epoxies take places at about 185 °C for 60min. Detailed description of the lamination process is given in literature [10]. New developments in RCC laminates can improve significantly the adhesion on the chip surface and the reliability of the embedded packages by new epoxy formulations with adjusted thermo-mechanical properties. For 50µm thick chips as in the case of QFN packages, RCCs with 90µm-100µm thickness are being used. The RCC thermal expansion is in the range of 60-70 ppm/°C and thereby leading to significant package warpage after lamination due to thermal expansion mismatch with the underyling silicon and copper structures. Therefore, alternative dual-layer RCC have been manufactured with one or both layers reinforced with very fine 1015 type glass fibers in order to stiffen the thin package and minimize warpage. The bottom layer which actually surrounds the chips after embedding can have a much lower elastic modulus for better absorption of the thermomechanical stresses. Another material combination, especially appropriate for thick chip embedding (~250μm) but not restricted to that, is the use of pre-cut prepregs with cavities around the chips and on top another ply of prepreg without cavity. Prepregs having a high Tg point (180-190°C) with low thermal expansion coefficient in the range of 12-14 ppm/°C can be selected in order to offer higher themal stability and robustness compared to RCC materials. Currently, new production QFN runs have used both prepreg materials with cavities and conventional RCC materials in order to compare their reliability performance.

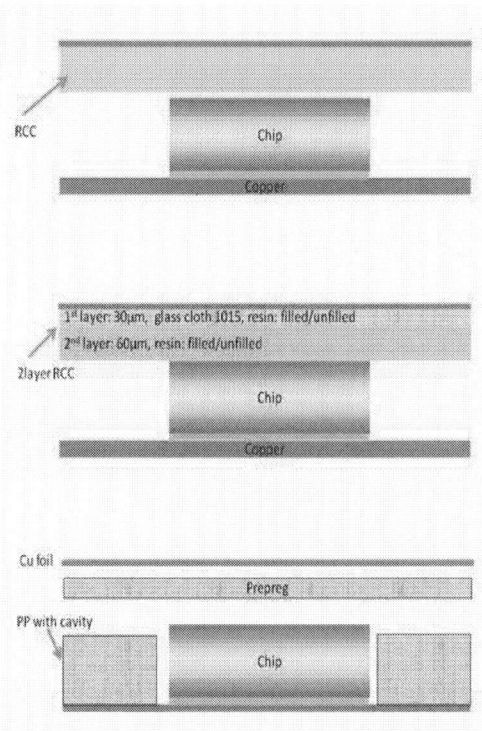

Figure 2. Different material combinations for chip lamination either with face-up or face-down embedding technologies.

Via opening and filling

In the face-up technology as shown in Figure 1(a), interconnections are achieved via microvia laser drilling to chip pads and subsequent metallization, similarly to the established formation of microvias on PCBs. For microvia drilling a pulsed 355 nm UV laser has been used. It can ablate Cu as well as the RCC dielectric (Figure 1 a). Accurate

alignment of the microvia drilling with respect to the underlying Cu pad pattern remains challenging for the yield of the interconnection. With I/O pad pitches between 120μm-150μm and enlarged Cu pads on the chips, alignment based on fiducials on each 10x10 cm² (sub-)panel are sufficient and a via diameter of 50μm can be feasible. For smaller pitches than 120μm, it is expected that local fiducials closer to each chip will be necessary. The I/O pitch for the QFN package is 100μm and that greatly challenges the via drilling with existing PCB UV laser drilling machines. Nevertheless, vias of 30μm could be successively opened on the enlarged Cu pads of 85μm. After drilling, a via cleaning (desmear) process is needed to remove remaining epoxy resin and sufficiently roughen the via wall to promote good adhesion of the subsequent electroplated copper. Sufficient conductive palladium is deposited on the epoxy surface which allows a direct copper via metallization without need of an extra 1μm electroless copper seed layer prior to electroplated copper. Figure 3 shows a 37μm filled via after desmear and copper electroplating. As evidenced in Figure 3, the achievement of the very small 37μm vias was crucial to meet the copper pads compensating for the slight misalignment of the vias with respect to the pads that has occurred.

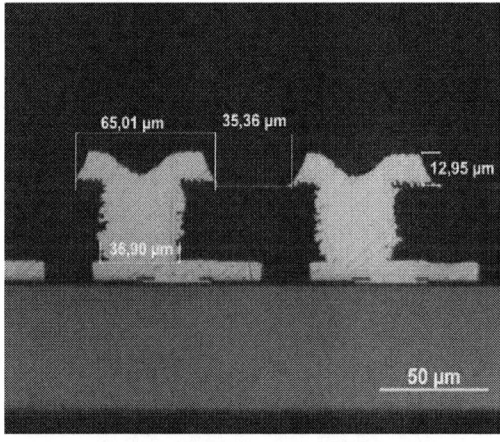

Figure 3. 37μm laser via at 100μm pitch after desmear and electroplating. (QFN package).

Fine line copper structuring for QFN package

A subtractive copper structuring is done for the realization of the QFN packages. The peripheral bond pad pitch of the daisy chain test chip of 100 μm results in a minimum 50μm L/S (line and space ratio) of the conductor lines. However, smaller spacing is required for the capture pads on the top of the package, which connect the micro via and the copper line, to guarantee a defect free connection to the chip. The structuring is done by the exposure of a negative resist mask using a laser direct imaging (LDI) followed by an acidic copper etching. Figure 4 illustrates the 50 μm line and space pattern after exposure and subtractive Cu etching. The inlet picture in Figure 4 shows in great detail that the capture pads should be enlarged and thus the spacing is reduced to 38μm.

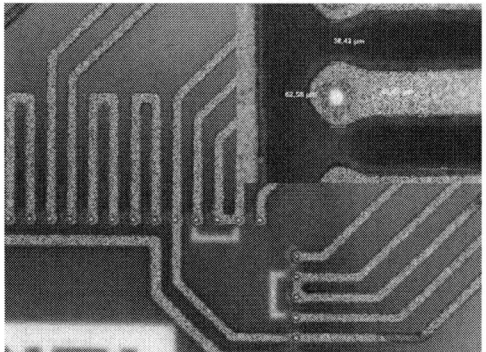

Figure 4. 50μm L/S for copper line structuring of the QFN-A package.

An electroless nickel and gold final finish is deposited on the QFN pads in order to preserve the solderability of the contact. Finally a pad defined solder mask is applied. A liquid solder mask material is applied by curtain coating followed by a lithographic structuring by LDI exposure and development. With this method it is possible to realize a solder mask thickness of about 15 μm. Figure 5 shows the final QFN package. The used die is 5mmx5mm in size and has a peripheral bond pad pitch of 100 μm. The QFN package has a standard pitch of 400μm and a size of 10mmx10mm.. The embedded die is 50μm thick and the final package thickness is 160μm. The QFN packages were produced in 10"x14" panel formats.

Figure 5. QFN-A package of 160μm thickness. (Package:10mmx10mm,Chip:5mmx5mm).

ULTRA FINE LINE STRUCTURING

Laser Direct Imaging (LDI) system enables the exposure of fine structures, by the use of a dry film photo resist and the exposure by direct writing with the UV laser beam. Currently available systems provide a resolution down to 15 μm lines and space on large substrate sizes with the possibility of local alignment. For embedding chips with pitches <100μm, very fine copper structuring is needed. The goal of the ultra fine line patterning is to produce structures down to 15 μm lines and spaces, which is also the maximum resolution of the LDI system. Such ultra fine line (UFL) structures can not obtained by subtractive technology and therefore a semi-additive technology is developed.

The adhesion of the resist necessitates the careful roughening of the copper foil top surface. The copper foil has a very smooth profile (~0.7μm) roughness. Figure 6 shows the 2μm foil developed in Hermes project for semi-additive technology. The inlet picture shows a 3μm resin adhesion promoter layer of which is coated in order to enhance the adhesion on the laminate material used as carrier for the development as well as the 2μm functional copper and the 70μm carrier foil which is peeled off after lamination.

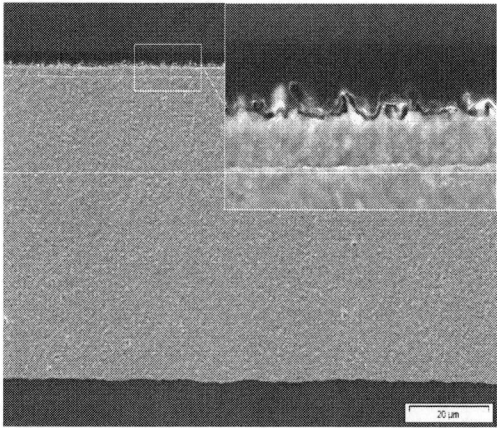

Figure 6. Very thin copper foil used for semi-additive technology(3μm adhesion promoter/2μm copper/70μm copper carrier foil).

A dry film resist of 15μm is laminated on top of the thin copper and after LDI exposure and development, the resist apertures are copper electroplated. Figure 7 shows 15μm L/S resist structures after development. After electroplating a ~12μm electroplated copper is deposited on top of the 2μm base copper. Figure 8 shows the 15μm L/S structures after resist removal. In the last step of differential copper etching, the copper base is etched away whereas the electroplated copper is also etched at a much lower rate. The lateral and vertical etching of the copper should be well controlled in order to obtain 15μm L/S. The semi-additive processing was performed in 610mmx457mm panels (24"x18") and the final 15μm UFL structures after differential etching were restricted to 228mmx305mm (9"x12") due to equipment

506

limitations. Detailed description of the semi-additive processing at IZM can be found in literature [9].

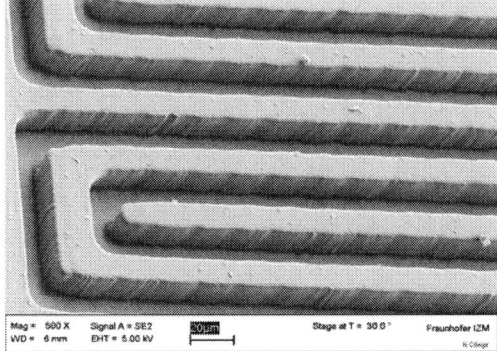

Figure 7. 15µm L/S resist structures after resist exposure and development.

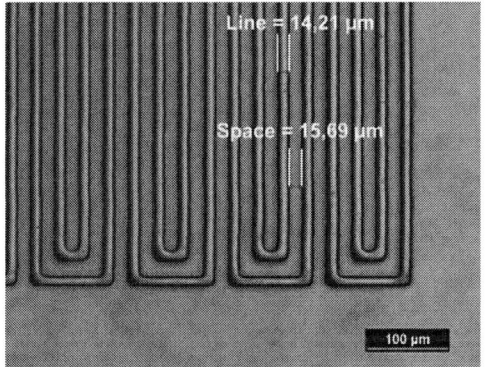

Figure 8. 15µm L/S copper structures after resist stripping.

FACE-DOWN EMBEDDING TECHNOLOGY

Ultra-fine-pitch embedding at 50µm pitch

The face-up technology with laser microvias can be used for embedding of chips with pitches up to 100µm due to limitations to very small via opening. For embedding of ultra-fine-pitch chips, IZM is in the process of developing a new "vialess" face-down technology where no vias will be needed for interconnection of the pads and thus it offers significant processing merits compared to the traditional face-down technology shown in Figure 1(b). The main idea is to reach "direct" contact of the chip copper pads with the copper foil by flip chip assembly using either glues like NCAs or even the B-stage resin of an RCC foil. IZM is currently working on alternative methodologies to assemble the chip face-down. One of these is described extensively in [9] where cavities in the thin 2µm foil should be etched until the carrier foil is reached and then the chip copper pads sit directly on the copper carrier foil. The assembly would be achieved using

NCA. Figure 9 describes another alternative where a very thin B-stage layer of 10µm is deposited on the 2µm copper foil as a normal RCC (stage1) and then an IZM 42 chip with 12µm thick copper pads at 50µm pitch is being positioned using temperature and pressure (stage 2). After prepreg lamination (stage 3), the carrier foil is peeled off (stage 4) and electroplated copper is deposited (stage 5). As a last step, direct semi-additive technology can be implemented for the final package structuring (stage 6). By using a B-stage RCC as the adhesive layer, the susbtrate is heated at about 100°C and the assembly head also at 130°C in order to reach the point where the B-stage resin becomes relatively fluid and then the chip is positioned with about 5 Kgr exerted pressure. The B-stage material can come in C-stage format by curing in the oven or later by lamination at 190° C. Figure 10 shows a cross sectional view of the embedded chip in prepreg layers. As it can be seen, the copper pad is only 1.7 µm away from the 2µm functional copper and experimental trials are currently underway to verify if chemical etching can remove the 1.7µm resin and in this way reach the chip copper pad. On the other hand, experimental work still focuses on achieving direct contact using a number of NCAs and assembly machines. The "vialess" face-down technology is still under development at IZM.

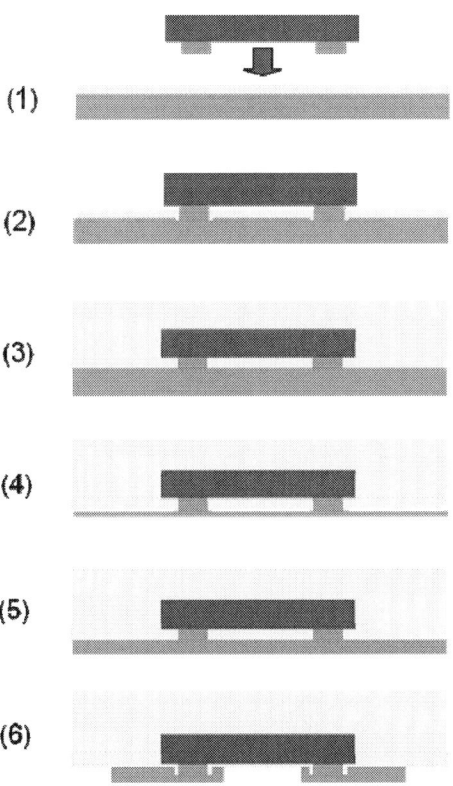

Figure 9. Process sequence for ultra fine line embedding using "vialess" face-down technology.

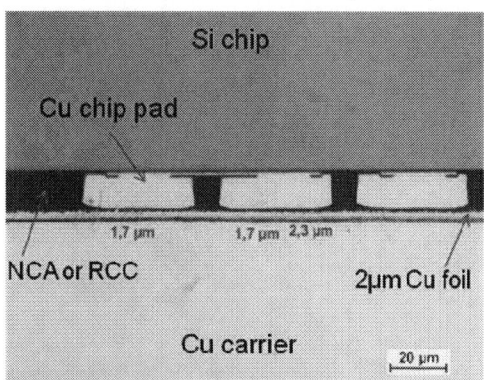

Figure 10. Ultra fine pitch embedding (50μm pitch) with face-down assembly using a B-stage resin or a NCA.

QUALITATIVE ANALYSIS ON EMBEDDED PACKAGES

Embedded packages with face-down technology

Embedded packages both produced with face-up and face-down technologies are subject to preconditioning and extensive reliability testing especially for examining the integrity of all constituent interfaces. For the face-down technology, the adhesion of the silicon backside/prepreg interface is tested before and after reliability testing whereas the delamination resistance is examined using acousto microscopy. Embedded face-down chips are subject to preconditioning according to JESD22-A113-E standard to reveal the effect of humidity on package stability. A gradual environmental loading was performed and thus samples: as produced, after baking, after MSL3 test (192h, 30°C/60%RH) and after MSL3 plus 3 x lead-free reflow cycles were tested in order to monitor any possible delamination evolutionary effect. All samples were examined by acousto microscopy. No delamination areas were detected for all samples even for the samples after MSL3 plus 3xreflows. Figure 11 shows indicatively an acousto micrograph of the silicon backside/prepreg interface of the sample after MSL3 plus 3xreflows. Weaving structure in Figure 11 comes is attributed to the glass fiber web in the prepreg. In addition, shear testing of the silicon/prepreg interface was performed and resulted in silicon breakage under 17 Kgr implying that the silicon/prepreg interface adhesion is very robust even after exposure of the samples on environmental loading.

Embedded packages with face-up technology (QFN)

The QFN packages presented in this study were preconditioned according to JESD22-A113-E standard. In the test frame, both MSL 1 (168h, 85°C, 85%RH) and MSL 3 (192h, 30°C/60%RH) tests were performed to increase the likelihood of failure precipitation. All QFNs showed warpage especially after MSL 1 testing and therefore the QFN package edges are not clearly shown in the acoustic microscopy tomographs. Tomographic views were taken from QFN surface, chip and DAF levels. No delaminations were detected at RCC/silicon, silicon/DAF interfaces. Figure 12 shows a typical acoustic tomograph from the resin/silicon interface. After MSL1 consitions only the DAF/FR4 interface shows locally DAF conglomeration but not obvious DAF/FR4 delamination.

Shear testing of the QFN samples was also conducted. For the samples without preconditioning, shearing at resin/chip interface results in local breakage of the resin over the chip surface under a total force of 8 kg whereas shearing at chip/DAF shows a final detachment of DAF/FR4 interface at 12 kg which evidences a very strong DAF/FR4 interface before preconditioning.

The samples after MSL3 preconditioning have exhibited lower strength as would be expected due to the MSL 3 testing. Shear testing at resin/chip interface has caused breakage of the resin only over the chip passivation area under 2 kg. Shearing at chip/DAF interface has resulted in chip breakage under a load of 4.5 kg. After MSL3 preconditioning, a partial DAF/FR4 delamination has been observed, however it does not extend on the entire chip bottom area.

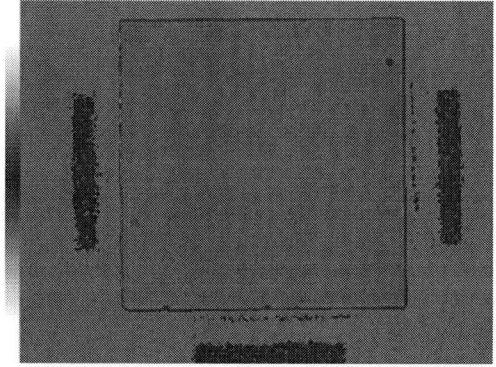

Figure 11. Acoustic tomograph of the Si/prepreg interface subject to MSL3 plus 3xlead-free reflows. No delamination is seen. Chip embedded with face-down technology.

508

Figure 12. Acoustic tomography of 2 QFNs after MSL1 test at chip/resin interface shows good adhesion of resin. Small white spots at edges are tiny voids on the chip edges.

Conclusions

This paper highlights the two main embedding technologies which essentially differ on the way the chip is assembled, either face-up or face-down, and how the interconnection to the pad is achieved. The face-up embedding technology has led to the emergence of a new prototype Embedded chip-QFN package with contact pads at 400µm pitch and a total number of 84I/Os with dimensions of 10mmx10mm. The total thickness of the package is 160µm. The embedded chip in the QFN package is 5mmx5mm in size and has a peripheral pad configuration at 100µm pitch. All Embedded chip-QFN packages have been manufactured in 10"x14" panels at prototype level. Subtractive copper structuring has been used for the QFN-A packages resulting in 50µm L/S. However, embedding of chips with smaller pitches than ~100µm will require also ultra fine routing at packaging level. Therefore, ultra fine line (UFL) technology has been developed on 610mmx450mm panels up to 15µm L/S copper structures. The optimisation on UFL technology continues focusing especially on resist stripping and differential etching. The face-down embedding is under development based on the concept of "vialess" embedding where direct contact of the chip pads with the copper foil is attempted using NCAs or B-stage RCCs. The final structuring is achieved with UFL package structuring. Qualitative analysis on QFN packages made by face-up technology and of embedded packages made by face-down technology has shown excellent reliability performance of all packages without evidence for delaminations at all constituent interfaces.

Acknowledgments

The authors would like to thank the European Commission for the financial support of HERMES project (FP7-ICT-224611).

References

[1] S.Norlyng, "Integrated component technologies: Opportunities, economics and trends", IMAPS Nordic, October 2003.

[2] Yole Developpment report, "Embedded Wafer-Level-Pacakges", 2010.

[3]. R. Schönholz, "Laser cavity, next step in evolution", Ruwel Symposium-PCB technologies, Kleve, 2010.

[4] Embedded active components and integrated passives: technologies and markets, TechSearch report, October 2007.

[5] T.Waris, R. Tuominen, J. Kivilahti, "Panel-sized integrated module board manufacturing", Proc. Polytronics Conference, Oct. 21- 24, 2001, Potszam.

[6] A.Ostmann, A. Neumann, P. Sommer, H. Reichl, "Buried components in printed circuit boards", Advancing Microelectronics, May/June 2005, pp. 13-18.

[7] L.Boettcher, D.Manessis, A. Ostmann, S. Karaszkiewicz, H.Reichl, "Embedded Chip Packages –Technology and Applications", Proc. International Surface Mount Technology Association Conference-SMTA 2008, Orlando, FL, August 2008, ISBN 978-0-9789465-7-9, pp. 43 – 48.

[8] D.Manessis, L.Boettcher, A. Ostmann, R. Aschenbrenner and H. Reichl, "Chip embedding technology developments leading to the emergence of miniaturized system-in-packages", Proceedings in the 60th Electronics Components & Technology Conference (ECTC) 2010, Las Vegas, NV, USA, June 1-4, 2010, ISBN: 978-1-4244-6412-8, ISSN: 0569-5503, pp. 803-810.

[9] L. Boettcher , D. Manessis , S. Karaszkiewicz, A. Ostmann, "Next Generation System in a Package Manufacturing ", Proceedings in International Surface Mount Technology Association Conference-SMTA 2010, Orlando, FL, USA, October 24-28, 2010, ISBN 978-0-9840562-3-1, pp. 222-228.

[10] D. Manessis, S-F.Yen, A. Ostmann, R. Aschenbrenner and H. Reichl, "Technical Understanding of Resin Coated Copper (RCC) lamination processes for realisation of reliable chip embedding technologies", Proc. Electronics Components & Technology Conference (ECTC) 2007, Reno, NV, May 29-June 1, 2007, pp. 278-285.

Technology development for a low-cost, roll-to-roll chip embedding solution based on PET foils

Maarten Cauwe [1], Bjorn Vandecasteele [1], An Gielen [1,] Johan De Baets [1],
Jeroen van den Brand [2], and Roel Kusters [2]

[1] imec - Cmst, Technologiepark 914a, B-9052 Zwijnaarde, Belgium
[2] Holst Centre, HTC31, Postbus 8550, 5605 KN Eindhoven, the Netherlands.

Tel.: +32-9-2645353, Fax: +32-9-29645374, Maarten.Cauwe@imec.be

Abstract

The aim of the research described in this paper is to develop a low-cost, roll-to-roll compatible process for the realization of electronic systems in foil using chip embedding. The small cost makes these systems suitable for disposable applications as food labels, medicine packages or smart bandages. Surface mount attaching of components on foils is a well-known process for building systems-in-foil. When using low-cost films like PEN and PET, there are serious restrictions on the maximum temperatures that can be used for the surface mounting process (soldering, adhesive bonding). Surface mounting has the additional disadvantage that the components are on the surface of the foil and are therefore not well protected mechanically and physically. The proposed process flow for embedding thin chips in PET foils overcomes these limitations. A key aspect of this technology is the application of a suitable adhesive to encapsulate the chips. The resulting product is based on full-metal copper which has a good thermal and electrical conductivity and allows for fine pitches. The process is compatible with several metal foils (Cu, Al ...), offering further possibilities in cost reduction, and does not rely on bumping of the chips or plating of the interconnections to the chips.

Key words: chip embedding, low-cost, roll-to-roll, PET

Introduction

Chip embedding is in essence a three-dimensional packaging technology that offers an alternative to wire bonding and flip-chip interconnects. Nowadays, two different categories of chip embedding technologies are commercially available. Fan-out wafer level packaging (WLP) enlarges the area available for interconnect routing on top of the die by embedding the chip in a polymer matrix [1], [2]. More system level approaches, such as the direct embedding of chips into the printed circuit boards (PCB), result in a System-in-Package concept [3], [4].

Flexible electronic products are mostly realized on polyimide with copper tracks and traditional pad finishes (NiAu, Ag, Sn...) because of the compatibility with "normal" assembly technology as e.g. soldering. The resulting flex foils are often stacked or folded to form three-dimensional constructions. These technologies and applications are still based on established printed circuit board (rigid and flex) processing. While the thickness of the dies embedded in rigid boards is usually around 100 um to 200 um, the embedding of active components inside a flexible circuit board relies on chips thinned down to 30 um and below. The ultra-thin chip package (UTCP) [5] is such a technology, where the chips are embedded in polyimide foils. The resulting package is only 60 μm thick. The chips are placed face-up on a polyimide layer and covered with a spin-on, photo definable polyimide. Vias to the chip contacts are realized using photo lithography and metallized by sputtering and galvanic plating. The embedded die can be used as a package, e.g. solder balls can be placed on the contacts and the package can be solder assembled on interconnection substrates, or further embedded between the layers of a flexible circuit board [6].

Apart from this polyimide-based flex technology, there are also developments on low-cost substrate materials. A very important application in this area is RFID and smart card technology. The chips used here are typically around 100 μm thick, have a small number of I/Os (2 – 8) and a large pitch (> 0.2 mm). They are connected using wire bonding or flip-chip to printed or plated conductors. The Holst Centre in Eindhoven focuses mainly on these low-cost substrates, where the use of polyester-based materials imposes additional challenges as the temperature budget for the interconnection process is limited. Current research topics include the assembly of packaged chips and bare dies on PET or PEN films with silver or copper interconnections (Figure 1), low-cost interconnections for OLEDs and the embedding of chips in PET foil, which is discussed in this paper.

Figure 1: Thin die assembled on Cu/PET foil

The most quoted advantages of chip embedding include the reduced interconnection length and freeing up board space. Due to the limited frequency range employed on polyester-based substrates and the low material cost, these topics are less of a concern for Systems-in-Foil. The possibility of thin and flat systems with improved reliability thanks to the better mechanical and physical protection of the die, as opposed to surface mounting where the die protrudes from the foil, are much more of a driving force for chip embedding in foil.

The aim of this research is not just to develop a chip embedding process based on PET films, but to realize a roll-to-roll compatible process with maximum cost reduction. Using low-cost material is a first step towards lower cost, but combining these materials with expensive processing steps is counterproductive. Thin-film processing and chemical or galvanic deposition should be avoided. If possible, bumping of the chips will be avoided, although the cost for bumping is small compared to the chip cost.

The process flow for chip embedding in PET foil is described in the next section, followed by a detailed overview of possible adhesives for encapsulating the chips. After a short discussion on chip assembly, this paper concludes with some cost considerations.

Process Flow

To avoid thermo-mechanical restrictions during die placement, the process starts from a single copper foil. The first step of this embedding technology is the application of the die bonding adhesive. One option is screen printing of isotropic conductive adhesive bumps and dispensing of non-conductive adhesive in between the bumps. Uniformity of the bump height and the minimal attainable pitch are important parameters of the printing process. During placement of the thinned chips, the pads on the chip are aligned to the bumps on the foil, realizing the interconnection between the

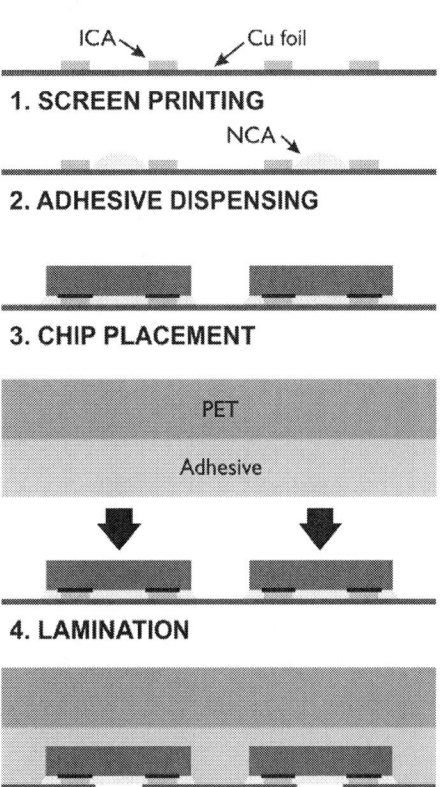

Figure 2: Process flow for chip embedding in PET foil

copper foil and the chip pads without the need for bumping the chip. As a backup solution for ultrafine-pitch applications, an anisotropic conductive adhesive could be used in combination with bumps on the chips. The actual embedding of the chips is performed in a lamination step using PET film and a suitable adhesive[1]. Void-free lamination, ensuring a good encapsulation of the chips, is a crucial step in the chip embedding process. In a final step, the copper is structured using conventional PCB processes or alternative patterning technologies to define the circuitry. The end result is a thin foil with embedded components and copper interconnects. Figure 2 shows a schematic overview of the process flow. Typical dimensions are a chip thickness of 30 μm, a 20 μm die bond adhesive layer, and a 50 μm thick PET film. The adhesive surrounding the chips will be about 70 μm to 100 μm in total, resulting in an overall thickness of less than 200 μm. For increased mechanical reliability, an additional cover layer can be laminated onto the copper side after structuring, moving the circuit closer to the neutral axis.

---

[1] Unless explicitly mentioned otherwise, the word "adhesive" in the following always refers to the adhesive used to encapsulate the chips.

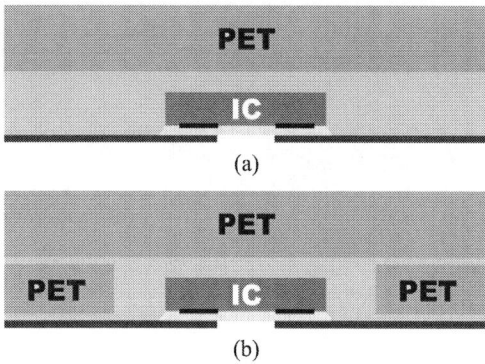

Figure 3: (a) Embedding using thick adhesive (b) Cavity-based embedding

The use of a full-metal foil not only provides better thermal and electrical conductivity, but also allows for finer pitches than an additive printing process. Surface mount assembly on PET or PEN is severely restricted by the maximum temperatures that can be tolerated by the materials. Die placement on a full-metal foil does not suffer from these restrictions, enlarging the selection of possible die bond adhesives. The process flow is compatible with other metals, as for example lower-priced aluminum, and other polymer substrates (PEN, PI, PUR …). A disadvantage for mass manufacturing is that just like in conventional surface mount technology each component needs to be placed and interconnected individually.

The adhesive used to encapsulate the chips is a critical aspect of the technology. Due to the combination with PET film, the processing temperature is limited to a maximum of 120 °C for a few minutes. The need for good adhesion to copper and PET is evident, whereas a good flow behavior is vital to successfully enclose the chips. Since the adhesive cost is a dominant factor in the overall material cost, primarily general-purpose industrial adhesives are considered. A this point in the process development, no additional specifications (thermal conductivity, ionic content, moisture permeation) requiring a higher grade adhesive are encountered. The resulting gamut of possible candidates is very broad, ranging from pressure sensitive tapes, over thermoplastic or thermosetting film adhesives, to UV curing liquid adhesives.

Depending on the available thickness and the cost of the adhesive, two options for embedding the chips in between the copper foil and the PET film are presented. The first possibility relies on the use of a thick adhesive layer, which is simply pressed onto the chips mounted on the copper foil (Figure 3a). The advantage of this approach is that the process is simple and straightforward. Since a thick layer of adhesive is needed to cover the height difference of 50 μm formed by the chip and the die bond adhesive layer, the bulk material cost of the adhesive needs to be very low. The alternative is to use thin layers of adhesive coated onto PET film and create cavities at the intended locations of the chips (Figure 3b). A second coated PET film is laid on top of the first and during lamination both adhesive layers flow together to fill the cavity. While this approach saves on adhesive cost, or allows for the use of a more robust adhesive at the same price, the process flow is more complex and a certain alignment is needed during layup or lamination. A third option, where only an adhesive layer without additional PET film is used to embed the chips, imposes additional requirements on the adhesive, as for example all the mechanical strength needs to be carried by the adhesive. The top PET film in option one or two can also contain certain functionality, resulting in a higher integration density for more complex Systems-in-Foil.

Material Evaluation

Since the preferred class for a suitable adhesive for chip embedding is general-purpose, industrial adhesives, instead of adhesives tailored for electronic applications, and no detailed set of specifications is available, the possible candidates are numerous. The chosen approach is to evaluate different types of adhesives based on processability, cost, performance and reliability.

Due to the restrictions in processing temperature, pressure sensitive double-sided tapes are a logical choice. A void-free encapsulation of the chips with these tapes requires a thick adhesive layer which is sufficiently soft to "flow" around the chips. A drawback of this type of adhesive is that it remains sticky, which might be an issues for the areas that are exposed after copper structuring.

The ideal adhesive for this technology would be a film adhesive which becomes liquid at a temperature below 80 °C, flows around the chips and subsequently solidifies. This could be either a thermoplastic or thermosetting polymer, although the latter would accommodate a low processing temperature without restricting the operational temperature of the circuitry. Alternatively, a liquid adhesive can be coated over the chips and cured during or after lamination of the PET film.

A total of seven different adhesives are evaluated: two pressure sensitive tapes with different softness ("PSA 1" and "PSA 2"), two thermoplastic

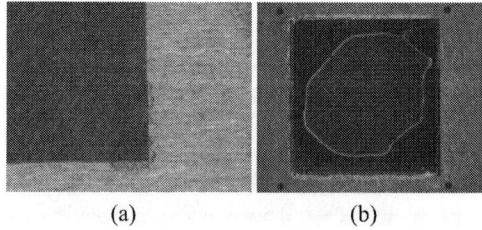

(a)   (b)

Figure 4: (a) Air bubbles entrapped between the edge of the chip and the pressure sensitive tape; (b) Insufficient filling of cavity using thermoplastic adhesive

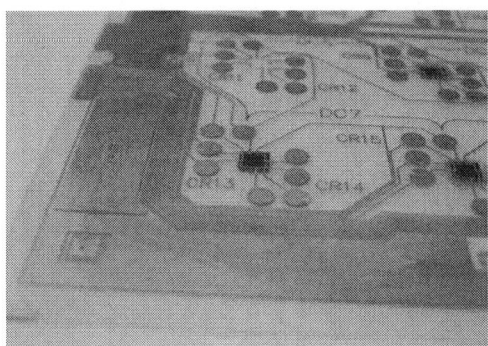

Figure 5: Chip embedding test vehicle (PSA 1)

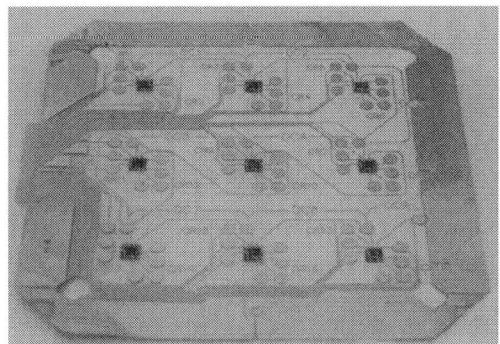

Figure 6: Chip embedding test vehicle (TP 1)

adhesives coated onto PET film ("TP 1" and "TP 2"), two heat-activated film adhesive with and without non-woven carrier ("HA 1" and "HA 2") and finally, a liquid adhesive that becomes pressure sensitive after UV exposure ("UV 1"). The processability and performance of these adhesives is evaluated by embedding dummy chips mounted onto copper foil. While roll lamination would be the method of choice for a roll-to-roll compatible process, all material evaluation experiments are performed using sheet lamination in a vacuum lamination press (*Lauffer RLKV 25*).

PSA 1 and the softer PSA 2 are double-sided tapes with an adhesive thickness of 130 μm, necessitating a two-step lamination process. In the first step, the tape is laminated to the PET film using a pressure of 1 bar at 40 °C for 5 minutes. After removing the second release liner, the PET + adhesive is laminated onto the copper foil with assembled chips. For this second lamination a slightly higher pressure (5 bar) and longer dwell time (10 min) are applied to assure a good encapsulation of the chips. As is the case for all film adhesives, the sheet lamination makes it difficult to avoid entrapped air. Contrary to the expectations, the use of the softer PSA 2 leads to more air bubbles, both around the chips as across the foil (Figure 4a). Cracking of the chips did occur when areas underneath the chip were not completely filled by the die bond adhesive, predominantly at the corners of the chips. Further optimization of the die bonding process will mitigate this issue. Structuring the copper reveals that the stickiness of the exposed area of the adhesive is noticeable, but not critical.

The heat-activated adhesives HA 1 and HA 2 follow a similar process flow, although the first lamination step is replaced by a manual application of the adhesive to the PET film using a hand roller. The thickness of the baseless HA 1 is only 50 μm, so two layers of adhesive are sequentially laminated to the PET film. HA 2 contains a non-woven cellulose carrier and has a total thickness of 90 μm. These adhesives show a reasonable amount of flow, but are designed to start curing in seconds when reaching temperatures above 100 °C, giving the adhesive

insufficient time to flow around the chips. A higher pressure and a slightly reduced temperature result in a void-free embedding without too much increase in processing time.

The low coating thickness of the thermoplastic adhesives TP 1 and TP 2 (25 μm and 14 μm, respectively) requires the use of cavities to overcome the height difference between the assembled chips and the copper foil. This complicates the process, since additional steps need to be introduced to cut out the cavities and to assure a good alignment of the cavities to the chips. The cutting of the adhesives coated on 50 μm PET film is done using a CO_2 laser (10.6 μm). On each side, the cavities are 100 μm larger than the chips and the parameters of the laser are optimized to obtain a well-defined cavity edge without affecting the flow behavior of the adhesive. The latter is very important for a successful embedding (Figure 4b). During the first trials, the fillet caused by the die bond adhesive at the edge of the chip appeared to be wider than 100 μm. As a result, the cavities would not fit around the chip and needed to be enlarged. With a margin of 200 μm and a better control of the die bonding process, a sufficient filling of the cavities was obtained. A dwell time of 20 minutes at 120 °C was needed to obtain these results, making the process not only more complex but also very slow. Due to the lengthy exposure of the PET film to elevated temperatures, a noticeable shrinkage occurred, resulting in severe warpage of the samples. The advantage of the thermoplastic materials is that insufficient filling of the cavities can be fixed by again heating the adhesive to above its melting point.

The last material under evaluation is a liquid, UV sensitive adhesive, which is applied over the chips by doctor blading. Care needs to be taken to allow sufficient spacing between the blade and the top of the chips. After applying the adhesive, the samples are exposed to a UV dose of 2 mJ/cm^2 and become tacky. A PET film is laminated on top using a pressure of 5 bar at room temperature for 5 minutes. First tests revealed that when the adhesive layer is too thin (100 μm or less), the UV-initiated

Table 1: Summary of material evaluation results

Mat.	Proc.	Cost	R2R	Perf.
PSA 1	+	++	++	+
PSA 2	+	++	++	+
HA 1	+	+	++	+
HA 2	++	+	++	+
TP 1	--	+	-	++
TP 2	--	+	-	++
UV 1	-	-	+	+

reaction does not continue to completion, resulting in insufficient adhesion to the PET film. Again, the exposed areas of the adhesive after structuring remain sticky.

Each of these adhesive types has its own advantages and drawbacks, making it difficult to identify a clear winner at this point. A single-sheet film adhesive with a one or two-step lamination would offer the best compromise between processability, cost and performance. Table 1 gives a summary of the behavior of the adhesives, evaluated in different categories: processability, including handling and complexity of the process flow, material cost, roll-to-roll compatibility of the process, and embedding performance. No lifetime or mechanical testing has been performed, so the reliability of these materials cannot be evaluated at this point.

Overall, the pressure sensitive tapes and the heat-activated films score the best. The embedding performance of the thermoplastic adhesives is very good, but the slow and complex process makes these materials less appropriate. The liquid adhesive does not have a clear advantage over the film adhesives due to the thickness requirements for this particular adhesive; although other types of liquid adhesives (e.g. thermosetting) could show better performance. Figure 5 and Figure 6 show chip embedding test vehicles using pressure sensitive tape and thermoplastic adhesive, respectively.

Assembly test run

To verify the contact resistance between the aluminum bond pads on the chip and the lands on copper foil, a dedicated test vehicle was designed containing daisy chain and four-point measurement test structures. This test design is realized on a single-sided polyimide flex (50 um PI and 18 um ED copper). The choice for PI was made to circumvent any temperature restrictions during die bonding, making it possible to use the same parameters as for bonding onto the copper foil. Two series of tests are performed. The goal of the first series is to determine the contact resistance at the copper side. To assure a good contact at the chip side, chips with Ni/Au bumps are used. The chips for the second series of experiments are not bumped, focusing on the contact between the isotropic conductive adhesive and the aluminum bond pads.

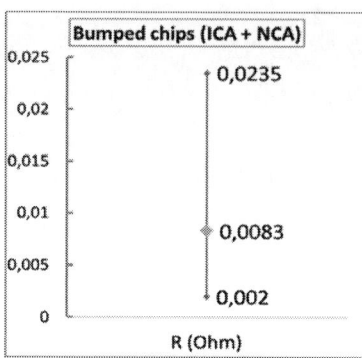

Figure 7: Contact resistance for bumped chips. (18 contacts measured: 6 showed no contact, 1 was non-ohmic, 1 had a resistance of 0.4 Ω, and the remaining 10 are included in the graph)

As expected, the series with bumped chips shows a very low contact resistance (Figure 7), while the non-bumped chips demonstrate non-ohmic contact resistance due to the presence of aluminum oxide on the chip pads. Some extra measures can be taken to remove this oxide prior to chip placement, although the long-term reliability without some form of bumping will remain a concern.

A complete filling of the die bond adhesive underneath the chip with a minimum fillet at the edge and a uniform thickness are essential to avoid damage to the chips during embedding.

Cost Considerations

The technology for embedding active components in foils is still under development, so a complete cost modeling cannot be performed at this time. Some general considerations concerning material cost can give an idea if this technology can truly be labeled "low-cost". For the calculation, a high-volume production line with a yield Y, manufacturing over 1 million units per year, is assumed. Each unit is tested after production and only good units are shipped.

A possible application for this technology is the smart blister, a medicine package which is meant to improve therapy compliance. The smart blister detects if and when a dosage is taken out of the blister, processes this information, and communicates the action to the environment (patient, doctor, nurse etc.), thus supporting a well-controlled medicine taking process. Typical dimensions for these circuits are 5 cm by 10 cm, including a single chip (1 mm x 1 mm) to collect and transmit the data. The material costs for a 50 µm thick PET film is around € 1.0/m^2 and € 3.0 for one square meter of untreated 20 µm rolled-annealed copper foil. The cost for the adhesive used for embedding is targeted to be between € 1.0/m^2 and € 5.0/m^2. This results in an overall material cost of € 5.0/m^2 to € 9.0/m^2. The smart blister contains one chip per module and with the given size, 200 modules can fit in one square meter. Even when the price of the chip would be as low as € 0.1 per chip, the chip cost per square meter quickly rises to € 20.0/m^2. Bumping and thinning would add another € 0.0025 and € 0.006 per chip, respectively, or € 0.5/m^2 and € 1.2/m^2. Combining the different cost contributions, excluding the processing cost, leads to a total cost of € 26.7/m^2 to € 30.7/m^2.

Expressing the chip-related cost in euro per square meter, albeit unusual, helps to compare the different cost factors. The cost of the chip is clearly the dominant factor, while the cost of chip thinning for this application is similar to the material cost of the PET. The cost calculation is based on 200 modules per square meter, resulting in a price per unit of between € 0.13 and € 0.15 divided by the yield Y. As mentioned before, the cost reduction from developing a process without the need for chip bumping would be minimal. On the other hand, lowering the cost of the adhesive from € 5.0/m^2 to € 1.0/m^2 reduces the overall cost by more than 13 %.

A comparable process using spin-on polyimide to coat a layer of 100 μm over the chips would cost € 192.7/m^2. For this competing technology, the material cost (€ 171.0/m^2) would be far greater than the chip cost (€ 21.7/m^2). A more correct comparison would need to take into account the difference in processing cost and the integration density of both technologies.

The cost of more advanced modules with several embedded dies is completely dominated by the chip cost. For a body area network patch of 3 cm by 5 cm including three chips, the silicon cost would be more than 99 % of the total material cost. As such, the low-cost aspects of the proposed technology become more apparent for large substrates with one or two small chips. Changing to a lower cost material, however, will only lead to an overall cost reduction if the yield and reliability are not reduced in the same way.

Conclusion

Developing a chip embedding solution based on PET foil requires a completely different approach than traditional chip embedding in rigid or flexible substrates. A simple and cost-efficient process is crucial. Requirements concerning electrical and mechanical reliability will depend strongly on the intended application. These would not only be situated in medicine packages and smart bandages, but also in areas where chip integration is not considered at the moment.

Future work will include further optimization of the process using the current materials, but also investigating alternatives as hot melt adhesives and pressure sensitive adhesives that continue to crosslink under heat exposure. Increasing the reliability without raising the cost will prove the main challenge for the future.

References

[1] M. Brunnbauer, E. Furgut, G. Beer, and T. Meyer, "Embedded wafer level ball grid array (eWLB)," in Proc. 8th Electron. Packag. Technol. Conf., Singapore, Dec. 2006, pp. 1–5.

[2] B. Keser, C. Amrine, T. Duong, S. Hayes, G. Leal, W. Lytle, D. Mitchell, and R. Wenzel, "Advanced packaging: The redistributed chip package," IEEE Trans. Adv. Packag., vol. 31, no. 1, pp. 39–43, Jan. 2008.

[3] P. Palm, J. Moisala, A. Kivikero, R. Tuominen, and A. Iihola, "Embedding active components inside printed circuit board (PCB)—A solution for miniaturization of electronics," in Proc. 10th Int. Symp. Adv. Packag. Mater.: Processes, Properties and Interfaces, Irvine, CA, Mar. 2005, pp. 1–4.

[4] A. Ostmann, J. De Baets, A. Kriechbaum, H. Kostner, and A. Neumann, "Technology for embedding active dies," in Proc. 15th Eur. Microelectron. Packag. Conf., Brugge, Belgium, Jun. 2005, pp. 107–110.

[5] Govaerts, J.; Bosman, E.; Christiaens, W.; Vanfleteren, J.; , "Fine-Pitch Capabilities of the Flat Ultra-Thin Chip Packaging (UTCP) Technology," IEEE Trans. Adv. Packag., vol.33, no.1, pp.72-78, Feb. 2010.

[6] W. Christiaens, T. Loeher, B. Pahl, M. Feil, B. Vandevelde, and J. Vanfleteren, "Embedding and assembly of ultrathin chips in multilayer flex boards," Circuit World, vol. 34, pp. 3-8, 2008.

Novel Approach for Integrating Electronics into Textiles at Room Temperature using a Force-Fit Interconnection

Erik P. Simon[1a], Christine Kallmayer[2], Rolf Aschenbrenner[2], Klaus-Dieter Lang[1]

[1]Technische Universität Berlin
Straße des 17. Juni 135
10623 Berlin, Germany

[2]Fraunhofer IZM
Gustav-Meyer-Allee 25
13355 Berlin, Germany

[a]simon@tu-berlin.de
+49 (0) 30 46403-652

Abstract

Integration of electronics into textiles needs two connection steps. On the one hand, the mechanical connection to a textile material and on the other hand, the electrical connection to integrated conductive structures. Both connections have to be reliable. In the last ten years different technologies were used for this task. Almost all of them need a temperature step either for melting or curing. This work presents an alternative method that can be applied at room temperature. It is simple, fast, and cost-efficient. The conducted experiments analyze the force-dependent contact resistance between three different conductors and metal strip lines of different widths. As a result conductive yarns need much higher contact forces to gain low and stable contact resistances. This is due to their particular configuration and thin peripheral conductive layers. Wider strip lines will enhance their contact resistances. Cross-section analyses of loaded yarns together with the measured data allow for these interpretations. A design of a force-fit interconnection as well as a prototype is presented. Two plates are clamped to a piece of fabric with integrated conductive yarn. The developing elastic force is directed to the contact members. A final thermal cycling test between −40 °C and +85 °C for 1000 hours confirms the applicability of force-fit interconnections for smart textiles.

Key words: Smart Textiles, Force-Fit Interconnection, Contact Resistance, Conductive Yarn

Introduction

The integration of electronics into textiles for various applications is gaining more and more interest by research facilities and companies, alike. The Fully Integrated EKG-Shirt [1] or the "Klight" [2] are examples for smart wearable textiles that attracted attention early on. Moreover, functional textiles have high potential for smart textile applications. The added value will largely increase the profit margin of these products. One example is a large area monitoring textile for dikes. A large woven fabric is equipped with a number of sensor nodes, which record movement, forces, and humidity, for example. These electronic modules have to be fixed onto the fabric and electrically connected to integrated conductive structures. The integration of electronics into textiles is mainly dealing with this interconnection task.

In recent years, a number of different integration and interconnection technologies have been presented. These have either been evolved from the textile field like embroidery of conductive yarns, or from electronics like soldering or the use of conductive and non-conductive adhesives. However, most of the technologies have one major drawback that is the need for a temperature step.

Mechanical connections can be made at room temperature. Hence, the objective of this work was to analyze the applicability of force-fit interconnections in the field of smart textiles.

State of the Art

The textile integration consists of two interconnection steps. The mechanical connection fixes the modules to the fabric and provides stability. The electrical connection allows for the flow of current. Certainly, it is desirable to combine both interconnection steps into a single one. Additionally, this processing step should be simple and use low-cost equipment to manufacture reliable contacts. An integration of the process into the textile production chain is favored.

The first interconnection technologies for textile integrated electronics were shown by M. Orth in the year 2000. The conductive yarn was knotted to an electronic circuit board by hand. [3] Later T. Linz presented the first machine embroidered electronic modules using conductive yarn. The embroidery needle pierced the conductive pad and the tightening of the yarn loop established an electrical and mechanical connection at the same time. [4] However, later it was shown that an embroidered yarn will not withstand thermal reliability tests without additional protection. The interconnection will lose its contact force and the resistance will rise up to an interrupt. [5] In 2010, T. Linz presented the use of a thermoplastic non-conductive adhesive to bond electronic modules to textile integrated conductive structures. [6] This technology is based on an adhesive layer that will melt at a certain temperature. During the process the applied pressure establishes the electrical contact. The process proved to be very reliable up to temperatures close to the glass temperature of the used adhesive.

Besides these technologies different people used standard soldering techniques as well as isotropic conductive adhesives. Soldering is performed at temperature levels that are critical for polymer based yarns. Furthermore, wetting was found to be bad on these yarns. Conductive adhesives provide good conductivity but also need a temperature step for curing. In the cured state they are brittle and cannot follow bending of flexible textiles and conductive yarns.

The following figure shows a sketch of the interconnection technology that is presented in this work.

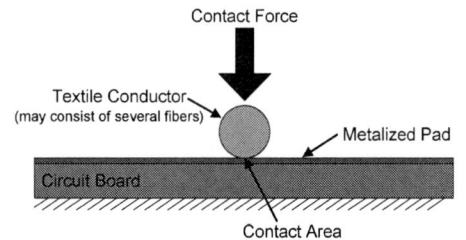

Figure 1: Sketch of an electrical contact between a textile conductor and a metalized pad due to an applied contact force

Contact Theory

From contact theory it is known that current is allowed to flow from one contact member to another as soon as their distance is in the order of the lattice spacing. This, however, is only true for perfectly smooth and clean surfaces. In reality surfaces are rough on the microscale and will be covered with tarnishing films like contaminations, oxides, sulfides, etc. These two effects lead to an increased resistance value at the contact member's interface – the total contact resistance R_t. [8]

$$R_t = R_c + R_f \qquad (1)$$

The rough surfaces only touch at a few certain spots distributed over the whole area. Hence, current density is inhomogeneous and the current is constricted in these small areas called a-spots. The size of the spots and their number depends on the contact force, i.e. the force which acts on the contact members. Plastic deformation leads to larger a-spots and merging of neighboring ones. The constriction resistance R_c considers this effect.

All metal surfaces will be covered with films except in an ultra-high vacuum. These films can be very thin and only slightly change the resistance. However, they can also be very thick and insulating. In that case they have to be removed or displaced to allow a successful electrical contact. This can be done by foregoing cleaning of the contact members or by self-cleaning during the interconnection due to plastic deformation and abrasive forces between the contact surfaces. [7] This effect is embodied in R_f.

Holm [8] found the following rule to be true for typical metal contact members. The constriction resistance of a single a-spot of two contact members of the same material is given by:

$$R_c = \frac{\rho}{2a} \qquad (2)$$

This equation, however, only contains the resistivity and size of a single a-spot. As has been said, the resistance also depends on the applied force as the number and size of a-spots changes. To account for that the following hyperbolic relation was found in [7]:

$$R_c = \frac{(\rho_1 + \rho_2)\sqrt{\pi H}}{4L(n/L)_0} F_{contact}^{-n} \qquad (3)$$

For typical contacts the exponent n lies between 0.9 and 1. Smaller values than 0.9 indicate surface contaminations of the contact areas.

However, the presented rules were found theoretically for solid metal bodies. In reality smart textiles will contain conductive yarns, which are primarily made of much softer polymers. The percentage of metal is low compared to the total volume hence the yarn will have the mechanical properties of a polymer. Additionally, the special configuration of the yarn incorporates lots of clearance, which makes the whole system even more compressible.

The resistivity is also much different from a solid conductor. The conductive part of the yarn is a thin peripheral silver layer of a few hundred nanometers. Firstly, the current needs to flow from one fiber surface to another in order to reach the contact pad. This represents a series resistance of fiber-to-fiber contact resistances, which also changes with the applied force. And secondly, the value of the constriction resistance will be different as the silver layers are very thin and non-uniform.

Hence, it can be expected that the relation between resistance and force will deviate from the analytical equations (2) and (3). However, the authors believe that the hyperbolic relationship between both values is also found here.

A SEM image and a cross section view of a conductive yarn are shown in figure 2 and in figure 3, respectively. The ratio of the silver layer to the polyamide core, the twisting, and the clearance between fibers is visible.

Figure 2: SEM overview of a conductive yarn

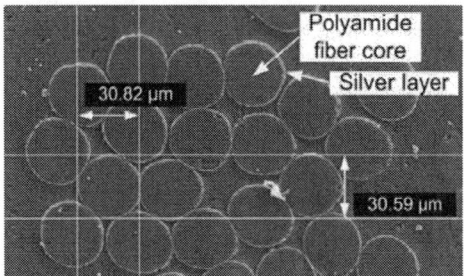

Figure 3: SEM cross-section view of a conductive yarn; the thin silver layers and clearance between fibers is visible

Contact resistance measurements of some assemblies were conducted in the past. The main interest, however, lay on the relative change of the contact resistance. It is a good indicator for the reliability of a smart textile during various tests, e.g. temperature cycling. Furthermore, it helps identifying potential failure mechanisms.

This work analyzes the contact resistance independently from the used technology. The results can be used for designs of embroidered-, force-fit-, adhesive-, as well as crimped-interconnections. It helps to define conductors, geometries, and forces to be used in the system.

Experiments

Figure 1 shows a sketch of the experiment conducted in this study, which has been derived from Holm's crossed rod experiment [8] to measure contact resistances of solid metal rods. Here, the focus was on the contact resistance between conductors for smart fabrics and contact pads of electronic circuit boards.

The test matrix consists of three conductor types and four different strip line widths of the contact pad. The widths are 50 µm, 100 µm, 250 µm, and 500 µm. The strip lines are made of 18 µm thick Cu, Ni, Flash-Au on a 500 µm thick FR4 circuit board. The widths are varied to study the area dependence of the contact resistance. A batch of test boards is shown in figure 4. A new board is used for each measurement. For the tested conductors two conductive yarns and one stranded wire are selected. As has been said, conductive yarns have the mechanics of ordinary yarns but also provide electrical conductivity, although it is much lower compared to metal conductors. They are widely used in the field of smart textiles. To be able to compare

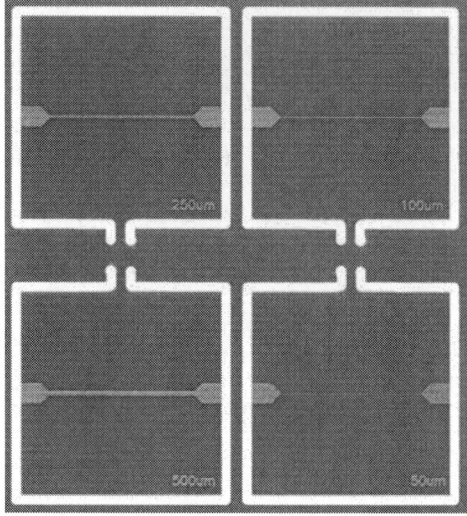

Figure 4: Test structures with four different line widths of 50 µm, 100 µm, 250 µm, and 500 µm

them in terms of contact resistance a metal conductor is selected as well.

Following, the three conductors are specified. Both conductive yarns are silver metalized nylon yarns. It should be mentioned that there are many other types of conductive yarns available. For more information it is referred to [9].

Shieldex 117/17 2-ply by *Statex* is a silver metalized spun nylon yarn. It consists of two twisted strands of 17 fibers each. The single fibers are about 30 µm in diameter and equipped with a thin silver layer of a few hundred nanometers. The conductivity is 349 Ohm/m with a standard deviation of 15 Ohm/m [9]. This particular yarn is shown in figures 2 and 3.

Shieldex 235/34 2-ply HC + B is an advanced version of that yarn with two strands of 34 fibers each and increased conductivity. The letter *B* stands for an additional layer to protect the silver metallization against wear. The conductivity was measured with 90 Ohm/m with a standard deviation of 4 Ohm/m.

The third conductor is a silver plated stranded copper wire from *AlphaWire*. It is an AWG 32 (Ø 202 µm) type with seven AWG 40 (Ø 80 µm) strands. The conductivity is 0.6 Ohm/m. The wire has a PTFE insulation that was removed prior to the measurements.

The experiment is conducted with a *Zwick* tensile tester that operates in compression mode. A circuit board is placed on a steel chuck and mechanically fixed. The conductor is placed perpendicular to the strip line onto the board. It is also fixed but without applying tension. A steel die of 15 mm in diameter is applying the load. All machine parts are electrically insulated to avoid short-circuiting. The die moves with 5 µm/s until the force limit of 100 N is reached. The test assembly is shown in figure 5.

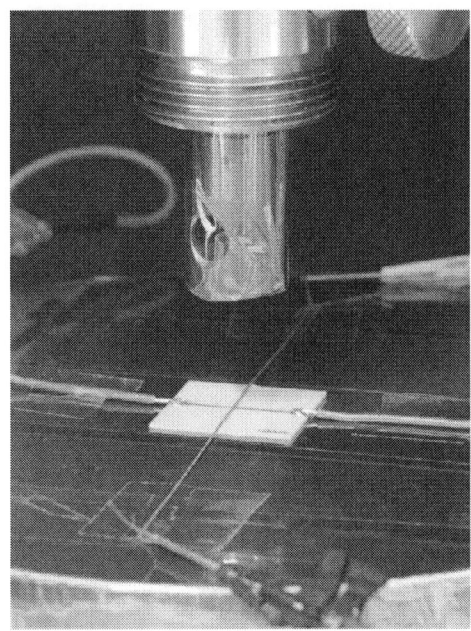

Figure 5: Contact resistance test assembly; Cu-Ni-Au strip line on FR4 and crossing conductor; movable steel die on top

The data recorded are the contact resistance and the force signal of the tensile tester. To be able to only measure the contact resistance without parasitic resistances a four-wire measurement is applied by using a constant current source with 10 mA output and a 1 MHz transient recorder (*TransCom-RackX*) to measure the voltage. The force signal is also recorded by the transient recorder.

The following data processing step consists of averaging 1000 readings to reduce the sampling rate to 1 kHz and generate a smoothened and strictly increasing force signal. Thereafter, the resistance can be plotted over the applied force data or the used strip line widths.

In order to generate meaningful results always five measurements for each of the 12 combinations are run resulting in 60 data sets.

Results

All data is presented in figures 6 to 8. These allow for an easy comparison of the curves of the three conductors. The mean contact resistances of five readings are plotted over the logarithmic contact force for the four strip line widths. Additionally, table 1 is generated from the three plots and contains specific data sets. The mean values from all measurements for each conductor are shown for the forces of 1 N, 10 N, and 100 N. Additionally, the relative changes to the 1 N and 10 N values are given. Finally, figure 9 shows the influence of the strip line width for the maximum applied force of 100 N.

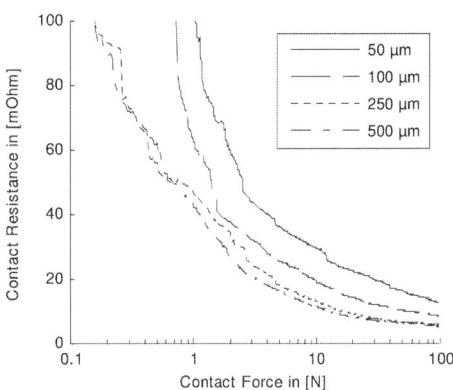

Figure 6: Contact resistance of *Shieldex 117/17 2-ply* over contact force for four strip line widths

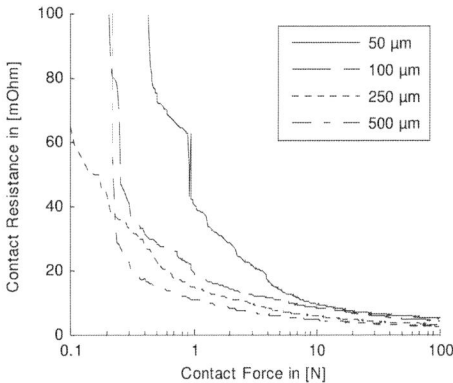

Figure 7: Contact resistance of *Shieldex 235/34 2-ply* over contact force for four strip line widths

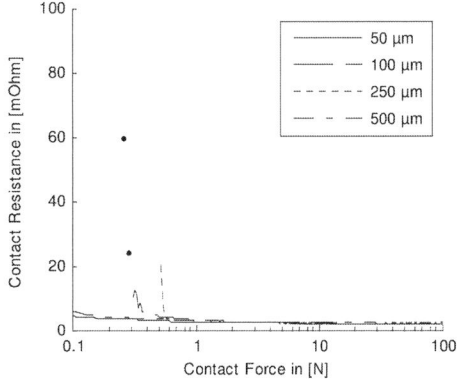

Figure 8: Contact resistance of stranded AWG 32 wire over contact force for four strip line widths

The first thing to note in all three plots is the expected hyperbolic behavior. For low forces the electrical contact is open although both members are in mechanical contact. Hence, the resistance is infinite but drops rapidly as soon as a certain minimum force is reached. It then decreases asymptotically to a saturated value with increasing force.

Table 1: Mean of contact resistances in [mOhm] of conductors for three force values. In brackets is the relative change to the 1 N value and the 10 N value, respectively.

	1 N	10 N	100 N
Shieldex 117/17	63.3	18.3 (71 %)	7.9 (88 %, 57 %)
Shieldex 235/34	21.2	7.1 (67 %)	3.8 (82 %, 47 %)
AlphaWire	2.9	2.2 (23 %)	2.1 (28 %, 6 %)

Giving a value for a "good contact resistance" is delicate. Of course, the lowest achievable value will be best. However, even higher values may be sufficient depending on the application. What can be stated is the following. The contact resistance should not change the overall resistance significantly and the change of contact resistance due to changes of contact force should be small. In a real application force values will be subject to change by mechanical motion, thermo-mechanical mismatch, relaxation, etc. In addition to the electrical characteristics of the contact it should also be noted that too high mechanical loads may even weaken the reliability of the contact. This is due to plastic deformation of the conductor, a profile decrease due to extrusion of material, and even a pinch-off of the conductor in the worst case.

At first, the contact behavior of the stranded wire is analyzed. The hyperbolic curvature is only hardly visible in the plots of mean values because a low contact resistance is already reached at very small forces. The first samples make contact for forces smaller than 0.1 N. From 1 N on all samples have contact with an already low value of 2.9 mOhm. Applying higher forces only slightly changes the value to 2.1 mOhms, i.e. by 28 %. When considering the plot in figure 9 it can also be concluded that even a small area provides for such low values.

The behavior of the conductive yarns is much different. First of all, the minimum forces to make the initial contact are much higher compared to the metal wire. First contacts are already made for small forces but lie in the range of 200 mOhm and are extremely unstable. In the worst case 1 N has to be applied to make the first electric contact. For 1 N the mean values for *Shieldex 117/17 2-ply* and *Shieldex 235/34 2-ply* are 63 mOhm and 21 mOhm, respectively. However, both yarns show large deviations for varying strip line widths. The *117 Shieldex* varies from 100 mOhm for the 50 µm line to 43 mOhm for the 500 µm line and the *235 Shieldex* varies from 40 mOhm to 12 mOhm, respectively. Furthermore, the slope is very steep, describing a high sensitivity of the resistance depending on relatively small changes in applied force. To overcome this sensitive region a force of about 10 N is necessary to stabilize the contacts and bring the resistance values in an acceptable range of

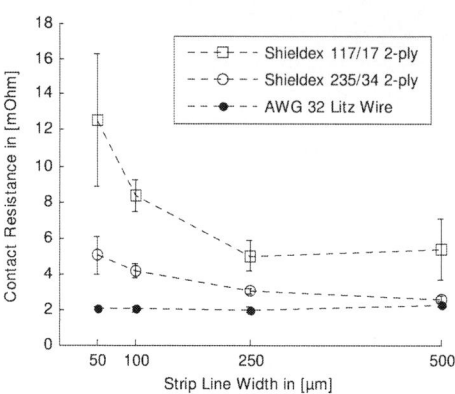

Figure 9: Minimum contact resistance values over strip line width for the maximum applied force of 100 N

10 to 20 mOhm. Nevertheless, the area dependence is still visible. Even if the maximum force is applied a large decrease by 50 % in contact resistance is visible if the line width is multiplied by the factor 10. This indicates that for the specific contact resistance of conductive yarns the apparent contact area plays an important role. A larger contact area potentially leads to a higher number of conducting a-spots. Additionally, from the graphs one can see that for wider strip lines the first contacts are made at smaller forces. It is assumed that an increase of the area of single a-spots plays a minor role in conductive yarns. Due to their soft material and their constitution the applied forces cannot be transferred to the contact interface as well as in full metal conductors. Nevertheless, higher applied forces and mechanical movements will support the displacement of tarnishing films and of insulators such as silver sulfide (Ag_2S).

To receive an impression of the quantitative results figure 10 shows cross-section views of three *Shieldex 235/34 2-ply* yarns loaded with 1 N, 10 N, and 100 N. The image for the lowest applied force looks very similar to the uncompressed state in figure 3. Only a few fibers are in mechanical contact with each other and clearance is visible. Hence, only

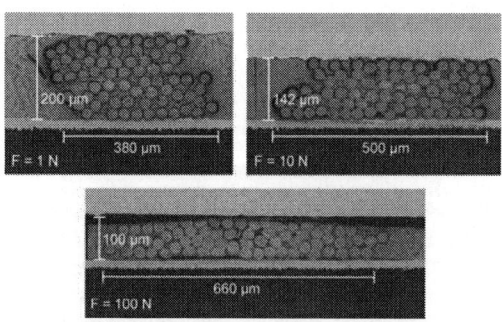

Figure 10: Cross-section views of compressed *Shieldex 235/34 2-ply* yarns for 1 N, 10 N, and 100 N

small contact forces seem to be acting between the fiber-to-fiber interfaces and the fiber-to-pad interface. The second image already shows a compressed and broadened yarn. More potential current paths are visible and the whole matrix is condensed. The last picture shows only small clearance and various potential current paths between fibers. The contact forces between all interfaces can be expected to be high. Additionally, the number of fibers making contact to the pad has doubled, which increases the number of a-spots. The images support the earlier theoretical assumption.

Design of a Force-Fit Interconnection

From the previous experiment one can estimate the resulting contact resistance for different combinations of conductor, strip line width, and applied force. This value stays constant if corrosion, elevated temperatures and high currents are avoided. However, electronics in the field will almost always experience varying temperatures, humidity changes, and mechanical stress. A common reliability test for microelectronics is temperature cycling between −40 °C and +85 °C for 1000 hours. Heat will of course have an effect on the resistivity of conductive materials. Even more important is the mechanical strain of contact systems due to the coefficient of thermal expansion (CTE). The majority of materials will expand with increasing temperature and contract with decreasing temperature.

For a force-fit interconnection this effect can be critical if the parts of the system change their dimensions and consequently the applied contact force with varying temperature. Depending on the initial contact force, the change of resistance can be exponentially as the graphs show.

Figure 11 shows a drawing of a force-fit interconnection using standard FR4 circuit boards. Clamping of the plates leads to bending due to the in-between die. The elastic stress of the plates applies force to the fabric integrated conductor, which presses onto the contact pad. This force depends on the dimensions and mechanical properties of the materials. Relaxation effects

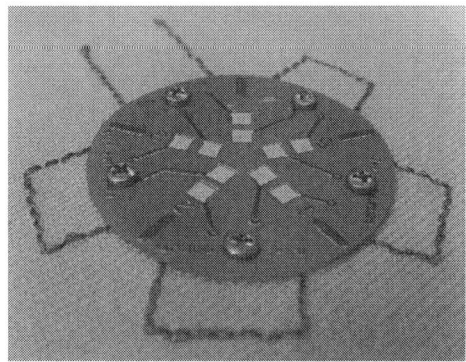

Figure 12: Example of a screwed force-fit interconnection of an FR4 circuit board and fabric integrated conductive yarn

especially for elevated temperatures need also be taken into account as they will lower the contact force. Most important for the contact behavior during thermal cycling is the CTE and mechanical stability of the fasteners. The z-component should be similar to the plate's CTE but preferably smaller.

Simulations will help identify optimum design values for the highest contact stability during different environmental loads.

Application

To validate the application of force-fit interconnections in smart textiles and their reliability a prototype is built. Figure 12 shows the assembly, which consists of a circular circuit board with bottom side contact pads. These are placed on an embroidered conductive structure of *Shieldex 117/17 2-ply*. Similar to figure 11 a second circular plate is positioned on the other side of the fabric. Both plates are screwed together using standard bolts and nuts. The contact force is applied by dies and due to elastic deformation of the plates. The application process is very fast and straightforward. Using a torque screwdriver ensures that the same forces are applied.

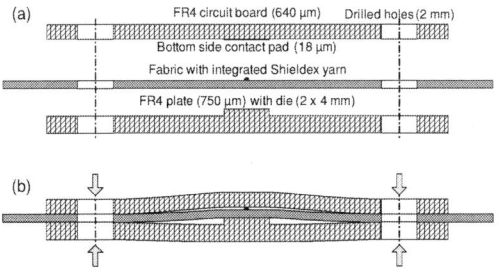

Figure 11: Drawing of a force-fit interconnection prior to assembly (a) and after assembly (b). The applied forces deform both plates elastically. The bottom plate's die applies force to the contact members. Compression of the fabric is neglected.

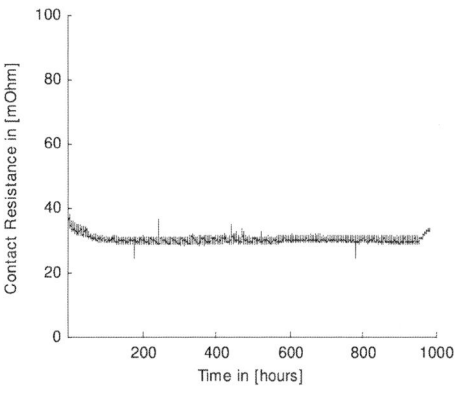

Figure 13: Contact resistance change of a screwed interconnection during temperature cycling between −40 °C and +85 °C for 1000 hours

The module was exposed to a standard thermal cycling test and all five contacts were measured with an in-situ system during 1000 hours. Figure 13 shows one of the five contacts of which none have failed. The measured resistance decreases in the first hours due to a change of the yarn resistance and stays stable over 1000 hours. Even during the minimum and maximum temperatures no resistance change is visible except due to the temperature dependent resistivity. That means that the contact force is keeping up the minimum contact resistance over all cycles.

Conclusion

The presented work proves the successful application of force-fit interconnections for integrating electronic systems in textile structures. They provide mechanical fixation and establish electrical contacts at the same time. However, a large difference between ordinary metal conductors and polymer-based conductive yarns is visible concerning contact resistance. In summary, it can be concluded that for conductive yarns much higher forces are necessary to allow stable and reliable contacts. This is due to their configuration, mechanical properties, and the thin conductive layers. Elastic stress can be used to apply force to contact members by using different types and materials of fasteners. The presented application of a screwed interconnection is an easy and fast alternative to other integration technologies. It proved to be very reliable in thermal cycling tests over 1000 hours.

Future work will include extensive mechanical simulations and redesigns as well as finding alternatives to the metal screws.

Acknowledgements

We thank the European Commission for funding part of this work under contract ICT-258724 (PASTA – Integrating platform for advanced smart textile applications).

References

[1] T. Linz et al., "Fully Integrated EKG Shirt based on Embroidered Electrical Interconnections with Conductive Yarn and Miniaturized Flexible Electronics", presented at the BSN, Cambridge, MA, Apr. 2006.

[2] R. Vieroth et al., "Stretchable Circuit Board Technology and Application", Presented at the ISWC, Linz, Austria, Sep. 2009.

[3] E. R. Post, M. Orth, et al., "E-broidery: Design and fabrication of textile-based computing", IBM Systems Journal, vol. 39, 2000.

[4] T. Linz et al., "New Interconnection Technologies for the Integration of Electronics on Textile Substrates", in Ambience 2005, Tampere, Finland, Sep. 2005.

[5] T. Linz, E. P. Simon, H. Walter, "Fundamental Analysis of Embroidered Contacts for Electronics in Textiles", Electronic System-Integration Technology Conference (ESTC), Berlin, September 13-16, 2010.

[6] T. Linz, M. von Krshiwoblozki, H. Walter, "Novel Packaging Technology for Body Sensor Networks Based on Adhesive Bonding", presented at the BSN, Singapore, Jun. 2010.

[7] A. Keil, "Elektrische Kontakte und ihre Werkstoffe", Springer, Berlin, Chapter 1, pp. 1-155, 1984.

[8] R. Holm, "Electric Contacts", Springer, fourth edition, Berlin, Chapter 1 and 8, p. 4, p. 43, 1967.

[9] E. P. Simon, "Analysis of Contact Resistance Change of Embroidered Interconnections", Diploma thesis, TU-Berlin, 2009.

Microsoldering with Short-Pulsed IR Laser

H. Hieber

ICR, Wilhelm-Kuelz-Str. 18, 99423 Weimar, Germany

+49(0)-3643-771650, +49(0)-3643-770397, hieber@icrjena.de

Abstract

A theoretical and experimental study investigates the minimum obtainable volume of soft soldered contacts to be made with an IR laser source. The parameters include the heat transfer and the surface energy of the liquid solder pool. The process is separated into a preheating step for the flux activation in the fine-grained solder paste, and into a melting and an overheating step. After switching-off the laser power a constitutional undercooling with the formation of intermetallics takes place. The optimization scheme considers the sensitivity of the calculated solder geometry with respect to the thermodynamic parameters for the state of equilibrium between power input, heat conduction and reflection of radiation.

Key words: Micro-soldering, IR laser, soft-solder process data

Introduction

Microsoldering with use of an IR laser as the heat source can be used for manufacturing and repair of electronic packages. A thorough understanding of the process is crucial for a good control of the properties of the produced parts. In this paper a first investigation is presented concerning the parameters which are relevant for the kinetics of preheating with extrusion of the volatile additives of the solder paste, the melting step with overheating the solder bath, and the cooling down step with constitutional undercooling of the melt and formation of the microstructure during the solidification.

Due to the fact that IR laser soldering with one laser source and without beam splitting is a sequential process the time steps required for positioning the wiring and the contact pads, the preheating, and the melting with subsequent cooling down must be summarized and determine the obtainable process rate. Previous investigations deal with the heat flow at surface-mounted chips on ceramic and printed circuit boards [1-4]. In this investigation special emphasis is given to the question which are the lower limits of solder mass, dimension of the solder joint, and time required for the soldering process.

Calculation of the heat transfer during soldering

Before starting the solder experiments a rough estimation is made of the power required for melting up a given amount of solder alloy, and for the heat losses in heat conductive wires and solder pads.

The theoretical approach divides the solder process into 3 steps. During the 1st step the volatile additives of the solder paste must be evaporated,

and the flux must be activated. During the 2nd step the melting up of the solder material and the overheating of the solder bath take place. Finally during the 3rd step after switching off the laser the liquid solder cools down to the melting interval of the alloy, and during the solidification the microstructure of the solder contact will be formed.

The heat sinks include not only the evaporation of volatile material and heat conduction but also the backscattered radiation due to partial absorption of the focussed laser beam. The kinetics of heat absorption of the powder is given by the energy flow [5]

$$\Phi = \frac{\eta a}{(4\eta-3)D}\{(1-\eta^2)e^{-\lambda}\left[(1-a)e^{-2a\zeta}+(1+a)e^{-2a\zeta}\right]$$
$$-(3+\eta e^{-2\lambda})\times\left[1+a-\eta(1-a)\right]e^{-2a(\lambda-\rho)}\}$$
$$-\frac{3(1-\eta)(e^{-\zeta}-\eta e^{\zeta-2\lambda})}{4\eta-3}$$

with λ the optical thickness of the powder bed, ζ the dimensionless depth coordinate. It is

$$a = \sqrt{1-\eta}$$

and

$$D = (1-a)\left[1-a-\eta(1+a)\right]e^{-2a\lambda}-(1+a)\left[1+a-\eta(1-a)\right]e^{-2a\lambda}$$

η is the absorption coefficient.

During the 2nd step the heat conduction is considered in the assembly of small solder balls inside the paste. It is estimated in a similar way as in a porous material bed or in a foam [6].

The melting up of the solder balls in the paste to the formation of the solder bath requires the heat flow into the solder pool. For practice the effective heat input [7] amounts to

$$N = \frac{\dfrac{P\eta}{\pi r_b^2}\,\Delta t}{\rho C_s (T_L - T_o) + \rho L}$$

It is determined by P the laser power, ρ the density of the solder bath, r_b the laser beam radius, T_L the melting temperature, and T_o the ambient temperature, C_s is the specific heat of the solid, and L the latent heat of fusion. Finally the solidification releases the heat of melting back. A part of the heat will be removed by conduction into the component and along the wiring.

Fig. 1 shows the calculated thermal characteristics for two exemplary cases with (a) low laser power and (b) high laser power both applied to $\approx 1 \dots 1.5$ mg solder mass. It is experienced that the laser power of e.g. 8 W needs a relatively long time interval of preheating to reach the flux activation temperature. With the laser power of 16 W the time intervals of preheating and of the melting step are markedly shorter. In the figs. the steps are indicated as P = pre-heating, H = full power heating, M = melting starts, O = overheating of the melt, R = switch off the laser power, S = solidification starts with constitutional undercooling.

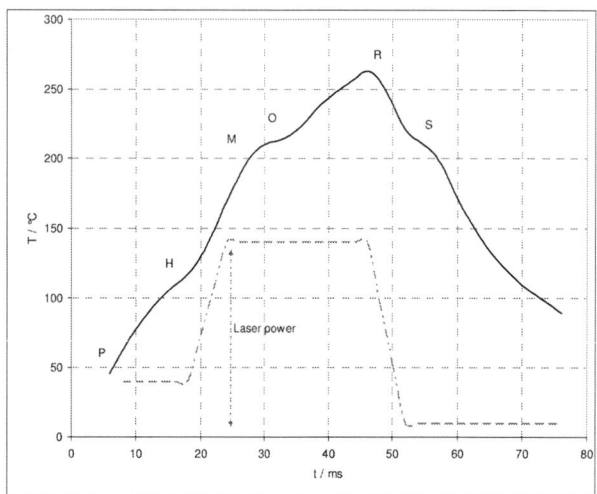

a) Laser pulse with low power 8 W on a solder mass ≈1 mg of Sn65.5Ag3Cu0.5

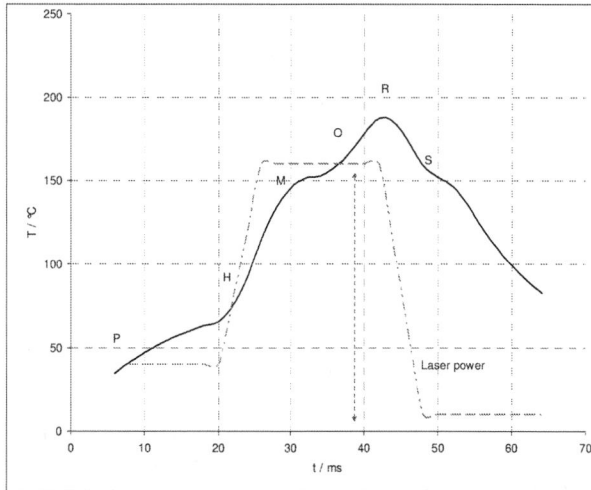

b) Laser pulse with increased power 12 W on a solder mass ≈1.5 mg of Sn42Bi58

Fig. 1: Calculation of the heating and cooling characteristics for different solder alloys and masses

Experimental solder process

The tested solder pastes consist of Sn96.5 Ag3Cu0.5 and Sn42 Bi58 alloys, respectively, and of additives. The ball sizes range between 25 and 45 µm. The filling degree is 85 %. The paste is extruded from nozzles with a diameter up to 200 µm. After positioning an amount of solder paste in the volume range between ≈0.01 and ≈0.06 mm^3 the extruder is shifted a small distance away, and the laser focus is put into correct position. The focus of the laser is adjusted to 50 ... 200 µm Gaussian width with use of local melting in a thin layer of solder paste. It is experienced that the wavelength of the laser $\lambda = 900 ... 915$ nm provides a sufficient absorption in the solder paste. The real heat affected area should be held in the lateral measure of the solder depot. The power of the laser is controlled electronically with pulse width modulation.

In the first step low power is applied to evaporate the paste additives and to activate the flux resin in a temperature range below the melting temperature of the solder alloy. After this step the power is increased for melting up the powder which forms the liquid solder bath. During this step the wetting of the contact material takes place. The melting of the solder bath is recorded with a high-speed camera with the required IR absorbing filter. The sequence of the video frames gives the exact times from the start of melting until the solidification which can be extracted due to characteristic changes in the surface of the solder bath.

The table shows some representative intervals of the physical parameters. The bold numbers indicate the intervals in which good solder quality can be obtained.

Table: Optimization of parameter windows for the IR laser soldering for the example of a low-melting Sn58Bi solder alloy, with use of a 25 µm solder paste

Dimension of the solder joints		Laser power 8 W		Laser power 12 W		Laser power 16 W	
Charact-eristic length [µm]	Solder mass [mg]	Heating interval [ms]	Tempe-rature peak [°C]	Heating interval [ms]	Tempe-rature peak [°C]	Heating interval [ms]	Tempe-rature peak [°C]
200 ... 250	0.29 ... 0.56	**9 ... 17**	**220 ... 160**	**6 ... 12**	**180 ... 220**	*4.5 ... 8*	*280 ... 240*
300 ... 350	0.97 ... 1.55	*30 ... 48*	*255 ... 190*	**20 ... 32**	**240 ... 175**	*15 ... 24*	*260 ... 220*
400 ... 450	2.31 ... 3.29	*72 ... 102*	*155 ... 175*	**48 ... 64**	**180 ... 160**	**36 ... 51**	**220 ... 175**

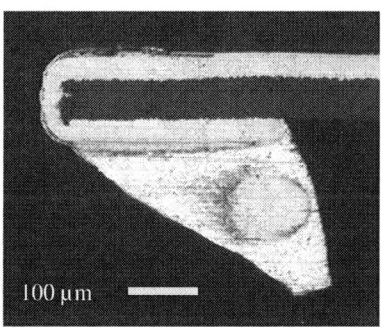

a) Cu wire at a solder ball

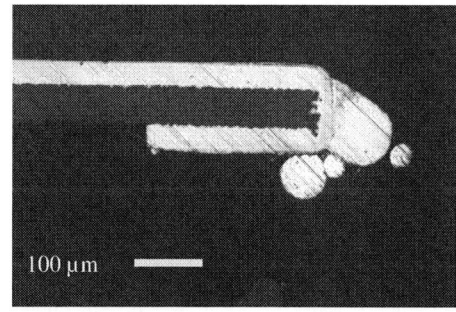

b) With solder balls

Figure 2: Soldered contacts without and with separate solder balls

Morphological results

Fig 2 shows exemplary cross sections of contacts soldered at solder pads of a component. Fig. 3 shows the melting behaviour of a metallization on a polymer fibre. It is seen that only proper adjustment of the geometry and of the physical parameters lead to good results.

The formation of metallic contacts requires a specific surface free energy $\Delta G(r) \sim \Delta G_o / r$. For ball radii e.g. $r = 25 \dots 45$ µm an increased force inhibiting the wetting process must be overcome. This is the reason why small solder droplets do not wet the contact pads properly and separate solder balls are formed. The Gibbs free energy must exceed a value of ≈ -10 kJ mol^{-1} [8]. On the other hand there are good conditions for the nucleation of intermetallic compounds, e.g. the η phase Cu_6Sn_5 with a thickness of 50 nm during reaction times down to ≈ 10 ms [9]. The subsequent growth of the phase is diffusion controlled [10].

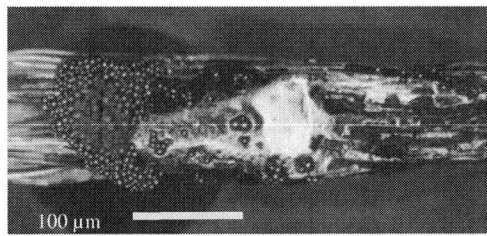

a) Partially molten solder paste

b) Fibre with soldered alloy

Figure 3: Soldered contacts at a metallized PE fibre

Conclusions

Microcontacts are soldered with use of the IR laser in the range of $0.001 \dots 0.06$ mm^3 volume, <100 ms time interval, and $8 \dots 16$ W laser power.

A phenomenological heat transfer and fluid flow model estimates the uncertain parameters necessary for solder pool modeling. Due to the short time intervals required for the soldering process the thickness of the intermetallics does not exceed few µm in the case of interaction between Sn-rich alloy and Cu wire. This is advantageous with respect to ageing kinetics and long term stability. Typical applications include the soldering of 40 and 80 µm Cu wires and a sequential process for soldering components with contact sizes of 200 µm up to 400 µm on a textile fabric to be used as a flexible electronic substrate.

The critical issue is the sensitive relationship between the solder mass and the applied laser power multiplied by the step duration of preheating and melting. Certain limits must not be exceeded. When the power applied for preheating is too high the volatile paste constituents explode. When the power applied during the melting step is too high or the time step is too short the solder forms small droplets which do not wet the contact pads.

The IR laser microsoldering is a sequential process with the disadvantage that the time steps for positioning the tools, the paste extrusion and the solder steps are added to the total process time which is proportional to the number of solder contacts to be manufactured. The advantage is the rapid local heating of the solder contacts with neglectable heat flow into the components and into the wiring. In particular, electronic constructions can be made with use of cheap but heat sensitive monofilament and fibre materials.

This technology can be considered for lightweight electronic constructions e.g. for sensorics and for MEMS devices which are exposed in harsh environments and subject to acceleration and fatigue. The IR laser microsoldering technology is also considered to be an effective method for repair in complex microelectronic and microoptical constructions.

References

[1] Semerad E, L Musiejovsky, J. Nicolics, "Laser soldering of surface mounted devices for high reliability applications", pp. 5065-5069, J. Mater. Sci., Vol. 28, 1993.

[2] Moeller, W, K U Vayhinger, C M Val, M Leroy, "Surface mount technology of 3D multichip modules by leadframe laser soldering", IEPS, Proc. of the Tech. Conf., 1994 Internat. Electronics Packaging Conf., Atlanta, USA, Sep 25-28, 1994, pp. 835-839, Wheaton, Internat. Electronics Packaging Soc., 1994.

[3] Herrmann A, Fertigungsorientierte Verfahrensentwicklung des Weichlötens mit Diodenlasern, Herbert-Utz-Verlag (2002), ISBN 978-3831600861.

[4] Naveed S M, R L Woods, "Diode laser soldering: a lumped parameter mathematical model and comparison of

different optical soldering technologies", pp. 142-157, Proc. SPIE Vol. 4973 (2003-06-19) High-Power Diode Laser Technologies and Applications, M S Zediker, editor.

[5] Verhaeghe F, T Craeghs, J Heulens, L Pandelaers, "A pragmatic model for selective laser melting with evaporation", No. 20, pp. 6006-6012, Acta Materialia 57, 2009.

[6] Conquard R, D Baillis, "Numerical investig-ation of conductive heat transfer in high-porosity foams", No. 18, pp. 5466-5479, Acta Materialia, Vol. 57, 2009.

[7] De A., T DebRoy, "A smart model to estimate effective thermal conductivity and viscosity in the weld pool", pp. 5220-5229, J. Appl. Phys. Vol. 95, 2004.

[8] Tu K N, T Y Lee, J W Jang, L Li, D R Frear, K Zeng, J K Kivilahti, "Wetting reaction versus solid state ageing of eutectic SnPb on Cu", No. 9, pp.4843-4849, J. Appl. Phys., Vol. 89, 2001.

[9] Park M S, R Arróyave, "Early stages of intermetallic compund formation and growth during lead-free soldering", Acta Materialia, pp. 4900-4910,Vol. 58, No. 14, 2010.

[10] Bader S, W Gust, H Hieber, "Rapid formation of intermetallic compounds interdiffusion in the Cu-Sn and Ni-Sn systems", No. 1, pp 329-337, Acta Metallurgica et Materialia, Vol. 43, 1995.

High Temperature Electronics for Harsh Environments

Bob Hunt, Andy Tooke

C-MAC MicroTechnology, Fenner Road, South Denes, Great Yarmouth, Norfolk NR30 3PX
Tel: +44 (0)1493 743100, E-mail RobertHunt@cmac.com

Abstract

A high performance, high operating temperature electronic module technology has been researched and developed that achieves reliable operation for 1000 hours at 225 Deg C.
The technology has been qualified with a unique approach to ensure reliable operation at maximum temperature and under high mechanical stress. The qualification model developed is somewhat unique and different from that prescribed for the aerospace and defence markets under MIL-STD-883/MIL-PRF-38534. Although specifically representative of the harsh conditions experienced at the drill tip for deep well oil and gas drilling, it is readily transferable to other market sectors.
This new technology brings significant value. It enhances system performance by enabling precise electronic sensing and control at 'Point of Use' in locations of extreme environment.
The paper covers some of the technologies, the qualification techniques adopted and results for a DC-DC converter module constructed with the technology developed.

Key words: 225 °C, Microelectronics, Module, Qualification, DC-DC converter

Introduction

The Aerospace and Defence market sector has historically pioneered research and development of high temperature, harsh environment, high reliability electronic systems.

This 'high temperature' of the past has been comprehensively established within the military specifications, specifically in MIL-STD-883/MIL-PRF-38534, and widely accepted as 125 °C. This being the temperature monitored on the system housing, heatsink or module case and typically translates to 150°C - 175°C semiconductor junction temperature. Although higher operating junction temperatures are possible, this is towards the upper limit of operation for single element (Silicon) semiconductor technology and so has been a logical standard to adopt.

Increasingly we are seeing the expansion of electrical and electronic systems in many industries. This is driven by cost benefit but also to enhance performance and achieve more effective and efficient operation. These requirements can often place sensors and conditioning electronics in remote and harsh environment locations where there are extreme temperatures coupled with high shocks and vibrations.

The Oil and Gas drilling industry utilises conventional industrial grade electronic components and integration technologies, those being surface mount, soldered components on printed circuit board (FR4). But, the industry is seeking hydrocarbons in ever increasing extreme locations and is establishing deep well drilling techniques where temperatures can reach 225°C. These deep well drills employ complex electronic sensing and control electronics requiring reliability and predictable life.

Unusually, most electronic environmental qualification test regimes and particularly those within the military standards require 'single condition' testing to be performed rather than 'concurrent condition' testing whereby, for example, shock or vibration testing is performed at high temperature. The value of 'concurrent conditions' with environmental testing is recognised as being critically important in drilling applications as it replicates the true operating conditions.

This paper describes the development of a 225 °C operating temperature electronic module and subsequent environmental testing with shock and vibration at maximum operating temperature.

Test Module

The approach adopted was to initially evaluate the range of technologies available. These technologies included the thick film materials for conductors, dielectrics and resistors, wire bonds, and adhesives for component attachment and the actual components themselves.

Initially, two test modules, or 'Capability Qualifying Circuits' (CQCs) were developed and subjected to an extensive range of representative electrical and environmental tests. The main test module is shown in Figures 1 and 2.

Figure 1. CQC Test Module – With welded lid

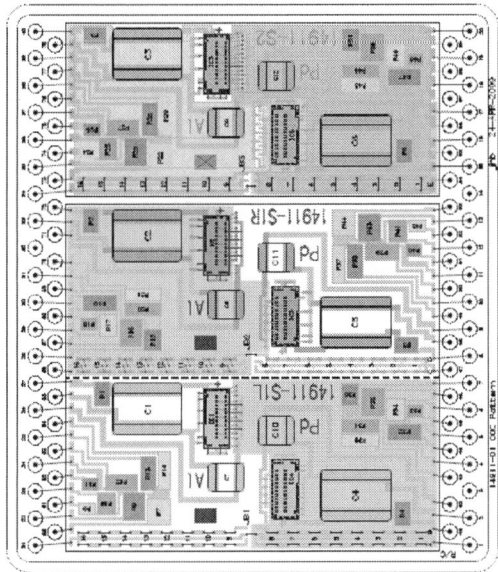

Figure 2. CQC Test Module – Internal Plan View

The test matrix for the thick film material sets and wire bonds is shown in Fig 3. This exercise enabled appropriate selection of the technology and proved the feasibility of producing a module capable of operating at 225 °C

Environmental Test Plan

The test modules were subjected to the following environmental tests. In all cases, the mechanical tests were performed at elevated temperatures. It should be noted, that due to lack of availability of suitable 'off-the-shelf' equipment, it was only possible to perform temperature cycle and thermal shock up to a maximum temperature of 200 °C.

- Life test 1000 hours @ 225°C case
- Temperature cycle 100 cycles, 200°C/-40°C
- Thermal shock 500 cycles, 200°C/-40°C
- Mechanical shock, 500g-peak sine, 1ms. 10 shocks, 3 axes at 225°C
- Vibration 15grms, 10Hz to 1kHz, 1 hour, 3 axes at 225°C

Figure 4. Vibration Test System

Shock and vibration at elevated temperature was performed with a unique arrangement of customised, self contained miniature ovens which encased the module and were mounted onto the equipment test bed.

	Type A Wirebonds	Type B Wirebonds		Type C Wirebonds		Link wire bonds	Link wire bonds	Link wire bonds	Link wire bonds	Link wire bonds	Numbers in cells are total number of wirebonds on this configuration
	Package Pins	Package Pins	25um Die Pads	Package Pins	25um Die Pads						
	38um	38um	25um	25um	25um	25um	38um	25um	25um	38um	
Substrate S1 - (Thick Film Materials with printed resistors)											
(Material A) - on ceramic	13		10		14	16	0.5				
(Material A) - on Dielectric	12		14		10	16	0.5				
Substrate S2 (Thick Film Materials with printed resistors)											
Material A - on ceramic	13										
Material A - on Dielectric	12										
(Material B) - on ceramic			14		10			16 wires	17 wires	One wire bond	
(Material B) - on dielectric			10		14						
Substrate S3 (Thick Film Materials with printed resistors)											
(Material C) on ceramic - Post plated	3	7	10	3	14				8]One 38u across		
(Material C) on dielectric - Post plated	3	6	14	3	10				8]		

Figure 3 – Material test matrix

Figure 5 Oven for shock and vibration testing at temperature extremes

The test plan included stress testing by three basic methods :-

A. An endurance or life test where parts were subjected to constant extreme operating temperature whilst being subjected to electrical bias. This demonstrated the ability of the products to perform their function whilst exposed to the temperatures that will be encountered in the field.

B. The application of mechanical stress by the use of vibration and shock testing. These tests were performed on product at maximum temperatures. This demonstrated the ability of the products to survive high levels of mechanical stress whilst being subjected to elevated temperatures.

C. Temperature cycling and thermal shock testing in air to evaluate the reliability of the components and the adhesives used to attach the components when subjected to the disparate thermal expansions and contractions that could be experienced in use.

There are a large number of testing sequences that could be adopted. For this exercise the main test sequences used were as follows:

SEQUENCE 1 :-

1. Endurance – 1000 hrs at 225°C
2. Hermeticity test
3. Residual Gas Analysis
4. Die shear / Wire bond pull

SEQUENCE 2:-
1. Vibration - at 225°C
2. Electrical test
3. Hermeticity test
4. Mechanical shock - at 225°C
5. Electrical Test
6. Hermeticity test
7. Die shear / Wire bond pull

SEQUENCE 3:-

1. Temperature cycle - at 200°C
2. Electrical test

Test Results

The data collected ranged widely, however, it was possible to select a combination of material sets and process that met the criteria.

The hermetic seals survived all testing conditions, and residual gas analysis showed level of between 119 and 304 ppm. Significantly below the MIL STD limit of 5000 ppm.

Wire bond pull and ball shear strengths were within the MIL STD specification, as was the component shear strengths.

Thick film materials also showing varying performance, particularly with adhesion and resistor stability. However, a suitable combination of materials was demonstrated which provided absolute resistor tolerances of +/-2.5% and Temperature Coefficient of Resistance (TCR) of less than 100ppm/°C over the range -55°C to + 225°C.

Having established suitable technologies, a specific functional module was designed and subjected to the same environmental testing to reinforce the findings.

Circuit Configuration

The circuit architecture developed by CISSOID S.A. for the module was a voltage mode, buck converter topology that uses duty-ratio modulation at a constant frequency. The advantage of selecting this circuit for the development is that it employs a wide range of passive and active components, and in particular some large body ceramic capacitors. In addition the circuit is sensitive to the impedances of connections and requires management of relatively high currents.

The DC-DC converter was constructed in a controlled manner in accordance with the materials and processes qualified with the CQC test circuit.

Technologies (See Fig 9)

Housing
A conventional Kovar (Iron:Nickel:Cobolt) package with a welded lid and glass to metal, electrical feedthroughs. This is a widely used and proven technology.
To meet the requirements of the deep well drilling, the package has a high aspect ratio envelope (~ 150mm long by 22mm wide) to enable it to fit within a 25.4 mm diameter drill tube.

Interconnections and Support Tile

Ceramic tile supporting printed and fired, multilayer thick film. With specially formulated materials for the printed thick film resistors.

Components - Actives

High operating (225°C) semiconductor chipset from CISSOID.

Components - Passives

High operating temperature, passive components were selected for surface mounting and electrical connection to the ceramic thick film interconnect. A toroidal inductor and high value ceramic capacitors (Body size 6560) used as filtering components on the output.

Environmental Testing

The DC-DC converter module built with the selected technologies and was tested in accordance with the same test regime as performed on the CQC test module above.

Results in general matched those established previously and demonstrated the suitability of the technology for operation at + 225°C, under shock and vibration for a minimum of 1000 hours operating lifetime.

Electrical performance was measured over the full temperature range -55°C to + 225°C and found to meet the objective specification.

Performance

Characterised for continuous use at Tcase = -55 °C / -67 °F to +225 °C / 437 °F
Robust technology allows operation to case temperatures of +250 °C / 482 °F
Output voltages factory tuneable from +2.5 to +25 V
Operating efficiencies up to 85 %
Load currents from 10 mA to 1.0 Amp
Input voltages from 7 V to 30 V
Typical output voltage changes of 0.25 mV / K
Low intrinsic standby current of less than 25 mA

Conclusions

The microelectronic module technology for reliable operation at 225 Deg.C was successfully developed and qualified.

The technology is highly suited to electronics for harsh environments and in particular Oil and Gas drilling equipment

A DC-DC converter was constructed utilising these technologies and fully electrically and environmentally tested and qualified

Figure 6. DC-DC Converter Module

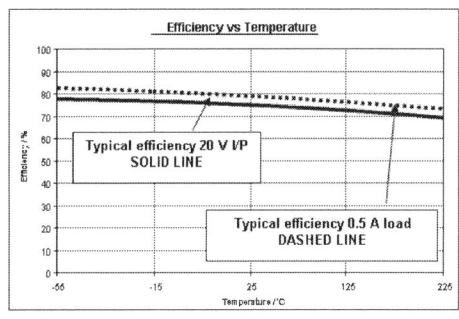

Figure 7. DC-DC Converter Efficiency

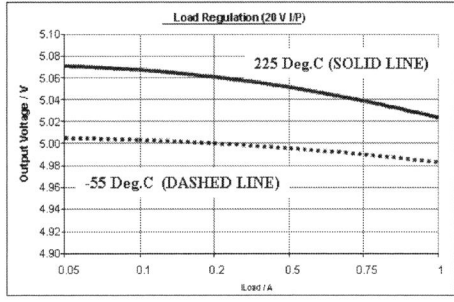

Figure 8. DC-DC Converter Load Regulation

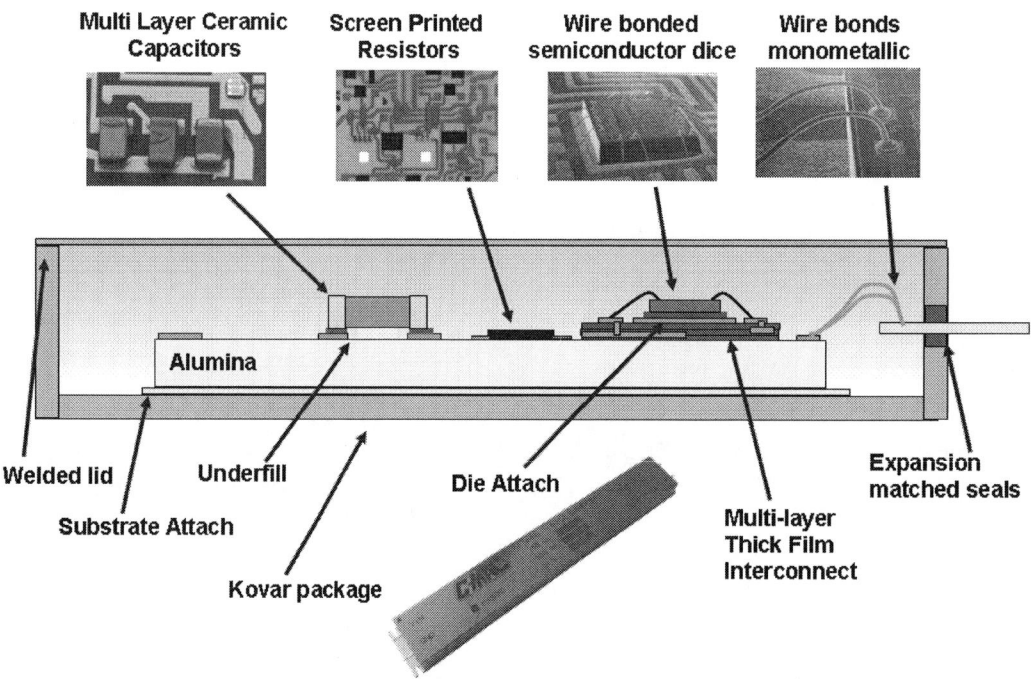

Figure 9. Cross section - Module technologies

References

[1] 'ETNA' DC-DC converter design, CISSOID
S.A.

Fine Pitch Flip Chip Chip Scale Packaging

Bernd K Appelt, Harrison Chung*, Chienfan Chen*, Raymond Wang*

ASE Group, 1255E Arques Ave, Sunnyvale, CA, USA
*ASE Group, Nantze Export Zone, Kaohsiung, Taiwan

1.408.768.8533 p, 1.408.636.9485 f, bernd.appelt@aseus.com

Abstract

Flip chip packaging used to be the domain of micro processors and ASICs. Application processors were next to adopt flip chip packaging in high volume. Both types of packaging were based on solder interconnects, initially PbSn eutectic solder on organic substrates and then later Pb-free solders. In order to solve electro migration issues, micro processors adopted Cu pillars. Flip chip packages are now also adopting Cu pillars but primarily as a fine pitch solution because pillars can be formed as slender columns which are maintain their initial shape during assembly unlike solder. The invariant shape of the Cu pillars brings a number of other benefits such as increased, effective spacing between pillars and invariant pillar height which enable simpler substrate design and molded underfill. Those benefits result ultimately in a lower cost flip chip chip scale package. Cu pillars are stiffer than solder, especially eutectic solder, and are a concern when assembly on ELK die. Finite element modeling is very effective in identifying solution spaces which lead to reliable Cu pillar based flip chip packages. Development efforts of Cu pillars, their assembly and modeling will be described here.

Key words: Cu pillars, fcCSP, modeling, assembly

Introduction

Flip chip (fc) packaging used to be the domain of high end applications like CPUs, MPUs, Chipsets, Graphics Processors, Gaming Processors, etc. Over the last few years, fc packaging, specifically fc chip scale packaging (fcCSP, see Figure 1), has also been introduced into mobile applications like application and baseband processors to take advantage of the superior electrical performance when compared to wire bond packaging [1]. The next step in interconnect evolution was the introduction of copper pillars (CuP) replacing the solder bump connections to overcome electro-migration. Enhanced thermal conductivity was a welcome side benefit. Again this was introduced on high end packages first. Now, CuP fc packaging is being introduced for mobile applications (Figure 2) but the motivation is as a solution for fine pitch requirements with a side benefit of a potential for packaging cost reduction. The issue with solder bump fc technology is that during reflow bumps collapse due to surface tension which lowers the die stand-off and reduces the space to adjacent bumps. The former complicates the underfill process by decreasing cleaning process efficiencies and by requiring small particulate underfill compounds of low viscosity to achieve void-free underfill. The later increases the propensity of solder bridging and settling or blocking of filler particles. These effects get progressively worse as pitches and solder bump diameters shrink which in turn is synonymous with advancing wafer node technology. CuPs on the other hand are dimensionally fixed members i.e.

diameter and height will not change during assembly of the die to the substrate. The dimensions can be selected over a wide range to optimize pillar height for reasonable stand-off and for pillar to pillar spacing. Hence above assembly problems can be circumvented. Because of the relatively small diameter of the pillars, the substrate pads can also be shrunk. With the proper optimization, the typical circular pads can be replaced by rectangular pads or traces which in turn increases the space between pillar traces. This additional spacing is sufficient to route an additional trace between pillars and thereby increasing the wireability of the substrate. This is in fact the biggest cost benefit of this CuP fc technology. The increased wireability can be utilized for substrate simplification: wider trace pitch i.e. cheaper substrate process and simpler substrate cross-section i.e. cheaper substrate.

Figure 1 depicts front and back side of a typical flip chip chip scale package.

CuPs are formed on the wafer by 'simply' replacing fine pitch solder plating with copper plating. The CuPs are capped with Pb-free solder. The assembly process has been optimized for direct reflow on pad/trace without pre-solder which does require better control of solder volume to ensure that all pillar caps will reflow and placement accuracy has to be controlled as well because the small solder volume has a lesser propensity for self-alignment. Instead of underfilling, the bumps are secured by molding only which has been enabled by the higher stand-off and the wider spacing between pillars.

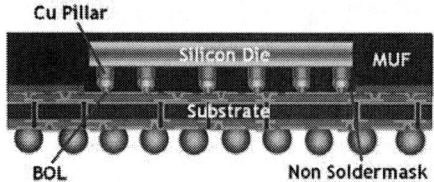

Figure 2 shows a schematic cross section of a Cu pillar based fcCSP.

For advanced nodes, FEM has provided the guide lines for pillar and UBM design on the fragile die dielectrics to avoid white bumps and chip cracking. Here, the FEM results will be presented for appropriate CuP design. Design rules for CuPs on die, substrate and assembly for a successful CuP fc package will be discussed. Successful qualification results will be shown to demonstrate the viability of this new technology.

Package on package (PoP) solutions with CuP are already being shipped in high volume. Here, the pitch is currently about 50 μ peripheral i.e. a wirebond die can be converted into a flip chip without redistribution. By contrast, solder flip chip die typically will require redistribution which is another cost savings for CuP. Another benefit of Cu pillars is that they are a green solution and fulfill the requirements of WEE and RoHS [2].

Low cost solution

Cu pillars are electroplated on under bump metallurgy (UBM) as a cylinder, sequentially electroplating Cu, Ni barrier and SnAg cap. Since the pillars photo-defined, virtually any pillar shape or orientation is possible. The SnAg cap is reflowed during soldering of the substrate. Hence, the height of Cu pillars can maintain more height and space at the same pitch. Table 1 shows the advantages of Cu pillars as a low cost solution as compared to solder bumps. A simpler and lower cost substrate, such as 4 layer or 2 layer instead of the 1+2+1 layer and non-solder mask defined (NSMD) instead of solder mask defined (SMD) or solder mask-less (SM-less) substrate may be used. The higher stand-off facilitates the encapsulation by molding only with cheaper, coarser filler instead of slow, expensive fine-filler capillary underfilling. Such a change in encapsulation method simplifies the process flow to only one step that can greatly enhance the throughput (10 to 20 times) and without capital expenditure for capillary underfilling equipment. The benefits are even greater when a two step process: underfilling plus molding is replaced. Furthermore, the layout of the passive devices can

be closer to the die and relax the keep-out zone required for capillary underfilling.

Table 1: This a comparison of substrates for solder and Cu pillar interconnects

Type	Standard practice	Low cost solution
Substrate layer	1+2+1 layer 4 layer	2 layer
Solder-mask	SMD	NSMD SM-less
Encap-sulation	Capillary underfill + overmold	Under + overmold

Engineering Development

The typical practice of capillary underfilling, Figure 4 (a), consist of die pick and place to the corresponding substrate pads, solder reflow to melt and join the solders to the substrate pads, capillary underfilling, curing and finally overmolding. Besides the aforementioned challenge, the other is that solder reflow can cause significant global warpage on thin substrates and can induce solder cap misalignment between the substrate pad and die pad. A thick core is typically necessary to maintain acceptable warpage. This counter measure limits the application of low-cost substrates for solder fcCSP.

The advantages, from the point of view of design flexibility, Table 2, are certainly the space saving as bumped die from peripheral wire bond to peripheral or full array flip chip, and the relaxed trace pitch and/or the routing escape traces on a layer between the slim pillars. A higher number and density of I/O can hence be achieved. The continuing development to finer pitch (<100 μ) and smaller gap from die to substrate (<45 μ), however, eliminates the under molding feasibility due to the filler size and pushes even the limit of capillary under filling.

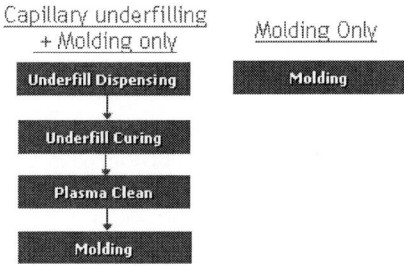

Figure 3: Comparison of process flow for capillary underfill + molding (left) and molding only (right)

Table 2: Comparison of design flexibility for wire bond copper pillar flip chip

Type	Standard practice	Design flexibility
1st level interconnection	Wire bond	Flip chip
Trace between pads		

Nevertheless, an alternative solution makes use of non-conductive paste (NCP) dispensing and thermal compression (TC) bonding, as shown in Figure 4(b). The characteristic of NCP dispensing and TC bonding is the localized heating of the bonding area. Heat-induced warpage is significantly decreased and thus improves the bondability. In addition, the NCP dispensing before bonding enables finer and finer pitch pillar encapsulation. Both process scenarios have been developed at ASE to cope with different substrate thickness, pillar pitches and die to substrate gaps. TC bonding is much lower in throughput and is therefore used only for pillar pitches below 100 μ. Development activities are under way to decrease the pillar pitch at which the more cost effective reflow can be used.

The need for Nickel (Ni) barrier plating on the pillar has long been discussed. To evaluate the need, assembled packages for both Ni barrier and without Ni barrier have been subjected to reliability testing. All packages went through preconditioning with moisture sensitivity level of 2a [6] and subsequently separated, half from each cell, for 1000-cycles temperature cycling test (TCT) of condition -55 °C to 125 °C [7] and the other half for 1000-hours high temperature storage test (HTST) maintained at 150 °C [8]. No electrical opens or shorts failure nor cracks and delamination by scanning acoustic tomography (SAT) were found for both features after TCT and HTST [9].

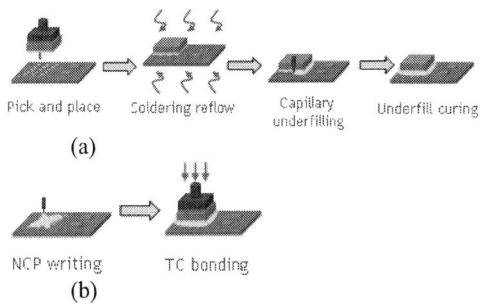

(a)

(b)

Figure 4: Comparison of process flows of (a) Reflow + capillary underfill and (b) NCP dispensing + TC bonding

Figure 5 shows the cross sections for both cells after HTST 1000 hours. It can be sees that the Cu pillar without Ni plating has very serious solder consumption forming thick intermetallic compounds (IMC) at the Cu pillar surfaces compared to the one with Ni barrier. This is especially severe at the downward surface of the pillar which is also consumed by the substrate bond pad. Thick Cu_3Sn is formed at the closest surfaces of the pillar and bond pad and secondary Cu_6Sn_5 is seen between the Cu_3Sn layers. At any rate, both of the features can live for these industrial standard reliability tests, i.e. useful life duration for the users. Reliability difference in electro-migration will be tested in the future.

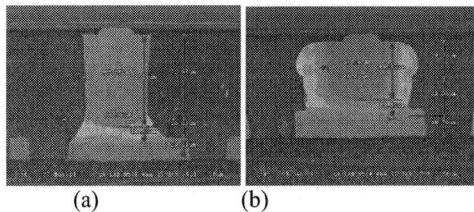

(a) (b)

Figure 5: Cross sections of Cu pillar (a) with Ni barrier and (b) without Ni barrier after HTST at 150C for 1000 hours

Cracks and delamination at the dielectric layer, Figure 6 (left), always are a concern since the implementation of fragile LK dielectrics with stiffer Cu damascene stacking structure in the die. To lower the electrical noise further nowadays, more porous extreme LK (ELK), i.e. more fragile dielectric materials are employed on 40 nm and 28 nm wafer technology nodes. The higher stiffness of Cu pillars may further aggravate the risk of damaging the LK and ELK layers. Structural re-design to decrease the stress reliability is necessary. Features close to the LK/ELK, polyimide (PI) and Cu redistribution layer

(RDL) are chosen to study the dimensional effect to the stress via finite element modeling, as shown in Figure 6 (middle). The analysis shows structural preferences for small PI opening, thick RDL and PI+RDL+PI structural stacking can minimize the ELK stress. These design rules have been validated by passing the pre-conditioning and 1000 cycles of TCT.

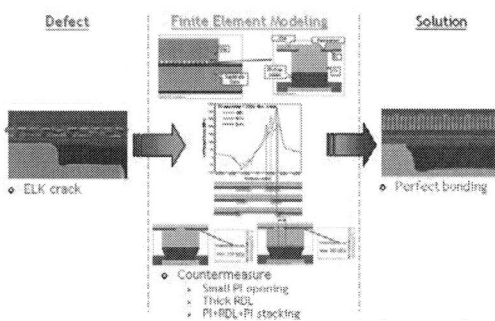

Figure 6: Structural design solution to ELK cracking via finite element modeling

The development of Cu pillars at ASE is categorized to normal pitch and fine pitch, applied on area array and peripheral distributions, respectively. The higher pillar and cap of the normal pitch has been qualified by package-level reliability tests on wafer technology nodes of 95 nm and 65 nm LK, and 40 nm and 28 nm ELK. Some of the fine-pitch prototypes have been built and are under reliability evaluations (Table 3).

Table 3: Current capability and qualification status

Type	Normal pitch	Fine pitch
Minimum pillar pitch (Distribution)	100 um (Area array)	45 um (Peripheral)
Wafer size	200 mm / 300 mm	300 mm
Repassivation type / thickness	Polyimide / 5 um	
Minimum repassivation opening	10 um	
UBM type	TiCu	
Structure	Cu pillar (40~60 um) + Ni (3 um) + SnAg2.4 (<30 um)	Cu pillar (20~30 um) + Ni (3 um) + SnAg2.4 (<25 um)
Qualified wafer technology node	95 nm LK, 65 nm LK, 40 nm ELK, 28 nm ELK	Under reliability tests

Summary

fcCSP based on Cu pillar interconnects exhibit superior thermal and electro-migration performance. The height of the pillar allows more space between die and substrate as well as between pillars at the same pitch enabling a low cost substrate consisting of 2 or 4 layers, NSMD or SM-less solder mask with relaxed registration, and mold only encapsulation. Great warpage on thin substrate packages induced by solder reflow can be mitigated by TC bonding minimizing the heat-affected zone. Ni barrier and no barrier on the Cu pillar show no defects after reliability, however, extensive diffusion and IMC formation was found on pillars without a Ni barrier. Finite element analysis provides an effective way to optimize structural designs avoiding ELK cracking which was validated by substantial test vehicle builds and analysis.

Acknowledgement

The authors would like to thank Chien-Fan Chen of the Bumping team, Spencer Tsai and Chien Liu of the Assembly team, and Tsung-Yueh Tsai and Meng-Kai Shih of the Central Product Solutions team at ASE for their technical consulting and discussion.

References

[1] L.F. Miller, "Controlled collapsed reflow chip joining," IBM Journal of Research and Development, Vol. 13, pp. 239-250, 1969.

[2] J. Cannis, "Green IC packaging," Advanced Packaging, Vol. 8, pp. 33-38, 2001.

[3] D. Mallik, K. Radhakrishnan, J. He, C.-P. Chiu, T. Kamgaing, D. Searls, J. D. Jackson, "Advanced package technologies for high-performance systems," Vol. 9, Issue 4, ISSN 1535-864X, pp. 259-271, 2005.

[4] A. Yeoh, M. Chang, C. Pelto, T.-L. Huang, S. Balakrishnan, G. Leatherman, S. Agraharam, G. Wang, Z. Wang, D. Chiang, P. Stover, P. Brandenburger, "Copper die bumps (first level interconnect) and low-K dielectrics in 65 nm high volume manufacturing," Electronic Components and Technology Conference, pp. 1611-1615, 2006.

[5] IPC/JEDEC, "Moisture/reflow sensitivity classification for nonhermetic solid state surface," J-STD-020D.1, March 2008.

[6] JEDEC, "Temperature cycling," JESD22-A104D, March 2009.

[7] JEDEC, "High temperature storage life," JESD22-A103-B, August 2001.

[8] P. J. Cheng, C. M. Chung, T. M. Pai, D. Y. Chen, "A challenge of 45 nm extreme low-k chip using Cu pillar bump as 1st interconnection," Electronic Components and Technology Conference, pp. 1618-1622, 2010.

Copper Wire Bond investigation on Multiple Surface Finishes - Enabling Wire Bond Packages without Gold

Mustafa Oezkoek
Atotech Deutschland GmbH
Erassmusstrasse 20, 10553 Berlin/Germany
Phone: +49 (0) 30 - 349 85 845   E-mail: mustafa.oezkoek@atotech.com

Nigel White
Atotech UK.,
William Street West Bromwich England
Phone:+44(0)1216067173   E-mail: nigel.white@atotech.com

Horst Clauberg
Kulicke and Soffa Industries, Inc. Fort Washington, PA
1005 Virginia Drive, PA 19034 Fort Washington
Phone: +1 215-784-6733 E-mail: hclauberg@kns.com

ABSTRACT

During the past two years, fine pitch copper wire bonding has finally entered high volume production. It is estimated that nearly 15% of all wire bonders used in production are now equipped for copper wire bonding. Most of these are used exclusively for copper wire bonding. In terms of pitch, copper wire is only barely lagging behind the most advanced gold applications. The most commonly used copper wire is 20um in diameter and 18um copper wire is entering final qualification. Evaluations with even finer wire are underway.

Although some technical challenges remain, many years of research have now resolved most of the problems associated with copper wire bonding and attention is beginning to shift from merely ensuring reliable manufacturing processes to optimizing processes for efficiency and throughput.   The most advanced wire bonders now have pre-configured processes specifically designed for copper. In addition to throughput optimization, further cost reductions are being sought. Among these is the desire to eliminate the high-cost gold not just from the wire, but also from the substrate.

On the substrate side the electronics packaging industry still works with electrolytic nickel / electrolytic (soft) gold (Ni/Au) for copper wire bond applications. This surface finish works with copper wire bonding but includes some disadvantages, such as:

- Thick expensive Au layers of 0.1 to 0.4μm
- Electrically connected pads (bussing for the plating) which require added space on the substrate.
- Pd-coated copper wire often delivers better results on gold covered finishes, but is two to three times more expensive as pure copper wire.

Furthermore electrolytic Ni/Au was not chosen as a result of in-depth investigations for the most effective surface finish. The selection was made because it was the surface finish with the highest distribution in the market for wire bond packages. This paper is offering the results of a two company joint work regarding an alternative copper wire bondable surface finish for substrates mainly with palladium as the final copper wire bondable layer. This offers further cost reduction possibilities. Furthermore, copper palladium intermetallics are regarded as very reliable.

Key words: ENEPIG, ENEP, wire bonding, gold wire bonding, copper wire bonding, pure Palladium.

SCOPE

This joint project (between Atotech Deutschland GmbH and Kulicke & Soffa) was initiated with three primary objectives in mind:

1. To screen all available surface finishes to determine their suitability for copper wire bonding
2. To determine a process window for copper wire bonding on the surface finishes that passed the first feasibility screening
3. To compare the copper wire bonding capability with that of the copper-palladium wire on selected finishes.
4. Pull test of stitch wire bonds over different temperature aging conditions.

In order to compare the results to a copper wire bondable reference, the surface finish of electrolytic Ni/Au (Soft gold) was selected. This finish is already proven in production to be copper wire bondable.
This paper focuses on the test results of the electroless Surface Finishes, there will be another paper where the results of the electrolytic surface finishes will be discussed.

EXPERIMENTAL

Parameter	Value/description
Cu-wire	20μm diameter Heraeus Maxsoft copper
Wire bonder	Kulicke and Soffa IConn^{PS} wire bonder equipped with the latest accessory kit for Cu wire bonding
Bonding tool	K&S CuPRAplus^{TM} tool (PN CU-FF-1099-P37, hole 1.0 mil, chamfer diameter 1.25mil, face angle 11°, inner chamfer angle 60°) with a granular surface finish. The surface finish of the bonding tool causes a slight granular structure on the copper wedge, as shown in Figure 1. This granular impression on the wedge enhances the bond adhesion of the copper wedge.
Cover gas	forming gas (5/95 H2/N2)
Bond temperature	165°C
Plasma treatment prior to bonding	Devices were cleaned in a March plasma cleaner for 5 min with Ar gas

Bonding windows were found by varying the ultrasonic bonding energy while maintaining all other parameter fixed in a simple thermosonic bonding process. The data in tables 1 and 4 were obtained by this simple bonding process. This simple thermosonic process was supplemented with a few microns of linear skid for the data in tables 3 and 4. For each ultrasonic energy value, 20 X-direction and 20 Y-direction wires were bonded. Results were averaged over the two directions.

It should be noted that more complex bonding process would likely result in stronger bonds, but the aim of this work was to compare surface finishes by the most straight-forward method. Fully optimized process would likely be different for each of the surface finishes. Also, no attempt was made to equalize the bonding results for X- and Y-direction wires by applying separate bonding parameters to each direction.

Wires were pull tested on a Dage 4000 with the hook placement 1/4 of the wire length from the stitch bond.

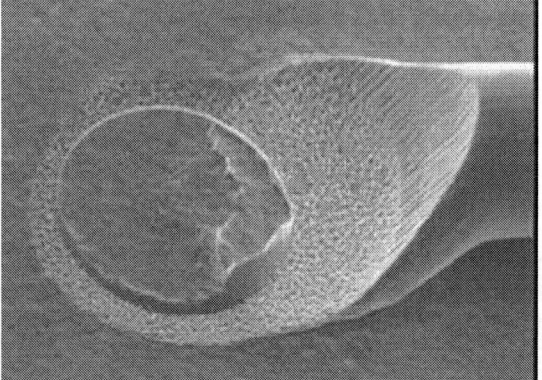

Figure 1: Cu-wedge with granular structure for advanced copper wire bonding

TEST RESULTS: SCREENING OF COPPER WIRE BONDABLE SURFACE FINISHES

A total of 21 different Atotech surface finishes were screened for copper wire bonding capability. Immediately, it was determined that seven of the surface finishes did not appear to be bondable with pure Cu wire. The surface finishes which failed in this first step were:
- Organic solderability preservative (OSP)
- Immersion Sn
- Electrolytic Ni/Sn
- Electrolytic Ni/Sn/Au
- Electroless Ni/immersion Au (ENIG)
- Electrolytic white bronze
- Electrolytic Ni/PdNi.

The OSP finish was discussed in the past as a copper wire bondable finish, but the test results show clearly that a copper wire bond could not be achieved. The OSP under the high temperature in the copper wire bonder begins to decompose and copper oxidation occurs on the substrate surface, which then prevents wire bonding.

In the first screening the surface finishes which showed limited potential for Cu-wire bonding with pure copper wire were:
- Electroless Ni / electroless Pd / immersion Au (ENEPIG) with a palladium layer co-deposited – with phosphorus (Pd-P)
- ENEPIG with a pure Pd layer
- Electrolytic Ni/Pd
- Electrolytic Pd/Au
- Electroless Ni / electroless Pd (ENEP) with Pd-P

Again it should be noted that more complex bonding processes could likely improve on these results

The surface finishes with the most potential for copper wire bonding were:
- ENEP (Electroless Nickel, Electroless Palladium with pure Palladium)
- Electroless pure Pd over Cu
- Electrolytic Ni/Pd/Au

Also, the reference surface finish, electrolytic Ni/Au with soft gold was found to be fully copper wire bondable.

TEST RESULTS: COPPER WIRE BOND WINDOW OF THE POTENTIAL SURFACE FINISHES

The ENEPIG surface finishes with different thickness variations were screened for copper wire bonding capability. Suprisingly, the surface finish of ENEPIG (proven for Au wire bonding) did not perform so well with pure Cu wire in all tested thickness variations (see Table 1, Sample Number 1-10). Also, different settings of ultrasonic energy (USG) did not help to improve the pure Cu wire bond on the ENEPIG finish.

As expected, the reference electrolytic Ni/soft gold samples performed well under different USG settings.(Table 1, Sample 11). Electrolytic Ni/soft gold also delivers good results in both the x- and y-directopn of copper wire bonding (Table 2 and 3, Sample Numer Λ,F).

Electroplated silver (Table 1, Sample Numer 13) also shows potential for copper wire bonding. However, silver is known to be less corrosion resistant [1] therefore, this finish was already phased out by several OEMs and packaging companies by 2008. In addition, because of current distribution during plating with the process has limitations when it comes to plating the same metal thickness on different pad sizes. Therefore, we did not focus any further attention on this finish.

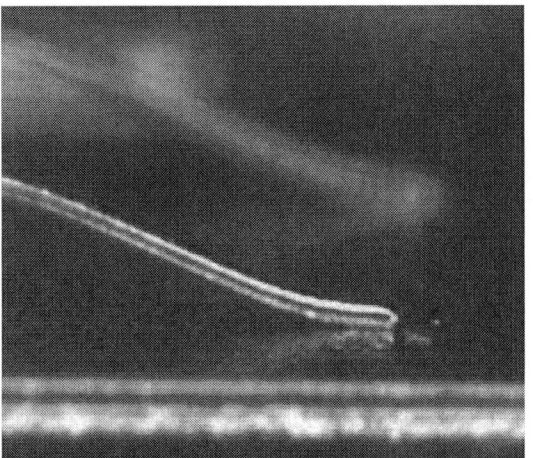

Figure 2: Cu wire bonding on ENEP with pure palladium

ENEP with pure Pd also performed very well in this test with a very good wire bonding window (Table 2 and 3, Sample Number G, H). This surface finish was evaluated as a potential copper wire bondable surface finish for the future. Its pull strengths are comparable to electrolytic Ni/Soft gold with pure copper wire manufactured at much lower material cost..

ENEP with Pd-P (4-6 wt %) offers a smaller process window compared to ENEP with pure Pd. Furthermore the variations of ultrasonic energy show inconsistent pull test results (Table 2 and 3, Sample Number G, H).

Excellent copper wire bond results were achieved with pure copper wire on electroless pure palladium over copper. (Table 1 and 3; Sample Number 12 and Table 2, 3; Sample Number L). Also this finish was identified as a surface finish for the future with excellent Cu-wire bonding capability in x and y direction with pure Cu wire.

TEST RESULTS: COPPER WIRE BOND WINDOW WITH PALLADIUM COATED CU-WIRE

In this investigation, it was unexpected to find that bare Cu-wire did not perform well on ENEPIG surface finishes. However, these results change when using Pd-coated Cu wire.

Electrolytic Ni/soft gold also shows very good results whith this type of wire (Table 4 Sample 3).
ENEPIG is copper wire bondable with this type of wire, but the pull test results show lower values compared to electrolytic plated Ni/soft gold. With respect to

electrolytic Ni/soft gold, the thicker gold is supposed to be the reason for the better results and an increasing of the Au thickness on ENEPIG to a similar level might improve results.

ENEP with pure Pd and electroless pure Pd over copper both either exhibit a narrower process window or offer even no bonding capability with Pd-coated Cu wire.

The test results with the Pd-coated copper wire shows clearly that the wire bonding characteristics of the wire are completely different compared to a pure copper wire. The pure copper wire attaches well on pure Pd layers, whereas Pd-coated wire types attach better on thicker Au Layers. This can be seen by comparing the results of Table 1,2,3 with those of Table 4.

COST COMPARISION OF DIFFERENT WIRE AND FINISH COMBINATIONS.

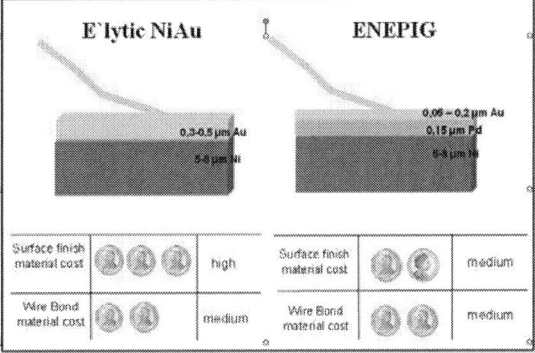

Figure 3: Au wire bonding on ENEPIG and electrolytic Ni/Au

Gold wire bonding on electrolytic Ni/soft gold has been a standard for many years, but the high cost of gold forced the packaging industry to consider ENEPIG finishes as a Au-wire bonding finish alternative. ENEPIG first became established in the packaging industry in 2006 and its use continues. Further cost reduction is possible by using electroless pure Pd in ENEPIG finishes, allowing a reduction in the Pd thickness to 0.05 μm or less [2].

Average Pull Strength - Cu wire					Plating Process	Ultrasonic Energy (mA)							
Sample	Ni [µm]	Pd [µm]	Au [µm]	Ag [µm]		70	80	90	100	110	120	130	140
1	7	0.05	0.05		ENEPIG (pure Pd)	2.6	NSOL	NSOL	2.7	2.5	2.5	3.0	2.8
2	7	0.15	0.05			NSOL	NSOL	2.4	2.6	3.2	4.0	3.8	3.7
3	7	0.15	0.1			NSOL	NSOL	1.6	2.3	3.0	2.8	3.1	4.0
4	7	0.25	0.05			NSOL	NSOL	2.0	2.8	3.2	3.5	3.5	4.2
5	7	0.35	0.05			NSOL	NSOL	1.3	1.8	2.1	2.3	2.8	3.1
6	7	0.05	0.05		ENEPIG (Pd-P)	NSOL	NSOL	NSOL	2.4	2.2	3.0	3.0	3.1
7	7	0.15	0.05			NSOL	NSOL	1.8	3.2	3.1	3.3	3.4	3.2
8	7	0.15	0.1			NSOL	NSOL	NSOL	1.4	1.5	2.6	3.1	2.4
9	7	0.25	0.05			NSOL	2.5	2.4	2.7	2.8	3.9	3.7	3.9
10	7	0.35	0.05			NSOL	NSOL	2.3	2.9	3.5	3.9	3.8	4.9
11	7	-	0.3		Electroplated Ni/Au Softgold	6.3	6.0	5.9	5.5	5.6	5.6	5.8	5.9
12		0.3			E'less pure Pd over Cu	NSOL	NSOL	7.0	6.4	6.5	5.7	5.6	5.3
13	7			1,0 -1,5	Electroplated Silver	4.4	4.7	4.6	4.5	4.5	4.9	4.9	5.1

Table 1. First test round results: average pull strength with pure Cu wire

Average Pull Strength (g) - Y Wires

	Plating				Cell:	9	8	7	6	5	4	3	2	1
Sample	Process	Ni	Pd	Au	USG:	110	120	130	140	150	160	170	180	190
A (Y)	Electrolytic Ni/soft Au	5,4	-	0,11		5,18	5,71	6,72	7,00	7,00	7,33	7,65	7,75	7,14
F (Y)	Electrolytic Ni/soft Au	7,2	-	0,31		5,91	6,03	6,97	7,12	6,82	7,70	8,20	8,17	8,18
G (Y)	ENEP (pure Pd)	7	0,15			7,27	6,58	6,67	6,54	6,26	6,17	6,87	7,44	7,66
H (Y)		7	0,3	-		7,04	6,70	6,64	6,56	6,94	6,69	6,39	6,94	7,38
I (Y)	ENEP (PdP)	8,1	0,1	-		NA	6,25	6,02	5,64	6,22	6,46	6,97	6,82	7,33
J (Y)		7,5	0,3	-		NA	6,23	6,04	5,84	5,35	6,08	5,96	5,66	6,35
K (Y)	Palladium Phosphor over Cu	-	0,2	-		NA	8,11	7,90	7,72	7,08	7,64	7,77	7,66	7,43
L (Y)	pure Palladium over Cu	-	0,2	-		8,45	7,97	7,49	7,53	7,04	6,72	6,52	6,84	7,17

Table 2. Second test round results: average pull strength in y direction with pure Cu wire

Average Pull Strength (g) - X Wires

	Plating				Cell:	9	8	7	6	5	4	3	2	1
Sample	Process	Ni	Pd	Au	USG:	110	120	130	140	150	160	170	180	190
A (X)	Electrolytic Ni/soft Au	5,4	-	0,11		4,28	3,44	4,57	5,02	5,04	5,39	5,20	5,78	5,75
F (X)	Electrolytic Ni/soft Au	7,2	-	0,31		6,20	6,31	6,61	6,28	6,15	6,70	6,64	6,74	NA
G (X)	ENEP (pure Pd)	7,5	0,12	-		5,17	7,72	7,26	7,10	6,61	5,95	4,83	5,03	4,64
H (X)		8	0,34	-		5,24	7,54	7,02	6,40	6,46	5,97	5,01	5,27	4,58
I (X)	ENEP (PdP)	8,1	0,1	-		NA	NA	7,05	6,02	6,09	5,57	5,27	4,71	3,96
J (X)		7,5	0,3	-		NA	8,14	6,39	6,07	NA	4,96	4,38	NA	4,05
K (X)	Palladium Phosphor over Cu	-	0,2	-		NA	NA	8,13	8,16	7,55	6,62	5,82	5,58	4,96
L (X)	pure Palladium over Cu	-	0,2	-		NA	8,70	8,53	7,39	7,44	6,73	5,27	4,95	4,28

Table 3. Second test round results: average pull strength in x direction with pure Cu wire

Average Pull Strength - Pd-Cu wire					Plating Process	Ultrasonic Energy (mA)					
Sample	Ni [µm]	Pd [µm]	Au [µm]	Ag [µm]		70	80	90	110	130	140
4	7	0.25	0.05		ENEPIG (pure Pd)	4.29	4.39	4.55	4.89	4.34	4.44
9	7	0.25	0.05		ENEPIG (Pd-P)	NSOL	4.71	4.32	4.80	5.27	5.29
11	7	-	0.3		Electroplated Ni/Au Softgold	7.96	8.39	7.90	7.56	7.37	7.72
13	7	0.15			ENEP (pure Pd)	NSOL	NSOL	3.27	4.11	4.66	4.29
14	7	0.3			ENEP (pure Pd)	NSOL	NSOL	NSOL	2.62	3.14	3.48
12		0.3			E'less pure Pd over Cu	NSOL	NSOL	NSOL	NSOL	NSOL	NSOL
13	7			1,0 -1,5	Electroplated Silver	6.92	6.00	6.35	5.56	5.69	5.76

Table 4. Third test round results: average pull strength with Pd coated Cu-wire

Legend:

NSOL or pull avg < 3g	3g < pull avg < 4.0g	Pull avg > 4.0g
Wedge or Ball lift off	Wire break	Wire break

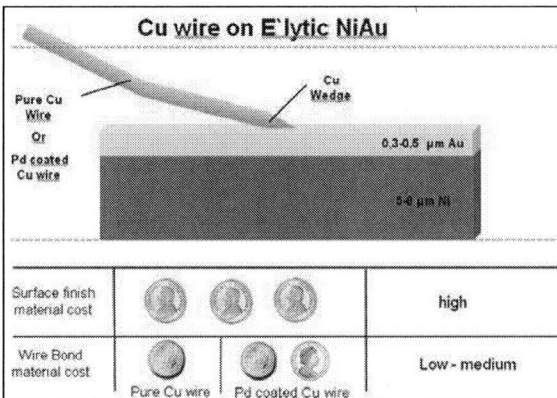

Figure 4: Cu wire bonding on electrolytic Ni/Au

Copper wire bonding on electrolytic Ni/soft gold was established in mass production in 2010. However, the high gold cost of the surface finish still remains in the electronic package. Furthermore, the Pd-coated copper wire is two to three times more expensive compared to pure Cu wire.

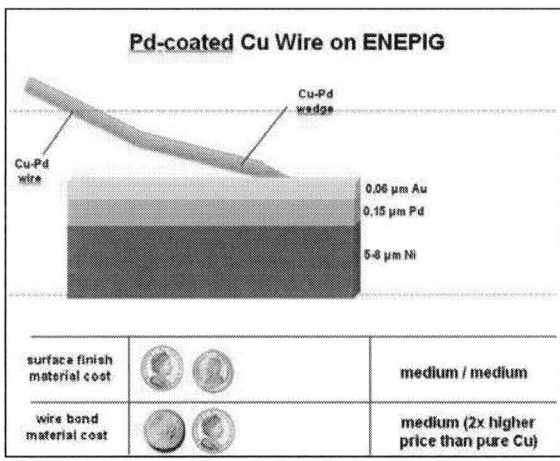

Figure 5: Pd coated Cu wire bonding on ENEPIG

Pure copper wire bonding on ENEPIG is more difficult and generally requires more complex, substantially slower bonding processes. Therefore, Pd-coated Cu is often preferred to achieve acceptable results. However, the results in this investigation show that this combination delivers worse results compared to electrolytic Ni/soft gold. Further increases of Au thickness in ENEPIG might bring similar results but at a higher cost level. Futhermore, the palladium coated Cu wire is harder which does bring further challenges for the ball bond on the Al pad.

Figure 6: Pure Cu wire bonding on ENEP with pure Pd

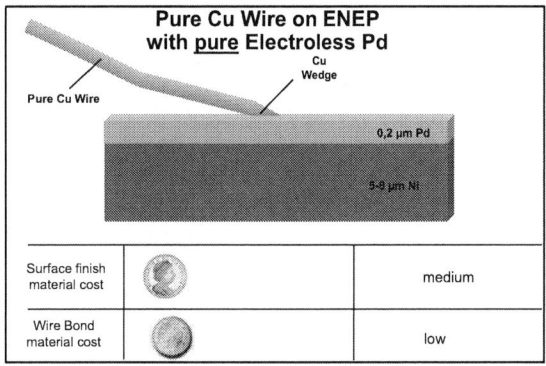

This investigation has shown that ENEP with pure Pd achieves pure copper wire bonding pull test results similar to those for electrolytic Ni/soft gold. As such, the electronics industry has an opportunity to (1) eliminate the Au costs from the surface finish and (2) utilize the pure copper wire, which performs well and is only half the cost of Pd-coated wire. Furthermore, if a barrier (i.e. nickel) is needed to prevent the consumption of underlying copper during soldering, the ENEP finish with pure Pd is the most cost-effective solution.

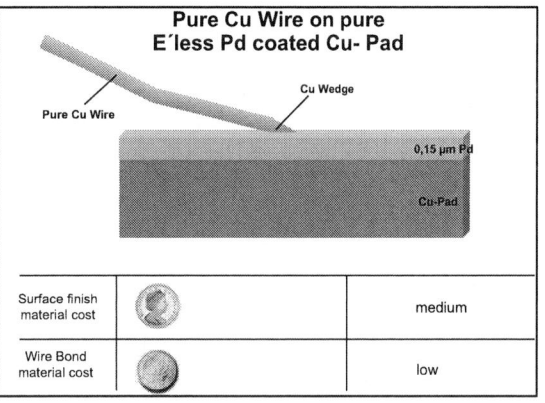

Figure 7: Pure Cu wire on pure electroless Pd on Cu-pad

Electroless pure Pd deposited directly on the copper pad is also very well suited for wire bonding with the pure Cu wire. The results are even better in comparison to pure Cu wire on electrolytic Ni/soft gold, as performed in this investigation. If no nickel barrier is needed (to stop the diffusion of copper into the solder) this surface finish is an alternative with excellent Cu-wire bonding results with pure Cu wire..

SUMMARY

A large variety of surface finishes were investigated to identify cost-effective surface finishes that are functionally suitable for Cu-wire bonding. When using pure Cu wire for bonding, two surface finishes were determined to be cost-effective alternatives to the standard process of electrolytic Ni/soft gold, namely:

- Electroless nickel / electroless palladium (ENEP) with a pure Pd deposit.
- Electroless pure Pd directly over copper

These two surface finishes have the potential to eliminate the Au cost from the electronic package, thus enabling lower manufacturing cost, without sacrificing performance or reliability.

FURTHER OUTLOOK

The Cu wire bonding performance of ENEP and electroless pure Pd over copper was determined in this study. There are further aspects currently being investigated that will be the topic of the next study, including:

1. Intermetallic growth between Cu wires bonded on different surface finishes

2. Study of the Pd-Au interconnection between the Cu-Pd wire and the pure Cu wire with Palladium covered finishes

3. Effect of temperature aging on performance

4. Pull test of stitch wire bonds over different temperature aging conditions.

Initial results from these latest investigations also appear very pomising. Furthermore, studies have already been performed on the 1st bond from the semiconductor side [3,4,5], which already show very good results for ENEP with pure Pd in terms of its reliability with pure Cu wire ball bonds over high-temperature storage testing.

REFERENCES:

[1] "Resistance of Common PWB Surface Finishes Against Corrosion in Harsh Environments" Mustafa Özkök , Mario Gensicke, Hugh Roberts, Joe McGurran, Guenter Heinz, SMTA 2010 Florida Orlando

[2] "The benefits of ENEPIG with pure Pd for Gold wire bonding" Mustafa Oezkoek, Gustavo Ramos, Dieter Metzger, Hugh Roberts

[3] „Nickel-Palladium Bond Pads for Copper and Gold Wire Bonding" Horst Clauberg, Asaf Hashmonai, Tom Thieme, Jamin Ling and Bob Chylak

[4] „Next Generation Nickel-Based Bond Pads Enable Copper Wire Bonding" Bob Chylak, Jamin Ling, Horst Clauberg, and Tom Thieme

[5] „Nickel-Palladium Bond Pads for Copper Wire Bonding " Horst Clauberg, Petra Backus and Bob Chylak

Low Temperature Wafer Bonding Technologies

Haubold, Marco (1); Baum, Mario (1); Schubert, Ina (1); Leidich, Stefan (1);
Wiemer, Maik (1); Gessner, Thomas (1,2);

Organization(s): 1: Fraunhofer ENAS, Germany; 2: Chemnitz University of Technology

Fraunhofer ENAS, Technologie-Campus 3, 09126 Chemnitz, GERMANY

Phone: +49-371-45001-274, Fax: +49-371-45001-374, E-mail: marco.haubold@enas.fraunhofer.de

Abstract

This paper will give an overview on up to date research results for wafer bonding technologies using heterogeneous substrate materials and advanced activation treatments. The following descriptions give a short introduction to certain bonding technologies focusing on low processing temperature. As examples direct and anodic bonding technologies were chosen and presented as part of the fabrication process for RF-MEMS. Furthermore the paper will show the theoretical aspects of the single bonding procedure as well as the experiments and the results of the development.

Key words: Low temperature, direct anodic bonding, packaging

Introduction

Low temperature wafer bonding technologies for joining two or more substrates with process temperature below 400 °C to avoid thermal and mechanical stress as much as possible are getting more and more important. Especially for smart systems integration the multitude of materials have to be considered within the field of packaging at zero level. Low temperatures open new possibilities for joining different substrate materials with different thermal expansion properties. So bonding technologies like silicon direct bonding, anodic bonding, eutectic bonding, adhesive and the glass frit bonding with low bonding temperatures have emerged. Especially the bonding of glass wafers offer new opportunities for fabricating Micro-Opto-Electro-Mechanical Systems (MOEMS) with a high internal volume resulting in new applications and more sophisticated products. The transparent encapsulation offers the unique opportunity to monitor the thermal impact and mechanical stress of single process steps such as bonding or grinding. For this purpose white light interferometry was used and the effect of low temperature anodic bonding on metallic cantilever beams was investigated.

Direct Bonding

Theoretical Aspects

Direct bonding is spread widely in MEMS fabrication due to its capability to fabricate stable compounds of silicon and glass stacks or heterogeneous compound wafers. Therefore the main fields of application can be seen in substrate fabrication (i.e. SOI, SOS), the production of bulk sensors for pressure monitoring or the measurement of acceleration and declination [l]. More recently, the technique has become a promising process step for the production of surface acoustic wave filters (SAW filter), where the Lithiumniobate is bonded to a silicon wafer [2].

The main physical driving force of the bonding technology is the formation of hydrogen bonds and attraction by Van der Waals forces when two, hydrophilic, sufficiently smooth and flat surfaces without any organic and inorganic contaminations are brought into atomic contact. The bonding strength is directly related to the density of chemical bindings between the touching surfaces. For silicon substrates, Silanol groups (Si-OH) provide the basic for bonding and in order to maximize their amount, chemical etching and plasma treatment prior to bonding has frequently been used [3]. The pre-bond is commonly performed either under atmospheric conditions or in a controlled gaseous atmosphere. The latter allows the adjustment of the internal pressure, supports the function of the device and reduces the probability for particles to be trapped inside the package most efficiently. Subsequently to pre-bonding, the wafer stack is annealed for promoting the formation of Siloxane bindings (Si-O-Si) and the bonding strength rises in consequence.

For the purpose of bonding glass substrates homogeneously with a sufficiently high bonding strength, several approaches have emerged. Some exploit high temperature, where the glass substrates are heated close to or above the transition temperature T_g. [4]. Other groups have employed a plasma treatment before bonding, resulting in ideally clean surfaces and the formation of Si-O-Si bonds as soon as the surfaces are in atomic contact [5].

Technological Aspects and Experiments

Within this work the direct bonding process was used for the fabrication of thick glass caps for the purpose of RF-MEMS packaging. Due to the out-of-plane movement of the structure, the cap demands for a cavity, covering the amplitude of the device's displacement. Because of monitoring the quality of the device during and after the fabrication processes by means of optical techniques, a sufficiently transparent housing for visible light is necessary. Several technologies exist for producing the desired pattern in glass substrates like mechanical milling, wet etching by hydrofluoric acid or reactive ion etching. All those technologies share the disadvantage that either the resulting bottom will be opaque or the process is rather unsuitable for the machining of hundreds of structures on wafer level. To overcome these issues sand blasting was chosen as alternative process on wafer level where either technology marks and through holes were fabricated within one cycle. The thickness of the blasted 4 inch wafer was set to 300 μm or 400 μm respectively. A characteristic property of holes, fabricated by sand blasting is a slope angle of 70 °. Hence, the dimension of the backside geometry is smaller than initially designed what has to be taken into consideration for the design process. During wafer fabrication, laminated resist served as mask layer which got locally removed by photolithography, allowing the glass to be exposed for perforation. The substrate's opposing surface was protected by a removable polymer film during the subsequent blasting procedure, ensuring a sufficiently high surface quality, allowing for direct bonding with the second glass wafer in order to seal the perforation hermetically (see figure 1).

As introduced, the direct bonding process requires a surface without contaminations and a roughness below 1 nm. The combination of the treatment by chemical mechanical polishing (CMP) with a subsequent RCA cleaning procedure fulfilled these criteria and was applied to the brittle and perforated glass substrate. The subsequent bond of the bare glass wafer with the perforated glass wafer was done under ambient conditions for monitoring the propagation of the bond wave. This qualitative characterization method allows the assessment of the surface quality. One can expect a high level of activation and cleanliness and a sufficiently low surface roughness in the case the bond wave propagates quickly and continuously. Due to the absence of temperature sensitive elements, no limitations were defined and the annealing at 400 °C served for the realization of a mechanically stable

and hermetically sealed cap compound. For other applications, the annealing temperature can be reduced significantly.

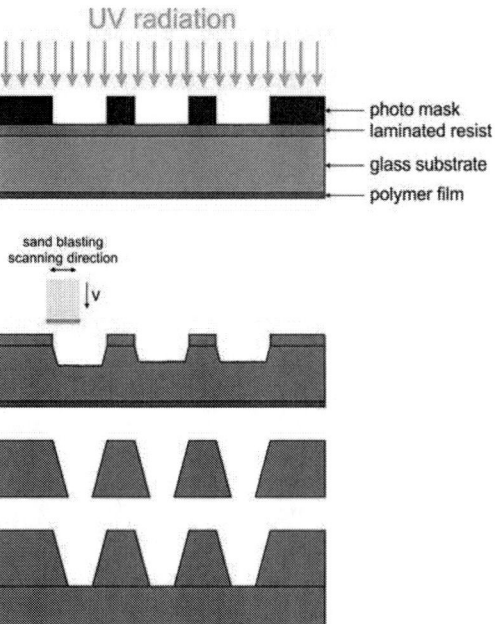

Figure 1: Schematic of the wafer preparation and bonding process for glass cap fabrication with cavities (1) UV radiation of laminated resist via photo mask, 2) sand blasting process for pattern fabrication, 3) removal of photo resist and polymer film, 4) direct bonding process with second glass substrate (from top to bottom)).

Results and Characterization

Before the first glass substrates could be bonded successfully, the CMP process needed to be adapted in respect to the through holes. Usually, structures and especially edges are getting rounded off during polishing, resulting in a bond void, surrounding the structure and reducing the bonded area therefore. To overcome this issue, the parameters down force, backpressure and polishing duration were optimized in order to find a minimum load, which still satisfied the requirements for direct bonding on wafer level. The Maszara blade test was chosen to evaluate the surface energy after bonding of two glass substrates. The method allows the calculation of interfacial energy by the measurement of the crack length, initiated by an inserted razor blade [6]. The results showed no significant decrease of the energy level when pressure and duration were reduced. The minimum observable crack length amounted up to 9.8 mm, corresponding to 0.81 J/m². The energy is seen to be sufficiently high and the polishing recipe got applied to the actual cap wafer.

In consequence of the conditioning prior to bonding, the bond wave spread fast and continuously even at the through holes' edges, showing the high suitability of the chosen technologies. In order to increase the bond strength, the wafer stack got annealed at 400 °C afterwards.

Figure 2 depicts the final result. A stable and almost void free bond could be achieved after bonding in ambient atmosphere and annealing at 400 °C. Hence, the direct bonding process offers a high potential to fabricate stable glass compound wafer stacks with high volume cavities and a perfect optical transparency.

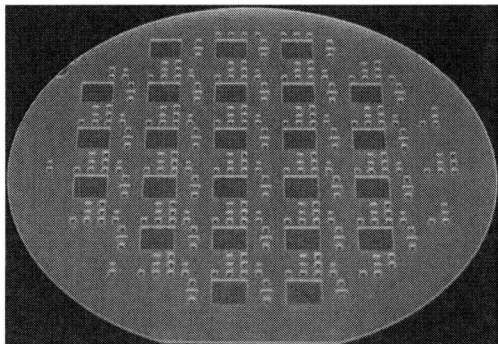

Figure 2: Photography of directly bonded glass wafers after preparation by sand blasting. The upper surface was protected by UV sensitive foil and the window showed transparency > 99 % for a wide spectrum.

Low Temperature MEMS Packaging by Anodic Bonding

Theoretical Aspects

The anodic bonding process of glass and silicon is a technology with a high level of maturity und has extensively been used for the fabrication of MEMS and MOEMS [7-9]. The technology involves the application of an electrical field to a silicon-glass stack with a voltage ranging from 200 V to 2000 V. The sodium ions in Borosilicate glasses serve as positive charge carrier and drift towards the cathode whereas electrons in the silicon wafer are accelerated towards the anode. Hence, an electrical field is formed at the wafers' interface, creating a high electrostatic force, resulting in the reduction of the gap in between and allowing the adjacent atoms to form chemical bindings (Si-O-Si). Usually the mobility of sodium ions in glass substrates is rather low, wherefore external energy demands to be provided by the increase of process temperature higher than 300 °C. The actual bonding process is limited by the maximum current (e.g. 10 mA) and lasts for 5 min to 40 min [10].

Beside the traditional anodic bonding process of glass and bare silicon substrates, successful experiments have been reported using aluminium or oxide covered glass wafers [11, 12]. The involved mechanisms at the glass-aluminium interface have been investigated in [13] and found to be a combination of directed growth of aluminium oxide into the glass substrate, accompanied with the oxidation of aluminium atoms at the bond interface and the migration of mobile ions in the glass.

For a large number of devices, the commonly applied temperature of 400 °C during bonding causes critical damage to the element. Typical failure mechanisms involved, are the generation of intrinsic stress due to differences in the coefficients of thermal expansion (CTE), the initiation of diffusion processes at material interfaces and in doped regions or the relaxation of pre-stressed structures. Hence, the reduction of bonding temperature is necessary . Two main techniques can be used without degrading the bonding strength to fulfil that demand. On the one hand, the voltage is raised to 1000 V or more, causing a higher electrostatic field which is capable to accelerate more sodium ions at lower temperatures. Hence a drift of positively charged ions occurs. But even though, the drift current of higher temperatures cannot be realized. In addition, layers being sensitive to high voltage will get damaged or destroyed. On the other hand, intermediate layers are used to lower the process temperature during anodic bonding. Past research projects have taken the deposition of thin layers of amorphous silicon or aluminium into consideration [14]. Hata et al. have shown that sufficiently strong bonds can be achieved even at 210 °C when aluminium as intermediate layer had been used [15]. The thickness of the aluminium layer was chosen to be 1 µm.

Technological Aspects and Experiments

An RF MEMS switch was chosen as application example for anodic bonding below 250 °C. The device is composed of a pre-stressed aluminium cantilever beam, showing a deflection in Z direction, designed to perform an out of plane movement when being actuated. A detailed description of the fabrication process of the switch and the corresponding electrical characterization can be found in [16]. The demand on low temperature packaging process is given by the relaxation of the cantilever beam when being exposed to a high temperature. Hence, the packaging process needs to realize: i) low thermal impact during the bonding process, ii) a stable housing and strong bonding with an insulating substrate and iii) a cavity surrounding the device with an ideal optical transparency, ensuring a proper deflection. The last requirement assures a highly reliable process monitoring of the active component by the method of white-light interferometry. The unique measurement principle involved the characterization of the beam deflection before and after critical processes such as the bonding procedure. The relaxation was monitored by the decrease of maximum displacement.

The above mentioned demand concludes the application of glass substrates as capping wafer. Because of the required cavity, the fabrication process of a glass compound wafer by sand blasting and direct bonding, as described in section 1, was used. Substrates of four inch diameter with a thickness of 300 μm were chosen to serve as bare capping substrate and as perforated wafer respectively. Either direct bonding or anodic bonding is suitable for the packaging at wafer level and meets the demand on low process temperatures. Both techniques were evaluated for its capabilities to encapsulate processed silicon wafers. Therefore, silicon dummy substrates were prepared by thermal oxidation, aluminium sputter deposition and patterning in phosphoric acid, covered by photo resist. In the case of low temperature anodic bonding, the actual bonding occurred between the glass cap and 200 μm narrow metal frames, surrounding the aluminium test structure. For direct bonding, no frames were present and the cap showed a gap for the compensation of the aluminium structures, ensuring the atomic contact between glass and the oxidized silicon surface. With reference to the results of those preliminary studies, the more proper method was used for bonding of the actual device wafer (thickness 200 μm) to the glass compound.

Results and Characterization

Preliminary Direct Bonding Investigations

Because of the MEMS-first fabrication procedure, the highly sensitive cantilever structures cannot be conditioned prior to bonding like usually applied substrates (i.e. plasma treatment and DI water flushing, CMP process). Hence, the bond surface got affected by all processes like metallization and patterning. The sole approach to secure a high bond yield could be seen in pre-treatment of the glass cap. In order to evaluate its effect on the bonding quality, the direct bonding was performed with and without a CMP pre-treatment of the glass substrates. The results of the pre-bond are shown in figures 3 and 4. Bonding voids occurred at the glass-silicon interface after pre-bonding for both recipes.

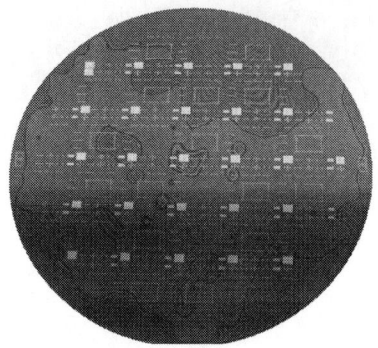

Fig. 3: Direct bond of glass to oxidized silicon substrate with aluminium test pattern. A high fraction of interference fringes could be observed, representing unbonded areas.

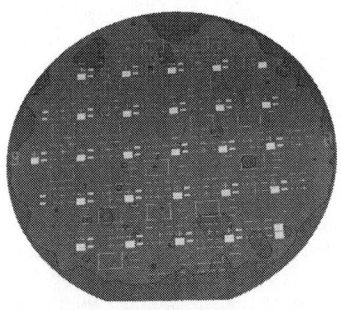

Fig. 4: Direct bond of glass to oxidized silicon substrate with aluminium test pattern. Bonding voids were reduced in comparison to none pre-treatment but were too high for a reliable packaging.

The main difference could be found in a faster and more stable propagation of the bond wave across the interface and in the reduction of the density of unbonded areas when the polishing process was applied to the glass wafer. The number of trapped particles, causing concentrical interference fringes will be reduced when bonding in vacuum atmosphere or inside a cleaning tool (e.g. SUSS CL200). Beside the high potential of direct bonding, the experiments have shown the importance of the device wafer's surface quality in respect to organic / inorganic contaminations and particles.

Low Temperature Anodic Bonding

Experiments analogue to the direct bonding investigations proved anodic bonding at low temperatures to be the most favourable bonding technology and got finally applied to the packaging process of the actual device, which is schematically shown in figure 5.

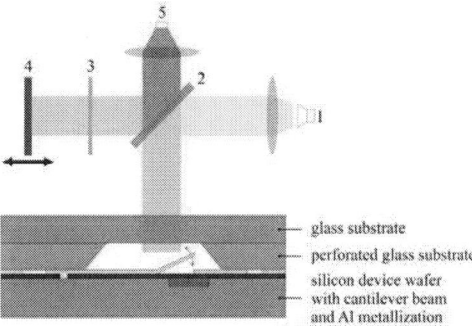

Fig. 5: Schematic of the packaged MEMS device. The cap composes of two directly bonded glass wafers in order to ensure a proper movement of the cantilever beam. Process monitoring was performed by white light interferometry through the transparent glass window, while a glass sample with same specifications was inserted into the reference path - 1) light source, 2) beam splitter, 3) glass chip for compensation of dispersion, 4) movable mirror, 5) detector.

In order to validate the impact of temperature on the MEMS switch, its deflection was measured before and after bonding by white light interferometry. The method's unique capability to measure vertical geometries up to several millimetres combined with a high lateral resolution affords reliable results with high accuracy. The compensation of dispersion of the light when measuring through the glass cap was realized in the reference path of the interferometer, where a glass sample of same specifications was inserted (compare to figure 5). Hence the interference fringes could be observed on the metal layer and the difference in height between surface and tip of the structure was measured. Figures 6 and 7 illustrate and table 1 summarizes the change in deflection due to bonding.

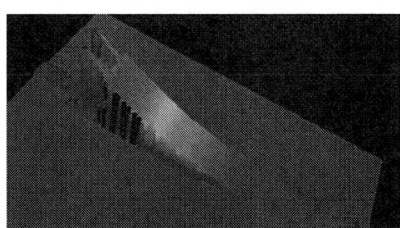

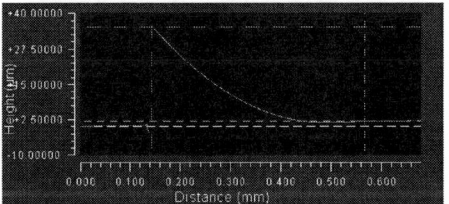

Figure 6: 3-dimensional image of the cantilever deflection *before* bonding by white light interferometry. The average deflection amounted to 33.4 µm.

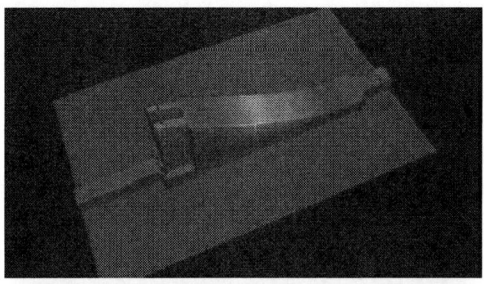

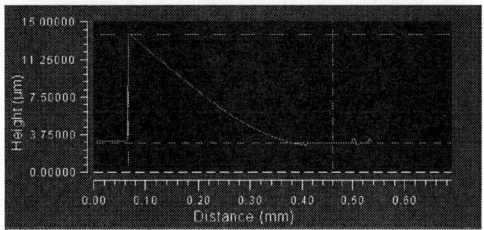

Figure 7: 3-dimensional image of the cantilever deflection *after* bonding by white light interferometry. The average deflection amounted to 11.2 µm.

Table 1: Summary of the measured values for metal layer deflection before and after the anodic bonding process at 210 °C.

	Deflection	
	before bonding	after bonding
Maximum	35.4 µm	12,4 µm
Minimum	27.9 µm	7,6 µm
Average	33.4 µm	11,2 µm

The deflection got reduced by almost 20 µm which is ascribed to recrystallization processes in the metallic lattice of the aluminium layer. Figure 8 shows the silicon-glass-glass wafer stack after anodic bonding. The MEMS switch can be recognized through the glass cap, surrounded by the aluminium bonding frames. For further testing of the bond, the stack got grinded on wafer level, in order to realize a total thickness of the silicon of 100 µm. The low temperature anodically bonded interface withstood that harsh process and neither damage nor voids could be observed after grinding.

The results proved the feasibility of low temperature bonding processes basing on anodic bonding for MEMS fabrication and emphasised its influence on the performance and maximum lifetime of a miniaturized device.

Fig. 8: Photography of the silicon-glass-glass triple stack after anodic bonding and grinding of silicon substrate to final thickness of 100 μm. The bonding was realized by 200 μm wide aluminium frames at 210 °C.

SLID Bonding

Theoretical Aspects

SLID bonding as a wafer bonding technology is based on a transient liquid phase of one bonding partner and the immediate isothermal solidification through diffusion and mixing with a second bonding partner (Solid Liquid Interdiffusion – SLID) and, additionally the bonding temperatures are dependent from the low melting metallic partner.

Solid liquid interdiffusion bonding was firstly described by Bernstein in 1966, who investigated the material combinations of Au/In, Ag/In, and Cu/In at a processing temperature around 156 °C, defined by the melting point of In. He described the following phases of reaction during the SLID process [17]:

    1. Wetting
    2. Alloying
    3. Liquid Diffusion
    4. Gradual Solidification
    5. Solid Diffusion

Today the focus lies on materials like Cu and Sn because there is a better availability of these materials in MEMS and microelectronics technologies. Munding published the investigation of CuSn SLID bonding in 2008 and he mentioned the advantage that CuSn forms a stable ε-Phase resulting in Cu₃Sn [18] (compare fig. 9).

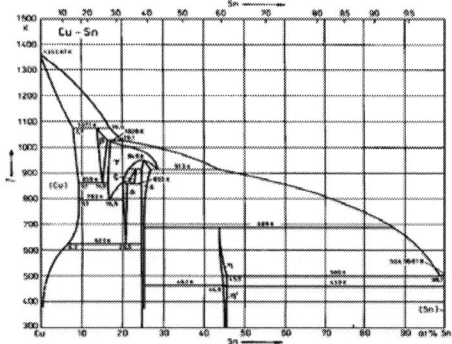

Fig. 9: Equilibrial phase diagram of CuSn binary system [19]

Technological Aspects and Experiments

Next to the state of the art this article describes the experimental investigation of the CuSn and AuSn SLID bonding at wafer level. For the experimental setup 4 inch silicon wafers were used. Cu and Au were chosen with Sn as SLID partner because these combinations are widely used for chip and component integration. For the characterization of the bond strength special chevron notches were designed that realize the bonding interface (see fig. 10). Onto oxidized Si wafer a 10 nm/ 15 nm Ti/TiN adhesion and a 150 nm Cu or Au seed layer were deposited by sputtering. After lithographic patterning of the test structures the Cu or Au were electro chemical grown to 3 μm after a native oxid was removed by using 10 % sulfuric acid. On the Cu or Au a certain layer of tin was deposited and the variation of the thickness should show the influence to the bond strength. Afterwards the resist was removed by a wet process and the wafers were finished for bonding test. Because of a liquid phase the requirements for surface roughness are not as high as for direct or anodic bonding.

 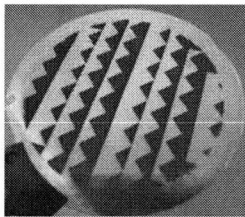

Fig. 10: Micro chevron notch lithographic mask and wafer after depositing Au and Sn andremoving the resist.

The bonding experiments were performed in a Suss SB6 substrate bonder under vacuum applying a tool pressure of 150 kPa and more. The temperature was set to T_{AuSn}=280 °C and T_{CuSn} = 300 °C. The process time for CuSn was 10 min and cool down [20]. In case of AuSn the process time was shortened to 5 min and cool down because there is no need for phase building diffusion.

Results and Characterization

The mechanical quality of the bond was tested with a razor blade first, performing a maszara blade test and measuring the crack between the substrates. If there is no crack a dicing process was made showing the yield of bonded chips after dicing (see also fig. 11). At least a pull test or micro chevron test could be applied as well as a shear test. In both cases the thinner layers showed the worse bonding properties and the weaker bonding strength after testing.

Regarding the microstructural analysis that was worked out using SEM, EDX and EBSD microscopy the main difference was the intermixing behavior between Cu and Sn and Au and Sn respectively (see also fig. 12).

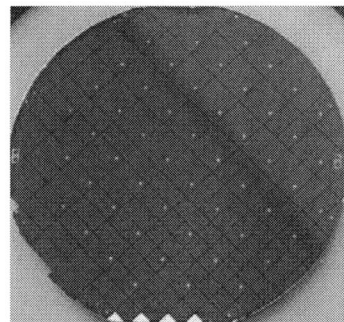

Fig. 11: SLID bonded wafer (AuSn) after mechanical treatment (dicing) showing a 100 % yield

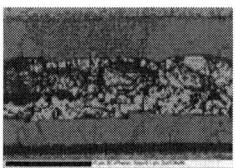 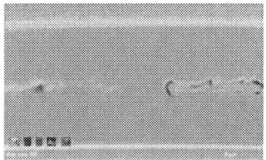

Fig. 12: EBSD analysis of CuSn left showing the Cu₃Sn Phase (light blue) and EDX analysis of AuSn right showing the complete intermixing of the materials and some remaining Sn spots (light blue) [19].

Conclusion and Outlook

Direct bonding of perforated glass substrates and anodic bonding of glass compound wafer to silicon substrates were successfully investigated within this work at 210 °C in order to fabricate a transparent package for an RF MEMS device. The transparency of the cap promotes the process monitoring by white light interferometry in respect to the reduced deflection of a pre-stressed cantilever beam after bonding process. Results have shown that a fraction of intrinsic tension got relieved on the one hand, but a particular high value of deflection remained after wafer level packaging when low temperature is applied By the increasing number of applied materials for MEMS/NEMS fabrication along with the demand on reducing the energy consumption and process time, low temperature bonding processes will gain further importance within the production procedure and the exploitation of metallic diffusion processes, nano scale effects as well as localized heating will seriously be taken into consideration for tomorrow's device fabrication.

Furthermore it could be shown that there is promising possibility to perform low temperature bonding using metals like CuSn and AuSn reaching a reliable bonding strength. Actual tests show Si fracture during shearing and pulling so the test structure must be adapted. Additional tests will be worked out to consider the hermeticity and the longterm quality of the bonded devices. With usual micro system technologies it is possible to perform the site-selective layer deposition of the bond interface materials as well as the bonding at wafer level. The technologies seem to be applicable for 3D integration processes.

Acknowledgements

The authors would like to thank the German Federal Ministry of Education and Research (BMBF) for funding the projects (contract 50YB0910 and 16SV3675). Furthermore the colleagues at the Center for Microtechnologies (ZfM) of Chemnitz University of Technology shall be acknowledged for their work.

References

[1] C. Jia, M. Wiemer, and T. Gessner, "Direct bonding with on-wafer metal interconnections", Microsystems Technologies, Vol. 12, No. 5, pp. 391-396, 2004.

[2] H. Takagi, R. Maeda, and T. Suga, "Room-temperature wafer bonding of Si to LiNbO3, LiTaO3 and Gd3Ga5O12 by Ar-beam surface activation", J. Micromech. Microeng., Vol. 11, pp. 348-352, 2001.

[3] V. Dragoi, G. Mittendorfer, C. Thanner, P. Lindner: "Wafer-level plasma activated bonding: new technology", Microsystems Technology for MEMS fabrication, 2002, pp. 509-515.

[4] N.F. Raley, J.C. Davidson and J.W. Balch, "Examination of glass-silicon and glass-glass bonding techniques for microfluidic systems", Proc. SPIE Vol. 2639, pp. 40-45; 1995

[5] M.M.R. Howlader, S. Suehara, T. Suga: "Room temperature wafer level glass/glass bonding", Sensors and Actuators A 127, pp. 31–36, 2006

[6] W.P. Maszara, G. Goetz, A. Caviglia, J.B. McKitterick: "Bonding of silicon wafers for silicon-on-insulator", J. Appl. Phys. Vol. 64, No. 10, pp. 4943-4350, 1988

[7] M. A. Schmidt, "Wafer-to-Wafer Bonding for Microstructure Formation", Proceedings of the IEEE, Vol. 86, No. 8, August 1998

[8] H. Henmi and S. Shoji and Y. Shoji and K. Yoshimi and M. Esashi, "Vacuum packaging for microsensors by glass-silicon anodic bonding", Sensors and Actuators A: Physical, Vol 43, No. 1-3, pp. 243 – 248, 1994.

[9] D.C. Abeysinghe, S. Dasgupta, J.T. Boyd, H.E. Jackson, "A novel MEMS pressure sensor fabricated on an optical fiber", Photonics Technology Letters, IEEE, Vol. 13, No. 9, pp. 993-995, 2001

[10] Q.-Y. Tong and U. Goesele; "Semiconductor Waferbonding"; John Wiley & Sons, New York, first edition, 1998.

[11] M. Neese, A. Hannborg: "Anodic bonding of silicon to silicon wafers silicon oxide, polysilicon or silicon nitride", Sensors and Acruators A, 37-38, 61-67, 1993.

[12] K. Schjølberg-Henriksen et al.:"Anodic bonding of glass to aluminium", Microsyst Technol Vol. 12, pp. 441–449, 2006.

[13] A.T.J. van Helvoort and K.M. Knowles: "Nanostructures at Electrostatic Bond Interfaces", J. Am. Ceram. Soc., Vol. 86, No. 10, pp. 1773–76, 2003.

[14] J. Wei, C.K. Wong, L.C. Lee, "Improvement of anodic bond quality with the assistance of sputtered amorphous film", Sensors and Actuators A, Vol. 113, pp. 218–225, 2005.

[15] S. Hata, J. Froemel, T. Gessner: "The improvement of the bonding strength of low temperature anodic bonding" Transactions Japan Institute of Electronics Packaging: Packaging Workshop, Tokyo, Mar 16-18, 2005.

[16] C. Siegel, V. Ziegler, B. Schoenlinner, U. Prechtel, H. Schumacher: "Simplified RF-MEMS switches using implanted conductors and thermal oxide", 36th European Microwave Conference, Manchester UK, Sep 10-15, 2006, pp. 1735-1738.

[17] L. Bernstein: "Semiconductor Joining by the Solid-Liquid-Interdiffusion (SLID) Process", Journal of Electrochemical Society, Vol. 113, No. 12, pp. 1281-1288, 1966.

[18] A. Munding et al: "Cu/Sn Solid-Liquid Interdiffusion Bonding", Wafer Level 3-D ICs Process Technology, chapter 7, pp. 131-170, Springer Verlag New York, 2008.

[19] Raynor, G.V., "The Cu-Sn Phase Diagram", Annoted Equilibrium Diagram Series, No. 2, The Institute of Metals, London, 1944.

[20] M. Baum et al: "Solid-Liquid-Interdiffusion Bonding für 3D-Integration und Wafer-Level-Packaging", Proceedings of Mikrosystemtechnik-Kongress, Berlin, Oct 12-14, 2009.

Effect of thermal shock conditions on reliability of chip ceramic capacitors

Alexander Teverovsky

Dell Services Federal Government, Inc.
Code 562, NASA GSFC, Greenbelt, MD 20771
Alexander.A.Teverovsky@nasa.gov

Abstract

Different size X7R MLCCs have been subjected to three types of thermal shock testing: terminal solder pot dip test, ice water test, and liquid nitrogen drop test. Electrical characteristics of the parts were measured through various test conditions to determine critical temperatures that result in fracturing and electrical failures. Optical examinations and cross-sectional analysis were used to confirm the presence of cracks. Mechanisms of fracturing and the effectiveness of different thermal shock methods are discussed. It is shown that the heat conduction and direction of temperature changes (heating or cooling) are critical factors of thermal shock testing of MLCCs. The probability of fracturing depends also on the level of residual mechanical stresses, mostly in terminal areas of ceramic capacitors.

Key words: ceramic capacitors, thermal shock, failure

Introduction

Soldering related thermal shock (TS) is one of the major causes of fracturing in MLCCs that might result in latent defects and cause failures with time during application. The probability of cracking generally increases with the size of capacitors, and is especially high for manual soldering (see literature review in [1]).

Various thermal shock techniques to assess the susceptibility of ceramics to fracture are described in military and industry specifications and in technical literature. The most popular techniques are water quenching, that is typically used for ceramic materials, and solder dip test that is widely used for high-reliability MLCCs and is described in relevant military specifications. Immersion in liquid nitrogen is often used to assess thermo-mechanical reliability of the parts intended for cryogenic applications.

Solder dip test provides thermal shock conditions that closely simulate stresses related to manual soldering. However, clamping of the parts or mounting in a fixture for plunging into molten solder might have a strong effect on test results [2]. GEIA-STD-0006 (2008) that was developed with the purpose of setting industry-wide requirements for solder dip to replace finishing on electronic parts emphasizes that the controls needed for the dip process are not well enough understood to be included in an industry standard [3].

The water quenching technique has been used for ceramic capacitors by several authors. Koripella [4] studied fracture toughness, modulus of rupture, and thermal shock resistance on COG, X7R, and Z5U type materials. The 1206 size COG and X7R samples showed a sudden drop in the strength at a temperature difference of 150 °C and Z5U samples at a temperature difference of 125 °C. These values were close to the predicted critical temperatures (thermal shock resistance parameter).

Chang and co-authors [5-7] used ice water test (IWT) to study fracturing in multilayer ceramic capacitors of different size. It was shown that the thermal shock resistance decreases in a row 0402, 0603, 0805, and 1206 capacitors, and the corresponding critical temperatures are 400 °C, 300 °C, 200 °C, and 100 °C. TS resistance of Y5V and Z5U 0.1 uF 0805 size capacitors was shown to be inferior compared to similar size and value of BT parts. Yeung et al. [8] confirmed that smaller capacitors have higher TS resistance than the larger ones. However, the dependence of thermal shock resistance on the thickness differ substantially from the one predicted by calculations, likely due to the presence of preexisting flaws.

Considering a wide use of manual soldering for space applications, there is a need for better understanding of the effect of TS on MLCCs and in development of adequate test methods that would assure the robustness of MLCCs at soldering conditions. The purpose of this work is to get more insight into the mechanism of the effect of TS on MLCCs and to compare the effectiveness of different thermal shock methods. For this purpose various types of large-size MLCCs were subjected to the terminal solder pot testing, ice water testing, and liquid nitrogen drop testing. Measurements of electrical characteristics of the parts, optical examinations, and cross-sectioning were used to assess the effect of different TS conditions.

Experiment

Sixteen types of multilayer ceramic chip capacitors from five vendors were used in this study (see Table 1). All parts employed X7R dielectric materials, nine parts were rated to 1 µF 50 V, five to 0.1 µF 100 V, and two to 10 µF 50V.

Table 1. Capacitors used for thermal shock testing

Gr	MLCC, uF/V	Mfr	Size	L (mm)	W (mm)	H (mm)
1	1/50	A	1210	3.29	2.61	2.02
2	1/50	A	1812	4.38	3.13	1.14
3	1/50	A	2220	5.76	5.36	1.72
4	1/50	B	1210	3.1	2.51	1.54
5	1/50	B	1812	4.3	3.1	1.28
6	1/50	B	2225	5.57	6.24	0.92
7	1/50	C	1206	3.29	1.74	1.66
8	1/50	C	0805	2.02	1.27	1.28
9	1/50	C	2220	5.78	4.82	1.89
10	10/50	A	2220	5.78	4.94	2.52
11	10/50	T	2220	5.87	5.34	2.26
12	0.1/100	A	1825	4.38	6.31	1.11
13	0.1/100	A	1812	4.43	3.11	0.92
14	0.1/100	P	1825	4.65	6.53	1.47
15	0.1/100	A	1210	3.2	2.5	1.5
16	0.1/100	A	1210	3.2	2.5	1.5

The parts were subjected to terminal solder dip test (TSD) at solder pot temperatures varying from 300 °C to 350 °C; ice-water quenching test at preheat temperatures from 150 °C to 250 °C, and liquid nitrogen (-197 °C) drop test. Details of the techniques used are described in the sections below. Vicinal optical illumination microscopy [9], measurements of capacitance, dissipation factor and leakage currents were used before and after each test.

Terminal solder dip test

Four types of large-size capacitors from Gr.3, Gr.6, Gr.12, and Gr.14 with ten samples each were used for this test. One terminal of each capacitor was dipped for 5 seconds into a pot of molten solder heated initially to 300°C and then the part left to cool at room temperature for 5 minutes. This solder dip and cooling procedures were performed 10 times for each part, after which all capacitors were tested for leakage currents. To enhance moisture sorption in possible cracks, post TS measurements were repeated after allowing samples to store at room conditions for one week. Then the solder dip test was repeated at molten solder temperatures of 325°C and 350°C.

Terminal solder pot test that have been used in this study provides more adequate test conditions compared to conventional solder dip testing. During this test one terminal of the capacitor is clamped in a fixture while the other is touching molten solder. This eliminates local pressure to ceramic and unpredictable changes in temperature distribution caused by application of tweezers, stabilizes temperature at the clamped terminal, and creates a significant temperature gradient at another terminal.

Figure 1 shows results of the testing for Gr. 3 and Gr. 6 of capacitors. No significant difference in leakage currents was observed before and after the solder shock test at 300°C and the currents remained stable after one week of storage. Ten additional solder dip cycles after solder temperature was increased to 325°C also did not result in degradation. For three part types, Gr.6, Gr.12, and Gr.14 no change in currents was observed even after 10 solder dip cycles at 350 °C. Microscopic examinations did not reveal any cracks in capacitors from Gr.6 and Gr.12. Some variations of capacitance and dissipation factors were consistent with the de-ageing process due to the temperature of the parts exceeding the Curie point.

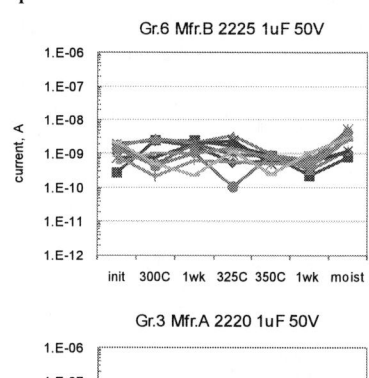

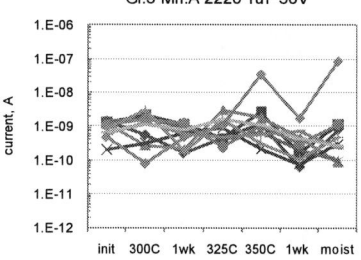

Figure 1. Variations of leakage currents during TSD testing in two groups of capacitors.

After 350 °C solder dip test one out of ten samples in Gr.3 increased DCL almost by two orders of magnitude (see Figure 1b), whereas all other parts in this group remained stable and did not degrade even after storage in humid environments (3 days at room temperature and 98% RH). Microscopic examinations confirmed the presence of cracks near terminals in the failed sample.

Results of this study show that large-size capacitors (2220 and 1825) have high resistance against thermal stresses developed during solder dip testing. No failures were observed in 40 parts after 10 cycles at 300 °C and 325 °C, and only one failure occurred after testing after 350 °C cycling. This is in agreement with the results of our previous study [1] that showed no failures in 70 samples from seven lots of large, 2220 size, parts tested by 100 terminal solder dip cycles at a solder temperature of 300 °C.

Ice water test.

Ten samples from each of the 16 groups shown in Table 1 were placed on a hot plate, covered with a Teflon cover, heated to a desired temperature and stabilized for 10 minutes. After stabilization, capacitors were rapidly quenched into water by sweeping the cover and dropping the parts from a hot plate into a nearby bath with ice water. Leakage currents were measured within one hour after IWT. The temperature of the hot plate was set to 150°C initially and then increased by 25°C increments for consecutive test cycles.

Examples of variations of leakage currents through IWT for three types of capacitors are shown in Figure 2 and clearly indicate the presence of a critical temperature when DCL increases substantially, two to five orders of magnitude. Most of the parts in 14 groups, except two parts in Gr.13, could withstand IWT at 150 °C without any significant changes in leakage currents, but none of the parts withstood TS at more than 225 °C.

To evaluate for how long this high-leakage-current state remains, the parts were stored at room conditions, and leakage currents were measured after one day, one week, two weeks, and one month of storage. In most cases a significant decrease in DCL was observed already after one day, but for some parts the currents gradually decreased during several weeks. In six groups (Gr. 1, 3, 4, 11, 12, 13) from 10% to 70% of the samples remained at a relatively low insulation resistance (IR) state even after one month of storage. The highest level of leakage currents after 1 month was in Gr.3, G.12, and Gr.13 that all had silver/palladium inner electrodes. All lots except for Gr.7, Gr.8, and Gr.14 had DCL values at least an order of magnitude greater than initially even after a month of storage.

Median values of IR after crack formation and water exposure are typically in the megohm range. Minimal values are in the hundredths of kiloohm to megohm range, and additional measurements showed that none of the parts failed at IR below one kiloohm.

Approximately 5% to 10% increase in capacitance that was observed after the first test and decrease after one week of storage are due to de-

aging/re-aging processes in barium titanate ceramics. However, no substantial variation in C and/or DF occurred in most of the parts even when the preheat temperature reached the critical level resulting in high DCL. These confirms that AC characteristics are much less sensitive to TS testing compared to leakage currents, which is in agreement with results of Chan and co-workers [10].

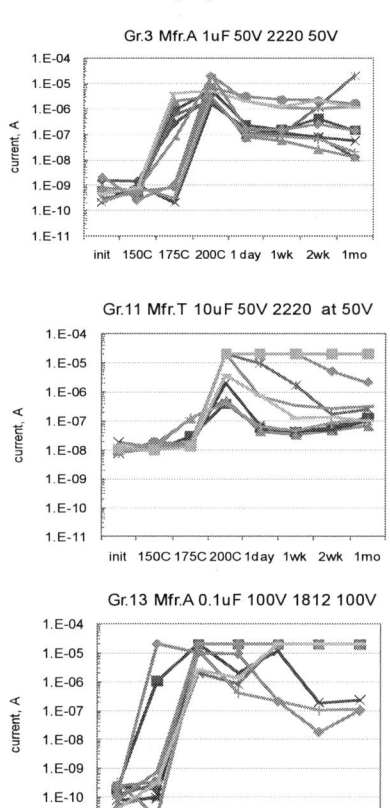

Figure 2. Variations of leakage currents through the ice-water-testing at different temperatures and storage conditions for three lots of MLCCs

Average values of the critical temperature, ΔT_c, and standard deviations determined based on results of IWT for ten parts are shown in Table 2. The minimal value of $\Delta T_c = 170$ °C was observed for Gr.13 parts, and maximum, 222.5 °C, for Gr.2. The standard deviations in all cases were relatively small varying from 0 to 22.3 (average value is 9.6 °C) thus indicating a relatively small spread and good reproducibility of test results. Repeat testing of 19 samples from Gr.1 capacitors resulted in average $\Delta T_c = 209.2$ °C and SDT = 18.5, which is close to the data presented in Table 2 and indicate good reproducibility of the IWT. Decreasing temperature increments during the testing and increasing sample size would provide more distinctive results and allow for better assessment of the critical temperature.

Values of capacitors, the type of metal electrode used and/or type of termination are apparently not critical factors for ice water thermal shock testing.

Table 2. Critical temperatures (in deg.C) per IWT.

Gr.	1	2	3	4	5	6	7	8	9	10	11	12	13	14	15	16
T_crit. avr.	210	222.5	185	197.5	196.1	194.4	200	218.8	192.5	197.2	196.3	175	170	175	208.3	195
STD	11.5	7.8	12.6	7.8	9.4	10.7	0	11.1	11.8	8.1	9.2	0	10.3	0	20.9	22.3

Analysis shows that none of the geometrical factors (size, thickness, and surface area of the parts) have a strong correlation with the critical temperature. In particular, no correlation between the thickness of the capacitors and ΔT_c as it is predicted by a simple TS model was observed. Figure 3a shows a trend of decreasing of the thermal shock resistance with the overall size of parts.

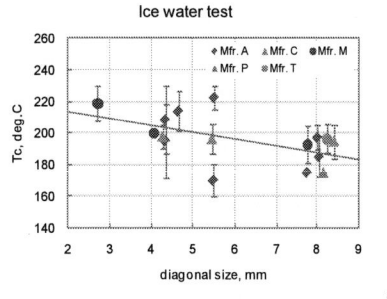

a)

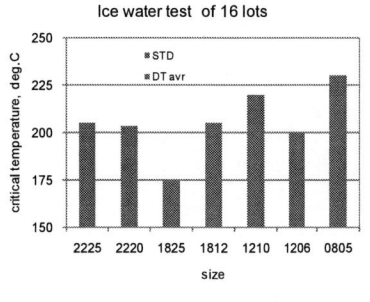

b)

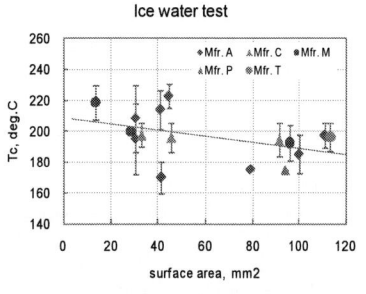

c)

Figure 3. Correlation between the critical temperature of IWT and standard size (a), diagonal size (b), and the surface area of the parts (c).

However, the correlation with the standard size (see Figure 3b) is not as good as reported in [7]. A somewhat better correlation exists between ΔT_c and the surface area of the parts, which is in agreement with the results presented by Yeung at al. [8] and likely indicates the presence of surface flaws in the capacitors.

Optical examinations showed that the cracks originated mostly at the corners/edges of the capacitors, often at the termination edge and there was a trend of increasing the number of cracks with the surface area. This is in agreement with the correlation between ΔT_c and the surface area and with a trend of decreasing values of post-IWT IR for large ceramic capacitors.

Results of cross-sectioning confirmed the presence of cracks crossing inner electrodes of the parts. In most cases the cracks were located at the terminal areas (see Figure 4).

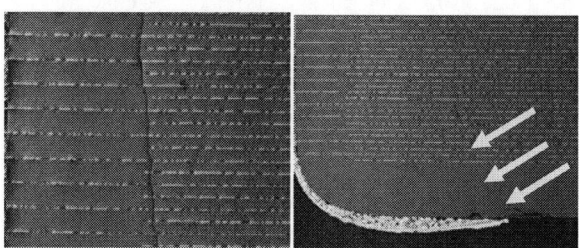

Figure 4. Cross-sectioning views of samples after IWT at critical conditions showing locations of cracks.

Liquid nitrogen drop test.

Four types of multilayer ceramic chip capacitors from Gr.1, Gr.3, Gr.6, and Gr.7, with 10 samples each were used in this experiment. Capacitors that were thermally stabilized at room temperature were dropped into a Dewar with liquid nitrogen (LN) and then removed into a sealed plastic bag to avoid moisture condensation and icing. After stabilizing at room temperature the parts were removed from the bag for electrical characterization. Because the difference between LN (-197 °C) and room temperature is ~220 °C, based on results of IWT, it was expected that the LN drop test is sufficient to cause fracturing in majority of the capacitors. However, results (see an example in Figure 5) showed no significant variations of DCL and some decrease in capacitance (~2%) that was likely due to the effect of mechanical stresses [11] associated with exposure to extremely low temperature.

One of the reasons that LN drop test did not increase leakage currents, might be related to the lack of exposure to moisture. In the absence of moisture DCL does not increase substantially even in fractured capacitors. To reveal possible cracks, the samples were stored in a humidity chamber at 85°C and 85% relative humidity for 10 days. However, post-moisture-exposure measurements showed no notable changes in leakage currents and restored initial values of capacitance.

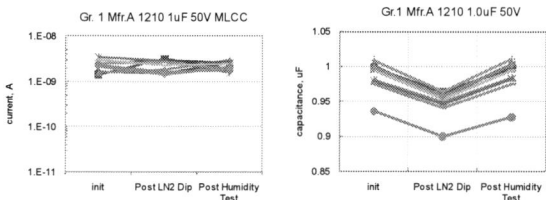

Figure 5. Variations of leakage current and capacitance in Gr.1 ceramic capacitors after liquid nitrogen drop test and exposure to 85 °C/85% RH for 10 days.

Vicinal illumination microscopy did not reveal cracks in capacitors from Gr.1. One tiny crack was found in one out of 10 parts in Gr.3. Two parts from Gr.7 had one small crack each. However, 9 out of ten parts had one or more cracks in Gr.6. Considering that Gr.6 capacitors have rather large margins (~ 300 µm), a relatively shallow cracks that formed in these parts would not cause any degradation of leakage currents even after moisture exposure.

The results show that thermal shock test caused by quenching in liquid nitrogen resulted in much less severe damage in ceramic capacitors compared to ice-water testing performed at conditions with similar temperature variations.

Discussion

Terminal solder dip testing did not produce failures up to 325 °C in four lots of capacitors tested in this study. This is consistent with our previous results [1] and with the work by Chan et al. [6] where no failures were observed in parts stressed by solder dip testing at 275 °C and some decrease in breakdown voltage appeared after testing at 425 °C only.

During solder dip, a ceramic capacitor experiences a hot thermal shock with an instant temperature increase that results in large transient compressive stresses at the surface and relatively small tensile stresses in the bulk of the specimen. Although this test provides good heat transfer condition and results in formation of significant

thermo-mechanical stresses, the probability of fracture and crack formation is low. This is mostly due to a high compressive strength of ceramic materials that is typically in the range from 1 GPa to 4 GPa and is approximately an order of magnitude greater that the tensile strength [12].

Another factor that likely mitigates the effect of solder dip stress and makes MLCCs generally robust enough to soldering induced TS stresses is increased thermal conductivity of capacitors along the metal plates. Our estimations showed that the for BME capacitors with the thickness of the dielectric layer of 10 µm, the effective thermal conductivity increases 2 to 6 times. For PME parts this increase is larger, from 9 to 20 times, and substantially reduces temperature gradients and stress transients in the parts.

An obvious explanation for the increase of DCL during IWT at a critical temperature is that it is caused by fracturing of the parts and formation of a conduction path by moisture in cracks crossing internal electrodes of the capacitor. Contrary to solder dip testing, IWT creates tensile stresses at the surface area of the parts and considering that ceramic materials have a relatively low strength under tensile conditions, this enhances fracturing during cold TS testing compared to hot TS test.

Analysis of mechanical characteristics of barium titanate ceramics and X7R type MLCCs reported in literature [4, 6, 13-14] shows that average values of the tensile strength, σ_c, are in the range from 70 MPa to 250 MPa, a typical values of the Young's modulus, $E \sim 100$ GPa, coefficient of thermal expansion, $\alpha \sim 1e-5$ 1/K, and Poisson's coefficient, $\nu = 0.33$.

The critical temperature during water-quenching test can be described by the relation [15]:

$$\Delta T_c = \frac{\sigma_c (1-\nu)}{\alpha \times E} \times \left[1.5 + \frac{3.25}{Bi} - 0.5 \times \exp\left(-\frac{16}{Bi}\right) \right],$$

where Bi is the Biot number that characterizes transient heat transfer processes in solids,

$$Bi \equiv \frac{h \times H}{\lambda},$$

λ is the thermal conductivity of ceramic, H is the half thickness of the capacitor, and h the coefficient of heat transfer.

A possible range of heat transfer coefficients, h, is likely in the range from 1,000 W/m^2_K to 30,000 W/m^2_K [16]. At the thickness of ceramic capacitors from 0.06 mm to 4 mm these values of h correspond to Biot numbers from 0.01 to 22.

Using typical mechanical characteristics for X7R ceramic, variations of the critical temperature with Bi for are shown in Figure 6. For the expected range of *Bi* during IWT, from 3 to 10 [16], ΔTc varies from 78 °C to 390 °C which overlaps the experimentally observed values from 170 °C to 222 °C. Estimations showed that the expected tensile strength of different X7R MLCCs used in this study varies in the range from 110 MPa to 200 MPa, which is consistent with literature data.

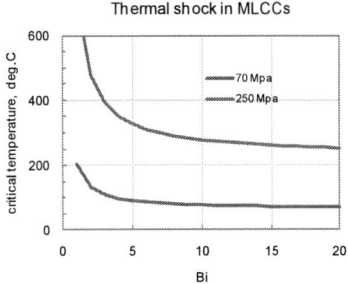

Figure 6. Estimated range of ΔTc for ice-water testing of MLCCs.

One of the factors that complicate analysis of thermal shock conditions on fracturing in MLCCS is the presence of residual or built-in mechanical stresses that are especially large at the edge, termination areas of the parts. These stresses are responsible not only for structural failures such as cracks, delaminations, and deformation, but also for modification of the dielectric behavior of ceramic materials [11]. Because the CTE of barium titanate ceramics is less than that of nickel, during the cooling from sintering (~1300 °C) to room temperature, the nickel electrode contracts more significantly than the barium titanate layer. This results in tensile stresses in the nickel layer and compressive stresses in the barium titanate layer. The latter might play a positive role by blocking crack propagation caused by cold thermal shock conditions [17]. Stresses in the terminal areas are attributed to different contraction of the terminal region compared to the active layer region in the MLCC. A significant variation of residual stresses from ~ -200 MPa (compressive) in the middle area to ~+15 MPa (tensile) at the edge, terminal areas of capacitors, has been demonstrated by finite element analysis and confirmed experimentally by Toonder at al. [18].

Figure 7 shows a trend of decreasing of ΔTc with the periphery of capacitors. This is consistent with the location of increased tensile stresses and with the fact that the majority of cracks are generated at the terminal areas. This indicates also that residual mechanical stresses are playing an important role in fracturing of MLCCs under thermal shock stress.

Although cold TS conditions during IWT do not simulate directly thermal shock stresses specific for initial phases of soldering, it allows assessing the probability of fracturing during the cooling, post-

soldering phase of the process. It is also useful to assess mechanical strength of the parts and to reveal lots with excessive residual stresses.

A cold thermal shock in MLCCs caused by LN drop test resulted in a temperature decrease of capacitors that exceeded ΔT_c values determined using IWT for most of the part types. However, this test did not cause significant fracturing, and none of the tested samples had degraded leakage currents even after moisture exposure. This result can be explained by much worse heat transfer conditions in liquid nitrogen compared to cold water quenching. Measurements of the heat transfer coefficients by Lee at al. [19] showed that for the liquid nitrogen quench, the value of *h* is an order of magnitude lower than for the water quench. This is due to a rapid formation of a gaseous N2 film that immediately shields the surface of the sample immersed in liquid nitrogen, reduces transient stresses, and the probability of fracturing.

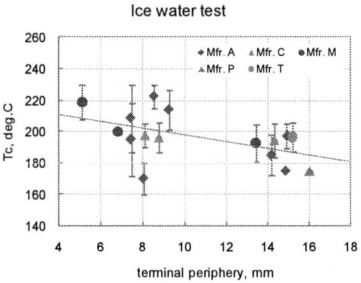

Figure 7. Effect of the length of terminal's periphery on critical temperature during IWT.

Conclusion

1. No degradation of electrical characteristics was observed after terminal solder dip testing up to 325°C, and only one sample out of forty had increased leakage currents after 350°C testing. A high robustness of MLCCs toward solder dip testing is mostly due to a high compressive strength of ceramic materials. Also, an increased thermal conductivity of capacitor along the metal plates mitigates the effect of solder dip stresses.

2. Ice water test is a reproducible and effective technique for evaluation of the susceptibility of MLCCs to fracturing during cold thermal shock conditions. Majority of the parts withstood IWT at 150 °C and the average critical temperature varied from 170 °C to 222 °C. The values of ΔT_c correspond to the tensile strength of the X7R capacitors in the range from 110 MPa to 200 MPa. Most of the TS-induced cracks formed at the terminal areas of the capacitors.

3. None of the geometrical factors of capacitors has a strong correlation with the critical temperature

determined by IWT. However, there is a trend of decreasing of the thermal shock resistance with the terminal periphery of capacitors. This indicates that a combination of residual and TS-related stresses is the major cause of fracturing and failures of MLCCs.

4. In spite of significant temperature difference that capacitors experienced during cold thermal shock caused by the liquid nitrogen drop test, this testing resulted in much less severe damage in ceramic capacitors compared to the ice-water testing performed at conditions with similar temperature difference. This result is due to a rapid formation of a gaseous N2 film that shields the surface of the capacitors and significantly reduces the heat transfer from the part and confirms that the heat transfer conditions are of critical importance for TS testing and for the assessment of the thermo-mechanical reliability of ceramic capacitors.

Acknowledgment

This work was sponsored by the NASA Electronic Parts and Packaging (NEPP) Program. The author is thankful to the Program Manager Michael Sampson for support and discussions, and appreciates the help of Susana Douglas, MEI, for assistance with electrical testing and microscopy.

References

[1] A. Teverovsky, "Effect of Manual-Soldering-Induced Stresses on Ceramic Capacitors," NASA Electronic Parts and Packaging (NEPP) Program, Part I, 2008, http://nepp.nasa.gov/.

[2] B. Rawal, *et al.*, "Reliability of multilayer ceramic capacitors after thermal shock," *Proceedings of the 38th Electronics Components Conference*, pp. 371 - 375, 9-11 May 1988 1988.

[3] ANSI/GEIA, GEIA-STD-0006, "Requirements for Using Solder Dip to Replace the Finish on Electronic Piece Parts", 2008

[4] C. Koripella, "Mechanical behavior of ceramic capacitors," *IEEE Transactions on Components, Hybrids, and Manufacturing Technology*, vol. 14, pp. 718 - 724, 1991.

[5] C. Chan and F. Yeung, "Electrical failure of multilayer ceramic capacitors subjected to environmental screening testing," *IEEE Transactions on Components, Packaging, and Manufacturing Technology, Part C*, vol. 19, pp. 138 - 143, 1996.

[6] Y. C. Chan, *et al.*, "Thermal effects on the dielectric and electrical properties of relaxor ferroelectric ceramic-based MLCs," in *Proceedings of 1995 Japan International, 18th IEEE/CPMT Electronic Manufacturing Technology Symposium, International* Omiya, 1995, pp. 328-333.

[7] Y. C. Chan and F. Yeung, "Failure analysis mechanisms of miniaturized multilayer ceramic capacitors under normal service conditions," in *43rd Electronic Components and Technology Conference*, 1993, pp. 1152 - 1155.

[8] F. Yeung, *et al.*, "Thermal shock resistance of miniaturized multilayer ceramic capacitors," *Journal of material science: materials in electronics*, vol. 5, pp. 339-343, 1994.

[9] S. Hull, "Nondestructive Detection of Cracks in Ceramics Using Vicinal Illumination," *ASM International, ISBN 0-87170-646-6*, Nov.1999 1999.

[10] Y. C. Chan, *et al.*, "Failure analysis of miniaturized multilayer ceramic capacitors in surface-mount printed-circuit board assemblies," *Journal of Materials Science-Materials in Electronics*, vol. 5, pp. 25-29, Feb 1994.

[11] G. Yang, *et al.*, "Evaluation of Residual Stress in a Multilayer Ceramic Capacitor and its Effect on Dielectric Behaviors Under Applied dc Bias Field," *Journal of American Ceramic Society*, vol. 91, pp. 887-892, 2008.

[12] R. C. Buchanan, Ed., *Ceramic materials for electronics*. Marcel Dekker, Inc, 2004, p.^pp. Pages.

[13] J. Bergenthal, "Mechanical strength properties of multilayer ceramic capacitors," in *Capacitor and Resistor Technology Symposium, CARTS'91*, 1991.

[14] P. Supancic, "Mechanical stability of BaTiO3-based PTC thermistor components: experimental investigation and theoretical modelling," *Journal of the European Ceramic Society*, vol. 20, pp. 2009-2024, Nov 2000.

[15] T. J. Lu and N. A. Fleck, "The thermal shock resistance of solids," *Acta Materialia*, vol. 46, pp. 4755-4768, Aug 1998.

[16] T. Nishikawa, *et al.*, "Heat transmission during thermal shock testing of ceramics," *Journal of Materials Science*, vol. 29, pp. 213-217, Jan 1994.

[17] G. De With, "Structural integrity of ceramic multilayer capacitor materials and ceramic multilayer capacitors," *Journal of the European Ceramic Society*, vol. 12, pp. 323-336, 1993.

[18] J. d. Toonder, *et al.*, "Residual stresses in multilayer ceramic capacitors: measurement and computation," *Journal of Electronic Packaging*, vol. 125, pp. 506-511, 2003.

[19] W. J. Lee, *et al.*, "The effect of quenching media on the heat transfer coefficient of polycrystalline alumina," *Journal of Materials Science*, vol. 28, pp. 2079-2083, Apr 1993.

IEEE
445 Hoes Lane
Piscataway, NJ 08854-4141

ISBN 978-1-4673-0694-2